Nanotechnology
Therapeutic, Nutraceutical, and Cosmetic Advances

Nanotechnology
Therapeutic, Nutraceutical, and Cosmetic Advances

Edited by
Bhaskar Mazumder, Subhabrata Ray, Paulami Pal, and Yashwant Pathak

CRC Press
Taylor & Francis Group
Boca Raton London New York

CRC Press is an imprint of the
Taylor & Francis Group, an **informa** business

CRC Press
Taylor & Francis Group
6000 Broken Sound Parkway NW, Suite 300
Boca Raton, FL 33487-2742

First issued in paperback 2022

ISBN-13: 978-0-815-36254-8 (hbk)
ISBN-13: 978-1-03-233855-2 (pbk)
DOI: 10.1201/9781351111874

Contents

Editors

Yashwant Pathak holds a Doctor of Philosophy (PhD) in Pharmaceutical Technology from Nagpur University, India and an Executive Master of Business Administration (EMBA) and a Master of Science (MS) in Conflict Management from Sullivan University. He is Professor and Associate Dean for Faculty Affairs at the College of Pharmacy, University of South Florida, Tampa, Florida. With extensive experience in academia as well as industry, he has more than 100 publications, two patents and two patent applications, and 21 edited books published, including seven books in nanotechnology and six in nutraceuticals and drug delivery systems. He has published several books on cultural studies and conflict management. He has received several National and International awards. Dr Yashwant Pathak is also an Adjunct Professor at the College of Public Health, Airlangga University, Surabaya, Indonesia.

Bhaskar Mazumder, M. Pharm, PhD, FIC, Professor in the Department of Pharmaceutical Sciences, Dibrugarh University, Assam, is a former Overseas Associate in the University of South Florida (Health). He is a former visiting fellow of the Department of Pharmaceutical Sciences, Assam University, Assam, India. He has more than 90 publications and 11 book chapters. He has nearly eight years of Industrial experience in Formulation and Development in various companies and more than twenty years of experience in academic and research.

Subhabrata Ray, Principal, Dr. B. C. Roy College of Pharmacy and AHS Durgapur, WB has more than eighteen years of Research and teaching experience. He completed B. Pharm (Hons) and M. Pharm (1st Class) from Jadavpur University in 1996 and 1998 respectively. He was awarded Doctor in Philosophy (PhD) in Pharmacy from Jadavpur University in year 2003. His field of Research Work is QSPR applications in Novel Drug Delivery Systems.

Paulami Pal is the Principal Investigator of projects sponsored under Women Scientist Scheme-A by the Department of Science & Technology, Govt. of India and she is conducting her research work in the Department of Pharmaceutical Sciences at Dibrugarh University, Assam, India. With more than two years of teaching experience, her research interest lies in the field of developing novel drug delivery systems with special attention to non-invasive drug delivery.

Contributors

Kazi Asraf Ali
Dr. B.C. Roy College of Pharmacy and Allied
    Health Sciences
Durgapur, India

Nilutpal Sharma Bora
Division of Pharmaceutical Technology
Defence Research Laboratory
Tezpur, India

and

Department of Pharmaceutical Sciences
Dibrugarh University
Dibrugarh, India

Srijita Chakrabarti
Division of Pharmaceutical Technology
Defence Research Laboratory
Tezpur, India

and

Department of Pharmaceutical Sciences
Dibrugarh University
Dibrugarh, India

Pronobesh Chattopadhyay
Division of Pharmaceutical Technology
Defence Research Laboratory
Tezpur, India

Anup Kumar Das
Konark Herbals and Health Care
Nani Daman, India

and

Department of Pharmaceutical Technology
Adamas University
Barasat, India

Pranab Joyti Das
Department of Pharmaceutical Sciences
Assam University
Assam, India

and

Department of Pharmaceutical Sciences
Dibrugarh University
Dibrugarh, India

Ratna Jyoti Das
Department of Pharmaceutical Sciences
Dibrugarh University
Dibrugarh, India

Sanjoy Kumar Das
Institute of Pharmacy, Jalpaiguri
Government of West Bengal
Jalpaiguri, India

Sandipan Dasgupta
Department of Pharmacy
NSHM Knowledge Campus
Kolkata Group of Institutions
Kolkata, India

and

Department of Pharmaceutical Sciences
Dibrugarh University
Dibrugarh, India

Supriya Datta
Bengal College of Pharmaceutical Sciences
    and Research
Durgapur, India

Sheba Rani N. David
PAPRSB Institute of Health Sciences
Universiti Brunei Darussalam
Brunei Darussalam

Sanjay Dey
Department of Pharmacy
Techno India University
Kolkata, India

Kumud Joshi
Division of Pharmaceutical Technology
Defence Research Laboratory
Tezpur, India

and

Department of Pharmaceutical Sciences
Dibrugarh University
Dibrugarh, India

Bibhuti Bhusan Kakoti
Department of Pharmaceutical Sciences
Dibrugarh University
Dibrugarh, India

Manjir Sarma Kataki
Department of Pharmaceutical Sciences
Dibrugarh University
Dibrugarh, India

Daphisha Marbaniang
Department of Pharmaceutical Sciences
Dibrugarh University
Dibrugarh, India

Bhaskar Mazumder
Department of Pharmaceutical Sciences
Dibrugarh University
Dibrugarh, India

Bankim Chandra Nandy
Department of Pharmacy
Techno India University
Kolkata, India

Lalduhsanga Pachuau
Department of Pharmaceutical Sciences
Assam University
Silchar, India

Paulami Pal
Department of Pharmaceutical Sciences
Dibrugarh University
Dibrugarh, India

Yashwant Pathak
College of Pharmacy
University of South Florida
Tampa Bay, Florida

and

Faculty of Public Health
Airlangga University
Surabaya, Indonesia

Manash Pratim Pathak
Division of Pharmaceutical Technology
Defence Research Laboratory
Tezpur, India

and

Department of Pharmaceutical Sciences
Dibrugarh University
Dibrugarh, India

Probin Kr Roy
Division of Pharmaceutical Technology
Defence Research Laboratory
Tezpur, India

Rajan Rajabalaya
PAPRSB Institute of Health Sciences
Universiti Brunei Darussalam
Brunei Darussalam

Ananya Rajkumari
Department of Pharmaceutical Sciences
Dibrugarh University
Dibrugarh, India

Subhabrata Ray
Dr. B.C. Roy College of Pharmacy and Allied
    Health Sciences
Durgapur, India

Sanjit Roy
Department of Pharmaceutical Technology
Jadavpur University
Kolkata, India

Ramkrishna Sen
Infectious Diseases and Immunology Division
C.S.I.R.: Indian Institute of Chemical Biology
Kolkata, India

Vipin Kumar Sharma
Department of Pharmaceutical Sciences
Faculty of Medical Science and Health
Gurukul Kangri Vishwavidyalaya
Haridwar, India

1

Potential of Nanostructures for Drug Delivery: With a Special Reference to Polymeric Nanoparticles

Pranab Jyoti Das, Laldusanga Pachuau, Ramkrishna Sen, and Bhaskar Mazumder

CONTENTS

ABSTRACT: The impact of nanomaterials on drug delivery is significant and the emergence of polymeric nanoparticles provides a unique platform for the delivery of drug molecules of a diverse nature. The conventional drug delivery system is widely changing with the development of nanoparticulate drug delivery systems. Because of their unique physicochemical properties, nanomaterials can be used for the strategic development of new drug delivery systems and reformulating existing drugs and therapeutics to escalate the effectiveness, patent protection, patient compliance, and safety of drugs. Nanoscale drug delivery systems such as nanoparticles, nanoliposome, nanoemulsions, nanosuspensions, dendrimers, nanopores, nanotubes, nanocrystals, quantum dots, and nanosponge are believed to have the potential to revolutionize the drug delivery strategy. Nanocarriers can be conjugated with a ligand such as an antibody to favor a targeted therapeutic approach. In the present chapter we will mainly highlight the current status of the nanoscale drug delivery systems along with the progress of research in the field. Furthermore, we will throw some light on the development of polymeric nanoparticles and how multi-functionality can be engineered into polymeric nanoparticles for tumor specific treatment. In short, we will provide an update on how nanotherapeutics may revolutionize the entire drug therapy strategy and bring it to a new look in the near future.

KEY WORDS: *Nanostructures, polymeric nanoparticles, drug delivery, polymer, nanoscale*

1.1 Introduction

Nanotechnology is a revolutionary area of science and technology which involves the creation and utilization of materials, devices, or systems on the nanometer scale. Mathematically a nanometer is equal to one thousand millionth of a meter. Nanotechnology is sparking innovation and plays a crucial role in various biomedical applications, especially in drug delivery. Delivering a therapeutic compound to the target site is a major problem in the treatment of many diseases. The conventional application of drugs is characterized by limited effectiveness, poor biodistribution, and lack of selectivity. These limitations and drawbacks can be overcome by controlling drug delivery (Wilczewska et al., 2012). In the past three decades, the explosive development of nanotechnology has been considered to be revolutionizing development in the field of drug targeting. The systems are exploited for therapeutic function to take the drug in the body in a controlled manner from the site of administration to the therapeutic target. Accumulation of therapeutic compounds in the target site increases and, consequently, the required doses of drugs are lower. Cancer stands out as a disease most likely to benefit from targeted drug delivery. Tumor cells express many molecules on their surface that distinguish them from normal cells. Therefore, nanoparticles have been considered as appropriate vehicles to provide an ideal platform for personalized approaches to cancer diagnosis and therapy in cancer disease management (Ruoslahti et al., 2010).

Advances in nanotechnology that enable drugs to preserve their efficacy while being delivered to precise therapeutic targets are creating a host of opportunities for drug developers. In addition, by combining nanotechnology-based, target-specific drug therapy with methods for early diagnosis of pathologies, we are getting closer to creating the ultimate functional drug carrier. Many researchers attach ethylene glycol molecules to nanoparticles that deliver therapeutic drugs to cancer tumors. The ethylene glycol molecules stop white blood cells from recognizing the nanoparticles as foreign materials, allowing them to circulate in the blood stream long enough to attach to cancer tumors. Researchers are also continuing to look for more effective methods to target nanoparticles carrying therapeutic drugs directly to diseased cells by using two types of nanoparticles. The first type of nanoparticle locates the cancer tumor and the second type of nanoparticle (carrying the therapeutic drugs) hones in on a signal generated by the first type of nanoparticle. Other researchers are using a photosensitizing agent to enhance the ability of drug carrying nanoparticles to enter tumors. First, they let the photosensitizing agent accumulate in the tumor, then illuminate the tumor with infrared light. The photosensitizing agent causes the blood vessels in the tumor to be more porous, therefore more drug-carrying nanoparticles can enter the tumor. One research group has found that a disk-shaped nanoparticle (nanodisk) will stick to the surface of a tumor longer than a spherical-shaped nanoparticle, providing more efficient transfer of therapeutic drugs to the tumor. Another group of researchers have found that rod-shaped nanoparticles are more effective at delivering chemotherapy drugs to breast cancer cells than spherical nanoparticles. Thus, the pharmaceutical scientists are attracted towards newer, novel nanoscale drug delivery systems for various reasons. The most important reason is that due to the increase in number of surface atoms or molecules to the total number of atoms or molecules, the surface area increases and thus helps in improving drug delivery strategies. This in turn helps to bind, adsorb, and allow passage of other bioactive compounds such as drugs, probes, and proteins. Sometimes the drug particle itself can be engineered to form nanoscale-size materials too (Hadjipanayis et al., 2010). Due to their unique size (smaller than eukaryotic or prokaryotic cells) nano-materials can eventually reach generally inaccessible areas, such as cancer cells, inflamed tissues, etc., in much higher amounts due to their enhanced permeability and retention effect (EPR) and can impair lymphatic drainage. Thus, this unique principle can be used for the administration of genes and proteins through the peroral route of administration (De Jong and Borm, 2008).

Safety is an important matter in the proper use of nanomaterials in drug delivery. Nanomaterials used for drug delivery should be easily soluble, safe, and biocompatible, as well as bioavailable at the site of treatment. They should not block blood vessels and should be less invasive. The toxicity associated with the nanomaterials for drug delivery should be negligible so that such material can be used to target the specific diseased tissue in a safe concentration (Webster et al., 2013). Nanomaterials help protect from the degradation of drug substances (both enzymatic as well as hydrolytic) in the gastrointestinal environment and they also help in bypassing "fast-pass" metabolism in the liver. The nanoscale drug delivery strategy increases the circulation time, especially those coated with hydrophilic polymers, and is hence suitable for increasing the efficacy of bioactives with short half-lives. This principle is again utilized by the formulation scientist for sustained release of drug molecules from nanomaterials, as well as for the delivery of deoxyribonucleic acid (DNA) (Chakroborty et al., 2013). The mass transfer rate of the drug to the biological fluid is enhanced, onset of therapeutic action is prolonged, and thus the therapeutic dose as well as dosing frequency is reduced. The premature loss of drug substances through rapid clearance, opsonization, and macrophageal uptake and metabolism can also be prevented (Sahoo et al., 2007). Nanomaterials also increase retention and accumulation due to their bio-adhesion property.

Thus, the nanoscale drug delivery vehicles have shown their uniqueness in encapsulating a wide range of bioactives such as small molecules (hydrophilic and/or hydrophobic), peptides, protein-based drugs, and nucleic acids. By encapsulating or attaching these molecules inside or on the surface of a nanocarrier, the solubility and stability of the drugs can be improved, providing an opportunity to reevaluate some potential drugs which were previously ignored because of their poor pharmacokinetics (Langer, 1998). Encapsulated or surface-adhered molecules can be released from nanocarriers in a controlled manner for a prolonged time to maintain a drug concentration within a therapeutic window or the release can be triggered by some stimulus unique to the delivery site (Moghimi, 2006). Nanomaterials can also be engineered to be multifunctional with the ability to target diseased tissue, carry imaging agents for detection, and deliver multiple therapeutic agents for combination therapy (Nasongkla et al., 2006). The multifunctionality of nanomaterial-based delivery systems offer the opportunity to develop novel approaches to deliver drugs that may result in alternative or complementary therapeutic options for the treatment of disease.

Nanoscale drug delivery systems such as nanoparticles, nanoliposomes, dendrimers, fullerenes, nanopores, nanotubes, nanoshells, quantum dots, nanocapsules, nanospheres, nanovaccines, nanocrystals, etc., are believed to have the potential to revolutionize drug delivery systems. Furthermore, nanomaterials on chips, nanorobotics, magnetic nanoparticles attached to specific antibodies, nano-sized empty virus capsids, and magnetic immunoassay are new dimensions for their use in drug delivery. Thus, nanomaterials can be used for the strategic development of new drug delivery systems and for reformulating existing drugs to enhance their effectiveness, patent protection, patient compliance, and safety, decreasing the cost of health care (Couvreur, 2013).

In this chapter we will mainly focus on therapeutic nanostructures with a particular emphasis on the development of nanocarrier drug delivery systems. We will also throw some light on polymeric nanoparticles and their various applications in novel drug delivery.

1.2 Various Nanostructures for Drug Delivery

The diversity of delivery system, as discussed hereunder, allows nanomaterials to be developed with a diverse array of shapes, sizes, and surface properties that enables them to be tailored for specific applications.

1.2.1 Nanoliposomes

The nano-sized vesicle made up of a phospholipid membrane, generally unilamellar, with an aqueous interior is known as a nanoliposome. Hydrophilic molecules can be encapsulated in the inner core while hydrophobic molecules can be carried in the hydrophobic domains of the lipid bilayer. Physicochemical properties of liposomes can be precisely modified to control surface charge, functionality, and size by simply mixing commercially available lipid molecules. The United States Food and Drug Administration (FDA) approved lipids used for the preparation of liposomal vesicles are 1,2-distearoyl-sn-glycero-3-pho sphoethanolamine (DSPE), hydrogenated phosphatidylcholine from soybean lecithin (HSPC), egg yolk phosphatidylglycerol (EggPG), and 1,2-distearoyl-glycero-3-phosphocholine (DSPC), etc. Among all other techniques, the thin film hydration technique is the most suitable and feasible for the preparation of nanoliposomes. Lipid vesicles have been investigated as carriers for anticancers, antifungals, analgesics, and gene therapies, as well as for vaccine delivery (Lian and Ho, 2001). Surface functionalization is also possible to produce stealth liposomes – liposomes that avoid Mononuclear Phagocyte System (MPS) uptake, thus having increased circulation times (Mufamadi et al., 2011). Polyethylene glycol (PEG), chitosan, silk-fibroin, and polyvinyl alcohol (PVA), etc., are generally used for surface functionalization and to avoid opsonization of the lipid vesicles. Targeted nanoliposomal drug delivery is more efficacious than the non-targeted drug delivery systems. The C6-ceremide ligand-induced nanoliposome used to treat blood cancer directly targets the over-expressed leukemic cells and decreases the high expression of survivin proteins in those leukemic cells. The future avenues that can be exploited using lipid vesicle platforms include association with imaging or tracing elements to track the fate of nanoliposomes *in vivo*, co-delivery of synergistic elements, and association of bio-responsive elements such as temperature sensitive elements or pH sensitive elements. Thus, liposome technologies are expected to bring lots of change in sophisticated drug delivery methods. A sample of literature indicating generalized advantages and disadvantages and the variety of ways liposomal formulations are currently being used is highlighted in Table 1.1.

Wang et al. (2015) successfully developed the dual-ligand liposomes modified with the specific ligand T7 motif and non-specific TAT. This liposomal delivery system possessed increased cellular uptake efficiency and targeting specificity in A549 cells, whose transferrin receptor (TFR) expression levels were high, and achieved an efficient, synergistic, targeted delivery of payload into tumor cells in A549 tumor-bearing nude mice, ultimately achieving excellent therapeutic efficacy on tumor-bearing mice. The findings of this study suggest that the T7- and TAT-modified liposomes are a potential antitumor drug delivery system (Wang et al., 2015). Li et al. (2015) developed PEGylated vinorelbine (VRB) plus quinacrine cationic liposomes for treating non-small cell lung cancer (NSCLC). The PEGylated VRB plus quinacrine cationic liposomes were able to increase cellular uptake and accumulate in the A549 cells. Furthermore, they exhibited significant inhibitory effects to A549 cells, and showed the strongest effects on inhibiting vasculogenic mimicry (VM) channels. In addition, studies on tumor-bearing mice confirmed that the PEGylated VRB plus quinacrine cationic liposomes did show an enhanced anticancer efficacy. Action mechanisms showed that the enhanced efficacies in treating NSCLC were related to activate apoptotic enzymes (caspases 9 and 3), pro-apoptotic proteins (Bax), and tumor suppressor genes (P53), and to suppress anti-apoptotic proteins (Bcl-2 and Mcl-1) (Li et al., 2015). Qu et al. (2014) formulated docetaxel (DTX) and BCL-2 small interfering ribonucleic acid (siRNA) incorporated PEGylated liposomes to systemically deliver in a lung cancer model (A549). The lipo-DTX/siRNA exhibited sustained-release kinetics, effectively inhibited cell proliferation (A549 and H226) and modified cell apoptosis and cell cycle analysis. An *in vivo* antitumor study on an A549 cell-bearing xenograft tumor model exhibited a remarkable tumor regression profile for lipo-DTX/siRNA with a 100% survival rate. The favorable tumor inhibition response was attributed to the synergistic effect of DTX potency and the multidrug resistance (MDR) reversing ability of siRNA in the tumor mass. Their unique findings support the hypothesis that siRNA treatment combined with a classical anticancer drug could represent a new approach in the treatment of lung cancer with MDR (Qu et al., 2014).

1.2.2 Polymersomes

Polymersomes are polymeric, tiny, hollow spheres that enclose a solution. They are amphiphilic block copolymers, forming a shell that can encapsulate aqueous-soluble compounds inside the aqueous core and entrap organic-soluble components within the shell itself. Although liposomes also work on the same principle, polymersomes have advantages over liposomes, with a longer half-life (20–30 hours) and

TABLE 1.1

Liposomal Nanostructures

General Advantages	General Disadvantages	Component Loaded	Size	Target Moieties	Outcome of Research	References
1. Research is well established	1. Tend to be more water permeable	siRNA	150 nm	Sigma receptor expressing cancer cells (anisamide)	Develop a new drug release mechanism for targeted siRNA delivery (LCP (Liposome/Calcium/Phosphate) core calcium phosphate dissolved in endosome causing swelling and endosomal rupture to delivery siRNA to cytosol)	Li et al. (2010)
2. Surface charge can be modified to improve circulation and cellular adhesion and uptake	2. Have a low circulation time	DOX	147 nm	Targeted release via localized ultrasound exposure	Evaluation of DOX-loaded liposome lipid gas-filled microbubble system as an anticancer therapeutic delivery system (complex was more effective at reducing cancer cell viability than free DOX or DOX-loaded liposomes, cavitation and implosion of microbubbles may enhance DOX uptake)	Lentacker et al. (2010)
		ITZ (Intraconazol)	264.5 nm	No active targeting	Biodistribution of ITZ-loaded liposomes (ITZ-loaded liposomes produced longer circulation half-life as well as greater ITZ levels in the liver and spleen than current commercial formulations)	Tang et al. (2010)
		Calcein and doxorubicin	46.43 nm	Antibody KN2/NRY or human haptoglobin	Stealth liposomes modified to recognize CD163 as a potential way to target drugs to macrophages	Etzerodt et al. (2012)

increased stability, for practical applications (Discher et al., 1999). Polymersomes are used for loading and protecting sensitive molecules, such as drugs, enzymes, other proteins and peptides, and DNA and RNA fragments. The polymersome membrane provides a physical barrier that isolates the encapsulated material from external materials, such as those found in biological systems. The term "polymersome" for vesicles made from block copolymers was coined in 1999. The use of synthetic polymers enables researchers to modify the physicochemical characteristics of the membrane and thus control permeability, release rates, stability, and other properties of the polymersome. There are many techniques available that can be used to prepare polymersomes by self-assembly of amphiphilic block copolymers. Typical methods are polymer rehydration techniques, which are based on the hydration of amphiphilic block copolymer films to induce self-assembly. Other important preparation methods are solvent-switching techniques. Lomas et al. (2007) have investigated the encapsulation of plasmid DNA by a biocompatible, pH-sensitive block copolymer poly(2-(methacryloyloxy)ethyl phosphorylcholine-co-poly(2-(diisopropyl amino)ethyl methacrylate) (PMPC-PDPA), polymersome formation, and delivery of the DNA-copolymer complex to the cell cytosol.

Polymersomes can also encapsulate diagnostic payloads. To date, polymersomes have been used in optical imaging, magnetic resonance imaging (MRI), and ultrasound imaging. By using polymersomes, useful information can be extended to obtain higher resolution and monitor biological pathways and cellular functions *in vivo*. The surface charge density plays a key role in the biodistribution and pharmacokinetics of polymersomes. Interactions with the opsonins can be reduced by the introduction of a slightly negative or positive charge on the surface of the polymersomes, yielding prolonged blood circulation times. Polymersomes based on poly(ethylene glycol)-b-poly(D,L-lactide) (PEG-PDLLA) with zeta potentials (−7.6 to −38.7 mV) were investigated for the effect of surface charge on blood circulation time and tissue distribution in tumor-bearing mice. PEG-PDLLA polymersomes with a low zeta potential (−7.6 mV) and a diameter of approximately 100 nm had a much longer half-life time and a reduced liver uptake (28% injected dose (ID) after 3 days) as compared to stealth liposomes (Lee et al., 2011). This implied that the charge density of anionic polymersomes effects circulation kinetics and biodistribution and showed that polymersomes with a slightly negative surface charge are most suited for *in vivo* administration. Table 1.2 highlights the properties and application of copolymers forming polymersomes.

1.2.3 Dendrimers

Dendrimers are branched polymers, resembling the structure of a tree. These nanostructured macromolecules show their potential abilities in entrapping and/or conjugating the high molecular weight hydrophilic as well as hydrophobic entities by host–guest interactions and covalent bonding respectively. They are in high demand because of their defined structure and versatility in drug delivery. Since Vogtle first used them in 1978, dendrimers have provided a novel, and one of the most efficient, nanotechnology platform for drug delivery (Buhleier et al., 1978). They are three-dimensional, branched, well-organized, nanoscopic macromolecules (typically 5,000–500,000 g/mol), possess a low polydispersity index, and are gaining more importance in the emerging field of nanomedicine. The name is actually derived from

TABLE 1.2

Properties and Application of Copolymers Forming Polymersomes

Copolymer	Properties	Application
Poly(ethylene oxide)-polystyrene	Ability to form large compound vesicular aggregates	Potential use in controlled release, biomedical applications
Poly(acrylic acid)-polystyrene	Ability to form large compound vesicular aggregates	Potential drug delivery vector
Poly(ethylene oxide)-poly(butadiene)	Mechanical toughness, biological inertness, crosslink ability	NIR (Near-infrared) imaging contrast agent
Poly(ethylene oxide)-poly(butylene oxide)	Selective permeability	Drug delivery
Poly(2 vinylpyridine)-poly(ethylene oxide)	pH sensitivity	Controlled release

the Greek word "dendron", meaning "tree", which refers to their unique tree-like branching structure. They are characterized by layers between each cascade point, popularly known as "generations". The shapes of dendrimers range from spheres to flattened spheroid-like (or disk-like) structures, especially in cases where surface charges exist and give the macromolecule a "starfish"-like shape. The branching of dendrimers depends on the synthesis processes.

There are two methodologies for the synthesis of dendrimers. One is known as the divergent methodology, first used by Tomalia et al. (1985) in the 1980s. Here the coupling of monomeric molecules containing reactive and protective groups with multifunctional core moiety results in the stepwise addition of generations around the core followed by the removal of protecting groups. The advantage of this technique is that it produces modified dendrimers by changing the end groups, and thus their physicochemical properties can be altered as per the required application. This method suffers from the limitation of structural deformities due to the incomplete reaction of groups, but this can be overcome by the addition of an excessive amount of monomer units. The second approach is the the convergent methodology and was pioneered by Hawker and Frechet (1990) in the 1990s. Here the strategy behind the formation of the dendrimer is inward growth by gradually assembling surface units with reactive monomers. The convergent methodology is used for the production of lower generation dendrimers. The other approaches available are the "hypercore and branched monomers", "double exponential", "lego chemistry", and "click chemistry" methodologies.

Dendrimers are popularly used for the transfer of genetic materials in cancer therapy or to treat other viral diseases in different organs because of their unique properties like monodispersity, high functional groups density, well-defined shape, and multivalency. In the case of gene delivery of polyamidoamines (PAMAM), dendrimers are widely used. Some other types of dendrimers are peptide dendrimers, glyco-dendrimers, PolypropylenimineDendrimers, polyethylene imine (PEI) dendrimers, etc. The versatility of the dendrimer is not limited to drug delivery. Other applications for dendrimers include photodynamic therapy (PDT), boron neutron capture therapy, and bioimaging applications. In spite of their various advantages, dendrimers are limited in use because of their high manufacturing cost and they are not subjected to "generally recognized as safe" (GRAS) status due to the inherent toxicity issues associated with them. Table 1.3 lists the commercially available dendrimers and dendrimer-based products.

Kaminskas et al. (2014) explored the utility of a 56 kDa biodegradable PEGylated polylysine dendrimer, conjugated to doxorubicin, to promote the controlled and prolonged exposure of lung-resident cancers to cytotoxic drugs. Twice-weekly intratracheal instillation of the dendrimer led to a >95% reduction in lung tumor burden after two weeks in comparison to intravenous (IV) administration of doxorubicin solution which reduced lung tumor burden by only 30–50%. Intratracheal instillation of an equivalent dose of doxorubicin solution led to extensive lung-related toxicity and death within several days of a single dose (Kaminskas et al., 2014). Liu et al. (2011) developed a universal tumor-targeting drug carrier; the NSCLC-targeting peptide (LCTP, sequence RCPLSHSLICY) and fluorescence-labeled molecule (FITC, a fluorescence labeling agent used as a tracer) were conjugated to an acetylated PAMAM dendrimer to generate a targeted drug delivery carrier (PAMAM–Ac–FITC–LCTP). The performance of this drug carrier was evaluated by *in vitro* culturing of NCI-H460 and 293T cells and *in vivo* using athymic mice with lung cancer xenografts. The specific LCTP-modified PAMAM dendrimer could be easily taken up by NCI-H460 cells *in vitro* and by tumors *in vivo* (Liu et al., 2011).

TABLE 1.3

List of Commercially Available Dendrimers

Brand Name	Dendrimer	Company	Application	Status
VivaGel®	Poly-L-lysine	Starpharma	Transmission prevention of HIV and STD	Clinical trial (Phase III)
Stratus® CS	PAMAM	Dade Behring	Cardiac assay diagnostic	Marketed
Superfect®	PAMAM	Qiagen	Transfection agent	Marketed
Priofect®	PAMAM	Starpharma	Transfection agent	Marketed
DEP® docetaxel	Not defined	Starpharma	Breast cancer treatment	Preclinical

1.2.4 Cyclodextrin Inclusion Complex

Cyclodextrins (CDs) have been used both in natural and modified (α-CD and β-CD) forms and some have gained US FDA approval or achieved GRAS status. CDs, grafted or crosslinked with polymers, are usually being developed into "smart" systems for efficient, targeted drug delivery, especially for water insoluble drugs. Amphiphilic CDs have the ability to form nanospheres or nanocapsules via a simple nanoprecipitation technique. The simplest way of incorporating CDs within a polymer structure is to covalently attach the CDs to a polymer backbone as a side group. Another approach is the creation of "star polymers" containing a CD core with polymer "arms" extending from the CD molecule. CDs are also being utilized for drug delivery in non-polymeric systems by either taking advantage of the hydrophobic portion in the inner region of CDs or modifying the CD ring by attaching ionic groups for further interactions. These types of carriers can be found in the literature for the delivery of various therapeutics such as paclitaxel (PTX) entrapped in the hydrophobic region of CD. In their study, Maciollek et al. (2010) investigated non-covalent incorporation of adamantyl-modified 5-flurocytocine (FC) into copolymers and release by enzymatic hydrolysis (the CD structure successfully interacts with adamantyl-modified 5-FC and can be released by a-amylase hydrolysis). In another study, Ren et al. (2009) studied Poly(N-vinyl-2-pyrrolidone-co-mono-6-deoxy-6-methacrylate-ethylamino-b-cyclodext rin) (PnvpCD) micelle formation via non-covalent interactions with adamantane end-capped polymer (resulting in successful formation of both single and multicore micelles based on preparation conditions). The limitation of CD-containing polymer is the location and amount of CD can limit loading, targeting, and hydrophilicity if using CD–AD (Adamantane) interactions for these purposes.

1.2.5 Micelles

Micelles are composed of lipids or other amphiphilic molecules. Micelles have been developed for the delivery of hydrophobic drugs. Genexol®-PM is the first non-targeted polymeric micellar formulation approved for cancer therapy. It is composed of block copolymer PDLLA–mPEG, forming micelles with a size of approximately 60 nm and PTX loading of approximately 15% (w/w). The maximum tolerated dose (MTD) of Genexol®-PM is threefold higher than Taxol® (PTX) (60 mg/kg versus 20 mg/kg, respectively) and the median lethal tolerated dose (LD_{50}) using Sprague–Dawley rats was reported to be approximately 20 times higher than Taxol®. Interestingly, the area under the plasma concentration (AUC) was similar for both formulations. However, PTX had more significant accumulation in tissues such as the liver and tumor than the Genexol®-PM formulation, leading to differential tumor cytotoxicity and reduction of tumor volume (Kim et al., 2001). Genexol®-PM was approved in Korea in 2006 as a first-line therapy for metastatic breast and NSCL cancer (currently in Phase III). It is currently being evaluated in a clinical phase II trial on "Nanoparticle Technologies for Cancer Therapy" in the US for metastatic pancreatic cancer therapy.

Various manufacturers are developing micelle systems. Among them Labopharm and Intezym are developing micelle systems for the delivery of a myriad of anticancer agents using polyamino acids and synthetic polymers formulations with sizes ranging from 10 to 200 nm. Krishnamurthy et al. (2014) successfully developed phenformin loaded into PEG/acid-functionalized polycarbonate diblock copolymer (PEG-b-PAC) and PEG/urea-functionalized polycarbonate diblock copolymer (PEG-b-PUC) mixed micelles. The phenformin-loaded micelles were characterized for drug loading level, particle size, size distribution, zeta potential, *in vitro* drug release, and stability in serum-containing solution. Cytotoxicity of phenformin and phenformin-loaded micelles against H460 cells was evaluated by MTS assay (Cell proliferation), and their non-specific cytotoxicity was also examined against human dermal fibroblasts (HDFs) and WI-38 human fibroblasts like the fetal lung cell line. The anticancer efficacy of micelles was studied in an H460 xenograft mouse model. The population of cancer stem cells in the tumor tissue after treatment with free phenformin and phenformin-loaded micelles was also analyzed (McMahon et al., 2011). Kim et al. (2014) developed a multifunctional siRNA-loaded polyion complex (PIC) micelle by elaborating a block copolymer PEG-poly[N‴-(N″-{N′-[N-(2-aminoethyl)-2-aminoethyl]-2-aminoethyl}-2-aminoethyl)aspartamide] (termed PAsp(TEP)). The ω-end of PAsp(TEP) was modified with cholesteryl moiety for enhanced core stability because of its hydrophobic interactions, and also the α-end of

PEG was installed with cyclic Arginylglycylaspartic acid (RGD) peptide ligand for targeting cancer cells as well as neovascular endothelial cells in tumor tissues. The targeted/stabilized micelle was systemically administered into subcutaneous lung tumor (A549)-bearing mice, demonstrating its therapeutic potential for siRNA-based cancer therapy through the gene silencing of polo-like kinase 1 (Plk1). These results demonstrate the strong potential of the targeted/stabilized PIC micelle for siRNA-based cancer therapy (Kim et al., 2014). Wu et al. (2014) formulated a dual drug delivery system for use in the treatment of lung cancer. An *in situ* gel-forming hydrogel composite was prepared by mixing monomethoxy poly(ethylene glycol)-poly(ε-caprolactone) (MPEG–PCL)/PTX micelles with cisplatin (DDP)-loaded poly(ethylene glycol)-poly(ε-caprolactone)-poly(ethylene glycol) (PECE) hydrogel by solid dispersion and produced a compound with high drug-loading ability, high encapsulation efficiency, small particle size, and a good *in vitro* drug release profile. The *in vivo* antitumor mouse model of lung cancer showed that the micelles can significantly inhibit tumor growth and prolong the survival time of tumor-bearing mice. Analysis of tissue biomarkers showed that the hydrogel composite suppressed tumor cell proliferation and inhibited angiogenesis in tumor tissue (Wu et al., 2014). Chen et al. (2013) developed methotrexate (MTX)-loaded pluronic P105/F127 mixed polymeric micelles (PF-MTX). These were prepared using a thin film hydration method. The optimized formulation displayed suitable particle size and distribution, high drug-loading, and pH-dependent drug release. *In vitro* and *in vivo* evaluation of PF-MTX was performed through *in vitro* cytotoxicity and cellular uptake in the sensitive and MDR tumor cells studies, biodistribution studies using fluorescein-labeled polymeric mixed micelles, and studies into antitumor activity pharmacological effects in a KBv xenograft tumor-bearing nude mice model (Chen et al., 2013).

1.2.6 Protein Nanoparticles

The protein-based drug delivery system (albumin-bound drug nanoparticles (~130 nm)) has had a high impact in the novel drug delivery approach. Recently Abraxane® (albumin-bound PTX) has gained approval from the FDA for metastatic breast cancer therapy, and multiple clinical trials based on albumin-bound nanoparticles are in progress for other specific types of cancer too. Albumin is a natural non-covalent physiological transporter of molecules across endothelial barriers through a transcytosis-mediated mechanism. Albumin is also tested as a platform of delivery for various types of molecules which are insoluble in aqueous medium. An example is rapamycin. Albumin-bound rapamycin (ABI-009) has been in clinical phase trials for the treatment of non-hematologic malignancies since January 2008.

1.2.7 Biological Nanoparticles

The use of bacterial cells to encapsulate hydrophobic and hydrophilic molecules for drug delivery led to the development of biological nanoparticles. EnGeneIC Pty Ltd developed a biological nanoparticle-based drug delivery system for cancer therapy. This system is known as the "nanocell", which consists of anucleate globular bacteria (~400 nm). Molecules with different solubilities and charge such as doxorubicin, PTX, and siRNA can be loaded efficiently in nanocells by diffusion into the bacterial cell. The limited application of biological nanoparticles is due to the immunological response to the carrier due to the presence of lipopolysaccharides (LPS).

1.2.8 Inorganic Nanoparticles

Inorganic nanoparticles are metal-based nanoparticles which have the potential to synthesize with near monodispersity. The use of nanometals (e.g., gold, silver, copper, palladium, platinum, molybdenum, titanium, and zinc) as a component of pharmaceuticals has been increasing remarkably due to their ability to act as a reducing agent, bind proteins and denature enzymes, and for their bactericidal properties in topical formulation. The demand for silver nanoparticles is due to their multiple uses in the treatment of traumatic or diabetic wounds, burns, tonsillitis, syphilis, and chronically-infected wounds, as well as in antiviral and antifungal treatments. Today, silver nanoparticles are used in consumer products and nutraceuticals more than any other metal. This is due to its ability to fight against bacterial growth. Colloidal silver is nowadays consumed as a dietary supplement also. Iron oxide nanoparticles coated with aminosilane (NanoTherm® M01) are also used for brain cancer therapy and recurrent prostate cancer therapy using hyperthermia as well as thermoablation methods. Similarly, silica nanoparticles coated with gold that absorbs near-infrared laser energy and converts it into heat to kill solid tumors are currently under investigation in a pilot study for head and neck cancer therapy. Their surface can be easily modified for attachment of drug molecules. These attachments are possible via ionic or covalent bonding, or by physical adsorption. The increase in circulation half-life and enhancement of stability of gold nanoparticles are achieved by surface functionalization. The attachment of PEG is an example of such a modification. Several examples exist for the attachment of PEG, including PEGylation attachment of PEG to the anticancer platinum-based therapeutic oxaliplatin (Brown et al., 2010),covalent attachment of PEG and PEG-tamoxifen for breast cancer therapy (Dreaden et al., 2009), covalent attachment of PEG and a Pt (IV) oligonucleotide conjugate for various cancer therapeutics (Dhar et al., 2009), attachment of PEG and an anticancer drug doxorubicin via amino group (Mirza et al., 2011), and many others. The main advantage of gold nanoparticles over other nanocarriers is the:

1. Increased drug-loading per particle.
2. Easy formation of smaller nanoparticles.
3. Natural uptake by cells.

However, gold nanoparticles also possess limitations, such as therapeutics moieties must be added via surface modification only, they are not biodegradable, and they have no controlled-release properties. Research is in high demand for the application of gold nanoparticles in drug delivery. A few examples of relevant research investigations are the evaluation of biomimetic HDL-AuNp as nucleic acid delivery vehicles (McMahon et al., 2011), the development of a AuNp-based delivery carrier for platinum-based anticancer therapeutics (Brown et al., 2010), and selective targeting/uptake of TAM-AuNp conjugates (Dreaden et al., 2009).

1.2.9 Hybrid Nanoparticles

Hybrid nanoparticles are composed of at least two different materials to form the core and the corona structure. Generally, metallic and polymeric materials form the core and are coated with single or multiple lipid layers to form a protective membrane (corona) similar to a liposome or micelle. An example of this is the study by Sengupta et al. (2005) where they reported the application of poly(lactic-co-glycolic acid) (PLGA)-core nanoparticles coated with a bi-phospholipid layer to carry multiple drugs. In their study, doxorubicin is conjugated to PLGA to form the core of the nanoparticles and combretastatin is conjugated with phospholipids and encapsulated in the lipid bilayer during the self-assembly process to form nanoparticles (~200 nm).

1.2.10 Viral Nanoparticles

Viral nanoparticles are spherical, unilamellar vesicles with a mean diameter of around 150 nm. The influenza virus is most commonly used for viral nanoparticle production. Viral nanoparticles cannot replicate but are pure fusion-active vesicles. In contrast to liposomes, viral nanoparticles contain functional viral envelope glycoproteins; influenza virus hemagglutinin (HA) and neuraminidase (NA) are intercalated within the phospholipid bilayer membrane. Further characteristics of viral nanoparticles depend on the choice of bilayer components. Viral nanoparticles can be optimized for maximal incorporation of the drug or for the best physiological effect by modifying the content or type of membrane lipids used. It is even possible to generate carriers for antisense-oligonucleotides or other genetic molecules, depending on whether positively- or negatively-loaded phospholipids are incorporated into the membrane. Various ligands, such as cytokines, peptides, and monoclonal antibodies (MAbs), can be incorporated into the virosome and displayed on the virosomal surface. Even tumor-specific monoclonal antibody fragments (Fab) can be linked to virosomes to direct the carrier to selected tumor cells (Almeida et al., 1975).

1.2.10.1 Advantages of Viral Nanoparticles as Drug Delivery Carriers

- Virosomal technology is approved by the FDA for use in humans and has a high safety profile (Helenius et al., 1977).
- Virosomes are biodegradable, biocompatible and non-toxic (Huckriede et al., 2005).
- No autoimmunogenity or anaphylaxis (Glück et al., 1994).
- Broadly applicable with almost all important drugs (anticancer drugs, proteins, peptides, nucleic acids, antibiotics, fungicides) (Helenius et al., 1977).
- Enables drug delivery into the cytoplasm of target cell (Helenius et al., 1977).
- Promotes fusion activity in the endolysosomal pathway (Helenius et al., 1977).
- Protects drugs against degradation (Helenius et al., 1977).

1.2.11 Fullerenes and Nanotubes

Fullerenes are ellipsoid tubes composed of carbon. Sometimes they also possess the shape form of a hollow sphere. These are also known as "buckyballs" because of their resemblance to the geodesic dome design of Buckminster Fuller. Fullerenes are being investigated for the drug delivery of anticancer agents, antiviral drugs, and antibiotics (Reilly, 2007). Soluble derivatives of fullerenes such as C_{60} have shown great utility as pharmaceutical agents.

Nanotubes are nanometer scale tube-like nanostructures. There are many types of nanotubes, like carbon nanotube, inorganic nanotube, DNA nanotube, membrane nanotube, etc. (Prato et al., 2008). Carbon nanotubes can be made more soluble by the incorporation of carboxylic or ammonium groups to their structures and can be used for the transport of peptides, nucleic acids, and other drug molecules. The ability of nanotubes to transport DNA across the cell membrane is used in studies involving gene therapy. DNA can be attached to the tips of nanotubes or can be incorporated within the tubes (McDevitt et al., 2007).

1.2.12 Quantum Dots

Quantum dots (QD) are semiconductor nanocrystal particles, whose size is generally not greater than 10 nm. The biomolecule conjugation of the QD can be modulated to target various biomarkers (Gangrade, 2011). QD may provide new insights into understanding the pathophysiology of cancer and real time imaging and screening of tumors. Bioconjugated QD are collections of variable-sized nanoparticles embedded in tiny beads made of polymeric material. In the "multiplexing" process they can be finely tuned to a myriad of luminescent colors that can tag a multitude of different protein biomarkers or genetic sequences in cells or tissues (Mulder et al., 2006). The new class of QD conjugate contains an amphiphilic triblock copolymer layer for *in vivo* protection and multiple PEG molecules for improved biocompatibility and circulation, making it highly stable and able to produce bright signals.

The advancement in the surface chemistry of QD has extended their use in biological applications, reduced their cytotoxicity, and rendered them a powerful tool for the investigation of distinct cellular processes, like uptake, receptor trafficking, and intracellular delivery. Another application of QD is for viral diagnosis. Rapid and sensitive diagnosis of Respiratory Syncytial Virus (RSV) is important for infection control and development of antiviral drugs. Antibody-conjugated nanoparticles rapidly and sensitively detect RSV and estimate relative levels of surface protein expression. A major development is the use of dual-color QD or fluorescence-energy-transfer nanobeads that can be simultaneously excited with a single light source (Amiot et al., 2008). QD linked to biological molecules, such as antibodies, have shown promise as a new tool for detecting and quantifying a wide variety of cancer-associated molecules. Thus, in the field of nanotechnology, QD give a worthy contribution to the development of new diagnostic and delivery systems as they offer unique optical properties for highly sensitive detection, they are well defined in size and shape, and they can be modified with various targeting principles.

1.2.13 Carbon Nanotubes

Carbon nanotubes are gaining importance in the field of nanomedicine. To increase their solubility and circulation time, carbon nanotubes can be treated with acids to create a carboxylic acid at the terminal position or a hydrophilic group can be attached by covalent bonding via functionalization chemistry. Excretion occurs mainly via the biliary pathway and secondarily via the renal pathway. PEGylated carbon nanotubes usually show low reticuloendothelial uptake which thus increases circulation time. The application of carbon nanotubes in drug delivery has been investigated by various groups of researchers. Some examples of such studies are:

- Targeted cancer cell killing with single-walled carbon nanotube (SWNT) bioconjugate (showed selectivity and significant regression of tumor growth compared to control) (Bhirde et al., 2009).
- Targeted delivery of anticancer platinum agents (shown to be greater than 8 times more toxic to folate receptor tumor cells) (Dhar et al., 2009).
- Controlled, targeted release via SWNT encapsulated by folate-chitosan (achieved sustained, controlled release along with the incorporation of folic acid targeting molecules) (Huang et al., 2011).
- Viability of SWNT conjugates for anticancer therapeutic delivery (showed no toxic effects of the SWNTs, tumor growth inhibition of 60% after 22 days on an *in vivo* murine breast cancer model, and circulation half-life of 1.1 hours with the most accumulation occurring in the liver and spleen) (Liu et al., 2009).

Although preliminary studies show some promising results for carbon nanotubes, much more biocompatibility information is needed to see the full extent of effects of carbon nanotubes throughout the body. The other limitations include:

1. It is not well established as a nanomedicine carrier.
2. There are currently no clinical trials underway.
3. Therapeutics must be added via surface modifications.
4. It needs to have surface functionalization to become soluble.

1.2.14 Solid Lipid Nanoparticles (SLNp)

The development of solid lipid nanoparticles (SLNp) is one of the emerging fields of lipid nanotechnology with several potential applications in drug delivery. SLNp are a nanoscale delivery system prepared from physiological lipids, primarily triglycerides, and phospholipids. Due to their unique size-dependent properties, lipid nanoparticles offer the possibility to develop new therapeutics. As the SLNp are based on using physiological components, they are less toxic and, as a result, more acceptable for pulmonary drug delivery. Phospholipid-based surfactant proteins present at the alveolar surface are essential for maintaining optimal surface tension and reducing friction in the lung tissue (Beck-Broichsitter et al., 2011; Paranjpe et al., 2014). Videra et al. (2012) showed the results obtained with a novel formulation of PTX encapsulated in a solid lipid matrix used for the treatment of lung metastases in an experimental mouse mammary carcinoma. They evaluated the effectiveness of the PTX-loaded SLNp (SLNp-PTX) treatment based on the decrease of the number and size of lung metastases in comparison with the treatment by IV administration of the same drug using the conventional formulation (Videira et al., 2012). In a study by Sahu et al. (2015), SLNp-PTXs were prepared using mannose as a lectin receptor ligand conjugate to target lung cancer and to increase the anticancer activity of PTX against A549 lung epithelial cancer cells. An *in vitro* cytotoxicity study on A549 cell lines showed an enhanced efficacy for the drug-loaded SLNp than the free drug. An *in vivo* biodistribution study of the drug revealed that SLNp delivered a higher concentration of PTX as compared to PTX-SLNp in the alveolar cell site (Sahu et al., 2015). Han et al. (2014) developed surface-modified, co-encapsulated SLNpcontaining enhanced green fluorescence protein plasmid (pEGFP) and doxorubicin in order to create a multifunctional delivery system that targets lung cancer cells. In addition, the coating of active targeting ligand transferrin (Tf) can improve the efficacy of the carriers at targeting lung cancer cells. They developed stable, solid, lipid, nano-sized particles by optimizing several critical process variables and carried out their physicochemical evaluation, such as particle size and surface charges, and an *in vitro* cytotoxicity study displayed a remarkable therapeutic effect both in drug delivery and gene therapy (Han et al., 2014). Paranjpe et al. (2013) observed the toxicological effect of sildenafil-loaded SLNp in *in vitro* and *ex vivo* models using lung and heart tissues from murine models for pulmonary administration for the treatment of pulmonary arterial hypertension. For *in vitro* models, human alveolar epithelial cell line (A459) and mouse heart endothelium cell line (MHEC5-T) were used. In this study, high LD_{50} values were observed in heart tissue models in comparison with the lung tissue models. Considering the toxicological aspects, sildenafil-loaded SLNp could have potential in the treatment of pulmonary hypertension via the inhalation route (Paranjpe et al., 2013).

1.3 Polymeric Nanoparticles for Drug Delivery

Nanoparticulate drug delivery has experienced significant development over the last few decades. This is probably due to the huge advancement in polymer science. Biodegradable polymers have gained significant attention from formulation scientists throughout the world due to their easy degradation and disposition from the biological system. Degradation of these biodegradable, long polymer chains into oligomers and individual monomers in the enzymatic environment of biological fluid helps in the removal of wastes from the body via various biochemical pathways. An example is the synthetic, biodegradable polymer PLGA, which metabolizes into lactic acid and glycolic acid and is washed out from the biological system via the Krebs Cycle. The degradation rate of PLGA in water is a function of the molecular weight and the lactide to glycolide ratio. Some polymers which are most commonly exploited for nanoparticulate formulation include PLGA, polylactic acid, dextran, hydroxypropyl methyl acrylate (HPMA), and chitosan. The encapsulated bioactive molecule is often released by the mechanism of erosion, diffusion, or both erosion and diffusion. Drug release from the polymer matrix follows Fickian kinetics due to the diffusion of the drug through the polymeric matrix, or is triggered in response to environmental stimuli or in the course of chemical degradation. The unique characteristics of polymeric backbone, such as molecular weight, polydispersity, architecture, charge, and hydrophobicity, have a potential impact on the pharmacokinetics and pharmacodynamics of the drug. Other crucial factors which are generally taken into serious consideration are chemical inertness, lack of leachable impurities, and biodegradability. The solubility of this potentially active, hydrophobic, anticancer drug can be enhanced by incorporating it into the polymer, thereby improving its bioavailability and biodegradability. Polymeric nanoparticles after administration have to cross several biological barriers and are attacked by the cells of the reticuloendothelial system, which in turn leads to the quick removal of the polymeric nanoparticles from circulation with less or no therapeutic effects. Thus the *in vivo* fate of polymeric nanoparticles plays an important role in eliciting the proper pharmacological effect of the encapsulated biomolecules. Studies are in urgent need in the areas of pharmacokinetics and biodistribution of polymeric nanoparticles.

Due to the small size and unique physicochemical properties of polymeric nanoparticles, the pharmacokinetic profile is quite different from free drugs or other conventional dosage forms. Moreover, the conventional approach to delivery of potent drugs may lead to an unwanted accumulation in various organs such as the kidney, liver, spleen, etc. Thus, the key issues in developing a polymer-based nanoparticulate system lie in overcoming the biological barrier, specific accumulation at the target site, and preventing rapid clearance.

MDR is the biggest hurdle towards any treatment. MDR has a detrimental effect on cancer treatment as a high dose or high frequent doses are often required for the treatment of cancer. To have an effective therapeutic strategy combating MDR is crucial. Nanoparticulate delivery systems are one of the best approaches among other possible drug delivery approaches. Most of the MDR reversal agents are hydrophobic in nature with additional unfavorable characteristics. Therefore, selective and elegant delivery of such molecules to the target site is only possible by means of polymeric nanoparticles. The various categories of such approaches are direct encapsulation of the drug in the polymeric core, multidrug-encapsulated polymeric nanoparticles conjugated with the targeting entity, and conjugation of the drug in the polymeric backbone before nanoparticle synthesis. Cisplatin-loaded poly(L-glutamic acid)-*g*-methoxy poly(ethylene glycol 5K) (PLG-*g*-mPEG5K) nanoparticles (CDDP-NPs) were characterized and exploited for the treatment of NSCLC. *In vitro* metabolism experiments were studied with HeLa cells and this suggests that the poly(glutamic acid) backbone of the PLG-*g*-mPEG5K is biodegradable. *In vivo* experiments with a Lewis lung carcinoma (LLC) model showed that the CDDP-NPs suppressed the growth of tumors. In addition, LLC tumor-bearing mice treated with the CDDP-NPs (5 mg/kg cisplatin eq.) showed much longer survival rates (median survival time of 51 days) as compared with mice treated with free cisplatin (median survival time of 18 days), due to the acceptable antitumor efficacy and low systemic toxicity of CDDP-NPs (Shi et al., 2015). In another study, Majumdar et al. (2014) formulated a polymer-encapsulated luteolin nanoparticle (nano-luteolin), which is water soluble, and studied its anticancer properties against squamous cell carcinoma of the head and neck (SCCHN, Tu212) and NSCLC (H292) cells. *In vitro* studies have shown that like luteolin, nano-luteolin can also inhibit the growth of H292 and Tu212 cells. *In vivo* studies using tumor xenograft mouse models have demonstrated that nano-luteolin has considerable inhibitory effect on tumor growth of SCCHN (Majumdar et al., 2014). Maya et al. (2014) developed chitosan cross-linked γ-poly(glutamic acid) (γ-PGA) nanoparticles loaded with DTX and decorated with cetuximab (CET) to target epidermal growth factor receptor (EGFR)-overexpressing NSCLC cells. EGFR-positive A549 cells significantly increased the uptake of CET-DTXL-γ-PGA nanoparticles. CET-DTXL-γ-PGA nanoparticles showed superior antiproliferative activity over non-targeted nanoparticles. CET-DTXL- γ-PGA nanoparticles inhibited cell division by the G2/M cell cycle in A549 cells. Targeted nanoparticles showed EGFR-specific accumulation in NSCLC cells.

1.3.1 Polymeric Nanoparticles for Targeted Therapy

The concept of targeted therapy appeared in the late 1970s with the development of antibodies (Schrama et al., 2006), and the application of targeted nanoparticles appeared later using immunoliposomes (Heath et al., 1980; Leserman et al., 1980). Advances in proteomics and bioinformatics in relation to cancer biology have allowed the development of targeted therapies. There are two main approaches of targeting: active targeting and passive targeting.

1.3.1.1 Passive Targeting

Passive targeting exploits the principle of the EPR effect. The EPR effect is time dependent and often takes more than 6 hours to accumulate the drug in the tumor tissue, where it resides for an extended period of time ranging from hours to several days. Tumor tissues are characterized by defective and chaotic hypervascularity, resulting in hyperpermeability and leakiness. This specific nature allows nano-carriers, especially polymeric nanoparticles, to diffuse into the tumor tissue. Further studies demonstrate that the EPR effect is potentiated by various factors such as vascular endothelial growth factor (VEGF), bradykinin, nitric oxide (NO), prostaglandin, peroxynitrite, and matrix metalloproteinase. The EPR effect was seen in the majority of human solid tumors, with the exception of poorly vascularized tumors, such as in prostate and pancreatic cancer. The metastatic nature of tumors favors high macrovascular density. VEGF and other inflammatory mediators also influence the penetration and accumulation of nanoparticles inside the tumor. Therefore, the design and development of a polymeric nanoparticle-based drug delivery system should be established based on the macrovascular density, cancer type, and secretion of permeability factors such as VEGF in order to exploit passive targeting through the EPR effect. This phenomenon is best described in Figure 1.1.

1.3.1.2 Active Targeting

Active targeting works on the principle that tumor tissues are composed of a wide variety of cells and those cell surfaces carry unique molecular markers. These markers are either not expressed in normal cells or are expressed at much lower levels than in tumor cells. Biomarkers are a variety of molecules that include mutant genes, RNAs, proteins, lipids, and even small metabolite molecules.

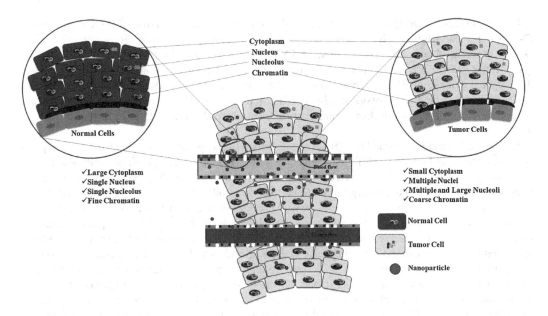

Cytoplasm
Nucleus
Nucleolus
Chromatin

Normal Cells

Tumor Cells

✓Large Cytoplasm
✓Single Nucleus
✓Single Nucleolus
✓Fine Chromatin

Blood flow

✓Small Cytoplasm
✓Multiple Nuclei
✓Multiple and Large Nucleoli
✓Coarse Chromatin

Normal Cell

Tumor Cell

Nanoparticle

FIGURE 1.1 Passive targeting by nanoparticles exploiting the EPR effect.

There are numerous widely-accepted methodologies and technologies available for screening potential tumor markers. Sophisticated technologies such as complementary (c)DNA microarrays, tissue microarrays, and immunohistochemical validations play a significant role in the molecular analysis of tumors and the identification of biomarkers enriched with valuable information on the behavior of the tumor, along with the stage, grade, clinical course, prognosis, and treatment efficacy. This helps in the successful targeting of drug moieties to the cancer cells by identifying the unique molecular targets. Various targeting approaches include:

1. Tyrosine kinase receptor targeting: HER2 is an example of a tyrosine kinase receptor belonging to the class ErbB/HER receptor family, which is highly expressed in numerous types of cancer.

2. EGFR targeting nanoparticles: Similar to HER2, these are also representative of the HER family. Surface functionalization of nanoparticles with various ligands, such as epidermal growth factor (EGF) and anti-EGFR antibodies (CET, Erbitux®), is the basis for designing the receptor-targeted approaches for tumors expressing EGFR.

3. Prostate cancer specific membrane antigen (PSMA) and polymeric nanoparticles: Aptamer-conjugated nanostructures are more studied than anti-PSMA antibodies for targeted delivery.

4. Folate receptor (FR) targeting: The FR is overexpressed in several epithelial malignancies, especially in gynecological cancers such as breast cancer. Surface modification of drug-loaded polymeric nanoparticles (PNPs) with folate, having a high binding affinity for FR (Kd ~ 10–10 M), may actively target the tumor cells and thus is likely to enhance the therapeutic potential of potent drugs.

Thus, pharmacokinetic properties of potent drugs can be altered by encapsulating them in the polymeric core. A few examples of research outcomes of polymeric nanoparticles containing anticancer drugs are highlighted hereunder.

In a research finding, Hu et al. (2010) reported a longer elimination half-life and an improved pharmacokinetic profile of Endostar® (a novel recombinant human endostatin, used for the treatment of NSCLC) when encapsulated in PEG-PLGA nanoparticles. Similarly, Park et al. (2009) investigated free doxorubicin and PEGylated polymeric nanoparticles loaded with doxorubicin and reported a several times increase in efficacy as well as a marked reduction in dose-related cardiac toxicity and side effects.

In another study, Yu and his team (2013) reported that the blood circulation time of DTX poly(lactide)-D-tocopheryl polyethylene glycol succinate (PLA-TPGS) nanoparticles was significantly extended compared to that of Taxotere® (a conventional DTX formulation), with a 13.2-fold longer half-life, 8.51-fold Mean Residence Time (MRT), 2.23-fold higher AUC, and substantially lower value of CL (41.9%). The targeted delivery of PTX to tumor neo-vasculature by peptide-conjugated PTX-loaded PNPs was investigated by Yu and his co-workers (2013). They reported the AUC–time curve of the polymeric nanoparticle is 30 times more than compared to the commercially available PTX formulations (i.e., Taxol®) (Yu et al., 2013).

Similar studies were performed by Milane and his group to encapsulate ionidamine and PTX in polymeric nanoparticles targeting EGFR. The bioactives were assessed by using high pressure liquid chromatography (Milane, 2011). The targeted PNPs demonstrated a superior drug pharmacokinetic profile relative to drug solution and non-targeted particles. Another study demonstrated that the uptake of folic acid conjugated doxorubicin by HeLa cells showed greater cytotoxicity when compared to non-folate-mediated nanoparticles. Folate-conjugated nanoparticles of doxorubicin exhibited higher uptake and cytotoxicity in FR-overexpressed ovarian cancer cells (SKOV3) in comparison with the pure doxorubicin and unmodified nanoparticles. Thus, a careful, systematic understanding of the properties of polymeric nanoparticles plays a significant role in the design development and *in vivo* performance of experimental nanoparticles. In recent laboratory experiments performed by our group, we tried to prepare antifungal drug-loaded PLGA nanoparticles and to deliver them to the lungs. The nanoparticles were prepared by using the multiple emulsion solvent evaporation technique with some modifications. The experimental nanoparticles showed their average particle diameter in the range of 232–500 nm with negative zeta potential values of −6.91 mV to −13.4 mV and voriconazole-loading capacities of 1.26–4.3% (w/w). Morphological study by scanning electron microscopy and transmission electron microscopy images (shown in Figure 1.2) of the nanoparticles further supports the data obtained from the Differential Light Scattering (DLS) experiment. The nanoparticles and the free drug were radiolabeled with technetium-99 m with 90% labeling efficiency, and the radiolabeled particles were administered to investigate the effect on their blood clearance, biodistribution, and *in vivo* gamma imaging. The *in vivo* deposition of the drug in the lobes of the lung was studied by LC-MS/MS study. The free drug was found to be excreted more rapidly than the nanoparticle-containing drug following the inhalation route as assessed by a gamma scintigraphy study. Thus, by rationally designing nanoparticles based on our improved knowledge of cancer biology and the tumor microenvironment, improved efficacy can be achieved.

1.3.2 Multifunctional Polymeric Nanoparticles

The versatility of polymeric nanocarriers opens up possibilities to incorporate combinations of drugs into a single drug delivery system. Concurrent administration of multiple drugs with a varied mechanism of actions will combat the development of drug resistance, which is the main cause of treatment failure

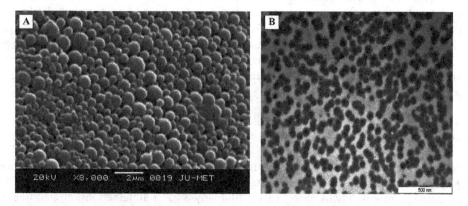

FIGURE 1.2   Scanning electron microscopic (A) and transmission electron microscopic (B) images of an antifungal drug-loaded PLGA nanoparticle, used for pulmonary delivery.

in most cases. The initial success of combination therapy using multifunctional polymeric nanoparticles caused a widespread acknowledgement for it to become a common consideration in current cancer therapy. The idea has also been exploited beyond combination chemotherapy to combine drugs with entirely distinct pharmacological targets (e.g., combinations of chemotherapeutic agents with angiogenesis inhibitors, protease inhibitors, immunotherapeutics, hormone therapeutics, and modulators of MDR); therapies largely stemming from advances in cancer molecular and cell biology leading to the identification of alternate therapeutic targets.

1.3.2.1 Simple Multiple Functionality of Polymeric Nanoparticles

Polymeric nanoparticles can be easily manipulated to improve the function of the nanoparticle through the attachment of tumor-specific target moieties such as antibodies or target ligands directed towards the cell surface markers unique to the tumor cell. Alternatively, simultaneous delivery of two drugs to the cancer cell is also possible to improve the biological properties of polymeric nanoparticles. However, there are several factors that must be addressed properly for the successful translation of nanoparticulate drug delivery systems from the laboratory to the clinic. One such factor is ratiometric drug loading. Nanoparticle-based drug delivery systems unify the pharmacokinetics of different drugs by administrating for the target site. This will maintain the correlation between *in vitro* and *in vivo* studies. Another factor is the sequential drug release. Many combination therapies show synergistic effects or adverse effects. This can be overcome by sequential delivery as well as scheduling of combinatorial drugs. An example is the release of ibandronate followed by tamoxifen in the treatment of breast cancer. Coadministration of antiangiogenic and cytotoxic drugs greatly enhances the cancer therapy. Examples include the combination of VEGF receptor 2 (VEGFR2) inhibitor JNJ-17029259 with PTX and doxorubicin (Emanuel et al., 2004), the use of monoclonal antibodies against VEGFR2 (ZD-6474, SU-6668, IMC-1121), or anti-VEGF (bevacizumab) with cisplatin and gemcitabine, and the combination of COX-2 inhibitors, which indirectly inhibit VEGF function, with temozolomide. The administration of bevacizumab in combination with standard chemotherapeutics like fluorouracil increases the survival rate of patients suffering from metastatic colon cancer. Similarly, drug delivery is also possible by encapsulating chemotherapeutics and multidrug resistance modulators (e.g., verapamil, ciclosporin A). Soma et al. (2000, p388) demonstrated the success of this strategy by successfully reversing MDR in monocytic leukemia cell lines by coadministrating doxorubicin and ciclosporin A in poly alkyl cyanoacrylate nanoparticles. These reports encourage the need for the development of multifunctional nanoparticles to successfully combat cancer that has attained a resistance to chemotherapeutic agents.

1.3.2.2 Complex Multiple Functionality of Nanoparticles

Complex multiple functionality includes imaging and treating tumor cells for the treatment of cancer. Superparamagnetic iron oxide (magnetite) nanoparticles are generally used for drug delivery because of their biocompatibility. One more advantage of this material is that it exhibits superparamagnetism. Superparamagnetism is the property that allows for stability and individual dispersion of the particles after the external magnetic field has been removed. Harisinghani et al. (2003) reported that unmodified iron oxide nanoparticles allowed for 90.5% detection of lymph node metastasis in patients with prostate cancer, as opposed to 35.4% detection using conventional MRI.

Successful exploitation of certain stimuli-responsive polymeric nanocarriers undergo thermodynamically reversible lower critical solution temperature (LCST) phase transition, also known as inverted phase transition (Meyer et al., 2001). Poly(*N*-isopropyl acrylamide) and certain elastin-like peptides are ideally suited for such thermally-targeted drug delivery in cancer.

1.4 Challenges for Nanoparticle-Based Drug Delivery

Nanoparticle-based medicine has infinite potential with novel applications continuously being developed for use in cancer diagnosis, imaging, detection, and treatment. Numerous nanoparticle-based experimental therapeutics for lung cancer utilize different approaches, balancing the design with targeting and imaging moieties and anticancer agents. It is expected that the current research work efforts in nanomedicine will continue to move towards safe, efficient, and feasible drug delivery, and highly sensitive and improved imaging agents for diagnostic and disease monitoring applications. However, nanomedicine research is facing a lot of challenges during the development of novel ideas and in translating them into clinical practice. Particle size distribution, structure, biocompatibility, and surface chemistry are the possible risk factors in the biological environment. A number of obstacles including immune reaction, efficiency in targeting, rate of clearance from circulation, and ability to cross biological barriers will follow when these nanoparticle systems enter the preclinical and clinical testing arenas. It is very difficult to determine the particle–particle interaction within a biological environment and the intracellular trafficking of nanoparticles. Particle sizes of more than 500 nm are not recommended for IV administration because these particles are rapidly eliminated from circulation. Particle sizes of less than 200 nm with a spherical shape and smooth texture can be easily transported through tumor vasculature and into tumor cells. Such physical characteristics are favorable to nanoparticles in utilizing the EPR effect associated with solid tumors. In lung cancer, the EPR effect plays an important role in determining the efficacy of the nanoparticle-based drug delivery system (Babu et al., 2013). Moreover, the poor lymphatic flows in the tumor tissue are advantageous to this EPR effect and result in enhanced retention of nanoparticles within the tumor site. Particle–particle interactions and aggregation tendencies are mostly dependent on the zeta potential of the nanoparticles. Positively charged nanoparticles have an increased affinity for the negatively charged cellular membranes of all cells in the body. Poor stability of nanoparticle systems has been attributed to their aggregation tendencies in the physiological environment. Once aggregated, it is almost impossible to redisperse the particles into their original distribution pattern; while shear forces can be used to redisperse the particles, this may affect the drug loading and therapeutic efficiencies.

Nanotoxicology research has shown that the interactions between nanomaterials and cells, animals, humans, and the environment are remarkably complex. As a matter of interest, increased scientific contributions to the field of nanomedicine have resulted in the emergence of a new area of research: nanotoxicology (Hubbs et al., 2011). Even nanoparticles having strong therapeutic properties towards various types of cancer in preclinical studies carry the risk of inducing toxicity to normal cells. Here mentioned are some inorganic and metal-based nanoparticles that induce toxicity to normal cells. In the past few years, several studies have investigated the toxic effects of silica nanoparticles *in vitro* and *in vivo*. For example, Napierska et al. (2009) observed that 15 nm silica nanoparticles could cause cytotoxic damage and decrease cell survival in human endothelial cells, McCarthy et al. (2012) found that silica nanoparticles in their amorphous state have the potential to cause inflammatory reactions in target organs resulting in apoptotic cell death, and Chang et al. (2007) observed that high dosages of silica nanoparticles induce cell membrane damage. Therefore, the use of silica-based nanoparticles for cancer therapy is limited to low concentrations (0.1 mg/mL) *in vitro*. There is still limited knowledge about the effects of silica nanoparticles on tumor cells. The toxicity of titanium oxide (TiO_2) nanoparticles towards healthy cells is also a matter of concern considering its biological applications (Trouiller et al., 2009). Research evidence suggests that TiO_2 nanoparticles may possess higher toxicity potential than their bulk materials (Long et al., 2007; Magaye et al., 2012; Zhao et al., 2009). Zhao et al. (2009) found that TiO_2 nanoparticles caused higher cytotoxicity than fine particles in cell culture. Due to their very small size, nanoparticles can penetrate basic biological structures, which may, in turn, disrupt their normal function (Buzea et al., 2007; Zhao et al., 2009). Li et al. (2008) observed that the erythrocytes treated with TiO_2 nanoparticles underwent abnormal sedimentation, hemagglutination, and hemolysis, which was totally different from those treated with TiO_2 fine particles. Carbon nanotubes have also been reported to exhibit toxicity to normal cells. Muller et al. (2008) and Ye et al. (2009) reported similar findings with SWNTs on the proliferation of A549 lung cancer cells and rat lung epithelial cells respectively. Carbon nanotubes upon interaction with live cells generate reactive oxygen species causing mitochondrial dysfunction and lipid peroxidation (Manke et al., 2013). Therefore, stringent *in vitro* and *in vivo* toxicity studies for each of the novel nanoparticle systems must be conducted to ensure safety prior to their application in humans.

1.5 Future Aspects of Nano-Size Drug Delivery Systems

A new paradigm is being established in the field of pharmaceutical sciences due to the advancement of nano-size drug delivery systems. The additive effect of science and technology leads to a new era of hope for humanity where medicines will elicit therapeutic effect with low doses, impose good bioavailability, and show stability towards all environmental conditions. Despite the huge progress in the area of nano drug delivery, few numbers of nanoformulation are approved by the various FDA departments. Although nanoformulations possess lots of advantages because of their unique physicochemical properties, there are many clinical, toxicological, and regulatory aspects which are matters of concern too. The knowledge of biocompatibility of nanomaterials is an important issue because of the effect of the nanomaterials in the body, ranging from cytotoxicity to hypersensitivity (Chakroborty et al., 2013). The host response to a specific nanomaterial should also be clinically transparent (Couvreur et al., 2013). Therefore, it is of prime importance to introduce more cost effective, better, and safer nano-biomaterials which will provide efficient drug loading and controlled drug release of some challenging drug moieties for which there is no other suitable delivery available yet.

1.6 Conclusion

Nanoparticle drug delivery systems are already helping to address key issues with traditional anticancer agents, such as low therapeutic efficiencies, poor bioavailability, nonspecific targeting, untoward side effects, and drug resistance, as well as surpassing their predecessors with the ability to detect early metastasis. Nanoparticle-based therapy must overcome the same hurdles faced by any new drug: optimal design of components and properties, reproducible manufacturing processes, institution of analysis methods for sufficient characterization, favorable pharmacology and toxicity profiles, and demonstration of safety and efficacy in clinical trials. Many clinical trials are ongoing to define the roles of targeted therapies along with relevant biomarkers, and to optimize regimens of multimodality therapies in different settings in lung cancer. It will be exciting to see the results from these trials during the next few years and whether truly significant improvements in overall and progression-free survival can be achieved.

Research continues regarding the development and optimization of PNPs for the treatment of cancer. While many non-tumor-targeted PNPs have already made their significant presence felt in today's commercial market, more focus has now been shifted to tumor-targeted PNPs. Though in recent years many works have been published on pharmacokinetics and the *in vivo* fate of PNPs, the definitive reports that establish the correlation of size, shape, and surface characteristics to their biodistribution, excretion, and adverse effects are still not enough. The exact mechanisms of the endocytosis and degradation pathways of PNPs are still under a cloud. Again, there is no specific research article available which has quantified and validated the actual amount of drug distribution in the cancer tissue and in the normal tissue between targeted and non-targeted PNPs. As the challenge for the production and optimization of tumor-targeted PNPs in industry needs huge expenditure, the detailed data of the *in vivo* journey of targeted PNPs relative to non-targeted PNPs may further help to open the doors for their large-scale production.

REFERENCES

Almeida JD, Brand CM, Edwards DC, Heath TD. Formation of virosomes from influenza subunits and liposomes. Lancet. 1975;2(7941):899–901.

Amiot CL, Xu S, Liang S, Pan L, Zhao JX. Near-infrared fluorescent materials for sensing of biological targets. Sensors. 2008;8(5):3082–105.

Babu A, Templeton AK, Munshi A, Ramesh R. Nanoparticle-based drug delivery for therapy of lung cancer: Progress and challenges. J Nanomater [Internet]. 2013;863951:11. Available from: http://dx.doi.org/10.1155/2013/863951.

Beck-Broichsitter M, Ruppert C, Schmehl T, Guenther A, Betz T, Bakowsky U, Seeger W, Kissel T, Gessler T. Biophysical investigation of pulmonary surfactant surface properties upon contact with polymeric nanoparticles in vitro. Nanomedicine. 2011;7(3):341–50.

Bhirde AA, Patel V, Patel J, Zhang G, Sousa AA, Masedunskas A, Leapman RD, Weigert R, Gutkind JS, Rusling JF. Targeted killing of cancer cells in vivo and in vitro with EGF-directed carbon nanotube-based drug delivery. ACS Nano. 2009;3(2):307–16.

Brown SD, Nativo P, Smith JA, Stirling D, Edwards PR, Venugopal B, et al. Gold nanoparticles for the improved anticancer drug delivery of the active component of oxaliplatin. J Am Chem Soc. 2010;132(13):4678–84.

Buhleier E, Wehner W, Vögtle F. Cascade-chain-link and nonskid-chain-link syntheses of molecular cavity topologies. Synthesis. 1978;2:155–8.

Buzea C, Pacheco II, Robbie K. Nanomaterials and nanoparticles: Sources and toxicity. Biointerphases. 2007;2(4):17–71.

Chakroborty G. Nanoparticles and nanotechnology: Clinical, toxi-cological, social, regulatory and other aspects of nanotechnology. JDDT. 2013;3(4):138–41.

Chang JS, Chang KL, Hwang DF, Kong ZL. In vitro cytotoxicity of silica nanoparticles at high concentrations strongly depends on the metabolic activity type of the cell line. Environ Sci Techonol. 2007;41(6):2064–8.

Chen Y, Sha X, Zhang W, Zhong W, Fan Z, Ren Q, Chen L, Fang X. Pluronic mixed micelles overcoming methotrexate multidrug resistance: *In vitro* and *in vivo* evaluation. Int J Nanomedicine. 2013;8:1463–76.

Couvreur P. Nanoparicles in drug delivery: Past, present and future. Adv Drug Deliv Rev. 2013;65(1):21–3.

De Jong WH, Borm PJA. Drug delivery and nanoparticle: Applications and hazards. Int J Nanomedicine. 2008;3(2):133–49.

Dhar S, Daniel WL, Giljohann DA, Mirkin CA, Lippard SJ. Polyvalent oligonucleotide gold nanoparticle conjugates as delivery vehicles for platinum (IV) warheads. J Am Chem Soc. 2009;131(41):14652–3.

Discher BM, Won YY, Ege DS, Lee JC, Bates FS, Discher DE, Hammer DA. Polymersomes: Tough vesicles made from diblock copolymers. Science. 1999;284(5417):1143–6.

Dreaden EC, Mwakwari SC, Sodji QH, Oyelere AK, El-Sayed MA. Tamoxifen_poly(ethylene glycol)_thiol gold nanoparticle conjugates: Enhanced potency and selective delivery for breast cancer treatment. Bioconjug Chem. 2009;20(12):2247–53.

Emanuel S, Gruninger RH, Fuentes-Pesquera A, Connnolly PJ, Seamon JA, HaZel S,et al. A vascular endothelial growth factor receptor-2 kinase inhibitor potentiates the activity of the conventional chemotherapeutic agents paclitaxel and doxorubicin in tumor xenograft models. Mol Pharmacol. 2004;66(3):635–47.

Etzerodt A, Maniecki MB, Graversen JH, Møller HJ, Torchilin VP, Moestrup SK. Efficient intracellular drug-targeting of macrophages using stealth liposomes directed to the hemoglobin scavenger receptor CD163. J Control Release. 2012;160(1):72–80.

Gangrade SM. Nanocrystals: A way for carrier free drug delivery. Pharma Buzz. 2011;6:26–31.

Glück R, Mischler R, Finkel B, Que JU, Scarpa B, Cryz SJ Jr. Immunogenicity of new virosome influenza vaccine in elderly people. Lancet. 1994;344(8916):160–3.

Hadjipanayis CG, Machaidze R, Kaluzova M, Wang L, Schuette AJ, Chen H, Wu X, Mao H. EGFRvIII antibody-conjugated iron oxide nanoparticles for magnetic resonance imaging-guided convection-enhanced delivery and targeted therapy of glioblastoma. Cancer Res. 2010;70(15):6303–12.

Han Y, Zhang P, Chen Y, Sun J, Kong F. Co-delivery of plasmid DNA and doxorubicin by solid lipid nanoparticles for lung cancer therapy. Int J Mol Med. 2014;34(1):191–6.

Harisinghani MG, Barentsz J, Hahn PF, Deserno WM, Tabatabaei S, Van de Kaa CH, de la Rosette J, Weissleder R. Noninvasive detection of clinically occult lymph-node metastases in prostate cancer. N Engl J Med. 2003;348(25):2491–9.

Hawker CJ, Frechet JM. Preparation of polymers with controlled molecular architecture. A new convergent approach to dendritic macromolecules. J Am Chem Soc. 1990;112:7638–47.

Heath TD, Fraley RT, Papahdjopoulos D. Antibody targeting of liposomes: Cell specificity obtained by conjugation of F(ab')2 to vesicle surface. Science. 1980;210(4469):539–41.

Helenius A, Fries E, Kartenbeck J. Reconstitution of smliki forest virus membrane. The Journal of cell biology. 1977;75:866–80.

Hu S, Zhang Y. Endostar-loaded PEG-PLGA nanoparticles: *In vitro* and *in vivo* evaluation. Int J Nanomedicine. 2010;5:1039–48.

Huang H, Yuan Q, Shah JS, Misra RDK. A new family of folate-decorated and carbon nanotube-mediated drug delivery system: Synthesis and drug delivery response. Adv Drug Deliv Rev. 2011;63:1332–9.

Hubbs AF, Mercer RR, Benkovic SA, Harkema J, Sriram K, Sahwegler-Berry D, et al. Nanotoxicology- a pathologist's perspective. Toxicol Pathol. 2011;39(2):301–24.

Huckriede A, Bungener L, Stegmann T, Daemen T, Medema J, Palache AM, Wilschut J. The virosome concept for influenza vaccines. Vaccine. 2005;23(S1):S26–38.

Kaminskas LM, McLeod VM, Ryan GM, Kelly BD, Haynes JM, Williamson M,et al. Pulmonary administration of a doxorubicin-conjugated dendrimer enhances drug exposure to lung metastases and improves cancer therapy. J Control Release. 2014;183:18–26.

Kim HJ, Ishii T, Zheng M, Watanabe S, Toh K, Matsumoto Y, Nishiyama N, Miyata K, Kataoka K. Multifunctional polyion complex micelle featuring enhanced stability, targetability, and endosome escapability for systemic siRNA delivery to subcutaneous model of lung cancer. Drug Deliv Transl Res. 2014;4(1):50–60.

Kim SC, Kim DW, Shim YH, Bang JS, Oh HS, Wan Kim S, Seo MH. *In vivo* evaluation of polymeric micellar paclitaxel formulation: Toxicity and efficacy. J Control Release. 2001;72(1–3):191–202.

Krishnamurthy S, Ng VW, Gao S, Tan MH, Yang YY. Phenformin- loaded polymeric micelles for targeting both cancer stem cells in vitro and in vivo. Biomaterials. 2014;35(33):9177–86.

Langer R. Drug delivery and targeting. Nature. 1998;392(6697 suppl):5–10.

Lee JS, Ankone M, Pieters E, Schiffelers RM, Hennink WE, Feijen J. Circulation kinetics and biodistribution of dual-labeled polymersomes with modulated surface charge in tumor-bearing mice: Comparison with stealth liposomes. J Control Release. 2011;155(2):282–8.

Lentacker I, Geers B, Demeester J, De Smedt SC, Sanders NN. Design and evaluation of Doxorubicin-containing microbubbles for ultrasound-triggered Doxorubicin delivery: Cytotoxicity and mechanisms involved. Mol Ther. 2010;18(1):101–8.

Leserman LD, Barbet J, Kourilsky F, Weinstein JN. Targeting to cells of fluorescent liposomes covalently coupled with monoclonal antibody or protein A. Nature. 1980;288(5791):602–4.

Li J, Chen YC, Tseng YC, Mozumdar S, Huang L. Biodegradable calcium phosphate nanoparticle with lipid coating for systemic siRNA delivery. J Control Release. 2010;142(3):416–21.

Li SQ, Zhu RR, Zhu H, Xue M, Sun XY, Yao SD, Wang SL. Nanotoxicity of TiO_2 nanoparticles to erythrocyte in vitro. Food Chem Toxicol. 2008;46(12):3626–31.

Li XT, Zhou ZY, Jiang Y, He ML, Jia LQ, Zhao L, Cheng L, Jia TZ. PEGylated VRB plus quinacrine cationic liposomes for treating non-small cell lung cancer. J Dug Target. 2015;23(3):232–43.

Lian T, Ho RJ. Trends and developments in liposome drug delivery systems. J Pharm Sci. 2001;90(6):667–80.

Liu J, Liu J, Chu L, Wang Y, Duan Y, Feng L, Yang C, Wang L, Kong D. Novel peptide–dendrimer conjugates as drug carriers for targeting non-small cell lung cancer. Int J Nanomedicine. 2011;6:59–69.

Liu Z, Tabakman S, Welsher K, Dai H. Carbon nanotubes in biology and medicine: In vitro and in vivo detection, imaging and drug delivery. Nano Res. 2009;2(2):85–120.

Lomas H, Canton I, MacNeil S, Du J, Armes SP, Ryan AJ, Lewis AL, Battaglia G. Biomimetic pH sensitive polymersomes for efficient DNA encapsulation and delivery. Adv Mater. 2007;19:4238–43.

Long TC, Tajuba J, Sama P, Saleh N, Swartz C, Parker J, Hester S, et al. Nanosize titanium dioxide stimulates reactive oxygen species in brain microglia and damages neurons in vitro. Environ Health Perspect. 2007;115(11):1631–7.

Maciollek A, Munteanu M, Ritter H. New generation of polymeric drugs: Copolymer from NIPAAM and cyclodextrin methacrylate containing supramolecular-attached antitumor derivative. Macromol Chem Phys. 2010;211:245–9.

Magaye R, Zhao J. Recent progress in studies of metallic nickel and nickel-based nanoparticles' genotoxicity and carcinogenicity. Environ Toxicol Pharmacol. 2012;34(3):644–50.

Majumdar D, Jung KH, Zhang H, Nannapaneni S, Wang X, Amin AR, et al. Luteolin nanoparticle in chemoprevention: *In vitro* and *in vivo* anticancer activity. Cancer Prev Res. 2014;7(1):65–73.

Manke A, Wang L, Rojanasakul Y. Mechanisms of nanoparticle-induced oxidative stress and toxicity. BioMed Res Int [Internet]. 2013;942916:15 . Available from: http://dx.doi.org/10.1155/2013/942916.

Maya S, Sarmento B, Lakshmanan VK, Menon D, Seabra V, Jayakumar R. Chitosan cross-linked docetaxel loaded EGF receptor targeted nanoparticles for lung cancer cells. Int J Biol Macromol, 2014;69:532–41.

McCarthy J, Inkielewicz-Stepmiak I, Corbalan JJ, Radomski MW. Mechanisms of toxicity of amorphous silica nanoparticles on human using submucosal cells in vitro: Protective effects of fisetin. Chem Res Toxicol. 2012;25(10):2227–35.

McDevitt MR, Chattopadhyay D, Kappel BJ, Jaggi JS, Schiffman SR, Antczak C, et al. Tumor targeting with antibody functionalized, radiolabelled carbon nanotubes. J Nucl Med. 2007;48(7):1180–9.

McMahon KM, Mutharasan RK, Tripathy S, Veliceasa D, Bobeica M, Shumaker DK, et al. Biomimetic high density lipoprotein nanoparticles for nucleic acid delivery. Nano Lett. 2011;11(3):1208–14.

Meyer DE, Shin BC, Kong GA, Dewhirst MW, Chilkoti A. Drug targeting using thermally responsive polymers and local hyperthermia. J Control Release. 2001;74(1–3):213–24.

Milane L, Duan Z, Amiji M. Therapeutic efficacy and safety of Paclitaxel/Ionidamine loaded EFGR-targeted nanopaticles for the treatment of multi-drug resistant cancer. PLoS One. 2011;6(9):e24075.

Mirza AZ, Shamshad H. Preparation and characterization of doxorubicin functionalized gold nanoparticles. Eur J Med Chem. 2011;46(5):1857–60.

Moghimi SM. Recent developments in polymeric nanoparticle engineering and their applications in experimental and clinical oncology. Anticancer Agents Med Chem. 2006;6(6):553–61.

Mufamadi MS, Pillay V, Choonara YE, Du Toit LC, Modi G, Naidoo D, Ndesendo VM. A review on composite liposomal technologies for specialized drug delivery. J Drug Deliv. 2011;939851.

Mulder WJ, Koole R, Brandwijk RJ, Storm G, Chin PT, Strijkers GJ, et al. Quantum dots with a paramagnetic coating as a bimodal molecular imaging probe. Nano Lett. 2006;6(1):1–6.

Muller J, Decordier I, Hoet PH, Lombaert N, Thomassen L, Huaux F, et al. Clastogenic and aneugenic effects of multiwalled carbon nanotube in epithelial cells. Carcinogenesis. 2008;29(2):427–33.

Napierska D, Thomassen LC, Rabolli V, Lison D, Gonzalez L, Kirsch-Volders M, et al. Size-dependent cytotoxicity of monodispersed silica nanoparticles in human endothelial cells. Small. 2009;5(7):846–53.

Nasongkla N, Bey E, Ren J, Ai H, Khemtong C, Guthi JS, et al. Multifunctional polymeric micelles as cancer-targeted, MRI-ultrasensitive drug delivery systems. Nano Lett. 2006;6(11):2427–30.

Paranjpe M, Muller Goymann CC. Nanoparticle-mediated pulmonary drug delivery: A review. Int J Mol Sci. 2014;15(4):5852–73.

Paranjpe M, Neuhaus V, Finke JH, Richter C, Gothsch T, Kwade A, et al. *In vitro* and *ex vivo* toxicological testing of sildenafil loaded solid lipid nanoparticles. Inhal Toxicol. 2013;25(9):536–43.

Park J, Fong PM, Lu J, Russell KS, Booth CJ, Saltzman WM, Fahmy TM. PEGylated PLGA nanoparticles for the improved delivery of doxorubicin. Nanomedicine. 2009;5(4):410–8.

Prato M, Kostarelos K, Bianco A. Functionalized carbon nanotubes in drug design and discovery. Acc Chem Res. 2008;4(11):60–8.

Qu MH, Zeng RF, Fang S, Dai QS, Li HP, Long JT. Liposome-based co-delivery of siRNA and docetaxel for the synergistic treatment of lung cancer. Int J Pharm. 2014;474(1–2):112–22.

Reilly RM. Carbon nanotubes: Potential benefits and risks of nanotechnology in nuclear medicine. J Nucl Med. 2007;48(7):1039–42.

Ren S, Chen M, Jiang M. Noncovalently connected micelles based on a bcyclodextrin containing polymer and adamantane end-capped poly (ecaprolactone) via host–guest interactions. J Polym Sci A Polym Chem. 2009;47:4267–78.

Ruoslahti E, Bhatia SN, Sailor MJ. Targeting of drugs and nanoparticles to tumors. J Cell Biol. 2010;188(6):759–68.

Sahoo SK, Parveen S, Panda JJ. The present and future of nanotechnology in human healthcare. Nanomedicine. 2007;3(1):20–31.

Sahu PK, Mishra DK, Jain N, Rajoriya V, Jain AK. Mannosylated solid lipid nanoparticles for lung-targeted delivery of Paclitaxel. Drug Dev Ind Pharm. 2015;41(4):640–9.

Schrama D, Reisfeld RA, Becker JC. Antibody targeted drugs as cancer therapeutics. Nat Rev Drug Discov. 2006;5(2):147–59.

Sengupta S, Eavarone D, Capila I, Zhao G, Watson N, Kiziltepe T, et al. Temporal targeting of tumour cells and neovasculature with a nanoparticle delivery system. Nature. 2005;436(7050):568–72.

Shi C, Yu H, Sun D, Ma L, Tang Z, Xiao Q, Chen X. Cisplatin-loaded polymeric nanoparticles: Characterization and potential exploitation for the treatment of non-small cell lung carcinoma. Acta Biomater. 2015;18:68–76.

Soma CE, Dubernet C, Bentolila D, Benita S, Couvreur P. Reversion of multidrug resistance by co-encapsulation of doxorubicin and cyclosporin A in polyalkylcyanoacrylate nanoparticles. Biomaterials. 2000;21(1):1–7.

Tang J, Wei H, Liu H, Ji H, Dong D, Zhu D, Wu L. Pharmacokinetics and biodistribution of itraconazole in rats and mice following intravenous administration in a novel liposome formulation. Drug Deliv. 2010;17(4):223–30.

Tomalia DA, Baker H, Dewald J, Hall M, Kallos G, Martin S, et al. A new class of polymers: Starburst-dendritic macromolecules. Polym J. 1985;17:117–32.

Trouiller B, Reliene R, Westbrook A, Solaimani P, Schiesti RH. Titanium dioxide nanoparticles induce DNA damage and genetic instability in vivo in mice. Cancer Res. 2009;69(22):8784–9.

Videira M, Almeida AJ, Fabra A. Preclinical evaluation of a pulmonary delivered paclitaxel-loaded lipid nanocarrier antitumor effect. Nanomedicine. 2012;8(7):1208–15.

Wang RH, Cao HM, Tian B, Wang Q, Ma H, Wu J. Efficacy of dual-functional liposomes containing paclitaxel for treatment of lung cancer. Oncol Rep. 2015;33(2):783–91.

Webster DM, Sundaram P, Byrne ME. Injectable nanomaterials for drug delivery carriers, targeting moieties, and therapeutics. Eur J Pharm Biopharm. 2013;84(1):1–20.

Wilczewska AZ, Niemirowicz K, Markiewicz KH, Car H. Nanoparticles as drug delivery systems. Pharmacol Rep. 2012;64(5):1020–37.

Will H De Jong, Paul JA Borm. Drug delivery and nanoparticle applications and hazards. Int J Nanomedicine, 2008;3(2), 133–149.

Wu Z, Zou X, Yang L, Lin S, Fan J, Yang B, et al. Thermosensitive hydrogel used in dual drug delivery system with paclitaxel-loaded micelles for in situ treatment of lung cancer. Colloids Surf B Biointerfaces. 2014;122:90–8.

Ye SF, Wu YH, Hou ZQ. ROS and NF-kappaB are involved in upregulation of IL-8 in A549 cells exposed to multi-walled carbon nanotubes. Biochem Biophys Res Commun. 2009;379(2):643–8.

Yu Y, Tan S, Zhao S, Zhuang X, Song Q, Wang Y, Zhou, et al. Antitumor activity of docetaxel-loaded polymeric nanoparticles fabricated by Shirasu porous glass membrane- emulsification technique. Int J Nanomed. 2013;8:2641–52.

Zhao J, Bowman L, Zhang X, Vallyathan V, Young SH, Castranova V, Ding M. Titanium dioxide (TiO$_2$) nanoparticles induce JB6 cell apoptosis through activation of the caspase-8/Bid and mitochondrial pathways. J Toxicol Environ Health. 2009;72(19):1141–9.

2

Characterization Techniques of Nanoparticles Applied in Drug Delivery Systems

Vipin Kumar Sharma and Daphisha Marbaniang

CONTENTS

ABSTRACT: Nanoscale materials are having direct and indirect impacts on the wellbeing of society. Nanotechnology is an emerging interdisciplinary area that is expected to have a wide range of implications in all fields of science and technology, such as material science, mechanics, electronics, optics, medicine, plastics, energy, and aerospace, etc. Nanoparticles are of current interest because of an emerging understanding of their possible effects on human health and environmental sustainability, and also owing to the increased output of man-made nanoparticles into the environment.

In the case of drug delivery systems, nanoparticle delivery systems have been proposed as colloidal drug carriers. The key advantages of nanoparticles are (1) improved bioavailability by enhancing aqueous solubility, (2) increasing resistance time in the body (i.e., increasing half-life for clearance, increasing specificity for its cognate receptors), and (3) targeting drugs to specific locations in the body (i.e., their site of action). This results in concomitant reduction in the quantity of the drug required and dosage toxicity, enabling the safe delivery of toxic therapeutic drugs and the protection of non-target tissues and cells from severe side effects. It is increasingly used in different applications, including drug-carrier systems and to pass organ barriers such as the blood–brain barrier, cell membrane, etc. They are based on biocompatible lipids and provide sustained effects by either diffusion or dissolution.

Nanoparticles are used in many different applications and are created by many different processes. The fundamental issue of nanotechnology lies in the fact that the properties of materials change dramatically when their size is reduced to the nanometer range. However, research is going on to synthesize nanostructured and nanophasic materials, and characterizing these nano-sized materials is also an emerging field posing lots of challenges to scientists and technologists.

The formal definition of a nanoparticle is a "nano-object with all three external dimensions in the nanoscale". Nanoparticles are a type of colloidal drug delivery system comprising particles with a size range from 10 to 1,000 nm in diameter. Their origins and properties are highly varied, making their study a rich branch of analytical science. Chemical properties of interest for nanoparticles include total chemical composition, mixing state (internal/external), surface composition, electrochemistry, and oxidation state, etc. Physical properties of interest include number and mass concentration, size, surface area, total mass, morphology, and optical properties, etc. Because of their very high surface area to mass ratio and high surface curvature, nanoparticles may be particularly chemically active.

The origins and properties of nanoparticles are highly varied, making their study a rich branch of analytical science. It is important to have robust analytical approaches for characterizing nanoparticles to maximize the benefits from them while mitigating their impact. The development and characterization of nanoscale entities have motivated an upsurge in research activities on the discovery and invention of techniques to allow a better regulation on morphology, size, and dimensions in the nano-range. The present chapter represents the basics of various nanoparticles applied as drug delivery devices with their benefits, as well as the properties of the materials applied in their fabrication. It also encompasses the applications and principles of different analytical techniques used in nanoparticle characterization.

KEY WORDS: *Drug delivery system, microscopy, nanoparticles, spectroscopy*

2.1 Introduction

Nanoparticles are used in many different applications and are created by many different processes. Their measurement and characterization pose interesting analytical challenges. The formal definition of a nanoparticle is "a nano-object with all three external dimensions in the nanoscale" (ISO, 2008), although in practice the term is often used to refer to particles larger than 100 nm (Gilham and Brown, 2010). The reason for this is that the behavior of nanoparticles and the applicability of measurement techniques vary with size and environment to the extent that 500 nm particles can either be considered very large or very small depending on the frame of reference.

Nanoparticles are produced by a wide range of natural (biogenic) and man-made (anthropogenic) processes. Some of the main biogenic sources of nanoparticles include mineral erosion, atmospheric chemistry processes, and evaporation of sea spray. Anthropogenic sources include combustion, mining/quarrying, and industrial processes, including the deliberate manufacture of engineered nanoparticles.

Their measurement and characterization pose interesting analytical challenges. The properties of nanoparticles often bridge the microscopic and macroscopic regimes, meaning that conventional theories do not necessarily allow us to predict their behavior. It is this uncertainty that lies at the heart of concerns surrounding the health and environmental impact of nanoparticles, but also of the excitement around opportunities for their application in new areas of science and technology. Therefore, it is important to have robust analytical approaches for characterizing nanoparticles, to maximize the benefits from them while mitigating their impact.

2.2 Properties of Materials Applied for Development of Nanostructure-Based Drug Delivery

Chemical properties of interest for those studying nanoparticles include total chemical composition, mixing state (internal/external), surface composition, electrochemistry, and oxidation state. Physical properties of interest include number and mass concentration, size, surface area, total mass, morphology, and optical properties. Because of their very high surface area to mass ratio and high surface curvature, nanoparticles may be particularly chemically active. Particle size has a dramatic effect on the physical properties of a collection of particles. Mass-based measurements are heavily weighted towards the largest particles whereas smaller particles have a much larger surface area per unit mass.

Nanoparticle characterization methods are required to cover a range of requirements from long term environmental monitoring campaigns over entire continents, where only basic properties are measured, to one-off laboratory measurements on specially prepared samples where a full chemical and physical analysis is performed (Novotny and Hecht, 2006). Also, the properties of the materials applied for fabrication of nanoparticles change dynamically on conversion into the nano-range. However, measuring the nano-range precisely is not an easy task. Hence, the research into the opportunities and challenges continues, as run by scientists and technologists.

2.3 Characterization Techniques

The characterization techniques can be subdivided by both general measurements and the phase in which the nanoparticles reside. Measurements of each type present their own difficulties and often have subtly different interpretations. Moreover, comparison of results between phases is very difficult, and matrix effects can be significant due to the high surface area to mass ratio of nanoparticles. The techniques presented in Table 2.1 give a general overview of common measurements made on nanoparticles for a range of applications (Table 2.1). Nanoparticles are generally characterized by their size, morphology, and surface charge, using such advanced microscopic techniques as scanning electron microscopy (SEM),

TABLE 2.1

Analytical Techniques Applied for the Evaluation of Physicochemical Characteristics of Solid Lipid Nanoparticles (SLNs)/Polymeric Nanoparticles (PNPs) (Ganesan and Narayanasamy, 2017; Carina and Barros 2017)

Analytical Methods/Instrumentation	Characterization Parameters
Scanning electron microscopy (SEM) Transmission electron microscopy (TEM) Atomic force microscopy (AFM) Freeze fracture microscopy (FFM) Phase contrast optical microscopy (PCM)	Surface and surface morphology
Atomic force microscopy (AFM) Phase contrast optical microscopy (PCM) Scanning electron microscopy (SEM) Transmission electron microscopy (TEM) Photon correlation spectroscopy (PCS) Scanning tunneling microscope (STM)	Size and size distribution, shape, structure, aggregation, surface properties
Differential scanning calorimetry (DSC)	Physiochemical state and possible interactions of the drug and the polymer
Zeta potential measurement pH sensitive probe	Electrical surface and surface pH, stability referring to surface charge
Laser light/dynamic light scattering (DLS) technique	Surface charge and electrophoretic mobility, hydrodynamic size distribution
Fluorescence spectroscopy	Critical association concentration (CAC) determination, drug content, *in vitro* drug release
X-ray photoelectron spectroscopy (XPS)	Elemental and chemical composition at the surface
Nuclear magnetic resonance (NMR)	Structure, composition, purity, conformational change
Near-field scanning optical microscopy (NSOM)	Size and shape of nanomaterials
Mass spectroscopy (MS)	Molecular weight, composition, structure, surface properties (secondary ion MS)
Infrared spectroscopy (IRS)	Structure and conformation of bioconjugates, functional group analysis
High performance liquid chromatography (HPLC)	Drug content, *in vitro* drug release
Hydrophobic interaction chromatography Two-phase partition Radiolabel probe Contact angle measurement X-ray photoelectron spectroscopy (XPS) Synchrotron radiation X-ray (SAX)	Surface hydrophobicity
Gas pycnometer	Density
Gel permeation chromatography (GPC)	Molecular weight
Viscometer	Rheology
Dialysis membrane dissolution test apparatus	*In vitro* release

transmission electron microscopy (TEM), and atomic force microscopy (AFM). The average particle diameter, their size distribution, and their charge affect the physical stability and the *in vivo* distribution of the nanoparticles. Electron microscopy techniques are very useful in ascertaining the overall shape of polymeric nanoparticles (PNPs), which may determine their toxicity. The surface charge of the nanoparticles affects the physical stability and redispersibility of the polymer dispersion as well as their *in vivo* performance (Kaminskyl and Dahms, 2008; Meier et al., 2004; Mitra et al., 2010).

2.3.1 Solid Phase

Nanoparticles in the solid phase exist either as a powder or encapsulated in a solid medium. The former can take several forms including loose powders and wet or dry "powder cakes" for convenience of handling. As such, any analysis must take into account how the particles will eventually be used because this will affect their final agglomeration state and other properties (Hinds, 1999).

- Size: There are many methods of measuring particle size (or more correctly, size distribution) but electron microscopy is widely used in conjunction with other measurements. Laser diffraction is a common technique for measuring bulk samples under ambient conditions and powder X-ray diffraction may also be used by examining peak broadening.
- Surface area: The most common technique is the nitrogen adsorption technique based on the BET isotherm and is routinely carried out in many laboratories.
- Pour density: This requires the weighing of a known volume of freshly poured powder.
- Composition: Suitable surface techniques include X-ray photoelectron spectroscopy (and other X-ray spectroscopy methods) and secondary ion mass spectrometry. Bulk techniques generally use digestion, followed by conventional wet chemical analyses such as mass spectrometry, atomic emission spectroscopy, and ion chromatography.
- Crystallography: Powder X-ray or neutron diffraction may be readily used to determine the crystal structure of simple lattice structures. It can also be applied to crystalline organic solids, but the data analysis is much more challenging.
- Dustiness: This is the property of a powder to become aerosolized by mechanical agitation. The dustiness of a particular sample is very dependent on its moisture content and static electrical properties.
- Morphology: Particle shape and aspect ratio are most readily determined by image analysis of electron micrographs.

2.3.2 Liquid Phase

Nanoparticles suspended in the liquid phase are often mixed with surfactants to moderate their agglomeration state. Furthermore, a range of chemical or even biological species may be present and affect the results obtained especially in samples taken from the environment. Therefore, care must be taken to prepare the sample to prevent unwanted matrix effects or changes to the sample (Hester and Harrison, 2007).

- Size: Dynamic light scattering (DLS) (or photon correlation spectroscopy (PCS)) and centrifugation are commonly used techniques. Other methods include image-tracking instruments.
- Surface area: Simple titrations may be used to estimate surface area but are very labor intensive. Nuclear Magnetic Resonance (NMR) experiments may also be used, and dedicated instrumentation has recently been developed.
- Zeta potential: This gives information relating to the stability of dispersions. It cannot be measured directly but is often measured indirectly using an electrophoretic method.
- Composition: Chemical digestion of the particles allows a range of mass spectroscopy and chromatography methods to be used. Interference from the matrix needs to be carefully managed.

- Morphology: It is difficult to measure the morphology of particles moving freely in a fluid. Deposition onto a surface for electron microscopy is most commonly used.

2.3.3 Gas Phase (Aerosols)

Nanoparticles in the gas phase may be monitored using a range of commercially available and relatively low-cost equipment. Generally, these instruments are quite robust and can be used for prolonged periods with little attention. They are also generally resistant to matrix effects. However, relative humidity and volatile organic species can sometimes affect measurements. Unlike the condensed phases, aerosols cannot be stored for later analysis and so reproducible sampling is very important. A range of inlets has been designed for different applications to reduce inconsistencies in this respect (Rubahn, 2008).

- Concentration: Usually expressed as a number concentration or a mass concentration. The latter will strongly weight the distribution curve in favor of larger particles – a single 10 μm diameter particle weighs the same as 1 million 100 nm particles!
- Size: There are many methods for measuring particle size but comparability between them is a problem. They include optical and aerodynamic methods, but these give no information about variations in morphology.
- Surface area: There are few routine surface area techniques available. The most common involves charging the aerosol using a corona discharge and measuring the charge concentration. Such methods are usually calibrated against size distribution measurements.
- Charge: Collisions of air ions with particles result in a steady state charge distribution. This distribution is Boltzmann-like, with small particles being much less likely to carry any charge than larger particles. This distribution is usually measured in the laboratory and subsequently utilized by other measurement techniques.
- Morphology: There are few instruments available for measuring morphology. The most common technique is to capture particles either electrostatically or by filtration for subsequent imaging using electron microscopy.
- Composition: Measuring aerosol composition is very challenging owing to the small amounts of matter present. Most speciation methods require the particles to be collected and then subjected to spectrometric or wet chemical techniques.

Besides these, some of other techniques have also been employed for particle analysis such as acoustic attenuation mobility absorption, quasi-elastic light scattering (QELS), laser doppler velocimetry (LDV), laser light diffraction or static light scattering, optical imaging, microelectrophoresis, sedimentation (gravitational and centrifugal), sieving, and tapered element oscillating microbalance (TEOM), etc. (Vasenka et al., 1993).

2.4 Advantages of Nano Drug Delivery Systems

The main objectives of nano drug delivery devices are more specific drug targeting and delivery, reduction in toxicity while maintaining therapeutic effects, greater safety and biocompatibility, and faster development of new safe medicines. The aims for nanoparticle entrapment of drugs are enhanced delivery to, or uptake by, target cells and/or a reduction in the toxicity of the free drug to non-target organs. In all situations, the therapeutic index of drugs is increased. The polymers that coat the nanoparticles mediate the release of the drug by controlling diffusion or erosion from the core. The release of the drug occurs either from the polymer membrane or from the polymeric matrix. When the release is regulated from the polymeric membrane, it acts as a barrier to release and solubility, and the diffusivity of the drug acts as the determining factor in drug release. Also, the auxiliary ingredients used in developing nanoparticle-based drug delivery devices have an impact on the drug release pattern. For an effective nanoparticulate system, drug release and biodegradation are important considerations. The factors governing drug release are (1) the solubility of the drug, (2) the desorption of the surface bound/adsorbed drug, (3) drug diffusion through the nanoparticle matrix, (4) nanoparticle matrix erosion/degradation, and (5) the combination of the erosion/diffusion process (Mohanraj and Chen, 2006). Keeping in view these factors governing drug release from nanoparticles, various devices have been developed as effective drug delivery devices such as fullerenes, solid lipid nanoparticles (SLNs), liposomes, nanostructured lipid carriers (NLC), nanoshells, quantum dots, superparamagnetic nanoparticles, dendrimers, etc. (Mudshinge et al., 2011).

A fullerene is any molecule composed of carbon in the form of a hollow sphere, ellipsoid, or tube. Spherical fullerenes are also called buckyballs and cylindrical ones are called carbon nanotubes or buckytubes. Buckyball clusters, or buckyballs composed of less than 300 carbon atoms, are commonly known as endohedral fullerenes and include the most common fullerene buckminsterfullerene, C_{60}. Nano "onions" are spherical particles based on multiple carbon layers surrounding a buckyball core, which are proposed for lubricants (Sano et al., 2001). Various categories of fullerenes have their applications in delivering different therapeutic agents and in checking the symptoms of different diseases, e.g. human immunodeficiency virus (HIV) protease (Friedman et al., 1993; Sijbesma et al., 1993), HIV-1 and HIV-2 replication (Marchesan et al., 2005), free radicals and oxidative stress (Krusic et al., 1991), liver toxicity and diminished lipid peroxidation (Slater et al., 1985), apoptosis of neuronal cells (Dugan et al., 1997), apoptosis of hepatoma cells (Huang et al., 1998), neurological disease including Parkinson's disease (Dugan et al., 1997), leukemia, and bone cancer (Thrash et al., 1999).

SLNs comprise mainly lipids that are available in solid phase at room temperature and surfactants for emulsification, the mean diameters of which range from 50 nm to 1,000 nm for colloidal drug delivery applications (zur Mhlen and Mehnert, 1998). Solid lipids utilized in SLN formulations include fatty acids (e.g., palmitic acid, decanoic acid, and behenic acid), triglycerides (e.g., trilaurin, trimyristin, and tripalmitin), steroids (e.g., cholesterol), partial glycerides (e.g., glyceryl monostearate and glyceryl behenate), and waxes (e.g., cetyl palmitate). Several types of surfactants are commonly used as emulsifiers to stabilize lipid dispersion, including soybean lecithin, phosphatidylcholine, poloxamer 188, sodium cholate, and sodium glycocholate (Zhang et al., 2010). The advantages of SLNs include improved bioavailability, protection of sensitive drug molecules from the environment (e.g., water and light), controlled and/or targeted drug release (Mehnert and Mäder, 2001; Müller et al., 2000; 2002), improved stability of pharmaceuticals, feasibility of carrying both lipophilic and hydrophilic drugs, and biodegradability of most lipids (Jenning et al., 2000; Müller and Runge, 1998). Various drugs like rifampicin, isoniazid (Pandey and Khuller, 2005), ibuprofen (Panga et al., 2009), tobramycin (Cavalli et al., 2002), doxorubicin (Wong et al., 2006), doxorubicin and paclitaxel (Serpe et al., 2004), and methotrexate and camptothecin (Ruckmani et al., 2006; Yang et al., 1999), etc., have been incorporated in different categories of SLNs for the potential of their respective therapeutic effects.

Nanostructured Lipid Carriers (NLCs) are produced from blends of solid and liquid lipids, but particles are in solid state at body temperature. Drug release from lipid particles occurs by diffusion and, simultaneously, by lipid particle degradation in the body. NLCs accommodate the drug because of their highly unordered lipid structures. A desired burst of drug release can be initiated by applying the trigger impulse to the matrix to convert it into a more ordered structure (Radtke and Müller et al., 2001). The major application areas are in topical, oral, subcutaneous, intramuscular, or intravenous routes of administrations. Various drugs amphotericin B (Takemoto et al., 2004), polymyxin B (Omri et al., 2002), xiprofloxacin (Magallanes et al., 1993), benzyl penicillin (Kim and Jones, 2004), streptomycin (Gangadharam et al., 1991), zidovudine (Kaur et al., 2008), and daunorubicin and doxorubicin (Park, 2002), etc., have been incorporated in different categories of NLCs for their potential effects in different diseases.

Liposomes are vesicular structures with an aqueous core surrounded by a hydrophobic lipid bilayer, created by the extrusion of phospholipids. Solutes, such as drugs, in the core cannot pass through the hydrophobic bilayer, however hydrophobic molecules can be absorbed into the bilayer, enabling the liposome to carry both hydrophilic and hydrophobic molecules. The lipid bilayer of liposomes can fuse with other bilayers such as the cell membrane, which promotes the release of its contents, making them useful for drug delivery and cosmetic delivery applications. Liposomes that have vesicles in the range of nanometers are also called nanoliposomes (Cevc, 1996; Zhang and Granick, 2006). The unique structure of liposomes, a lipid membrane surrounding an aqueous cavity, enables them to carry both hydrophobic and hydrophilic compounds without chemical modification. In addition, the liposome surface can be easily functionalized with "stealth" material to enhance their *in vivo* stability or target ligands to enable preferential delivery of the liposomes. Different drugs like daunorubicin and doxorubicin have been incorporated in liposomes (Park, 2002). Anti-GD2 immunoliposomes, liposomes entrapping fenretinide, and gold-containing liposomes have also been prepared for their effective applications (Di Paolo et al., 2009). Hepatically targeted liposomes containing insulin have been developed for effective use in diabetes mellitus (Spangler, 1990).

Nanoshells are also notorious as core-shells; nanoshells are spherical cores of a particular compound (concentric particles) surrounded by a shell or outer coating of a thin layer of another material, which is 1–20 nm thick (Davies et al., 1998; Liz-Marzan et al., 2001; Templeton et al., 2000; Xia et al., 2000). Nanoshells possess highly favorable optical and chemical properties for biomedical imaging and therapeutic applications. A few applications in the area of imaging and diagnostics are the detection of deoxyribonucleic acid (DNA) (Thaxton et al., 2005), detection of analytes through immunoassay (Hirsch et al., 2003a), detection of cancer cells (Loo et al., 2004), detection of tumors (Hirsch et al., 2003b), detection of antibodies (Kalele et al., 2005), detection of microorganisms (Kalele et al., 2006), and diseases (Sparnacci et al., 2002), etc.

Quantum dots are semiconductor nanocrystals and core shell nanocrystals containing an interface between different semiconductor materials. The size of quantum dots can be continuously tuned from 2 to 10 nm, which, after polymer encapsulation, generally increases to 5–20 nm in diameter. Particles smaller than 5 nm are quickly cleared by renal filtration (Choi et al., 2007a; 2007b). The quantum dot core can serve as the structural scaffold, and the imaging contrast agent and small molecule hydrophobic drugs can be embedded between the inorganic core and the amphiphilic polymer coating layer. Hydrophilic therapeutic agents, including small interfering ribonucleic acid (siRNA) and antisense oligodeoxynucleotide (ODN), and targeting biomolecules, such as antibodies, peptides, and aptamers, can be immobilized onto the hydrophilic side of the amphiphilic polymer via either covalent or noncovalent bonds.

Superparamagnetic molecules are those that are attracted to a magnetic field but do not retain residual magnetism after the field is removed. Nanoparticles of iron oxide with diameters in the 5–100 nm range have been used for selective magnetic bioseparations. Typical techniques involve coating the particles with antibodies to cell-specific antigens, for separation from the surrounding matrix. The main advantages of superparamagnetic nanoparticles are that they can be visualized in magnetic resonance imaging (MRI) due to their paramagnetic properties, they can be guided to a location by the use of a magnetic field, and they can be heated by a magnetic field to trigger the drug release (Irving, 2007). Various applications are associated with superparamagnetic nanoparticles, such as use as MRI contrast agents for detecting liver tumors (Smith et al., 2007), identifying dangerous arteriosclerotic plaques by MRI (zur Mühlen et al., 2007; Smith et al., 2007), use as enhanced MRI contrast agents in breast cancer xenografts and metastases in the lungs (Meng et al., 2009), measuring macrophage burden in atherosclerosis (Morishige et al., 2010), magnetic fluid hyperthermia (MFH) in cancer treatment (Jordan et al., 1999), and purification of plasmid DNA from bacterial cells (Chiang et al., 2005), etc.

Dendrimers are unimolecular, monodispersive, micellar nanostructures, around 20 nm in size, with a well-defined, regularly-branched symmetrical structure and a high density of functional end groups at their periphery. Dendrimer vectors are most commonly used as parenteral injections, either directly into the tumor tissue or intravenously for systemic delivery (Tomalia et al., 2007). Dendrimers used in drug delivery studies typically incorporate one or more of the following polymers: polyamidoamine (PAMAM), melamine, poly L-glutamic acid (PG), polyethylene imine (PEI), polypropylene imine (PPI), polyethylene glycol (PEG), and chitin. There are several potential applications for dendrimers in the field of imaging, drug delivery, gene transfection, and non-viral gene transfer. Some of the applications of dendrimers are to diagnose certain disorders of the heart, brain, and blood vessels (Wiener et al., 1994), strep throat (*Streptococcus*), staph infection (*Staphylococcus aureus*), and the flu (*Haemophilus influenza*) (Abeylath et al., 2008; Ma et al., 2007), to use as an antifungal against *Candida albicans*,

Aspergillus niger and *Sachromyces cerevasae* (Khairnar et al., 2010), to use as an antiviral against HIV, herpes simplex virus (HSV), and sexually transmitted infections (Rupp et al., 2007), to induce a systemic antitumor immune response against residual tumor cells (Culver, 1994), and to use as a diagnostic tool for arteriosclerotic vasculatures, tumors, infarcts, kidney complications, or efferent urinary tract complications (Schumann et al., 2003), etc.

2.5 Instrumental Techniques for Nanoparticle Characterization

2.5.1 Scanning Electron Microscopy (SEM)

The morphological examination can be performed by direct visualization of the SEM. There are numerous areas of advantage associated with this technique for an estimation of sizing analysis and morphological characterization. However, limited information can be obtained regarding size distribution and the true population average. For SEM studies, the solution containing the nanoparticles should be converted into dry powder and then mounted on a sample holder. If the material is non-conductive, it is changed into conductive material by metallic coating, with such as gold, in a sputter coating chamber. The nanomaterial should have the capacity to withstand the vacuum as in many cases the applied electron beam can damage the polymeric materials. The mean particle size data obtained by this technique is then compared with data obtained from DLS. Generally, this technique is time consuming, costly, and sometimes requires basic information regarding size-distribution (Molpeceres et al., 2000).

In this technique, the images of the samples are acquired by scanning with the help of high energy beams of electrons. Where a conventional microscope is used, a series of glass lenses are applied to bend the light waves and a magnified image is produced. But on the other hand, in SEM, the magnified image is created by high energy electrons.

2.5.1.1 Basic Principle

The beam of high energy electrons strikes the surface of the specimen. It interacts with the atoms of the sample and creates a signal in the form of secondary electrons, or back-scattered electrons. By these high energy electrons, the characteristic X-rays are generated that have the information regarding the sample's morphology and composition, etc. By using this technique, very high-resolution images of a sample surface can be obtained that may reveal details about particles 1–5 nm in size in its primary detection mode. The X-rays generated during the striking of the sample with high energy electrons are the second most common imaging mode. These X-rays are applied as a diagnostic tool in analyzing the elemental composition of the sample in energy dispersive X-rays (EDX). The back-scattered electrons coming from the sample are also used in generating the image. These images are used in elemental composition determination of the samples by applying analytical SEM with the spectra made from the characteristics of the X-rays.

The electron beam passes through the pairs of scanning coils or pairs of deflector plates in the electron column and then, as it reaches the final lens, it reflects the beam horizontally and vertically. Afterwards, it scans in a raster fashion over a rectangular area of the sample surface. The electronic devices used in SEM instrumentation are applied to detect and amplify the signals. These signals are displayed as an image on a cathode ray oscilloscope in which the raster scanning is synchronized with that of the microscope. The image displayed in SEM is just like the distribution map of the intensity of the electronic signals being emitted from the scanned area of the specimen (Joshi and Vishwanathan, 2006; Joshi et al., 2006; Joshi et al., 2008; Phillip and Alexander, 2008;; Zhang et al., 2005).

The pre-requisite of sample imaging is to prepare the sample to be electronically conductive. Non-conductive solid samples are made conductive by coating with a layer of conductive material by low vacuum sputter coating or high vacuum evacuation. Environmental SEM (ESEM) and in field emission gun (FEG) SEM operated at low voltage/high vacuum or at low vacuum/high voltage are used for imaging uncoated, non-conductive materials.

2.5.1.2 Applications

Detailed, three dimensional images can be obtained at many high resolutions (up to ×300,000) as compared to light microscope (up to ×10,000). The images obtained by SEM are black and white as they are produced without light waves (Joshi et al., 2006). The surface morphology of polymer nanocomposites, fracture surfaces, nanofibers, nanoparticles, and nanocoating can be imaged. Images with a resolution up to 1–5 nm can be obtained (Zhang et al., 2005). This technique is most suitable for studying the coating of nanofibers on polymeric substances. Field emission SEM has also been applied to determine the size, morphology of the surface, and information such as homogeneity of synthesized Fe_3O_4, Fe_3O_4@basil seed mucilage (BSM), and Fe_3O_4@BSM-cephalexin (CPX) monohydrate nanocarriers. The mean size and size-distribution of the particles were determined to be 12 and 6 nm with 0.28 and 0.25 polydispersity index (PDI) values respectively (Rayegan et al., 2018). This study proved the nature of synthesized nanoparticles in nano-range. The limited and uniform size distribution of these nanoparticles resulted in a similar amount of drug absorption, and more accurate characterization with equal probability of being magnetically trapped in the target tissue. They also had uniform physical and biochemical properties resulting in homogenous colloidal properties (Kayal and Ramanujan, 2010; Mahmoudi et al., 2011; Varadan et al., 2008).

2.5.2 Transmission Electron Microscopy (TEM)

This technique is also used for the characterization of the shape and size of PNPs.

2.5.2.1 Basic Principle

TEM is two dimensional and is formed from the electrons transmitted through the sample. The resolution of this instrumentation is of sub-nanometer range that provides the information on inner-particle structures of the nanocarriers. It is less valuable for measuring the polymeric wall of the nanocapsules. Membrane thickness of 1–2 nm range in PCL nanocapsules has been measured by TEM. This technique has some limitations (Ma et al., 2006). For characterization by TEM, the ideal specimen thickness should be less than 100 nm. Also, the organic samples can suffer structural damage due to local heating from the kinetic energy (KE) absorbed by the sample (Kaminskyl and Dahms, 2008). The heating problem of the samples can be circumvented by applying cryo-TEM, in which the samples are analyzed in a frozen-hydrated state, with minimal sample modifications (Stewart, 2016).

To apply this technique for sample analysis in metallurgy and biological sciences, the specimens should be very thin and able to withstand the high vacuum present inside the instrument. The maximum thickness of the specimen should be 1 μm. The biological samples are typically held at liquid nitrogen temperatures after embedding them in vitreous ice; these are fixed by using a negative staining material such as uranyl acetate or by plastic embedding. These processes are adopted for the samples to withstand the vacuum.

The properties of specimens such as nanocomposites depend to a large extent on successful nano-level dispersion or intercalation/exfoliation of nanocrystals.

2.5.2.2 Applications

TEM images may reveal the distribution and dispersion of nanoparticles in polymer matrices of nanocomposites, fibers, and nanocoatings, etc. The extent of the exfoliation, intercalation, and orientation of the nanoparticles can be visualized using TEM micrographs. For determining the size, shape, and uniformity of the particles, TEM images of Fe_3O_4@BSM-CPX were obtained (Rayegan et al., 2018). The images revealed relatively uniform size and limited size-distribution. The size of the coated nanoparticles measured by TEM was also compared with uncoated nanoparticles as observed by FESEM (Field Emission Scanning Electron Microscopy). The TEM images showed that putting basil mucilage coating on the magnetic nanoparticles reduced the interaction between the particles, and the size of the nanoparticle cores was also decreased. The core shell structure of the magnetic nanoparticles had additional advantages such as good dispersion, high stability against oxidation, and it could load a considerable amount of drug on the shell.

2.5.3 High Resolution Transmission Electron Microscopy (HRTEM)

This technique allows imaging of the crystallographic structure of a sample at an atomic scale. Besides this, this technique is an imaging mode of the transmission electron microscope that allows the imaging of the crystallographic structure of a sample at an atomic scale.

2.5.3.1 Basic Principle

HRTEM does not use absorption by the sample for image formation, but instead the contrast arises from the interference in the image plane of the electron wave with itself. Each imaging electron interacts independently with the sample. As a result of the interaction with the sample, the electron wave passes through the imaging system of the microscope where it undergoes further phase changes and interferes with the image wave in the imaging plane. Actually, the recorded image is not a direct representation of the sample's crystallographic structures.

2.5.3.2 Applications

As it has high resolution, the nanoscale properties of crystalline materials can be precisely studied using this technique. At present, the possible highest resolution is 0.8 Å. At these small scales, individual atoms and crystalline defects can be imaged (Sano et al., 2002).

2.5.4 Scanning Tunneling Microscopy (STM)

STM is an instrument for producing surface images with atomic scale, lateral resolution in which a fine probe tip is scanned over the surface of a conducting specimen with the help of a piezoelectric crystal at a distance of 0.5–1 nm and the resulting tunneling current, or the position of the tip required to maintain a constant tunneling current, is monitored.

2.5.4.1 Basic Principle

This technique is based on the principle of quantum tunneling. When a conducting tip is brought very near to a metallic or semi-conducting surface, a bias between the two can allow electrons to tunnel through the vacuum between them. For low voltages, this tunneling current is a function of the local density of states at the Fermi level of the sample. Variations in current as the probe passes over the surface are translated into an image. In this technique, good resolution is considered to be 0.1 nm lateral resolution and 0.01 nm depth resolution. The images are normally generated by holding the current between the tip of the electrode and the specimen surface while the tip is piezoelectrically scanned in a raster pattern over the region of the specimen surface being imaged, by holding the force rather than the electric current between the tip and the specimen at a set-point value. When the height of the tip is plotted as a function of its lateral position over the specimen, an image results looking very much like the surface topography. This technique can be applied not only in an ultra-high vacuum (UHV) but also in air and various other liquid or gas states. However, extremely clean surfaces and sharp tips are required, which makes it challenging as a technique.

2.5.4.2 Applications

STM has been applied for the characterization and modification of a variety of materials. It has been used to detect and characterize materials like carbon nanotubes (single-walled and multi-walled). It has also been used to move single nanotubes or metal atoms and molecules on smooth surfaces with high precision (Mejer et al., 2004; Terrones et al., 2004).

2.5.5 Atomic Force Microscopy (AFM)

This technique is also known as scanning probe microscopy (SPM). It is a strong tool for studying the surface morphology with nanometer resolution as well as for sensitive force measurements (Binnig et al., 1986;

Ducker et al., 1991; Dufrêne, 2002; Russel and Batchelor, 2001). In many biological processes, molecular interaction studies are also required, and by applying this technique single molecular-level measurements of the forces on colloidal interactions can be studied (Drelich et al., 2004; Hodgesm et al., 2002).

2.5.5.1 Basic Principle

The imaging in this technique is performed by sensing the attractive/repulsive forces between a very sharp probing unit and the sample surface (Rao et al., 2007). The force is measured by attaching the probe to a pliable cantilever and monitoring the deflection of the cantilever through a laser-photodiode system, which detects the difference in the photo-detector output voltage. The scanning motion is controlled by a piezoelectric scanner that controls the position of the sample and moves it with respect to the tip. Several AFM imaging modes are available; the three most widely used are contact mode, non-contact mode, and tapping mode AFM (Gu et al., 2005; Tiede et al., 2008). In contact mode, one simply records the cantilever deflection while the sample is scanned horizontally, i.e., at a constant height. In contrast, tapping mode consists of oscillating the cantilever at its resonance frequency and lightly "tapping" on the surface during scanning. The repulsive force on the tip is set by pushing the cantilever against the specimen's surface with a piezoelectric positioning element. The deflection of the cantilever is measured, and the AFM images are created. In non-contact mode, the scanning tip floats about 50–150 Å above the specimen's surface. In contrast, tapping imaging is implemented in ambient air by oscillating the cantilever assembly at its resonant frequency (often hundreds of kilo hertz) using a piezoelectric crystal (Crucho and Barros, 2014, 2017; Kaminskyl and Dahms, 2008). The piezo motion causes the cantilever to oscillate when the tip is not in contact with the surface of a material. The oscillating tip is then moved towards the surface until it begins to tap the surface. During scanning, the vertically oscillating tip alternately contacts the surface and lifts off, generally at a frequency of 50,000–5,00,000 Hz. As the oscillating cantilever begins to intermittently contact the surface, the cantilever oscillation is reduced due to energy loss caused by the tip contacting the surface. The reduction in oscillation amplitude is used to measure the surface characteristics (Celasco et al., 2014).

Similar to SEM, AFM can be used to study the size and size distribution, shape, and aggregation of PNPs. Additionally, AFM is a relatively nondestructive technique, meaning, in other words, a large range of substrates can be studied (Russel and Batchelor, 2001). The sample preparation is perhaps the most important step in the analysis because it can affect the concentration, aggregation, and the cleanliness of the sample prior to further analysis. For instance, the water evaporation entailed in the AFM sample preparation influences the PNPs' size. Aggregation may be caused by the hydrophilicity of the mica substrate, which could have traces of water even after evaporation, leading to nanoparticle coalescence (Rao et al., 2007). One way to avoid that is through lyophilization of the sample on the mica wafer (Crucho and Barros, 2014). AFM has the advantage to image a variety of biomaterials in aqueous fluids and it is also possible to observe macromolecular motion on a surface of the same sample in real time (Gu et al., 2005). The main limitation of AFM for PNP visualization is that the size of the cantilever tip is often larger than the particles being probed, which leads to overestimations of the PNPs' lateral dimensions (Tiede et al., 2008). Therefore, lateral dimension measurements should be used with great caution and accurate size measurements should only be taken on the height (z-axis). SEM has a larger depth of focus than AFM, but AFM is more accurate in the z dimension, which is not available in SEM (Kaminskyl and Dahms, 2008). AFM equipment, which has comparable resolution to SEM apparatus, costs much less, requires substantially less laboratory space, and is much simpler to operate. Even though SEM and AFM appear very different, it is possible to find some similarities (Drelich et al., 2004; Rao et al., 2007). Their images have comparable lateral resolution (5–100 nm) and both techniques have image artifacts (Binnig et al., 1986; Celasco et al., 2014; Russel and Batchelor, 2001).

2.5.6 Dynamic Light Scattering (DLS)

A number of techniques can be applied for the measurement of particle size and its distribution, such as sieving, optical counting, electroresistance counting, sedimentation, laser diffraction, DLS, acoustic spectroscopy, etc. Among these, DLS is applied mostly for particle size determination and its distribution of nanoparticles. This technique is also known as PCS or QELS. It is a non-invasive, well-established technique for measuring the submicron region and, with the latest technology, can measure lower than 1 nm.

2.5.6.1 Basic Principle

For experimentation of DLS, a colloidal suspension is illuminated by a monochromatic laser light and this light is scattered into a photon detector. During this, Brownian motion of the particles occurs and the detected scattered light intensity fluctuates in time. This fluctuated light is related to the particle size with the help of an auto-correlation function. By applying the Stokes–Einstein equation, the hydrodynamic diameter can be obtained. The above said relation can be expressed as follows for spherical particles:

$$D_h = \frac{k_B T}{3\pi\eta D_t}$$

In this equation, D_h is the hydrodynamic diameter, η is the relative viscosity of the solvent, k_B is Boltzmann's constant, T is the temperature, and D_t is the translational diffusion coefficient (this is what is measured by DLS). DLS experimentation provides the inclusive information of the whole particulate population in a short span of time, but the basic limitation is that it considers the whole sample as all particles being spherical in nature. As larger particles scatter more light than smaller ones, even a small amount of aggregates or dust particles could shift the particle size distribution to the larger value. Hence, the information collected for particle size distribution is interpreted with caution.

The Stokes–Einstein equation is also being applied in a new emerging technique called nanoparticle tracking analysis (NTA) (Kaszuba et al., 2008; Lim et al., 2013). In this technique, both the laser light scattering microscopy and a charge-coupled device camera are used simultaneously. It helps in concurrent visualization and for the recording of individual nanoparticles moving under Brownian motion in solution. Also, both the nanoparticle distributions and nanoparticle concentration can be assessed at the same time. However, NTA is a time-consuming technique and requires skilled man-power for the setting and adjustment of the software. However, the major advantage of NTA over DLS is that it has the ability to resolve multiple size populations with more accuracy.

2.5.6.2 Applications

The nanoparticle size as ascribed in DLS data is an intensity-based diameter (Z-average) and it is the value that is obtained directly from the correlation function. By applying Mie theory, this data can be converted into a volume or number distribution if there is no error in the intensity distribution. As these are number-based metrologies, the corresponding size distribution for comparison is the number distribution. The nanoparticle data obtained from the DLS technique should be supplemented with imaging techniques such as AFM, SEM, etc., to resolve the associated ambiguities. When the nanoparticle size of the preparation composed of PEG-*g*-PLA fabricated by emulsification solvent evaporation method was measured by DLS, the size was found to be in the range of 178–192 nm. However, on applying the resuspension process for the formulation in spite of using surfactant or sonication, the size of the particles was increased (Sant et al., 2005).

2.5.7 X-Ray Photoelectron Spectroscopy (XPS)

XPS is a quantitative, spectroscopic, surface chemical analysis technique and is used to estimate the empirical formula or elemental composition, chemical state, and electronic state of the elements on the surface (up to 10 nm of the material). It is also known as electron spectroscopy of chemical analysis (ESCA).

2.5.7.1 Basic Principle

X-ray irradiation of a material under UHV leads to the emission of electrons from the core orbitals of the top 10 nm of the surface elements being analyzed. Measurement of the KE and the number of electrons escaping from the surface of the material gives the XPS spectra. From the KE, the binding energy (BE) of the electrons to the surface atoms can be calculated. The BE of the electrons reflects the oxidation state of the specific surface elements. The number of electrons reflects the proportion of the specific elements on the surface. As the energy of a particular X-ray wavelength used to excite the electron from a core orbital is a known quantity, the electron BE of each of the emitted electrons can be determined by applying the following expression based on Ernest Rutherford's research work (Joshi et al. 2008):

$$E_{binding} = E_{photon} - E_{kinetic} - \Phi$$

In this equation, $E_{binding}$ is the energy of the electron emitted from the electron configuration within the atom, E_{photon} is the energy of the X-ray photons being used, $E_{kinetics}$ is the KE of the emitted electron as measured by the instrument, and ϕ is the work function of the spectrophotometer (not the material).

2.5.7.2 Applications

This technique is used to determine the elements and the quantity of those elements that are present within approximately 10 nm of the sample surface. It also detects the contamination, if any exists, in the surface or the bulk of the sample. If the material is free of excessive surface contamination, XPS can generate the empirical formula of the sample and the chemical state of one or more of the elements can be identified. It is generally applied to determine the thickness of one or more thin layers (1–8 nm) of different materials within the top 10 nm of the surface. It can also measure the uniformity of elemental composition of surfaces after nano-level etching, finishing, or coating of the surfaces. The major limitation is that it cannot detect hydrogen ($Z=1$) or helium ($Z=2$), because these two elements do not have any core electron orbitals, but only valence orbitals.

The thickness of the copolymer layer in PEG nanocapsules prepared by the solvent displacement method has been measured by XPS. By tuning the synthesis parameters, the thickness of the shell varied from 1 to 10 nm. This technique also revealed that PEG chains were more pronounced on the surface of PEG nanospheres than that of nanocapsules. The polymeric wall thickness in nanocapsules can be estimated by this more direct and accessible technique.

2.5.8 Wide Angle X-Ray Diffraction (XRD)

XRD are electromagnetic radiations that are similar to light but with a much shorter wavelength (a few Angstrom). They are produced when electrically-charged particles of sufficient energy are decelerated. In an X-ray tube, the high voltage maintained across the electrodes draws electrons toward a metal target (the anode). X-rays are produced at the point of impact and radiate in all directions.

2.5.8.1 Basic Principle

When an incident X-ray beam encounters a crystal lattice, the general scattering occurs. Although most scattering interferes with itself and is eliminated (destructive interference), diffraction occurs when scattering in a certain direction is in phase with scattered rays from other atomic planes. Under this condition, the reflections combine to form new enhanced wave fronts that mutually reinforce each other (constructive interference). The relation by which diffraction occurs is known as Bragg's law (Newbury and Ritchie, 2013). As each crystalline material, including semi-crystalline polymers, metal and metal oxide nanoparticles, and layered silicate nanoclays, has a characteristic atomic structure, it diffracts X-rays in a unique characteristic diffraction order or pattern.

2.5.8.2 Applications

X-ray diffraction data from polymers generally provides information about crystallinity, crystallite size, orientation of the crystallites, and phase composition in semi-crystalline polymers. With appropriate accessories, X-ray diffraction instrumentation can be used to study the phase change as a function of stress or temperature, to determine lattice strain, to measure the crystalline modulus, and, with the aid of molecular modeling, to determine the structure of the polymer.

Besides the above characterization, this technique can be used to characterize PNPs, polymer layered silicate nanocomposites, etc. The crystal structure of magnetite nanoparticles and the change in crystal structure after coating and deposition of drugs has been studied by this technique. When the nanoparticles of Fe_3O_4 (magnetite), magnetite coated with BSM (Fe_3O_4@BSM), and coated formulation with deposited drug (Fe_3O_4@BSM-CPX) were analyzed for X-ray diffraction patterns, seven characteristic peaks were obtained for each sample. All three samples studied were found to have approximately the same consistency, such as magnetic reference sample standard card number JCPDS=00–003-0863 (Rayegan et al., 2018). It revealed the presence of the inverse spinel structure of Fe_3O_4 in all three samples. The absence of any additional peak in spectrogram confirmed the absence of any impurity. It was also observed that after coating of magnetite nanoparticles with BSM (Fe_3O_4@BSM) and deposition of the drug, no structural or phase changes were observed in the spectrogram of the nanoparticles. However, the Fe_3O_4@BSM sample showed the peaks with half-widths (the peak width at half the maximum peak intensity) bigger than those of the Fe_3O_4 nanoparticles and it was speculated that this was due to the amorphousness of the BSM and the reduction of its crystalline structure. It indicated the successful coating of the nanoparticles.

2.5.9 Energy Dispersive X-Ray Analysis (EDX)

By applying this technique, near surface elements and their proportion at different positions can be analyzed.

2.5.9.1 Basic Principle

EDX's basic use is associated with SEM. An electron beam having energy in the range of 10–20 keV is used to strike the surface of a conducting sample. The X-rays are emitted by this phenomenon and the energy of these emitted radiations depends upon the properties of the materials under study. However, the X-rays are generated in a region about 2 μm in depth. The image of each element preset in the material can be obtained by moving the electron beam all around the material. As the energy of the emitted X-rays is of very low extent, it takes a lot of time to acquire the images.

2.5.9.2 Applications

The presence of heavy metals in the nanoparticles or their composition in the nanoparticles can be estimated by applying this technique. The presence of Au, Pd, and Ag nanoparticles has been estimated by the EDX technique (Newbury and Ritchie, 2013). However, the elements with a low atomic number are difficult to detect. The Si-Li detector, protected by a beryllium window, cannot detect elements below an atomic number of 11 (i.e., sodium (Na)). However, the elements with as low an atomic number as 4 (beryllium (Be)) can be detected by using windowless system. EDX spectra have to be taken by focusing the beam at different regions of the same sample to verify the spatially uniform composition of the bimetallic materials.

2.5.10 Fourier Transform Infrared Spectroscopy (FTIR)

This technique is applied to investigate the structural properties of PNPs and to assess the molecular interactions between drugs and the encapsulating polymer. The FTIR technique is categorized under vibrational spectroscopies.

2.5.10.1 Basic Principle

This technique is concerned with the absorption of infrared (IR) radiation, which causes vibrational transition in the molecule, due to which it is also termed "vibrational spectroscopy". It is used mainly in structure elucidation to determine functional groups (Elliot and Ambrose, 1950). In the electromagnetic radiation spectra, the IR region extends from 800 nm to 1,000 μm in which near-IR, middle-IR, and far-IR regions are available in the region 0.8–2 μm, 2–15 μm, and 15–1,000 μm respectively (Anastasopoulou et al., 2009). Most of the analytical studies are confined to the middle-IR region because absorption of organic molecules is high in this region. The atoms of molecules are connected with different types of bondings, and these have spring and ball like vibrations. These characteristic vibrations are called the natural frequency of vibration. When energy in the form of IR radiation is applied, it causes the vibration between the atoms of the molecules; at equal frequency of applied IR frequency to the natural frequency of vibration, the absorption of IR radiation causes a peak in the spectra. Different functional groups absorb characteristic frequencies of IR and give characteristic peak values. Therefore, the IR spectrum of a chemical substance is a finger print of a molecule for its identification. A molecule can absorb IR radiation if the natural frequency of vibrations of some part of a molecule is the same as the frequency of the incident radiation. Also, the molecule can only absorb IR radiation when its absorption causes a change in its electric dipole. An electric dipole property remains in the molecule when there is a slight positive and a slight negative charge on its component of atoms (Colthup et al., 1990). Due to incident IR radiations, various types of stretching and bending vibrations occur in molecules. These vibrations generate characteristic peaks in the IR spectra of a molecule.

2.5.10.2 Applications

By applying FTIR studies, the presence of known compounds in the fingerprint region can be investigated. The presence of impurities or unexpected interactions can also be detected by the presence of characteristic functional groups in the spectra. This technique has also been applied to identify the existing molecules and functional groups, to ensure the formation of bindings, and to prove the coating and loading of the drug on nanocarriers. For such assessments, the FTIR spectrum of BSM, Fe_3O_4, CPX, and Fe_3O_4@BSM were recorded and studied (Rayegan et al., 2018). It was observed by comparing the IR spectrum obtained from the synthesized magnetite with the spectra obtained from similar particles that a reasonable consistency was maintained between them. It revealed the successful synthesis of magnetite nanoparticles (Dorniani et al., 2012; Varadan et al., 2008). For determining the effective coating of magnetite nanoparticles by BSM, the FTIR spectra of Fe_3O_4, BSM, and Fe_3O_4@ BSM were recorded. The stretching vibration of the hydroxyl (OH) group binding was observed in the form of increased transmittance intensity in the peak of the 3,423 cm^{-1}. The 1,629 cm^{-1} zone in the Fe_3O_4-BSM spectrum compared to the peaks of the Fe_3O_4 spectrum and the slight displacement of their sites were also observed. These peculiar changes in spectral study indicated a bond in magnetite nanoparticles and BSM. The bonds that have appeared in the 441 cm^{-1} zone due to the stretching vibration of Fe-O in the nanoparticles also appeared in the Fe_3O_4@BSM spectrum. However, these bonds were not observed in the BSM spectrum. The peaks that appeared in the 1,030 cm^{-1} zone were related to the C-O and C-O-C stretching groups of the mucilage and these also appeared in the Fe_3O_4@BSM spectrum. However, these bonds were not present in the Fe_3O_4 spectrum. The comparative study of all these spectra defined the effective coating operation of BSM on magnetite nanoparticles. Besides this, the deposition of CPX on magnetite nanoparticles has also been ascertained by FTIR spectroscopy. For this, the FTIR spectra of Fe_3O_4@BSM and Fe_3O_4@BSM-CPX were compared (Rayegan et al., 2018). The apparent difference in spectra was observed in the form of the increased transmittance intensity of the Fe_3O_4@BSM-CPX spectrum at 1,386 cm^{-1} and it was indicative of symmetrical bending vibrations of C-H and symmetrical stretching vibrations of carboxyl existing in CPX. It confirmed the successful loading of the drug in the magnetite nanoparticles. Another indication of successful drug loading in magnetite nanoparticles was observed due to emergence of a peak in the 719 cm^{-1} bond range showing the presence of the stretching vibrations of C-S in the Fe_3O_4@BSM-CPX sample. A slight increase in the intensity of the two peaks in the 2,900 cm^{-1} range was also observed and that was considered due to the presence of the stretching vibrations of C-H related to the drug. These observations in the FTIR spectrum of the Fe_3O_4@BSM-CPX sample indicated the successful presence and loading of CPX.

2.5.11 Matrix-Assisted Laser Desorption/Ionization Time-of-Flight Mass Spectrometry (MALDI-TOF-MS)

Matrix-assisted laser desorption/ionization (MALDI) has been applied as an indispensable tool for high molecular-weight compounds measurement (Korytár et al., 2005; Tanaka et al., 1988; Wetzel et al., 2004).

2.5.11.1 Basic Principle

The samples for MALDI are mixed uniformly into a large amount of matrix. During the operation, the matrix absorbs the ultraviolet light, usually nitrogen light of wavelength 337, and coverts it to heat energy. Afterwards, a small amount of the matrix, about 100 nm down from the top outer surface of the analyte, is heated rapidly. This process takes several nanoseconds and is vaporized together with the sample.

However, in the case of MALDI time-of-flight mass spectroscopy (MALDI-TOF-MS), charged ions of various sizes are generated from the samples. A potential difference, V_0, between the sample and the ground attracts the ions towards the detector. The velocity of the attracted ions (v) is determined by the law of conservation of energy. As the V_0 is constant with respect to all ions, ions with smaller mass to charge (m/z) value (lighter ions) and more highly-charged ions move faster through the drift space to reach the detector. Consequently, the time of ion flight differs in accordance to the m/z value of the ion.

With the advances in matrix modifications and the development of novel materials, MALDI-TOF-MS has been successfully applied in the analysis of low-molecular-weight compounds (Crank and Armstrong, 2009; Mullens et al., 2011; Prabhakaran et al., 2012; Tseng et al., 2010; Zhang et al., 2006). By applying soft ionization techniques, protonated analyte is observed in MALDI-TOF-MS. Extensive fragments from the typical organic matrix are a serious obstacle for measuring analytes in the low mass region. Alkali metal ions exist abundantly in the environment and samples are dominant in mass spectra, giving rise to poor peak intensity of protonated analytes (Montaudo et al., 2006), especially in utilization of the inorganic matrix and surface assisted laser desorption/ionization (SALDI). In addition, some low-molecular-weight compounds such as normal alkanes, organic acids, sugars, and biologically active substances are unable to be observed in the protonated state. To obtain a non-destructive ionization process for short-chain n-alkanes, soft ionization mass spectrometry techniques such as MALDI and electrospray ionization (ESI) could be a better choice. Owing to the non-polar saturated hydrocarbon, short-chain n-alkanes are unsuitable for conventional MALDI and ESI techniques. In light of this difficulty, new ionization methods have been created, including chemical ionization (Dehon et al., 2011) and laser desorption/ionization (Wallace et al., 2010). Porous materials are commonly used in modification of the matrix for MALDI-TOF-MS analysis of low-molecular-weight compounds (Li et al., 2009). Zeolites, which are widely used as catalysts and sorbents, are porous aluminosilicates which consist of a three-dimensional framework of SiO_4 and AlO_4 tetrahedron linked by oxygen bridges. As a consequence of the isomorphous replacement of Si^{4+} by Al^{3+} in the crystal, negative charges on the Si-O-Al bridge sites are balanced by cations such as H^+ and Na^+ that are able to be exchanged with other cations. Because of their porous structures and Brönsted acid sites, zeolites were recently found to be applicable to MALDI-TOF-MS analysis (Yang et al., 2018). They were successfully used for eliminating interference from fragments and alkali metal ion adducts in conventional MALDI-TOF-MS analysis (Komori et al., 2010; Yamamoto and Fujino, 2012). Moreover, alkali-cation substituted zeolites were also developed as a cationization reagent for ionization of low-molecular-weight compounds (Suzuki and Fujino, 2012; Suzuki et al., 2012). In our previous research, it was found that loading nanoparticles on zeolite could prevent aggregation and that the improved spatial distribution of nanoparticles could lead to the high reproducibility of MALDI-TOF-MS measurements (Yang and Fujino, 2014a).

2.5.11.2 Applications

Silver nanoparticles loaded on ammonium-exchanged mordenite zeolite have been used as a matrix for MALDI-TOF-MS analysis, with an Ag^+ ionization probe for the detection of the short-chain n-alkane. It has also been studied that using zeolite as a supporter provides the more uniform and homogenous distribution of silver nanoparticles and diminishes the aggregation effect of the nanoparticles (Yang and Fujino, 2013). Also, the active site on zeolite surfaces influence the electronic state of cluster and silver cations due to the decreased excitation threshold. Zeolite as a matrix provides the more sufficient use of laser energy for silver nanoparticles. The studies have also shown that silver nanoparticles loaded on zeolite with NH_4^+ termination of Brönsted acid sites provide a more efficient MALDI matrix (Yang and Fujino, 2013; Yang and Fujino, 2014b). Also, the ammonium termination generated a high amount of Ag^+ ions, which were used as an ionization source of small molecules. By using this technique, acetyl salicylic acid has been detected with high intensity (Yang et al., 2018). The ammonium-ion-terminated mordenite prevented the dissolution of the silver nanoparticles (Yang et al., 2018).

The protonation of n-alkanes is very hard due to their saturated structures. When conventional organic matrix α-cyano-4-hydroxycinnamic acid (CHCA), with a strong absorption at 337 nm wavelength, was applied to assist in ionization of long-chain n-alkane ($C_{23}H_{48}$), many noisy peaks generated from the conventional matrix were found in spectra without the presence of the native peak of the alkane having a concerned m/z ratio. However, a concerned peak of alkane (m/z) was obtained using silver nanoparticles loaded with NH4+-terminated mordenite (Yang et al., 2018). Similarly, various categories of metallic nanoparticles have been applied to develop the matrices for the MALDI-TOF-MS analysis of alkanes.

2.5.12 Ultraviolet-Visible Spectroscopy

The ultraviolet-visible spectrophotometers consist of a light source, reference and sample beams, a monochromator, and a detector. The ultraviolet spectrum for a compound is obtained by exposing a sample of the compound to ultraviolet light from a light source such as a xenon lamp.

2.5.12.1 Basic Principle

The reference beam in the spectrophotometer travels from the light source to the detector without interacting with the sample. The sample beam interacts with the sample exposing it to ultraviolet light of a continuously changing wavelength. When the emitted wavelength corresponds to the energy level which promotes an electron to a higher molecular orbital, energy is absorbed. The detector records the ratio between the reference and sample beam intensities. The computer determines at what wavelength the sample absorbed a large amount of ultraviolet light by scanning for the largest gap between the two beams. When a large gap between intensities is found, where the sample beam intensity is significantly weaker than the reference beam, the computer plots this wavelength as having the highest ultraviolet light absorbance when it prepares the ultraviolet absorbance spectrum (Mock et al., 2002).

2.5.12.2 Applications

In certain metals such as silver and gold, the plasmon resonance is responsible for their unique and remarkable optical phenomenon. Metallic (silver or gold) nanoparticles, typically 40–100 nm in diameter, scatter optical light elastically with remarkable efficiency because of a collective resonance of the conduction electrons in the metal, which is known as surface plasmon resonance. The surface plasmon resonance peak in the ultraviolet absorption spectra is known by these plasmon resonant nanoparticles. The magnitude, peak wavelength, and spectral bandwidth of the plasmon resonance associated with a nanoparticle are dependent on the particle's size, shape, and material composition, as well as the local environment (Bruning et al., 1974).

The surface plasmon resonance peak in the absorption spectra of silver particles is shown by an absorption maximum at 420–500 nm. The surface peaks vary with size, shape, and concentration of the metallic nanoparticles. It has been studied to confirm that the value of λmax is shifted towards a higher wavelength with increasing Ag content ($\lambda_{max} = 398$ nm at 5×10^{-4} M Ag sol and $\lambda_{max} = 406$ nm at 5% Ag/kaolinite) in silver nanoparticles/kaolinite composites (Patakfalvi et al., 2003). It has also been reported that the truncated plane as the basal plane displayed the strongest biocidal actin compared with spherical, rod-shaped nanoparticles or with Ag^+ (in the form of $AgNO_3$) (Patakfalvi et al., 2003). The shape of the silver nanoparticles can be identified by observing the corresponding peak.

2.5.13 Raman Spectroscopy

Raman spectroscopy is a spectroscopic technique used to study vibrational, rotational, and other low-frequency modes in a system. It is based on inelastic scattering, or Raman scattering, of monochromatic laser light. The laser light interacts with the photons or other excitations in the system resulting in the energy of the laser photons being shifted up or down. The shift in energy gives information about the photons mode in the system.

2.5.13.1 Basic Principle

The Raman effect occurs when light impinging upon a molecule interacts with the electron cloud of the bonds of that molecule and an incident photon excites one of the electrons into a virtual state. For the spontaneous Raman effect, the molecule will be excited from the ground state to a virtual energy state, and then relax into a vibrational excited state, which generates Stokes Raman scattering. If the molecule was already in an elevated vibrational energy state, the Raman scattering is then called anti-Stokes Raman scattering. A molecular polarizability change or the amount of deformation of the electron cloud with respect to the vibrational coordinate is required for the molecule to exhibit the Raman effect. The amount of polarizability change determines the Raman scattering intensity whereas the Raman shift is equal to the vibrational level that is involved.

2.5.13.2 Applications

As vibrational information is very specific for the chemical bonds in molecules, Raman spectroscopy plays a major role in studying the chemical bonding of drugs and polymers. It provides a fingerprint by which the molecule can be identified in the range of 500–2,000 cm^{-1}. Raman gas analyzers are use in medicine for real time monitoring of anesthetic and respiratory gas mixtures during surgery.

2.5.14 Dual Polarization Interferometry (DPI)

By applying this technique, simple, phase-sensitive nanoparticle detection can be performed with no active optical elements. Two measurements are taken simultaneously, allowing amplitude and phase to be decoupled. However, the limitations of various optical techniques are associated with their inability to disclose the difference between changes in sample layer thickness and density. Due to this, the better analysis of membrane interactions as well as molecular aggregation cannot be performed to a higher sensitivity. This technique analyzes thin films to acquire unique combinations of multiple opto-geometrical properties such as density, mass, thickness, and refractive index (RI).

2.5.14.1 Basic Principle

The dual polarization interferometry (DPI) instrument is based on a dual slab wave guide sensor chip that has an upper sensing wave guide as well as a lower optical reference wave guide lit up with an alternating, orthogonal polarized laser beam. Two differing wave guide modes are created — specifically, the transverse magnetic (TM) mode and the transverse electric (TE) mode. Both modes generate an evanescent field at the top sensing wave guide surface and probe the materials that make contact with this surface.

As material interacts with the sensor surface it leads to phase changes in interference fringes. Then, the interference fringe pattern for each mode is mathematically resolved into RI and thickness values. Because of this important capability, the sensor is able to measure extremely subtle molecular changes on the sensor surface. In the dual-phase detection interferometer, four components are associated (Deutsch et al., 2010):

1. Configuration for immobilized particles: A 1.4 NA oil-immersion objective is used to focus light on a particle, and the particle is scanned through the focus.
2. Nanoscale channel configuration: Particles in solution are loaded into reservoirs at the ends of a 15 µm channel of cross-section (500 nm by 400 nm), and a microscope objective focuses light on the channel. The black arrow indicates the direction of electro-osmotic flow.
3. Numerical aperture increasing lens (NAIL) configuration: A NAIL is water-bonded to a cover-slip and is used to focus light on an immobilized particle, which is scanned through the focus.
4. Schematic of dual-phase interferometer: A linear polarizer sets the polarization state of the signal beam to 45°, and a combination of a polarizer and quarter-wave plate (QWP) set the reference polarization to be circular. After reference and signal are combined with a non-polarizing beam splitter (NPBS), a polarizing beam-splitter (PBS) sends the two orthogonal polarization states to detectors. The difference in relative phase between signal and reference at the two detectors is $\pi/2$.

2.5.14.2 Applications

The detection of 25 nm Au particles in liquid in $\Delta t \sim 1$ ms with a signal-to-noise ratio of 37 has been performed with this technique. Such a performance makes it possible to detect nanoscale contaminants or larger proteins in real time without the need of artificial labeling.

2.5.15 Nuclear Magnetic Resonance (NMR)

The analysis of nanoparticles still represents a considerable challenge for common methods of NMR spectroscopy. In addition, motional inhomogeneity complicates the spectral analysis; while the fluid constituents exhibit rapid rotational diffusion, the solid contributions mostly lack the rapid isotropic tumbling which could lead to a desired motional averaging of the chemical shift tensors. Finally, spectroscopists who deal with dispersed nanoparticles have to be concerned about the long-term stability of their samples, as agglomeration and precipitation may occur especially under experimental conditions such as magic angle spinning (MAS). This method in general is completely non-destructive and allows for long-term studies on sensitive particle systems. At the same time, NMR spectroscopy offers the unique benefit of rendering simultaneous information on the chemical identity as well as the molecular mobility of individual components in a complex inhomogeneous mixture. Further, it yields data which gives access to domain sizes, chemical exchange, proximity to paramagnetic centers, and reaction kinetics. Unlike common techniques for nanoparticle characterization, such as electron and scanning probe microscopy, it does not require any sample preparation procedures which always include the risk of causing artefacts or structural deterioration. With all these advantages given, NMR may prove to be a valuable asset in the analysis of dispersed nanoparticles. By applying this technique, rotational diffusion of particles, assignment to structural elements of particle dispersions, phase transitions of the nanoparticle matrix, molecular exchange on the nanoparticle surface, molecular exchange through nanocapsule walls, local phase separations, and particle degradation, etc., can be studied (Mayer, 2018).

2.5.16 Nanoparticle Tracking Analysis (NTA)

This technique uses the Stokes–Einstein equation in NTA (Filipe et al., 2010). This technique combines laser light scattering microscopy with a charge-coupled device camera that enables the visualization and recording of individual nanoparticles moving under Brownian motion in solution. Nanoparticle size distributions and nanoparticle concentrations can then be determined. Although NTA is time consuming and requires some operational skills for the adjustment of the software settings, one major advantage over DLS is the ability to resolve multiple size populations with more accuracy.

2.5.17 Zeta Potentiometry

Zeta (ζ)-potential analysis is routinely used to determine the surface charge and stability of a PNP colloidal suspension, i.e., to what degree aggregation will occur over time. Besides, it is also a key factor influencing PNPs' *in vivo* fate, for example in interactions with cell membranes. The particle surface charge is determined by measuring the ζ-potential of a suspension, which is generally done by the well-known electrophoresis method (Tantra et al., 2010). The basic principle involves measuring the electrophoretic mobility of charged particles under an applied electrical potential. ζ-potential is related to the electrophoretic mobility (μ) by the Henry equation:

$$\mu = \frac{2\zeta\varepsilon}{3\eta_0} f(kr)$$

In this equation, ε is the dielectric constant, η_0 is the medium's viscosity, and f(kr) refers to Henry's function. ζ-potential represents a measure of the stability of a colloidal suspension. Values between –30 mV and +30 mV indicate a condition towards instability, aggregation, coagulation, or flocculation (Sapsford et al., 2011). Thus, particle aggregation is less likely to occur for high ζ-potential values due to electric repulsion. Ultraviolet spectroscopy can be used to determine the aggregation state by turbidity measurements. In an interesting work, the influence of one or two charged groups per polymer chain on the size of PLGA PNPs obtained by nanoprecipitation were studied (Reisch et al., 2015). The authors were able to reduce the size from over 100 nm to less than 25 nm. The successful introduction of several charged end groups on the polymer chain was determined by ζ-potential measurements. For example, PLGA modified with carboxy and sulfonate end groups afforded negative values (–40 mV), while positive values were obtained for PLGA-NMe$_3$ (+15 mV). PNPs in aqueous solutions can be stabilized by electrostatic stabilization and/or steric stabilization using surfactants. The stability and hence the ζ-potential can be influenced by many factors such as pH, concentration, ionic strength of the solution, and the nature of the surface ligands (Sapsford et al., 2011). The uptake of PNPs usually increases with increasing the ζ-potential values. However, cationic particles are believed to be more toxic than their anionic or neutral counterparts (Elsabahy and Wooley, 2012).

2.5.18 Differential Scanning Calorimetry (DSC)

Differential scanning calorimetry (DSC) is commonly used to study the physicochemical state and possible interactions of the drug(s) loaded in PNPs. By applying this technique, the phase transitions such as glass transition, crystallization, and melting can also be assessed. The glass transition temperature (Tg) is the reversible transition in amorphous materials or in amorphous regions in semi-crystalline materials from a hard and brittle state into a molten or rubber-like state.

2.5.18.1 Basic Principle

In this technique, the amount of heat required to increase the temperature of reference and the sample are measured as a function of temperature. A similar temperature for the sample and the reference is maintained throughout the study. The program of temperature control is maintained in such a way that the temperature of the sample holder increases linearly as a function of time. The characteristic of the reference sample is that it should have a well-defined heat capacity over the range of temperatures to be scanned. The phase transition studies such as melting, glass transition, or exothermic decompositions are studied by this technique. In all these phase transitions, energy changes or heat capacity is involved, which is detected by DSC with more sensitivity.

When the sample undergoes the physical transformation, i.e., phase transition, heat is required in amounts of more or less to maintain the sample and reference at the same temperature. The need of heat in terms of less or more depends upon the processes, such as exothermic or endothermic. Simply saying, when a solid sample converts into liquid during melting, it requires more flow of heat to increase its temperature at the same rate as the reference. This occurs due to the absorption of heat (endothermic process) by the sample as it undergoes phase transition from solid to liquid. On the other hand, in some exothermic processes, like crystallization, heat is required to raise the temperature. The difference in the flow of heat between the sample and the reference enables the assessment of the amount of heat absorbed or released during such types of phase transitions.

In DSC experimentations, the data is represented in the form of a curve of heat flux versus temperature or time. The enthalpies of phase transitions are also measured by these curves and it is performed by integrating the peak corresponding to a given phase transition. The enthalpies of phase transitions can be represented by the following expression:

$$\Delta H = KA$$

In above equation, ΔH is the enthalpy of transition, K is the calorimetric constant, and A is the area under the curves.

2.5.18.2 Applications

Numbers of applications are associated with this technique, like melting points, glass transitions, crystallization times and temperatures, heat of melting and crystallization, percent crystallinity, oxidative stabilities, compositional analysis, heat capacity, heats of cure, completeness of cure, percent purity, purities, thermal stabilities, polymorphism, heat set temperatures, recyclates or regrinds, etc. The glass transition occurs when the temperature of an amorphous solid is increased. In recorded DSC spectrograms, these transitions appear as a step in the baseline and this occurs due to a change in heat capacity of the sample. When the temperature of an amorphous solid increases, it becomes less viscous and, at some points, the molecules obtain freedom of motion due to this temperature rise. These molecules also arrange themselves into a crystalline form. The temperature at which this phenomenon occurs, is called the crystalline temperature (Tc). The transition of amorphous solid to crystalline solid is an exothermic process and produces a peak in the DSC spectrogram. On gradually increasing the temperature of the sample, the melting point (Tm) is reached and this melting point process results in an endothermic peak in the DSC curve. The estimation of phase transition temperatures and enthalpies helps in producing the phase diagrams of various chemical systems. During transition from solid to liquid state, some of the materials enter into a third state which displays properties of both phases. This anisotropic liquid is known as liquid crystalline or mesomorphous state. As this phenomenon results because of small energy changes, the matter shows transitions from a solid to a liquid crystal and from a liquid crystal to an isotropic liquid. In such studies for liquid crystal, DSC is an important technique.

The oxidative stability of drugs and their products has also been studied by DSC. Generally, an airtight sample chamber is required to study the stability of drug and their formulations. These experimentations are performed in isothermal conditions by changing the atmosphere of the sample. In the initial stage of the experimentation, the sample is maintained at the desired test temperature under an inert atmosphere such as nitrogen. Afterwards, oxygen is added to the system. If the oxidation process takes place in the sample, a deviation in baseline of the DSC spectrogram of the sample is observed. These type of studies may be used to determine the stability and optimum storage conditions for drugs and their formulations.

For the design of polymers applied in dosage forms, the curing processes of these materials can be studied by DSC. The process of crosslinking is an exothermic process and it occurs in most of the fabrication techniques used to develop nanoparticles as drug delivery devices. The crosslinking phenomenon results in a positive peak in the DSC curve and it usually appears soon after the glass transition. The molecular weight of the polymers can also be assessed, along with composition analysis. The melting point and the glass transition temperature of polymers with known molecular weight are constant and the degradation of the polymers is expected from lowering of the melting point. As the melting point depends upon molecular weight, the lower grades developed during degradation will have lower melting points than expected. The percentage crystallinity of the polymer can also be determined by DSC. It can be calculated from the crystallization peak from the DSC graph, since the heat of fusion can be calculated from the area under an absorption peak. The impurities in polymers can also be determined by examining thermograms for anomalous peaks and plasticizers can be detected at their characteristic boiling points.

In dosage forms, a well-defined and characterized drug is required to show its desirable effect. For example, if the amorphous form of a drug is to be required for proper delivery, it becomes necessary to process the drug at a temperature below those at which crystallization can occur. In such types of drug analysis, DSC plays an important role. The purity analysis of drugs and their products can be performed by DSC. The less pure compound produces a broadened melting peak in the DSC spectrogram, starting at lower temperature than a pure compound.

The polymers applied in PNPs, if available in their glassy state, can provide mechanical strength to the formulations and avoid particle aggregation. Corrigan and Li (2009) studied the thermal characteristics of ketoprofen-loaded PLGA nanoparticles. The absence of a melting peak from the DSC thermogram of the loaded nanoparticles indicated that the drug was present in a non-crystalline state.

2.6 Instrumental Characterization of Drug–Polymer Interactions

Drug–polymer interactions have an impact on drug-loading efficiency and, to an extent, on drug release. Hence, these interactions, types of polymers, and their physicochemical properties are taken into consideration. The amorphous state of the active component of nanostructured drug delivery devices has an impact on formulation characteristics, as this state of drug has a higher internal energy which leads to greater solubility and bioavailability (Hancock and Parks, 2000; Yu, 2001). Being physically unstable in nature, the amorphous form of a drug tends to convert spontaneously into crystalline form and hence both the amorphous and crystalline forms of the drug are possible. Structural stabilization can be achieved by incorporating the drug into an amorphous polymer with concomitant formation of an amorphous solid solution (Raula et al., 2007). The interactions between the molecules of the drug and the polymer control the solubility of these materials in each other and this phenomenon is accomplished by van der Waals' forces and hydrogen bonding. When the drug–polymer molecular interactions are comparable to the drug–drug and the polymer–polymer molecular interactions, the polymeric component is able to solubilize the drug material and large amounts of the drug can be incorporated in the polymer matrix without drug crystallization (Raula et al., 2007). On the other hand, when the drug–polymer interactions are weaker, drug crystallization occurs because the drug and the polymer have a preference to interact with the molecules of their own kind (Raula et al., 2007). In PNPs, DSC can be used to study the physicochemical state and the possible interactions of the drug and the polymers. It is also able to detect the phase transitions such as glass transitions, crystallization, and melting (Kerč and Srčič, 1995). The glass transition temperature is the reversible transition in amorphous materials or in the amorphous region in semi-crystalline materials, from a hard and brittle state into a molten or rubber-like state. The polymers in their glassy state can provide mechanical strength to the polymeric nanoparticles (Yang et al., 2014).

2.7 Toxicity Aspects of Nanoparticles

The ultimate goals of nanoparticles having entrapped drugs or the drug in itself are for either enhanced delivery to, or uptake by, target cells and/or reduction in toxicity of the free drug to non-target organs. To achieve these, target specific delivery with prolonged retention is required. The retention of drugs within the therapeutic range results in therapeutic efficacy, e.g., tumor cell death, as well as toxicity in other organ systems. The applied biodegradable polymers result in drug release after degradation in the body. The major problem associated with particulate drug carriers is their entrapment in mononuclear phagocytic systems, as available in the liver and spleen (Moghmi et al., 2001). However, this property is favorable in treating liver diseases, like tumor metastasis or hepatitis. Also, the cytotoxicity of the nanomaterials occurs due to the interaction between the nanomaterial surface and cellular components. The surface area of the particle increases exponentially on decreasing the diameter. Therefore, when the particles have the same composition, the significant difference in their cytotoxicity depends upon their particle size and surface reactivity. Braakhuis et al. (2016) studied the effect of different sizes of silver nanoparticles on cytotoxicity after inhalation. Silver nanoparticles prepared in particle size ranges of 18, 34, 60, and 160 were exposed to rats. After a specified time of exposure, the rats were sacrificed and the amounts of the nanoparticles were determined. It was observed that silver nanoparticles in sizes of 18 nm and 34 nm after exposure of 24 hours resulted in induced lactate dehydrogenase (LDH) expression, which is a marker of cell damage, in a dose-dependent manner. However, no dose-dependent cell damage was observed during exposure of 60 nm and 160 nm nanoparticles. More of the nanoparticles having the particle size of 60 nm and 160 nm were found in the lungs overall and more of the 18 nm and 34 nm nanoparticles were found in the alveoli. The study revealed that increased surface area of the nanoscaled particles was the key factor contributing to the toxicology of the silver nanoparticles. It has also been determined that the nanoparticles having a diameter greater than 6 nm accumulate in specific organs such as the liver and spleen and cannot be excreted by the kidney. The accumulation of these formulations causes serious side effects, e.g., cadmium selenide quantum dots remain in the tissue for up to eight months and cause hepatotoxicity (Ballou et al., 2004). The particle size and the surface chemistry affect the pharmacokinetic properties of nanoparticles.

In the rat model, the distribution of gold nanoparticles has been assessed for *in vivo* distribution. The range of nanoparticles under study were 10–205 nm. The study revealed that the distribution pattern of 10 nm nanoparticles was different than their counterparts. These were spread in most organs, including the blood, brain, lungs, heart, thymus, spleen, liver, kidneys, and testes. However, the particles having a size larger than 50 nm were detected only in the blood, liver, and spleen. In reference to their size, it has also been determined that various pathways for internalization into the cells are adopted (Kou et al., 2013; Zhao et al., 2011). The best suitable size for uptake ranges from 10 to 500 nm, with an upper limit of 5 mm. The large particles are engulfed by macropinocytosis. The size of a vesicle involved in clathrin-mediated endocytosis is about 100 nm while the size involved in caveolae-mediated endocytosis is usually about 60–80 nm. It can also be concluded that by decreasing the particle size, its biological activity increases substantially. The smaller particles occupy less volume and result in pathophysiological toxicity mechanisms such as oxidative stress, ROS generation, and mitochondrial perturbation, etc. (Duffine et al., 2002; Oberdorster et al., 2005). However, it is presumed that the size of the nanoparticle alone may not be responsible for toxicity but that the total number per unit volume may be important. It has also been demonstrated that pulmonary toxicity assessed after treatment with several particles, including carbon black, titanium dioxide, etc., was induced by ultrafine and fine materials that have large surface areas (Sager et al., 2008; Sager and Castranova, 2009; Stoeger et al., 2005). The particles' aggregation properties, termed as particle surface reactivity, also play a significant role in cytotoxicity (Monteiller et al., 2007; Lin et al., 2006; Rabolli et al., 2011; Warheit et al., 2009). The *in vivo* biodistribution of nanoparticles is also affected by changes in surface charges. The particles also reveal different degrees of toxicity depending on their surface charges (Asati et al., 2010; Frohlich, 2012; Jiang et al., 2008; Schaeublin et al., 2011). The formulations having positive surface charges tend to have much higher toxicity. Agglomeration of dosage forms in nano-ranges could be a potent inducer of inflammatory lung injury in humans (Bantz et al., 2014; Li et al., 2007).

Besides size and surface charges of the nanoparticles, the morphology also has impacts on toxicological aspects of these formulations (Shin et al., 2015). Nanoscale fibers (e.g., carbon nanotubes) have been reported as a potent causative agent of inflammation and prolonged exposure may result in several cancers (Cha and Myung, 2007; Firme and Bandaru, 2010; Huczko et al., 2001). Some of the studies have also revealed that carbon nanotubes are more toxic than other ultrafine carbon black or silica dusts (Chalupa et al., 2004). The exposure of single-walled carbon nanotubes (SWCNTs) beyond the current permissible exposure limit (PEL) caused the development of lung lesions (Chalupa et al., 2004). Also, the carbon nanotubes have been shown to cause the death of targeted kidney cells via inhibited cell growth, induced by decreased cell adhesiveness (Handy et al., 2008). Serve lung damage has also been reported in humans after exposure to fullerene (also termed as buckyballs) in addition to the destruction of fish brains and water flea death (Lovern and Klaper, 2006; Lyon et al., 2005; Lyon et al., 2006; Oberdörster, 2004).

2.8 Conclusion

In medicine, the applicability of nanoparticles should also be assessed thoroughly for their safety in biological systems. All the concerned physicochemical properties of nanoparticles should be evaluated for elucidation of their interaction with the cellular organelles, cell tissues, and organisms. Such investigations will surely provide strategies to develop new generations of nontoxic products containing nanoparticles. Also, these fundamental studies will help in generating new criteria for acceptance and rejection for *in vivo* applications.

The present scenario of research and development activities in advanced materials and polymers indicates a shift to nanomaterials as the new tool to improve properties and gain multifunctionalities. However, many of the challenges are also associated with nanoparticles and one of the most important is the transition from lab-scale proof of concept research to reproducible, with precise physicochemical properties, and high-yielding production of useful nanomaterials. Regarding this, several methods are being extensively applied for fabrication of PNPs and techniques are being applied for scaling-up production. Nanomedicines provide vast opportunities in improving the therapeutics and development of new treatment options in all diseases in general, and for diseases previously thought difficult or impossible to treat in particular. However, some of the nanomedicines do not even reach clinical trials. To some extent, this can be overcome if the regulatory requirements for clinical trials and the key features that make a polymer suitable for biological applications (biodegradable, stable, nontoxic, and well-characterized) are taken into account.

During the last few years, research has been extensive. Many of the nanomedicines are now available in the market, and it has been possible by adopting remarkable techniques and instrumentation for nanoparticle fabrication and characterization. Furthermore, the quality and extent of information derived through instrumental characterization techniques such as SEM, TEM, AFM, and Raman spectroscopy also depends to a large extent on the level of understanding of the user, their expertise, and the right sample preparation. Thus, it can be summarized that nanotechnology research advancements in drug delivery devices have a lot of potential as a futuristic approach but would be governed largely by simultaneous progress in newer, simpler, and more efficient characterization techniques for nanomedicines.

REFERENCES

Abeylath SC, Turos E, Dickey S, Lim DV. Glyconanobiotics: Novel carbohydrated nanoparticle antibiotics for MRSA and Bacillus anthracis. Bioorg Med Chem. 2008;16(5):2412–8.

Anastasopoulou J, Theophanides TH. Chemistry and symmetry. In: Greek National Technical University of Athens, NTUA, 94p.

Asati A, Santra S, Kaittanis C, Perez JM. Surface-charge-dependent cell localization and cytotoxicity of cerium oxide nanoparticles. ACS Nano. 2010;4:5321–31.

Ballou B, Lagerholm BC, Ernst LA, Bruchez MP, Waggoner AS. Noninvasive imaging of quantum dots in mice. Bioconjug Chem. 2004;15:79–86.

Bantz C, Koshkina O, Lang T, Galla HJ, Kirkpatrick CJ, Stauber RH, Maskos M. The surface properties of nanoparticles determine the agglomeration state and the size of the particles under physiological conditions. Beilstein J Nanotechnol. 2014;5:1774–86.

Binnig G, Quate CF, Gerber C. Atomic force microscope. Phys Rev Lett. 1986;56:930–3.

Braakhuis HM, Cassee FR, Fokkens PH, de la Fonteyne LJ, Oomen AG, Krystek P, de Jong WH, van Loveren H, Park MV. Identification of the appropriate dose metric for pulmonary inflammation of silver nanoparticles in an inhalation toxicity study. Nanotoxicol. 2015.

Bruning JH, Herriott DR, Gallagher JE, Rosenfeld DP, White AD, Brangaccio DJ. Digital wavefront measuring interferometer for testing optical surfaces and lenses. Appl Optics. 1974;13:2693–703.

Cavalli R, Gasco MR, Chetoni P, Burgalassi S, Saettone MF. Solid lipid nanoparticles (SLN) as ocular delivery system for tobramycin. Int J Pharm. 2002;238(1–2):241–5.

Celasco E, Valente I, Marchisio DL, Barresi AA. Dynamic light scattering and X-ray photoelectron spectroscopy characterization of PEGylated polymer nanocarriers: Internal structure and surface properties. Langmuir. 2014;30:8326–35.

Cevc G. Transfersomes, liposomes and other lipid suspensions on the skin: Permeation enhancement, vesicle penetration, and transdermal drug delivery. Crit Rev Ther Drug Career Syst. 1996;13(3–4): 257–388.

Cha KE, Myung H. Cytotoxic effects of nanoparticles assessed in vitro and in vivo. J Microbiol Biotechnol. 2007;17:1573–8.

Chalupa DC, Morrow PE, Oberdörster G, Utell MJ, Frampton MW. Ultrafine particle deposition in subjects with asthma. Environ Health Perspect. 2004;112:879–82.

Chiang CL, Sung CS, Wu TF, Chen CY, Hsu CY. Application of superparamagnetic nanoparticles in purification of plasmid DNA from bacterial cells. J Chromat B. 2005;822(1–2):54–60.

Choi AO, Cho SJ, Desbarats J, Lovric J, Maysinger D. Quantum dot-induced cell death involves Fas upregulation and lipid peroxidation in human neuroblastoma cells. J Nanobiotechnol. 2007a;5:1–4.

Choi HS, Liu W, Misra P, Tanaka E, Zimmer JP, Ipe BI, Bawendi MG, Frangion JV. Renal clearance of quantum dots. Nat Biotechnol. 2007b;25(10):1165–70.

Colthup NB, Daly LH, Wiberley SE. Introduction to infrared and Raman spectroscopy. 3rd ed. London: Academic Press Ltd. p. 547.

Corrigan OI, Li X. Quantifying drug release from PLGA nanoparticulates. Eur J Pharm Sci. 2009;37:477–85.

Crank JA, Armstrong, DW. Towards a second generation of ionic liquid matrices (ILMs) for MALDI-MS of peptides, proteins, and carbohydrates. J Am Soc Mass Spectrom. 2009;20:1790–1800.

Crucho CIC, Barros MT. Surfactant-free polymeric nanoparticles composed of PEG, cholic acid and a sucrose moiety. J Mater Chem B. 2014;2:3946–55.

Crucho CIC, Barros MT. Polymeric nanoparticles: A study on the preparation variables and characterization methods. Mater Sci Eng C. 2017;80:771–784.

Culver KW. Clinical applications of gene therapy for cancer. Clin Chem. 1994;40(4):510–2.

Davies R, Schurr GA, Meenam P, Nelson RD, Bergna HE, Brevett CAS, Goldbaum RH. Engineered particle surfaces. Adv Mater. 1998;10(15):1264–70.

Dehon C, Lemaire J, Heninger M, Chaput A, Mestdagh H. Chemical ionization using CF^{3+}: Efficient detection of small alkanes and fluorocarbons, Int J Mass Spectrom. 2011;299:113–9.

Deutsch B, Beams R, Novotny L. Nanoparticle detection using dual-phase interferometry. Appl Opt. 2010;49(26):4921–5.

Di Paolo D, Loi M, Pastorino F, Brignole C, Marimpietri D, Becherini P, Caffa I, Zorzoli A, Longhi R, Gagliani C, Tacchetti C, Corti A, Allen TM, Ponzoni M, Pagnan G. Liposome-mediated therapy of neuroblastoma. Methods Enzymol. 2009;465:225–49.

Dorniani D, Hussein MZB, Kura, AU, Fakurazi S, Shaari AH, Ahmad Z. Preparation of Fe_3O_4 magnetic nanoparticles coated with gallic acid for drug delivery. Int J Nanomed. 2012;7:5745–56.

Drelich J, Tormoen GW, Beach ER. Determination of solid surface tension from particle–substrate pull-off forces measured with the atomic force microscope. J Colloid Interface Sci. 2004;280:484–97.

Ducker WA, Senden TJ, Pahley RM. Direct measurement of colloidal forces using an atomic force microscope. Nature. 1991;353:239–41.

Duffine R, Tran CL, Clouter A, Brown DM, MacNee W, Stone V, Donaldson K. The importance of surface area and specific reactivity in the acute pulmonary inflammatory response to particles. Ann Occup Hyg. 2002;46:242–5.

Dufrêne YF. Atomic force microscopy: A powerful tool in microbiology. J Bacteriol. 2002;184(19): 5205–13.

Dugan LL, Turetsky DM, Du C, Lobner D, Wheeler M, Almli CR, Shen CK, Luh TY, Choi DW, Lin TS. Carboxyfullerenes as neuroprotective agents. Proc Natl Acad Sci USA. 1997;94(17):9434–9.

Elliot A, Ambrose E. Nature, structure of synthetic polypeptides. Nature. 1950;165:921. Woernley DL. Infrared absorption curves for normal and neoplastic tissues and related biological substances. Curr Res. 1950;12:516.

Elsabahy M, Wooley KL. Design of polymeric nanoparticles for biomedical delivery applications. Chem Soc Rev. 2012;41:2545–61.

Filipe V, Hawe A, Jiskoot W. Critical evaluation of Nanoparticle Tracking Analysis (NTA) by nanosight for the measurement of nanoparticles and protein aggregates. Pharmaceut Res. 2010;27:796–810.

Firme CP, Bandaru PR. Toxicity issues in the application of carbon nanotubes to biological systems. Nanomed Nanotechnol Biol Med. 2010;6:245–56.

Friedman SH, DeCamp DL, Sijbesma RP, Srdanov G, Wudl F, Kenyon GL. Inhibition of the HIV-1 protease by fullerene derivatives: Model building studies and experimental verification. J Am Chem Soc. 1993;115(15):6506–9.

Frohlich E. The role of surface charge in cellular uptake and cytotoxicity of medical nanoparticles. Int J Nanomed. 2012;7:5577–91.

Ganesan P, Narayanasamy D. Lipid nanoparticles: Different preparation techniques, characterization, hurdles, and strategies for the production of solid lipid nanoparticles and nanostructured lipid carriers for oral drug delivery. Sustainable Chem Pharm. 2017;6:37–56.

Gangadharam PR, Ashtekar DA, Ghori N, Goldstein JA, Debs RJ, Duzgunes N. Chemotherapeutic potential of free and liposome encapsulated streptomycin against experimental Mycobacterium avium complex infections in beige mice. J Antimicrob Chemother. 1991;28(3):425–35.

Gilham RJJ, Brown RJC. The characterization of nanoparticles. In: Thompson M, editor. Analytical Methods Committee Technical Brief. Royal Society of Chemistry; 2010 . p. 1–3. Available from: http://www.rsc.org/images/characterisation-nanoparticles-technical-brief-48_tcm18-214815.pdf.

Gu X, Nguyen T, Oudina M, Martin D, Kidah B, Jasmin J, Rezig A, Sung L, Byrd E, Martin JW. Microstructure and morphology of Amine-Cured Epoxy coatings before and after outdoor exposures—An AFM study. J Coat Technol Res. 2005;2:547–56.

Hancock BC, Parks M. What is the true solubility advantage for amorphous pharmaceuticals? Pharm Res. 2000;17:397–404.

Handy RD, Henry TB, Scown TM, Johnston BD, Tyler CR. Manufactured nanoparticles: Their uptake and effects on fish—A mechanistic analysis. Ecotoxicol. 2008;17:396–409.

Hester R, Harrison RM. Nanotechnology: Consequences for human health and the environment. RSC Publishing; 2007.

Hinds WC. Aerosol technology: Properties, behaviour and measurements. 2nd ed. Wiley-Blackwell; 1999.

Hirsch LR, Jackson JB, Lee A, Halas NJ, West JL. A whole blood immunoassay using gold nanoshells. Anal Chem. 2003a;75(10):2377–81.

Hirsch LR, Stafford RJ, Bankson JA, Sreshen SR, Rivera B, Price RE, Hazle JD, Halas NJ, West JL. Nanoshell mediated near-infrared thermal therapy of tumours under magnetic resonance guide. Proc Natl Acad Sci USA. 2003b;100(23):13549–54.

Hodgesm CS, Cleaver JAS, Ghadiri M, Jones R, Pollock HM. Forces between polystyrene particles in water using the AFM: Pull-off force vs particle size. Langmuir. 2002;18:5741–8.

Huang YL, Shen CKF, Luh TY, Yang HC, Hwang KC, Chou CK. Blockage of apoptotic signaling of transforming growth factor-b in human hepatoma cells by carboxyfullerene. Eur J Biochem. 1998;254:38–43.

Huczko A, Lange H, Całko E, Grubek-Jaworska H, Droszcz P. Physiological testing of carbon nanotubes: Are they asbestos-like? Fuller Sci Technol. 2001;9:251–4.

Irving B. Nanoparticle drug delivery systems. Inno Pharm Biotechnol. 2007;24:58–62.

International Organization for Standardization (ISO). Standard ISO/TS27687:2008 – Nanotechnologies – Terminology and definitions for nano-objects – Nanoparticle, nanofibre and nanoplate. 2008. Available from: ttps://www.iso.org/standard/44278.html.

Jenning V, Gysler A, Schafer-Korting M, Gohla S. Vitamin A loaded solid lipid nanoparticles for topical use: Occlusive properties and drug targeting to the upper skin. Eur J Pharm Biopharm. 2000;49(3):211–8.

Jiang J, Oberdörster G, Biswas P. Characterization of size, surface charge, and agglomeration state of nanoparticle dispersions for toxicological studies. J Nanopart Res. 2008;11:77–89.

Jordan A, Scholz R, Wust P, Fähling H, Felix R. Magnetic fluid hyperthermia (MFH): Cancer treatment with AC magnetic field induced excitation of biocompatible superparamagnetic nanoparticles. J Magn Magn Mater. 1999;201(1–3):413–9.

Joshi M, Vishwanathan V. High-performance filaments from compatibilized polypropylene/clay nanocomposites. J Appl Polym Sci. 2006;102(3):2164–74.

Joshi M, Butola BS, Simon G, Kukalevab N. Rheological and viscoelastic behavior of HDPE/octamethyl-POSS nanocomposites. Macromolecules. 2006;39(5):1839–49.

Joshi M, Bhattacharya A, Wazed Ali S. Characterization technique for nanotechnology applications in textiles. Indian J Fibre Textile Technol. 2008;33:304–17.

Kalele SA, Ashtaputre SS, Hebalkar NY, Gosavi SW, Deobagkar DN, Deobagkar DD, Kulkarni SK. Optical detection of antibody using silica–silver core-shell particles. Chem Phys Lett. 2005;404(1–3):136–41.

Kalele SA, Kundu AA, Gosavi SW, Deobagkar DN, Deobagkar DD, Kulkarni SK. Rapid detection of Echerischia coli using antibody conjugated silver nanoshells. Small. 2006;2(3):335–8.

Kaminskyl SGW, Dahms TES. High spatial resolution surface imaging and analysis of fungal cells using SEM and AFM. Micron. 2008;39:349–61.

Kaszuba M, McKnight D, Connah MT, McNeil-Watson FK, Nobbmann U. Measuring sub nanometer sizes using dynamic light scattering. J Nanopart Res. 2008;10:823–9.

Kaur CD, Nahar M, Jain NK. Lymphatic targeting of zidovudine using surface-engineered liposomes. J Drug Target. 2008;16(10):798–805.

Kayal S, Ramanujan RV. Doxorubicin loaded PVA coated iron oxide nanoparticles for targeted drug delivery. Mat Sci Eng C. 2010;30:484–90.

Kerč J, Srčič S. Thermal analysis of glassy pharmaceuticals. Thermochimica Acta. 1995;248:81–95.

Khairnar GA, Chavan-Patil AB, Palve PR, Bhise SB, Mourya VK, Kulkarni CG. Dendrimers: Potential tool for enhancement of antifungal activity. Int J Pharm Tech Res. 2010;2(1):736–9.

Kim HJ, Jones MN. The delivery of benzyl penicillin to Staphylococcus aureus biofilms by use of liposomes. J Liposome Res. 2004;14(3–4):123–39.

Komori Y, Shima H, Fujino T, Kondo JN, Hashimoto K, Korenaga T. Pronounced selectivity in matrix-assisted laser desorption-ionization mass spectrometry with 2, 4, 6-trihydroxyacetophenone on a zeolite surface: Intensity enhancement of protonated peptides and suppression of matrix-related ions. J Phys Chem C. 2010;114:1593–600.

Korytár P, Parera J, Leonards PEG, Santos FJ, de Boer J, Brinkman UAT. Characterization of polychlorinated n-alkanes using comprehensive two-dimensional gas chromatography–electron-capture negative ionization time-of-flight mass spectrometry. J Chromatogr A. 2005;1086:71–82.

Kou L, Sun J, Zhai Y, He Z. The endocytosis and intracellular fate of nanomedicines: Implication for rational design. Asian J Pharm Sci. 2013;8:1–10.

Krusic PJ, Wasserman E, Keizer PN, Morton JR, Preston KF. Radical reactions of C60. Science. 1991;254(5035):1183–5.

Li X, Wu X, Kim JM, Kim SS, Jin M, Li D. MALDI-TOF-MS analysis of small molecules using modified mesoporous material SBA-15 as assisted matrix. J Am Soc Mass Spectrom. 2009;20:2167–73.

Li Z, Hulderman T, Salmen R, Chapman R, Leonard SS, Young SH, Shvedova A, Luster MI, Simeonova PP. Cardiovascular effects of pulmonary exposure to single-wall carbon nanotubes. Environ Health Perspect. 2007;115:377–82.

Lim J, Yeap SP, Che HX, Low SC. Characterization of magnetic nanoparticle by dynamic light scattering. Nanoscale Res Lett. 2013;8:381.

Lin W, Huang YW, Zhou XD, Ma Y. In vitro toxicity of silica nanoparticles in human lung cancer cells. Toxicol Appl Pharm. 2006;217:252–9.

Liz-Marzan LM, Correa-Duarte MA, Pastoriza-Santos I, Mulvaney P, Ung T, Giersig M, Kotov NA. Core-shell and assemblies thereof. In: Nalwa HS, editor. Handbook of surfaces and interfaces of materials. Amsterdam: Elsevier; 2001. p. 189–237.

Loo C, Lin A, Hirsch L, Lee M, Barton J, Halas N, West, J, Drezek R. Nanoshell-enabled photonics-based imaging and therapy of cancer. Technol Cancer Res Treat. 2004;3(1):33–40.

Lovern SB, Klaper R. Daphnia magna mortality when exposed to titanium dioxide and fullerene (C60) nanoparticles. Environ Toxicol Chem. 2006;25:1132–7.

Lyon DY, Adams LK, Falkner JC, Alvarez PJJ. Antibacterial activity of fullerene water suspensions: Effects of preparation method and particle size. Environ Sci Technol. 2006;40:4360–6.

Lyon DY, Fortner JD, Sayes CM, Colvin VL, Hughes JB. Bacterial cell association and antimicrobial activity of a c60 water suspension. Environ Toxicol Chem. 2005;24:2757–62.

Ma J, Qi Z, Hu Y. Synthesis and characterization of polypropylene/clay nanocomposites. J Appl Polym Sci. 2006;82(14):3611–7.

Ma M, Cheng Y, Xu Z, Xu P, Qu H, Fang Y, Xu T, Wen L. Evaluation of polyamidoamine (PAMAM) dendrimers as drug carriers of antibacterial drugs using sulfamethoxazole (SMZ) as a model drug. Eur J Med Chem. 2007;42(1):93–8.

van der Maas JH. Basic infrared spectroscopy. 2nd ed. London: Heyden & Son Ltd; 1972. p. 105.

Magallanes M, Dijkstra J, Fierer J. Liposome-incorporated ciprofloxacin in treatment of murine salmonellosis. Antimicrob Agents Chemother. 1993;37(11):2293–7.

Mahmoudi M, Sant S, Wang B, Laurent S, Sen T. Superparamagnetic iron oxide nanoparticles (SPIONs): Development, surface modification and applications in chemotherapy. Adv Drug Deliver Rev. 2011;63:24–46.

Marchesan S, Da Ros T, Spalluto G, Balzarini J, Prato M. Anti-HIV properties of cationic fullerene derivatives. Bioorg Med Chem Lett. 2005;15(15):3615–8.

Mayer, C. Annual reports on NMR spectroscopy. Volume 55, ISSN. 0066. p. 4103, 2005.

Mehnert W, Mäder K. Solid lipid nanoparticles: Production, characterization and applications. Adv Drug Deliver Rev. 2001;47(2–3):165–96.

Meier J, Schiǿtz J, Liu P, Nǿrskov JK, Stimming U. Nano-scale effects in electrochemistry. Chem Phys Lett. 2004;390(4–6):440.

Meng J, Fan J, Galiana G, Branca RT, Clasen PL, Ma S, Zhou J, Leuschner C, Kumar CSSR, Hormes J, Otiti T, Beye AC, Harmer MP, Kiely CJ, Warren W, Haataja MP, Soboyejo WO. LHRH-functionalized superparamagnetic iron oxide nanoparticles for breast cancer targeting and contrast enhancement in MRI. Mat Sci Eng C. 2009;29(4):1467–79.

Mitra A, Deutsch B, Ignatovich F, Dykes C, Novotny L. Nano-optofluidic detection of single viruses and nanoparticles. ACS Nano. 2010;4:1305–12.

Mohanraj VJ, Chen Y. Nanoparticles: A review. Trop J Pharm Res. 2006;5(1):561–73.

Molpeceres J, Aberturas MR, Guzman M. Biodegradable nanoparticles as a delivery system for cyclosporine: Preparation and characterization. J Microencapsul. 2000;17(5):599–614.

Montaudo G, Samperi F, Montaudo MS. Characterization of synthetic polymers by MALDI-MS. Prog Polym Sci. 2006;31:277–357.

Monteiller C, Tran L, MacNee W, Faux S, Jones A, Miller B, Donaldson K. The pro-inflammatory effects of low-toxicity low-solubility particles, nanoparticles and fine particles, on epithelial cells in vitro: The role of surface area. Occup Environ Med. 2007;64:609–15.

Morishige K, Kacher DF, Libby P, Josephson L, Ganz P, Weissleder R, Aikawa M. High-resolution magnetic resonance imaging enhanced with superparamagnetic nanoparticles measures macrophage burden in atherosclerosis. Circulation. 2010;122(17):1707–15.

Müller RH, Mader K, Gohla S. Solid lipid nanoparticles (SLN) for controlled drug delivery – A review of the state of the art. Eur J Pharm Biopharm. 2000;50(1):161–77.

Müller RH, Radtke M, Wissing SA. Solid lipid nanoparticles (SLN) and nanostructured lipid carriers (NLC) in cosmetic and dermatological preparations. Adv Drug Deliver Rev. 2002;54:S131–55.

Müller RH, Runge SA. Solid lipid nanoparticles (SLN) for controlled drug delivery. In: Benita S, editor. Submicron emulsions in drug targeting and delivery. Amsterdam: Harwood Academic Publishers; 1998. p. 219–34.

Mudshinge SR, Deore AM, Patil S, Bhalgat CM. Nanoparticles: Emerging carriers for drug delivery. Saudi Pharm J. 2011;19:129–41.

Mullens CP, Anugu SR, Gorski W, Bach SBH. Modified silica-containing matrices towards the MALDI-TOF-MS detection of small molecules. Int J Mass Spectrom. 2011;308:311–5.

Newbury DE, Ritchie NW. Is scanning electron microscopy/energy dispersive Xray spectrometry (SEM/EDS) quantitative? Scan. 2013;35:141–68.

Novotny L, Hecht B. Principles of nano-optics. Cambridge: Cambridge University Press; 2006. p. 404.

Oberdörster E. Manufactured nanomaterials (fullerenes, C60) induce oxidative stress in the brain of juvenile largemouth bass. Environ Health Perspect. 2004;112:1058–62.

Oberdorster G, Maynard A, Donaldson K, Castranova V, Fitzpatrick J, Ausman K, Carter J, Karn B, Kreyling W, Lai D, et al. Principles for characterizing the potential human health effects from exposure to nanomaterials: Elements of a screening strategy. Part Fibre Toxicol. 2005;2:8–43.

Omri A, Suntres ZE, Shek PN. Enhanced activity of liposomal polymyxin B against Pseudomonas aeruginosa in a rat model of lung infection. Biochem Pharmacol. 2002;64(9):1407–13.

Pandey R, Khuller GK. Solid lipid particle-based inhalable sustained drug delivery system against experimental tuberculosis. Tuberculosis (Edinb). 2005;85(4):227–34.

Panga X, Cui F, Tian J, Chen J, Zhou J, Zhou W. Preparation and characterization of magnetic solid lipid nanoparticles loaded with ibuprofen. Asian J Pharm Sci. 2009;4(2):132–7.

Park JW. Liposome-based drug delivery in breast cancer treatment. Breast Cancer Res. 2002;4(3):95–9.

Phillip DB, Alexander EB. Electrical conductivity study of carbon nanotube yarns, 3-D hybrid braids and their composites. J Compos Mater. 2008;42(15):1533–45.

Prabhakaran J, Yin B, Nysten H, Degand P, Morsomme T, Mouhib S, Yunus P, Bertrand A. Metal condensates for low-molecular-weight matrix-free laser desorption/ionization. Int J Mass Spectrom. 2012;315(2012):22–30.

Rabolli V, Thomassen LC, Uwambayinema F, Martens JA, Lison D. The cytotoxic activity of amorphous silica nanoparticles is mainly influenced by surface area and not by aggregation. Toxicol Lett. 2011;206:197–203.

Radtke M, Müller RH. Novel concept of topical cyclosporine delivery with supersaturated SLN_ creams. Int Symp Control Rel Bioact Mater. 2001;28:470–1.

Rao A, Shoenenberger M, Gnecco E, Glatzel TH, Meyer E, Brändlin D, Scandella L. Characterization of nanoparticles using Atomic Force Microscopy. J Phys Conf Ser. 2007;61:971–6.

Raula J, Eerikäinen H, Lähde A, Kauppinen EI. Aerosol flow reactor method for the synthesis of multicomponent drug nano- and microparticles. In: Thassu D, Deleers M, Pathak YV, editors. Nanoparticulate drug delivery systems. CRC Press; 2007. p. 111–28.

Rayegan A, Allafchian A, Abdolhosseini I, Kameli P. Synthesis and characterization of basil seed mucilage coated Fe_3O_4 magnetic nanoparticles as a drug carrier for the controlled delivery of cephalexin. Int J Biomacromol. 2018;113:317–328.

Reisch A, Runser A, Arntz Y, Mély Y, Kylmchenko AS. Charge-controlled nanoprecipitation as a modular approach to ultrasmall polymer nanocarriers: Making bright and stable nanoparticles. ACS Nano. 2015;9:5104–16.

Rubahn HG. Basics of nanotechnology. 3rd ed. Wiley VCH; 2008.

Ruckmani K, Sivakumar M, Ganeshkumar PA. Methotrexate loaded solid lipid nanoparticles (SLN) for effective treatment of carcinoma. J Nanosci Nanotechnol. 2006;6(9–10):2991–5.

Rupp R, Rosenthal SL, Stanberry LR. VivaGel (SPL7013 Gel): A candidate dendrimer–microbicide for the prevention of HIV and HSV infection. Int J Nanomed. 2007;2(4):561–6.

Russel P, Batchelor D. SEM and AFM: Complementary techniques for surface investigations. In: Microscopy and analysis. John Wiley & Sons Ltd.; 2001. p. 9–12.

Sager TM, Castranova V. Surface area of particle administered versus mass in determining the pulmonary toxicity of ultrafine and fine carbon black: Comparison to ultrafine titanium dioxide. Part Fibre Toxicol. 2009;6:15–46.

Sager TM, Kommineni C, Castranova V. Pulmonary response to intratracheal instillation of ultrafine versus fine titanium dioxide: Role of particle surface area. Part Fibre Toxicol. 2008;5:17–32.

Sano N, Wang H, Alexandrou I, Chhowalla M, Teao KBK, Amaratunga GAJ. Properties of carbon onions produced by an arc discharge in water. J Appl Phys. 2002;92(51):2783.

Sano N, Wang H, Chhowalla M, Alexandrou I, Amaratunga GAJ. Synthesis of carbon 'onions' in water. Nature. 2001;414(6863):506–7.

Sant S, Nadeau V, Hildgen P. Effect of porosity on the release kinetics of propafenone-loaded PEG-g-PLA nanoparticles. J Control Rel. 2005;107:203–14.

Sapsford KE, Tyner KM, Dair BJ, Deschamps JR, Medintz IL. Analyzing nanomaterial bioconjugates: A review of current and emerging purification and characterization techniques. Anal Chem. 2011;83:4453–88.

Schaeublin NM, Braydich-Stolle LK, Schrand AM, Miller JM, Hutchison J, Schlager JJ, Hussain SM. Surface charge of gold nanoparticles mediates mechanism of toxicity. Nanoscale. 2011;3:410–20.

Schumann H, Wassermann BC, Schutte S, Velder J, Aksu Y, Krause W. Synthesis and characterization of water-soluble tin-based metallodendrimers. Organometallics. 2003;22(10):2034–41.

Serpe L, Catalano MG, Cavalli R, Ugazio E, Bosco O, Canaparo R, Muntoni E, Frairia R, Gasco MR, Eandi M, Zara GP. Cytotoxicity of anticancer drugs incorporated in solid lipid nanoparticles on HT-29 colorectal cancer cell line. Eur J Pharm Biopharm. 2004;58(3):673–80.

Shin SW, Song IH, Um SH. Role of physicochemical properties in nanoparticle toxicity. Nanomaterials. 2015;5:1351–65.

Sijbesma R, Srdanov G, Wudl F, Castoro JA, Wilkins C, Friedman SH, DeCamp DL, Kenyon GL. Synthesis of a fullerene derivative for the inhibition of HIV enzymes. J Am Chem Soc. 1993;115(15):6510–2.

Slater TF, Cheeseman KH, Ingold KU. Carbon tetrachloride toxicity as a model for studying free-radical mediated liver injury. Philos Trans R Soc Lond B Biol Sci. 1985;311(1152):633–45.

Smith BR, Heverhagen J, Knopp M, Schmalbrock P, Shapiro J, Shiomi M, Moldovan NI, Ferrari M, Lee SC. Localization to atherosclerotic plaque and biodistribution of biochemically derivatized superparamagnetic iron oxide nanoparticles (SPIONs) contrast particles for magnetic resonance imaging (MRI). Biomed Microdevices. 2007;9(5):719–27.

Spangler RS. Insulin administration via liposomes. Diabetes Care. 1990;13(9):911–22.

Sparnacci K, Laus M, Tondelli L, Magnani L, Bernardi C. Core-shell microspheres by dispersion polymerization as drug delivery systems. Macromol Chem Phys. 2002;203(10–11):1364–9.

Stewart PL. Cryo-electron microscopy and cryo-electron tomography of nanoparticles. Wiley Interdiscip. Rev. Nanomed. Nanobiotechnol. 2016;9(2):e1417–1433.

Stoeger T, Reinhard C, Takenaka S, Schroeppel A, Karg E, Ritter B, Heyder J, Schulz H. Instillation of six different ultrafine carbon particles indicates a surface area threshold dose for acute lung inflammation in mice. Environm Health Perspect. 2005;114:328–33.

Suzuki J, Fujino T. Matrix-assisted laser desorption ionization mass spectrometry of maltohexaose and acetyl-salicylic acid using alkali metal cation-substituted zeolites. Anal Sci. 2012;28:901–4.

Suzuki J, Sato A, Yamamoto R, Asano T, Shimosato T, Shima H, Kondo JN, Yamashita KI, Hashimoto K, Fujino T. Matrix-assisted laser desorption ionization using lithium-substituted mordenite surface. Chem Phys Lett. 2012;546:159–63.

Takemoto K, Yamamoto Y, Ueda Y, Sumita Y, Yoshida K, Niki Y. Comparative studies on the efficacy of AmBisome and Fungizone in a mouse model of disseminated aspergillosis. J Antimicrob Chemother. 2004;53(2):311–7.

Tanaka K, Waki H, Ido Y, Akita S, Yoshida Y, Yoshida T. Protein and polymer analyses up to MLZ 100 000 by laser ionization time-of-flight mass spectrometry. Rapid Commun Mass Spectrom. 1988;2: 151–3.

Tantra R, Schulze P, Quincey P. Effect of nanoparticle concentration on zetapotential measurement results and reproducibility. Particuology. 2010;8:279–85.

Templeton AC, Wuelfing WP, Murray RW. Monolayer protected cluster molecules. Acc Chem Res. 2000;33(1):27–36.

Terrones M, Jorio A, Endo M, Rao AM, Kim YA, Hayashi T, Terrones H, Charlier JC, Dresselhaus G, Dresselhaus M. New direction in nanotube science. Materials Today. 2004;7 (10):30.

Thaxton CS, Rosi NL, Mirkin CA. Optically and chemically encoded nanoparticle materials for DNA and protein detection. MRS Bull. 2005;30(5):376–80.

Thrash TP, Cagle DW, Alford JM, Ehrhardt GJ, Wright K, Mirzadeh S, Wilson LJ. Toward fullerene-based radiopharmaceuticals: High-yield neutron activation of endohedral 165Ho metallofullerenes. Chem Phys Lett. 1999;308(3–4):329–36.

Tiede K, Boxall ABA, Tear SP, Lewis J, David H, Hassellöv M. Detection and characterization of engineered nanoparticles in food and the environment. Food Addit Contam A. 2008;25:795–821.

Tomalia DA, Reyna LA, Svenson S. Dendrimers as multipurpose nanodevices for oncology drug delivery and diagnostic imaging. Biochem Soc Trans. 2007;35(1):61–7.

Tseng MC, Obena R, Lu YW, Lin PC, Lin PY, Yen YS, Lin JT, Huang LD, Lu KL, Lai LL. Dihydrobenzoic acid modified nanoparticle as a MALDI-TOF MS matrix for soft ionization and structure determination of small molecules with diverse structures. J Am Soc Mass Spectrom. 2010;21:1930–9.

Varadan VK, Chen L, Xie J. Nanomedicine: Design and applications of magnetic nanomaterials, nanosensors and nanosystems. John Wiley & Sons; 2008.

Vasenka J, Manne S, Giberson R, Marsh T, Henderson E. Colloidal gold particles as an incompressible AFM imaging standard for assessing the compressibility of biomolecules. Biophys J. 1993;65:992–7.

Wallace WE, Lewandowski H, Meier RJ. Reactive MALDI mass spectrometry of saturated hydrocarbons: A theoretical study. Int J Mass Spectrom. 2010;292:32–7.

Warheit DB, Reed KL, Sayes CM. A role for nanoparticle surface reactivity in facilitating pulmonary toxicity and development of a base set of hazard assays as a component of nanoparticle risk management. Inhal Toxicol. 2009;21:61–7.

Wetzel SJ, Guttman CM, Girard JE. The influence of matrix and laser energy on the molecular mass distribution of synthetic polymers obtained by MALDI-TOF-MS. Int J Mass Spectrom. 2004;238:215–25.

Wiener EC, Brechbiel MW, Brothers H, Magin RL, Gansow OA, Tomalia DA, Lauterbur PC. Dendrimer-based metal chelates: A new class of MRI contrast agents. Magn Reson Med. 1994;31(1):1–8.

Wong HL, Rauth AM, Bendayan RA, Manias JL, Ramaswamy M, Liu Z, Erhan SZ, Wu XY. New polymer-lipid hybrid nanoparticle system increases cytotoxicity of doxorubicin against multidrug-resistant human breast cancer cells. Pharm Res. 2006;23(7):1574–85.

Xia Y, Gates B, Yin Y, Lu Y. Monodispersed colloidal spheres: Old materials with new applications. Adv Mater. 2000;12(10):693–713.

Yamamoto R, Fujino T. 2,4,6-Trihydroxyacetophenone on zeolite surface: Correlation between electronic relaxation and fragmentation on mass spectra. Chem Phys Lett. 2012;543:76–81.

Yang H, Yuan B, Zhang X, Scherman OA. Supramolecular chemistry at interfaces: Host-guest interactions for fabricating multifunctional biointerfaces. Acc. Chem Res. 2014;47:2106–15.

Yang M, Hashimoto K, Fujino T. Silver nanoparticles loaded on ammonium exchanged zeolite as matrix for MALDI-TOF-MS analysis of short-chain n-alkanes. 2018;706:525–532.

Yang M, Fujino T. Silver nanoparticles on zeolite surface for laser desorption/ionization mass spectrometry of low molecular weight compounds. Chem Phys Lett. 2013;576:61–4.

Yang M, and Fujino T. Gold nanoparticles loaded on zeolite as inorganic matrix for laser desorption/ionization mass spectrometry of small molecules. Chem Phys Lett. 2014a;592:160–3.

Yang M, and Fujino T. Quantitative analysis of free fatty acids in human serum using biexciton auger recombination in cadmium telluride nanoparticles loaded on zeolite. Anal Chem. 2014b;86:9563–9.

Yang SC, Lu LF, Cai Y, Zhu JB, Liang BW, Yang CZ. Body distribution in mice of intravenously injected camptothecin solid lipid nanoparticles and targeting effect on brain. J Control Rel. 1999;59(3):299–307.

Yoo MK, Kim IY, Kim EM, Jeong HJ, Lee CM, Jeong YY, Akaike T, Cho CS. Superparamagnetic iron oxide nanoparticles coated with galactose-carrying polymer for hepatocyte targeting. J Biomed Biotechnol. 2007;10:94740.

Yu L. Amorphous pharmaceutical solids: Preparation, characterization and stabilization. Adv Drug Deliv Rev. 2001;48:27–42.

Zhang L, Granick S. How to stabilize phospholipid liposomes (using nanoparticles). Nano Lett. 2006;6(4):694–8.

Zhang L, Pornpattananangkul D, Hu CMJ, Huang CM. Development of nanoparticles for antimicrobial drug delivery. Curr Med Chem. 2010;17(6):585–94.

Zhang M, Shi Z, Bai Y, Gao Y, Hu R, Zhao F. Using molecular recognition of β-cyclodextrin to determine molecular weights of low-molecular-weight explosives by MALDI-TOF mass spectrometry. J Am Soc Mass Spectrom. 2006;17:189–93.

Zhao F, Zhao Y, Liu Y, Chang X, Chen C, Zhao Y. Cellular uptake, intracellular trafficking, and cytotoxicity of nanomaterials. Small. 2011;7:1322–37.

Zhang S, Gelain F, Zhao X. Designer self-assembling peptide nanofiber scaffolds for 3D tissue cell cultures. Semin Cancer Biol. 2005;15(5):413–20.

zur Mühlen A, Mehnert W. Drug release and release mechanism of prednisolone loaded solid lipid nanoparticles. Pharmazie. 1998;53:552–5.

zur Mühlen A, von Elverfeldt D, Bassler N, Neudorfer I, Steitz B, Petri-Fink A, Hofmann H, Bode C, Peter K. Superparamagnetic iron oxide binding and uptake as imaged by magnetic resonance is mediated by the integrin receptor Mac-1 (CD11b/CD18): Implications on imaging of atherosclerotic plaques. Atherosclerosis. 2007;193(1):102–11.

3

Therapeutic Nanostructures in Antitubercular Therapy

Paulami Pal, Subhabrata Ray, Anup Kumar Das, and Bhaskar Mazumder

CONTENTS

ABSTRACT: Nano-biomaterials have immense potential as carriers in nanotherapeutic therapy of chronic and resurgent tuberculosis that can affect several human organs, including the brain, the kidneys, the bones, and, primarily, the lungs. World Health Organization (WHO) recommended antitubercular drugs are well known for their toxicity (especially hepatotoxicity), drug interactions, and multidrug resistance problems – a perfect case for the application of nanobiotechnology. In this chapter, the strategies of utilizing nanostructures in addressing lacunae in the therapy of tuberculosis will be discussed. Targeted use of nanotechnology will help to overcome the biological barriers and finally develop an effective dose regimen for the treatment of this deadly disease. The different nanometer scale technologies, materials, systems, and their routes of administration will be discussed, as well as possible toxicity concerns and future prospects. Primarily, pulmonary tuberculosis is the main focus here, however, the potential of targeted delivery to other sites of the infection has also been addressed.

KEY WORDS: *Therapeutic nanostructures, tuberculosis, antitubercular therapy, lung targeted therapy, nanotoxicity*

3.1 Introduction

Tuberculosis (TB) is one of the oldest diseases known to mankind, though its true existence was never established until the early 1980s with its increasing incidence and the development of new resistant forms. Airborne transmission of TB is a result of incompletely characterized host, bacterial and environmental factors. TB is a leading cause of death in youthful, healthy adults globally and the universal curse of multidrug resistant (MDR)-TB has reached epidemic proportions. It is endemic in most of the developing countries, whereas resurgence in developed and developing countries continues with a high rate of infection with the human immunodeficiency virus (HIV). TB is currently the second largest killer worldwide, after AIDS under the head of infection/microbe-caused diseases.

TB infection may spread with the inhalation of airborne particles contaminated with *Mycobacterium tuberculosis* (of particles less than 5 µm in size) (Edwards and Kirkpatrick, 1986). Once the bacilli reach the alveolar network, they are supposed to be ingested by the alveolar macrophage system (lungs), the primary refuge of the tubercular bacilli. The remaining bacilli multiply within the macrophage and sooner or later spread to other organs of the body. In the case of patients infected with HIV, the macrophage system malfunctions in response to simultaneous infection with *M. tuberculosis*, which increases susceptibility to TB (Patel et al., 2009). In spite of this, there is no convincing data that HIV-seropositive people are more likely to get a TB infection than HIV-seronegative individuals, where they are known to have the same degree of exposure (Meltzer et al., 1990; Whalen et al., 2011).

3.2 Biomaterial and Its Present Status in Drug Delivery and Therapeutics

A nano-biomaterial is any engineered entity demonstrating nanoscopic characteristics. They may be natural or synthetic constructs, developed using a variety of physicochemical approaches utilizing metals, biological materials, polymers/composites, or ceramics, etc. Nano-biomaterials have immense potential as nanotherapeutic systems in antimicrobial therapy, as they can be targeted to specific sites of infection, thereby reducing systemic toxicity. Different biodegradable biomaterials that are generally used or can be used to develop therapeutic nanoparticles are listed in Table 3.1.

3.3 Tuberculosis in the Twenty-First Century

3.3.1 Tuberculosis and Its Status

This highly ubiquitous, contagious, chronic, recurrent, granulomatous bacterial disease is caused by the *M. tuberculosis*. It affects several vital organs of the human body, including the brain, kidneys, liver, bone marrow, lymph nodes, and spleen, but most commonly it affects the lungs, in which case it is clinically termed as pulmonary TB. Tubercular bacilli always target the macrophage cells deep inside the alveolar sac within the lungs. Researchers have proved that TB is becoming a bigger threat with patients suffering from AIDS/HIV, making the situation even worse. In this context, it is worth mentioning that TB is a major cause of concern in the African and Indian subcontinents due to the increasing incidence of AIDS in these regions. The emergence of TB among children is also a major point of concern in society when medical practitioners are yet to standardize the ideal pediatric dose regimen for its treatment. Drug resistance, diabetes, smoking, poor hygiene, and malnutrition makes the TB scenario worse and more difficult to control.

In 2012, the World Health Organization (WHO) reported that an estimated 8.6 million people developed TB and 1.3 million died from the disease (including 320,000 deaths among HIV-positive people) worldwide. Although most TB cases and deaths occur among men, burden of the disease is also high among women and children. In 2012, an estimated 410,000 women died from TB (250,000 among

TABLE 3.1

List of Biomaterials/Nano-Biomaterials That Can Be or Are Generally Used as Carriers in Targeted Drug Delivery

Biomaterials		
Lipids	Triacylglycerols	Tricaprin, trilaurin, trimyristin, tripalmitin, tristearin
	Acylglycerols	Glycerol monostearate, glycerol behenate, glycerol palmitostearate
	Fatty acids	Stearic acid, palmitic acid, decanoic acid, behenic acid
	Waxes	Cetyl palmitate, yellow carnauba wax
Polymers	Cyclic complexes	Cyclodextrin
	Natural biodegradable polymer	Chitosan, fibrin, collagen, gelatin, hyaluronan, chitin, starch, cellulose
	Semisynthetic biodegradable polymer	Microbial polymers: poly(β-hydroxyalcanoate) (PHA), Poly(hydroxybutyrate) (PHB), Poly(hydroxybutyrate-co-hydroxyvalerate) (PHBV)
		Other Sources: starch-poly(ethylene-co-vinyl alcohol) (EVOH), starch-polyvinyl alcohol, starch-PVA, starch-PHB, starch-PBS
	Synthetic biodegradable polymer: aliphatic polyesters	Polylactic acid (PLA), Polyglutamic acid (PGA), Poly(lactide-co-glycolic) acid (PLGA), Polycaprolactone (PCL), polyorthoester, poly(dioxanone), poly(anhydride), poly(trimethylene carbonate), polyphosphazene, poly(butylene succinate) (PBS), poly(p-dioxanone) (PPDO)
Ceramics, metals, proteins, peptides, DNA, RNA		
Surfactants and co-surfactants	Sorbitan ethylene oxide	Polysorbate 20, polysorbate 60, polysorbate 80
	Phospholipids	Soy lecithin, egg lecithin, phosphatidylcholine
	Bile salts	Sodium cholate, sodium glycocholate, sodium taurocholate, sodium taurodeoxycholate
	Alcohols	Ethanol, butanol
	Ethylene oxide	Poloxamer 188, poloxamer 182, poloxamer 407, poloxamine 908, poloxamer 181
	Alkylaryl polyether alcohol polymers	Tyloxapol

HIV-negative women and 160,000 among HIV-positive women). There were also an estimated 74,000 reported cases of child death due to TB. Apart from this, in India alone 1,289,000 TB cases were reported in 2012 in which 106,000 were relapse cases of TB. Also 2.2% of cases with MDR-TB were reported in the 21,000 notified cases with pulmonary TB (WHO, 2012).

The pathogen *M. tuberculosis* belongs to the *Mycobacteriaceae* family. The family is comprised of several slow-growing, acid-fast bacilli, residing mostly in soil and water, which play a vital role in the degradation of organic materials. Five species are capable of causing TB in human beings, three of which are extremely pathogenic. *M. tuberculosis*, *M. africanum*, and *M. bovis* are the three high-risk, pathogenic bacilli that are capable of causing TB in humans. Among these, *M. tuberculosis* is the most popular causative agent of TB in the vast majority of human beings.

M. canettii and *M. microti* can also cause TB in humans, but these are rare etiologic microbes. This group of five intimately-related bacteria capable of causing TB in humans is referred to as the "mycobacterium complex" collectively. Due to the fact of dominance in human infection and the global disease burden, our discussion will focus on infections caused by *M. tuberculosis*.

Thin, rod-shaped mycobacteria are about 4 μm by 0.3 μm in size. The cell wall of these aerobic bacilli contains a high concentration of heavy molecular weight lipids (Figure 3.1). With effect to this, these organisms behave in a hydrophobic manner and hence are resistant to water-based bactericidal agents. Furthermore, due to their hydrophobic nature, they do not dry out, which is very important for the infectivity of *M. tuberculosis*. They show very slow growth in culture, which can obstruct diagnosis for patients with active TB. It requires almost four to six weeks as a growing period on solid media and 9 to 16 days when cultured in rapid liquid cultures.

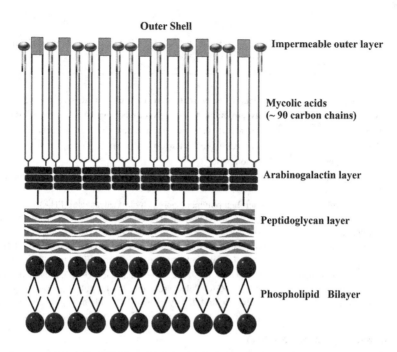

FIGURE 3.1 Structure of the cell wall of *M. tuberculosis*.

3.3.2 Types and Resistant Forms of Tuberculosis

Infection with *M. tuberculosis* originates with dormant infection that slowly steps forward to form an active disease. The primary infection is usually asymptomatic in the majority of cases, although a small number go on to have symptomatic haematological dissemination which may result in miliary TB. Only 5% of patients, usually those with impaired immunity, go on to have progressive primary TB.

Latent tuberculosis infection (LTBI) is mostly asymptomatic and does not represent an active disease state. The majority of people infected with LTBI are unaware of their clinical state of infection. After primary infection, the immune system of the infected person controls as well as restrains the infection, however, classically, does not abolish the infection. Around 5% to 10% of infected individuals develop primary TB as they are not capable of controlling the initial infection, while the rest of the population are left with LTBI. It is an amazing capacity of *M. tuberculosis* to remain in this dormant state within the macrophages and other cells for decades. These dormant bacteria may again become active in about 5% to 10% of individuals having LTBI as a result of various factors, such as a compromised immune system. Active TB is often termed as reactivation TB amid this group of infected individuals (Figure 3.2).

TB can be both chronic and acute, with symptoms like coughing, with sputum with or without blood, night sweats, fever with chills, anorexia, weight loss, fatigue, and sometimes chest pain. Similar symptoms are also observed in cases of extra pulmonary TB, but prominent symptoms will primarily influence the affected organs.

Clinical cases with TB mostly show presence of infection in the lungs and respiratory tract. Nevertheless, the tubercular pathogens are potent enough to affect almost any organ of the human body. Active TB can be evident as pulmonary or extra pulmonary disease irrespective of whether the affected person is a primary or reactivation case. While extra pulmonary TB can be seen more frequently in immune-compromised people, almost 80% of tubercular cases are pulmonary among individuals with superior immune function.

Extra pulmonary TB is mostly observed with the gastrointestinal tract, pericardium, lymph nodes, genitourinary tract, bones, liver, kidney, meninges, adrenal glands, eyes, and skin (Nancy C E., 1992).

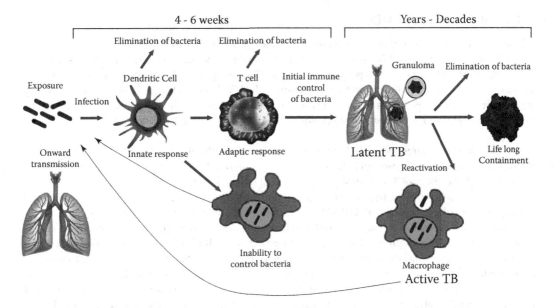

FIGURE 3.2 Pathogenesis of *M. tuberculosis.*

Also, systemic TB may occur when *M. tuberculosis* is circulated throughout the body by the lymph or blood circulation systems, with diminutive bruising appearing in major organs. Disseminated TB was initially termed as "miliary" TB because of the fact that its lesions showed up looking like grains of millet; it is most commonly seen in children and immune-compromised patients.

The primary mode of transmission of TB is by airborne microorganisms. It is to be noted here that although people with LTBI are reservoirs for the bacilli, they are not transmitters of the disease. The organism is transmitted from a host with active pulmonary TB through the release of tiny droplets into the air when they communicate or cough. Transmission of infection between hosts and susceptible persons is governed by several factors like immune-competence of the susceptible person, intimacy and frequency of contact, severity of disease state in the host, and other social conditions. Apart from airborne transmission, these bacilli can also be transmitted through the genitourinary tract, gut, and compromised skin. Such modes of transmission are relatively rare, but they are familiar in areas of high prevalence of the disease.

The usual record of TB directly affects its transmission among a populace.

Stage 1: In a given community in a definite time duration, all the inhabitants with LTBI act as a reservoir for the pathogen.

Stage 2: A certain percentage of individuals bearing latent TB develop secondary TB each year.

Stage 3: Cases of new mycobacterium infections are cited when the carriers of active TB transmit the bacteria to liable contacts.

Stage 4: A small percentage of the patients freshly infected by this secondary infection are susceptible to developing primary TB, and thus become new potential transmitters to their fellow contacts. The remaining newly-infected individuals remain with LTBI and continue to serve as pathogen reservoirs.

The most significant attribute of the TB epidemiology surfaces at a vital point where the physical and social backgrounds unite. Overcrowding of population density is often noticed in areas where there is a TB epidemic. For TB to become endemic, regular close communication and contact between infected and susceptible individuals is compulsory. Other factors such as dwelling circumstances, like poor ventilation due to the dense construction of houses, and extensive proximity between family members and neighbours are also responsible.

3.3.3 Tuberculosis and AIDS

It is worth mentioning here that simultaneous infection with HIV is doubtless the most significant risk factor for initiation of active TB from latent TB. The risk of re-emergence of active TB ranges from 3% to 14% per year, and averages 10% per year (Walsh, 2013), among individuals who were susceptible and also a recessive host of LTBI prior to their infection with HIV. People who are HIV-positive are highly susceptible to active TB. That's because TB benefits from a damaged immune system, which is why it's called an opportunistic infection. Worldwide, TB is the leading cause of death for people infected with HIV.

The detrimental progression of HIV infection is synergistically affected by TB mainly because of the increased immune activation and also the overexpression of CXCR4 and CCR5 coreceptors on CD4 cells, although the effect of TB on HIV ribonucleic acid (RNA) is not as certain.

This noxious disease is known to have demoralizing effects on the socioeconomic development of a country because of its alliance with HIV/AIDS, malnutrition, and poor hygiene in major sectors of society, where poverty reigns over education and health awareness. Non-compliance with the present choice of tubercular treatment regimen along with co-infection with HIV/AIDS led to the appearance of a noxious form of tubercular strain which is resistant to all forms of antitubercular drugs. This extensively drug resistant form of TB (known as XDR-TB) is not widely curable, but some TB control programs have shown that a cure is possible for an estimated 30% of affected people. Successful outcomes of treatment depend greatly on the extent of the drug resistance, the severity of the disease, and whether the patient's immune system is weakened (Martin and Portaels, 2007; MedicineNet, 2007). TB may occur in any stage of HIV infection, even in the early stages. There is a chance of increased contamination with bacilli along with the first instance of HIV in the host and of dispersed infection as the CD4 cell count decreases. Even with effective immune reconstitution with antiretroviral therapy, the risk of TB remains generally high in patients infected with HIV, above the background risk of the entire population, even with high numbers of CD4 cells (Gupta et al., 2012; Moore et al., 2007; Van Rie et al., 2011).

WHO estimates that one-third of the global population is infected with *M. tuberculosis*, among which 14.8% of these TB patients have AIDS. Now 50–80% of TB-AIDS co-infected patients originate from parts of Sub-Saharan Africa. The frequency of TB associated with HIV/AIDS reached 1.39 million in 2005, however it is decreasing now. Still, looking at the bigger picture, TB kills 1 in 3 patients suffering from AIDS (Raviglione et al., 1995; Zwang et al., 2007). Additionally, in the United States an increase in TB cases was observed in the mid-1980s (Luetkemeyer, 2013).

3.3.4 Current Mode of Treatment and Its Challenges

The current treatment of pulmonary TB involves long term oral treatment with high doses of antitubercular drugs, which are associated with unwanted side effects like hepatotoxicity, drug–drug incompatibility, and poor patient compliance, apart from the emergence of drug resistance. The five "first-line" antitubercular agents used at present are: isoniazid, rifampicin, pyrazinamide, ethambutol, and streptomycin. Treatment of TB is an ongoing process. The initial six months of treatment is mandatory due to the existence of the assorted population of both active and dormant tubercular bacilli in an infected human being. Effective medication against active mycobacteria are not always effective against latent mycobacteria, thus extended treatment ensures complete destruction of the whole population of *M. tuberculosis*. Prolonged treatment with antibiotics helps in the emergence of drug resistance in *M. tuberculosis* due to genetic mutation. Thus, to reduce the possibility of developing drug-resistant bacilli, at least two effective drugs must be coadministered. Adherence to treatment with the complete schema is essential for the success of the treatment. Non-compliance may lead to therapeutic failure in the individual, but also to the development of *M. tuberculosis* forms that are antibiotic resistant. To limit the spread of resistance to the antibiotic in the population, short durations for the therapy are recommended under direct observation treatment (DOT). The therapy is composed of four actives, namely isoniazid, rifampicin, ethambutol, and pyrazinamide, for two months, following which isoniazid and rifampicin need to be further administered for an additional four months. DOT requires that a health worker must closely follow every TB patient and observe the patient take every dose of drugs against TB.

However, antibiotic resistance has been a difficult problem at hand for several years, now reaching a state of crisis whereby certain strains of *M. tuberculosis* are resistant to the two most powerful drugs, namely isoniazid and rifampicin. These strains lead to a disease condition which can be referred to as MDR-TB. To make the situation even worse, the development of resistance by other strains leads to inadequate management of TB. These MDR strains are also resistant to isoniazid and rifampin, as well as being resistant to at least one of the quinolones, capreomycin, kanamycin, or amikacin. The disease caused by such strains is known as extensively drug-resistant TB (XDR-TB).

DOT is at the center of the "Stop TB Strategy" agenda by WHO. The five fundamentals of the DOT approach (as outlined in Figure 3.3) need special attention to enable known constraints to be addressed and new challenges met (WHO, 2012).

3.3.4.1 Prevention and Control

There are several key factors that need to be implemented to achieve an effective TB prevention and control program. Treatment and case-first, rigorous research is obviously crucial to save the individual as well as to interrupt the transmission of the infection to the contacts. First, the case identification must combine microscopy and clinical symptoms, and treatment should consist of short duration therapy under DOT. Second, comprehensive contact tracing for contacts of each active TB case must proceed in the field so that new cases of infection can be identified and treated before becoming active. Third, a good monitoring system is fundamental to the control of any infectious disease. A system of registration of cases and the monitoring of outcomes is necessary to estimate the onset of the disease and identify temporal trends and spatial clusters. In addition, molecular epidemiology surveillance instruments are needed to monitor resistance to the antibiotic in *M. tuberculosis*. Fourth, an adequate supply of TB drugs must be available for population endemic TB. This may seem obvious, and it is, but unfortunately the lack of a steady supply of drugs has hindered control in many particular programs in poor areas of the third world. Fifth, there must be a commitment from the relevant countries' governments to the fight against TB. Important resources and infrastructure for public health and sustainability, including the

Political commitment with increased and sustained financing: Government attention to increase awareness and contestant financing: Planning, management, training and field work along with human resources.

Case detection through quality-assured bacteriology: Strengthening TB laboratories, drug resistance surveillance.

Standardized treatment with supervision and patient support: TB treatment and programmers management guidelines, International. (ISTC), PPM, Practical Approach to Lung Health (PAL), community-patient involvement

An effective drug supply and management system: Availability of antitubercular drugs, dose regimen and management, Global Drug Facility (GDF), Green Light Committee (GLC)

Monitoring and evaluation system and impact measurement: TB recording and reporting systems, Global TB Control Report, data and country profiles, TB planning and budgeting tool, WHO epidemiology and surveillance online training.

FIGURE 3.3 The five components of DOT.

staff required to apply and maintain control of the TB programs, is essential because the fight against this disease requires a long-term effort. Therefore, a strong commitment by government agencies who can mobilize the necessary resources and infrastructure is essential to the regional fight against TB.

3.3.5 Short Comings in Present Treatment Methodology

The present treatment of TB, as discussed earlier, is with the long-term administration of high doses of antitubercular drugs like isoniazide, rifampicin, pyrazinamide, and ethambutal. But these drugs suffer from drug–drug interaction, first-pass metabolism, high hepatotoxicity, and poor patient compliance, apart from the emergence of drug resistance.

The limitations associated with the oral delivery of these drugs include a slower onset of action and/or considerable first-pass effect by the live drugs being poorly absorbed due to gastrointestinal chemicals. In contrast with the oral route, the parenteral route shows the highest bioavailability and is not subject to first-pass metabolism or degradation due to harsh gastrointestinal environments. This route of drug administration also provides the bioavailable dose at the site of infection (Pham et al., 2015). Nevertheless, parenteral drug administration is often alleged as a painful route of administration, leading to discomfort for the patient, and also requires trained hands for administration. It is important to note that both of these conventional routes may lead to subtherapeutic levels of antitubercular drugs at the site of infection due to poor pulmonary distribution. As a result, the chance of drug-resistant strain appearance may be rather high (Conte et al., 2002). Even if therapeutic drug concentrations are achieved clinically in the regions of infection, conventionally administered antitubercular drugs may lack the potential to penetrate into the granulomas-protected mycobacterium bacilli. With a narrow therapeutic window and a prolonged treatment regimen, antitubercular drugs have less patient compliance, increasing the chances of the emergence of drug-resistant strains. The complete duration of treatment is also limited by constraints on drug dosage, systemic adverse drug reaction, which is exacerbated in HIV patients, along with insufficient drug dose at the pathological site (Sacks et al., 2001).

Multiple drug treatment is restricted in many cases due to the degradation of rifampicin in the presence of isoniazid due to the drug–drug interaction (as rifampicin is reduced by isoniazid) and increasing the dose or through the incorporation of antioxidants. There is also evidence of degradation of rifampicin and isoniazid in the gastric environment, although pyrazinamide has been found to be stable. Hence, a segregated delivery of rifampicin and isoniazid for improved rifampicin bioavailability could be a step in the right direction to address the issue of treatment failure. However, this increases patient incompliance, as these medicines are suggested to be taken along with and in between food intake.

But successful lung-targeting with these antibiotics will help to combat the incompatibility problems associated with the model antitubercular drugs proposed to be used in the investigation, avoid instability in the blood stream and uptake by the macrophages polymorphonuclear system (MPS), and also bypass the fast-pass metabolism in the liver, resulting in better drug utilization at the site of infection.

3.4 Need and Use of Targeted Nanotechnology in Treatment of Tuberculosis

In general, TB drugs have a good oral absorption profile with a bioavailability of about 80%, for except rifampicin which has a 50% bioavailability, approximately. Rifampicin is unstable in the acidic environment of the stomach and its degradation is accelerated in the presence of isoniazid, which reduces the bioavailable dose for absorption of rifampicin. This degradation of rifampicin in such *in vivo* conditions is clinically significant and a decrease of up to 33% in its bioavailability can be observed. In the gastrointestinal tract, the solubility of the antitubercular drugs is another issue directly reflecting on their bioavailability. However, the main limitation of the drugs for this particular therapy appears to be their half-life. Rifampicin has a plasma half-life of about 2 to 5 hours whereas the half-life of isoniazid ranges between 1.06 to 6.99 hours, depending on whether the patient on the medication is a slow or rapid acetylator of the drug. The drugs used in drug-resistant TB also exhibit short half-lives, for example, ethionamide exhibits a half-life between 1.7 and 2.1 hours. Some authors suggest that current administration of short half-life drugs (e.g., rifampicin) together with longer half-life drugs (e.g., isoniazid, pyrazinamide) can be referred to as a pharmacokinetic (PK) mismatch, which is a significant cause of bacterial drug resistance (Dube et al., 2013).

Nanotechnology overcomes several lacunae in the conventional dosage forms. Developing a nanoparticle-based system and targeting it to the site of colonization of the microorganisms in the lungs will help to increase the efficacy of the dose regimen and will reduce the toxicity compared to the oral route through optimal drug utilization. Solid lipid nanoparticles (SLNs) combine the advantage of both polymeric nanoparticles and liposomes with the possibility of controlled drug release and drug targeting, increased drug stability, incorporation of both lipophilic and hydrophilic drugs, etc. The proposed approach for formulation development will not only help to combat the associated problems with these model antitubercular drugs, but also will result in better drug utilization at the site of infection.

In spite of the emergence of new drugs and therapy regimens with antitubercular drugs, the combination therapy of isoniazid, rifampicin, and pyrazinamide remains the golden standard for TB treatment. Though it has hepatotoxicity and drug–drug interactions to some extent, having a reasonable efficacy/toxicity ratio has led it to be one of the most acceptable antitubercular combination agents.

Several authors have developed nanoformulations and lipid-based formulations with the combination of these three drugs, but these formulations suffer from problems like drug incompatibility, drug resistance, increased toxicity due to non-targeted delivery through the oral route, partial gastric degradation of the drugs, etc. The problem becomes severe when the chance of TB incidence increases with HIV/AIDS victims. Developing a lipid-based nanoparticle system and delivering it to the lungs by inhalation will help to release the drug at the site of infection and will reduce the toxicity compared to the oral route (Figure 3.4). Though a few scientists have tried to develop a nanoparticle system for antitubercular drugs, poor loading and combining the drugs in the same formulation owing to drug incompatibility still remain problems. Moreover, there has not been much work done on nanoparticle delivery by inhalation.

3.5 Various Nanoscale Technologies for Antitubercular Therapy

The clinical management of mycobacterial diseases, with a special emphasis on TB, through anti-mycobacterial chemotherapy remains a tricky assignment. It is a classical treatment with a long-term therapy where the drugs do not reach the site of mycobacterial infection in the macrophages in sufficient amounts and/or do not persist for a long enough time to develop the desired effect. Also, the available therapeutic agents induce a severe hepatotoxicity effect.

Nanotechnology has shown considerable impact on the pharmacodynamics of a drug candidate by re-engineering the drug delivery systems for their potential to target the phagocytic cells infected by intracellular pathogens, such as mycobacteria. The application of nanotechnology helps in increasing the therapeutic index of antimicrobacterial drugs, improving the solubility of the hydrophobic agents, and reducing dosing frequency, along with allowing the administration of higher doses. Many researchers have formulated liposomes, polymeric nanoparticles, SLNs, NLCs (Nanostructured Liquid Carriers), nanosuspensions, etc., with antitubercular drugs and tried to target them to the lungs or the primary site of infection.

Pyrgiotakis et al. (2014) have tried to inactivate *M. parafortuitum* through engineered water nanostructures (EWNS). The EWNS will interact with airborne mycobacteria and inactivate them by significantly reducing their concentration levels. Moreover, EWNS has also been shown to inactivate *M. parafortuitum* by effectively delivering reactive oxygen species, encapsulated during the electrospray process, to the bacteria, oxidizing their cell membrane which results in inactivation.

Spray-dried capreomycin inhalable powders were prepared for pulmonary delivery for the targeted treatment of TB (Schoubben et al., 2014). The researcher developed two capreomycin sulfate powders with suitable properties for inhalation by using the nanospray-dryer B-90.

Nanostructured, particulate, antitubercular drug carriers for suitable inhalable delivery, including liposomes (Justo and Moraes, 2003; Kurunov et al., 1995; Pandey et al., 2004; Vyas et al., 2004) and nanoparticles (Pandey et al., 2003a; Pandey et al., 2003b; Pandey and Khuller, 2005a ; Sharma et al., 2004), have been established by various researchers. From their studies, it was revealed that these nano

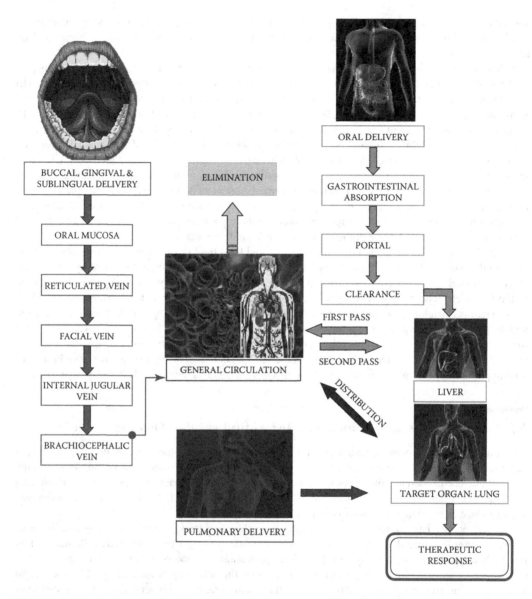

FIGURE 3.4   Need for lung-targeting delivery systems for the treatment of TB.

drug-loaded systems have the potential to lower the high drug doses and reduce toxicity, along with an effective change in dosing frequency, which would be beneficial to patients with pulmonary TB for prolonged treatment.

Sustained release formulation based on a zinc-layered, hydroxide nanocomposite of an antitubercular drug, 4-amino salicylic acid, was formulated and these developed nanocomposites showed reduced cell viability in a dose- and time-dependent manner when treating normal 3T3 cells (Saifullah et al., 2013).

3.6 Methods of Preparation for Nanoparticles

The various methods of development for both polymeric and lipoidal nanoparticles that are commonly used have been discussed below.

3.6.1 High Pressure Homogenization

The most effective and powerful technique in modern times for formulation of nanoparticles is high pressure homogenization, although it has been used in developing nanoemulsions for several years. In this method, the liquid premix is passed through a narrow orifice of a few microns under high pressure (around 100–2,000 bar) at a high velocity (over 1,000 km/hr) for a very short distance. The optimization of the particle size mainly depends on shear stress and cavitation force. There are two methods of nanoparticle production by this process, namely "hot homogenization" and "cold homogenization" (Mehnert and Mader, 2001).

Thermolabile substances cannot be subjected to the hot homogenization method as initially the drug and the lipid are melted together before being dispersed in a hot, aqueous, surfactant mixture. The high temperature results in a decrease of particle size due to the lower viscosity in the inner phase, but at the same time may cause drug and lipid degradation. The number of homogenization cycles and the high pressure must be optimized during formulation development, or else an increase in either or both might result in an increase in particle size due to particle coalescence as a result of the high kinetic energy of the particle. The initial product of hot homogenization is always a nanoemulsion, as until that time the lipid is in a liquid form. SLN is formed only after subsequent cooling of this nanoemulsion to room temperature or below, and the resultant product remains as a supercooled melt for several years by the virtue of its small size and the presence of the surfactant.

Cold homogenization, on the other hand, is carried out at room temperature with solid lipids, followed by high pressure milling of the cold premix. The quality of the premix greatly affects the quality and nature of the final formulation. Naturally, in comparison with the hot homogenization technique, cold homogenization yields dispersion in a polydispersed formulation with a larger particle size. Still it helps to bypass the problems related to hot homogenization, like temperature induced drug degradation, drug distribution in the aqueous phase, chances of crystallization, etc.

3.6.2 High Shear Homogenization and/or Ultrasound

High shear homogenization and ultrasound are together effective emancipating techniques which are well used to make nanoparticles, especially SLNs. But usually the obtained nanodispersion remains in a state of polydispersed formulation, with a co-existence of microparticles. Moreover, metal contamination might take place if the ultrasound is used for the reduction of the particle size. It is reported in many cases that an increase in shear rate does not always reflect in smaller particles but it plays a role in improving the polydispersity state of the dispersion. The particle size and zeta potential controlling parameters for this process are emulsification time, shear rate in terms of stirring rate, and, most importantly, the cooling conditions; all these factors are experiment specific depending upon the drug candidate chosen, the other excipients used, and what needs to be optimized (Mehnert and Mader, 2001).

3.6.3 Solvent Diffusion Method

The polymeric nanoparticle was first developed by the solvent diffusion method by Leroux et al. (1995) and the method was further developed and patented by Quintanar-Guerrero et al. (1996). In this method, the carrier is solubilized in partially-soluble organic solvent containing the drug in the same or in a different volatile solvent. Commonly used water-miscible organic solvents are isobutyric acid, butyl lactate, benzyl alcohol acid, and tetrahydrofuran, etc. The organic solvent is first saturated with water to attend to the thermodynamic equilibrium, followed by solubilization of the carrier, i.e., polymer or lipid. With this, the solvent-saturated aqueous phase containing the surfactant is emulsified to form the primary emulsion. The obtained primary emulsion is again emulsified within an aqueous phase, so that the nanoparticles can be obtained by extraction of the solvent into the external water phase.

Antitubercular formulation containing isoniazide, rifampicin, and pyrazinamide was developed by Rajesh Pandey and his group by using the previously mentioned method (Pandey et al., 2005a).

3.6.4 Solvent Evaporation Technique

Solvent evaporation is a well-established technique used in the preparation of pseudolatex (Vanderhoff et al., 1979). Nanoparticles can also be obtained by this method by dissolving the lipid or polymer in a suitable organic solvent, with or without drugs, in the same organic phase, followed by emulsification with aqueous surfactant solution. Later, complete removal of the organic phase is necessary under low pressure with constant stirring.

Maraicar and his group formulated an antitubercular SLN formulation of isoniazid by this method (Maraicar et al., 2014).

3.6.5 Double Emulsion Technique

In double emulsion technique, the hydrophilic active ingredient is solubilized in the internal water phase of a water-in-oil-in-water (w/o/w) emulsion with the addition of stabilizers. This is required to prevent drug loss to the external aqueous phase during solvent evaporation. In this method, an aqueous surfactant solution of the drug is emulsified in molten lipids to give a primary water-in-oil (w/o) emulsion. Thereafter, the primary emulsion is dispersed in a second aqueous stabilizer solution under constant stirring at an optimized rpm to obtain the final w/o/w emulsion. Constant stirring for longer periods leads to precipitation of the nanoparticles.

This method is also applicable for preparation of polymeric nanoparticles, when the drug and polymer is dissolved in organic solvents before making the primary emulsion, then the rest of the process remains the same. During the process, the organic solvent gets evaporated and the dispersion is kept for overnight stirring for the purpose of complete removal of the organic phase from the formulation. Nanoparticles thus obtained by the double emulsion technique are reported to have a particle size of less than 300 nm (and maybe as low as about 100 nm also) with a uniform range of size distribution.

Antitubercular nanoparticulate formulation was developed and patented by Kalombo and the double emulsion technique was used to get optimal drug loading of isoniazid-loaded poly(lactide-co-glycolic) acid (PLGA) nanoparticles (Kalombo, 2010).

3.6.6 Solvent Displacement Method

Solvent displacement, better known as the solvent injection technique, is a modification of the solvent diffusion method. It is mainly used for the preparation of microspheres, liposomes, and polymeric nanoparticles. In this technique, the drug-dissolved organic phase is spontaneously added to an aqueous surfactant mixture under continuous stirring. The aqueous phase also contains lipids solubilized in a semi-polar, water-miscible solvent or water-soluble solvent mixture. The nanoparticles are obtained by precipitation as the solvent is distributed in the continuous aqueous phase. The general particle size of nanoparticles obtained by this method mostly ranges from 100 to 250 nm. The particle size depends on the rate at which the organic solvent diffuses through the lipid-solvent barrier.

A modified method known as "solvent injection lyophilization" has been reported by Wang et al. (2010). In this method, the usual procedure of the classical solvent injection technique is followed, however, the organic solvent is directly injected into the lyoprotectant-rich aqueous medium with stirring. This yields lipid nanoparticles dispersed in the organic solvent/aqueous medium co-solvent system. This system may be lyophilized to form free-flowing nanoparticulate powder that is easy to rehydrate into an aqueous lipoidal dispersion.

Nair et al. (2011) developed an antitubercular formulation with isoniazid by using the solvent injection method.

3.6.7 Microemulsion-Based SLN Preparation Technique

The microemulsion technique is a simple yet low energy-consuming technique for the preparation of SLNs, as well as other lipoidal dispersions like NLC and nanoemulsion. A good and reproducible quality of nanoparticles with good particle size distribution can be obtained by this method. The lipid was heated at 41°F–50°F above its melting point or oil mixture is made ready under continuous stirring with an aqueous surfactant and co-surfactant mixture at the same elevated temperature to obtain the hot clear microemulsion. This microemulsion is dispersed in cold water (35.6°F—37.4°F) under stirring at a constant rate, the volume ratio (microemulsion:cold water) being in a range of 1:25 to 1:50. The dilution ratio is subject to the composition of the microemulsion itself. As the particles are already in a submicron range, the cooling step only helps in the solidification of the lipid particles and no further external energy is required to decrease the particle size. The quality of the dispersion largely depends on the temperature gradient and the pH value. This technique has one disadvantage: the rapid crystallization of the lipids as it is subjected to high temperature conditions (Mehnert and Mader, 2001).

This method was used to develop antitubercular nano-formulation by Bhandari and Kaur in 2013.

3.6.8 Microwave-Aided Microemulsion Technique

The novel and modified microwave-aided microemulsion technique for the production of SLNs was developed by Shah et al. (2014). Instead of separate heating as in the conventional system, here the lipid is heated above its melting point along with the aqueous surfactant mixture under stirring in a controlled microwave environment. The resultant hot microemulsion is again dispersed in cold water as per the conventional method. By this method, the yield of the formulation was found to increase substantially, along with an increase in the drug loading and entrapment efficiency of the stearic acid formulation (Shah et al., 2014). This method can be adopted to increase the drug loading of antitubercular drugs in nanoparticulate formulations.

3.6.9 Phase Inversion Temperature Technique

Heurtault et al. (2002) developed a novel, solvent-free formulation technique for lipiodol nanostructured capsules, which can be considered to be an intermediate between polymeric and lipid nanoparticles. This method was based on the phase inversion of an emulsion at a particular phase inversion temperature (PIT). Lipid, surfactant, drug, and water are thoroughly mixed through stirring. The mixture is then subjected to successive cycles of heating and cooling (e.g., from room temperature to the PIT, to 140°F to the PIT, to 140°F to the PIT, to room temperature), functional at a constant rate of 39.2°F/min. Finally, the emulsion is diluted with stirring under cooling conditions (35.6°C–39.2°F).

3.7 Route of Nanoparticle Delivery in Tuberculosis

For the purpose of tubercular treatment, the drugs are conventionally given by the oral route. But they can be given through the intravenous (IV) route also. With a wide variety of available polymers, lipids, and oils, nanoparticles can be developed and applied through oral, IV, or even inhalation routes (like nebulization, dry powder inhalation, and aerosol delivery).

3.8 Lung-Targeting Delivery System for the Treatment of Pulmonary Tuberculosis

Lung-targeted delivery systems are a non-invasive route of drug delivery for both local and systemic action. The lungs are perhaps one of the oldest routes of drug delivery so far known. In early 1500 BC, Egyptian therapists introduced the first inhalable vapors for the treatment of various diseases. However, the lung-specific delivery was soon obsolete, and it was not until the early 1950s that it was given serious consideration again with the appearance of the metered dose inhaler (DaSilva et al., 2013). But the current decade has perhaps taken clinicians closest to really useful lung-specific drug delivery systems. With the rapid growth in popularity and the technical sophistication of lung-targeting chemotherapy for the treatment of pulmonary TB, there is a growing demand for inhalable drug particles, as it is able to provide the most effective delivery of drugs to the lungs to obtain the most optimal therapeutic results. For the purpose of in-demand formulation, a wide variety of new particle technologies have emerged in the last decade. For inhalable drug products, formulations are given by nebulization, dry powder inhalation, or aerosol systems (Chow et al., 2007).

The aerodynamic diameter (dA) is defined as the diameter of a sphere of unit density, which comes to the same velocity air stream as a non-spherical particle's arbitrary density. This diameter defines the mechanism particle deposition in the respiratory system. The particle size and morphology have an impact on all aspects of drug delivery to the lungs, including deposition, dissolution, and the winding mechanism. Particles less than 5 mm can be distributed deep into the smallest airways and this penetration correlates well with good clinical management for local treatment responses (Malcolmson and Embleton, 1998). The fraction of particles with dA in the range 1–2 μm is probably the most efficient deposition in the alveolar capillary-rich airspaces. It is a target for systemic drug delivery that is less efficient or less convenient when delivered by other means. For this size range, the other main particle mechanisms like deposition and diffusion are most important, sedimentation being significant due to the very low air velocity in this area of the lungs (Shekunov et al., 2007). Particles in the submicron range (dA of less than 0.5 mm) can be exhaled out if not accumulated and/or if there is insufficient time available for their movement towards the lung walls. Therefore, slow, deep breathing and holding the breath helps the deposition of such small particles to exhibit different interactions with the trachea, bronchi, and alveolar epithelium.

Now, it is well known that *M. tuberculosis* resides in the deep lung tissue inside the alveolar macrophage system. Nanostructured particles are engineered to deliver the drugs inside the MPS to provide better drug utilization and to reduce the systemic side effects of the high dose antibiotics used for the treatment of TB. By the virtue of lung-targeting the dose of the drug will also automatically reduce.

When the lungs are targeted through the nasal route, it adds to the advantages of easy accessibility and good permeability of lipophilic drugs and low molecular weight drugs. Now, when antitubercular drugs are delivered by SLNs it becomes a lipophilic unit, thereby improving alveolar permeability. The main challenges of this route of drug delivery are ciliary movement and enzymatic activity. But in the case of lung-targeting through the nose, ciliary movement with help SLNs to reach the target organ and the SLNs will not be affected by enzyme activity as they do not stay in the nasal track (Figure 3.5).

3.9 Nanoparticles to Target Other Sites of Tubercular Infection

Nanoparticles are being extensively used in targeted drug delivery for treating diseased conditions with optimal drug utilization and zero or negligible side effects. As discussed earlier in the chapter, TB can spread to the brain, bone, spleen, and other major organs in the later stages if it remains untreated, or in cases of XRD-TB, or with patients infected with HIV. Polymeric nanoparticles, SLNs, metallic nanoparticles, and liposomes, etc., have been front-line research areas for targeted drug delivery to the above mentioned organs (Justo and Moraes, 2003; Kurunov et al., 1995; Pandey et al., 2003a; Pandey et al., 2004; Pandey and Khuller, 2005b; Sharma et al., 2004; Vyas et al., 2004). If antitubercular drugs can be targeted to the site of colonization of the causative organism inside the specific organs, then their systemic side effects can be avoided. Most importantly, the chances of hepatotoxicity can be minimized, as they will bypass the liver. Nanoparticles can be triggered to release antitubercular drugs to the site of the infection in the form of liposomes or other nanostructures.

Pandey and Khuller (2006a) developed a successful brain-targeted drug delivery system for the treatment of TB in a later stage, when the infection spreads to the brain by the oral route. In *Mycobacterium* TB H37Rv infected mice they found no detectable bacilli in the meninges after five oral doses of the nanoparticle formulation were administered every tenth day, as assessed on the basis of colony forming units (CFU) and a histopathology study. In another study they also found that in the administration of two doses there was a significant reduction of CFU count both in the lungs and the spleen (Pandey and Khuller, 2006b).

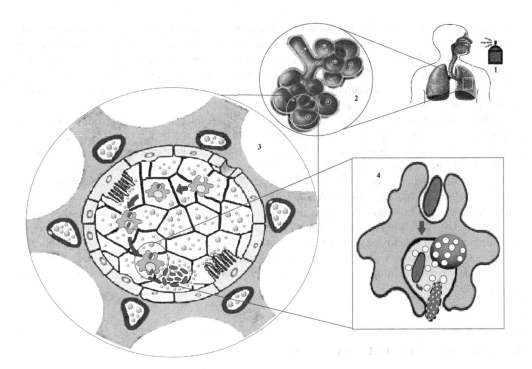

FIGURE 3.5 Fate of nanoparticles within the body upon inhalation: (1) Nanospray, (2) Inside the alveoli where the nanoparticles reach, (3) Inside of an alveolus depicting the growth and simultaneous killing of mycobacterium bacilli by the drugs loaded in nanoparticles, (4) Inside the macrophage where the nanoparticle-encapsulated drug is released in the infected macrophage to inhibit the mycobacterium growth and accumulation.

3.10 Possible Toxicities of Nano-Biomaterials in Tuberculosis Therapy

Since nanotechnology is still relatively at its infancy, health research scientists and clinicians are more engrossed in exploring its potential horizons, and much less work is being directed at looking at the cons of the technology. As with any human endeavor, nanotechnology is also a two-edged sword. Focus is now steadily shifting towards the banes of this technology, so that more judicious applications can be formulated (Ansary and Daihan, 2009).

Naturally, therapeutic nanoparticles and nano-biomaterials used therein are also being scrutinized for safety issues. Quite a few reports now point out that all may not be well with nanoparticles. It has been well known for some decades that environmental pollution and the associated growth of nanoparticles in the atmosphere are taking their toll on the human body. Several researchers have noted the plethora of toxic effects on earth flora and fauna, including human beings, due to the increase in particulate levels of nanoscale materials in the atmosphere. By virtue of their extremely small size, the nanomaterials interact with the deepest of body tissues and components after gaining entry through dermal, mucosal, and respiratory routes (Ansary and Daihan, 2009; Li et al., 2010; Sosnik et al., 2010).

Therefore, therapeutic nanoparticles, too, need thorough investigation as regards to their acute and chronic toxicities. Indeed, certain categories of therapeutic nanoparticles composed of metal, ceramic, and carbon materials have been reported to be cytotoxic. Their toxicities are not necessarily restricted to the portals of entry to the human body, but extend far beyond that into other organs and tissues. However, literature comparison between studies reporting beneficial and toxic effects suggest that at this point in time, huge lacunae exist in our knowledge of the extent to which these nanostructures can affect our body systems and their normal working (Boisselier and Astruc, 2009; Wan et al., 2012).

But we are confident that this is going to change, since a number of nanoparticulate formulations have appeared in the market in the last 20 years, and, eventually, we are going to learn a lot regarding the nanotoxicity of these therapeutic nanostructures. The clinically-used and marketed nanostructures are mainly the hard nanoparticles (like the metal nanoparticles), though about 40% are soft nanoparticles (like the liposomes). However, a huge number of potential candidates are in the pipelines of major pharmaceutical houses in various clinical stages of development (particularly Phases I and II). These figures exclude the cosmetic nanoformulations, which also, if systematically probed, could lead us to valuable nanotoxicity information (Jones et al., 2009).

It is important to note that of the huge amount of data that is being generated worldwide on nanotherapeutics through studies, only about one-third of the studies cover the toxicity aspects experimentally. However, in the authors' opinion, these studies are not always extensive or dependable, or they simply lack long-term data. Another point of concern is the non-availability of toxicological data from industrial research and development, since most of the literature, including patents, is from academic or governmental investigations. Industry usually likes to withhold such sensitive data.

It is encouraging to note that toxicity concerns among nanotech researchers are increasing spectacularly. Now-a-days, regulatory or purely clinical concerns are leading to the integration of toxicological evaluations (such as genotoxicity, mutagenicity, cytotoxicity, etc.) in the developmental agenda of the industry and academia that are actively engaged in this field of study. Maybe, in not so distant future, we will see a clear picture coming out regarding the toxicity potential of therapeutic nanoparticles (Kolter et al., 2015).

3.11 Conclusion and Future Prospects

TB is a bane for developing nations, though even developed nations are also witnessing the resurgence of the disease. Conventional treatments available can cure TB, but its toxicity and resurgence are of primary concern. Nanotechnology is revolutionizing the potentials of existing drugs to treat several diseases, including TB. Nanostructures, particularly those made of biocompatible natural materials, have great promise in overcoming these toxicity and recurrence problems. They may be manufactured through several well-established techniques, then evaluated and scaled up for industrial production. Nanotechnology has shown good *in vivo* potential as antitubercular delivery systems through several routes. The nasal and inhalation routes may be best suited for pulmonary TB treatment, though nanostructures have penetrated the other poorly accessible sites of TB. Although most research on antitubercular nanoplatforms has not yet reached the advanced clinical stages, the next few years may bring new hope if proper budgetary allocations are made by governments and industries to explore this vista.

REFERENCES

Ansary AE, Daihan SA. On the toxicity of therapeutically used nanoparticles: An overview. J Toxicol. 2009;2009:754810.

Bhandari R, Kaur IP. A method to prepare solid lipid nanoparticles with improved entrapment efficiency of hydrophilic drugs. Curr Nanosci. 2013;9(2):211–20.

Boisselier E, Astruc D. Gold nanoparticles in nanomedicine: Preparations, imaging, diagnostics, therapies and toxicity. Chem Soc Rev. 2009;38:1759–82.

Chow AHL, Tong HHY, Chattopadhyay P, Shekunov BY. Particle engineering for pulmonary drug delivery. Pharm Res. 2007;24(3):411–37.

Conte Jr. JE, Golden JA, McQuitty M, Kipps J, Duncan S, McKenna E, Zurlinden E. Effects of gender, AIDS, and acetylator status on intrapulmonary concentrations of isoniazid. Antimicrob Agents Chemother. 2002;46:2358–64.

DaSilva AL, Santos RS, Xisto DG, Alonso SDV, Morales MM, Rocco PRM. Nanoparticle-based therapy for respiratory diseases. An Acad Bras Cienc. 2013;85(1):137–46.

Dube A, Lemmer Y, Hayeshi R, Balogun M, Labuschagne P, Swai H, Kalombo L. State of the art and future directions in nanomedicine for tuberculosis. Expert Opin Drug Deliv. 2013;10(12):1725–34.

Edwards D, Kirkpatrick CH. The immunology of mycobacterial diseases. Am Rev Respir Dis. 1986;134:1062–71.

Gupta A, Wood R, Kaplan R, Bekker LG, Lawn SD. Tuberculosis incidence rates during 8 years of follow-up of an antiretroviral treatment cohort in South Africa: Comparison with rates in the community. PLoS One. 2012;7(3):e34156.

Heurtault B, Saulnier P, Pech B, Proust JE, Benoit JP. A novel phase inversion-based process for the preparation of lipid nanocarriers. Pharm Res. 2002;19(6):875–80.

Jones MAM, Bantz KC, Love SA, Marquis BJ, Haynes CL. Nanomedicine. 2009;4(2):219–41.

Justo OR, Moraes AM. Incorporation of antibiotics in liposomes designed for tuberculosis therapy by inhalation. Drug Deliv. 2003;10:201–7.

Kalombo L. Nanoparticle carriers for drug administration and process for producing same. US Patent US8518450 B2. 2010.

Kurunov IN, Ursov IG, Krasnov VA, Petrenko TI, Iakovchenko NN, Svistelnik AV, Filimonov PA. Effectiveness of liposomal antibacterial drugs in the inhalation therapy of experimental tuberculosis. Probl Tuberk. 1995;1:38–40.

Leroux J, Allemann E, Doelker E, Gurny R. New approach for the preparation of nanoparticles by an emulsification-diffusion method. Eur J Pharm Biopharm. 1995;41(1):14–8.

Li JJ, Muralikrishnan S, Ng CT, Yung LYL, Bay BH. Nanoparticle-induced pulmonary toxicity. Exp Biol Med. 2010;235:1025–33.

Luetkemeyer, A. Tuberculosis and HIV: HIV InSite Knowledge Base Chapter [Internet]. San Francisco, CA: University of California San Francisco (UCSA); 2013 [cited 2014 Dec 12]. Available from: http://hiv insite.ucsf.edu/insite?page=kb-05-01-06.

Malcolmson RJ, Embleton JK. Dry powder formulations for pulmonary delivery. PSTT. 1998;1:394–8.

Maraicar KSH, Narayanan T. Design and characterization of solid lipid nanoparticle by solvent evaporation method followed by homogenization. Int J Biopharm. 2014;5(3):190–6.

Kolter M, Ott M, Hauer C, Reimold I, Fricker G. Nanotoxicity of poly(n-butylcyano-acrylate) nanoparticles at the blood–brain barrier, in human whole blood and in vivo. J Control Release. 2015;197:165–79.

Martin A, Portaels F. Drug resistance and drug resistance detection. In: Palomino JC, Leão SC, Ritacco V, editors. Tuberculosis 2007: From basic science to patient care. 1st ed: TuberculosisTextbook.com; 2007. p. 635–60. Available from: pdf.flyingpublisher.com/tuberculosis2007.pdf.

MedicineNet. Extensively Drug-Resistant Tuberculosis (XDR TB) [Internet]. San Clemente, CA: MedicineNet; 2007 [cited 2014 Dec 26]. Available from: http://www.medicinenet.com/extensively_drug-resistant_ tuberculosis_xdr_tb/page3.htm#can_xdr_tb_be_treated_and_cured.

Mehnert W, Mader K. Solid lipid nanoparticles: Production, characterization and application. Adv Drug Deliv Rev. 2001;47:165–96.

Meltzer MS, Skillman DR, Gomatos PJ, Kalter DC, Gendelman HE. Role of mononuclear phagocytes in the pathogenesis of human immunodeficiency virus infection. Annu Rev Immunol. 1990;8:169–94.

Moore D, Liechty C, Ekwaru P, Were W, Mwima G, Solberg P, Rutherford G, Mermin J. Prevalence, incidence and mortality associated with tuberculosis in HIV-infected patients initiating antiretroviral therapy in rural Uganda. AIDS. 2007;21(6):713–9.

Nair R, Vishnupriya K, Arun Kumar KS, Badivaddin TM, Sevukarajan M. Formulation and evaluation of solid lipid nanoparticles of water soluble drug: Isoniazid. J Pharm Sci & Res. 2011;3(5):1256–64.

Nancy C E. Extrapulmonary tuberculosis: A review. Fam Med. 1992;1:91–8.

Pandey R, Khuller GK. Antitubercular inhaled therapy: Opportunities, progress and challenges. J Antimicrob Chemother. 2005a;55:430–5.

Pandey R, Khuller GK. Solid lipid particle-based inhalable sustained drug delivery system against experimental tuberculosis. Tuberculosis. 2005b;85:227–34.

Pandey R, Khuller GK. Oral nanoparticle-based antituberculosis drug delivery to the brain in an experimental model. J Antimicrob Chemother. 2006a;57:1146–52.

Pandey R, Khuller GK. Nanotechnology-based drug delivery system(s) for the management of tuberculosis. Indian J Exp Biol. 2006b;44(5):357–66.

Pandey R, Sharma A, Zahoor A, Sharma S, Khuller GK, Prasad B. Poly (DL-lactide-co-glycolide) nanoparticle-based inhalable sustained drug delivery system for experimental tuberculosis. J Antimicrob Chemother. 2003a;52:981–6.

Pandey R, Zahoor A, Sharma S, Khuller GK. Nanoparticle encapsulated antitubercular drugs as a potential oral drug delivery system against murine tuberculosis. Tuberculosis. 2003b;83:373–8.

Pandey R, Sharma S, Khuller GK. Nebulization of liposome encapsulated antitubercular drugs in guinea pigs. Int J Antimicrob Agents. 2004;24:93–4.

Pandey R, Sharma S, Khuller GK. Oral solid lipid nanoparticle-based antitubercular chemotherapy. Tuberculosis. 2005:85(5–6):415–20.

Patel NR, Swan K, Li X, Tachado SD, Koziel H. Impaired M. tuberculosis-mediated apoptosis in alveolar macrophages from HIV+ persons: Potential role of IL-10 and BCL-3. J Leukoc Biol. 2009;86(1):53–60.

Pham DD, Fattal E, Tsapis N. Pulmonary drug delivery systems for tuberculosis treatment. Int J Pharmaceutics. 2015;478(2):517–29.

Pyrgiotakis G, McDevitt J, Gao Y, Branco A, Eleftheriadou M, Lemos B, Nardell E, Demokritou P. Mycobacteria inactivation using Engineered Water Nanostructures (EWNS). Nanomedicine. 2014;10(6):1175–83.

Quintanar-Guerrero D, Fessi H, Allemann E, Doelker E. Influence of stabilizing agents and preparative variables on the formation of poly(d,l-lactic acid) nanoparticles by an emulsification-diffusion technique. Int J Pharm. 1996;143(2):133–41.

Raviglione MC, Snider DE, Kochi A. Global epidemiology of tuberculosis: Morbidity and mortality of a worldwide epidemic. JAMA. 1995;273:220–6.

Sacks LV, Pendle S, Orlovic D, Andre M, Popara M, Moore G, Thonell L, Hurwitz S. Adjunctive salvage therapy with inhaled aminoglycosides for patients with persistent smear-positive pulmonary tuberculosis. Clin Infect Dis. 2001;32(1):44–9.

Saifullah B, Hussein MZ, Hussein-Al-Ali SH, Arulselvan P, Fakurazi S. Sustained release formulation of an anti-tuberculosis drug based on para-amino salicylic acid-zinc layered hydroxide nanocomposite. Chem Central J. 2013;7:72.

Schoubben A, Giovagnoli S, Tiralti MC, Blasi P, Ricci M. Capreomycin inhalable powders prepared with an innovative spray-drying technique. Int J Pharm. 2014;469:132–9.

Shah R, Malherbe F, Eldridge D, Palombo E, Harding I. Physicochemical characterization of solid lipid nanoparticles (SLNs) prepared by a novel microemulsion technique. J Colloid Interf Sci. 2014;428:286–94.

Sharma A, Sharma S, Khuller GK. Lectin-functionalized poly(lactide-co-glycolide) nanoparticles as oral/ aerosolized antitubercular drug carriers for treatment of tuberculosis. J Antimicrob Chemother. 2004;54:761–6.

Shekunov BY, Chattopadhyay P, Tong HHY, Chow AHL. Particle size analysis in pharmaceutics: Principles, methods and applications. Pharm Res. 2007;24:203–27.

Sosnik A, Carcaboso AM, Glisoni RJ, Moretton MA, Chiappetta DA. New old challenges in tuberculosis: Potentially effective nanotechnologies in drug delivery. Adv Drug Del Rev. 2010;62:547–59.

Van Rie A, Westreich D, Sanne I. Tuberculosis in patients receiving antiretroviral treatment: Incidence, risk factors, and prevention strategies. J Acquir Immune Defic Syndr. 2011;56(4):349–55.

Vanderhoff J, El-Aasser M, Ugelstad J. Polymer emulsification process. US Patent 4177177. 1979.

Vyas SP, Kannan ME, Jain S, Mishra V, Singh P. Design of liposomal aerosols for improved delivery of rifampicin to alveolar macrophages. Int J Pharm. 2004;269:37–49.

Walsh M. Tuberculosis [Internet]. Brooklyn, NY: Infection Landscapes; 2013 [cited 2014 Sep 8]. Available from: http://www.infectionlandscapes.org/2013_04_01_archive.html.

Wan R, Mo Y, Feng L, Chien S, Tollerud DJ, Qunwei Zhang Q. DNA damage caused by metal nanoparticles: Involvement of oxidative stress and activation of ATM. Chem Res Toxicol. 2012;25:1402–11.

Wang T, Wang N, Zhang Y, Shen W, Gao X, Li T. Solvent injection-lyophilization of tertbutyl alcohol/water cosolvent systems for the preparation of drug-loaded solid lipid nanoparticles. Colloid Surface B. 2010;79(1):254–61.

Whalen CC, Zalwango S, Chiunda A, Malone L, Eisenach K, Joloba M, Boom WH, Mugerwa R. Secondary attack rate of tuberculosis in urban households in Kampala, Uganda. PLoS One. 2011;6(2):e16137.

World Health Organization (WHO). Global tuberculosis report [Internet]. France: WHO; 2012. Available from: http://apps.who.int/iris/handle/10665/75938.

Zwang J, Garenne M, Kahn K, Collinson M, Tollman SM. Trends in mortality from pulmonary tuberculosis and HIV/AIDS co-infection in rural South Africa (Agincourt). Trans R Soc Trop Med Hyg. 2007;101(9):893–8.

4

Therapeutic Nanostructures for Improved Wound Healing

Lalduhsanga Pachuau, Pranab Jyoti Das, and Bhaskar Mazumder

CONTENTS

ABSTRACT: The fusion of nanotechnology with medicine has offered a plethora of opportunity to control, manipulate, study, and manufacture structures and devices at the nanometer size range for biomedical applications, including drug delivery. Encapsulation of drugs into nanoparticles offers a great opportunity to improve the standard care and prognosis for challenging healthcare issues such as impaired wound healing. In fact, the use of nanoparticulate drug delivery vehicles for wound healing has been predicted to revolutionize the future of diabetic therapy. The slow and sustained release of the encapsulated drug from the nanoparticle can increase the safety of the drug for topical delivery, as the whole amount of the encapsulated drug is never in direct contact with the skin at one time. Controlled-release wound healing formulations are also highly beneficial in management of chronic wounds, as they eliminate the need for frequent dressing changes thus increasing patient compliance. Reduced dose, along with higher localized concentration and prolonged delivery at the wounds, can be achieved avoiding high dose, systemic exposure to the drugs. In addition, susceptible drugs such as growth factors can also be delivered safely while there is also a scope for tailoring their release profiles.

KEY WORDS: *Wound healing, solid lipid nanoparticles (SLNs), liposomes, nanoemulsions, nanofiber mats, nanosponge*

4.1 Introduction

Nanotechnology is science, engineering, and technology conducted at the nanoscale, which is about 1 to 100 nm (NNI, 2017). It is the study and application of extremely small things across all the scientific fields such as chemistry, biology, physics, material science, and engineering. When particles are reduced to such nanometer-size dimensions, quantum physics begins to take over and particles begin to show entirely different physicochemical properties (Bawarski et al., 2008). At this size, particles possess novel optical, electronic, and structural properties that are not manifested by either individual molecules or bulk solids. Nanotechnology thus offers a plethora of opportunities across the different scientific fields to engineer and manipulate many structural properties to produce novel or superior functional materials.

Nanomedicine is the application of nanotechnology in the development of a safer and more effective delivery system for medicine. It has rapidly expanded across the biomedical fields over the past decade. In fact, these novel approaches to the delivery and formulation of therapeutics using nanotechnology are revolutionizing the future of medicine (Bawarski et al., 2008). A large variety of drugs, including both hydrophilic and hydrophobic, as well as vaccines and other large molecule protein drugs, can also be effectively delivered by nanoparticles over a prolonged period of time. The drug may be either dissolved, entrapped, encapsulated, or attached to a nanoparticulate matrix and, depending on the method of preparation, nanoparticles, nanospheres, or nanocapsules can be obtained (Soppimath et al., 2001). Due to their varying composition, structure, and surface characteristics, nanocarriers can be constructed and fine-tuned to allow for specific applications. Intravenous administration of nanoparticles usually results in rapid clearance from the body, exhibiting a very short blood circulation time, as they are easily recognized by the body's immune systems (Soppimath et al., 2001). However, it is now well understood that a stealth property can be imparted to the nanoparticles by coating them with hydrophilic polymers such as polyethylene glycol (PEG). Such coating would avoid the rapid uptake and optionization of nanoparticles by the mononuclear phagocytes, leading to prolonged blood circulation. The targeting ability of nanoparticles can also be greatly enhanced through bioconjugation with relevant affinity ligands, such as antibodies, antibody fragments, peptides, sugars, and small molecules (Kamaly et al., 2012). This surface modification of nanoparticles is yet another exciting application for nanoparticles in medicines for targeting and enhancing drug accumulation to tumors or organs. As a result, nanotherapeutic structures thus offer a great opportunity in increasing the therapeutic efficacy of the drugs while also improving the pharmaceutical and pharmacological properties of the drugs.

Several investigations on various types of nanostructures have been carried out for the therapeutic drug delivery and treatment of a variety of diseases and disorders. Due to several reasons, much of the research work on nanomedicines has been dedicated especially to cancer therapy. However, in recent times, therapeutic nanoconstructs, such as polymeric nanoparticles, liposomes, SLNs, and others, have been demonstrated to improve the currently available treatments for challenging healthcare issues such as impaired wound healing (Tocco et al., 2012). Such encapsulation of drugs into nanoparticles is found to be an advantageous means to overcome several of the problems of free drugs, such as poor aqueous solubility, limited biodistribution, and quick degradation and clearance (Pan et al., 2011). Due to their small size and high surface-to-volume ratio, nanostructures provide an ease of intracellular access and passage through the skin barrier which is ideal for topical drug delivery. The slow and sustained release of the encapsulated drug from these nanoparticulate constructs is also considered to reduce the toxicity while increasing the safety of drugs for topical delivery, as the whole amount of the encapsulated drug is never in direct contact with the skin at one time (Krausz et al., 2015).

4.2 Wounds and Wound Healing

4.2.1 Wounds

The skin is the largest organ of the body, whose primary function is to serve as a protective barrier against the environment. Intact skin is vital to the preservation of body fluid homeostasis, thermoregulation, and protection against infection (Church et al., 2006). Damage to the skin results in a breach through which bacteria can enter to cause inflammation, local or systemic infection, and loss of tissue fluids (Percival, 2002), therefore any break on the skin must be rapidly and efficiently mended (Martin, 1997).

The loss of the integrity of large portions of the skin as a result of injury or illness may lead to major disability or even death (Singer and Clark, 1999).

Wounding of the skin occurs when the epidermal layer is breached, thus exposing the underlying dermis to air (Heng, 2011). A wound is defined as a disruption of normal anatomic structure and function, which results from pathophysiologic processes beginning internally or externally to the involved organs (Lazarus et al., 1994). Wounds are one of the oldest afflictions associated with mankind and its history is as old as humanity (Singh et al., 2013). Even though there is no standard classification system, wounds are generally categorized as acute or chronic depending on the nature of the repair process. Wounds, such as burns, other traumatic injuries, and surgical wounds, that normally heal through an orderly and timely reparative process with rapid establishment of skin barrier function are termed as acute wounds. In contrast to acute healing, impaired and delayed healing is encountered in chronic wounds due to certain local factors, including contamination or the underlying pathophysiological conditions such as venous leg ulcers, diabetic foot ulcers, and pressure ulcers. Such healing fails to proceed through the orderly and timely process of reparation, preventing proper closure of wounds and may even re-occur requiring a longer and more intensive treatment (Boateng et al., 2008; Lazarus et al., 1994; Li et al., 2008; Moore et al., 2006). Acute and chronic wounds have been reported to affect millions of people around the world and their incidence, along with the treatment burden, is growing rapidly and becoming a major health burden and a drain on resources (Harding et al., 2002). Acute wounds affected 11 million people in the United States alone, resulting in hospitalization of approximately 300,000 people every year, and the annual expenditure on the treatment of chronic wounds is also in excess of 25 billion US dollars (Brem et al., 2007; Rice et al., 2012; Singer and Dagum, 2008).

4.2.2 Wound Healing

When a wound is inflicted on the body, either accidentally or by design, the body responds to such an injury through a well-orchestrated series of complex biological and molecular events called the healing process. This ability to heal appears to be universal throughout the animal kingdom (Dyson, 1997). The healing process, which is a response to tissue injury, can be divided into four overlapping phases of coagulation, inflammation, migration-proliferation (including matrix deposition), and remodeling (Figure 4.1) (Falanga, 2005). Whereas the initial responses are similar, subsequent events depend on the type of tissue as well as the degree of damage (Halloran and Slavin, 2002).

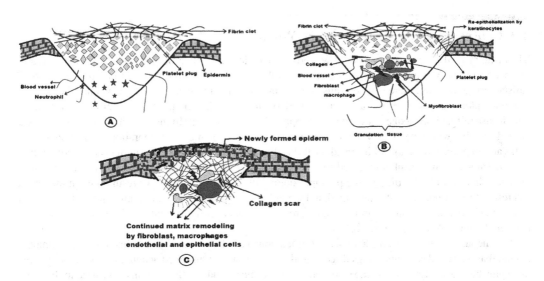

FIGURE 4.1 The phases of the wound healing process: (A) inflammation, (B) proliferation, and (C) scar maturation. (Reproduced with permission from Gunasekaran et al. (2012), Copyright©2012 Elsevier Inc.)

4.2.2.1 Inflammation and Hemostasis

Bleeding is the body's first response to an injury, mainly to flush out bacteria and other harmful antigens from the injured site. The first step in wound healing is hemostasis, which takes place through the formation of a fibrin clot to stop the bleeding. Simultaneously, inflammation occurs which involves both cellular and vascular responses, including the release of histamine and other cell mediated factors. This results in vasodilation, enhanced capillary permeation, and stimulation of pain receptors. Phagocytes that were recruited into the wound sites digest and kill the bacteria and scavenge tissue debris. This inflammatory phase occurs within a few minutes to 24 hours of the time of injury.

4.2.2.2 Migratory

Epithelial cells and fibroblasts, which are the key cells involved in the formation of the extracellular matrix, migrate to the injured site, replacing the damaged and lost tissue. Keratinocytes also migrate across the wound bed. This migration takes place from the margin and lasts for 2–3 days. Migration coupled with wound contraction results in re-epithelialization and wound closure.

4.2.2.3 Proliferation

The migration phase is followed immediately from the third day onwards by the proliferative phase. As a result of the proliferation of capillaries and lymphatic vessels into the wound, granulation tissue is formed as well as fibroblast-synthesized collagen, which gives tensile strength to the skin. As continuous proliferation takes place, further epithelial cell migration provides closure and visible wound contraction, which continues until the collagen bridges the wound.

4.2.2.4 Maturation

After wound closure has been achieved through the proliferation phase, the maturation phase, also known as the remodeling phase, takes place, which is the final phase in wound healing. This remodeling phase involves a reduction in blood flow to the scar tissue and the enlargement of collagen fibers. Remodeling may take place over a month or years during which the tensile strength of the repair reverts back to as close as possible to the pre-injured tissue.

4.3 Controlled Drug Delivery to the Wounds

The primary goal in the management of wounds is to achieve rapid healing with optimal functional and aesthetic results. A major economic and ethical problem is posed by the fact that not all wounds heal satisfactorily. However, there is much cause for optimism in the treatment of chronic wounds, because of the tremendous strides taken in our scientific understanding of the repair process over the past decades and the use of that knowledge to develop new approaches to treatment. In the past, the main function of a wound dressing was to keep the wound as dry as possible by allowing the evaporation of the wound exudates and to prevent the entry of harmful bacteria (Boateng et al., 2008). However, it is now realized that a warm, moist environment is required for achieving faster and more successful wound healing. These advances in our understanding of the pathophysiology of various wounds have resulted in an increased interest in the development of multifunctional wound dressings that provide not only physical protection but also help in maintaining an optimal moisture environment for the wound to achieve faster and more efficient healing (Adhirajan et al., 2009).

Over the last few years, controlled-release formulations have been the subject of interest in developing a system that delivers therapeutic drugs directly to the wound site for sustained action. Such delivery systems are highly beneficial, especially in the management of chronic wounds, as they eliminate the need for frequent administration of drugs with conventional dosage forms, thus improving patient compliance to the treatment regime. Also, topical administrations of antimicrobials in the form of controlled-release formulations are reported to be more beneficial than systemic administration in the management of chronic or heavily-colonized wounds. As the surface of the chronic wounds usually lack sufficient blood supply, potentially high doses of antibiotics are often required to be administered by systemic routes to control the infection, while a higher localized concentration with prolonged delivery can be achieved by topical administration, avoiding high dose, systemic exposure to the antibiotics. Encapsulation of the active drug for sustained-release topical delivery is also considered to enhance the safety of the drug, as the whole amount of the encapsulated drug is never in direct contact with the skin at one time (Krausz et al., 2015). In addition, controlled-release wound healing formulations are also highly beneficial in improving the stability of susceptible drugs such as growth factors (Losi et al., 2013), while there is also scope for tailoring their drug release profiles.

4.4 Nanostructures for Wound Healing

An increasing number of investigations have been devoted towards the development of nanoconstructs for the therapeutic drug delivery and treatment of impaired wound healing. Innovations in this area over the last few years have resulted in the development of drug delivery systems for wound healing, including polymeric nanoparticles, liposomes, SLNs, silver and gold nanoparticles, nanoemulsions, hydrogels, nanosponges, and nanofiber mats. This chapter will provide an insight into the latest developments in these areas of research.

4.4.1 Metallic Nanoparticles

The synthesis and development of metallic nanoparticles is an area of active research in applied nanotechnology. Particularly, metals such as silver (Ag), zinc (Zn), and gold (Au) have received much attention for their medicinal applications (Thakkar et al., 2010). It is a fact that Ag has a long history of being used as an antimicrobial agent for managing different kinds of wounds and infections. However, the discovery of apparently more powerful and cost-effective antibiotics has significantly declined their existence and opportunities in the biomedical field (Holbrook et al., 2014). Yet, the rising issues on antibiotic resistance due to the widespread and irrational consumption of antibiotics have mooted for the renewed interest in Ag and other metallic nanoparticles (Rai et al., 2009). Ag nanoparticles are easily prepared into various shapes and sizes (2–500 nm), possessing excellent antibacterial and anti-inflammatory properties with a potential for surface modification to suit their applications (Gunasekaran et al., 2012).

Treatment of wounds using Ag nanoparticles was shown to significantly accelerate the wound healing *in vivo*. The healing time for a thermal burn injury was reduced by more than a week when BALB/C mice were treated with Ag nanoparticles (from 35 days with control to 26 days with Ag nanoparticles), while it took 37 days with Ag sulfadiazine treatment (Tian et al., 2007). Ag nanoparticles were also found to be highly effective in reducing the wound healing time in diabetic animals. Another desirable property of the Ag nanoparticles is their aesthetic appeal, as there was less hypertrophic scarring and the hair growth on the wound surface was reported to be almost normal. To achieve a controlled and sustained therapy, Ag nanoparticles may also be incorporated into dressings based on chitosan or gelatin nanofiber mats prepared through electrospinning (Lu et al., 2008; Rujitanaroj et al., 2008). These kinds of formulations imbibe the beneficial effects of providing a moist, warm, and nutritious environment for wound healing, as contributed by the dressings. Due to the slow and continuous release of the entrapped Ag nanoparticles, the Ag levels in the blood and tissues are lower than those of Ag sulfadiazine, making it a safer wound dressing product and significantly enhancing the healing in partial-thickness wound models.

The two most important factors that hinder and impede efficient wound healing are diabetic hyperglycemia and microbial infections (Ahmadi and Adibhesami, 2017; Singla et al., 2017). Therefore, these issues should be clearly addressed by a satisfactory and functional wound healing preparation. Clinical trials in nursing clients with leg ulcers compromised by bacterial burden showed that nanocrystalline Ag was found to be associated with a quicker healing rate during the first two weeks of treatment as compared to cadexomer iodine preparation, which is another commonly used agent (Miller et al., 2010).

Au nanoparticles (AuNPs) are another safe and effective option in wound-healing metallic nanoparticles. Au and Ag nanoparticles can be synthesized through various means, utilizing different materials.

When AuNPs were prepared with antioxidant epigallocatechin gallate (EGCG) and α-lipoic acid (ALA), and applied topically *in vivo* in mice, the anti-inflammatory and antioxidant actions of the mixture demonstrated enhanced healing of the wound area (Leu et al., 2012). Ag nanoparticles and AuNPs prepared using *Brassica oleracea* L. var. *capitata* f. *rubra* also promoted topical wound healing without any toxic effect (Sivakumar et al., 2017) and modifications such as conjugation of antimicrobial peptides to AuNPs are also a promising platform for increasing the pro-regenerative properties (Comune et al., 2017).

Innovations in metallic wound dressing nanostructures include the use of a combination of metallic nanoparticles for improved healing and safety. In this manner, lower quantities of Ag nanoparticles are needed to be incorporated while maintaining its antibacterial property. When chitosan was loaded with Au-Ag nanoparticles, a safer and better promotion of wound healing was observed as compared to chitosan loaded with only Ag nanoparticles (Li et al., 2017). The antibacterial properties of nanofibers incorporated with a ZnO-Ag nanoparticle combination against *Staphylococcus aureus* and *Escherichia coli* were also significantly improved than their individual metal nanomaterial loaded into the same nanofibers. Apart from their enhanced activity, the cytotoxicity of ZnO-Ag bimetallic nanofibers against fibroblasts was significantly reduced (Hu et al., 2017).

4.4.2 Liposomes

Liposomes are vesicular nanostructures made up of hydrated lipid bilayers that resemble the lipid cell membrane of the human body. The unique make-up of the phospholipid complex offers vesicular carriers such as liposomes an ideal vehicle for topical drug delivery (Jangde and Singh, 2016). If the drug is lipophilic, it can be suitably accommodated in the lipid bilayer, and if it is water soluble, the aqueous core is still available for accepting the water-soluble drug. Such a unique construct thus makes it possible to enhance the bioavailability and stability of a wide range of drugs, produce the prolonged release of the entrapped drug, and even improve the membrane permeability of polar compounds with large molecular weight (Rahman et al., 2010; Wang et al., 2014). The versatility of liposomes is exemplified by the fact that liposomal products for wound healing extend from those that incorporate plant extracts to those with growth factors (Alemdaroglu et al., 2008; Moulaoui et al., 2015; Xiang et al., 2011). The typical release characteristics observed in liposomes are the initial burst release (about 40%) as a result of the drug presence on the surface of the liposomes and this will be followed by 24 hours of sustained release of the entrapped drug (Alemdaroglu et al., 2008; Jangde and Singh, 2016).

Polyphenols such as quercetin, curcumin, and resveratrol are promising topical wound healing agents due to their free-radical scavenging and anti-inflammatory properties. However, their efficient clinical applications have been hampered by their extreme water insolubility and rapid clearance (Prasain and Barns, 2007). Formulations of natural polyphenols into topical nanovesicular systems have been reported to overcome these limitations, resulting in significant improvement in their wound healing ability and protection from oxidative damage. An astonishingly high entrapment, approaching 90%, can also be achieved with such poorly aqueous soluble drugs such as quercetin when formulated into liposomes (Castangia et al., 2014; Jangde and Singh, 2016; Pando et al., 2013). Liposomal nanovesicles can be applied not only for the treatment of various wounds and their related complications, but also for preventing skin lesions and damages. Various biochemical processes that cause epithelial loss and skin damage by the induction of cutaneous ulceration by 12-O-tetradecanoylphorbol 13-acetate (TPA) treatment are abrogated by quercetin and curcumin liposomes (Castangia et al., 2014). Such phytochemical-loaded liposomes and penetration-enhancer-containing vesicles (PEVs) were demonstrated to be superior over the simply dispersed phytodrugs in reducing myelo-peroxidase (MPO) activity in damaged tissues. The MPO inhibition value was higher, especially for quercetin liposomes (59%) and curcumin liposomes and PEG-PEVs (about 68%). This technology has also been extended to encapsulate a polyphenolic phytocomplex obtained from the extract of medicinal plants used in folk medicine (Moulaoui et al., 2015). Efficient entrapment of the plant extract (40%) as well as a good aggregation of the phospholipid (75%) was obtained, which exhibits enhanced *in vitro* antioxidant activity as well as antioxidant activity against oxidative stress in human keratinocytes. In an animal model, the novel PEG-PEVs improved the local bioavailability and intracellular antioxidant activity, ultimately promoting wound healing.

Growth factors such as epidermal growth factors (EGF) and basic fibroblast growth factors (bFGF) possess the ability to stimulate fibroblast, vascular endothelial cell, and keratinocyte division *in vitro* and granulation tissue formation and epidermal regeneration *in vivo*, apart from stimulating collagen synthesis to alleviate the scarring observed in wound healing (Alemdaroglu et al., 2008; Xiang et al., 2011). However, the clinical applications of the growth factors have been limited by their instability to environmental change and their inherent short shelf-life. Liposome preparations are reported to be possessed of most of the biological cell properties needed for a successful drug delivery system in general, and for growth factors in particular, for the treatment of wounds (Alemdaroglu et al., 2008).

A bFGF liposome with high bFGF-entrapment efficiency (80%) was formulated to improve its properties and prolong its effects *in vivo*, which otherwise has a very short *in vivo* half-life (Xiang et al., 2011). Depending on the dose administered, the wound healing time of the liposomes in a deep second-degree burns model was reduced by five days. The expression of proliferating cells nuclear antigen (PCNA) TGF-β along with hydroxyproline was significantly increased by the liposomal treatment, indicating its ability to improve the quality of the healing. When multilamellar-type liposomes containing EGF were prepared, the stability was reported to be lower, which was temperature dependent, and encapsulation efficiency was also found to be lower at 58.1% (Alemdaroglu et al., 2008). A drug release study showed an initial burst release followed by a sustained release of up to 85.6% over the 24 hours. In a second-degree uniform deep burn wounds model, animals treated with the liposome formulations for 14 days showed a significant increase in the average numbers of proliferated cells, enhanced epidermal thickness, and fibroblast nucleus areas when compared to other groups. Madecassoside, a triterpenoid saponin from *Centella asiatica*, exhibits a powerful potency against several skin disorders such as wound healing and psoriasis, etc., however it possesses very low permeability to the skin. A madecassoside liposomal formulation was prepared with 41% encapsulation efficiency, with improved physical stability and a preparation favorable for application to the skin (Wang et al., 2014).

The broadening application of liposomes resulted in the development of textile-based, biocompatible, controlled-release wound dressing systems functionalized with liposomes (Ferreira et al., 2013). As the vesicles represent a negative surface charge, cationization of the surfaces of the non-woven gauzes resulted in significantly enhanced attachment of the liposomes to the gauzes. The preparation also leads to better drug release and improves the process of wound healing. A liposomal-complex containing an antioxidant flavonoid of plant origin that promotes the reduction of the inflammatory response to thermal burn wound injuries was also reported (Naumov et al., 2010). There was an intense regeneration of the skin along with reparation of the hair follicles and sebaceous gland, signifying the high quality of the healing.

Persistent ulcers have been found to be the result of topical malperfusion which causes ischemia or hypoxia, and this condition may be treated efficiently with hyperbaric oxygen therapy. A liposome formulation was applied to encapsulate hemoglobin with high oxygen affinity to accelerate skin wound healing in a mice model (Fukui et al., 2012). To test the wound healing property of the formulation, skin wounds were created on the back of anesthetized mice using a circular skin tome of 6 mm in diameter. The results suggest that liposome-encapsulated hemoglobin may accelerate skin wound healing in BALB/C mice via mechanisms involving reduced inflammation and enhanced metabolism (Fukui et al., 2012). Clodronate liposomes were also shown to reduce the excessive formation of scars in a burn injury wound model in mice by reducing collagen deposition and the expression of TGF-β1 (Lu et al., 2014).

4.4.3 Polymeric Nanoparticles

Nanoparticulate structures based on both natural and synthetic polymers such as chitosan, fibrin, poly (D,L-lactide-co-glycolide) acid (PLGA), and others, along with their nanocomposites, have been evaluated for controlled delivery of a wide range of therapeutics to wounds. Encapsulation of curcumin using PLGA was shown to enhance its stability and solubility, as required for wound healing applications, while lactic acid from the PLGA provides an add-on effect on wound healing (Cheredy et al., 2013). The nanoparticles were prepared by modified oil-in-water (O/W) emulsion solvent evaporation technique and the *in vitro* drug release study showed a biphasic pattern of curcumin release from the PLGA nanoparticles, which included the initial burst release within the first 24 hours followed by a gradual and sustained release over an eight day period. Drug release was found to take place through diffusion and degradation/erosion mechanisms. The PLGA-Curcumin nanoparticles were found to significantly promote re-epithelialization and granulation of tissue while inhibiting myeloperoxidase activity and the expression of inflammatory mediators such as glutathione peroxidase and NFκB (Figure 4.2).

The application of exogenous vascular endothelial growth factor (VEGF) and bFGF to the wound site is also known to stimulate cell proliferation and improve wound healing. However, the local application of such growth factors in diabetic foot ulcers was shown to have poor efficiency due to the rapid leakage and short half-life of the growth factors (Hanft et al., 2008; Uchi et al., 2009). To improve the therapeutic efficiency and enhance the duration of action, PLGA-based nanoparticulate formulations were prepared by the modified solvent diffusion technique and evaluated for their wound healing properties in genetically diabetic mice (db/db) (Losi et al., 2013). To prevent the burst release of the nanoparticles and help in controlling the drug release, the nanoparticles can be further incorporated into a scaffold construct. Such scaffolds loaded with growth factor nanoparticles significantly enhanced wound closure without any adverse reaction, and the scaffold can also be removed from the skin without any bleeding or disruption of the newly-formed granulation tissue.

Recently, fibrin-based nanoparticles were evaluated as a platform for delivering antibiotics to microbe-infected wounds, as they are also reported to stimulate the body's wound healing response (Alphonsa et al., 2014). An extended release of up to 144 hours was achieved with these nanoformulations. The formulated nanoparticles were also shown to possess good *in vitro* antimicrobial activity against *E. coli*, *S. aureus*, and *Candida albicans*.

The infection of burn wounds with *Candida spp.* leads to an invasive disease with a 14–70% mortality rate (Sanchez et al., 2014). The amphotericin B (AmB) encapsulated in chitosan-based nanoparticles (Amb-np) containing PEG was shown to be a highly effective topical therapeutic against such *Candida spp.* infections of wounds, inhibiting greater than 82% of *C. albicans* growth in an animal burn wound model as compared to the untreated control group. The chitosan-based AmB nanoparticles formulation exhibited a quicker efficiency in fungal clearance when compared to the AmB solution.

Often, therapeutic nanoparticles are embedded with a polymeric matrix for controlled release of the therapeutic drug for wound healing (Archana et al., 2013; Vasile et al., 2014). Such a system also utilizes the inherent antibacterial activity of composites such as chitosan. A chitosan matrix incorporated with zinc oxide nanoparticles is coated with gentamicin to obtain a hybrid inorganic–organic material with a consistency suitable for application as a wound dressing (Vasile et al., 2014). The incorporation of gentamicin did not cause any change in the crystalline structure of the zinc oxide or in the initial burst release or sustained release of the encapsulated gentamicin. The nanocomposite was found to show great antimicrobial properties, inhibiting *S. aureus* and *Pseudomonas aeruginosa* growth in both planktonic as well as surface-attached conditions. The developed wound dressing system also maintains a moist environment at the wound interface, providing a cooling sensation and soothing effect, apart from slowly releasing the antibiotic gentamicin. Chitin and chitosan contain N-acetyl glucosamine, which is a major component of dermal tissue essential for the repair of scar tissue (Singh and Ray, 2000), and also possess broad antibacterial activity and low toxicity to mammalian tissue (Archana et al., 2013). A blend of chitosan, poly(N-vinylpyrrolidone) (PVP), and titanium dioxide (TiO_2) nanocomposite dressing with an excellent antimicrobial property and biocompatibility was also developed.

A silane composite nanoparticle incorporated with curcumin through sol-gel-based polymerization was reported to effectively treat burn wounds infected with methicillin-resistant *S. aureus* (MRSA) and *P. aeruginosa* (Krausz et al., 2015). Such encapsulation of curcumin in a nanocarrier extends the inherent antimicrobial properties of curcumin by facilitating interaction with the pathogen surfaces unattainable by free curcumin, thereby reducing bacterial load and enhancing wound healing activity. It was also found to accelerate the formation of granulation tissue, collagen deposition, and the formation of a new blood vessel.

4.4.4 Nano and Microemulsions

Nano and microemulsions are thermodynamically stable, isotropic, and transparent or translucent dispersions of oil and water stabilized by surfactants (Ghosh et al., 2013). Their droplet size varies between

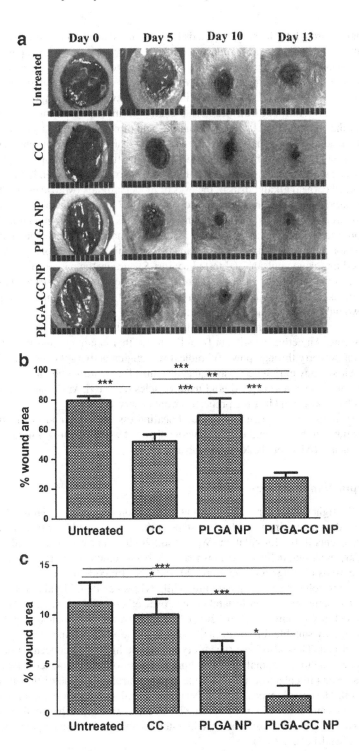

FIGURE 4.2 Effect of PLGA-CC nanoparticles on wound healing (wound closure over time was measured as a percentage of the original area, with ruler units in mm and data as mean ± SD): (A) representative images of the wounds of four test groups: untreated, CC, PLGA nanoparticles, and PLGA-CC nanoparticles, (B) dermal wound area at day 5 (n = 10), (C) dermal wound area at day 10 (n = 7). (Reproduced with permission from Chereddy et al. (2013), Copyright©2013 Elsevier BV.)

10–100 nm prepared through high energy methods such as ultrasonication and high shear homogenizations. As a result of their nanostructure, construct, and efficient stability, with superior solubilization efficiency, many wound healing preparations have been formulated based on nano and microemulsions. Such qualities have established them to be the favored delivery system for plant-based wound healing formulations (Honorio-Franca et al., 2008). The interaction between the components of the emulsion and the lipid layers of stratum corneum leads to a change in its structural integrity, which enhances the permeation of the entrapped drugs even without the help of permeation enhancers (Aggarwal et al., 2013). Therefore, nano and microemulsions offer an added advantage over conventional emulsions for delivering drugs through the skin. Microemulsion has been fine-tuned for the topical application of plant extracts, which was found to accelerate wound healing while also reducing the inflammatory response to silver nitrate-induced corneal alkali burns in the eyes of model animals (Martin et al., 2013).

Cinnamon oil-based nanoemulsions with tiny droplet sizes (5.7 nm) were demonstrated to significantly reduce the healing time of *S. aureus* infected wounds by six days in experimental animals (Ghosh et al., 2013). When appropriately optimized, treatment with the stable nanoemulsion formulation results in better wound contraction and epithelialization than standard antibiotics such as neomycin. When eucalyptus oil was formulated into nanoemulsion and tested against *S. aureus*, the pathogen was found to lose its viability within 15 minutes of interaction, signifying the magnified antibacterial activity of the nanoconstruct (Sugumar et al., 2014). These preparations are non-irritant to the skin and they reportedly exhibit a higher rate of wound contraction than certain antibiotics in animal models.

Seed oil from *Nigella sativa*, a black cumin, has been used traditionally as a therapy for several diseases and ailments. An active constituent from the plant, thymoquinone, has been reported to possess an antioxidant property through powerful radical scavenging activity (Guler et al., 2014). When *N. sativa* emulsion without any enrichment was equated against the *N. sativa* oil enriched with *Calendula officinalis* extract and lipoic acid-capped gold nanoparticles (AuNP–LA) through nanoemulsification, improved antioxidant and wound healing properties were observed.

Nano and microemulsions, as carriers for wound healing essential oils and therapeutic drugs, have also been demonstrated to be safe and nontoxic, with enhanced wound healing, improved spreadability, and therapeutic efficacy (Alam et al., 2017; Chhibber et al., 2015).

4.4.5 Solid Lipid Nanoparticles

The selection of the right dosage form and the right formulation is highly crucial for the efficient management of heavily damaged skin, as the healing steps of the wound should not be impeded by the vehicles and other excipients in the formulation (Wolf et al., 2009). Nanostructured lipid carriers (NLCs), including SLNs, are emerging delivery systems that have been demonstrated to be safe and effective in delivering diverse kinds of drugs (Gainza et al., 2014). SLNs are colloidal particles containing highly purified lipids that are solid at room temperature, while NLCs use liquid lipids (oil) for their preparations. SLNs provide better protection to labile drugs over other colloidal carriers due to their higher physicochemical stability (Bonifacio et al., 2014). The non-irritant and nontoxic lipid components of SLNs make them highly suitable for use in the topical application of therapeutic drugs where the skin is inflamed or damaged. They also have the ability to provide high drug concentrations in the treated skin area. Moreover, due to their small size and lipid composition, close contact between the skin and the nanoparticles results in enhanced resident time and a controlled release of the encapsulated drugs (Gainza et al., 2014). The only drawback with SLNs is that such dispersions are not sufficiently viscous to facilitate topical applications on skin, necessitating its incorporation into other semi-solid dosage forms such as hydrogels to promote the consistency and long-term stability of the final nanoparticulate products (Chen et al., 2013; Sandri et al., 2013).

Lipid nanoparticles are an efficient system for controlled delivery of angiogenic proteins such as recombinant human EGF (rhEGF) for wound healing with high entrapment (95%) (Gainza et al., 2014). Both lipid nanoparticle types, SLNs and NLCs, exhibit an initial burst release phase, a fast release phase (8–24 hours), and slower release phase (24–72 hours). Between the NLCs and SLNs, similar viscosity or thickness was obtained, but SLNs were found to be more occlusive to the skin than the NLCs. Their topical administrations significantly improved healing in terms of wound closure, restoration of the inflammatory process, and re-epithelization. However, based on the entrapment efficiency and lack of solvent utilization, NLC-based nanoparticles were found to be a better alternative than SLN-based nanoparticles for the topical wound healing delivery of rhEGF.

Encapsulation of poorly water-soluble drugs such as astragaloside IV, the chief component of a popular Traditional Chinese Medicine (TCM) Radix Astragali (obtained from the root of *Astragalus membranaceus*), into the SLN and its further incorporation into hydrogels was also reported to promote wound healing by enhancing the repair process while reducing the formation of scar tissue (Figure 4.3). This was mainly attributed to the enhanced proliferation and migration of keratinocytes, along with an increased uptake of the drug by fibroblasts facilitated by the SLN formulation (Chen et al., 2012; Chen et al., 2013). SLNs also present an added advantage of sustaining the release of the encapsulated drug. Such formulations offer great opportunities for delivering anti-inflammatory and analgesic drugs, including opioids, which can reduce and alleviate pain associated with severely damaged skin wounds (Wolf et al., 2009). They can be delivered efficiently without any undesired effects, while such systems were also reported to promote keratinocyte migration and wound closure (Kuchler et al., 2010).

The wide range of adaptability of lipid nanoparticles implies that a complex delivery system for better control and effective wound management can be developed. An active drug, such as silver sulfadiazine (AgSD), or plant extracts can be loaded into the SLNs system and this system can be further incorporated into different wound dressings, such as water-soluble polymeric bases like chondroitin sulfate and sodium hyaluronate, hydroxypropyl methyl cellulose (HPMC), or chitosan glutamate (CSglu). In this manner, the dressing materials are able to maintain and enhance the antimicrobial activity or therapeutic activity of their encapsulated drugs, suggesting their possible clinical application in the management of wounds (Arana et al., 2015; Sandri et al., 2013).

4.4.6 Nanoconstruct Fiber Mats and Matrices

Advances in nanotechnology have enabled the development of nano- to micrometer sized polymeric fibers from materials of diverse origin (Rujitanaroj et al., 2008). Electrospinning is one technique that can be used to produce materials with unique architecture that resembles an extracellular matrix, providing an excellent platform with high entrapment of the active drug for wound healing (Choi et al., 2015; Rath et al., 2015). Due to the high surface-to-volume ratio exhibited by these nanofibers, there is an enhanced porosity that allows the permeation of gas, promotes cell respiration, removes exudates, regenerates the skin, and promotes hemostasis while retaining moisture (Rieger et al., 2013). Different kinds of therapeutic or antimicrobial agents can be loaded to functionalize the electospun fiber mats for enhancing their adaptability to any kind of wound surface.

The incidence of chronic wounds is on the rise due to increasing diabetic and peripheral circulation disorders (Lai et al., 2014). Impaired angiogenesis and cellular migration in such cases results in increased wound healing time. This often necessitates the external supply of angiogenic factors such as VEGF that help in the formation of new blood vessels for complete regeneration at the wound site.

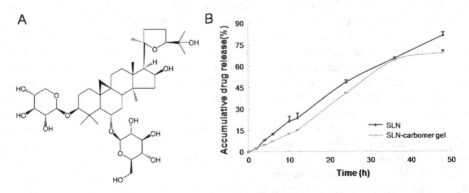

FIGURE 4.3 (A) Chemical structure of astragaloside IV and (B) cumulative astragaloside IV release profiles of astragaloside IV-loaded SLNs and SLN-gel. (Reproduced with permission from Chen et al. (2013), Copyright©2013 Elsevier BV.)

Nanoconstruct fiber mats are a promising delivery system for growth factors as they can also function as skin substitutes, with similar tensile strength and Young's modulus. Growth factors can be delivered in a controlled manner to effect dermal re-epithelialization and reconstruction. Delivery of rhEGF through poly (ε-caprolactone) (PCL)-PEG nanofibers was also reported to enhance the expression of EGF receptors and significantly increase wound closure in diabetic animals (Choi et al., 2008).

Several antibiotics have been effectively delivered through nanofiber constructs for the treatment of chronic and diabetic wounds. The improved hydrophilicity, higher tensile strength, and sustained release of antibiotics from the chitosan-based nanofiber construct resulted in efficient wound healing in a short period of time (Fazli and Shariatinia, 2017). Nanoscale composite ion-exchange fibers prepared with poly(styrene sulfonic acid-co-maleic acid) and polyvinyl alcohol (PVA) for controlled release of cationic antibiotic neomycin also produced satisfactory wound healing in the study animals (Nitanan et al., 2013). A porous, gelatin-based nanofiber scaffold was also reported to release the entrapped cefazoline in a controlled manner and, due to its unique construct, it promotes cell adhesion and other biological processes for improving wound healing (Rath et al., 2015).

Bioresorbable and biocompatible composite fibers have also been used for delivering wound healing plant extracts directly to the wound sites. Different polymers and their nanocomposites evaluated for wound healing preparations include PLGA, chitosan, gelatin, silk fibroin, PVA, PCL, sodium alginate, and poly(lactic acid). Loading plant extracts into biocompatible nanofibers could significantly enhance their antioxidant property and improve the survival rate of skin cells exposed to oxidative stress (Lin et al., 2016). Chitosan-based nanofiber mats loaded with *Garcinia mangostana* extracts were also found to provide a good alternative for accelerating wound healing (Charernsriwilaiwat, et al., 2013).

4.4.7 Nanohydrogels

In the past, traditional dressings were used, the main function of which was to keep the wound dry and prevent the entry of harmful bacteria to the wound (Boateng et al., 2008). However, it is now understood that a warm and moist wound bed, not dehydration, is essential for proper wound healing. Polymeric hydrogels offer a good candidate for wound healing drug delivery as they have the ability to absorb tissue exudates and prevent wound dehydration while allowing oxygen to permeate (Anumolu et al., 2011; Moritz et al., 2014). Apart from these factors, hydrogels possess porosity that is capable of creating a moist environment to deliver the therapeutic substances in a sustained manner at the wound site (Anumolu et al., 2011).

The most desirable wound healing hydrogels are those *in situ* gel forming systems which are converted into gels at the administration site while remaining as a sol (solution) prior to administration (Tran et al., 2011). *In situ* gel formation can be affected through different mechanisms such as temperature or pH-induced gel formation. Thermosensitive polymers such as PCL have been utilized in developing such nanoconstructs and, loaded with active compounds such as curcumin, were demonstrated to significantly enhance the re-epithelization of the epidermis and deposition of collagen within the wound tissue in animal models (Gong et al., 2013). The pH at wounds is reported to be ranging from 6.7 to 7.9. Therefore, *in situ* hydrogel forming systems are promising options for delivering a wide range of wound healing agents, especially growth factors such as VEGF and EGF, as their activity has been reported to be influenced by pH (Banerjee et al., 2012). In fact, the wound healing response was reported to be improved when compared to a non-pH sensitive delivery system. A multifunctional *in situ* forming hydrogel (MISG) was prepared from a mixture of poloxamers 407 and 188 as the matrix, which was designed to stop bleeding, inhibit inflammation, relieve pain, and improve healing (Du et al., 2012). The MISG was composed of several agents including aminocaproic acid (to stop bleeding), an anti-infective povidone iodine, lidocaine, and chitosan. It was found to significantly decrease the hemostasis time, possessing a strong bacteriostatic action, while also increasing the pain threshold and accelerating wound healing.

Nanohydrogel has also been extended to enhance the stability of sensitive wound healing agents, such as growth factor rhEGF, while also promoting their controlled release (Zhou et al., 2011). Increasing the stability of such cell proliferating agents promotes wound healing significantly. Depending on the kind of polymers used for hydrogel formation, additional functionality can be imparted into the system, including bioadhesiveness or injectable form formulation.

Cellulose-based nanohydrogels, such as bacterial nanocellulose (BNC), have also been proven to be perfectly suitable biomaterials that fulfill all the requirements of an ideal wound dressing (Bielecki et al., 2012). Even though the BNC does not possess antimicrobial activity by itself, it can be loaded with such drugs for treating acute and chronically infected wounds. The incorporation of antiseptics, such as octenidine, does not cause any change in the physicochemical properties of the BNC, however a biphasic sustained release of the drug based on diffusion and swelling was observed which sustained the release of the drug up to 96 hours (Moritz et al., 2014). An advanced wound healing dressing based on nanofibrillated cellulose was reported to trigger blood coagulation, resulting in the rapid formation of blood clots. The incorporation of kaolin into the nanohydrogel was also found to further improve clot formation (Basu et al., 2017). Cytocompatible and flexible nanocomposite hydrogels based on carboxymethyl cellulose have also been demonstrated to exhibit potential for wound dressings and sustained drug release (Rakhshaei and Namazi, 2017).

Nanohydrogels of inter-penetrating polymer networks (IPN) based on inorganic clays such as laponite (lithium magnesium silicate hydrate) are another nanocontruct that have been explored for their wound healing applications (Golafshan et al., 2017; Yang et al., 2017). pH and glucose dual-responsive and injectable nanohydrogels for diabetic wound healing, along with nanocomposite hydrogels capable of delivering silver, *Aloe vera*, and curcumin triple loaded are another extension of advanced wound dressings systems (Anjum et al., 2016; Zhao et al., 2017).

4.4.8 Nanoconstruct Wafers and Sponges

Wafers and sponges are extremely porous structures produced by freeze-drying polymer solutions and gels, which can be applied to an exuding wound surface with ease (Anisha et al., 2013; Pawar et al., 2014). As a result of their porosity and their ability to absorbed large amounts of water, they offer several advantages over other constructs, such as maintenance of a moist environment and the proper circulation of oxygen. Due to their nanoconstruct and components, efficient loading is possible with both water-soluble and insoluble drugs. Since they are directly applied into the wound, there is minimal human contact, thus maintaining sterility at the wound site to minimize infections (Matthews et al., 2008). One of the striking uniquenesses of wafers is that when they are applied to the wound, they revert back into a highly viscous fluid or gel. This property is highly beneficial for wound healing as it provides a system suitable for the sustained release of the wound healing agent for an extended period of time (Boateng et al., 2010; Labovitiadi et al., 2012). Further, the lyophilized system in wafers and sponges was found to provide protection to the entrapped drugs, thereby enhancing their stability when compared to semi-solid formulations of similar drugs (Matthews et al., 2008).

A wide range of nanocomposites have been utilized for preparing wafers and sponges for effective wound healing. Some of the polymers reported for preparing such composites include chitosan (Anisha et al., 2013; Mohandas et al., 2015), polyox, carrageenan and sodium alginate (Pawar et al., 2014), carboxymethyl cellulose (Ng and Jumaat, 2014), and hyaluronic acid and collagen (Muthukumar et al., 2014; Niiyama and Kuroyanagi, 2014). The excellent texture it provided, the sponginess, and the flexibility, along with a high drug loading ability, make them an excellent system for wound healing applications (Figure 4.4).

In a recent study, a hydrogel system that releases a toxin-absorbing nanosponge was developed for antivirulence treatment of local bacterial infections (Wang et al., 2015). The nanosponge consists of PLGA nanoparticles wrapped with Red Blood Cells (RBCs) membrane, which was then loaded into a hydrogel system. The nanoparticles were retained in the administration site following sub-cutaneous injection and in an MRSA-subcutaneous mouse model, mice treated with the nanosponge–hydrogel system exhibited markedly reduced development of MRSA skin lesions (Wang et al., 2015). Imiquimod, which has been observed to inhibit cell proliferation and induced apoptosis in normal skin and hypertrophic scar fibroblasts, has been incorporated into a nanosponge system based on β-cyclodextrin (Chiara et al., 2014). A high drug loading along with a slow and sustained *in vitro* release of the imiquimod was observed in the nanosponge system, which also exhibited a greater antiproliferative effect when compared to the free imiquimod.

Recent developments have extended wafers and sponges to functional dressings for treating pain and infections associated with chronic wounds (Catanzano et al., 2017). Multifaced lyophilized wafers that

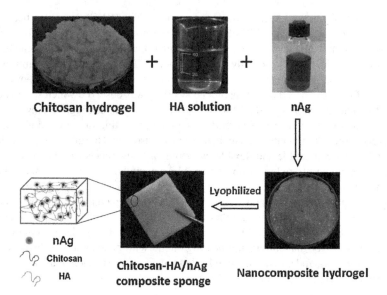

FIGURE 4.4   Fabrication of chitosan–HA/nAg composite sponges. (Reproduced with permission from Anisha et al. (2013), Copyright©2013 Elsevier BV.)

incorporate antibiotic-loaded liposomes have also been reported recently with controlled release and improved wound healing (Avachat and Takudage, 2017).

4.5 Conclusion

Traditionally, the main function of conventional wound dressings was to keep the wound dry and prevent the entry of harmful microorganisms. However, it is now well understood that effective wound management requires a warm, moist environment with proper circulation at the wound site. The application of nanotechnology in medicine has offered a plethora of opportunities for the efficient management of acute and chronic types of wounds. The ability to control, manipulate, and manufacture structures and devices at the nanometer size range has been predicted to revolutionize the future of nanomedicine. Depending on the need, several therapeutic nanostructures are available to choose for controlled delivery of wound healing drugs directly to the wound site. Several studies have also demonstrated the superior properties of these therapeutic nanostructures in the management of different types of wounds.

REFERENCES

Adhirajan N, Shanmugasundaram N, Shanmuganathan S, Babu S. Functionally modified gelatin microspheres impregnated collagen scaffold as novel wound dressing to attenuate the proteases and bacterial growth. Eur J Pharm Sci. 2009;36:235–45.

Aggarwal N, Goindi S, Khurana R. Formulation, characterization and evaluation of an optimized microemulsion formulation of griseofulvin for topical application. Colloids Surf B Biointerfaces. 2013;105:158–66.

Ahmadi M, Adibhesami M. The effect of silver nanoparticles on wounds contaminated with *Pseudomonas aeruginosa* in mice: An experimental study. Iran J Pharm Res. 2017;16:661–9.

Alam P, Ansari MJ, Anwer MK, Raish M, Kamal YKT, Shakeel F. Wound healing effects of nanoemulsion containing clove essential oil. Artif Cells Nanomed Biotechnol. 2017;45:591–7.

Alemdaroglu C, Degim Z, Celebi N, Sengezer M, Alomeroglu M, Nacar A. Investigation of epidermal growth factor containing liposome formulation effects on burn wound healing. J Biomed Mater Res A. 2008;85:271–83.

Alphonsa BM, Sudheesh Kumar PT, Praveen G, Biswas R, Chennazhi KP, Jayakumar R. Antimicrobial drugs encapsulated in fibrin nanoparticles for treating microbial infested wounds. Pharm Res. 2014;31:1338–51.

Anisha BS, Biswas R, Chennazhi KP, Jayakumar R. Chitosan–hyaluronic acid/nano silver composite sponges for drug resistant bacteria infected diabetic wounds. Int J Biol Macromol. 2013;62:310–20.

Anjum S, Gupta A, Sharma D, Gautam D, Bhan S, Sharma A, Kapil A, Gupta B. Development of novel wound care systems based on nanosilver nanohydrogels of polymethacrylic acid with *Aloe vera* and curcumin. Mater Sci Eng C. 2016;64:157–66.

Anumolu SNS, Menjoge AR, Deshmukh M, Gerecke D, Stein S, Laskin J, et al. Doxycycline hydrogels with reversible disulfide crosslinks for dermal wound healing of mustard injuries. Biomaterials. 2011;32:1204–17.

Arana L, Salado C, Vega S, Aizpurua-Olaizola O, Arada I, Suarez T, et al. Solid lipid nanoparticles for delivery of *Calendula officinalis* extract. Colloids Surf B Biointerfaces. 2015;135:18–26.

Archana D, Singh BK, Dutta J, Dutta P. In vivo evaluation of chitosan–PVP–titanium dioxide nanocomposite as wound dressing material. Carbohydr Polym. 2013;95:530–9.

Avachat AM, Takudage PJ. Design and characterization of multifaceted lyophilized liposomal wafers with promising wound healing potential. J Liposome Res. 2017.

Banerjee I, Mishra D, Das T, Maiti TK. Wound pH-responsive sustained release of therapeutics from a Poly(NIPAAm-co-AAc) hydrogel. J Biomater Sci Polym Ed. 2012;23:111–32.

Basu A, Hong J, Ferraz N. Hemocompatibility of Ca^{2+}-crosslinked nanocellulose hydrogels: Toward efficient management of hemostasis. Macromol Biosci. 2017:1700236.

Bawarski WE, Chidlowsky E, Bharali DJ, Mousa SA. Emerging nanopharmaceuticals. *Nanomedicine: NBM*. 2008;4:273–82.

Bielecki S, Kalinowska H, Krystynowicz A, Kubiak K, Kolodziejczyk M, de Groeve M. Wound dressings and cosmetic materials from bacterial nanocellulose. In: Gama M, Gatenholm P, Klemm D, editors. Bacterial nanocellulose: A sophisticated multifunctional material. Boka Raton, FL: Taylor and Francis (CRC Press); 2012. p. 157–74.

Boateng JS, Auffret AD, Matthews KH, Humphrey MJ, Stevens HN, Eccleston GM. Characterisation of freeze-dried wafers and solvent evaporated films as potential drug delivery systems to mucosal surfaces. Int J Pharm. 2010;389:24–31.

Boateng JS, Matthews KH, Stevens HN, Eccleston GM. Wound healing dressings and drug delivery systems: A review. J Pharm Sci. 2008;97:2892–923.

Bonifacio BV, Silva PB, Ramos MA, Negri KM, Bauab TM, Chorilli M. Nanotechnology-based drug delivery systems and herbal medicines: A review. Int J Nanomedicine. 2014;9:1–15.

Brem H, Stojadinovic O, Diegelmann RF, Entero H, Lee B, Pater I, et al. Molecular markers in patients with chronic wounds to guide surgical debridement. Mol Med. 2007;13:30–9.

Castangia I, Nacher A, Caddeo C, Valenti D, Fadda AM, Diez-Sales O, Ruiz-Sauri A, Manconi M. Fabrication of quercetin and curcumin bionanovesicles for the prevention and rapid regeneration of full-thickness skin defects on mice. Acta Biomater. 2014;10:1292–300.

Catanzano O, Docking R, Schofield P, Boateng J. Advanced multi-targeted composite biomaterial dressing for pain and infection control in chronic leg ulcers. Carbohydr Polym. 2017;172:40–8.

Charernsriwilaiwat N, Rojanarata T, Ngawhirunpat T, Sukma M, Opanasopit P. Electrospun chitosan-based nanofiber mats loaded with *Garcinia mangostana* extracts. Int J Pharm. 2013;452:333–43.

Chen X, Peng LH, Li N, Li QM, Li P, Fung KP, Leung PC, Gao JQ. The healing and anti-scar effects of astragaloside IV on the wound repair in vitro and in vivo. J Ethnopharmacol. 2012;139:721–7.

Chen X, Peng LH, Shan YH, Li N, Wei W, Yu L, Li QM, Liang WQ, Gao JQ. Astragaloside IV-loaded nanoparticle-enriched hydrogel induces wound healing and anti-scar activity through topical delivery. Int J Pharm. 2013;447:171–81.

Chereddy KK, Coco R, Memvanga PB, Ucakar B, Rieux A, Vandermeulen G, Preat V. Combined effect of PLGA and curcumin on wound healing activity. J Control Release. 2013;171:208–15.

Chhibber T, Wadhwa S, Chadha P, Sharma G, Katare OP. Phospholipid structured microemulsion as effective carrier system with potential in methicillin sensitive *Staphylococcus aureus* (MSSA) involved burn wound infection. J Drug Target. 2015;23:943–52.

Chiara B, Sara S, Daniela A, Irene C, Mara F, Stefania C, et al. Cyclodextrin-based nanosponges as a nanotechnology strategy for imiquimod delivery in pathological scarring prevention and treatment. J Nanopharm Drug Deliv. 2014;2:311–24.

Choi JS, Kim HS, Yoo HS. Electrospinning strategies of drug-incorporated nanofibrous mats for wound recovery. Drug Deliv Transl Res. 2015;5:137–45.

Choi JS, Leong KW, Yoo HS. In vivo wound healing of diabetic ulcers using electrospun nanofibers immobilized with human epidermal growth factor (EGF). Biomaterials. 2008;29:587–96.

Church D, Elsayed S, Reid O, Winston B, Lindsay R. Burn wound infections. Clin Microbiol Rev. 2006;19:403–34.

Comune M, Rai A, Chereddy KK, Pinto S, Aday S, Ferreira AF, et al. Antimicrobial peptide-gold nanoscale therapeutic formulation with high skin regenerative potential. J Control Release. 2017;262:58–71.

Du L, Tong L, Jin Y, Jia J, Liu Y, Su C, Yu S, Li X. A multifunctional in situ–forming hydrogel for wound healing. Wound Repair Regen. 2012;20:904–10.

Dyson M. Advances in wound healing physiology: The comparative perspective. Vet Dermatol. 1997;8:227–33.

Falanga V. Wound healing and its impairment in the diabetic foot. Lancet. 2005;366:1736–43.

Fazli Y, Shariatinia Z. Controlled release of cefazolin sodium antibiotic drug from electrospun chitosan-polyethylene oxide nanofibrous mats. Mater Sci Eng C. 2017;71:641–52.

Ferreira H, Matama T, Silva R, Silva C, Gomes AC, Paulo A. Functionalization of gauzes with liposomes entrapping an anti-inflammatory drug: A strategy to improve wound healing. React Funct Polym. 2013;73:1328–34.

Fukui T, Kawaguchi AT, Takekoshi S, Miyasaka M, Sumiyoshi H, Tanaka R. Liposome-encapsulated hemoglobin accelerates skin wound healing in mice. Artif Organs. 2012;36:161–9.

Gainza G, Pastor M, Aguirre JJ, Villullas S, Pedraz JL, Hernandez RM, Igartua M. A novel strategy for the treatment of chronic wounds based on the topical administration of rhEGF-loaded lipid nanoparticles: In vitro bioactivity and in vivo effectiveness in healing-impaired db/db mice. J Control Release. 2014;185:51–61.

Ghosh V, Saranya S, Mukherjee A, Chandrasekaran N. Antibacterial microemulsion prevents sepsis and triggers healing of wound in wistar rats. Colloids Surf B Biointerfaces. 2013;105:152–7.

Golafshan N, Rezahasani R, Esfahani TM, Kharaziha M, Khorasani SN. Nanohybrid hydrogels of Laponite: PVA-Alginate as a potential wound healing material. Carbohydr Polym. 2017;176:392–401.

Gong CY, Wu QJ, Wang YJ, Zhang D, Luo F, Zhao X, Wei Y, Qian Z. A biodegradable hydrogel system containing curcumin encapsulated in micelles for cutaneous wound healing. Biomaterials. 2013;34:6377–87.

Guler E, Barlas FB, Yavuz M, Demir B, Gumus ZP, Baspinar Y, Coskunol H, Timur S. Bio-active nanoemulsions enriched with gold nanoparticle, marigold extracts and lipoic acid: In vitro investigations. Colloids Surf B Biointerfaces. 2014;121:299–306.

Gunasekaran T, Nigusse T, Dhanaraju MD. Silver nanoparticles as real topical bullets for wound healing. J Am Coll Clin Wound Spec. 2012;3:82–96.

Halloran CM, Slavin JP. Pathophysiology of wound healing. Surgery. 2002;20:i–v.

Hanft JR, Pollak RA, Barbul A, Gils C, Kwon PS, Gray SM, et al. Phase I trial on the safety of topical rhVEGF on chronic neuropathic diabetic foot ulcers. J Wound Care. 2008;17(30):34–7.

Harding KG, Morris HL, Patel GK. Healing chronic wounds. BMJ. 2002;324:160–3.

Heng MCY. Wound healing in adult skin: Aiming for perfect regeneration. Int J Dermatol. 2011;50:1058–66.

Holbrook RD, Rykaczewski K, Staymates ME. Dynamics of silver nanoparticle release from wound dressings revealed via in situ nanoscale imaging. J Mater Sci Mater Med. 2014;25:2481–9.

Honorio-França AC, Marins CMF, Boldrini F, Franca EL. Evaluation of hypoglicemic activity and healing of extract from amongst bark of "Quina do Cerrado" (*Strychnos pseudoquina* ST. HILL). Acta Cir Bras. 2008;23:504–10.

Hu M, Li C, Li X, Zhou M, Sun J, Sheng F, Shi S, Lu L. Zinc oxide/silver bimetallic nanoencapsulated in PVP/PCL nanofibres for improved antibacterial activity. Artif Cells Nanomed Biotechnol. 2017.

Jangde R, Singh D. Preparation and optimization of quercetin-loaded liposomes for wound healing, using response surface methodology. Artif Cells Nanomed Biotechnol. 2016;44:635–41.

Kamaly N, Xiao Z, Valencia PM, Radovic-Moreno AF, Farokhzad OC. Targeted polymeric therapeutic nanoparticles: Design, development and clinical translation. Chem Soc Rev. 2012;41:2971–3010.

Krausz AE, Adler BL, Kabral V, Navati M, Doerner J, Charafeddine RA, et al. Curcumin-encapsulated nanoparticles as innovative antimicrobial and wound healing agent. Nanomedicine: NBM. 2015;11:195–206.

Kuchler S, Wolf NB, Heilmann S, Weindl G, Helfmann J, Yahya MM, Stein C, Schafer-Korting M. 3D-Wound healing model: Influence of morphine and solid lipid nanoparticles. J Biotechnol. 2010;148:24–30.

Labovitiadi O, Lamb AJ, Matthews KH. Lyophilised wafers as vehicles for the topical release of chlorhexidine digluconate-Release kinetics and efficacy against *Pseudomonas aeruginosa*. Int J Pharm. 2012;439:157–64.

Lai HJ, Kuan CH, Wu HC, Tsai JC, Chen TM, Hsieh DJ, Wang TZ. Tailor design electrospun composite nanofibers with staged release of multiple angiogenic growth factors for chronic wound healing. Acta Biomater. 2014;10:4156–66.

Lazarus GS, Cooper DM, Knighton DR, Margolis DJ, Percoraro RE, Rodeheaver G, Robson MC. Definitions and guidelines for assessment of wounds and evaluation of healing. Wound Repair Regen. 1994;2:165–70.

Leu JG, Chen SA, Chen HM, Wu WM, Hung CF, Yao YD, Tu CS, Liang YJ. The effects of gold nanoparticles in wound healing with antioxidant epigallocatechin gallate and α-lipoic acid. Nanomedicine: NBM. 2012;8:767–75.

Li J, Chen J, Kirsner R. Pathophysiology of acute wound healing. Clin Dermatol. 2008;25:9–18.

Li Q, Lu F, Zhou G, Yu K, Lu B, Xiao Y, et al. Silver inlaid with gold nanoparticle/chitosan wound dressing enhances antibacterial activity and porosity, and promotes wound healing. Biomacromolecules. 2017.

Lin S, Chen M, Jiang H, Fan L, Sun B, Yu F, et al. Green electrospun grape seed extract-loaded silk fibroin nanofibrous mats with excellent cytocompatibility and antioxidant effect. Colloids Surf B Biointerfaces. 2016;139:156–63.

Losi P, Briganti E, Errico C, Lisella A, Sanguinetti E, Chiellini F, Soldani G. Fibrin-based scaffold incorporating VEGF- and bFGF-loaded nanoparticles stimulates wound healing in diabetic mice. Acta Biomater. 2013;9:7814–21.

Lu S, Gao W, Gu HY. Construction, application and biosafety of silver nanocrystalline chitosan wound dressing. Burns. 2008;34:623–8

Lu S, Zhang X, Luo H, Fu Y, Xu M, Tang S. Clodronate liposomes reduce excessive scar formation in a mouse model of burn injury by reducing collagen deposition and TGF-β1 expression. Mol Biol Rep. 2014;41:2143–9.

Martin P. Wound healing – Aiming for perfect skin regeneration. Science. 1997;276:75–81.

Martin LFT, Rocha EM, Garcia SB, Paula JS. Topical Brazilian propolis improves corneal wound healing and inflammation in rats following alkali burns. BMC Complement Altern Med. 2013;13:337.

Matthews KH, Stevens HNE, Auffret AD, Humphrey MJ, Ecclesson GM. Formulation, stability and thermal analysis of lyophilized wound healing wafers containing an insoluble MMP-3 inhibitor and a non-ionic surfactant. Int J Pharm. 2008;356:110–20.

Miller CN, Newall N, Kapp SE, Lewin G, Karimi L, Carville K, Gliddon T, Santamaria NM. A randomized-controlled trial comparing cadexomer iodine and nanocrystalline silver on the healing of leg ulcers. Wound Repair Regen. 2010;18:359–67.

Mohandas A, Anisha BS, Chennazhi KP, Jayakumar R. Chitosan-hyaluronic acid/VEGF loaded fibrin nanoparticles composite sponges for enhancing angiogenesis in wounds. Colloids Surf B Biointerfaces. 2015;127:105–13.

Moore K, McCallion R, Searle RJ, Stacey MC, Harding KG. Prediction and monitoring the therapeutic response of chronic dermal wounds. Int Wound J. 2006;3:89–96.

Moritz S, Wiegand C, Wesarg F, Hessler N, Muller FA, Kralisch D, Hipler UC, Fischer D. Active wound dressings based on bacterial nanocellulose as drug delivery system for octenidine. Int J Pharm. 2014;471:45–55.

Moulaoui K, Caddeo C, Manca ML, Castangia I, Valenti D, Escribano E, et al. Identification and nanoentrapment of polyphenolic phytocomplex from Fraxinus angustifolia: In vitro and in vivo wound healing potential. Eur J Med Chem. 2015;89:179–88.

Muthukumar T, Prabu P, Ghosh K, Sastry TP. Fish scale collagen sponge incorporated with *Macrotyloma uniflorumplant* extract as a possible wound/burn dressing material. Colloids Surf B Biointerfaces. 2014;113:207–12.

Naumov AA, Shatalin YV, Potselueva MM. Effects of a nanocomplex containing antioxidant, lipid, and amino acid on thermal burn wound surface. Bull Exp Biol Med. 2010;149:62–6.

Ng SF, Jumaat N. Carboxymethyl cellulose wafers containing antimicrobials: A modern drug delivery system for wound infections. European J Pharm Sci. 2014;51:173–9.

Niiyama H, Kuroyanagi Y. Development of novel wound dressing composed of hyaluronic acid and collagen sponge containing epidermal growth factor and vitamin C derivative. J Artif Organs. 2014;17:81–7.

Nitanan T, Akkaramongkolporn P, Rojanarata T, Ngawhirunpat T, Opanasopit P. Neomycin-loaded poly(styrene sulfonic acid-co-maleic acid) (PSSA-MA)/polyvinyl alcohol (PVA) ion exchange nanofibers for wound dressing materials. Int J Pharm. 2013;448:71–8.

National Nanotechnology Initiative (NNI). What is nanotechnology? 2017 [cited 2017 November 6]. Available from: http://www.nano.gov/nanotech-101/what/definition.

Pan J, Chan SY, Lee WG, Kang L. Microfabricated particulate drug-delivery systems. Biotechnol J. 2011;6:1477–87.

Pando D, Caddeo C, Manconi M, Fadda AM, Pazos C. Nanodesign of olein vesicles for the topical delivery of the antioxidant resveratrol. J Pharm Pharmacol. 2013;65:1158–67.

Pawar HV, Boateng JS, Ayensu I, Tetteh J. Multifunctional medicated lyophilised wafer dressing for effective chronic wound healing. J Pharm Sci. 2014;103:1720–33.

Percival NJ. Classification of wounds and their management. Surgery. 2002;20:114–7.

Prasain JK, Barnes S. Metabolism and bioavailability of flavonoids in chemoprevention: Current analytical strategies and future prospectus. Mol Pharmaceutics. 2007;4:846–64.

Rahman Z, Zidan AS, Khan MA. Non-destructive methods of characterization of risperidone solid lipid nanoparticles. Eur J Pharm Biopharm. 2010;76:127–37.

Rai M, Yadav A, Gade A. Silver nanoparticles as a new generation of antimicrobials. Biotechnol Adv. 2009;27:76–83.

Rakhshaei R, Namazi H. A potential bioactive wound dressing based on carboxymethyl cellulose/ZnO impregnated MCM-41 nanocomposite hydrogel. Mater Sci Eng C. 2017;73:456–64.

Rath G, Hussain T, Chauhan G, Garg T, Goyal AK. Fabrication and characterization of cefazolin-loaded nano-fibrous mats for the recovery of post-surgical wound. Artif Cells Nanomed Biotechnol. 2015;44:1783–92.

Rice TN, Hamblin MR, Herman IM. Acute and impaired wound healing: Pathophysiology and current methods for drug delivery, part 2: Role of growth factors in normal and pathological wound healing: Therapeutic potential and methods of delivery. Adv Skin Wound Care. 2012;25:349–70.

Rieger KA, Birch NP, Schiffman JD. Designing electrospun nanofiber mats to promote wound healing – A review. J Mater Chem B. 2013;1:4531–41.

Rujitanaroj P, Pimpha N, Supaphol P. Wound-dressing materials with antibacterial activity from electrospun gelatin fiber mats containing silver nanoparticles. Polymer. 2008;49:4723–32.

Sanchez DA, Schairer D, Tuckman-Vernon C, Chouake J, Kutner A, Makdisi J, et al. Amphotericin B releasing nanoparticle topical treatment of *Candida spp.* in the setting of a burn wound. Nanomedicine: NBM. 2014;10:269–77.

Sandri G, Bonferoni MC, Autilia F, Rossi S, Ferrari F, Grisoli P, Caramella C. Wound dressings based on silver sulfadiazine solid lipid nanoparticles for tissue repairing. Eur J Pharm Biopharm. 2013;84:84–90.

Singer AJ, Clark RAF. Cutaneous wound healing. N Engl J Med. 1999;341:738–46.

Singer AJ, Dagum AB. Current management of acute cutaneous wounds. N Engl J Med. 2008;359:1037–46.

Singh DK, Ray A. Biomedical applications of chitin, chitosan and their derivatives. J Macromol Sci C: Polym Rev. 2000;40:69–83.

Singh MR, Saraf S, Vyas A, Jain V, Singh D. Innovative approaches in wound healing: Trajectory and advances. Artif Cells Nanomed Biotechnol. 2013;41:202–12.

Singla R, Soni S, Patial V, Kulurkar PM, Kumari A, Mahesh S, Padwad YS, Yadav SK. *In vivo* diabetic wound healing potential of nanobiocomposites containing bamboo cellulose nanocrystals impregnated with silver nanoparticles. Int J Biol Macromol. 2017;105:45–55.

Sivakumar AS, Krishnaraj C, Sheet S, Rampa DR, Kang DR, Belal SA, et al. Interaction of silver and gold nanoparticles in mammalian cancer: As real topical bullet for wound healing — A comparative study. In Vitro Cell Dev Biol. 2017;53:632–45.

Soppimath KS, Aminabhavi TM, Kulkarni AR, Rudzinski WE. Biodegradable polymeric nanoparticles as drug delivery devices. J Control Release. 2001;70:1–20.

Sugumar S, Ghosh V, Nirmala MJ, Mukherjee A, Chandrasekaran N. Ultrasonic emulsification of eucalyptus oil nanoemulsion: Antibacterial activity against *Staphylococcus aureus* and wound healing activity in Wistar rats. Ultrason Sonochem. 2014;21:1044–9.

Thakkar KN, Mhatre SS, Parikh RY. Biological synthesis of metallic nanoparticles. Nanomedicine: NBM. 2010;6:257–62.

Tian J, Wong KKY, Ho CM, Lok CN, Yu WY, Che CM, Chiu JF, Tam PKH. Topical delivery of silver nanoparticles promotes wound healing. ChemMedChem. 2007;2:129–36.

Tocco I, Zavan B, Bassetto F, Vindigni V. Nanotechnology-based therapies for skin wound regeneration. J Nanomater. 2012.

Tran NQ, Joung YK, Lih E, Park KD. In situ forming and rutin-releasing chitosan hydrogels as injectable dressings for dermal wound healing. Biomacromolecules. 2011;12:2872–80.

Uchi H, Igarashi A, Urabe K, Nakayama J, Kawamori R, Tamaki K, Furue M. Clinical efficacy of basic fibroblast growth factor (bFGF) for diabetic ulcer. Eur J Dermatol. 2009;19:461–8.

Vasile BS, Oprea O, Voicu G, Ficai A, Andronescu E, Teodorescu A, Holban A. Synthesis and characterization of a novel controlled release zincoxide/gentamicin–chitosan composite with potential applications in wounds care. Int J Pharm. 2014;463:161–9.

Wang F, Gao W, Thamphiwatana S, Luk BK, Angsantikul P, Zhang Q, et al. Hydrogel retaining toxin-absorbing nanosponges for local treatment of methicillin-resistant *Staphylococcus aureus* infection. Adv Mater. 2015;27:3437–43.

Wang H, Liu M, Du S. Optimization of madecassoside liposomes using response surface methodology and evaluation of its stability. Int J Pharm. 2014;473:280–5.

Wolf NB, Kuchler S, Radowski MR, Blaschke T, Kramer KD, Weindl G, Schafer-Korting M. Influences of opioids and nanoparticles on in vitro wound healing models. Eur J Pharm Biopharm. 2009;73:34–42.

Xiang Q, Xiao J, Zhang H, Zhang X, Lu M, Zhang H, et al. Preparation and characterisation of bFGF-encapsulated liposomes and evaluation of wound-healing activities in the rat. Burns. 2011;37:886–95.

Yang C, Xue R, Zhang Q, Yang S, Liu P, Chen L, Wang K, Zhang X, Wei Y. Nanoclay cross-linked semi-IPN silk sericin/poly(NIPAm/LMSH) nanocomposite hydrogel: An outstanding antibacterial wound dressing. Mater Sci Eng C. 2017;81:303–13.

Zhao L, Niu L, Liang H, Tan H, Liu C Zhu F. pH and glucose dual-responsive injectable hydrogels with insulin and fibroblasts as bioactive dressings for diabetic wound healing. ACS Appl Mater Interfaces. 2017;9:37563–74.

Zhou W, Zhao M, Zhao Y, Mou Y. A fibrin gel loaded with chitosan nanoparticles for local delivery of rhEGF: Preparation and in vitro release studies. J Mater Sci Mater Med. 2011;22:1221–30.

5

The Role of Nanotechnology in the Treatment of Drug Resistance Cancer

**Sandipan Dasgupta, Anup Kumar Das, Paulami Pal,
Subhabrata Ray, and Bhaskar Mazumder**

CONTENTS

5.1 Introduction

Carcinogenesis is a sequence of multistep phenomenon, developed due to an amalgamation of several factors, like environmental and genetic factors. It leads to a series of epigenetic and genetic changes that occurs with progression of the disease through different stages. Carcinogenic cells arise from a particular mutated cell that suffered from genetic and epigenetic changes. This diseased state of cancer is a result of quite a few chronological events triggered by diverse factors including virus-induced transformation, genetic predispositions along with radiation, and/or certain chemicals (Kanwar et al., 2012).

According to a survey conducted by the American Cancer Society, the most common forms of cancer in men were found to be prostate, lung, and colorectal cancers, whereas in women, it was lung, breast, colon, and rectum cancer. In the present time scenario, about 45% of all cancers are resistant to chemotherapy and the chief obstacle in cancer treatment is development of multidrug resistance (MDR) by the cancer cells. By definition, MDR is a phenomenon involving tumor cells that are resistant to the cytotoxic or cytostatic actions of the structurally-dissimilar and functionally-opposing drugs that are conventionally applied in cancer chemotherapy (Kapoor, et al, 2013).

Resistance to chemotherapeutics can be categorized into two broad classes: those intrinsically resistant to chemotherapy and tumors that acquire resistance to chemotherapy.

Intrinsic resistance denotes that before receiving chemotherapy, resistance factors pre-exist within the tumor cells that make the chemotherapy-mediated cancer treatment ineffective. Intrinsic or primary drug resistance does not have a particular cellular origin, but often arises from duct cells or the cells lining of excretory organs (e.g., digestive organs, respiratory system, and urinary organs). The cells lining the excretory organs, which normally function to detoxify, transport, and excrete many toxic compounds, may maintain this normal property when they turn into cancerous cells to detoxify, excrete, and eliminate chemotherapeutic agents (Fojo et al., 1987). Acquired or induced drug resistance is seen initially in responsive tumors, which over the course of time no longer respond to the chemotherapeutics to which they were initially sensitive. This latter type is commonly termed drug resistance and it may develop for one drug or it may develop simultaneously for multiple agents, as in multidrug resistance (MDR) (Longley and Johnston, 2005).

Generally, drug resistance due to pharmacokinetic (PK) factors, such as drug absorption, distribution, metabolism, and elimination (ADME), restricts the amount of a systemically administered drug that

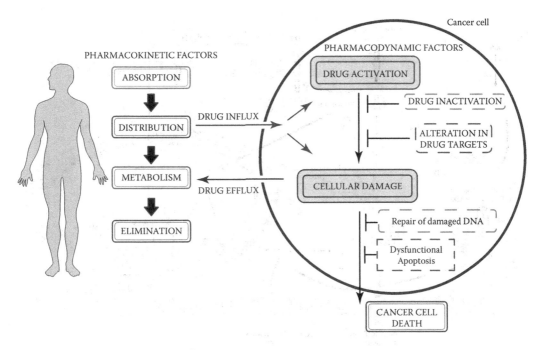

FIGURE 5.1 Pharmacokinetic (PK) pharmacodynamic (PD) factors affecting drug resistance.

reaches the tumor cells. In the tumor or cancerous cell, the effects of the drug on the cancer cell are collectively termed its pharmacodynamic (PD) properties as shown in Figure 5.1.

As seen from the figure, the anticancer activity of a drug can be limited by poor drug influx or disproportionate efflux, inactivation or lack of drug activation, changes in levels of expression of the targeted proteins, adaptive pro-survival response activation, and abnormal apoptosis, which is a trademark of cancer (Borst and Elferink, 2002; Fojo and Bates, 2003; Gottesman et al., 2002).

5.2 Mechanisms of Cancer Resistance

Even though many effective anticancer agents are available nowadays, their therapeutic potential is often compromised due to reasons encompassing tumor sensitivity, drug access, and PKs. The core of all these problems is the drug resistance towards chemotherapy. The mechanism by which cancerous cells or tumors develop resistance towards chemotherapy is one of the major hurdles in successful cancer management.

Although many types of cancers are initially vulnerable to chemotherapeutic agents, over the course of time they can develop resistance through the above mentioned or other mechanisms, namely deoxyribonucleic acid (DNA) mutations and changes in metabolic pathway that promote drug inhibition followed by gradual degradation. In this section, we outline how drug resistance through drug inactivation, drug target alteration, drug efflux, DNA damage repair, and apoptosis inhibition develops in cancer in response to current treatments; how these problems can be addressed has been discussed in the subsequent sections to follow.

5.2.1 Decreased Drug Entry: Role of Uptake Transporters

Beside pharmaceutical related factors such as drug administration, distribution, metabolism, and excretion, the primary obstacle which blocks a drug molecule from entering the intracellular compartment is the plasma membrane. As shown in Figure 5.2, anticancer agents can interact with many molecules resulting in a complex formation represented as a hexagon (D) which can enter cells either by passive

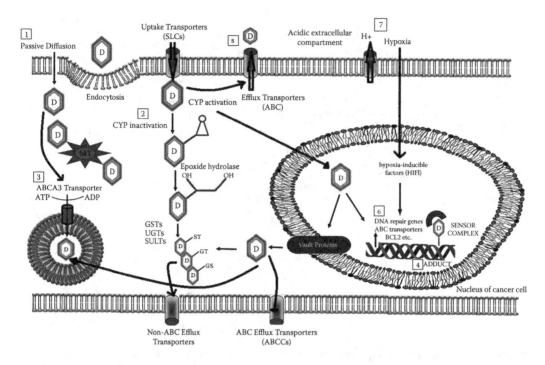

FIGURE 5.2   Mechanism of multidrug resistance (MDR) in cancer.

diffusion, endocytosis, or uptake transportation. Although the exact mechanisms of cellular uptake are not properly understood for most anticancer drugs, it has been well established that the decreased expression of reduced-folate carriers (SLC19A1) and polymorphisms in their genes significantly hampers a patient's response towards chemotherapeutic agents. Notable among these is the example of methotrexate (Assaraf, 2007).

5.2.2 Drug Inactivation: Role of Detoxifying Enzymes

Once inside the cells, drug metabolism enzymes are the second line of cellular resistance for anticancer drugs. This process involves the role of Phase I and II metabolizing enzymes.

Phase I metabolism occurs primarily through cytochrome P450 enzymes (CYPs) and epoxide hydrolases. CYPs are found in mitochondria and the endoplasmic reticulum (ER) and catalyze the endogenous monooxygenase reaction by introducing one atom of oxygen into the substrate (in this case anticancer drugs) and one into water. On the other hand, microsomal CYPs metabolize both endogenous and exogenous compounds. Anticancer drugs are therefore metabolized by microsomal CYPs and epoxide hydrolases, and in the process transform the highly mutagenic aromatic metabolites (epoxides) created from the CYP metabolism into a metabolite that can be easily conjugated by the Phase II enzymes, followed by efflux through the adenosine triphosphate (ATP) binding cassette subfamily C (ABCC) transporter family, thus ensuring less systemically available anticancer drugs in the body (Brown et al., 2008).

Phase II enzymes manage conjugation reactions through sulfation, glutathionylation, and glucuronidation. These enzymes include arylamine N-acetyltransferases (NAT), glutathione-S-transferase (GST), UDP glucuronosyltransferases (UGT), and sulfotransferases (SULT), which transform the parent drug or Phase I metabolite into a hydrophilic conjugate that favors excretion. The same thing happens with anticancer drugs when they enter the cells and come in close contact with these metabolizing enzymes. The Phase II conjugated metabolites thus formed are then effluxed out of the cancerous cells by members of the ABCC family of ATP-binding cassette (ABC) transporters, thus making the anticancer drugs inactive (Deeley et al., 2006).

The GST superfamily is group of detoxifying enzymes which protect cellular macromolecules from electrophilic compounds. The elevation of GST levels in cancer cells increases the rate of detoxification of the anticancer drugs, which results in less cytotoxic damage to the cancer cells (Manolitsas et al., 1997).

Another mechanism involves the conjugation of the anticancer drug to glutathione (GSH), a potent anti-oxidant that protects the cells against the damaging effects of reactive oxygen species (Wilson et al., 2006), in the process making radiation therapy ineffective.

GSH conjugation with platinum drugs, such as oxaliplatin and cisplatin used in the treatment of various types of cancers, makes them substrates for ABC transporters which boost drug efflux out of the cancer cells (Ishikawa and Ali-Osman, 1993; Meijer et al., 1992).

5.2.3 Drug Sequestration

Recent scientific studies have led us to believe that intracellular drug sequestration is also one of the important causes of MDR. Abnormalities in lysosomal and protein trafficking results in subtherapeutic concentration in the target compartment, thus leading to treatment failure (Duvvuri and Krise, 2005).

In a study it has been demonstrated that cisplatin, which is commonly employed in the treatment of many malignant tumors, is sequestered into lysosome, golgi, and secretory compartments and is then effluxed out from the cell (Chauhan et al., 2003).

Recent investigations have analyzed the role of an ABC efflux transporter, ABCA3, on intracellular drug sequestration.

The expression of this transport protein can be related to low survival rates in patients with acute myeloid leukemia. Intracellular drug sequestration caused by the ABCA3 transporter was found to be responsible for developing resistance towards chemotherapeutics such as daunorubicin, etoposide, imatinib, mitoxantrone. and vincristine (Chapuy et al., 2008; Chapuy et al., 2009; Cheong et al., 2006; Steinbach et al., 2006).

In addition to drug sequestration in intracellular vesicles, "scavenger" metallothioneins (MTs) play a major role in trapping anticancer drugs within a cell. MTs are cysteine-rich and have a high attraction for metal ions. This property, along with their ability to trap reactive oxygen species, strongly suggests them as significant contributors in developing resistance towards radiation treatments and metal-based therapeutic agents (Thirumoorthy et al., 2007).

5.2.4 Alteration of Drug Targets

The efficacy of a drug is often influenced by its molecular targets and mutation-induced alteration of these targets generally leads to drug resistance. As an example, certain anticancer agents like anthracyclines and epipodophyllotoxins act on topoisomerase II, an enzyme that prevents DNA from becoming supercoiled or under-coiled. The conjugation between DNA and topoisomerase II is usually very unstable, but these drugs destabilize it, leading to DNA damage, inhibition of DNA synthesis, and a cessation of mitotic processes in the cancer cells. In the course of time, cancer cells have modified themselves against these drugs by introducing mutations in the genes of topoisomerase II (Hinds et al., 1991; Stavrovskaya, 2000).

Another example has been observed in the case of androgen receptors. In cancers of the prostrate, the androgen receptor often gets genomically altered and, as a result, these cancers become resistant to androgen-deprivation therapy drugs leuprolide and bicalutamide (Holohan et al., 2013; Palmberg et al., 1997). These drugs become incapable of inhibiting all the molecular targets present, and thus these cancers are considered resistant to them.

An overall understanding of the previously mentioned methods of drug target alteration is important for diagnosing and developing new therapies to treat drug-resistant cancers.

5.2.5 Drug Efflux: Role of Transport Proteins

5.2.5.1 Role of ABC Efflux Transporters

One of the factors that plays a critical role in cancer patients is the prevention of intracellular accumulation of anticancer drugs by transport proteins that pump the drugs out of the cells (Figure 5.2).

In addition, they also block the entry of anticancer drugs to their cellular targets (Rajagopal and Simon, 2003).

These transporter proteins come under the mammalian ABC family and remain associated with the plasma membrane of cells. After entering through the plasma membrane, anticancer drugs are recognized by the transporters, where they use the energy provided by ATP hydrolysis to expel the drug molecules out of the cells, resulting in a low bioavailability, finally leading to resistance.

5.2.5.2 Role of Multidrug-Resistant Proteins (P-Glycoprotein)

The outward movement or efflux of drugs mostly governs the defense of cancer cells against anticancer drugs. In this connection, P-glycoprotein (P-gp/MDR1, also called ABCB1) is one of the major transmembrane transporters in humans and is encoded by the ABCB1/MDR1 gene (Gottesman et al., 2002).

It has been noticed that in carcinomas of pancreatic, renal, colorectal, adrenocorticoid, and hepatocellular origin, P-gp is often overexpressed and as a result actively transports several anticancer drugs (e.g., anthracyclines, vinca alkaloids, podophyllotoxins, and taxanes) out of the cells (Wang et al., 2003; Zhou, 2008).

5.2.5.3 Multidrug Resistance-Associated Proteins (MRPs)

The other type of pump which confers resistance to anticancer drugs is the multidrug resistance-associated proteins (MRPs). They remove drugs from within the cancerous cells with or without GSH conjugation. An overexpression of MRPs results in resistance towards epipodophyllotoxins, anthracyclines, vinca alkaloids, etoposide, methotrexate, mitoxantrone, cisplatin, and epirubicin, etc. (Borst et al., 2000; Choi, 2005; Toyoda et al., 2008; Zhang et al., 2012b).

5.2.5.4 Breast Cancer Resistance Protein (BCRP)

Another transporter protein that is a member of the G subfamily (G is the subfamily of the ABCG family) of ABC transporters is breast cancer resistance protein (BCRP). In tumors, BCRP promotes the efflux of methotrexate, mitoxantrone, irinotecan, and topotecan from cells, thereby causing resistance towards these chemotherapeutics (Doyle et al., 1998).

5.2.5.5 Lung Resistance-Related Protein (LRP)

Lung resistance-related protein (LRP), also known as vault protein, has been found to be involved in intracellular transport processes (Scheffer et al., 2000).

As shown in Figure 5.2, it is located in the cytoplasm and mediates the bidirectional distribution of cytotoxic drugs between the nucleus and the cytoplasm. Such an effect in drug distribution has been noticed with doxorubicin (DOX) and, finally, in its resistance (Kitazono et al., 1999).

5.2.6 Resistance Mechanism Activated after Drug Enters Cancer Cells

After a sufficient amount of anticancer drug has entered, accumulated, and inhibited its cellular targets, the outcome is dependent on how the cancer cell reacts. Ideally, anticancer drug-induced damage results in cancer cell death. However, several intrinsic adaptive responses are triggered that promote cancer cell survival. Additionally, the pathways that regulate cell death through apoptosis frequently become dysfunctional, which has become one of the hallmarks of cancer (Hanahan and Weinberg, 2000).

5.2.6.1 Evasion of Drug Induced Apoptosis by Cancer Cells

Defects in the apoptotic pathway can be one of the alternative modes of anticancer drug resistance (Dive and Hickman, 1991).

It is a well-known fact that the p53 protein plays a vital role in regulating normal cell cycles and is sensitive to any DNA damage which prompts gap 1 (G_1) (G is the subfamily and 1 is the member) phase arrest and/or apoptosis to prevent the production of defective cells.

In cancer patients, mutations in the p53 gene are frequently observed. This signifies a loss of p53 function and as a result cells with damaged DNA continue replicating, which means resistance to DNA-damaging drugs. In a variety of tumors, p53 deletion was reported to be associated with MDR (Grunicke, 1998).

In one study conducted by Johnson and Fan (2002) it was concluded that diminished expression of p53 in human breast cancer cells modified the response towards paclitaxel (PAX) and 5-fluorouracil (5-FU). Additionally, the vulnerability of cancer cells to apoptosis is influenced by a series of proto-oncogenes and tumor suppressor genes. The occurrence of mutations in such genes potentially increases the number of surviving mutant cells, thus leading to greater tumor heterogeneity.

Another way of evading the toxic effects of anticancer drugs is through the upregulation of anti-apoptotic proteins such as survivin and B cell lymphoma 2 (Bcl-2) family proteins (Kanwar et al., 2011; Zhang et al., 2012b). Survivin induces drug resistance by directly blocking apoptotic proteins like caspase and procaspase signaling mechanisms, resulting in the upregulation of the expression of MDR proteins like P-gp, MRP-1, and MRP-2.

5.2.6.2 Repairing of Damaged DNA by Cancer Cells

The repair of damaged DNA of cancer cells has a clear role in developing anticancer drug resistance. In response to anticancer drugs that either directly or indirectly damage DNA, cancer cells can reverse the drug-induced damage by triggering a number of DNA repair mechanisms.

DNA repair involves a complex network of repair systems, each of which targets a specific subset of wound. These pathways include (1) the mismatch repair (MMR) pathway, (2) the nucleotide excision repair (NER) pathway, (3) the base excision repair (BER) pathway, (4) the homologous recombination (HR) pathway, and (5) the nonhomologous end joining (NHEJ) pathway (Hakem, 2008).

Figure 5.3, will elaborately demonstrate the various DNA repair pathways involved in repairing toxic DNA lesions formed by cancer treatments.

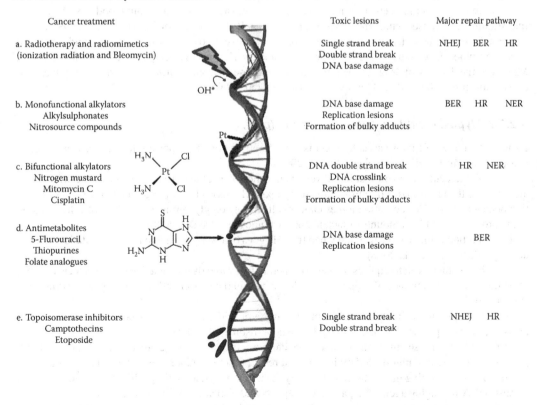

FIGURE 5.3 An overview of DNA repair pathway.

As seen from the figure, the NHEJ repair pathway is triggered to repair the double-strand breaks caused by radiomimetic drugs and ionizing radiation.

Anticancer drugs like alkylsulphonates, nitrosourea compounds, cisplatin, and nitrogen mustards cause DNA base damage, where the DNA double strands break apart, forming bulky DNA adducts. But these lesions are being healed by cancer cells through the BER and NER pathways, together with alkyl-transferases (ATs), which is another major repair pathway. Antimetabolite drugs like 5-FU and thiopurine interfere with nucleotide metabolism and DNA synthesis, causing replication lesions in cancer cells, but cancer cells have evolved themselves by triggering the BER pathway mechanism.

Topoisomerase inhibitors like camptothecin and etoposide form transient cleavage complexes with DNA, thus inducing DNA breaks and interfering with the replication of cancer cells. Through the HR pathway and the NHEJ pathway, cancer cells evade the damaging effects of the topoisomerase inhibitors (Helleday et al., 2008).

5.2.7 Other Physiological Factors

The physiological factors that are mainly held responsible for MDR are high interstitial fluid pressure (IFP), hypoxia, and low extracellular pH (pHe) (Cairns et al., 2006). The previously mentioned factors are interrelated and affect each another.

5.2.7.1 Interstitial Fluid Pressure (IFP)

Pro-angiogenesis and anti-angiogenesis are the two primary factors which regulate the process of blood vessel formation, apart from maintaining an efficient vasculature to meet nutritional and oxygen requirements for a normal and healthy cell.

During tumor growth, this delicate balance of regulation of angiogenesis is lost and unorganized vasculature away from the cells originates, causing the leakage of proteins from blood vessels into the interstitium. As a consequence, there is a rise in the IFP (IFP > 100 mm of Hg) (Less et al., 1992).

Apart from all these factors, the presence of convective transport, transcapillary fluid flow, and decreased blood flow actually creates a problem in delivering chemotherapeutic agents intravenously. Moreover, the delivery of drugs to the target cells sometimes becomes difficult due to the deep localization of cancer cells beyond the blood vessels (Heldin et al., 2004).

5.2.7.2 Hypoxia and Low Extracellular pH (pHe)

In more than 50% of tumor cases, hypoxia is considered one of the important factors for causing drug resistance in cancer cells (Vaupel and Mayer, 2007).

Due to the irregular formation of blood vessels in cancer cells, the oxygen demand for a cancer cell is always more than a normal cell, thus leading to hypoxia. Hypoxia also develops due to the deep localization of cancer cells. This condition forces cancer cells to rely on glycolysis for ATP supply, thus leading to the production of high amounts of lactate and carbonic acid (Denko, 2008).

The condition is further exaggerated when the pHe goes below 7.0, owing to the accumulation of these acidic products (Lee et al., 2008).

Radiation therapy, which requires oxygen to generate free radicals like reactive oxygen species (ROS), often fails due to the absence of proper oxygen levels within the cancer cells due to hypoxia (Brown and Wilson, 2002).

Second, drugs such as melphalan, bleomycin, and etoposide, etc., which depend on partial oxygen pressure, may not perform properly during hypoxia (Wilson et al., 2011).

Third, under the harsh environment of anoxia, cancer cell cycle arrest occurs either at G_1, G_2, or the S phase (S is the phase of mitotic cell division). As a result, there is an increase in the activity of the DNA repair enzymes, resulting in resistance towards cycle-selective cytotoxic drugs, like 5-FU and PAX, and towards DNA-damaging agents, like alkylating agents and cisplatin (Wilson et al., 2011).

Additionally, hypoxia has been found to be associated in the activation of the ABC transporters, as discussed in Section 5.2.2 (Jabr-Milane et al., 2008).

The presence of a pH gradient due to the accumulation of acidic products results in the "ion trapping" phenomenon, which causes a permeability difference between the ionized and non-ionized forms of an anticancer drug. This phenomenon plays a crucial role for weakly basic drugs like DOX and vincristine, which remain ionized and therefore get trapped in the acidic extracellular environment (Wojtkowiak et al., 2011).

5.3 Present Scenarios in the Treatment of MDR in Cancer Therapy

Cancer patients are invariably treated with conventional chemotherapeutic agents and/or radiation therapy. However, the response to such therapies varies widely in different subsets of patients. In the majority of cases, an initial response is encouraging, whereas in some cases the response fades away with the passage of time. Again, in some patients the initial positive response is also absent. Evidently, in those two categories of patients, the cancer cells demonstrate mechanisms of resistance to apoptotic stimuli, resulting in cancer development and metastasis and ultimately the death of the patients.

The development of novel therapies will primarily focus on demarcating the biochemical, molecular, and genetic mechanisms that control tumor cell resistance to anti-neoplastic drug-induced apoptosis. Such mechanisms have revealed gene products that directly regulate resistance and, thus, one may develop new drugs that target these resistance factors.

5.3.1 Modification of Chemotherapy Regimens

Non-cross-resistant multidrug chemotherapeutic regimens function on the principle of utilization of the largest number of active agents at the highest possible doses, on the assumption that mutations conferring drug resistance will not impart resistance to all of the agents in the regimen. This approach assumes that despite resistance to standard doses of anticancer drugs, a dose–response relationship still exists for these tumors and that high doses of chemotherapy might overcome this resistance.

5.3.1.1 Inactivation of MDR-Associated Genes by Targeting Specific mRNA for Degradation

Drug resistance cells are sensitized by reducing the expression of P-gp, MRP, and BCRP using antisense oligonucleotides (ASOs) and catalytic ribonucleic acids (RNAs) (Nadali et al., 2007, Ran et al., 2008; Stewart et al., 1996). The modern advancement targeted to messenger RNA (mRNA) degradation, which depends on the RNA interference post-transcriptional gene silencing mechanism. Both small interfering RNA (siRNA), which has a transient effect, and short hairpin RNA (shRNA), which has longer term effects, silence the gene expression of P-gp, MRP, and BCRP in MDR cancer cells. To limit the exposure of normal cells to the inhibitor (siRNA) and the anticancer drug and maximize synergy, both P-gp-targeted siRNA and PAX were co-encapsulated in various formulations.

This robust silencing effect of RNA interference (RNAi) makes it a valuable research tool both in cell culture and in living organisms, although introducing siRNA into cells *in vivo* remains a significant obstacle. The silencing of P-gp and the down regulation of MRPs could reverse the MDR phenotype, re-sensitizing MDR cells to anti-neoplastic agents.

5.3.1.2 Use of Monoclonal Antibodies for P-gp

A monoclonal antibody to P-gp has been shown to inhibit tumor growth in an athymic nude mouse model (Goda et al., 2006; Watanabe et al., 1996). However, there are some issues with the antibody approach, for example, MRK-16, a specific monoclonal antibody for P-gp, may target MDR1-expressing cells in normal tissues. Thus, the potential of this approach to lead to unacceptable toxicities could be overlooked in certain mouse models.

5.3.1.3 Development of New Anticancer Drugs That Are Not Substrates of P-gp

P-gp induced efflux of drugs can be terminated by modified drug analogues, which can affect the binding of P-gp and as a result the P-gp cannot identify the analogues. For example, taxane analogs DJ-927 (Phase I) (Ono et al., 2004; Shionoya et al., 2003) and ortataxel (Phase II) are considered to conquer drug resistance. Similarly, other taxane analogs, such as BMS-184476 (Phase I) (Altstadt et al., 2001; Rose et al., 2001) and RPR 109881A (Phase II) (Gelmon et al., 2000; Kurata et al., 2000), were also claimed to possess a broad spectrum of activity both in susceptible and resistant tumor cell lines.

Recently, PAX vitamin E emulsion (TOCOSOL), containing the P-gp inhibitor D-α-tocopherol polyethylene glycol 1000 succinate (TPGS) as an excipient, was developed as a P-gp substrate and subsequently investigated in a Phase II clinical trial to counter drug resistance. Promising efficacy of the drug was observed when TOCOSOL was administrated weekly in patients with refractory cancers (Lissianskaya et al., 2004). Table 5.1 shows compounds which interact with P-gp.

5.3.1.4 Photodynamic Therapy (PDT)

Photodynamic therapy (PDT) is a form of cancer treatment that involves the use of photosensitizers as therapeutic agents. In the presence of light, photosensitizers enter a triplet state of excitation. This triplet state of energy is readily transferred to oxygen molecules, which are subsequently converted into reactive oxygen species that are capable of causing cell damage (Konan et al., 2001; Samia et al., 2003; Tang et al., 2004). This method of treatment has high selectivity, since only the cells which are exposed to both the light and the photosensitizer are affected. Photofrin 2, which is a derivative of hematophorphyrin, is the only PDT drug that is approved for clinical use in Canada, the Netherlands, and Japan for the treatment of bladder, lung, and esophageal cancer, respectively (Konan et al., 2001).

5.3.1.5 ABC Transporter Inhibitors

This is an important approach to resist MDR and is receiving a great deal of attention. Usually ABC transporters are classified into two generations, such as inhibitors that can transport themselves and can act as competitive antagonists and the others that are not transported but affect transporter function. P-gp mediated drug resistance attracted various approaches for the treatment of MDR; one of them is the calcium channel blocker verapamil, used to reverse vincristine resistance in murine leukemia cells (Tsuruo et al., 1981, 1982).

TABLE 5.1

Compounds Which Interact with P-glycoprotein (P-gp)

MDR Drugs	P-gp Inhibitory Agents	Endogenous P-gp Substrates	Reference
Daunorubicin	Verapamil	Cortisol	Lehne, 2000
Doxorubicin	Nifedipin	Aldosterone	
Epirubicin	Azodipin	IL-2	
Mitoxantrone	Quinine	IL-4	
Etoposide	Flupentixol	Interferon-γ	
Teniposide	Progesteron	Sphingolipids	
Vinblastine	Megestrol acetate	–	
Vincristine	Tamoxifen	–	
Mitomycin C	Cyclosporin A	–	
Paclitaxel	FK 506	–	
Actinomycin D	PSC 833	–	
Topotecan	280–446	–	
–	LY 335979	–	
–	GF 120918	–	

However, the promising preclinical data failed to translate into therapeutic clinical outcomes as far as the role of first- and second-generation inhibitors of ABC transporters are concerned.

The different reasons these inhibitors fail to show their beneficial effects are

- They are nonspecific and less potent inhibitors.
- In clinical trials, it was found that the required concentration of inhibitor was not achieved. In case of potent inhibitors, associated toxicities of anticancer molecules resulting from the inhibition of transporters within normal tissues remains a major concern.
- Coadministration of anticancer drugs and inhibitors has also reported failure due to programmed and modest PK interaction. For example, when verapamil is administered along with DOX in humans, an unassuming PK interaction is observed (Kerr et al., 1986).
- In addition, some second-generation inhibitors of P-gp are substrates of CYP. Competition with these P-gp inhibitors for CYP-mediated oxidative reactions may lead to undesirable PK interactions (Bates et al., 2001). Increased toxicity of the anticancer drugs may occur due to both PK effects and inhibition of a protective function of P-gp in normal tissues.

To date, clinical trials of third generation inhibitors, such as LY335979-targeted P-gp, R101933-targeted P-gp, XR9576-targeted P-gp, GF120918-targeted P-gp, BCRP, and MRP1 are in process. The *in vitro* and *in vivo* associated toxicities of the XR9576 compound and several other compounds show high potency and better PK profile in Phase I trials, but the Phase III study was called off because of the increased toxicity of non-small cell lung cancer (NSCLC) (Nobili et al., 2006). As a result, the coadministration of such inhibitors had to be redefined and also redesigned to reverse clinically significant drug resistance.

MRP, another protein involved in the MDR mechanism, was also found to be a pump, particularly as a member of the ABC transmembrane transporter superfamily. Since then, both the P-gp and MRP have been significant targets for anticancer drug discovery. Table 5.2 shows the list of companies with programs in this area. The development of drugs through antisense therapy by the Isis Company is very successful in preventing the generation of MRPs and also its oligonucleotides. In another successful approach by Vertex, two compounds were developed, Incel (Biricodar Dicitrate, VX-710), intended for the intravenous route, and VX-853, to be given by the oral route, for blocking MRP and P-gp (MDR1). Incel and VX-853 are to be coadministered with anticancer chemotherapeutics, helping in the non-removal of these anticancer agents by the cancer cells from within. The MDR inhibitor product of Incel along with chemotherapy is in Phase II clinical trials targeting various cancers, namely ovarian cancer, small-cell lung cancer (SCLC), breast cancer, prostate cancer, and soft tissue sarcoma (STS). In another program, the CT-2584 compound was developed for cell therapeutics for the treatment of patients with chemorefractory, including prostate cancer and sarcomas (Cancer multidrug resistance, 2000; Krishna et al., 2000).

5.3.2 Clinical Status

Various trials on MDR to cancer chemotherapy intended with MDR1 and/or MRP as targets are successfully going on. Vertex's Incel in combination with other agents for the treatment of SCLC and ovarian cancer was in Phase II clinical trials. The compound will be tested for various drug resistance cancers such as ovarian, breast, SCLC, and prostate. STS caused by the overexpression of MDR1 and MRP is unresponsive to chemotherapy. BioChem Pharma is funding an STS trial and is aiming to market the developed product in Canada.

Incel, a compound under investigation, helps to regain or improve the activity of the DOX in STS patients suffering from an aggressive form of the disease that has also developed an acquired or inherent resistance to this anticancer drug, as evident from the Phase I/II data. In the United States and Canada, drugs like DOX are standard chemotherapeutic agents, which are becoming unresponsive in about 7,000 new patients annually. It is seen in many cases that patients' refractory to chemotherapy with a survival rate of five years is a low 10–30%. Frequent relapses of cases are observed with approximately 70% of patients initially remaining unresponsive to chemotherapy. As per Vertex's data, Incel blocks both MRP

TABLE 5.2

Selected Companies with Multidrug Resistance (MDR) Programs

Company	Program	Status	Reference
Aronex (The Woodlands, TX)	Annamycin against chemorefractory cancers	Phase II, 9/2000	Nature Biotechnology Volume 18
Avigen (Alameda, CA)	MDR gene therapy	Patent	Supplement 2000
Biochem Pharma (Quebec, Canada)/ Vertex (Cambridge, MA)	MDR in cancer	Phase II	
Cell Therapeutics (Seattle, WA)	Small molecule inhibitor of multidrug resistance-associated protein (MRP) for prostate cancer and sarcomas	Phase II	
CytRx (Norcross, GA)	MDR in acute leukemia	Patent	
Genelabs (Redwood City, CA)	MDR in small-cell lung and colorectal cancer	Preclinical	
Genetic Therapy (Gaithersburg, MD)	MDR1 gene for breast cancer	Phase I/II	
Genetix Pharmaceuticals (Cambridge, MA)	Bone marrow chemoprotection	Phase I/II	
Immunex (Seattle, WA)	Restoration of tumor sensitivity to anticancer drugs	Market	
Ingenex (Menlo Park, CA)	MDR gene therapy	Phase I/II	
Isis Pharmaceuticals (Carlsbad, CA)	Antisense oligonucleotides for MDR	Phase II/III	
Ixsys (San Diego, CA)	MDR monoclonal antibodies	Phase II	
SuperGen (San Ramon, CA)	MDR in cancer	Preclinical	
Titan Pharmaceuticals (San Francisco, CA)	Induction of MDR chemoprotection in stem cells	Phase I	
Xenova (Slough, UK)	MDR in cancer	Preclinical	

and MDR1 to regain the sensitivity of tumor cells towards anticancer therapy by physically increasing the concentration of anticancer drugs within the target cells.

Another molecule that is under trial is Aronex. It is basically an anthracycline, known as annamycin, which is under development to treat drug-resistant breast cancer. Like Incel, this compound also has the potential to treat a broad range of cancers. Toremifene is another key compound that can be used to reduce MDR *in vitro*, but it shows effects when tried on patients, almost certainly because in blood, toremifene remains bound to serum proteins. Nevertheless, a current study shows that the combination therapy of toremifene and vinblastine is well tolerated for short term therapy and can also reverse the MDR *in vitro* (Braybrooke et al., 2000).

A recent study shows that a new compound olivacine, labeled S16020-2, has shown noteworthy antitumor activity against cells that exhibit resistance through the MDR1 phenotype both *in vitro* and *in vivo*. The probable mechanism of this compound is its rapid rate of uptake and by-passing the P-gp leads to higher intracellular concentration (Pierre et al., 1998). Several other lead compounds are available with a similar functional profile against MDR. High throughput screening was applied to find the lead compounds. Development in the treatment of MDR1 and MRP related drug resistance leads to establishing a link between other mutants, like p53 expression. A recent study found an association between mutant p53 expression and MRPs, which can be preferentially used for diagnosis (Oshika et al., 1998). Some other such correlations were also found for the treatment of MDR cancer. Table 5.3 shows inhibitors of the PI3K/Akt pathway currently undergoing clinical trials.

5.3.3 Drug Delivery Challenges

Various biological barriers are found in every level, from the body to organisms and tissues to cells up to the sub-cellular level or molecular level. Due to that reason, successful drug delivery has become challenging. The main problem in cancer drug delivery is the localization of the drug molecule into

TABLE 5.3

Inhibitors of the PI3K/Akt Pathway Currently Undergoing Clinical Trials (Martin et al., 2014)

Drug	Target	Breast Cancer Selection Criteria	Combination Therapies	Phase	Trial Identifier
AZD5363	Akt	ER+	Paclitaxel / none	I	NCT01625286, NCT01226316
GSK2110183	Akt	Drug resistant	None	I	NCT 01476137
GSK 2141795	Akt	Not stated	None	I	NCT 00920257
MK2206	Akt	ER+	Lapatnib ditosylate/ paclitaxel / anastrozole / letrozole / exemestane / fulvestrant / none	II	NCT01245205, NCT01277757, NCT01776008, NCT01344031
		HER2+	Lapatnib ditosylate / trastuzumab	I	NCT01705340, NCT01281163
		Not stated	Paclitaxel	Ib	NCT01263145
Triciribine Phosphate, Monohydrate	Akt	Not stated	Paclitaxel / doxorubicin / cyclophosphamide	I/II	NCT01697293
BAY 80-6946	PI3K	Not stated	Paclitaxol	I	NCT01411410
BKM 120	PI3K	ER+	Fulvestrant / letrozole	I/III	NCT 01339442, NCT 01248494
		HER2	Lapatnib / trastuzumab / capecitabine	I/II	NCT01589861, NCT01132664
		Trastuzumab-resistant HER2+	Trastuzumab / paclitaxel	I/II	NCT01285466, NCT01816594
		HER2-	Paclitaxel	II	NCT01572727
		Triple-negative	Post-chemotherapeutics	I/II	NCT01629615
BYL719	PI3K	ER+/HER2-	Letrozole / fulvestrant	I	NCT01791478, NCT01219699
GDC-0941	PI3K	ER+	Trastuzumab / paclitaxel / bevacizumab	I	NCT00960960
XL147	PI3K	ER+/HER2+	Letrozole	I	NCT01082068
CC223	mTOR	ER+	Unresponsive tumors	I/II	NCT01177397
Everolimus	mTOR	ER+/HER2+/−	Endocrine therapies (tamoxifen) / bevacizumab / lapatnib	II/III	NCT01298713, NCT1805271
		ER+, AI-resistant	Fulvestrant / chemotherapeutics / exemestane	II/III	NCT01797120, NCT01088893, NCT00863655, NCT01626222
		HER2+	Paclitaxel / trastuzumab	III	NCT00876395
		HER2-	Vinorebine	II	NCT01520103
		Triple-negative	Gemcitabine / cisplatin	I	NCT01939418
Rapamycin	mTOR	HER2+	Trastuzumab	II	NCT00411788
Temsirolimus	mTOR	ER+	Letrozole	II	NCT00062751
		HER2+/ triple-negative	Neratinib	I/II	NCT01111825
Ridaforolimus	mTOR	ER+	Dalotuzumab / exemestane	II	NCT01234857
BEZ235	PI3K/mTOR Dual inhibitor	ER	Letrozole / everolimus	I	NCT01248494, NCT01482156
		HER2+	Paclitaxel / trastuzumab	I	NCT01285466
		HER2-	Paclitaxel	I/II	NCT01495247
	PI3K/mTOR Dual inhibitor	ER+	Fulvestrant	II	NCT01437566
	PI3K/mTOR Dual inhibitor	ER+ HER2-	Letrozole	II	NCT01082068

the intended site. The maximum drug response of a compound can only be possible if the compound will concentrate on the target site, otherwise it can damage the other healthy cells, resulting in toxicity and an adverse drug reaction. So, the main aim to treat cancer cells are targeted drug delivery systems. Although this is a simple aim, the process of achieving such a result will be costly and time consuming, and no guarantee of success can be assured.

An experiment reported in *Cell Reports* disclosed that to conquer the resistance of lapatinib, a novel experimental drug called a BET (Bromo- and Extra-Terminal domain) bromodomain inhibitor, whose role is to inhibit the expression of certain genes, was developed in HER2-positive breast cancer cell lines.

5.3.4 New Agents with Reduced Drug Efflux Properties

Apart from approaches to inhibit the activity drug efflux pumps, existing anticancer agents might be modified such that these active molecules do not act as substrates for P-gp or other MDR proteins, hence avoiding such efflux mechanisms (Nobili et al., 2011). Normally compounds like topoisomerase II inhibitors daunorubicin, DOX, and idarubicin are substrates for P-gp, which cause efflux from the tumor site leading to insufficient therapeutic concentration. One of the new alternative agents is amonafide, a novel topoisomerase II inhibitor with reduced drug efflux properties, which was successfully applied for the treatment for AML (Acute Myeloid Leukemia) (Allen and Lundberg, 2011; Chau et al., 2008). In another study it was reported that taxane cabazitaxel, a dimethyloxy derivative of docetaxel, can cross the blood–brain barrier and has no affinity for P-gp (Paller and Antonarakis, 2011). Other newly developed agents like GSH transferase inhibitor 6-(7-nitro-2,1,3-benzoxadiazol-4-ylthio) hexanol (Ascione et al., 2009) and a series of pyrrolo-1,5-benzoxazepine compounds (Nathwani et al., 2010) are not substrates for P-gp. Otherwise, agents could be made more lipophilic to undergo influx more readily. These approaches help to improve the compound's passive lipid permeability, increase passive diffusion, and also resist increasing concentration gradients. They may even eradicate resistance due to efflux transporters irrespective of the fact that the compounds are substrates or not (Raub, 2006). This concept can be established (in reverse) through a related work on uptake transporters performed with imatinib and nilotinib, used in the treatment of chronic myeloid leukemia (CML). Both of these active agents act as substrates for MDR efflux transporters and human organic cation transporter 1 (hOCT1) influx protein along with other solute carrier transporters (Minematsu and Giacomini, 2011). It is worth mentioning here that nilotinib is less hydrophilic than imatinib and hence can better enter cells. It has been observed that imatinib uptake decreases with the reduced activity of hOCT1 (Crossman et al., 2005; White et al., 2006), which in turn results in reduced responses among patients suffering from CML (Engler et al., 2011; White et al., 2007). But it should be noted that nilotinib uptake remained unaffected by hOCT1 activity (Davies et al., 2009; White et al., 2006). When DOX was linked to a hybrid cell-penetrating and drug-binding peptide, its influx and efflux kinetic were distorted to boost cytotoxicity in MDR KD30 leukemia cells (Zheng et al., 2010). Similarly, nanotechnology can be a capable approach to help bioactive agents escape efflux, which is discussed later in this chapter.

5.3.5 Modified MDR Drugs

Unfortunately, drugs such as antimetabolites, alkylating agents, and platinum compounds, which were successfully used in the treatment of malignancies, are suffering from MDR and are not easily substituted with non-MDR drugs. Drugs like anthracyclines and taxanes are presently unique in a range of chemotherapy regimens because of their exclusive anti-neoplastic activity. Thus, the affinity of these drugs for P-gp can be reduced by the chemical modification of their structures. In anthracyclines, the 9-alkyl substitution of the anthracene A ring and certain sugar modifications have been associated with a reduced affinity for P-gp and the maintenance of cytotoxic activity in certain MDR tumor cell lines (Coley et al., 1990; Scott et al., 1986).

The ability of anthracyclines to avoid MDR is shown to progress with increasing lipophilicity (Lampidis et al., 1997).

Recently it was found that verapamil-induced manipulation of the P-gp inhibitor does not affect annamycin's anti-neoplastic action (Consoli et al., 1996; Perez-Soler et al., 1994). Additionally, idarubicin has been shown to preserve considerable cytotoxicity in various selected MDR cell lines (Ross et al., 1995), in cells transfected with the MDR1 gene (Kuffel and Ames, 1995), and in P-gp-positive blasts recovered from leukemia patients treated with the drug (Nussler et al., 1997).

5.4 Strategies in General for Overcoming Cancer MDR Using Nanotechnology

One of the major causes of failure of cancer chemotherapy is the emergence of resistance of cancer cells due to the proliferation of several precursor genes and repair pathways.

Therefore, it is the need of the hour to explore strategies for eradicating the MDR of cancers as well as to make the chemotherapy more effective by utilizing the available drugs.

Strategies for eradicating MDR by using nanotechnology are presented in Figure 5.4 and are discussed here under. Such strategies have the potential to emerge as promising alternatives to conventional chemotherapeutics, by encapsulating and conjugating drug molecules within nanocarriers.

5.4.1 Delivery of Inhibitors of MDR Proteins

The cytotoxic activity of anticancer drugs that are present within the cells depends on their concentration and availability in the cell cytoplasm and in nuclei. Additionally, several proteins, such as P-gp, MRP, BCRP, and LRP, play a vital part in pumping out the internally accumulated drugs to the outer environment of the cell (Figure 5.2). Moreover, several detoxifying enzymes like GST detoxify the internalized drugs through their enzymatic activity (Figure 5.2). By inhibiting the expression of responsible proteins and enzymes, it will thus become possible for the drug molecules to be available for effective functioning. Research in the field of cancer has helped to reveal several inhibitors of these proteins or enzymes which could be utilized to suppress their expressions (Choi, 2005; Wang et al., 2003). To block the function of P-gp, several inhibitors like astemizole, itraconazole, verapamil, diltiazem, and cyclosporin A, etc., have been discovered (Choi, 2005; Wang et al., 2003).

Genistein and estrone, etc., inhibit the proper functioning of the BCRP (Choi, 2005). Drugs like ethacrynic acid, benastatin A, nitazoxanide, and aloe-emodin (Laborde, 2010) were found to modulate drug resistance by sensitizing tumor cells to anticancer drugs through GST inhibition.

However, even with the availability of such types of inhibitors to block the function of MDR proteins, they lack specificity towards the target receptor and can block the function of the target proteins in normal organs, leading to side effects like cardiotoxicity, nephrotoxicity, and neurotoxicity (Rezzani, 2004).

In order to avoid unwanted side effects in normal and healthy tissues, it is absolutely necessary to deliver the inhibitors specifically to the tumor sites by encapsulating them in nanoparticles (Figure 5.4A), which can be modified further by adding specific ligands to render them more tumor specific.

Delivering combinations of cytotoxic drugs and efflux pump inhibitors, such as cyclosporine, verapamil, and tariquidar, within a nanoparticle has shown promising results in terms of reversing MDR in cancer cells (Patil et al., 2009; Soma et al., 2000; Wu et al., 2007).

Nanoencapsulation of cyclosporin A and DOX result in about two-fold higher therapeutic efficacy in DOX-resistant leukemia cells as compared to free cyclosporin A or only DOX (Soma et al., 2000).

Patil et al. (2009) showed that biotin-conjugated poly lactic-co-glycolic acid (PLGA) nanoparticles loaded with PAX and tariquidar have improved therapeutic efficacy in breast cancer as compared to the non-targeted formulations (Patil et al., 2009).

Thus, the application of nanoparticles loaded with protein inhibitors and cytotoxic drugs would be a promising means for overcoming MDR, apart from being safe in treating cancers that are resistant to chemotherapy.

5.4.2 Delivery of Nucleic Acids Aimed to Target MDR Proteins

The gene delivery in order to control the activity of a specific gene by RNAi has become a powerful tool in cancer therapy. RNAi is a post-transcriptional gene silencing mechanism that is mediated by siRNAs (21–25 nucleotides (NT)) (Filipowicz et al., 2005).

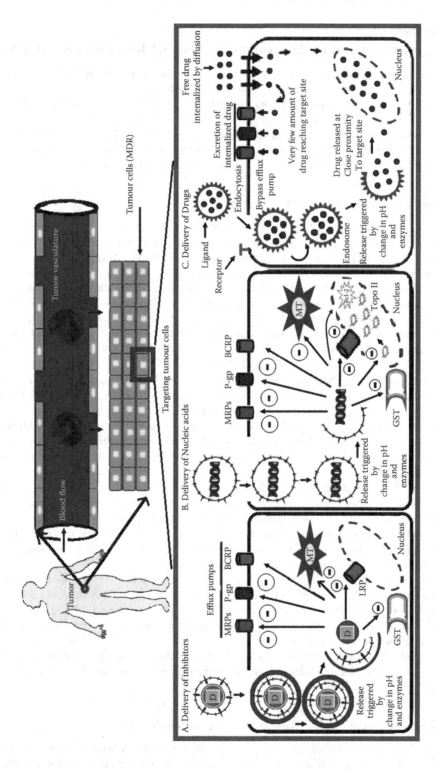

FIGURE 5.4   Schematic representation of application of nanocarriers.

The double-stranded RNA molecules are incorporated into a RNA-induced silencing complex (RISC), which causes the degradation of the target mRNA in a specific manner (Filipowicz et al., 2005). In solid tumors, the membrane transporters play an important role in the distribution and excretion of clinically relevant anticancer drugs. In recent years, several attempts to control the expression of ABC transporters by delivering nucleic acid-loaded nanoparticles (with siRNA, micro RNA (miRNA), etc.) to the site of the tumors have been made (Patil and Panyam, 2009; Wang et al., 2010), as illustrated in Figure 5.4B.

Nanoparticles loaded with P-gp and DOX siRNA were applied for the delivery encapsulated content at the site of tumors where the siRNA silences the expression of P-gp, which therefore increases the intracellular concentration of the DOX (Meng et al., 2010). Following the same order, modified liposomes containing specific ligands of P-gp siRNA or DOX are also used to treat drug resistant tumors (Jiang et al., 2010). Nanoparticles loaded with MDR siRNAs showed a higher gene transfection (Nakamura et al., 2011), owing to their systemic stability and target specificity, compared with the free siRNAs that are unstable in serum and show poor cellular absorption (Gao et al., 2009). It was also reported that coadministration of PAX- and Bcl-2-targeted siRNA nanoparticles showed greater activity in the cell lines of breast cancer (MDA-MB-231). Such a combination worked by down regulating Bcl-2 expression when compared to the individual therapeutic agents (Wang et al., 2006).

Other studies reported that when PAX is combined with P-gp siRNA-loaded PLGA copolymer, it resulted in higher PAX retention in cancer cells resistant to multiple drugs and showed improved *in vivo* activity (Patil et al., 2010). Furthermore, DOX-loaded liposomes and MRP-1 and Bcl-2 siRNAs caused cell apoptosis and the reversal of MDR in lung cancer (H69AR, human SCLC) (Saad et al., 2008).

It is well known that surviving, which is a negative regulator of apoptosis, generally shows high expression levels in MDR cancer cells. Therefore, inhibition of survivin expression can be considered as an effective means in increasing apoptosis in cancer cells (Kanwar et al., 2011).

In this context, the polyamidoamine (PAMAM), dendrimer-modified, magnetic nanoparticles were used to deliver antisense oligodeoxynucleotide (asODN) to suppress mRNA levels of survivin and protein levels in breast cancer cells (MCF-7 and MDA-MB 435) as well as in liver cancer cells. The above techniques resulted in the resensitization of cancer cells that were initially resistant to single or multiple drugs. Therefore, prior to the administration of cytotoxic drugs, it is of prime importance to lower the expression levels of the respective genes by administering the nucleic acids via nanoparticles (Figure 5.4B).

5.4.3 Delivery of Nanoparticles to Modulate the Uptake Route of Drugs

Drugs which are encapsulated in nanoparticles have different PK properties compared to free drugs. Free drug molecules generally enter cells by a diffusion process through cell membrane pumps and drug efflux pumps. These are also present on the cell membrane and can detect this movement (Figure 5.4C) and in response make the cells vulnerable to ABC transporters, leading to drug expulsion. Apart from increasing drug efficiency, delivering drugs into the inner organelles through nanocarriers would be an effective tool to overcome the outflow of free drugs (Figure 5.4C). Nanocarriers are internalized into cells via the endocytosis pathway and can cross the cell membrane surreptitiously, thus avoiding recognition of the drugs by the efflux pumps (Huwyler et al., 2002; Rejman et al., 2004).

This type of endocytosis is called "endocytosis stealth" (Figure 5.4C), and results in greater intracellular drug accumulation (Davis et al., 2008).

Under this process, drug particles get internalized into endosomes which eventually release the drug deep within the cytoplasm, away from the vicinity ABC transporters present on the cell membrane (Kunjachan et al., 2012; Shen et al., 2008). In the process the cytotoxic drugs remain protected from the detoxification enzyme, methionine synthase (Murakami et al., 2011). It was reported that taxol-containing liposomes exhibited antitumor effects on a resistant colon-26 tumor model (Sharma et al., 1993). In addition, a polymer-drug conjugate composed of a carboxymethyl dextran-PAX showed antitumor activity *in vivo* against colon-26 carcinoma cells resistant to PAX (Sugahara et al., 2007). Thus, by modulating the uptake or entry pathway as well as by targeting cellular compartments, drug delivery-utilizing nanocarriers would be an efficient tool for reversing MDR in cancer cells.

5.5 Different Nanocarriers in the Treatment of MDR Cancer

It was because of the limitations of the various drugs and their combinations used in the treatment of MDR cancer, as discussed in the previous sections, that novel nano-drug delivery systems came into force. Nanoparticles/nanomedicines are being extensively explored for their use in the treatment of cancer. Nano-delivery systems are applied to improve drug delivery, increase the efficacy of treatment, reduce side effects, overcome drug resistance, and improve localization of anti-neoplastic compounds in the tumor site. A variety of nanotechnology-based drug carriers, which come from various classes based on their nature, have been investigated for chemotherapeutic delivery in MDR cancer. An illustration of the different families of nanocarriers is given in Figure 5.5. Tables 5.4 and 5.5 show, respectively, nano-drug carriers with improved anticancer efficacy in chemotherapeutics against MDR and recent progress in overcoming tumor drug resistance by using nanomedicines.

5.5.1 Liposome

Liposomes are vesicular structures with an aqueous core surrounded by a lipid bilayer shell which encapsulate drugs either in the aqueous compartment or in the lipid bilayer, depending on their nature. Recently the development of liposomes such as mAb (Monoclonal antibody) 2C5 with DOX and anti-HER2 mAb with PAX have been tested in the preclinical phase, while at the same time many others are undergoing clinical trials (Torchilin, 2010; Yang et al., 2007a; Yang et al., 2007b). Novel approaches like using the burst release of the drug once introduced to the body have also been investigated in the treatment of hyperthermia, for example ThermoDOXO® which is currently in Phase III trials (Allen, 2013; Needham et al., 2000; Perche and Torchilin, 2013).

Recently, a neutral 1,2-dioleoylsn-glycero-3-phosphatidylcholine (DOPC) liposome-siRNA delivery system for siRNA-dependent silencing of the cisplatin resistance transporter mRNA of ATP7B was developed by Mangla et al. (2009). This finding reveals that the system is very effective in reducing ATP7B expression *in vivo*, leading to the control of tumor growth by using non-encapsulated cisplatin. These types of investigations support the encapsulation of important anticancer compounds such as PAX and DOX as DOPC liposome-siRNA delivery systems. The first Food and Drug Administration (FDA) approved preparation of DOX-entrapped liposomes (Doxil®), to treat AIDS-related Kaposi's sarcoma

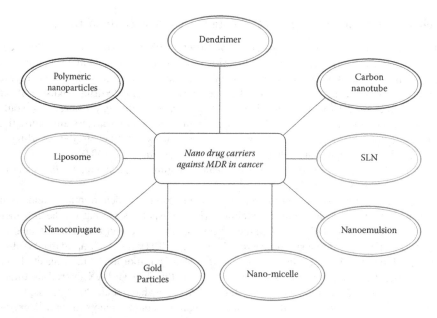

FIGURE 5.5 Illustration of the different families of nanocarriers.

TABLE 5.4

Examples of Nano-Drug Carriers That Showed Anticancer Improved Efficacy of Chemotherapeutics in *In Vivo* and *In Vitro* Studies against MDR

Nanomedicine	Drug	Types of Cancer	Model	Outcome	References
Liposome	Doxorubicin	Human colon cancer	HT 29 cells and HT 29-dx cells	Doxorubicin encapsulated liposome (Lipodox) more effective than free doxorubicin in HT 29-dx cells	Riganti et al., 2011
	Mitoxantrone	Hepatocellular carcinoma (HCC)	HCC Huh-7 cells and MDCK II/BCRP cells; HCC xenograft BALB/c mice	2.3-fold higher cytotoxicity in Huh-7 cells and 14.9-fold increase in mitoxantrone accumulation in MDCK II/BCRP cells; furthermore, improved antitumor activity *in vivo* in orthotopic HCC xenograft BALB/c mice	Zhang et al., 2012a
	Doxorubicin	Small-cell lung cancer (SCLC)	SBC 3/ADM cells	3.5 times more cytotoxicity than free doxorubicin	Kobayashi et al., 2007
Polymeric micelles	FG020326	Nasopharyngeal carcinoma	KB(v200) cells	Human KB(v200) cells treated with vincristine were ~5 times more resensitized when treated with folate-functionalized and FG020326-loaded polymeric micelles; moreover, the micelles significantly inhibited the P-gp drug efflux function	Yang et al., 2008
	Doxorubicin	Colorectal carcinoma	SW480 cells	The coadministration of DOX and TRAIL in P(MDS-co-CES) micelles resulted in increased cytotoxicity against resistant tumor cells	Lee et al., 2011
	Doxorubicin	Breast cancer	MCF-7/DOX(R) cells and MCF-7/DOX(R) cells xenografts model	Doxorubicin-loaded pH-sensitive micelles decorated with folate showed more than 90% cytotoxicity to DOX-resistant MCF-7 cells *in vitro*; moreover, in the MCF-7/DOX(R) xenografts model the accumulated doxorubicin levels in the solid tumors was 20 times higher than free doxorubicin	Lee et al., 2005
Nanoemulsion	Paclitaxel and curcumin	Human ovarian adenocarcinoma	SKOV3 and SKOV3(TR) cell	Co-therapy of paclitaxel and curcumin in nanoemulsion formulations was very effective in enhancing the cytotoxicity in wild-type and resistant cells by promoting the apoptotic response	Ganta and Amiji, 2009
Polymeric nanoparticles	Paclitaxel	Human colon adenocarcinoma	HCT-15 cells and HCT-15 mouse xenograft model	Significant inhibition in tumor growth was observed in HCT-15 mouse xenograft model	Koziara et al., 2006
	Doxorubicin	Osteosarcoma	KHOS and KHOSR2 cells; U-20S and U-20SR2 cells	Nanoparticle loaded with doxorubicin showed increased apoptosis in osteosarcoma cells as compared with doxorubicin alone	Susa et al., 2009
	Doxorubicin and curcumin	Chronic myeloid leukemia (CML)	K562 cells	Coformulation of doxorubicin and curcumin in nanoparticles inhibits the development of multidrug resistance in K562 cells by the enhancement of antiproliferative activity	Misra and Sahoo, 2011
	Doxorubicin	Ovarian and uterine carcinoma	SKOV-3 and MES-SA/ Dx cells	Cellular uptake of drug from non-conjugated and HER2 conjugated PLGA nanoparticles were higher compared with both free drug and non-conjugated PLGA nanoparticles	Lei et al., 2011

(Continued)

TABLE 5.4 (CONTINUED)

Examples of Nano-Drug Carriers That Showed Anticancer Improved Efficacy of Chemotherapeutics in *In Vivo* and *In Vitro* Studies against MDR

Nanomedicine	Drug	Types of Cancer	Model	Outcome	References
Dendrimers	Doxorubicin	Colon carcinomas	C-26 cells and C-26 tumor bearing BALB/c mice	Dendrimer-doxorubicin was 10 times more toxic than free doxorubicin toward C-26; moreover, its tumor uptake was 9-fold higher than intravenously-administered free doxorubicin in C-26 bearing BALB/c mice	Lee et al., 2006
Cyclodextrin nanoparticles	Doxorubicin	Breast cancer	MCF-7/ADR cells	There was a 3-fold decrease in IC 50 value by doxorubicin-loaded cyclodextrin nanoparticles as compared to free doxorubicin in resistant MCF-7/ADR cells	Qiu et al., 2010
Gold nanoparticles	Doxorubicin	HCC	HepG2R cells	Extent of intracellular drug uptake of gold particles was increased in contrast to free doxorubicin in resistant HepG2R cells as confirmed by confocal imaging and plasma mass spectrometry	Gu et al., 2012
	Doxorubicin	Breast cancer	MCF-7/ADR cells	Doxorubicin-tethered gold nanoparticles enhanced drug accumulation and retention in multidrug resistant MCF-7/ADR cells and induced elevated apoptosis of cancer cells	Wang et al., 2011
Magnetic nanoparticles	Doxorubicin	CML	K562 cells	Remarkable synergistic effect of $Fe(3)O(4)$ nanoparticles on drug uptake of daunorubicin leukemia K562 cells	Wang et al., 2007
Silica nanoparticles	Doxorubicin and siRNA	Nasopharyngeal carcinoma	KB-V1 cells	The dual delivery of doxorubicin and siRNA in drug resistant KB-V1 cells was capable of increasing the intracellular as well as the intranuclear drug concentration to levels exceeding that of free doxorubicin.	Meng et al., 2010
Carbon nanotube	Doxorubicin	CML	K562 cells	2.4-fold higher cytotoxicity and significant cell proliferation suppression toward multidrug resistant K562 leukemia cells as compared with free doxorubicin	Li et al. 2010
Solid lipid nanoparticles	Doxorubicin	Breast cancer	MCF –7/ADR cells	Efficiently enhanced apoptotic cell death through the higher accumulation of doxorubicin in resistant MCF-7/ADR cells in comparison with free doxorubicin	Kang et al., 2010
	Doxorubicin and mitomycin C	Breast cancer	MDA 435/LCC 6 cells	Hybrid systems, solid polymer–lipid nanoparticles were effective in killing MDR cells at 20–30-fold lower doses than the free drugs	Shuhendler et al., 2009
Polymer conjugate	Doxorubicin	Lymphoma	Mice bearing T cell lymphoma EL4 of B cell lymphoma 38 C13	The best antitumor effects were produced by conjugates with 10–13 wt. % of bound doxorubicin; free doxorubicin up to 4.6% relative to total drug content had no impact on the treatment efficacy	Sirova et al., 2010

TABLE 5.5

Recent Progress in Overcoming Tumor Drug Resistance by Using Nanomedicines

Tumor Type	Nanomedicines	Active Groups	Action Mechanism	References
Docetaxel (DTX)-resistant human ovarian A2780/T *in vitro* model	D-α-Tocopheryl polyethylene glycol 1000-block-poly (β-amino ester) containing micellar nanoparticles	D-α-Tocopheryl polyethylene glycol, docetaxel (DTX)	Inhibition of P-gp to decrease DTX efflux; DTX inhibition of cell division	Zhao, et al., 2013
H460/Tax R human non-small-cell lung cancer overexpressing P-gp, *in vitro* model	D-α-Tocopheryl polyethylene glycol 1000 succinate containing micellar nanoparticles	D-α-Tocopheryl polyethylene glycol 1000 succinate; paclitaxel (PTX); fluorouracil (5-FU)	Inhibition of P-glycoprotein by Tocopheryl polyethylene glycol 1000 succinate; inhibition of cell division by PTX; irreversible inhibition of thymidylate synthase; synergism of PTX/5-FU	Wang, et al., 2013
Human MCF 7/ADR tumor on BALB/c Nude mice, *in vivo* breast cancer model	Poly(bis2-hydroxylethyl)-disulfide-diacrylate-β-tetraethylene pentamine)-polycaprolactone copolymer(PBD-PCL) containing micelle nanoparticles	shRNA to survivin; PBD-PCL; doxorubicin	Inhibition of P-glycoprotein; inhibition of glutathione S-transferase interaction into DNA	Yin, et al., 2013
CD138-CD34- cells isolated from a human U266 multiple myeloma cell line inoculated in mice with non-obese diabetic/severe combined immunodeficiency (NOD/SCID), *in vivo* model	Polyoxypropylene chain and oleic acid coated iron oxide nanoparticles	Anti-ABCG2 antibody; paclitaxel (PTX)	Antibody blocking of ABCG2 to inhibit PTX resistance; PTX inhibition of cell division	Yang et al., 2013; Yang et al., 2014
Human lung adenocarcinoma A549-Bcl-2 cells, *in vivo* model	Micelleplexes	siRNA to Bcl-2; paclitaxel (PTX)	Downregulation of Bcl-2; PTX inhibition of cell division	Yu et al., 2014
CAI27 cisplatin-resistant human oral cancer (CAR) cells, *in vitro* model	PLGA nanoparticles	Curcumin; cisplatin	Pt-DNA crosslinks; MDR1 suppression; triggering of apoptosis	Chang, et al., 2013
Human breast cancer MDA-MB-231 cells inoculated into BALB/c nu/nu mice, xenogeneic *in vivo* model	PLGA nanoparticles conjugated to anti-CD 133	Anti-CD133; paclitaxel (PTX)	Targeting tumor initiating cells CD133+; PTX inhibition of cell division	Swaminathan et al., 2013
GS5 glioblastoma multiforme cells (obtained from human U87GM cells enriched by stem cells) injected intracranially in rats, *in vivo* model	PGLA nanoparticles treatment by convection-enhanced delivery (CED)	Dithiazanine iodide (DI)	DI displays toxicity towards brain cancer stem cells	Zhou et al., 2013; Günther, et al., 2008

(Continued)

TABLE 5.5 (CONTINUED)

Recent Progress in Overcoming Tumor Drug Resistance by Using Nanomedicines

Tumor Type	Nanomedicines	Active Groups	Action Mechanism	References
Rat F98 glioblastoma inoculated on Ficher 344 rats, orthotopic syngeneic *in vivo* model	PGLA–chitosan	Carmustine (BCNU); O(6)-benzylguanine (BG)	BCNU for DNA alkylating and crosslinking; BG for inhibition of O(6)-methylguanine-DNA-methyl transferase (MGMT)	Qian, et al., 2013
Chemotherapy and antiandrogen-resistant mAR+/GPRC6A+DU-145 human prostate carcinoma cells, *in vivo* model	Gold nanoparticles	Multiple α-Bicalutamide (Bic) and β-Bic antiandrogens	Multivalent binding to androgen receptor and to G-protein coupled receptor (GPCR6A); the antiandrogens inhibit binding of androgens	Dreaden et al., 2012; Borsellino et al., 1995
Human breast MDA-MB-231 and MDA-MB 468 cell lines and brain cancer cell lines U87 MG and T98G, *in vitro* model	Polymer-drug conjugate based on poly(β-L-malic acid) platform	Temozolomide (TMZ)	TMX is a DNA alkylating agent preventing cell division	Mohri et al., 2000
Human Lewis drug carcinoma A549 cells subcutaneously inoculated into C57BL/6N mice, *in vivo* model	Nanoliposomes in combination with radiation therapy	Cisplatin (CDDP); radiation therapy	Cisplatin alkylating and crosslinking DNA; sensation to radiation lesions	Zhang et al., 2011b
Human CW480 colorectal cancer, *in vitro* model	Human serum albumin-based anti-survivin siRNA delivery in combination with radiation therapy	Anti-survivin siRNA; radiation therapy	Knocking down of survivin promotes apoptosis	Gaca et al., 2012
Human melanoma cells HMV II radiation resistance under hypoxic conditions, *in vitro* model	Liposomes in combination with radiation therapy	Pimonidaole (Pmz); radiation Therapy	DNA fragmentation and crossing sensitizes for radiation damage	Kato et al., 2012
Human U251 glioblastoma intracranially grown on nu/nu rats, *in vivo* model	Magnetic ferric oxide nanoparticles in combination with radiation therapy	TRAIL, a type II transmembrane homotrimeric protein (TNF gene superfamily); gamma radiation	Radiation sensitizes for TRAIL induced apoptosis; sensitization by TRAIL by conjugation to the ferric oxide nanoparticle	Perlstein et al., 2013
HMLER (shE-cadherin) human breast cancer stem cells (BCSCs) inoculated into mice to treat triple-negative breast cancer, *in vivo* model	Multiwalled carbon nanotubes (MWCNTs) in combination with photothermal (laser) treatment	Nanotubes without active targeting but with specific permeation into BCSCs; photothermal treatment	Thermal therapy promotes rapid MWCNT membrane permeabilization resulting in necrosis of BCSCs and differentiated cancer cells	Burke et al., 2012; Gupta et al., 2009

and oncology related symptoms in 1995, made a remarkable change in the treatment of MDR cancer (Ganta et al., 2008).

On the other hand, liposome-induced modulation of P-gp marks another important means of enhancing the therapeutic efficacy of anticancer drugs. In this background, Riganti et al. (2011) considered an anionic liposomal formulation of DOX (Lipodox®) which is much more effective in resistant HT29-dx cells in comparison to free DOX.

In another approach, an MDR mechanism was treated with a multifaceted liposomal system consisting of an anticancer agent (DOX), ASOs targeting MDR1 mRNA, and ASOs targeting Bcl-2 mRNA (Pakunlu et al., 2006). From this development it was established that the prepared formulation was more lethal *in vitro* in resistant A2870/AD human ovarian carcinoma cells as compared to free DOX, DOX liposomes only, and DOX liposomes with ASOs.

In another similar type of study, Zhang et al. (2012b) prepared a mitoxantrone-loaded liposome with a double function from synthetic, polymeric nano-biomaterial (Gal-P123) for its efficiency in targeting cancer cells and reversing MDR in hepatocellular carcinoma (HCC) cells. The investigation revealed a significant increase in cytotoxicity by enhancing intracellular accumulation in MDCKII/BCRP cells by means of mitoxantrone-incorporated liposomes (MX-LPG) in comparison to free mitoxantrone. In addition, MX-LPG showed better activity and increased selectivity in BALB/c mice bearing orthotopic HCC xenograft tumors.

Kobayashia et al. (2007) found that DOX-loaded liposomes for targeting transferrin receptors (Tf-R) overcome MDR by avoiding P-gp-mediated drug efflux into MDR cells (SBC-3/ADM) through Tf-R-mediated endocytosis (Kobayashi et al., 2007). In this study, four types of liposomes were prepared, consisting of untargeted and Tf-R-targeted liposomes using either egg-PC/cholesterol (EPC) or hydrogenated EPC. The results showed a significant enhancement in cytotoxicity via targeted EPC-liposomes in comparison to free DOX. It was concluded that Tf-R-targeted EPC-liposomes possess a great potential as a drug delivery system to circumvent P-gp-mediated MDR in tumors.

5.5.2 Inhibition of MDR Using Peptides

Peptides can be used to reverse the effect of P-gp in MDR cancer. One of the synthetic P-gp-derived peptides coupled with polyethylene glycol (PEG) and entrapped into liposomes is reported to overcome MDR. Post-immunization with liposomes loaded with these peptides, along with simultaneous treatment with DOX, survival time was observed to increase in 83% of mice inoculated with P388R cells. Although complete eradication of the cancer was not observed, these mice didn't show any autoimmune responses. Hence, these results encourage a potential novel approach to rupture immune tolerance towards MDR1 proteins and therefore modulate the sensitivity of resistant tumors to anticancer chemotherapy (Gatouillat et al., 2007; Hamada and Tsuruo, 1986; Mechetner and Roninson, 1992).

5.5.3 Polymeric Micelles

A polymeric micelle block copolymer consists of a hydrophobic portion, which encapsulates the poorly soluble drug, and a hydrophilic portion, which faces outwards to form a shell.

In the early 1990s, Kataoka and associates designed DOX-conjugated block copolymer micelles which resulted in benchmark outcomes and compelled a great deal of interest of researchers in this area (Kataoka et al., 2000; Kwon et al., 1997).

Yang et al. formulated folate-functionalized polymeric micelles by encapsulating FG020326 and vincristine from di-block copolymers of PEG and biodegradable poly(ε-caprolactone) (PCL). This formulation was tested for blocking P-gp-mediated imidazole derivative FG020326 and vincristine resistant KB-V200 cells (Yang et al., 2008). The result showed the significant sensitization of KB-V200 cells *in vitro* by folate-functionalized, FG020326-loaded micelles as compared to their folate-free counterparts. Furthermore, the prepared folate-functionalized micelles were shown to obstruct the P-gp-dependent rhodamine 123 effluxes.

In another study, human TNF (tumor necrosis factor)-related, apoptosis-inducing ligand (Apo2L/TRAIL) and self-assembled micelles of DOX were coadministered with a cationic copolymer of poly{*N*-methyl di etheneaminesebacate)-co-[(cholesteryl oxo carbonyl amino ethyl) methyl bis (ethylene) ammonium bromide]sebacate} (P(MDS-co-CES)), resulting in increased cytotoxicity against resistant tumor cells (Lee et al., 2011).

5.5.4 Nanoemulsion

Nanoemulsions are a biphasic dosage form consisting of an aqueous continuous phase in which nano-range size oil droplets are enclosed with surfactant molecules (Acosta, 2009; McClements et al., 2007, 2009).

Different types of anticancer drugs with highly lipophilic natures, such as taxanes (PAX, docetaxel), etoposides, tamoxifen, and dacarbazine, have been successfully delivered using nanoemulsion as a carrier system (Ahmad, 2013).

The effect of PAX and curcumin was studied by Ganta and Amiji (2009) by coadministering them using nanoemulsion as a delivery system. They selected curcumin as it inhibits NFκB and causes down regulation in ABC transporters in wild-type SKOV3 and resistant SKOV3 TR (Taxol resistant) human ovarian adeno-carcinoma cells (Ganta and Amiji, 2009). Their findings established an efficient delivery of entrapped drugs within SKOV3 and SKOV3 TR cells. In addition, coadministration of curcumin resulted in the inhibition of NFκB activity and the down regulation of P-gp expression in resistant cells.

5.5.5 Particulate Nanocarriers

Particulate nanocarriers such as nanocapsules, polymeric nanoparticles, solid lipid nanoparticles (SLNs), inorganic nanoparticles, polymeric conjugates, carbon nanotubes, and dendrimers have been extensively studied to overcome MDR in cancer. These carriers have also been found to be very effective in overcoming the limitations of MDR in tumors.

5.5.5.1 Nanocapsules

Consisting of a liquid lipid core, nanocapsules are stabilized by surfactants and may sometimes be coated by polymers. Medium and long chain fatty acids containing triglycerides, usually vegetable oils, form the lipid cores. The active ingredients remain entrapped within this lipid core, behaving as a reservoir from where hydrophobic drugs are released at a controlled rate. Thus, nanocapsules are a promising prospect as a carrier for lipophilic and hydrophobic drugs.

5.5.5.2 Solid Lipid Nanoparticles (SLN)

SLNs evolved in the early 1990s as an attractive choice of nano-colloidal carrier system for controlled drug delivery. It is usually prepared with biodegradable lipids that remain in solid state at normal body temperature. Here, the drug is consistently and physically dispersed as a solid lipid. The mechanism of drug release from the lipid matrix attributes to the degradation of the particles by the lipases present at the site of delivery, leading to a sustained and extended release of drugs (Muller et al., 2000).

There are various methods of preparation for SLNs such as high-pressure homogenization, high shear homogenization, ultrasonication, microemulsification techniques, etc. (Subedi et al., 2009). SLNs overcome the general limitations of polymeric systems by exhibiting low toxicity due to the presence of biodegradable lipids, and by their extremely small size, which facilitates circumvention of the reticuloendothelial system (RES).

In one of the studies, Kang et al. (2010) prepared DOX-loaded SLNs using Capmul MCM C10 (glyceryl caprate) as the lipid phase, Solutol HS15 (PEG 660 hydroxy stearate) as the surfactant, and curdlan as the shell forming material (Kang et al., 2010; Subedi et al., 2009). The developed DOX SLNs showed a significant increase in cellular uptake at one hour and two hours, respectively.

In another study, Shuhendler et al. (2009) prepared a polymer–lipid hybrid SLN by co-incorporating DOX and mitomycin C using myristic acid, HPESO, pluronic F68, PEG100SA, and PEG40SA. The developed SLNs showed a 20- to 30-fold increase in toxicity in resistant MB435/LCC6/MDR1 cells when compared with free DOX.

5.5.5.3 Polymeric Nanoparticles

Nowadays, polymeric nanoparticles are widely investigated for their amazing potential as a drug delivery system for anticancer compounds. They are prepared either by encapsulation, dissolution, and entrapment of the drug in biodegradable polymers or by embedding the drug in the polymeric matrix. The binding of drugs to hydrophilic polymers increases their circulation time and minimizes toxicity to normal tissues (Hu and Zhang, 2012). Therefore, long circulating nanoparticles are more frequently formulated using PEG, which avoids opsonization.

Currently, polylactide and PLGA are mainly used as biodegradable polymers for the synthesis of FDA-approved nanomedicines, while many more are undergoing clinical trials (Mattheolabakis et al., 2012).

Recently, Koziara et al. (2006) prepared PAX-loaded nanoparticles for a PAX-resistant human colorectal tumor (HCT-15) xenograft model, which revealed a noticeable inhibition of tumor growth in mice treated with PAX-nanoparticles in contrast with free PAX. In another study, chitosan-based nanoparticles loaded with shRNA were prepared, targeting MDR1, and showed a significant reversal of PAX resistance in A2780/TS cells in a time-dependent manner (Yang et al., 2009).

In yet another study, DOX-loaded, stearyl amine-modified dextran nanoparticles were successfully prepared. From this study it was found that there was an increased accumulation of DOX-loaded nanoparticles in the nucleus of resistant osteosarcoma cells in contrast to free DOX, which was confined to the cytoplasm in resistant cells (Susa et al., 2009).

Misra and Sahoo (2011) developed PLGA nanoparticles by coadministering DOX with curcumin, which leads to the facilitation of the retention of DOX in the nucleus in addition to down regulating the expression of P-gp and Bcl-2 in K562 cells. Furthermore, Lei et al. (2011) formulated HER2 antibody-conjugated, DOX-loaded PLGA nanoparticles and compared their cellular uptake and cytotoxicity to free DOX and non-targeted nanoparticles in resistant ovarian SKOV-3 and uterine MES-SA/Dx5 cells (Lei et al., 2011). The details are given in Table 5.4.

In a further study, DOX- and photosensitizer in 4-armedporphyrinpolylactide was coadministered and the surface of the nanoparticle was coated with d-α-tocopheryl PEG 1,000 succinate (TPGS), which is a potential P-gp inhibitor (Shieh et al., 2011). The findings signify that this type of combination demonstrates a marked synergistic effect that resulted in the enhanced transport of DOX to the nucleus in resistant MCF-7/ADR cells.

5.5.5.4 Dendrimers

Dendrimers are a type of multi-branched polymer, which consists of a central core, branches of repeating units, and an outer layer of multivalent functional groups. These functional groups can electrostatically interact with charged polar molecules, whereas the hydrophobic inner cavities can encapsulate uncharged, nonpolar molecules through a number of interactions.

This complex structure consists of different functional groups that can allow for controlled delivery of the drug. This structure can be modified to control the release of the drug in a certain pH or when encountered by specific enzymes; targeting molecules, such as the RGD (Arginylglycylaspartic acid) peptide or mAbs, are also used.

Various clinical trials are in progress using the amphiphilic di-block copolymer to deliver PAX-forming micelles for treating breast, NSCLC, and advanced pancreatic cancer (Webster et al., 2013). In another study, DOX-dendrimers are prepared by hydrazone linkage (Lee et al., 2006). The study demonstrates that DOX-dendrimers exhibit controlled drug-loading via multiple attachment sites and a modulated solubility profile through PEGylation, along with characteristic drug release which is influenced by pH-sensitive hydrazone dendrimer linkages. The details of polyester dendrimer PEO-DOX conjugate are given in Table 5.4.

5.5.6 Cyclodextrin Polymers

Recently, cyclodextrin polymers with a transferrin-targeting moiety delivery system are frequently used in the treatment of MDR cancer (Davis, 2009). Cyclodextrin polycation is one of the components of this system, used for nucleic acid condensation along with transferrin-linked PEG adamantine. It is used to stabilize the particle and also to target the cell surface receptors (Davis, 2009). In another study, scientists developed a complex cyclodextrin-based formulation of DOX (sPEL/CD). In this formulation, methoxy PEG and poly lactic acid (PLA) was attached to DOX to obtain linear mPEG-PLA, which is a well-established structure like the arms of core β-cyclodextrin (Kolesnick and Kronke, 1998). The result showed a significant decrease in IC50 (Inhibitory concentration) value by DOX-loaded sPEL/CD as compared to free DOX in resistant MCF-7/ADR cells.

5.5.7 Metallic and Magnetic Nanoparticles

Gold nanoparticles (Au-NPs) are extensively used for biomedical imaging and biosensing. Likewise, anti-HER2 was attached to the nanoparticle surfaces to facilitate the cellular internalization of gold nanoparticles and gelatin or albumin (van Vlerken and Amiji, 2006).

Due to the photo-physical property of gold nanoparticles, they can conjugate with other materials easily through ionic and covalent bonds. They are considered nontoxic and the therapeutic constituents can be released in a controlled manner. Various drugs' small particles, like DNA and RNA, can also be delivered by attaching them to the gold particles. The anticancer agents can be formulated with Au-NPs by physical adsorption, ionic bonding, and/or covalent bonding (Chen et al., 2007; Podsiadlo et al., 2008).

In addition to other targeting agents, PEG can also be attached to the surface of metallic nanoparticles to increase their stability and circulation time (Webster et al., 2013).

More recently, sarcomas and melanomas treated with TNF α-integrated colloidal Au-NPs are in Phase I clinical trials (Libutti et al., 2010).

To reduce the intracellular drug intake, Gu et al. (2012) prepared Au-NPs of DOX by combining DOX in PEGylated Au-NPs via a disulfide bond (Au-PEG-SS-DOX), which resulted in increased intracellular drug uptake in contrast to free DOX in resistant HepG2-R cells. Wang et al. (2011) designed DOX Au-NPs using a hydrazone linker (DOX-Au-Hyd-NPs). The DOX-Au-Hyd-NPs resulted in a marked increase in DOX intracellular uptake and minimum efflux, in addition to a significantly enhanced cytotoxicity in comparison to free DOX in resistant MCF-7/ADR cells.

During the synthesis of gold, sodium citrate can be used as a reducing agent and stabilizer to avoid aggregation (Arvizo et al, 2010).

5.5.8 Super Paramagnetic Nanoparticles

Iron oxide (Fe_3O_4) nanoparticles are a newer approach to treat cancer. The drug can be delivered into targeted sites by local hyperthermia or by applying oscillation strategies. Magnetic fields can also be used to guide the drug to the intended target area within the body.

Different types of magnetic materials, like magnetite, iron, nickel, cobalt, neodymium iron- boron, and samarium-cobalt, have a large range of magnetic properties. Furthermore, some ferrofluids are strongly magnetized in the presence of a magnetic field. Ferrofluids are basically colloidal suspensions of nano-dimension ferromagnetic particles. Such types of nanoparticles, once introduced, are internalized within the lysosomes of the RES cells and dissociate to ferritin and/or hemosiderin (Pankhurst et al., 2003).

An *in vitro* study of the role of Fe_3O_4 magnetic nanoparticles in conjugation with daunorubicin for MDR in sensitive and resistant K562 cells was investigated (Wang et al., 2007). The nanoparticles were coated with tetraheptylammonium (THA) to purposely improve the interaction between the nanoparticles and the lipid portion of the cell membrane. From the study on the resistant K562 cells, it was found that THA-coated Fe_3O_4 nanoparticles increased the uptake of daunorubicin.

5.5.9 Silica Nanoparticles

Inorganic nanoformulations like silica are used in chemotherapeutics. In one study, DOX and mesoporous silica was covalently added by hydrazone bonding (DOX-Hyd-MNSP), as designed by Huang et al. (2011), which revealed considerable apoptosis *in vitro* and *in vivo* in resistant MES-SA/Dx-5 cells in contrast to the controls.

In another study, DOX and MDR1 siRNA was co-incorporated in MNSP (Mesoporous silica nanoparticle) (Meng et al., 2010). MNSP was activated with a phosphonate group which facilitates DOX binding by electrostatic action. After that, a cationic polyethylenimine (PEI) was used to coat the functional group which supports the complexation with the anionic MDR1 siRNA. From this study, it was found that this delivery system notably improved the intracellular and intranuclear DOX uptakes over free DOX or DOX MNSP without siRNA in resistant KB-V1 cells. In a more comprehensive study, Chen et al. (2009b) co-engineered DOX and Bcl-2 siRNA as MNSP by the entrapment of DOX inside MNSP pores and complexation of Bcl-2 siRNA with modified PAMAM dendrimers of MNSP. The study findings reveled an increase in perinuclear localization of DOX leading to cytotoxicity by MNSP as compared with the free DOX in resistant A2780/AD human ovarian cancer cells.

5.5.10 Carbon-Based Nanoparticles/Carbon Nanotubes

Carbon nanotubes have a needle-like structure which can penetrate through the tumor cell very easily and can deliver drug molecules into the intended site. Due to their special structure they have a large surface area, which leads to a high entrapment of drug molecules and a large number of attachment sites. These carbon nanotubes also have electrical and thermal conductivity, which may prove to be useful in future cancer therapy applications, such as thermal ablations.

Recently, carbon nanotubes were used to integrate various anti-neoplastic agents such as DOX and PAX, nucleic acids including ASOs, and siRNAs (Fabbro et al., 2012).

In a study, DOX loaded by physical adsorption into carbon nanotubes was prepared by Li et al. (2010) by linking a P-gp antibody by a diimide-activated amidation reaction. Due to the physical adsorption of DOX in nanotubes, the molecular integrity was kept intact by preventing chemical bonding. The release of DOX from this formulation was activated by exposure under near-infrared radiation of resistant human leukemia K562R cells. Furthermore, it was concluded that the coupling of the P-gp antibody with nanotubes offers enormous stereo hindrance for P-gp recognition of DOX, which results in the inhibition of its P-gp-mediated efflux.

5.5.11 Polymeric Nanoconjugates

The conjugation of hydrophilic polymers with proteins and anticancer drugs is one of the most vigorously explored approaches for polymer drug delivery. This process establishes polymer therapeutics as one of the first classes of anticancer nanomedicine (Duncan, 2006).

Drug conjugation with a synthetic polymer like PEG covalently increases residence time in the plasma, decreases protein immunogenicity, and widens their therapeutic index. Nowadays, many PEGylated enzymes such as L-asparaginase and cytokines that include interferon-α and granulocyte colony-stimulating factor are being utilized very frequently.

Poly(*N*-[2-hydroxypropyl]methacrylamide) (polyHPMA) and HPMA copolymers have been used to conjugate DOX to overcome drug resistance. HPMA happens to be a polar and non-immunogenic synthetic polymer (Kopecek et al., 2000; Nori and Kopecek, 2005; Omelyanenko et al., 1998).

DOX and HPMA conjugate was prepared for a multiple mechanism of MDR in addition to P-gp-mediated drug resistance. After the internalization of the HPMA–DOX conjugate, the connecting spacer between the HPMA–DOX was hydrolyzed by lysosomal enzyme leads to release the drug into the targeted site. It has been found that MDR is not induced after constant exposure of A2780 cells to HPMA–DOX conjugates, which was quantified through MDR1 gene expression. Also, it was evident that MPR gene expression was inhibited and resistance was decreased against taxol (Minko et al., 1999a).

Polymeric conjugates can be delivered either via passive targeting by the enhanced permeability of lysosomotropic delivery following the EPR (enhanced permeability and retention) effect or actively by binding cell-specific ligands for receptor-mediated targeting. One such example of polymeric conjugates in chemotherapeutics includes a polyglutamic acid–PAX which is undergoing Phase III trials for NSCLC in females.

The ability of HPMA–DOX conjugates to avoid MDR *in vivo* was established in solid tumor mouse models of sensitive human ovarian carcinoma A2780 and resistant A2780/AD tumors (Minko et al., 2000). HPMA–DOX conjugates were reported to remarkably decrease sensitive tumor size by 28-fold , whereas in resistant tumors it was by 18-fold while free DOX was capable of cutting sensitive tumor size by only 2.8-fold and showed no effect in the case of the resistant tumor model when compared with the control.

In a study, Sirova et al. (2010) investigated the *in vivo* efficacy and safety of HPMA-based copolymers of DOX through a spacer containing a pH-sensitive linkage in murine tumor models bearing T cell lymphoma EL4 or B cell lymphoma 38C13. Their findings revealed that conjugates with 10–13% weight of bound DOX produced remarkable antitumor effects. The various formulations used to overcome MDR and their anticipated mechanisms are given in Table 5.6.

5.5.12 pH-Sensitive Nanocarriers

Drug delivery through pH-sensitive nanocarriers is another innovative approach in the treatment of MDR cancer. According to researchers, the vehicles of pH-sensitive nanocarriers are accumulated into non-diseased cells characterized by the basal expression of antigens, carbohydrates, and receptors. This type of pH-sensitive nanocarrier releases the drug by the action of stimuli like pH.

To overcome the efflux dependent MDR, acidic pH is applied, which leads to the disruption of the endosomal membrane and burst release of the drug-loaded nanocarriers into the cytoplasm was reported (Yadav et al., 2009). Therefore, the difference of pH between the extracellular (not less than 5.4) and lysosomal surrounding (not greater than 5.7) can be investigated to develop pH-sensitive drug release nanocarriers targeting lysosomal compartments instead of endosomal compartments, hence avoiding undesirable drug release into the tumor stroma (Engin et al., 1995; Yamashiro and Maxfield, 1998).

TABLE 5.6

List of Formulations to Overcome Multidrug Resistance and Their Proposed Mechanisms

Formulations	Proposed Mechanisms	Status	References
Liposomes	Endocytosis	*In vitro*	Ford and Hait, 1990
	Interaction of liposomes with P-gp	*In vitro*	Rahman et al., 1992
	Co-encapsulation of a P-gp inhibitor (Verapamil)	*In vitro*	Wu et al., 2007
Polymethcrylate NPs	Endocytosis	*In vitro*	Astier et al., 1988
Polyisohexylcyanoacrylate NPs	No endocytosis, saturation of P-gp by high concentrations of the drug, formation of an ion pair between NP degradation product and the drug	*In vitro*	Colin et al., 1997; Pepin et al., 1997; Brigger et al., 2002
Polymer–lipid hybrid NPs	Phagocytosis	*In vitro*	Wong et al., 2006
AOT-alginate NPs	Not established	*In vitro*	Chavanpatil et al., 2007
Solutol HS-15-based lipid NPs	Interaction of the released intracellular free solutol HS-15 with MDR efflux pump	*In vitro*	Garcion et al., 2006
	Redistribution of intracellular cholesterol	*In vitro*	Garcion et al., 2006
Lipid-based NPs containing Brij 78	Endocytosis, inhibition of P-gp, ATP depletion	*In vitro*	Dong et al., 2009; Koziara et al., 2006
HPMA copolymer-doxorubicin conjugates	Endocytosis, inhibition of drug detoxification systems, inhibition of cellular defensive mechanisms	*In vitro*	Omelyanenko et al., 1998; Nori and Kopecek, 2005; Kopecek et al., 2000; Minko et al., 1999; Minko et al., 2000; Minko et al., 2001
Pluronic® block copolymer micelles	Change in membrane micro-viscosity, inhibition of drug efflux transporters, ATP depletion, influence of cell apoptosis signaling, inhibition of the GSH-GST detoxification system, inhibition of mitochondria respiratory chain and decrease chain and decrease of oxygen consumption	*In vitro*, Phase II	Alakhov et al., 1996; Batrakova and Kabanova , 2008; Kabanov et al., 2002; Alakhov et al., 1999; Batrakova et al., 2004; Pruitt et al., 2000; Yang et al., 2007; Batrakova, 2001; Batrakova EV, et al., 2001

However, pH-sensitive polymers cause the destruction of endosomal membranes most likely by proton absorption and by interacting with these membranes, which leads to osmotic swelling and rupturing of the membranes (Chen et al., 2009a; Lee et al., 2005).

Lee et al. (2005) developed pH-dependent, DOX-loaded polymeric micelles coated with folate (PHSM/f). PHSM/f were prepared using a mixture of two block copolymers of poly(L-histidine)-*b*-PEG-folate (75 wt.%) and poly(L-lactic acid)-*b*-PEG-folate (25 wt.%). The study shows significant cytotoxicity in DOX-resistant MCF-7 (MCF-7/DOX R) via PHSM/f. From this study it was assumed that various mechanisms are involved in enhanced cytotoxicity, such as active internalization of PHSM/f through folate-receptor-mediated endocytosis, ionization of histidine residues resulting in micelle destabilization, and disruption of endosomal membranes.

In the same way, He et al. (2011) prepared a pH-sensitive nano-multidrug delivery system via *in situ* co-self-assembly of DOX, surfactant micelles-CTAB (chemosensitizer), and silica (DOXO-CTAB-MSNs). This multidrug nanoformulation, DOXO-CTAB-MSNs, demonstrated exceptionally specific, pH-responsive drug release both *in vitro* and *in vivo* and significant anticancer response in overcoming MDR.

In a different study, scientists evaluated the DOX-loaded second generation of pH-sensitive micelles consisting of poly(L-histidine-*co*-L-phenylalanine (16 mol%, MW: 5 K))-*b*-PEG(MW: 2 K) and poly(L-lactic acid)(MW: 3 K)-*b*-PEG(MW: 2 K)-folate (80/20 wt/wt%) for the first endosomal pH-targeting (pH 6.0) using *in vivo* MDR ovarian tumor xenografted mouse models (Kim et al., 2009). Their study showed extended circulation of the drug carrier, higher tumor-selective accumulation, as well as increased intracellular drug delivery. From this study they concluded that micelle formulation was better than its first-generation formulation targeting pH 6.8 and the folate receptor.

5.5.13 Pluronic Micelles

Pluronic micelles are a thermodynamically stable, colloidal delivery system formed spontaneously. They consist of surfactant molecules and lipid digestion products, such as bile salt mixtures, which can facilitate the absorption of fatty acids and fat-soluble vitamins. They are normally in the size range of 5–100 nm. Amphiphilic copolymers were used recently to enhance the solubility of poorly soluble drugs as a substitute to lipid-based surfactant systems. These types of formulations are very flexible in terms of structure, stabile inside the biological lipid bilayer, and have low CMC (carboxy methyl cellulose) level leads to enhance their conjugation ability to attach the ligands to the surface of the colloidal system.

Pluronics are different from polymeric micelles, as they consist of inert block copolymers poly(EO) (hydrophilic) and poly(PO) (hydrophobic). Due to their amphiphilic nature, they are different from HPMA copolymers. Their surfactant properties allow them to self-assemble into micelles with a hydrophobic PO inner core and a hydrophilic EO outer shell. Though these two types of delivery system are used to overcome drug resistance, a wide range of clinical trials have been going on with SP1049C-loaded DOX in pluronic L61 and F127 to overcome the MDR limitations of metastatic adenocarcinoma in the upper GI tract (Alakhov et al., 1996; Batrakova and Kabanova, 2008; Kabanov et al., 2002). Furthermore, SP1049C has shown a high amount of accumulation in tumor cells as compared to free DOX (Alakhov et al., 1999). An *in vivo* study shows that SP1049C has action on both sensitive and resistant tumor models, such as P388 and P388/ADR murine leukemia, Sp2/0 and Sp2/0-Dnr murine myeloma, 3LL-M27 Lewis lung carcinoma, MCF-7 and MCF-7/ADR human breast carcinomas, and KBv human oral epidermoid carcinoma (Alakhov et al., 1999; Batrakova and Kabanova, 2008).

The compound mechanism of pluronic block polymers on MDR cancer was predicted to be the altering membrane micro-viscosity (membrane fluidization) (Batrakovaet al., 2001a, 2004). It was confirmed by investigation that single copolymers are only responsible for the biological modification of the cell membrane. In the case of pluronic copolymers, they alter membrane structure and decrease its micro-viscosity, leading to an inhibition of drug efflux transporters (such as P-gp and MRPs, through the inhibition of P-gp ATPase activity), and reduced intracellular ATP levels (Batrakova et al., 2001a, 2001b, 2004). The hydrophobic PO chains of pluronic unimers insert into the hydrophilic regions of the membrane; as these pumps are dependent on energy, the reduction of these pumps was related to energy deprivation and the eradication of pump-associated ATPase activity. Therefore, it can be concluded that pluronic micelle-dependent energy depletion can stop the efflux pumps and, as a result, cell lines become susceptible to chemotherapeutic agents.

5.5.14 Ceramide Combination Therapy

Ceramide is a very important molecule in the cell membrane. It also acts as a second messenger in different signaling pathways, including in apoptosis and immune response (Kolesnick and Kronke, 1998; Struckhoff et al., 2004). During radiation and chemotherapeutic treatment, ceramide activates the apoptotic pathway and is also involved in the clustering of the death receptor (CD95) (Gulbins and Grassme, 2002; Kolesnick and Kronke, 1998; Pettus et al., 2002; Schenck et al., 2007). Additionally, ceramide is available in the mitochondrial outer membrane in the permeable channels, which allow the secretion of pro-apoptotic factors such as cytochrome C (Elrick et al., 2006; Siskind, 2005; Siskind et al., 2006). In the mechanism of MDR, the overexpression of glucosylceramide synthase, an enzyme which alters active ceramide into an inactive glucosylceramide, leads to an increase in the threshold of apoptotic cells, a reduction in signaling potential, and a decrease in the amount of intracellular ceramide level (Itoh et al., 2003; Morjani et al., 2001; Senchenkov et al., 2001).

Presently, ceramide and PAX combination therapy in nanoform has been introduced to combat MDR cancer. These preparations have been vigorously explored *in vivo* and *in vitro* (van Vlerken et al., 2007, 2008). In this formulation, PAX was coupled with a pH-sensitive polymer poly(beta amino ester) (PbAE) for fast release and ceramide was tagged with PLGA for sustained release (van Vlerken et al., 2008). This combined nano-system is very effective in treating MDR *in vitro* as well as improving efficacy in MDR tumor xenografts in *nu/nu* mice (van Vlerken et al., 2007, 2008). Furthermore, the MDR phenotype in human breast cancer cell lines was successfully treated with glucosylceramide synthase antisense complementary DNA (cDNA) (Liu et al., 2001). In another approach to increase the level of ceramide inside the cell is to utilize siRNA for silencing glucosylceramide synthase, which leads to decreased P-gp expression in MDR cells. This will confirm the importance of ceramide in apoptotic modulation (Gouaze et al., 2005).

5.6 Physical Approaches to Overcome MDR Using Nanoparticles: Combination Drug and Thermal Therapies

Hyperthermia is a process in which the temperature of any organ rises between 105.8°F to 114.8°F. This process indirectly effects the MDR treatment as it can complement the efficacy of chemotherapy and can also increase radiation-induced tumor damage (Jordan et al., 1999). Due to the elevated temperature, tumor morphology changes, which leads to an increase in tissue perfusion, resulting in enhanced delivery of polymeric and liposomal nanoparticles (Kong et al., 2000). Various investigations show super paramagnetic Fe_3O_4 is used to induce hyperthermia. To deliver super paramagnetic Fe_3O_4, liposome nanoparticles are used since they have a confirmed higher accumulation in tumor cells. After the administration of the nanoparticles, a magnetic field was applied between 100–120 kHz for 30 minutes to reach temperatures between 104°F and 113°F. A study showing the combined therapy of hyperthermia with DOX-loaded liposomes in MDR cervical cancer cells reduced the IC50 value of the drug-loaded, targeted liposomes (Gaber, 2002). Temperature sensitive polymers are also being studied extensively. These polymers change their solubility in aqueous systems as the temperature changes. This temperature is known as the lower critical solution temperature (LCST). These polymeric properties can be utilized to synthesize self-assembling nanoparticles, such as poly(nisopropyl acrylamide) (pNIPAAm) which has a LCST of 89.6°F. pNIPAAm is used as the hydrophilic segment of a di-block copolymer micellar system, along with a hydrophobic segment of cholesterol or cholic acid (Kong et al., 2000). In addition to the variation in the structure of pNIPAAm, incorporating an amine tail can change the LCST value, which can control the solubility of the nanocarriers and their release profile (Ganta et al., 2008).

5.7 Conclusion

A new archetype in cancer management will be established with the aptitude to specifically target a particle type of cancer cell with functionally bio-relevant molecular signals, such that each type of cancer is identified and defined for its uniqueness with logical target prioritization using a computational process of algorithms.

The chronological development and eventual success of cancer chemotherapeutics based on nanotechnology is, however, dependent upon a number of parameters. A cumulative approach for the treatment of cancer using strategies with nanocarriers is most important for its success. Apart from the fundamental knowledge about the known targets and causes of MDR in cancer therapy, it is highly important to look for new targets. Knowing the recently developed targets helps to study the origin of the cancer and develop better MDR targets; such targets are significantly associated with the response of tumors and tumor vasculature to cancer chemotherapy.

Due to their identical size, nanomaterials and nanocarriers are similar to the cellular organelles such as mitochondria, lipoproteins, surface receptors, siRNA, miRNA, genes, and unknown proteins. Thus, nanocarriers can easily interfere with different bio-molecules. The flexibility of platforms or carriers developed with the help of nanotechnology has paved a fortunate and rapid invasion of these molecules into the epitome of basic cancer research and the specialty of MDR oncology. In this chapter, the approaches described to overcome MDR cancer therapy starts with liposomes that can be considered to be the simplest and most primary nanovectors for targeted delivery of cytotoxic active molecules. Each of the nanocarriers mentioned in this chapter are essential to achieve noteworthy progress in overcoming MDR-related oncology therapeutics and gain a better understanding of MDR development mechanisms.

REFERENCES

Acosta E. Bioavailability of nanoparticles in nutrient and nutraceutical delivery. Curr Opin Colloid Interface Sci. 2009;14(1):3–15.

Ahmad J, Kohli K, Mir SR, Amin S. Lipid based nanocarriers for oral delivery of cancer chemotherapeutics: An insight in the intestinal lymphatic transport. Drug Deliv Lett. 2013;3:38–46.

Alakhov VY, Klinski E, Li S, Pietrzynski G, Venne A, Batrakova E, Bronitch T, Kabanov A. Block copolymer-based formulation of doxorubicin: From cell screen to clinical trials. Colloids Surf B Biointerfaces. 1999;16:113–34.

Alakhov VY, Moskaleva E, Batrakova EV, Kabanov AV. Hypersensitization of multidrug resistant human ovarian carcinoma cells by Pluronic P85 block copolymer. Bioconjug Chem. 1996;7(2):209–16.

Allen SL, Lundberg AS. Amonafide: A potential role in treating acute myeloid leukemia. Expert Opin Investig Drugs. 2011;20:995–1003.

Altstadt TJ, Fairchild CR, Golik J, Johnston KA, Kadow JF, Lee FY, Rose WC, Vyas DM, Wong H, Wu MJ, Wittman MD. Synthesis and antitumor activity of novel c-7 paclitaxel ethers: Discovery of BMS-184476. J Med Chem. 2001;44(26):4577–83.

Arvizo R, Bhattacharya R, Mukherjee P. Gold nanoparticles: Opportunities and challenges in nanomedicine. Expert Opin Drug Deliv. 2010;7:753–63.

Ascione A, Cianfriglia M, Dupuis ML, Mallano A, Sau A, Pellizzari TF, Pezzola S, Caccuri AM. The gluta-thione S-transferase inhibitor 6-(7-nitro-2,1,3-benzoxadiazol-4-ylthio) hexanol overcomes the MDR1-P-glycoprotein and MRP1- mediated multidrug resistance in acute myeloid leukemia cells. Cancer Chemother Pharmacol. 2009;64:419–24.

Asaraf YG. Molecular basis of antifolate resistance. Cancer Metastasis Rev. 2007;26:153–81.

Astier A, Doat B, Ferrer MJ, Benoit G, Fleury J, Rolland A, Leverge R. Enhancement of adriamycin antitumor activity by its binding with an intracellular sustained-release form, polymethacrylate nanospheres, in U-937 cells. Cancer Res. 1988;48(7):1835–41.

Bates S, Kang M, Meadows B, Bakke S, Peter Choyke RN, Merino M, Goldspiel B, Chico I, Smith T, Chen C, Robey R, Bergan R, William FD, Fojo T. A Phase I study of infusional vinblastine in combination with the P-glycoprotein antagonist PSC 833 (valspodar). Cancer. 2001;92(6):1577–90.

Batrakova EV, Kabanov AV. Pluronic block copolymers: Evolution of drug delivery concept from inert nanocarriers to biological response modifiers. J Control Release. 2008;130(2):98–106.

Batrakova EV, Li S, Elmquist WF, Miller DW, Alakhov VY, Kabanov AV. Mechanism of sensitization of MDR cancer cells by Pluronic block copolymers: Selective energy depletion. Br J Cancer. 2001b;85(12):1987–97.

Batrakova EV, Li S, Li Y, Alakhov VY, Elmquist WF, Kabanov AV. Distribution kinetics of a micelle-forming block copolymer Pluronic P85. J Control Release. 2004a;100(3):389–97.

Batrakova EV, Li S, Li Y, Alakhov VY, Kabanov AV. Effect of Pluronic P85 on ATPase activity of drug efflux transporters. Pharm Res. 2004b:21(12);2226–33.

Batrakova EV, Li S, Vinogradov SV, Alakhov VY, Miller DW, Kabanov AV. Mechanism of Pluronic effect on P-glycoprotein efflux system in blood–brain barrier: Contributions of energy depletion and membrane fluidization. J Pharmacol Exp Ther. 2001a;299(2):483–93.

Borsellino N, Belldegrun A, and Bonavida B. Endogenous interleukin 6 is a resistance factor for cis-diammin-edichloroplatinum and etoposide-mediated cytotoxicity of human prostate carcinoma cell lines. Cancer Res. 1995;55:4633–9.

Borst P, Evers R, Kool M, Wijnholds J. A family of drug transporters: The multidrug resistance-associated proteins. J Natl Cancer Inst. 2000;92:1295–302.

Borst P, Elferink RO. Mammalian ABC transporters in health and disease. Annu Rev Biochem. 2002;71:537–92.

Braybrooke JP, Vallis KA, Houlbrook S, Rockett H, Ellmen J, Anttila M, Ganesan TS, Harris AL, Talbot DC, et al. Evaluation of toremifene for reversal of multidrug resistance in renal cell cancer patients treated with vinblastine. Cancer Chemother Pharmacol. 2000;46:27–34.

Brigger I, Dubernet C, Couvreur P. Nanoparticles in cancer therapy and diagnosis. Adv Drug Deliv Rev. 2002;54(5):631–51.

Brown CM, Reisfeld B, Mayeno AN. Cytochromes P450: A structure-based summary of biotransformations using representative substrates. Drug Metab Rev. 2008;40:1–100.

Brown JM, Wilson WR. Exploiting tumour hypoxia in cancer treatment. Nat Rev Cancer. 2002;4:437–47.

Burke AR, Singh RN, Carroll DL, Wood JC, D'Agostino RB, Ajayan PM, Torti FM, Torti SV. The resistance of breast cancer stem cells to conventional hyperthermia and their sensitivity to nanoparticle-mediated Photothermal therapy. Biomaterials. 2012;33:2961–70.

Cairns R, Papandreou I, Denko N. Overcoming physiologic barriers to cancer treatment by molecularly targeting the tumor microenvironment. Mol Cancer Res. 2006;4:61–70.

Chang PY, Peng SF, Lee CY, Lu CC, Tsai SC, Shieh TM, Wu TS, Tu MG, Chen MY, Yang JS. Curcumin-loaded nanoparticles induce apoptotic cell death through regulation of the function of MDR1 and reactive oxygen species in cisplatin-resistant CAR human oral cancer cells. Int J Oncol. 2013;43:1141–50.

Chapuy B, Koch R, Radunski U. Intracellular ABC transporter A3 confers multidrug resistance in leukemia cells by lysosomal drug sequestration. Leukemia. 2008;22:1576–86.

Chapuy B, Panse M, Radunski U. ABC transporter A3 facilitates lysosomal sequestration of imatinib and modulates susceptibility of chronic myeloid leukemia cell lines to this drug. Haematologica. 2009;94:1528–36.

Chau M, Christensen JL, Ajami AM, Capizzi RL. Amonafide, a topoisomerase II inhibitor, is unaffected by P-glycoprotein-mediated efflux. Leuk Res. 2008;32(3):465–73.

Chauhan SS, Liang XJ, Su AW. Reduced endocytosis and altered lysosome function in cisplatin-resistant cell lines. Br J Cancer. 2003;88:1327–34.

Chavanpatil MD, Khdair A, Gerard B, Bachmeier C, Miller WD, Shekhar MPV, Panyam J. Surfactant-polymer nanoparticles overcome P-glycoprotein-mediated drug efflux. Mol Pharm. 2007;4(5):730–8.

Chen AM, Zhang M, Wei D, Stueber D, Taratula O, Minko T, He H. Co-delivery of doxorubicin and Bcl-2 siRNA by mesoporous silica nanoparticles enhances the efficacy of chemotherapy in multidrug-resistant cancer cells. Small. 2009b;5:2673–7.

Chen R, Khormaee S, Eccleston ME, Slater NKH. The role of hydrophobic amino acid grafts in the enhancement of membrane-disruptive activity of pH-responsive pseudo-peptides. Biomaterials. 2009a;30:1954–61.

Chen YH, Tsai CY, Huang PY, Chang MY, Cheng PC, Chou CH, Chen DH, Wang CR, Shiau AL, Wu CL. Methotrexate conjugated to gold nanoparticles inhibits tumor growth in a syngeneic lung tumor model. Mol Pharm. 2007;4:713–22.

Cheong N, Madesh M, Gonzales LW. Functional and trafficking defects in ATP binding cassette A3 mutants associated with respiratory distress syndrome. J Biological Chem. 2006;281:9791–800.

Choi CH. ABC transporters as multidrug resistance mechanisms and the development of chemosensitizers for their reversal. Cancer Cell Int. 2005;5:30.

Coley HM, Twentyman PR, Workman P. 9-Alkyl, morpholinyl anthracyclines in the circumvention of multi-drug resistance. Eur J Cancer. 1990;26:665–7.

Colin DVA, Dubernet C, Nemati F, Soma E, Appel M, Ferté J, Bernard S, Puisieux F, Couvreur P. Reversion of multidrug resistance with polyalkylcyanoacrylate nanoparticles: Towards a mechanism of action. Br J Cancer. 1997;76(2):198–205.

Consoli U, Priebe W, Ling YH, Mahadevia R, Griffin M, Zhao S, Perez-Soler R, Andreeff M. The novel anthracycline annamycin is not affected by P-glycoprotein-related multidrug resistance: Comparison with idarubicin and doxorubicin in HL-60 leukemia cell lines. Blood. 1996;88(2):633–44.

Crossman LC, Druker BJ, Deininger MW, Pirmohamed M, Wang L, Clark RE. hOCT 1 and resistance to imatinib. Blood. 2005;106:1133–4.

Davies A, Jordanides NE, Giannoudis A, Lucas CM, Hatziieremia S, Harris RJ, Jørgensen HG, Holyoake TL, Pirmohamed M, Clark RE, Mountford JC. Nilotinibconcentration in cell lines and primary CD34 chronic myeloid leukemia cells is not mediated by active uptake or efflux by major drug transporters. Leukemia. 2009;23:1999–2006.

Davis ME. The first targeted delivery of siRNA in humans via a self-assembling, cyclodextrin polymer-based nanoparticle: From concept to clinic. Mol Pharm. 2009;6(3):659–68.

Davis ME, Chen ZG, Shin DM. Nanoparticle therapeutics: An emerging treatment modality for cancer. Nat Rev Drug Discov. 2008;7:771–82.

Deeley RG, Westlake C, Cole SP. Transmembrane transport of endo- and xenobiotics by mammalian ATP-binding cassette multidrug resistance proteins. Physiol Rev. 2006;86:849–99.

Denko NC. Hypoxia, HIF1 and glucose metabolism in the solid tumour. Nat Rev Cancer. 2008;8:705–13.

Dive C, Hickman JA. Drug-target interactions: Only the first step in the commitment to a programmed cell death? Br J Cancer. 1991;64:192–96.

Dong X, Mattingly CA, Tseng MT, Cho MJ, Liu Y, Adams VR, Mumper RJ. Doxorubicin and paclitaxel-loaded lipid-based nanoparticles overcome multidrug resistance by inhibiting P-glycoprotein and depleting ATP. Cancer Res. 2009;69(9):3918–26.

Doyle LA, Yang W, Abruzzo LV, Krogmann T, Gao Y, Rishi AK, Ross DD. A multidrug resistance transporter from human MCF-7 breast cancer cells. Proc Natl Acad Sci USA. 1998;95:15665–70.

Dreaden EC, Gryder BE, Austin LA, Tene DBA, Hayden SC, Pi M, Quarles LD, Oyelere AK, El SMA. Antiandrogen gold nanoparticles dual-target and overcome treatment resistance in hormone-insensitive prostate cancer cells. Bioconjug Chem. 2012;23:1507–12.

Duncan R. Polymer conjugates as anticancer nanomedicines. Nat Rev Cancer. 2006;6:688–701.

Duvvuri M, Krise JP. Intracellular drug sequestration events associated with the emergence of multidrug resistance: A mechanistic review. Front Biosci. 2005;10:1499–509.

Elrick MJ, Fluss S, Colombini M. Sphingosine, A product of ceramide hydrolysis, influences the formation of ceramide channels. Biophys J. 2006;91(5):1749–56.

Engin K, Leeper DB, Cater JR, Thistlethwaite AJ, Tupchong L, McFarlane JD. Extracellular pH distribution in human tumours. Int J Hyperthermia. 1995;11:211–6.

Engler JR, Hughes TP, and White DL. OCT-1 as a determinant of response to antileukemic treatment. Clin Pharmacol Ther. 2011;89:608–11.

Fabbro C, Ali BH, Da RT. Targeting carbon nanotubes against cancer. Chem Commun (Camb). 2012;48:3911–26.

Filipowicz W, Jaskiewicz L, Kolb FA, Pillai RS. Posttranscriptional gene silencing by siRNAs and miRNAs. Curr Opin Struct Biol. 2005;15:331–41.

Fojo AT, Ueda K, Slamon DJ. Expression of multidrug resistance gene in human tumors and tissues. Proc Natl Acad Sci USA. 1987;84:265–69.

Fojo T, Bates S. Strategies for reversing drug resistance. Oncogene. 2003;22:7512–23.

Ford JM, Hait WN. Pharmacology of drugs that alter multidrug resistance in cancer. Pharmacol Rev. 1990;42(3):155–99.

Gaber MH. Modulation of doxorubicin resistance in multidrug-resistance cells by targeted liposomes combined with hyperthermia. J Biochem Mol Biol Biophys. 2002;6(5):309–14.

Gaca S, Reichert S, Rödel C, Rödel F, Kreuter J. Survivin-miRNA-loaded nanoparticles as auxiliary tools for radiation therapy: Preparation, characterisation, drug release, cytotoxicity and therapeutic effect on colorectal cancer cells. J Microencapsul. 2012;29:685–94.

Ganta S, Amiji M. Coadministration of paclitaxel and curcumin in nanoemulsion formulations to overcome multidrug resistance in tumor cells. Mol Pharm. 2009;6:928–39.

Ganta S, Devalapally H, Shahiwala A, Amiji M. A review of stimuli-responsive nanocarriers for drug and gene delivery. J Control Release. 2008;126(3):187–204.

Gao S, Dagnaes-Hansen F, Nielsen EJ, Wengel J, Besenbacher F, Howard KA, Kjems J. The effect of chemical modification and nanoparticle formulation on stability and biodistribution of siRNA in mice. Mol Ther. 2009;17:1225–33.

Garcion E, Lamprecht A, Heurtault B, Paillard A, Aubert-Pouessel A, Denizot B, Menei P, Benoît JP. A new generation of anticancer, drug-loaded, colloidal vectors reverses multidrug resistance in glioma and reduces tumor progression in rats. Mol Cancer Ther. 2006;5(7):1710–22.

Gatouillat G, Odot J, Balasse E, Nicolau C, Tosi PF, Hickman DT, López-Deber MP, Madoulet C. Immunization with liposome-anchored pegylated peptides modulates doxorubicin sensitivity in P-glycoprotein-expressing p388 cells. Cancer Lett. 2007;257(2):165–71.

Gelmon KA, Latreille J, Tolcher A, Nicolau C, Tosi PF, Hickman DT, López-Deber MP, Madoulet C. Phase I dose-finding study of a new taxane, RPR109881a, administered as a one-hour intravenous infusion days 1 and 8 to patients with advanced solid tumors. J Clin Oncol. 2000;18(24):4098–108.

Goda K, Fenyvesi F, Bacsó Z. Complete inhibition of P-glycoprotein by simultaneous treatment with a distinct class of modulators and the UIC2 monoclonal antibody. J Pharmacol Exp Ther. 2006;320(1):81–8.

Gottesman MM, Fojo T, Bates SE. Multidrug resistance in cancer: Role of ATP-dependent transporters. Nature Rev Cancer. 2002;2:48–58.

Gouaze V, Liu YY, Prickett CS, Yu JY, Giuliano AE, Cabot MC. Glucosyl ceramide synthase blockade down-regulates P-glycoprotein and resensitizes multidrug-resistant breast cancer cells to anticancer drugs. Cancer Res. 2005;65(9):3861–7.

Grunicke H. Oncogenes and drug resistance. Paper presented at: Deutsche Krebsgesellschaft e.V. Materials, 23rd Congr; 1998 June 8–12; Berlin.

Gu YJ, Cheng J, Man CW, Wong WT, Cheng SH. Gold-doxorubicin nanoconjugates for overcoming multidrug resistance. Nanomedicine. 2012;8(2):204–11.

Gulbins E, Grassme H. Ceramide and cell death receptor clustering. Biochim Biophys Acta. 2002;1585(2–3):139–45.

Günther HS, Schmidt NO, Phillips HS, Kemming D, Kharbanda S, Soriano R, Modrusan Z, Meissner H, Westphal M, Lamszus K. Glioblastoma-derived stem cell-enriched cultures form distinct subgroups according to molecular and phenotypic criteria. Oncogene. 2008;27:2897–909.

Gupta PB, Onder TT, Jiang G, Tao K, Kuperwasser C, Weinberg RA, Lander ES. Identification of selective inhibitors of cancer stem cells by high-throughput screening. Cell. 2009;138:645–59.

Hakem R. DNA-damage repair; the good, the bad, and the ugly. Embo J. 2008;27:589–605.

Hamada H, Tsuruo T. Functional role for the 170-kDa to 180-kDa glycoprotein specific to drug resistant tumor cells as revealed by monoclonal antibodies. Proc Natl Acad Sci USA. 1986;83(20):7785–9.

Hanahan D, Weinberg RA. The hallmarks of cancer. Cell. 2000;100:57–70.

He Q, Gao Y, Zhang L. A pH-responsive mesoporous silica nanoparticles-based multidrug delivery system for overcoming multi-drug resistance. Biomaterials. 2011;32:7711–20.

Heldin CH, Rubin K, Pietras K. High interstitial fluid pressure – An obstacle in cancer therapy. Nat Rev Cancer. 2004;4:806–13.

Helleday T, Petermann E, Lundin C, Hodgson B, Sharma RA. DNA repair pathways as targets for cancer therapy. Nat Rev Cancer. 2008;8:193–204.

Hinds M, Deisseroth K, Mayes J, Altschuler E, Jansen R, Ledley F, Zwelling L. Identification of a point mutation in the topoisomerase II gene from a human leukemia cell line containing an amsacrine resistant form of topoisomerase II. Cancer Res. 1991;51:4729–31.

Holohan C, Van Schaeybroeck S, Longley DB, Johnston PG. Cancer drug resistance: An evolving paradigm. Nat Rev. 2013;13:714–26.

Hu CMJ, Zhang LF. Nanoparticle-based combination therapy toward overcoming drug resistance in cancer. Biochem Pharmacol. 2012;83:1104–11.

Huang IP, Sun SP, Cheng SH, Lee CH, Wu CY, Yang CS, Lo LW, Lai YK. Enhanced chemotherapy of cancer using pH-sensitive mesoporous silica nanoparticles to antagonize P-glycoprotein-mediated drug resistance. Mol Cancer Ther. 2011;10:761–9.

Huwyler J, Cerletti A, Fricker G, Eberle AN, Drewe J. By-passing of P-glycoprotein using immunoliposomes. J Drug Targets. 2002;10:73–9.

Ishikawa T, Ali-Osman F. Glutathione-associated *cis*-diamminedichloroplatinum(II) metabolism and ATP-dependent efflux from leukemia cells: Molecular characterization of glutathione-platinum complex and its biological significance. J Biol Chem. 1993;268:20116–25.

Itoh M, Kitano T, Watanabe M, Kondo T, Yabu T, Taguchi Y, Iwai K, Tashima M, Uchiyama T, Okazaki T. Possible role of ceramide as an indicator of chemoresistance: Decrease of the ceramide content via activation of glucosylceramide synthase and sphingomyelin synthase in chemoresistant leukemia. Clin Cancer Res. 2003;9(1):415–23.

Jabr-Milane LS, van Vlerken LE, Yadav S. Multi-functional nanocarriers to overcome tumor drug resistance. Cancer Treat Rev. 2008;34:592–602.

Jiang J, Yang SJ, Wang JC, Yang LJ, Xu ZZ, Yang T, Liu XY, Zhang Q. Sequential treatment of drug-resistant tumors with RGD-modified liposomes containing siRNA or doxorubicin. Eur J Pharm Biopharm. 2010;76:170–78.

Johnson KR, Fan W. Reduced expression of p53 and p21WAF1/CIP1 sensitizes human breast cancer cells to paclitaxel and its combination with 5-fluorouracil. Anticancer Res. 2002;22:3197–204.

Jonathan PM. Cancer multidrug resistance: Progress in understanding the molecular basis of drug resistance in cancer promises more effective treatments. Nat Biotechnol. 2000;18:IT18–20.

Kabanov AV, Batrakova EV, Alakhov VY. Pluronic block copolymers for overcoming drug resistance in cancer. Adv Drug Deliv Rev. 2002;54(5):759–79.

Kang KW, Chun MK, Kim O, Subedi RK, Ahn SG, Yoon JH, Choi HK. Doxorubicin-loaded solid lipid nanoparticles to overcome multidrug resistance in cancer therapy. Nanomedicine. 2010;6(2):210–3.

Kanwar JR, Mahidhara G, Kanwar RK. Nanotechnological based systems for cancer. In: Sebastian M, Ninan N, Elias E, Editors. Advances in nanoscience and nanotechnology, Volume 2: Nanomedicine and cancer therapies. New Jersey: Apple Academic Press, Inc.; 2012. p. 1.

Kanwar JR, Kamalapuram SK, Kanwar RK. Targeting survivin in cancer: The cell-signaling perspective. Drug Discov Today. 2011;16:485–94.

Kapoor K, Sim HM, Ambudkar SV. Multidrug resistance in cancer: A tale of ABC drug transporters. In: Bonavida B, editor. Molecular mechanisms of tumor cell resistance to chemotherapy: Targeted therapies to reverse resistance, Volume I: Resistance to targeted anti-cancer therapeutics. New York: Springer Science+Business Media; 2013.p. 2.

Kataoka K, Matsumoto T, Yokoyama M. Doxorubicin-loaded poly(ethylene glycol)-poly(beta-benzyl-L-aspartate) copolymer micelles: Their pharmaceutical characteristics and biological significance. J Control Release. 2000;64:143–53.

Kato S, Kimura M, Kageyama K, Tanaka H, Miwa N. Enhanced radiosensitization by liposome-encapsulated pimonidazole for anticancer effects on human melanoma cells. J Nanosci Nanotechnol. 2012;12(6):4472–7.

Kerr DJ, Graham J, Cummings J. The effect of verapamil on the pharmacokinetics of adriamycin. Cancer Chemother Pharmacol. 1986;18(3):239–42.

Kim D, Gao ZG, Lee ES, Bae YH. In vivo evaluation of doxorubicin-loaded polymeric micelles targeting folate receptors and early endosomal pH in drug-resistant ovarian cancer. Mol Pharm. 2009;6:1353–62.

Kitazono M, Sumizawa T, Takebayashi Y, Chen ZS, Furukawa T, Nagayama S, Tani A, Takao S, Aikou T, Akiyama S. Multidrug resistance and the lung resistance-related protein in human colon carcinoma SW-620 cells. J Natl Cancer Inst. 1999;91(19):1647–53.

Kobayashi T, Ishida T, Okada Y, Ise S, Harashima H, Kiwada H. Effect of transferrin receptor-targeted liposomal doxorubicin in P-glycoprotein-mediated drug resistant tumor cells. Int J Pharm. 2007;329(1–2):94–102.

Kolesnick RN, Kronke M. Regulation of ceramide production and apoptosis. Annu Rev Physiol. 1998;60:643–65.

Konan YN, Gurny R, Allemann E. State of art in the delivery of photo sensitizers for photo dynamic therapy. J Photochem Photobiol. 2001;66(2):89–106.

Kong G, Braun RD, Dewhirst MW. Hyperthermia enables tumor-specific nanoparticle delivery: Effect of particle size. Cancer Res. 2000;60(16):4440–5.

Kopecek J, Kopeckova P, Minko T, Lu Z. HPMA copolymer-anticancer drug conjugates: Design, activity and mechanism of action. Eur J Pharm Biopharm. 2000;50(1):61–81.

Koziara JM, Whisman TR, Tseng MT, Mumper RJ, et al. In vivo efficacy of novel paclitaxel nanoparticles in paclitaxel-resistant human colorectal tumors. J Control Release. 2006;112(3):312–9.

Kuffel MJ, Ames MM. Comparative resistance of idarubicin, doxorubicin and their C-13 alcohol metabolites in human MDR1 transfected NIH-3T3 cells. Cancer Chemother Pharmacol. 1995;36:223–6.

Kunjachan S, Blauz A, Mockel D, Theek B, Kiessling F, Erich T, Ulbricht K, Bloois LV, Storm G, Bartosz G, Rychlik B, Lammers T. Overcoming cellular multidrug resistance using classical nanomedicine formulations. Eur J Pharm Sci. 2012;45:421–8.

Kurata T, Shimada Y, Tamura T. Phase I and pharmacokinetic study of a new taxoid, RPR109881a, given as a 1-hour intravenous infusion in patients with advanced solid tumors. J Clin Oncol. 2000;18(17):3164–71.

Kwon G, Naito M, Yokoyama M. Block copolymer micelles for drug delivery: Loading and release of doxorubicin. J Control Release. 1997;48:195–201.

Laborde E. Glutathione transferases as mediators of signaling pathways involved in cell proliferation and cell death. Cell Death Differ. 2010;17:1373–80.

Lampidis TJ, Kolonias D, Podona T, Israel M, Safa AR, Lothstein L, Savaraj N, Tapiero H, Priebe W. Circumvention of P-GP MDR as a function of anthracycline lipophilicity and charge. Biochemistry. 1997;36:2679–85.

Lee AL, Dhillon SH, Wang Y, Shazib P, Weimin F, Yang YY. Synergistic anti-cancer effects via co-delivery of TNF related apoptosis-inducing ligand (TRAIL/Apo2L) and doxorubicin using micellar nanoparticles. Mol Bio Syst. 2011;7:1512–22.

Lee CC, Gillies ER, Fox ME, Guillaudeu SJ, Fréchet JM, Dy EE, Szoka FC. A single dose of doxorubicin-functionalized bow-tie dendrimers cures mice bearing C-26 colon carcinomas. Proc Natl Acad Sci USA. 2006;103(75):16649–54.

Lee ES, Na K, Bae YH. Doxorubicin loaded pH-sensitive polymeric micelles for reversal of resistant MCF-7 tumor. J Control Release. 2005;103:405–18.

Lee ES, Gao Z, Bae YH. Recent progress in tumor pH targeting nanotechnology. J Control Release. 2008;132:164–70.

Lehne G. P-glycoprotein as a Drug Target in the treatment of Multidrug Resistant Cancer. Curr Drug Targets. 2000;1:85–9.

Lei T, Srinivasan S, Tang Y, Fernandez FA, McGoron AJ. Comparing cellular uptake and cytotoxicity of targeted drug carriers in cancer cell lines with different drug resistance mechanisms. Nanomedicine. 2011;7:324–32.

Less JR, Posner MC, Boucher Y. Interstitial hypertension in human breast and colorectal tumors. Cancer Res. 1992;52:6371–4.

Li R, Wu R, Zhao L, Wu M, Yang L, Zou H. P-glycoprotein antibody functionalized carbon nanotube overcomes the multidrug resistance of human leukemia cells. ACS Nano. 2010;4:1399–408.

Liang GW, Lu WL, Wu JW, Zhao JH, Hong HY, Long C, Li T, Zhang YT, Zhang H, Wang JC. Enhanced therapeutic effects on the multi-drug resistant human leukemia cells in vitro and xenograft in mice using the stealthy liposomal vincristine plus quinacrine. Fundam Clin Pharmacol. 2008;22:429–37.

Libutti SK, Paciotti GF, Byrnes AA, Alexander HR, Gannon WE, Walker M, Seidel GD, Yuldasheva N, Tamarkin L. Phase I and pharmacokinetic studies of CYT-6091, a novel PEGylated colloidal gold-rhTNF nanomedicine. Clin Cancer Res. 2010;16:6139–49.

Lissianskaya A, Gershanovich M, Ognerubov N, Golubeva O, and Pratt J. Paclitaxel injectable emulsion: Phase 2a study of weekly administration in patients with platinum-resistant ovarian cancer. J Clin Oncol. 2004;22(14):5047.

Liu YY, Han TY, Giuliano AE, Cabot MC. Ceramide glycosylation potentiates cellular multidrug resistance. Faseb J. 2001;15(3):719–30.

Longley DB, Johnston PG. Molecular mechanisms of drug resistance. J Pathol. 2005;205:275–92.

Mangala LS, Zuzel V, Schmandt R, Leshane ES, Halder JB, Armaiz-Pena GN, Spannuth WA, Tanaka T, Shahzad MM, Lin YG, Nick AM, Danes CG, Lee JW, Jennings NB, Vivas-Mejia PE, Wolf JK, Coleman RL, Siddik ZH, Lopez-Berestein G, Lutsenko S, Sood AK. Therapeutic targeting of ATP7B in ovarian carcinoma. Clin Cancer Res. 2009;15:3770–80.

Manolitsas TP, Englefield P, Eccles DM, Campbell IG. No association of a 306 bp insertion polymorphism in the progesterone receptor gene with ovarian and breast cancer. Br J Cancer. 1997;75:1397–9.

Martin HL, Smith L, Tomlinson DC. Multidrug resistant breast cancer: Current perspectives. Breast Cancer (Auckl). 2014;6:1–13.

Mattheolabakis GB, Rigas B, Constantinides PP. Nanodelivery strategies in cancer chemotherapy: Biological rationale and pharmaceutical perspectives. Nanomedicine. 2012;7:1577–90.

McClements DJ, Decker EA, Park Y, Weiss J. Structural design principles for delivery of bioactive components in nutraceuticals and functional foods. Crit Rev Food Sci Nutr. 2009;49(6):577–606.

McClements DJ, Decker EA, Weiss J. Emulsion-based delivery systems for lipophilic bioactive components. J Food Sci. 2007;72(8):R109–24.

Mechetner EB, Roninson IB. Efficient inhibition of P-glycoprotein-mediated multidrug resistance with a monoclonal antibody. Proc Natl Acad Sci USA. 1992;89(13):5824–8.

Meijer C, Mulder NH, Timmer H, Sluiter WJ, Meersma GJ, DeVries EG. Relationship of to the cytotoxicity and resistance of seven platinum compounds. Cancer Res. 1992;52:6885–9.

Meng H, Liong M, Xia T, Li Z, Ji Z, Zink JI, Nel AE. Engineered design of mesoporous silica nanoparticles to deliver doxorubicin and P-glycoprotein siRNA to overcome drug resistance in a cancer cell line. ACS Nano. 2010;4:4539–50.

Merkel TJ, DeSimone JM. Dodging drug-resistant cancer with diamonds. Sci Transl Med. 2011;3(73):73ps8. (DOI:10.1126/scitranslmed.3002137).

Minematsu T, and Giacomini KM. Interactions of tyrosine kinase inhibitors with organic cation transporters and multidrug and toxic compound extrusion proteins. Mol Cancer Ther. 2011;10:531–39.

Minko T, Kopeckova P, Kopecek J. Chronic exposure to HPMA copolymer-bound Adriamycin does not induce multidrug resistance in a human ovarian carcinoma cell line. J Control Release. 1999a;59(2):133–48.

Minko T, Kopeckova P, Kopecek J. Comparison of the anticancer effect of free and HPMA copolymer-bound adriamycin in human ovarian carcinoma cells. Pharm Res. 1999b;16(7):986–96.

Minko T, Kopeckova P, Kopecek J. Efficacy of the chemotherapeutic action of HPMA copolymer-bound doxorubicin in a solid tumor model of ovarian carcinoma. Int J Cancer. 2000;86(1):108–17.

Minko T, Kopeckova P, Kopecek J. Preliminary evaluation of caspases-dependent apoptosis signaling pathways of free and HPMA copolymer-bound doxorubicin in human ovarian carcinoma cells. J Control Release. 2001;71(3):227–37.

Misra R, Sahoo SK. Coformulation of doxorubicin and curcumin in poly (D, L -lactide-coglycolide) nanoparticles suppresses the development of multidrug resistance in K562 cells. Mol Pharm. 2011;8:852–66.

Mohri M, Nitta H, Yamashita J. Expression of multidrug resistance-associated protein (MRP) in human gliomas. J Neurooncol. 2000;49:105–15.

Morjani H, Aouali N, Belhoussine R, Veldman RJ, Levade T, Manfait M. Elevation of glucosylceramide in multidrug-resistant cancer cells and accumulation in cytoplasmic droplets. Int J Cancer. 2001;94(2):157–65.

Muller RH, Mader K, Gohla S. Solid lipid nanoparticles (SLN) for controlled drug delivery – a review of the state of the art. Eur J Pharm Biopharm. 2000;50(1):161–77.

Murakami M, Cabral H, Matsumoto Y, Wu S, Kano MR, Yamori T, Nishiyama N, Kataoka K. Improving drug potency and efficacy by nanocarrier-mediated subcellular targeting. Sci Transl Med. 2011;3:64ra2. (DOI:10.1126/scitranslmed.3001385).

Nadali F, Pourfathollah AA, Alimoghaddam K, Nikougoftar M, Rostami S, Dizaji A, Azizi E, Zomorodipour A, Ghavamzadeh A. Multidrug resistance inhibition by antisense oligonucleotide against MDR1/mRNA in P-glycoprotein expressing leukemic cells. Hematology. 2007;12(5):393–401.

Nakamura K, Abu Lila AS, Matsunaga M, Doi Y, Ishida T, Kiwada H. A double-modulation strategy in cancer treatment with a chemotherapeutic agent and siRNA. Mol Ther. 2011;19(11):2040–7.

Nathwani SM, Butler S, Fayne D, McGovern NN, Sarkadi B, Meegan MJ, Lloyd DG, Campiani G, Lawler M, Williams DC, et al. Novel microtubule-targeting agents, pyrrolo-1,5-benzoxazepines, induce apoptosis in multi-drug-resistant cancer cells. Cancer Chemother Pharmacol. 2010;66:585–96.

Needham D, Anyarambhatla G, and Kong G. A new temperature sensitive liposome for use with mild hyperthermia: Characterization and testing in a human tumor xenograft model. Cancer Res. 2000;60:1197–201.

Nobili S, Landini I, Giglioni B, Mini E. Pharmacological strategies for overcoming multidrug resistance. Curr Drug Targets. 2006;7(7):861–79.

Nobili S, Landini I, Mazzei T, Mini E. Overcoming tumor multidrug resistance using drugs able to evade P-glycoprotein or to exploit its expression. Med Res Rev. 2011;32(6):1220–62.

Nori A, Kopecek J. Intracellular targeting of polymer-bound drugs for cancer chemotherapy. Adv Drug Deliv Rev. 2005;57(4):609–36.

Nussler V, Gieseler F, Zwierzina H, Gullis E, Pelka-Fleischer R, Diem H, Abenhardt W, Schmitt R, Langenmayer I, Wohlrab A, Kolb HJ, Wilmanns W. Idarubicin monotherapy in multiply pretreated leukemia patients: Response in relation to P-glycoprotein expression. Ann Hematol. 1997;74(2):57–64.

Omelyanenko V, Kopeckova P, Gentry C, Kopecek J. Targetable HPMA copolymer-adriamycin conjugates: Recognition, internalization and subcellular fate. J Control Release. 1998;53(1–3):25–37.

Ono C, Takao A, Atsumi R. Absorption, distribution, and excretion of DJ-927, a novel orally effective taxanein mice, dogs, and monkeys. Biol Pharm Bull. 2004;27(3):345–51.

Oshika Y, Nakamura M, Tokunaga T, Fukushima Y, Abe Y, Ozeki Y, Yamazaki H, Tamaoki N, Ueyama Y. Multidrug resistance-associated protein and mutant p53 protein expression in non-small cell lung cancer. Mod Pathol. 1998;11(11):1059–63.

Pakunlu RI, Wang Y, Saad M, Khandare JJ, Starovoytov V, Minko T. In vitro and in vivo intracellular liposomal delivery of antisense oligonucleotides and anticancer drug. J Control Release. 2006;114(2):153–62.

Paller CJ, Antonarakis ES. Cabazitaxel: A novel second-line treatment for metastatic castration-resistant prostate cancer. Drug Des Devel Ther. 2011;5:117–24.

Palmberg C, Koivisto P, Hyytinen E, Isola J, Visakorpi T, Kallioniemi O, Tammela T. Androgen receptor gene amplification in a recurrent prostate cancer after monotherapy with the nonsteroidal potent antiandrogen Casodex (bicalutamide) with a subsequent favorable response to maximal androgen blockade. Eur J Urol. 1997;31(2):216–9.

Pankhurst QA, Connolly J, Jones SK, Dobson J. Applications of magnetic nanoparticles in biomedicine. J Phys Appl Phys. 2003;36:R167–81.

Patil Y, Sadhukha T, Ma L, Panyam J. Nanoparticle-mediated simultaneous and targeted delivery of paclitaxel and tariquidar overcomes tumor drug resistance. J Control Release. 2009;136(1):21–9.

Patil Y, Panyam J. Polymeric nanoparticles for siRNA delivery and gene silencing. Int J Pharm. 2009;367(1–2):195–203.

Patil YB, Swaminathan SK, Sadhukha T, Ma L, Panyam J. The use of nanoparticle-mediated targeted gene silencing and drug delivery to overcome tumor drug resistance. Biomaterials. 2010;31(2):358–65.

Pepin X, Attali L, Domrault C, Gallet S, Metreau JM, Reault Y, Cardot PJ, Imalalen M, Dubernet C, Soma E, Couvreur P. On the use of ion-pair chromatography to elucidate doxorubicin release mechanism from polyalkylcyanoacrylate nanoparticles at the cellular level. J Chromatogr B Biomed Sci Appl. 1997;702(1–2):181–91.

Perche F, Torchilin VP. Recent trends in multifunctional liposomal nanocarriers for enhanced tumor targeting. J Drug Deliv. 2013;2013:1–32. (DOI:http://dx.doi.org/10.1155/2013/705265).

Perez-Soler R, Ling YH, Zou Y, Priebe W. Cellular pharmacology of the partially non-cross-resistant anthracycline annamycin entrapped in liposomes in KB and KB-V1 cells. Cancer Chemother Pharmacol. 1994;34(2):109–18.

Perlstein B, Finniss SA, Miller C, Okhrimenko H, Kazimirsky G, Cazacu S, Lee HK, Lemke N, Brodie S, Umansky F, Rempel SA, Rosenblum M, Mikklesen T, Margel S, Brodie C. TRAIL conjugated to nanoparticles exhibits increased anti-tumor activities in glioma cells and glioma stem cells in vitro and in vivo. Neuro Oncol. 2013;15(1):29–40.

Pettus BJ, Chalfant CE, Hannun YA. Ceramide in apoptosis: An overview and current perspectives. Biochim Biophys Acta. 2002;1585(2–3):114–25.

Pierre A, Léonce S, Pérez V, Atassi G. Circumvention of P-glycoprotein-mediated multidrug resistance by 16020-2: Kinetics of uptake and efflux in sensitive and resistant cell lines. Cancer Chemother Pharmacol. 1998;42(6):454–60.

Podsiadlo P, Sinani VA, Bahng JH, Kam NW, Lee J, Kotov NA. Gold nanoparticles enhance the anti-leukemia action of a 6-mercaptopurine chemotherapeutic agent. Langmuir. 2008;24(2):568–74.

Pruitt JD, Husseini G, Rapoport N, Pitt WG. Stabilization of Pluronic P-105 micelles with an interpenetrating network of N,N-diethylacrylamide. Macromolecules. 2000;33(25):9306–9.

Qian L, Zheng J, Wang K, Tang Y, Zhang X, Zhang H, Huang F, Pei Y, Jiang Y. Cationic core-shell nanoparticles with carmustine contained within O (6)-benzylguanine shell for glioma therapy. Biomaterials. 2013;34(35):8968–78.

Qiu LY, Wang RJ, Zheng C, Jin Y, Jin le Q. Beta-cyclodextrin-centered star-shaped amphiphilic polymers for doxorubicin delivery. Nanomedicine. 2010;5(2):193–208.

Rahman A, Husain SR, Siddiqui J, Verma M, Agresti M, Center M, Safa AR, Glazer RI. Liposome-mediated modulation of multidrug resistance in human HL-60 leukemia cells. J Natl Cancer Inst. 1992;84(24):1909–15.

Rajagopal A, Simon SM. Subcellular localization and activity of multidrug resistance proteins. Mol Biol Cell. 2003;14(8):3389–99.

Ran Y, Wang Y, Zhang Y, Wei D. Overcoming multidrug resistance in human carcinoma cells by an antisense-oligodeoxynucleotide doxorubicin conjugate in vitro and in vivo. Mol Pharm. 2008;5(4):579–87.

Raub TJ. P-glycoprotein recognition of substrates and circumvention through rational drug design. Mol Pharm. 2006;3(1):3–25.

Rejman J, Oberle V, Zuhorn IS, Hoekstra D. Size dependent internalization of particles via the pathways of clathrin- and caveolae-mediated endocytosis. Biochem J. 2004;377:159–69.

Rezzani R. Cyclosporine A and adverse effects on organs: Histochemical studies. Prog Histochem Cytochem. 2004;39:85–128.

Riganti C, Voena C, Kopecka J, Corsetto PA, Montorfano G, Enrico E, Costamagna C, Rizzo AM, Ghigo D, Bosia A. Liposome-encapsulated doxorubicin reverses drug resistance by inhibiting P-glycoprotein in human cancer cells. Mol Pharm. 2011;8(3):683–700.

Rose WC, Fairchild C, Lee FY. Preclinical antitumor activity of two novel taxanes. Cancer Chemother Pharmacol. 2001;47(2):97–105.

Ross DD, Doyle LA, Yang W, Tong Y, Cornblatt B. Susceptibility of idarubicin, daunorubicin, and their C-13 alcohol metabolites to transport-mediated multidrug resistance. Biochem Pharmacol. 1995;50(10):1673–83.

Saad M, Garbuzenko OB, Minko T. Co-delivery of siRNA and an anticancer drug for treatment of multidrug resistant cancer. Nanomedicine. 2008;3(6):761–76.

Samia AC, Chen X, Burda C. Semiconductor quantum dots for photodynamic therapy. J Am Chem Soc. 2003;125(51):15736–7.

Scheffer GL, Schroeijers AB, Izquierdo MA, Wiemer EA, Scheper RJ. Lung resistance-related protein/major vault protein and vaults in multidrug-resistant cancer. Curr Opin Oncol. 2000;12(6):550–6.

Schenck M, Carpinteiro A, Grassme H, Lang F, Gulbins E. Ceramide: Physiological and pathophysiological aspects. Arch Biochem Biophys. 2007;462(2):171–5.

Scott CA, Westmacott D, Broadhurst MJ, Thomas GJ, Hall MJ. 9-alkyl anthracyclines. Absence of cross-resistance to adriamycin in human and murine cell cultures. Br J Cancer. 1986;53(5):595–600.

Senchenkov A, Litvak DA, Cabot MC. Targeting ceramide metabolism- a strategy for overcoming drug resistance. J Natl Cancer Inst. 2001;93(5):347–57.

Sharma A, Mayhew E, Straubinger RM. Antitumor effect of taxol-containing liposomes in a taxol-resistant murine tumor model. Cancer Res. 1993;53(24):5877–81.

Shen F, Chu S, Bence AK, Bailey B, Xue X, Erickson PA, Montrose MH, Beck WT, Erickson LC. Quantitation of doxorubicin uptake, efflux, and modulation of multidrug resistance (MDR) in MDR human cancer cells. J Pharm Exp Ther. 2008;324(1):95–102.

Shieh MJ, Hsu CY, Huang LY, Chen HY, Huang FH, Lai PS. Reversal of doxorubicin resistance by multifunctional nanoparticles in MCF-7/ADR cells. J Control Release. 2011;152(3):418–25.

Shionoya M, Jimbo T, Kitagawa M, Soga T, Tohgo A. DJ-927, A novel oral taxane overcomes P-glycoprotein mediated multidrug resistance in vitro and in vivo. Cancer Sci. 2003;94(5):459–66.

Shuhendler AJ, Cheung RY, Manias J, Connor A, Rauth AM, Wu XY. A novel doxorubicin-mitomycin C coencapsulated nanoparticle formulation exhibits anti-cancer synergy in multidrug resistant human breast cancer cells. Breast Cancer Res Treat. 2009;119(2):255–69.

Sirova M, Mrkvan T, Etrych T, Chytil P, Rossmann P, Ibrahimova M, Kovar L, Ulbrich K, Rihova B. Preclinical evaluation of linear HPMA-doxorubicin conjugates with pH-sensitive drug release: Efficacy, safety, and immunomodulating activity in murine model. Pharm Res. 2010;27(1):200–8.

Siskind LJ. Mitochondrial ceramide and the induction of apoptosis. J Bioenerg Biomembr. 2005;37(3):143–53.

Siskind LJ, Kolesnick RN, Colombini M. Ceramide forms channels in mitochondrial outer membranes at physiologically relevant concentrations. Mitochondrion. 2006;6(3):118–25.

Soma CE, Dubernet C, Bentolila D, Benita S, Couvreur P. Reversion of multidrug resistance by co-encapsulation of doxorubicin and cyclosporine: A in polyalkylcyanoacrylate nanoparticles. Biomaterials. 2000;21(1):1–7.

Stavrovskaya AA. Cellular mechanisms of multidrug resistance of tumor cells. Biochem . 2000;65(1):95–106.

Steinbach D, Gillet JP, Sauerbrey A. ABCA3 as a possible cause of drug resistance in childhood acute myeloid leukemia. Clin Cancer Res. 2006;12:4357–63.

Stewart AJ, Canitrot Y, Baracchini E, Dean NM, Deeley RG, Cole SP. Reduction of expression of the multidrug resistance protein (MRP) in human tumor cells by antisense phosphorothioate oligonucleotides. Biochem Pharmacol. 1996;51(4):461–9.

Struckhoff AP, Bittman R, Burow ME, Clejan S, Elliott S, Hammond T, Tang Y, Beckman BS. Novel ceramide analogs as potential chemotherapeutic agents in breast cancer. J Pharmacol Exp Ther. 2004;309(2):523–32.

Subedi RK, Kang KW, Choi HK. Preparation and characterization of solid lipid nanoparticles loaded with doxorubicin. Eur J Pharm Sci. 2009;37(3–4):508–13.

Sugahara S, Kajiki M, Kuriyama H, Kobayashi TR. Complete regression of xenografted human carcinomas by a paclitaxel-carboxymethyl dextran conjugate (AZ10992). J Controlled Release. 2007;117(1):40–50.

Susa M, Iyer AK, Ryu K, Hornicek FJ, Mankin H, Amiji MM, Duan Z. Doxorubicin loaded polymeric nanoparticulate delivery system to overcome drug resistance in osteosarcoma. BMC Cancer. 2009;9:399. DOI:10.1186/1471-2407-9-399.

Swaminathan SK, Roger E, Toti U, Niu L, Ohlfest JR, Panyam J. CD133-targeted paclitaxel delivery inhibits local tumor recurrence in a mouse model of breast cancer. J Control Release. 2013;171(3):280–7.

Tang W, Xu H, Kopelman R, Philbert MA. Photodynamic Characterization and in vitro application of Methylene Blue containing nanoparticle platform. Photochem Photobiol. 2004;81(2):242–9.

Thierry A, Vige D, Coughlin S, Belli J, Dritschilo A, Rahman A. Modulation of doxorubicin resistance in multidrug-resistant cells by liposomes. FASEB J. 1993;7(6):572–9.

Thirumoorthy N, Manisenthil Kumar KT, Shyam Sundar A, Panayappan L, Chatterjee M. Metallothionein: An overview. World J Gastroenterol. 2007;13(7):993–6.

Torchilin VP. Antinuclear antibodies with nucleosome-restricted specificity for targeted delivery of chemotherapeutic agents. Ther Deliv. 2010;1(2):257–72.

Toyoda Y, Hagiya Y, Adachi T, Hoshijima K, Kuo MT, Ishikawa T. MRP class of human ATP binding cassette (ABC) transporters: Historical background and new research directions. Xenobiotica. 2008;38(7–8):833–62.

Tsuruo T, Iida H, Tsukagoshi S, Sakurai Y. Overcoming of vincristine resistance in p388 leukemia in vivo and in vitro through enhanced cytotoxicity of vincristine and vinblastine by verapamil. Cancer Res. 1981;41(5):1967–72.

Tsuruo T, Iida H, Yamashiro M, Tsukagoshi S, Sakurai Y. Enhancement of vincristine- andadriamycin-induced cytotoxicity by verapamil in p388 leukemia and its sublines resistant to vincristine and adriamycin. Biochem Pharmacol. 1982;31(19):3138–40.

van Vlerken LE, Amiji MM. Multi-functional polymeric nanoparticles for tumour-targeted drug delivery. Expert Opin Drug Deliv. 2006;3(2):205–16.

van Vlerken LE, Duan Z, Little SR, Seiden MV, Amiji MM. Biodistribution and pharmacokinetic analysis of Paclitaxel and ceramide administered in multifunctional polymer-blend nanoparticles in drug resistant breast cancer model. Mol Pharm. 2008;5(4):516–26.

van Vlerken LE, Duan Z, Seiden MV, Amiji MM. Modulation of intracellular ceramide using polymeric nanoparticles to overcome multidrug resistance in cancer. Cancer Res. 2007;67(10):4843–50.

Vaupel P, Mayer A. Hypoxia in cancer: Significance and impact on clinical outcome. Cancer Metastasis Rev. 2007;26:225–39.

Wang D, Tang J, Wang Y, Ramishetti S, Fu Q, Racette K, Liu F. Multifunctional nanoparticles based on a single-molecule modification for the treatment of drug-resistant cancer. Mol Pharm. 2013;10(4):1465–9.

Wang F, Wang YC, Dou S, Xiong MH, Sun TM, Wang J. Doxorubicin-tethered responsive gold nanoparticles facilitate intracellular drug delivery for overcoming multidrug resistance in cancer cells. ACS Nano. 2011;5(5):3679–92.

Wang X, Zhang R, Wu C, Dai Y, Song M, Gutmann S, Gao F, Lv G, Li J, Li X, Guan Z, Fu D, Chen B. The application of Fe3O4 nanoparticles in cancer research: A new strategy to inhibit drug resistance. J Biomed Mater Res A. 2007;80(4):852–60.

Wang RB, Kuo CL, Lien LL, Lien EJ. Structure activity relationship: Analyses of p-glycoprotein substrates and inhibitors. J Clin Pharm Ther. 2003;28(3):203–28.

Wang Y, Gao S, Ye WH, Yoon HS, Yang YY. Codelivery of drugs and DNA from cationic core-shell nanoparticles self-assembled from a biodegradable copolymer. Nat Mater. 2006;5:791–6.

Wang Z, Li Y, Ahmad A, Azmi AS, Kong D, Bannered S, Sarkar FH. Targeting miRNAs involved in cancer stem cell and EMT regulation: An emerging concept in overcoming drug resistance. Drug Resist Updates. 2010;13(4–5):109–18.

Watanabe T, Naito M, Kokubu N, Tsuruo T. Regression of established tumors expressing P-glycoprotein by combination of adriamycin, cyclosporine derivatives, and MRK16 antibodies. J Natl Cancer Inst. 1997;89(7):512–8.

Webster DM, Sundaram P, Byrne ME. Injectable nanomaterials for drug delivery: Carriers, targeting moieties, and therapeutics. Eur J Pharm Biopharm. 2013;84(1):1–20.

White DL, Saunders VA, Dang P, Engler J, Venables A, Zrim S, Zannettino A, Lynch K, Manley PW, Hughes T. Most CML patients who have a suboptimal response to imatinib have low OCT-1 activity: Higher doses of imatinib may overcome the negative impact of low OCT-1 activity. Blood. 2007;110(12): 4064–72.

White DL, Saunders VA, Dang P, Engler J, Zannettino AC, Cambareri AC, Quinn SR, Manley PW, Hughes TP. OCT-1-mediated influx is a key determinant of the intracellular uptake of imatinib but not nilotinib (AMN107): Reduced OCT-1 activity is the cause of low in vitro sensitivity to imatinib. Blood. 2006;108(2):697–704.

Wilson WR, Hay MP. Targeting hypoxia in cancer therapy. Nat Rev Cancer. 2011;11(6):393–410.

Wilson TR, Longley DB, Johnston PG. Chemoresistance in solid tumours. Ann Oncol. 2006;17(10):315–24.

Wojtkowiak JW, Verduzco D, Schramm KJ, Gillies RJ. Drug resistance and cellular adaptation to tumor acidic pH microenvironment. Mol Pharm. 2011;8(6):2032–8.

Wong HL, Rauth AM, Bendayan R, Manias JL, Ramaswamy M, Liu Z, Erhan SZ, Wu XY. A new polymer–lipid hybrid nanoparticle system increases cytotoxicity of doxorubicin against multidrug-resistant human breast cancer cells. Pharm Res. 2006;23(7):1574–85.

Wu J, Lu Y, Lee A, Pan X, Yang X, Zhao X, Lee RJ. Reversal of multidrug resistance by transferrin-conjugated liposomes co-encapsulating doxorubicin and verapamil. J Pharm Pharm Sci. 2007;10(3):350–7.

Yadav S, Van VLE, Little SR, Amiji MM, et al. Evaluations of combination MDR-1 gene silencing and paclitaxel administration in biodegradable polymeric nanoparticle formulations to overcome multidrug resistance in cancer cells. Cancer Chemother Pharmacol. 2009;63(4):711–22.

Yamashiro DJ, Maxfield FR. Regulation of endocytic processes by pH. Trends Pharmacol Sci. 1998;9:190–3.

Yang C, Wang J, Chen D, Chen J, Xiong F, Zhang H, Zhang Y, Gu N, Dou J. Paclitaxel-Fe3O4 nanoparticles inhibit growth of CD138(–) CD34(–) tumor stem-like cells in multiple myeloma-bearing mice. Int J Nanomedicine. 2013;8:1439–49.

Yang C, Xiong F, Wang J, Dou J, Chen J, Chen D, Zhang Y, Luo S, Gu N. Anti-ABCG2 monoclonal antibody in combination with paclitaxel nanoparticles against cancer stem-like cell activity in multiple myeloma. Nanomedicine. 2014;9(1):45–60.

Yang T, Choi MK, Cui FD, Lee SJ, Chung SJ, Shim CK, Kim DD. Antitumor effect of paclitaxel-loaded pegylated immunoliposomes against human breast cancer cells. Pharm Res. 2007b;24(12):2402–11.

Yang TF, Chen CN, Chen MC, Lai CH, Liang HF, Sung HW. Shell-crosslinked Pluronic 1121 micelles as a drug delivery vehicle. Biomaterials. 2007a;28(4):725–34.

Yang X, Deng W, Fu L, Blanco E, Gao J, Quan D, Shuai X. Folate-functionalized polymeric micelles for tumor targeted delivery of a potent multidrug-resistance modulator FG020326. J Biomed Mater Res A. 2008;86(1):48–60.

Yang Y, Wang Z, Li M, Lu S. Chitosan/pshRNA plasmid nanoparticles targeting MDR1 gene reverse paclitaxel resistance in ovarian cancer cells. J Huazhong Univ Sci Technolog Med Sci. 2009;29(2):239–42.

Yin Q, Shen J, Zhang Z, Yu H, Chen L, Gu W, Li Y. Multifunctional nanoparticles improve therapeutic effect for breast cancer by simultaneously antagonizing multiple mechanisms of multidrug resistance. Biomacromolecules. 2013;14(7):2242–52.

Yu H, Xu Z, Chen X, Xu L, Yin Q, Zhang Z, Li Y. Reversal of lung cancer multidrug resistance by pH-responsive micelleplexes mediating co-delivery of siRNA and paclitaxel. Macromol Biosci. 2014;14(1):100–9.

Zhang B, Liu M, Tang HK, Ma HB, Wang C, Chen X, Huang HZ. The expression and significance of MRP1, LRP, TOPOIIb, and BCL2 in tongue squamous cell carcinoma. J Oral Pathol Med. 2012b;41(2):141–8.

Zhang X, Guo S, Fan R, Yu M, Li F, Zhu C, Gan Y, et al. Dual-functional liposome for tumor targeting and overcoming multidrug resistance in hepatocellular carcinoma cells. Biomaterials. 2012a;33(29):7103–14.

Zhang X, Yang H, Gu K, Chen J, Rui M, Jiang GL. In vitro and in vivo study of a nanoliposomal cisplatin as a radiosensitizer. Int J Nanomedicine. 2011b;6:437–44.

Zhao S, Tan S, Guo Y, Huang J, Chu M, Liu H, Zhang Z. pH-sensitive docetaxel-loaded D-α-tocopheryl polyethylene glycol succinate-poly(β-amino ester) copolymer nanoparticles for overcoming multidrug resistance. Biomacromolecules. 2013;14(8):2636–46.

Zheng Z, Aojula H, Clarke D. Reduction of doxorubicin resistance in P-glycoprotein overexpressing cells by hybrid cell-penetrating and drug-binding peptide. J Drug Target. 2010;18(6):477–87.

Zhou J, Patel TR, Sirianni RW, Strohbehn G, Zheng MQ, Duong N, Schafbauer T, Huttner AJ, Huang Y, Carson RE, Zhang Y, Sullivan DJ, Piepmeier JM, Saltzman WM. Highly penetrative, drug-loaded nano-carriers improve treatment of glioblastoma. Proc Natl Acad Sci USA. 2013;110:11751–6.

Zhou SF. Structure, function and regulation of P-glycoprotein and its clinical relevance in drug disposition. Xenobiotica. 2008;38(7–8):802–32.

6

Nanoemulsions in Non-Invasive Drug Delivery Systems

Ratna Jyoti Das, Subhabrata Ray, Paulami Pal, Anup Kumar Das,
and Bhaskar Mazumder

CONTENTS

ABSTRACT: Nanoemulsions (often referred to as submicron, ultrafine, or miniemulsions) are kineti-cally stable, heterogeneous dispersions with a mean droplet size ranging between a few hundred nanome-ters. These water-oil-surfactant ternary systems seem to be virtually stable in the thermodynamic sense due to their long-term physical stability. The nano-size of the droplets allows them to deposit uniformly on skin surfaces, resist gravitational effects and coalescence, as well as impart unique properties including optical clarity. Nanoemulsions are suitable for the efficient delivery of active ingredients through topical and transdermal routes. The large surface area and the low surface and interfacial tension of the systems facilitates superior penetration of drugs through the rough skin surface. By virtue of these unique quali-ties, nanoemulsions are believed to be efficient drug delivery vehicles for a variety of molecules, especially BCS class II and IV drugs. These systems could be applied through other non-invasive routes, namely ocular, nasal, ungual, and pulmonary to overcome the inherent challenges of drug delivery. Throughout this chapter, the authors discuss the formulation aspects of nanoemulsions and how these complex systems might overcome biopharmaceutical and other challenges of non-invasive drug delivery.

KEY WORDS: *Nanoemulsion, miniemulsion, kinetic stability, non-invasive drug delivery, thermo-dynamics, BCS classification, biopharmaceutical challenges*

6.1 Introduction with Definition

Scientifically, it is prudent to say that this era in human evolution is the era of nanotechnology. We are deal-ing with nanotech products in our daily life, from computer microchips (or shall we say nanochips) to the nanopharmaceuticals and everything in between. The novelty in drug delivery research moved on many years back from "micro" systems to "nano" systems. Right from the liposomes to nanoparticles, coarse emulsions to microemulsions, the trend has always been to explore the smallest, as was aptly foreseen by the Nobel Laureate physicist, Sir Richard Feynman, way back in 1959 in *There's Plenty of Room at the Bottom* (Feynman, 1959). Maybe the culmination of that vision in the pharmaceutical field is the concept of nano-structured systems of drug delivery, especially nanoemulsions from a practical, translational point of view.

   Nanoemulsions are kinetically stable, isotropic, colloidal dispersions stabilized by surfactant molecules (Anton et al., 2008). Often referred to as "miniemulsions", "fine dispersed emulsions", "ultrafine emul-sions", or "submicron emulsions", nanoemulsions can be water-in-oil (w/o) or oil-in-water (o/w) types, whose particle size ranges between 20–500 nm (Anton et al., 2008; Kong et al., 2011). They are optically transparent, translucent, or occasionally milky white in appearance, and cannot be created spontaneously as they are non-equilibrium systems. Nanoemulsions are capable of entrapping lipophilic drugs and can resist collisions between droplets due to Brownian motion, resulting in excellent kinetic stability for the colloidal systems (Figure 6.1). Even though nanoemulsions are kinetically stable, practically they are thermodynamically unstable dispersions owing to the higher free energy of the formulation state than that in the phase-separated state. However, due to their nano-size, nanoemulsions acquire unique "virtual

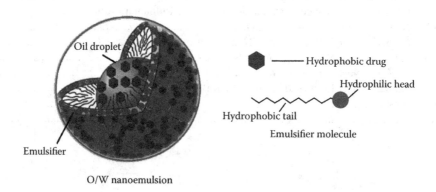

O/W nanoemulsion

FIGURE 6.1   Structure of Nanoemulsion.

stability", imparting long-term physical stability (Tadros et al., 2004). To get better control of particle size, different combinations of surfactants have been used during formulation, which in turn also improved the stability of nanoemulsions in terms of their zeta potential (Zhou et al., 2008).

Nanoemulsions are a useful means of encapsulation, protection, and delivery of poorly water-soluble, bioactive components for applications in both pharmaceuticals and functional foods (Chen et al., 2011; McClements, 2010). Nanoemulsions are capturing an increasing amount of attention from present-day researchers due to their inimitable structure and properties, such as an enormously smaller particle size, superior surface area to volume ratio, increasing drug solubility and entrapment efficiency, and also their high dissolution rate. This type of formulation helps to protect active compounds against degradation, improves the permeability of molecules across intestinal membranes, and, finally, improves their bioavailability (Bali et al., 2010; Ganta and Amiji, 2009; Kuo et al., 2008; Singh and Vingkar, 2008; Tagne et al., 2008; Vyas et al., 2008). They also help to reduce the dosing frequency of hydrophobic drugs and reduces their systemic toxicity.

About 40% of the recently discovered drugs or bioactive molecules proposed for oral, parenteral, or transdermal routes are lipophilic compounds that reduce bioabsorption, which is dissolution-rate limited (Sivakumar et al., 2014). For this reason, these drugs suffer from bioavailability related issues, hindering their clinical translations. Conventionally, BCS (Biopharmaceutical Classification System) III and IV drugs are formulated as tablets, capsules, and injectables, etc., but these techniques with good reproducibility suffer from several disadvantages, like poor drug absorption through the gastrointestinal (GI) tract due to the larger particle size, erratic drug release, non-specific drug delivery, and higher doses and dosing frequency. Anatomically, the GI tract is a nine-meter long, flexible, hollow tube and gastric fluid is basically a pale, digestive fluid, secreted from the mucous membrane of the abdomen, containing mainly hydrochloric acid along with mucin, pepsin, and rennin.

Nanoemulsions, contrary to conventional dosage forms, offer many crucial benefits. By virtue of their nanometric droplet size, nanoemulsions provide particularly low interfacial tension and exceptionally large surface areas, leading to a drastic enhancement of drug absorption in the GI tract (Doh et al., 2013). Thus, nanoemulsion technology improves the therapeutic index of active moieties with its higher payload, protecting the therapeutic agents from physiological barriers and also enabling the development and synthesis of novel classes of polysaccharide nanoparticles (Burapapadh et al., 2012; Ethirajan et al., 2008; Manchun et al., 2014). In nanoemulsion, the drug is encapsulated within the internal oil droplets to protect the therapeutic activity of the drug and also the particle integrity. It is well stabilized by suitable surfactant–co-surfactant mixes so as to prevent partitioning of the drug from the internal phase to the external phase during delivery. Due to their superior ratio of surface area to volume, nanoemulsions provide a high concentration gradient across the membrane, facilitating drug absorption and improved biodistribution and pharmacokinetics of therapeutics, thereby reducing the systemic toxicity of drugs by means of better drug targeting. They improve the solubility and stability of hydrophobic compounds and a variety of therapeutic agents, like peptides and oligonucleotides, and make them suitable for parenteral, per-oral, transdermal, and other non-invasive routes of administration. In particular, the bioavailability of encapsulated compounds in an emulsion increases by decreasing the emulsion droplet size to nanometric range (Acosta, 2009). It has been observed that with the reduction of emulsion droplet size below the size of cells, both drug absorption and particle uptake increases through the intestinal walls by the passive transport mechanism (Luo et al., 2006). Various studies have revealed that nanoemulsions with positive zeta potential successfully improved the permeation of anti-inflammatory drugs through the transdermal route in comparison to the conventional gels and emulsions; this may be accountable to their superior interaction with skin epithelia, which is negatively charged (Sandig et al. 2013; Shakeel et al., 2007). Therefore, the droplet charge may be an important determinant in the success of bioavailability enhancement. Overall, nanoemulsions seem to be an efficient, practical, lithe formulation strategy, projecting them as capable candidates for pharmaceutical applications.

6.2 Principles of Nanoemulsion Formation

In thermodynamics, free energy can be defined as the energy of a physical system that can be converted to do work, i.e., the internal energy of the system minus the energy that cannot be used to create work.

Nanoemulsion is considered to be thermodynamically unstable since the free energy of the nano-sized, colloidal dispersion is always greater than the free energy of oil and water co-existing as separate phases (Figure 6.2). Kinetic stability of nanoemulsions can be achieved only if a significant, large energy barrier exists between the two states; this state of kinetic stability of the nanoemulsion can be better termed as "metastable". Hence, with sufficient time, a nanoemulsion will always crack owning to its thermodynamic instability. However, the rate at which this breakdown will occur depends on two main factors: first, the specific mass process (rearrangement of molecules of a particle system to form a new configuration, i.e., change of a system from one state to another) and, second, the height of energy barriers (activation energy) between the nanoemulsion state and separated states (Figure 6.2) (McClements, 2012). Again, the height of the energy barrier directs the kinetic stability of this nano-sized colloidal system; the energy barrier is directly proportional to the stability of the nanoemulsion. Classically, this energy barrier is required to be greater than 20 kT to render a stable nanoemulsion (McClements, 2005). Factors that determine the height of this energy barrier are certain physicochemical phenomena that push the droplets of the nanoemulsion apart, such as colloidal interactions existing between droplets and repulsive hydrodynamics within the system. The rate at which a nanoemulsion breaks down is ascertained by the rate of recurrence of droplet collision with one another due to factors like gravitational forces, Brownian motion, or applied shear (McClements, 2012). A detailed argument of these factors is beyond the scope of the current chapter, and can be found by the interested readers in the given references Hunter (1989), Israelachvili (1992), and McClements (2005).

Based on thermodynamic theory, the colloidal dispersion of oil in water can be discussed and explained with the help of Gibbs free energy (ΔG) as expressed in the following equation (Equation 6.1) (Anton et al., 2008):

$$\Delta G = \mu_2 - \mu_1 \tag{6.1}$$

In Equation 6.1, μ_1 and μ_2 are the chemical potentials of state 1 and 2 respectively. Now, in cases where $\Delta G > 0$, it means the phase transformation is thermodynamically impossible. When $\Delta G = 0$, it can be said that the oil and water mixture is in equilibrium in terms of thermodynamics. But, when $\Delta G < 0$, the transformation from heterogeneous, multi-phase oil and water mixture to single phase emulsion is possible, which is spontaneous and also reversible. Therefore, when $\Delta G < 0$, one bulk phase could be dispersed into another bulk phase.

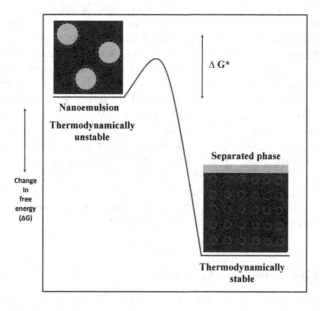

FIGURE 6.2 Nanoemulsion formation free energy.

Let us consider a system that is in equilibrium between the colloidal nanodispersion and the separate phases, i.e., oil and an aqueous surfactant solution (Hunter, 1989). Now, the free energy required for formation of a colloidal dispersion ($\Delta G_{formation}$) from the separate phases can be represented in terms of configuration entropy ($-T \Delta S_{conf}$) and interfacial free energy (ΔG_I), as in Equation (6.2) :

$$\Delta G_{formation} = \Delta G_I - T \Delta S_{conf} \tag{6.2}$$

When temperature, pressure, and interfacial chemical potential are kept constant:

$$\Delta G_I = \gamma \Delta A \tag{6.3}$$

Here, γ is the interfacial tension and ΔA is the increase in contact area between the oil and aqueous phases in Equation 6.3 and configuration entropy depends on the possible permutation of arrangement of the oil phase within the system.

The "ΔG_I" term always remains positive, as the increase in contact area and interfacial tension is always positive. Hence, this interfacial free energy forever hinders the formation of nanoemulsions. On the other hand, the term "ΔS_{conf}" is always considered to be negative because the number of possible arrangements of the oil phase in an emulsified state is always greater than that in a non-emulsified state, and consequently it constantly favors the formation of nanoemulsions.

The configurational entropy change, ΔS_{conf}, can be described by a statistical analysis of the number of configurations the oil droplets can develop within the nano-sized colloidal dispersion and separated states, as in Equation 6.4 (Overbeek, 1978):

$$\Delta S_{conf} = -nk_B \left[\ln\phi + \left\{ \left(\frac{1-\phi}{\phi} \right) \ln(1-\phi) \right\} \right] \tag{6.4}$$

Here, n is the number of droplets of the dispersed phase, k_B is the Boltzmann constant, and ϕ is the dispersed phase volume fraction. By combining the above two equations, i.e., Equation 6.3 and 6.4, Gibbs free energy of the nanoemulsion formation can be expressed as follows in Equation 6.5:

$$\Delta G_{formation} = \gamma \Delta A - T \Delta S_{conf} \tag{6.5}$$

In Equation 6.5, $\Delta G < 0$ is only possible when the value of the $\gamma \Delta A$ term is much lower than the $T \Delta S_{conf}$ term. Thus it is of utmost importance for a surfactant to produce an interfacial tension which is especially low. Here, it is assumed that the interfacial tension of a surfactant monolayer around an oil droplet is the same as that at a planar oil–water interface. Nevertheless, the interfacial tension is believed to depend on the curvature of a surfactant monolayer; it decreases as the monolayer advances towards its optimum curvature, which in turn depends on the molecular geometry of the surfactant molecules involved (McClements, 2012). With the help of the following simple phenomenological equation, it can be shown how interfacial tension depends on the curvature of droplet as in Equation 6.6:

$$\gamma = \gamma_0 + \left[(\gamma_\infty - \gamma_0) \frac{(R_0 - R)^2}{R_0^2 + R^2} \right] \tag{6.6}$$

Here
γ_0 = interfacial tension when the surfactant monolayer attends its optimum curvature
γ_∞ = interfacial tension at a planar oil–water interface
R_0 = radius of the droplet when the surfactant monolayer is at its optimum curvature
R = radius of droplet at planar oil–water interface.

It is worth mentioning here that the process of nanoemulsion formation requires a severe external energy, usually supplied by using a high-pressure homogenizer, high-speed homogenizer, Microfluidizer, or ultrasonication, making the process non-spontaneous. The high amount of external energy needed for the formation of nanodroplets of a particular average size can be described by the Young–Laplace Equation (Equation 6.7):

$$p = \gamma \left[\frac{1}{R_1} - \frac{1}{R_2} \right]$$

(6.7)

Here, p is the Laplace pressure, R_1 and R_2 are the principal radii of curvature of the droplets, and γ indicates interfacial tension. Theoretically this is possible, provided the subjected shear rate is suitably much higher than the Laplace pressure of tiny droplets. As a result, the need for a surfactant to significantly reduce the interfacial tension, γ, is evident. With the presence of this surfactant, applied stress can be minimized and the twist or deformation of tiny droplets can be evaded (Sivakumar et al., 2014). The Weber number (We), a dimensionless parameter, has been widely used to describe droplet deformation, which is given below in Equation 6.8:

$$We = \frac{G_\eta \eta_c R}{2\gamma}$$

(6.8)

Here, G_η is the velocity gradient, η_c is the viscosity of the continuous phase, R is the radius of the curvature of the droplet, and γ is interfacial tension. It can be inferred from here that with an increase in We, the possibility for the deformation of droplets also increases and the breakdown of droplets occurs when it reaches a specific value termed the "critical Weber number".

6.3 Synthesis and Basic Processes for the Fabrication of Nanoemulsions

The basic requirements for fabrication of nanoemulsions are four primary components: (1) oil phase, (2) aqueous phase, (3) surfactant, and (4) high energy. Figure 6.3 shows the components involved and their role in the generation of nanoemulsions.

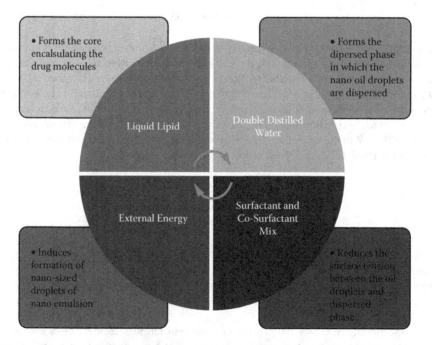

FIGURE 6.3 Components of Nanoemulsion.

As discussed in the previous section, nanoemulsions cannot be formed spontaneously in the absence of high external energy. The fabrication of tremendously fine emulsions is a major challenge, as the pharmacological properties of such miniemulsions can be mostly influenced by their droplet size and size distribution, surface morphology, zeta potential, and viscosity, etc. The chemical potential of nano-emulsions depends mainly on the choice of surfactants and co-surfactants used during formulation. But the drug encapsulation efficiency of a formulation depends on the solubility of the active molecule in the liquid lipids or oils used for the fabrication of the nanoemulsion. A list of the most commonly used oils and surfactants are tabulated in Table 6.1 for convenience.

As tremendous shear must be applied to overcome the Laplace pressure for the purpose of breaking the larger droplets into nano-size droplets, the formation of nanoemulsion is impossible with traditional emulsification systems. Typically, different homogenizing technologies such as ultrasonication, high speed agitation, colloidal mills, static mixers, and high shear mixers have been employed to produce nanoemulsions. Both the physicochemical and organoleptic properties of nanoemulsions are a function of the method used to produce it, such as their stability, surface morphology, texture, and rheological behavior, which are significantly affected by the droplet size and its distribution in the nanoemulsion. It has been observed that the smaller the particle size the better the formulation comes out, e.g., a creamier mouth feel and higher stability in terms of shelf-life (Overbeek, 1978).

Nanoemulsion preparation can be broadly classified under two heads, namely (1) low-energy emul-sification methods and (2) high-energy emulsification methods. Low-energy emulsification covers processes like the phase solvent-diffusion method, spontaneous emulsification method or microemul-sion technique, and phase inversion temperature technique (PIT), whereas high-energy emulsification includes high-pressure homogenization, double emulsion technique, Microfluidization, ultrasonica-tion, and high-speed homogenization due to the larger mechanical energy generated by them. Among the low-energy consuming methods, the most efficient is the PIT technique, followed by the micro-emulsion technique. Both these methods generate the chemical potential (zeta potential) required to stabilize the system by using a surfactant–co-surfactant mix. Nevertheless, quite a few limitations are associated with both these methods, such as the use of a large quantity of surfactant which may lead to cytotoxicity at a molecular level, the requirement for careful selection of both surfactant and co-surfactant for long-term stability of the system, a high polydispersity index (PDI) leading to insta-bility with long shelf-life, and the moderately heavy, time consuming, and tedious workload required to identify the inversion temperature of the system and develop of the phase diagram. In contrast, nanoemulsions prepared by high-energy involvement, like high-speed homogenization followed by ultrasonication, propose a higher stability of the formulation with a reduced Ostwald ripening rate. However, for the purpose of obtaining a formulation with smaller droplet sizes, higher energy input is required for further droplet break off (Shi et al., 2011), but the time duration used for the homogeniza-tion and sonication with a probe sonicator is crucial, as over doing it may lead to particle agglomera-tion resulting in larger particle size.

The best method to produce a good quality nanoemulsion with uniform particle size, reproducibility, and stability is high-pressure homogenization. In this method, the liquid pre-mix is passed through a narrow orifice of a few microns under high pressure (around 100–2,000 bar) at high velocity (over 1,000 km/hr) for a very short distance. The optimization of particle size mainly depends on the applied shear stress and cavitation force. There are two methods of nanoemulsion production by this process, namely hot homogenization and cold homogenization (Mehnert and Mader, 2001).

Thermolabile substances cannot be subjected to the hot homogenization method as the drug and the liquid lipid are heated together in this method, before dispersing it in a hot aqueous surfactant mixture at the same temperature. High temperature results in a decrease of particle size due to a lower viscosity in the inner phase, but at the same time may cause drug and lipid degradation. For the process, a number of homogenization cycles and high pressure thus applied must be optimized during formulation devel-opment, or else an increase in either or both might result in an increase in particle size due to particle coalescence as a result of the high kinetic energy of the globules. The product obtained is a hot nano-emulsion, which is subsequently cooled to room temperature.

Cold homogenization, on the other hand, is carried out at room temperature with liquid lipids, i.e., oil followed by high pressure milling of the cold pre-mix. The quality of the pre-mix greatly affects

TABLE 6.1

List of Commonly Used Excipients in the Development of Nanoemulsions

Liquid Lipids	Stabilizers	Aqueous Phase
Ajowan oil	Brij 72, 721	Double distilled water,
Almond oil	Caseinate	di-ionized water,
Arachis oil	Cremophor EL	Millipore water
Avocado oil	Deoxycholic acid	
Camputol MCM EP (glycerol monocaprylocaprate, type 1)	Dodecyltrimethylammonium bromide (DTAB)	
Captex 200P (mixed diesters of caprylic/capric acids on propylene glycol)	Egg lecithin	
Captex 355 EP (triglycerides of caprylic/capric acid)	Ethanol	
Castor oil	Ethylene glycol	
Coconut oil	Glycerine	
Corn oil	Gum arabic	
D-limonene	Hydroxypropyl-β-cyclodextrin	
Ethyl oleate	Isopropyl alcohol	
Fennel oil	Isostearyl isostearate	
Flaxseed oil	Labrasol	
Ginger oil	Labrasol	
IsoparM (Exxon)	Lipoid E80	
Isopropyl myristate	Modified starch	
Isopropyl palmitate	Organosilicones	
Isopropyl palmitate	Phospolipon	
Jojoba oil	Plurol oleique	
Labrafil	Poloxamar	
Lauroglycol 90	Polyethylene glycol	
Lauroglycol 90	Polyvinyl alcohol	
Lemongrass oil	Polyvinylpyrrolidone K25	
Linseed oil	Propylene glycol	
Liquid Paraffin	Sodium dodecyl sulfate (SDS)	
MCT oil	Sodium oleate	
Medium chain triglyceride	Sodium stearate	
Miglyol 812	Sorbitol	
Octyl palmitate	Soya lecithin	
Oil of eucalyptus	Span 20, 60, 80, 85	
Oleic acid	Stearylamine	
Olive oil	Stearylamine	
Pine nut oil	TPGS	
Sesame oil	Transcutol	
Silicone oil	Tween 20, 60, 65, 80, 85	
Soybean oil	Tween 80	
Sunflower oil	ULTROLs L70 (ethoxylated lauryl ether)	
Vegetable oil	Xanthan, pectin, carrageenan, sodium alginate	

the quality and nature of the final formulation. Naturally, in comparison with the hot homogenization technique, cold homogenization yields dispersion with poly-dispersed formulations with a larger particle size. Still, it helps to bypass problems related to hot homogenization, like temperature induced drug degradation, drug distribution in aqueous phase, and chances of crystallization, etc.

Nanoemulsions can be characterized by different *in vitro* evaluation parameters, such as particle size and distribution, zeta potential, surface morphology studies, rheological studies, drug loading, drug entrapment efficiency, and, lastly, drug release studies.

6.4 Application of Nanoemulsions in Pharmaceuticals

6.4.1 Cosmetics

The importance of nanoemulsions as good vehicles is increasing for the delivery of cosmetics in a controlled manner and to optimize dispersions of medicaments in targeted skin layers. Nanoemulsions, having a lipophilic interior, are suitably used for the transport of lipophilic drugs. Nanoemulsions like liposomes enhance the penetration of active ingredients through the skin and thus increases their concentration where required. Nanoemulsions allow better delivery of the active ingredients to the skin due to their small droplet size with high surface area. There is also an increasing interest in nanoemulsions because they have their own bioactive effects. Nanoemulsions strengthen the barrier function of the skin and thereby may lower trans-epidermal water loss. This type of formulation is more advantageous than microemulsion in cosmetics because of the absence of properties like sedimentation, creaming, flocculation, or coalescence. The use of high-energy equipment during the manufacturing of nanoemulsions helps to avoid or to reduce the incorporation of potentially irritating surfactants. Formulations of nanoemulsions for cosmetics which are polyethylene glycol (PEG)-free have been developed and show optimum stability (Charles and Attama, 2011; Rutvij et al., 2011; Subhashis et al., 2011; Surbhi and Kumkum, 2012; Yashpal et al., 2013).

6.4.2 Antimicrobials

Antimicrobial nanoemulsions (o/w droplets) range from 200–600 nm in size and are made up of oil and water stabilized by surfactants and alcohol. Nanoemulsions show a broad spectrum of activity against bacteria such as *Escherichia coli*, *Staphylococcus aureus*, salmonella linked viruses (e.g., human immunodeficiency virus (HIV)), fungi (like *Candida*), dermatophytes, herpes simplex, and spores (including anthrax) (Charles and Attama, 2011; Rutvij et al., 2011; Subhashis et al., 2011; Yashpal et al., 2013).

6.4.3 Prophylactic in Bioterrorism Attack

Nanoemulsions have antimicrobial activity, therefore research is ongoing to protect people exposed to biological attacks, like Ebola and anthrax, by using nanoemulsions as a prophylactic medicated dosage form in a human protective treatment. In December 1999, the US Army checked broad-spectrum nanoemulsions on surfaces for the decontamination of anthrax spores. They have also experimentally used it as a chemical decontamination agent. Nanoemulsions may be applied to contaminated heels and wounds to salvage limbs and are being tested on gangrene as well as *Clostridium botulinum* spores. These nanoemulsions are marketed as NANOSTAT™ (Nanobio Corp.) and can be formulated into a cream, liquid, foam, and spray, etc., for the decontamination of a huge variety of materials (Charles and Attama, 2011; Rutvij et al., 2011; Subhashis et al., 2011; Yashpal et al., 2013).

6.4.4 Vaccine Delivery

The delivery of vaccines can be done through nanoemulsions. It may be an effective formulation to fight against HIV by developing mucosal immunity, since HIV mainly infects the mucosal immune system (Rodriguez et al., 1999). It is administered via the nose, which is opposed to traditional routes of vaccination. Recent researchers reported that the vaccines administered into the nasal mucosa provide a genital mucosal immunity (Charles and Attama, 2011; Rutvij et al., 2011; Subhashis et al., 2011; Yashpal et al., 2013). Transportation of inactivated organisms to the mucosal surface to produce an immune response can be done by nanoemulsions; an influenza vaccine and a HIV vaccine are currently undergoing clinical trials. The application of nanoemulsion to the mucosal surface causes proteins to be adjuvant and helps in uptake by antigen-presenting cells due to the development of particular immunoglobin G (IgG) and

IgA antibodies and cellular immunity, as well as the transport of organisms which are not active to the surface of mucosa for the production of an immune response. Work on animals has showed that influenza can be prevented after mixing nanoemulsions with a single mucosal contact with the virus. Research on nanoemulsions for nasal mucosa demonstrate significant responses to HIV where animals are exposed to recombinant glycoprotein (gp) 120, and thus this nanoemulsion can be giving as a HIV vaccine. Other work on vaccines for anthrax and hepatitis B are ongoing to provide a clear concept in animal trials. The Michigan University has licensed the mentioned technology to NanoBio (Rutvij et al., 2011).

6.4.5 Nanoemulsions as Nontoxic Disinfectant Cleaner

Nanoemulsions have been employed as disinfectant cleaners. EnviroSystems developed a nontoxic disinfectant cleaner, marketed as EcoTru™, for day to day use in healthcare, travel, food processing, and military applications. Plenty of fungi, viruses, and bacteria have been found to be destroyed within 5–10 minutes. It does not pose any hazards, unlike other categories of disinfectants, and does not require any warning labels. It can be administered through the skin, inhaled, or swallowed without harm and it does not irritate the eyes. Nanoemulsions (the disinfectant formulation) are prepared by suspending nanospheres of oil droplets (less than 100 µm) in water with some active ingredient in minute quantity (i.e., parachlorometaxylenol). The presence of charge on the surface of nanospheres enables them to enter the microorganisms' membranes by breaking through an electrical barrier. The preparation may permit parachlorometaxylenol to target tissues and penetrate cell walls, instead of "drowning" the cells. Parachlorometaxylenol concentration (range of 1–2 times) is lower compared to other disinfectants, and thereby poses reduced harmful side effects in humans, animals, as well as the environment (Charles and Attama, 2011; Rutvij et al., 2011; Subhashis et al., 2011; Yashpal et al., 2013). Conventional disinfectants require large doses of active ingredients in order to surround the pathogen cell wall and disintegrate microbes, ideally "drowning" cells in the disinfectant solution. The disinfectant of nanospheres is safe, not flammable, and can be stored anywhere without any stability problems. It is poorly oxidizing, less acidic, and non-ionic, and thus will not corrode plastic, metals, or acrylic, which makes the product ideal for application to instruments and equipment, enhances the economy, and reduces health risks, which includes hazardous chemical disposal since it is environmentally safe. It may be applied on hard surfaces of equipment, counters, walls, fixtures, and floors, hence the preparation is considered a broad-spectrum disinfectant cleaner. Now-a-days this product can replace several others, thereby lowering product inventories, protecting storage capacity, reducing the cost of chemical disposal, and overall cleaning and disinfection costs can be reduced (Charles and Attama, 2011; Rutvij et al., 2011; Subhashis et al., 2011; Yashpal et al., 2013).

6.4.6 Nanoemulsions in Cell Culture Technology

For the production of biological actives, such as antibodies, recombinant deoxyribonucleic acid (DNA)/proteins, or for *in vitro* assays, cell cultures are used. A large number of molecules/blood serum can be added to the culture medium for the optimization of cell growth. It is one of the latest methods for the administration of lipophilic substances into human cell cultures. Nanoemulsions used for such purposes are transparent and in sterile, phospholipid-stabilized systems. The cells can take up the oil droplets easily and encapsulate them, therefore increasing the cells' bioavailability in culture.

The advantages of nanoemulsions in cell culture technology include the following:

- Enhanced uptake of oil-soluble additives in cell cultures
- Increased growth and vitality of cultured cells
- The possibility to carry out toxicity studies on oil-soluble drugs through cell culture techniques (Breitenbach, 2002; Venkatesan et al., 2005).

6.4.7 Nanoemulsions to Improve Oral Bioavailability of Poorly Soluble Drugs

Different formulations of nanoemulsions have been developed to enhance the oral bioavailability of drugs with solubility problems (like paclitaxel). In a recent study, o/w nanoemulsions were prepared with pine nut oil as the internal phase and egg lecithin as the primary emulsifier. The stearyl amine gives a positive charge while deoxycholic acid gives a negative charge to the droplets. The particle size ranged from 100–120 nm, with zeta potential varying between 34–245 mV. Oral administration of the nanoemulsions significantly increased the concentration of paclitaxel in the systemic circulation system compared to the aqueous control. The study indicated that it can be a promising vehicle to increase the oral bioavailability of hydrophobic drugs (Charles and Attama, 2011; Rutvij et al., 2011; Subhashis et al., 2011; Yashpal et al., 2013; Zhang et al., 2014).

Perhaps a huge potential exists for the improvement of oral bioavailability of poorly soluble drugs through nanoemulsion vehicles as exemplified by recent reports. A multitude of drugs such as trans-resveratrol (Singh and Pai, 2014), breviscapine (Ma et al., 2015), atorvastatin (Agrawal et al., 2015), simvastatin (Chavan et al., 2013), dapsone (Monteiro et al., 2012), curcumin (Yu and Huang, 2012), colchicine (Shen et al., 2011), amlodipine (Chhabra et al., 2011), and ezetimibe (Bali et al., 2010) have been formulated as nanoemulsions for improving and stabilizing the bioavailability profile of typically poor water soluble therapeutics. It is expected that as the experience of the formulators with the system improves, this list will expand quickly.

6.4.8 Nanoemulsions in Ocular and Otic Drug Delivery

Ophthalmic drug delivery is one of the most promising but challenging aspects of contemporary pharmaceutical technology (Hughes and Mitra, 1993). The demerits in conventional ophthalmic delivery systems (such as eye drops) include lachrymal secretion and nasolacrimal drainage, which results in reduced residence time in the eye and hence poor bioavailability and therapeutic response (Patton and Robinson, 1976; Sieg and Robinson, 1977). Repeated administration of concentrated solutions is required to reach the desired therapeutic levels, because the majority of the drug is washed away from the precorneal area very quickly (Chein et al., 1982). The administered drug can cause side effects when it is transported via the nasolacrimal duct into the intestinal tract and is systemically absorbed (Middleton et al., 1990). For better effect of the drug, the time of contact of dosage form to the eye should be increased. This may improve the bioavailability, lower systemic absorption, and decrease the need for frequent administration leading to improved patient compliance. There are numerous advantages in dilutable nanoemulsions (potent drug delivery vehicles) for ophthalmic use, such as sustained action and an increased ability in the drug to penetrate into the deeper layers of the ocular structure and aqueous humor. Hussein et al. (2009) formulated the antiglaucoma drug, dorzolamide hydrochloride, as an ocular nanoemulsion of high therapeutic efficacy and it showed prolonged efficacy. These nanoemulsions showed acceptable physicochemical properties with slow drug release. The systems showed lower irritability, a higher therapeutic index, and quicker onset of action when compared to other drug solutions or marketed products of the drug. It was concluded that dorzolamide hydrochloride nanoemulsion affords an improved treatment in glaucoma, decreases the number of daily applications, and enhances patient compliance when compared to conventional eye drops.

6.4.9 Nanoemulsions as Transdermal Drug Delivery Vehicles

Using nanoemulsions for the delivery of drugs through the skin to the systemic circulation system has become an interesting field for treatment of many clinical conditions, including cardiovascular conditions, anxiety, Parkinson's disease, Alzheimer's disease, and depression, etc. (Gaur et al., 2009; Muller Goymann, 2004). However, due to the skin structure, which acts as a barrier for effective penetration of the bioactive, this mode of administration is limited. It gives an efficient delivery of drugs because of its minute globule size, which enables them to easily penetrate through the pores of the skin layers and reach the systemic circulation system quickly (Ravi and Padma, 2011). Caffeine, which was earlier developed to treat different types of cancer by oral delivery, has also been formulated for the transdermal route (w/o nanoemulsion formulations of caffeine). Drug-loaded nanoemulsions showed a significant increase in permeability parameters in comparison to aqueous caffeine solutions (Shakeel and Ramadan, 2010). Therefore, the use of nanoemulsions to enhance therapeutic levels and drug bioavailability without adverse effects has become one challenging area of research in transdermal delivery. It is a promising formulation with an increased transdermal drug flux (due to supersaturation in the oil droplets), more storage stability, economy of production, lack of organic solvents, and, finally, improved scale-up feasibility, offering improved plasma profiles and reproducible drug bioavailability. Currently such systems are used to provide dermal and surface effects, and for deeper penetration into skin (Ravi and Padma, 2011). On numerous occasions, nanoemulsions have improved transdermal and dermal drug delivery characteristics *in vitro* (Delgado Charro et al., 1997; Dreher et al., 1997; Kreilgaard et al., 2000; Lee et al., 2003; Osborne et al., 1991; Schmalfus et al., 2001; Trotta et al., 1996) as well as *in vivo* (Kemken et al., 1992; Kreilgaard et al., 2001a, 2001b). The substitution of conventional topical products by nanoemulsion has improved the transdermal permeation of many drugs (Gasco et al., 1991; Kriwet et al., 1995; Ktistis et al., 1991; Sheela et al., 2012; Trotta, 1999).

6.4.10 Nanoemulsion in Cancer Therapy/Targeted Delivery

The treatment of cancer by submicron-sized emulsions has demonstrated targeting to tumors. Apart from vehicles for administering poorly soluble drugs, they have also recently been used as colloidal carriers for the targeted administration of anticancer drugs, diagnostic tools, photo sensitizers, and neutron capture therapeutic agents. The improvement of magnetic nanoemulsions for use as photo sensitizers (Foscan®) for deeper tissue layers across the skin, thereby enhancing hyperthermia for the subsequent generation of free radicals. As a photodynamic therapy, this technology can be used to treat cancer (Primo et al., 2007). There are several other reports of improved cytotoxicity and tumor internalization of anticancer drug-loaded, nanoemulsion-based delivery. Some notable recent reports of nanoemulsion-based delivery and targeting of neoplasia include the platinum-based theranostic nanoemulsions and phosphatidylinositol 3-kinase inhibitor-loaded nanoemulsions for the treatment of ovarian cancer (Ganta et al., 2014a; Talekar et al., 2012), and gene delivery in the treatment of solid tumors (Zhou, 2013). Interested readers are referred to reviews on the topic by Ganta et al. (2014b) and Zhou (2013) for further insights.

6.4.11 Nanoemulsions and Intranasal Drug Delivery

Recently, the intranasal drug delivery system has been recognized as a preferable route for drug administration, next to the oral and parenteral routes. Nasal mucosa is emerging as a therapeutically successful channel for drug administration into systemic circulation and also seems to be an interesting way to challenge the obstacles to the direct passage of drugs to the target site (Ugwoke et al., 2005). This route is painless, well tolerated, and non-invasive. The nasal cavity is rich in immunoactive sites, has a moderately permeable epithelium, and poor enzymatic activity (Pardridge, 1999). There are numerous challenges linked with drug targeting to the brain, particularly hydrophilicity and large molecular weight. Due to the impervious nature of the endothelium, it multiplies barriers in the systemic circulation system between the blood and the brain (Kumar et al., 2008a) Nasal mucosa in the olfactory region gives a direct line of contact from the nose with the brain, and, thereby, by using nanoemulsions loaded with drugs, Alzheimer's disease, migraines, depression, meningitis, schizophrenia, and Parkinson's disease, etc., may be treated (Csaba et al., 2009; Mistry et al., 2009). Nanoemulsions loaded with risperidone for brain delivery through the nose have been reported (Clark et al., 2001). It has been claimed that this emulsion is more effective via the nasal rather than the intravenous route. The next application of intranasal delivery of medicament in therapeutics is in their potential for the preparation of vaccines. Immunity is obtained by the application of mucosal antigens. The first intranasal vaccine has already been marketed (Bhanushali et al., 2009). Among the possible delivery systems, nano-based carriers have a great advantage in protecting the bioactives, enhancing nanocarrier interaction with the mucosal membrane and directly with antigens of the lymphoid tissues. Hence it is possible that the application of nanoemulsions in a nasal delivery system will bring about brilliant results in drug targeting to the brain in the treatment of disorders related to the central nervous system (Tamilvanan, 2004). Bhanushali et al. (2009) developed intranasal nanoemulsion and gel formulations for rizatriptan benzoate for prolonged action. Varying mucoadhesives have been tried out to form thermo-triggered, mucoadhesive nanoemulsions. The mucoadhesive gel of rizatriptan was formulated using different proportions of Hydroxypropyl methylcellulose (HPMC) and Carbopol 980. The comparative evaluation of nasal nanoemulsions and intranasal mucoadhesive gels reported that higher brain-targeting could be attained with nanoemulsions. Other drugs which were developed for nasal administration include insulin and testosterone (Tamilvanan, 2004).

6.4.12 Nanoemulsions and Parenteral Drug Delivery

The parenteral route is one of the most common and efficient routes of drug administration, which may be preferred for actives of reduced bioavailability and narrow therapeutic index. Their capability to dissolve huge amounts of hydrophobics with their compatibility and their capacity to prevent the medicaments from hydrolysis and enzymatic degradation has made nanoemulsions a promising vehicle for the needs of parenteral transport. In addition to this, the frequency and dosage time can be lowered in the medicament therapy period since nanoemulsions enable drug release in a sustained and controlled order over extended time periods. In addition, the lack of flocculation, sedimentation, and creaming, the large surface area, and free energy provides serious advantages over macroemulsions (Ravi and Padma, 2011). Large interfacial area influences the transport of drugs and their delivery, as associated with their targeting to particular sites. Clinical and pre-clinical trials have been carried out with parenteral nanoemulsion-based carriers. The advancement in novel parenteral drug delivery, including nanoemulsions, has been reported by Patel and Patel (2010). Nanoemulsions consisting of thalidomide have been synthesized with a dose as low as 25 mg leading to therapeutic plasma concentrations (Araujo et al., 2011). A significant lowering in the drug content of the nanoemulsion has been observed in 0.01% drug formulation after two months storage, but this can be overcome by the addition of polysorbate 80. Chlorambucil, a lipophilic anticancer agent, was reported to be used against breast and ovarian cancer. Its pharmacokinetics and anticancer activity were studied by loading it in emulsions for parenteral use and preparing it through the high energy ultrasonication method. Treatment of adenocarcinoma of the colon in an experimental mouse model was carried out using this nanoemulsion, which resulted in a higher tumor suppression rate than the plain drug solution treatment and demonstrated that the drug-loaded emulsion may be a beneficial carrier for its delivery in cancer treatment (Ganta et al., 2010). Carbamazepine, an anticonvulsant drug, had no parenteral treatment available for patients because of its low water solubility. Kelmann et al. (2007) developed a nanoemulsion for intravenous delivery, which demonstrated favorable *in vitro* release kinetics. Parenteral nanoemulsion formulations of diazepam, propofol, dexamethasone, etomidate, flurbiprofen, and prostaglandin E1 have also been documented (Brussel et al., 2012).

6.4.13 Nanoemulsions and Pulmonary Drug Delivery

The lung is a promising target for drug delivery because of its non-invasive administration via inhalation aerosols. It avoids first-pass metabolism, can be directly delivered to the target site for the treatment of respiratory disorders, and has the availability of a large surface area for both local and systemic drug absorption. Nanocarrier systems in pulmonary drug delivery provide several advantages, such as the ability to reach the systemic circulation system, relatively uniform drug distribution and dose among the alveoli, achieving improved solubility of the drug, a sustained drug release which deliberately lowers the dosing rate, enhances patient compliance, and lowers the incidence of adverse effects, and the ability of drug internalization by cells (Heidi et al., 2009). To date, the submicron emulsion system has not yet been fully popularized for pulmonary drug delivery and even less has been reported in this area (Heidi et al., 2009). Some recent studies have been reported by Nesamoy et al. (2014, 2013) on nanoemulsion mists for pulmonary delivery and on amphotericin B nanoemulsion aerosols (Nasr et al., 2012).

Bivas-Benita et al. (2004) reported that cationic submicron emulsions are advantageous carriers for DNA vaccines to the lung because they are capable of transfecting pulmonary epithelial cells, that possibly exert cross-priming of antigen-presenting cells and directly initiate dendritic cells, resulting in the stimulation of antigen-specific T-cells. Hence, the submicron emulsion nebulization will be a new and upcoming research area. However, extensive research is needed for the successful preparation of inhalable submicron emulsions because of the possible adverse reactions of lung surfactants with oils on the lungs, which may affect alveoli function. A novel pressurized aerosol system of salbutamol has been introduced for pulmonary delivery using lecithin-stabilized microemulsions formulated in trichloro trifluoro ethane.

6.4.14 Nanoemulsions as Gene Delivery Vectors

Nanoemulsion systems have been reported as alternative gene transfer vectors for liposomes (Liu et al., 1996). Studies on cationic nanoemulsion for gene delivery (via non-pulmonary routes) demonstrated that the binding of the emulsion/DNA complex was tighter than that of liposomal carriers (Yi et al., 2000). This stable emulsion system administered genes more effectively than liposomes (Liu and Yu, 2010). Silva et al. (2009) evaluated factors which influence DNA compaction in cationic lipids that contain a primary amine group when they are in solution (André et al., 2012). The effect of the stearyl amine incorporation phase (water or oil), complexation time, and incubation temperatures were also observed. The complexation rate was assessed by using the electrophoresis migration method on agarose gel, lipoplex, and nanoemulsion, and characterization was done by dynamic light scattering (DLS). The results demonstrated that the best DNA compaction process occurs after 120 minutes of complexation, at reduced temperature ($39.2 \pm 33.8°F$), and after addition of the cationic lipid into the aqueous phase. Though the zeta potential of lipoplexes was found to be lower than that for basic nanoemulsions, the granulometry did not change. As well as this, it was found that lipoplexes are more suitable vehicles for gene delivery.

6.4.15 Nanoemulsions for Phytopharmaceuticals

Presently, considerable emphasis has been given to the formulation of novel drug delivery systems for herbal medicaments (Ajazuddin and Saraf, 2010). However, some challenges with plant bioactives, such as instability in acidic pH and liver metabolism, led to a lower drug level than therapeutic concentration in the blood plasma, resulting in poor or no therapeutic effect (Goyal et al., 2011). Therefore, the encapsulation of plant exudates or their bioactives will lower their risk of degradation, metabolism, and serious hazardous effects, because of the accumulation of drugs in non-targeted areas, and increase their ease of administration in the pediatric/geriatric groups of patients (Uhumwangho and Okor, 2005). Waxy nanoemulsions with oil from medicinal plants or lipophilic drugs have been reported to increase drug solubility, lower the side effects of various potent drugs, enhance drug bioavailability, and extend the pharmacological effects over other conventional preparations, such as microemulsions (Youenang et al., 1999). The preparation of nanoemulsions with photoactives have been reported. The effect of nanoemulsion on the intestinal absorption of colchicine has been demonstrated *in vivo*. Colchicine nanoemulsion was formulated with isopropyl myristate, Tween 80, eugenol, water, and ethanol, using eugenol as the oil phase in the preparation. The results indicated that the intestinal absorption of colchicine was significantly increased by the nanoemulsion preparation (Shen et al., 2011). In different *in vitro* experiments genistein has shown anticancer activities, but the same effects could not be translated into clinical testing because of its low bioavailability. Researchers have tried various nano-approaches, including the incorporation of genistein into topical nanoemulsion preparations composed of egg lecithin, medium chain triglycerides, or octyldodecanol with water by using the spontaneous emulsification method, resulting in enhanced activity (Silva et al., 2009). O/w nanoemulsion formulation demonstrated the enhanced anti-inflammatory activity of curcumin (Wang et al., 2008).

6.5 Application of Nanoemulsion in Non-Invasive Drug Delivery

6.5.1 Transdermal Delivery

The delivery of drugs via the skin to the systemic circulation system is advantageous for several clinical conditions and hence there is considerable interest in this area (Gaur et al., 2009; Muller Goymann, 2004). It has the advantages of steady-state, controlled drug delivery for an extended period of time and self-administration is also possible. By removing the transdermal patch, drug administration can be terminated at any time. Their transparent nature and fluidity confers a pleasant skin feel on nanoemulsions. There is also an absence of the GI side effects which are generally associated with oral delivery. Several transdermal products are available for various diseases including Parkinson's and Alzheimer's disease, anxiety, cardiovascular conditions, depression, and so on. However, the skin also imposes a barrier for the penetration of bioactive molecules, which is a limitation of the route. Drugs penetrate the skin through hair follicles, sweat ducts, and sometimes directly across the stratum cornea, because of which there is a restriction in the absorption and thereby bioavailability. For the improvement of drug pharmacokinetics and targeting, the primary skin barriers need to be overcome. Nano-sized emulsions can easily penetrate the pores of the skin and reach the systemic circulation system, and hence they are advantageous for effective delivery. Literature is available on the use of caffeine for the treatment of different types of cancer by the oral route. W/o nanoemulsion formulations of caffeine have been developed for transdermal drug delivery. When the *in vitro* skin permeation profile of the prepared formulation was compared with aqueous caffeine solutions there was a significant increase in the permeability of nanoemulsion-loaded drugs (Shakeel and Ramadan, 2010). Nanoemulsions in transdermal drug delivery systems (TDDS) demonstrate a major area of research, which enhances therapeutic efficacy and bioavailability without having side effects. They are also regarded as a promising technique with several advantages like more storage stability, less preparation cost, practical thermodynamic stability, feasibility of loading both hydrophilic and lipophilic drugs, absence of organic solvents, and better production feasibility, as well as the reproducibility of plasma concentration profiles and the bioavailability of drugs. These systems are currently used for providing dermal and surface effects and for their deep skin penetration power. Several studies have also shown that nanoemulsion formulations possess improved transdermal and dermal delivery properties *in vitro* (Alvarez-Figueroa and Blanco-Mendez, 2001; Delgado Charro et al., 1997; Dreher et al., 1997; Kreilgaard et al., 2000; Lee et al., 2003; Osborne et al., 1991; Rhee et al., 2001; Schmalfus et al., 1997; Trotta et al., 1996) as well as *in vivo* (Kemken et al., 1992; Kreilgaard, 2001). Nanoemulsions afford enhanced transdermal permeation in several drugs over the traditional topical formulations (Gasco et al., 1991; Ktistis and Niopas, 1998) and gels (Kriwet et al., 1995; Trotta, 1999; Ribier et al., 1998; Rocha-Filho et al., 2014).

Two mechanisms to explain the advantages of nanoemulsions as TDDS according to Barakat (2011) are

1. The increased solubility effect of drugs in nanoemulsion systems which might lead to higher thermodynamic activity in the skin.
2. Nanoemulsion ingredients generally act as penetration enhancers, leading to enhanced transdermal flux.

Surfactants are present in nanoemulsions as one of their components, which help to enhance the permeability of the membrane and thereby increase transdermal transport. According to the literature, drug release can be controlled by nanoemulsion (NE) and the bioavailability of many drug compounds also improved by incorporating them in nanoemulsions (Barakat, 2011). The role of surfactants in these systems is of paramount importance, which is evident from the definition of nanoemulsions provided by Caraglia et al. (2011) where "nanoemulsion may be defined as an isotropic, thermodynamically stable, transparent or translucent dispersion of water and oil which may be stabilized by using an interfacial film of surfactant molecules having droplet size of 20–500 nm". Several features of nanoemulsions which have drawn the attention of researchers include their ease of preparation and scale-up, and increased bioavailability, solubility, and stability (Faiyaz, 2009; Kakumanu et al., 2012; Rachmawati and Haryadi, 2014; Silva, 2014). The various advantages of nanoemulsions, according to Faiyaz (2009), Kakumanu et al. (2012), Rachmawati and Haryadi (2014), and Silva (2014), include the following:

- Increased absorption rate
- Easy solubilization of lipophilic drugs
- Varying routes of drug administration
- Rapid and efficient penetration of bioactives

- Enhanced patient compliance
- Thermodynamically stable system and self-emulsification is possible due to the stability of the system (Table 6.2)

Baboota et al. (2007) have carried out an o/w nanoemulsion with 10% (m/m) oil (propylene glycol mono-caprylic ester and glycerol triacetate), 50% (m/m) surfactant (diethylene glycol mono ethyl ether and Tween 80), and 40% (m/m) water and suggested this as a vehicle for the improved transdermal delivery of celecoxib. *In vitro* skin permeation studies of the same showed an increased percutaneous uptake of celecoxib from the nanoemulsion in comparison to a simple gel. The nanoemulsion improved the inhibition of carrageenan-induced paw edema in rats (Baboota et al., 2007).

6.5.2 Topical Drug Delivery

Drug delivery via topical routes are non-invasive and have the capability of self-administration with the least side effects. There are enormous advantages in the topical delivery of biopharmaceuticals. These are the agents with a controlled size of droplet, the ability to dissolve lipophilic drugs efficiently, enhanced skin permeation, and improved release pattern in both lipophilic and hydrophilic drugs. The use of topical nanoemulsions is interesting due to their good sensorial and physical properties. They get completely dispersed on the skin. The ability of nanoemulsion formulations to lead topical drug delivery is reported for ketoprofen (Kim et al., 2008). It was claimed that the drug permeation rate can be altered by changing the quantity of the surfactants, oil, and co-surfactants. Similarly, a study of aceclofenac-loaded nanoemulsion reported enhanced permeation of the drug into the abdominal skin of rats, and concurrently increased the anti-inflammatory effect of carrageenan-induced paw edema in rats *in vivo*, when compared with a gel formulation (Shakeel et al., 2007). A different type of nanoemulsion for topical use has been developed for the introduction of an adrenergic β-blocking agent, adaprolol maleate, into the eye for the treatment of glaucoma without any systemic side effects. On the other hand, the drug shows an important clinical discouragement as it led to irritation of the eyes, some burning sensations, as well as local discomfort. But a nanoemulsion formulation, on the other hand, maintains the therapeutic efficacy of the drug while reducing the level of ocular irritation (Amselem and Friedman, 1996). Now-a-days, nanoemulsions are found in a variety of cosmetics including anti-wrinkle creams, bath oils, body creams, and anti-aging preparations since they have minute, uniform droplet size. Nanoemulsions are clear, fluid, and smooth to touch (Guglielmini, 2008; Sarker, 2005). In comparison

TABLE 6.2

Active Compounds Encapsulated in Nanoemulsions for TDDS

S.No.	Formulation	Active Ingredient	References
1	Nanoemulsion	Ibuprofen	Salim (2011); Behzad (2012)
2	Nanoemulsion gel	Carvedilol	Singh (2012)
3	Nanoemulsion gel	Etoricoxib	Lala and Awari (2014)
4	Nanoemulsion gel	Betamethasone valerate	Ali (2013)
5	Nanoemulsion	Lecithin	Huafeng (2010)
6	Nanoemulsion	Glibenclamide	Abdul (2011)
7	Nanoemulsion	Turmeric oil	Ali (2012)
8	Nanoemulsion	Dithranol	Gidwani and Singnurkar (2003)
9	Nanoemulsion gel	Beclomethasone dipropionate	Ali (2012)
10	Nanoemulsion gel	Aceclofenac gel	Jignesh et al. (2011); Ali (2012)
11	Microemulsion	Methoxalen	Shah (2013)
12	Nanoemulsion	Indomethacin	Barakat (2011)
13	Nanoemulsion	Clarithromycin	Stuti (2014)
14	Nanoemulsion	Pterostilbene	Yue (2014)
15	Nanoemulsion	Felodipine	Poluri (2011)
16	Nanoemulsion	Rice bran oil	Bernardi et al. (2011)

to traditional emulsions, they have more spreading ability on the skin. Due to their rare texture and rheological properties, they become very effective in cosmetic applications. Nanoemulsion formulations range from water-like fluids to semi-solid gels. Numerous patents resemble the active development of nanoemulsion formulations in cosmetics. For example, nanoemulsions having fluid, amphiphilic lipids, non-ionic, diglyceryl isostearate, sorbitan oleate, and α-butylglucoside caprate were found to be stable in storage from 32°F to 113°F (Guglielmini, 2008). They can contain a tremendous quantity of fragrance and led the entry of the drug into the skin. Nanoemulsions composed of anionic, amphiphilic lipids of phosphoric acid fatty esters and oxyethylenated derivatives also possess a transparent appearance and better cosmetic ability even after the addition of a large quantity of oil into the formulations (Simonnet et al., 2001). Another o/w nanoemulsion based on one or more non-ionic and/or anionic 8 amphiphilic lipids and one or more water-soluble neutral polymers (e.g., poly(ethylene oxide), polyvinyl alcohol, polyvinyl caprolactam) allowed the viscosity of the composition to be increased without influencing its transparency or influencing the level of the oil phase (L'alloret et al., 2006). For stable and translucent nanoemulsions as cosmetics, dermatological applications, or ophthalmological applications, the nano-emulsions consisting of a ternary surfactant system with ethoxylated fatty ester polymer, fatty acid ester of sorbitan, and alkali metal salts of cetyl phosphate or palmitoyl sarcosinate need not use gelling agents for stabilization (Quemin, 2005), and are deemed suitable for application even on sensitive skin. Other nanoemulsion formulations have been developed for numerous properties, including sun-protection, anti-wrinkling, and anti-aging of the skin and other cosmetic targets (Calderilla-Fajardo et al., 2006; Haake et al., 2007; Huglin et al., 2005; Yilmaz and Borchert, 2006; Yoo et al., 2006).

6.5.2.1 Patented Nanoemulsions

Certain notable patents on nanoemulsions primarily in the cosmetic field are

1. Method of preventing and treating microbial infection (Assignee: NanoBio Corporation (US), US Patent number: 6,506,803).
2. Nanoemulsion based on phosphoric acid esters and its uses in the cosmetics, dermatological, pharmaceutical, and/or ophthalmological fields (Assignee: L'Oreal (Paris, FR), US Patent number: 6,274,150) (Simonnet et al., 2001).
3. Nontoxic antimicrobial compositions and methods of use (Assignee: NanoBio Corporation (US), US Patent numbers: 6,559,189 and 6,635,676).
4. Nanoemulsion based on ethylene oxide and propylene oxide block co-polymers and its uses in the cosmetics, dermatological, and/or ophthalmological fields (Assignee: L'Oreal (Paris, FR), US Patent number: 6,464,990).
5. Nanoemulsion based on oxyethylenated or non-oxyethylenated sorbitan fatty esters, and its uses in the cosmetics, dermatological, and/or ophthalmological fields (Assignee: L'Oreal (Paris, FR), US Patent number: 6,335,022).

6.5.3 Transungual Drug Delivery

Transungual drug delivery systems are associated with drug application across the nail barrier to achieve targeted drug delivery for the treatment of diseases associated with nails. The hardness and permeability of the nail implies a challenge to drug administration. Drug delivery via the oral route has a variety of disadvantages in the form of systemic side effects, including drug interactions (Kobayashi et al., 1999). Because of the localized effect of topical drug delivery, it is needed for treating nail disorders to ensure minimal adverse effects, and possibly thereby improve adherence and compliance (Murdan, 2002). Advancements in drug delivery for diseases associated with nails, like fungal nail diseases (ony-chomycosis and nail psoriasis), can reduce the need for the systemic administration of drugs with their associated side effects (Walters and Flynn, 1983). Nail drug delivery/penetration is shown in Figure 6.4.

Research on transungal drug delivery mainly focuses on altering the nail plate barrier by the appli-cation of various chemical penetration enhancers, like sulfites, mercaptans, keratolytic agents (such as

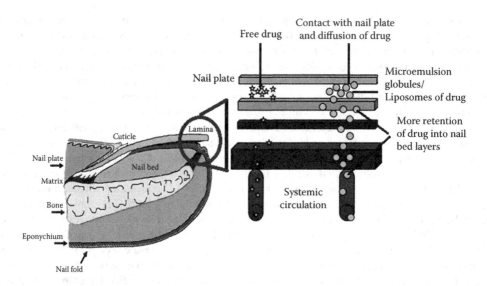

FIGURE 6.4 Nail drug delivery penetration.

enzymes), hydrogen peroxides, urea, and water, etc., the use of physical techniques, like iontophoresis, carbon dioxide laser, acid etching, occlusion and hydration, electroporation, photodynamic therapy, and sonophoresis/phonophoresis, etc., and mechanical means, like nail avulsion and nail abrasion, for drug penetration. Several reports exist for the ungual delivery of oxiconazole, ticonazole, econazole, miconazole, ketoconazole, sertaconazole, terbinafine, ciclopirox, and so on (Rubio et al., 2005). The increased attention of researchers in this particular area may be suitable due to the infancy of the field; the launch and success of several commercial, topical, antifungal nail lacquers like Curanil, Loceryl, and Penlac point out the absolute opportunity for the research and development of new products in this previously neglected area. Table 6.3 shows the clinical trial status of certain transungual nanoemulsions and Table 6.4 shows different formulations and ongoing research into ungual drug delivery.

The permeability of topically applied drugs via the keratinized nail plate is very poor and uptake of the drugs into the nail apparatus is severely low. Figure 6.5 shows the fate of the drug following topical application to the nail plate.

Nail lacquers composed of drugs are an innovative kind of dosage form. They have to be applied on to the nail plate using a brush. The field of ungual drug delivery following topical application has not been fully explored and more research is needed in this field to resolve conflicting findings on the physico-chemical parameters which influence ungual drug permeation and to find and characterize new penetration enhancers and delivery vehicles. Drug transport into the nail plate consists of filing the nail plate before drug formulations are applied topically, as well as by using chemical enhancers. Enhancing the ungual drug uptake following topical application may be categorized into three approaches: first, realizing the physicochemical factors which influence drug penetration into the nail plate, second, use of chemical enhancers which cause alterations in the nail plate leading to drug permeation, and, third, use of drug-containing nail lacquers which are brushed onto nail plates and act as a drug depot from where drugs can be released continuously into the nail. For optimal ungual permeation and uptake, active moieties must be smaller in size and uncharged. Nail drug delivery has a major drawback of deficiency

TABLE 6.3

Clinical Trials Status

Therapeutic Agents	Formulation and Dose	Clinical Status	Diseased Condition
NB-002	Nanoemulsion	Phase II	Onychomycosis
NB-00X	Nanoemulsion	Phase II	Onychomycosis

TABLE 6.4

Different Formulations and Ongoing Research

Route	Drugs Pharmacokinetically Compared			Reference
	Itraconazole	Terbinafine	Floconazole	
Nail penetration	7–14 days	7–14 days	<7	Gupta et al. (2001)

in understanding both the barrier of the nail and formulations to increase ungual delivery by increasing the efficiency of topical treatments for nail diseases. The topical delivery of systemic therapeutics offers several benefits but presents a greater technical challenge. Among the benefits, first-pass metabolism avoidance, convenience, and sustained release are most often cited. This avoids the oral toxicity of antifungal drugs and provides a longer contact time with the application site. For the diagnosis and resolution of onychomycosis both clinical and mycological parameters are important. So, the formulation should have a better therapeutic effect and less site effect, which is necessary for the treatment of nail diseases (Effendy, 1995).

Nail psoriasis and fungal infections are very common among patients, particularly those who are immune compromised. Nail diseases are generally treated by the administration of huge doses of drugs orally over a prolonged duration. Oral therapy is mainly associated with various adverse effects. However, the impermeability of the nail plate lowered the amount of drug entry from topical products into the deeper tissues of the nail, hence, topical therapy was poorly successful as a monotherapy. There is an immediate need for the development of technologies for the improvement of the topical delivery of drugs via and across the nail to improve its positive influence and to minimize the recurrence of nail disorders. The use of chemical nail permeability enhancers and biophysical technologies, such as iontophoresis and sonophoresis, have been investigated as potential approaches to improve drug delivery into the nail bed. The topical delivery of different drug formulations as nail lacquers, creams, lotions, and gels are hindered by the low penetration into the human nail plates and there is a need for repeated dosing for a longer period of time for effective treatment. This is where the potential of nanoemulsions as delivery vehicles for transungual delivery comes into focus.

Kumar et al. (2012) carried out research work on the effective delivery of ciclopirox olamine (CIC) via human nail plates by increasing the penetration and retention time of CIC in the skin layers. For this need, a nanoemulsion gel containing surfactant, oil, co-surfactant, and cabopol was developed by the

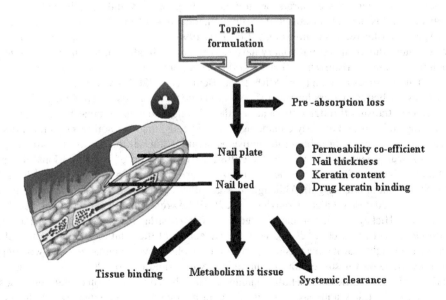

FIGURE 6.5 Fate of drug after topical formulation.

aqueous phase titration method and analyzed for different *in vitro* attributes. They successfully formu-lated a thermodynamically stable, antifungal nanoemulsion gel containing ciclopirox olamine, which can be retained for a prolonged period of time. Such a formulation achieves better local concentration with low systemic absorption of potentially toxic, anti-fungal drugs. Hence, the formulation has proved to be a promising approach for the treatment of onychomycosis.

Hamouda et al. (2008), using nanoemulsion technology, developed an antifungal nanoemulsion (NB-002) with potent activity against both fungal spores and hyphae. Using previously published principles of nanoemulsion formulation, they formulated NB-002 with constituents regarded as either "generally rec-ognized as safe" (GRAS) or with a history of safe use in humans. Stability was categorized by physical appearance, droplet size, pH, and potency. Stable emulsions were analyzed in an *in vitro* fungicidal assay for the determination of the minimum inhibitory concentration (MIC) and minimum lethal concentra-tion (MLC) against laboratory provided, clinical dermatophyte isolates consisting of onychomycosis, as well as several *Candida* species. An *in vivo* safety study was performed in appropriate animal models. Their conclusion is that NB-002 is a novel nanoemulsion with potent activity against the organisms that cause onychomycosis and other dermatomycoses. It seems to be safe for the application of topical use at doses 1,000-fold higher than the minimum fungicidal concentration. A Phase I study representing the tolerance, safety, and pharmacokinetics of NB-002 has been completed, and a Phase II study is on the way using over 400 subjects with distal subungual onychomycosis.

6.5.4 Pulmonary Drug Delivery

The lungs are increasingly accessible and are permeable by the means usual inhalation, hence the pulmonary route reflects a prospective medium for the systemic delivery of small molecules as well as macromolecules (proteins and peptides). In the last few years, there have been intensive research efforts made on pulmonary drug delivery (PDD) not only for local administration but also for systemic delivery, as well as for diagnostic purposes. This interest is primarily due to the several advantages the pulmonary route possesses over other routes of drug delivery. Macromolecular drugs undergoing extensive first-pass metabolism are generally preferable via the pulmonary route. Even though the lung is enzymatically active, there are certain reports of increased drug bioavailability using the pulmonary route (Michael et al., 2008). It is less invasive, which increases patient acceptance, and there are fewer incidences of side effects, especially for local therapies, which is generally needed due to the localized deposition of the drug and its decreased systemic and generalized exposure. Until now, nanoemulsion systems had not yet been fully explored for pulmonary drug delivery and very limited literature has been published in this field (Mansour et al., 2009). Nanoemulsion formulations are now regarded as alternative gene transfer vectors to liposomes (Liu et al., 1996). Other studies of nanoemulsion for gene delivery by routes other than non-pulmonary have shown that the binding of the emulsion complex was better than that of liposomal carriers (Yi et al., 2000). This stable emulsion system delivers genes more effectively than liposomes (Liu and Yu, 2010). Bivas-Benita et al. (2004) reported that cationic submi-cron emulsions are better carriers for vaccines of deoxyribonucleic acids to target lungs tissue as they are appropriate to transmit through pulmonary epithelial cells, and hence increasing cross priming for cells presenting antigens and directly enhancing dendritic cells, thereby motivating to the stimulation of antigen-specific T-cells. For submicron emulsions nebulization will be a novel research approach. But detailed research approach is needed for optimized formulation of nanoemulsions of inhalable route as there is chances of unwanted reactions between surfactants and oils may occur on lung alveoli functions (Surfactant of lung origin). Although several authors have suggested using nanoemulsion for-mulations for metered dose inhalers (Courrier et al., 2003; Lawrence and Rees, 2000; Patel et al., 2003; Sommerville and Hickey, 2003; Sommerville et al., 2002; Tenjarla, 1999), nebulizable nanoemulsion reports are far scarcer (Nasr et al., 2012). With the advantage of the solution-like properties of nano-emulsions, it is thought that nanoemulsions perform as a solution when nebulized and will improved the aerosol experience over suspension formulations. Amir Amani et al. (2010) carried out work to evaluate the *in vitro* properties of a nebulized nanoemulsion-based formulation containing budesonide as a model steroid with low aqueous solubility for administration into the lungs by nebulizers. They came to the conclusion that the nanoemulsion formulation containing budesonide exhibited a distinct improvement over the suspension preparation of budesonide to perform *in vitro* aerosolization, when evaluated using a jet and a vibrating mesh nebulizer.

6.5.4.1 Significance of Nanoemulsion for the Pulmonary Route

The pulmonary route offers local targeting for the treatment of respiratory diseases and effectively appears to be a viable option for the administration of drugs systemically. Pulmonary delivery is via a variety of ways, including aerosols, metered dose inhaler systems, powders (dry powder inhalers), and solutions (nebulizers) that may contain nanostructures such as micelles, liposomes, and nanoparticles, including nanoemulsions and microemulsions. Lung delivery research is driven by the potential for successful protein and peptide drug delivery, by the challenge for an efficient delivery mechanism for gene therapy, and by the urgency to substitute chlorofluorocarbon as propellants into inhaler systems (metered dose). But the success of the pulmonary delivery of protein drugs is hindered by proteases in the lung, which reduce overall bioavailability, due to the barrier between the capillary blood and alveolar air (the air–blood barrier). Hence, adjuvants like absorption enhancers and protease inhibitors are sometimes needed to increase the pulmonary absorption of similar macromolecular or poorly absorbable actives (Ugwoke et al., 2007). Drug penetration from nanoemulsion via the pulmonary route is represented in Figure 6.6.

Niven et al. (1990) reported that adjuvants including oleic acid, oleyl alcohol, and Span 85 can enhance the transfer rate of disodium fluorescein in isolated rat lungs.

Yamamoto et al. (1997) reported that the absorption of insulin was enhanced in the presence of various constituents including glycocholate, Span 85, surfactin, and nafamostat. They have also reported that n-lauryl β-D-maltopyranoside (LM), linoleic acid-HCO60 mixed micelles (MM), and sodium glycocholate (NaGC) are generally effective in increasing the pulmonary absorption of insulin and fluorescein isothiocyanate dextrans (FDs). They did not cause severe mucosal damage to rat lungs, when experimented with through the leakage of Evans blue from the systemic circulation system. These findings suggest that the use of protease inhibitors and absorption enhancers would be a useful approach for improving the pulmonary absorption of biologically active drugs like peptides. Besides the importance of the choice

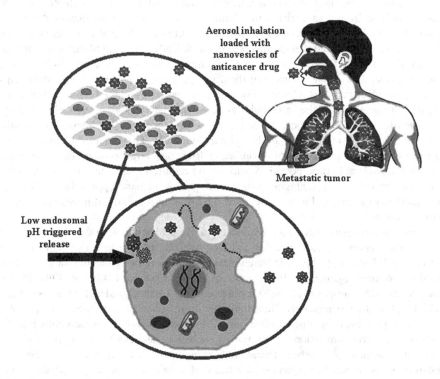

FIGURE 6.6 Drug penetration from nanoemulsion via pulmonary route.

of excipients to increase the pulmonary absorption of actives, the design of the nanoemulsion can also contribute to the improvement of the treatment.

Acute lung injury (ALI) is an adverse disease causing death if not treated quickly. Emergency medicament is required to treat ALI. Dimethyl silicone (DMS) is a suitable agent to defoam the bubble in the lung induced by ALI. However, DMS aerosols (a marketed product) affect the environment and their use will be limited in the future. Lifei et al. (2015) first reported dry nanoemulsion inhalation for pulmonary delivery. Recently, dry DMS nanoemulsion inhalations (DSNIs) were formulated. The optimized preparation of DMS nanoemulsions (DSNs) was stable and homogenous, constituting Cremophor RH40/PEG 400/DMS (4:4:2, w/w/w) and water. The DSNs showed a minute size of 19.8 nm, zeta potential of 9.66 mV, and low PDI of 0.37. This type of DSN was o/w. The DSNs were mixed with mannitol followed by freeze-drying to achieve DSNIs, which were white-colored powders, of good fluidity, and also capable of rapid reconstitution into DSNs. The DSNs could adhere to the surfaces of mannitol crystals which were being lyophilized. The aerodynamic diameter of DSNIs was 4.82 mm, suitable for pulmonary delivery. The defoaming rate of DSNIs was 1.25 mL/s, much faster than those of the blank DMS, DSNIs, and DMS aerosols. As per lung appearances, histological sections, and lung wet/dry weight ratios, the DSNIs are reported to have a higher anti-ALI effect on the ALI rat models than the blank DSNIs and the DMS aerosols, meaning they are effective anti-ALI nanomedicines. The novel DMS preparation is a promising replacement for DMS aerosols.

6.5.5 Nanoemulsion in Ocular Drug Delivery

Ocular drug delivery needs some special characteristics because of the peculiar physiological structure of the eye. The eye is said to be a challenging and unique organ for drug delivery on the surface as well as inside the ocular structure. Its anatomy and physiology interfere with the fate of the administered bioactives. Tears contain lysozymes and immunoglobulins, because of which they permanently wash the structure of the eye and produce anti-infectious activity. Drugs may have the chance to bind to tear proteins and conjunctival mucin. For the treatment of local ophthalmic disorder, an eye drop is the suitable dosage form because of its convenience for administration and patient compliance. Most of the conventional eye drops contain drugs in solution form and have reduced therapeutic effect due to its low bioavailability. Repeated and frequent instillations are often needed while using conventional eye drops to achieve the required therapeutic effect, which leads to increased inconvenience and side effects. Generally, not more than 5% of the drug introduced penetrates the cornea/sclera and reaches the intraocular tissue, with a major portion of the drug applied getting systemically absorbed via the conjunctiva and nasolacrimal duct. Corneal, conjunctival epithelia, and the tear film impose a barrier preventing drug absorption into the intraocular area, which gives results of poor bioavailability and undesired systemic side effects. Hence, the delivery of drugs for ocular therapeutics is a challenging concept, and is a matter of interest to multidisciplinary scientists. The major problem concerned with ocular therapeutics is the maintenance of an optimum drug concentration in eye tissue or at the application site for a considerable time period to achieve the required therapeutic response. Ophthalmic drug administration may get advantages from the uniqueness of nano-approaches to drug delivery (Sahoo and Labhasetwar, 2003). Figure 6.7 depicts the different routes of administration for ophthalmic disease treatment.

The use of nanosystems may provide a bright feature for topical ocular drug delivery, because of their capability to protect encapsulated molecules and facilitate transport to various chambers of the eye (Kayser et al., 2005; Losa et al., 1993). Different nanosystems have been adopted for corneal retention and bioavailability encouragement. Colloidal systems include liposomes (Bochot, 1998; Pleyer et al., 1993), nanoparticles (De Campos et al., 2003; Losa et al., 1993), nanocapsules (De Campos et al., 2004; Losa et al., 1993), and nanoemulsions. Nanoemulsions have achieved controlled and sustained drug release from the cornea and deeper penetration into the interior layers of the eye structure and the aqueous humor over the native conventional drug. Earlier successful nanoemulsions of pilocarpine, timolol maleate, and chloramphenicol have been formulated (Akhter et al., 2011). Extensive investigations were carried out over the last decade to support the view that cyclosporin A (CYA), a hydrophobic peptide with powerful immunosuppressive action, is effective to treat keratoconjunctivitis sicca, dry eye syndrome,

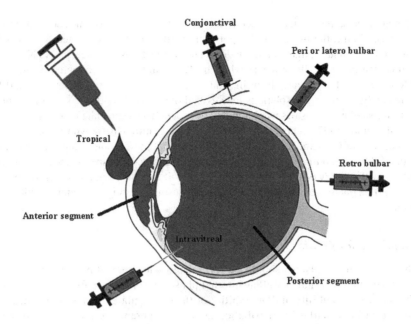

FIGURE 6.7 Routes of administration for ophthalmic disease treatment.

and for the prevention and treatment of corneal allograft rejection (Power et al.,1993). The poor aqueous solubility of CYA (6.6 mcg/mL) is a drawback for the preparation of solutions to be administered in the eye (Lallemand et al., 2003). Though there is evidence that for the treatment of these diseases the target sites are the cornea and conjunctiva (Acheampong et al., 1999; Gunduz and Ozdemir, 1994;), the CYA delivery systems investigated so far (i.e., emulsions, oils, collagen shields, nanocapsules, and liposomes) have not been successful. Collagen shields were found to provide sustained delivery of CYA onto the eye ball surface, however the use of such a system is restricted because it causes ocular irritation and blurring of vision (Dua et al., 1996). The oil-based vehicles possess serious limitations which include the slow partition rate of CYA into the epithelium of the cornea (Acheampong et al., 1999), the intraocular and/ or systemic drug absorption of CYA (Bellot et al., 1992; Foets et al., 1985), and the local adverse effects linked with the use of oils (symptoms of irritation, itching, blurred vision, transient epithelial keratitis, and toxic effects at the cornea) (Kaswan et al., 1989). In 2002, a CYA 0.05% lipid emulsion (Restasis™, Allergan, Irvine, California) received Food and Drug Administration (FDA) approval for the first time and it was the only therapy for patients inflicted with keratoconjunctivitis sicca, a lack of tear production assumed to be because of ocular inflammation. If the corneal concentration reached with dosing four times a day is not sufficient to prevent immunologic graft reactions, Restasis™ was not effective in preventing rejection after corneal allograft (Price and Price, 2006). An earlier attempt to improve the ocular penetration of CYA by developing CYA-loaded poly-o-caprolactone nanocapsules was effective for improving the transcorneal transport of CYA by lowering systemic absorption but could not produce significant CYA at the target tissue for a prolonged time (Calvo et al., 1996). Polymeric nanosystems release the drug entrapped in controlled manner over an extended time period, but in this situation it is not possible to get the therapeutic concentration instantly. Therefore, particularly for infectious as well as immunological diseases related to the eyes, it is desirable that the delivery system should release the bioactive molecule rapidly and then elicit the immunosuppressive action. Additionally, an increase in the tissue penetration and ocular retention further reduces the dose. The design of a drug administration system with enhanced properties in the ocular surface will be a challenging fact towards the management of external ocular disorders, including keratoconjunctivitis sicca or dry eye syndrome. From this information, and also from the evidence that cornea and conjunctiva possess a negative charge, it was thought that the use of mucoadhesive polymers, which might interact intimately with these extra-ocular structures, may enhance the concentration and residence time of the constituent drug. Nanoemulsion systems which are mucoadhesive and can act as carriers for drugs may become successful for effective ocular delivery. Until now, the mucoadhesive polymer mostly widely investigated is the cationic polymer chitosan (CH), which has gained attention due its unique features, like acceptable biodegradability, biocompatibility (Hirano et al., 1990; Knapczyk et al., 1989), and capacity to increase paracellular drug transport. Moreover, CH has recently been introduced as a material with a better potential for ocular therapy. Interestingly, whereas CH solutions may extend the corneal residence time of antibiotics (Felt et al., 1999), CH-coated nanocapsules were found to be more effective in enhancing the intraocular drug penetration (Calvo et al., 1997; Genta et al., 1997). Recent literature has shown the interaction and prolonged residence time of CH nanoparticles at the ocular mucosa after topical application to rabbits (De Campos et al., 2009). Depending upon these considerations, mucoadhesive nanoemulsions containing CH encapsulated with 3% w/v of CYA were produced as well as characterized to perform *in vitro*. An *in vivo* study was carried out to validate the corneal retention using γ-scintigraphy and exhibited a favorable biodistribution.

6.5.5.1 Excipients for Ocular Nanoemulsion

Topical ophthalmic nanoemulsions should be formulated with compatible vehicles and additives. The constituents of both external as well as internal phases of the nanoemulsions should be selected to enable the increased solubility and/or stability of the encapsulated ocular active drug. It should also be designed to infer ocular biodistribution as well as therapeutic index. Before the formulation design of the nanoemulsions, the solubility of the drug in the oil vehicle should be determined. In addition to this, other prerequisite information is needed for oil vehicle compatibility with the other preparation ingredients and with the already present ocular tissues should be compiled before the dosage form can be prepared (Anton et al., 2008; El-Aasser et al., 1986; Pouton, 1997; Shah et al., 2007; Sole et al., 2006). Table 6.5 shows the nanoemulsions used in ocular delivery (Nagariya, 2013).

TABLE 6.5

Nanoemulsion in Ocular Delivery

S.No.	Drug	Result	Reference
1	Pilocarpine	i. Physically stable for six months at 39.2°F ii. Bioavailability is pH dependent	Nagariya (2013)
2	Cyclosporine	i. The viscosity of the formulation is increased upon dilution with tear fluid ii. Prolonged precorneal residence time	
3	Adaprolol maleate	Safe and effective in human studies	
4	Indomethacin	Improved ocular bioavailability	
5	Piroxicam	Positively charged submicron emulsion shows a pronounced effect on the ulceration rate as well as epithelial defects in the management of corneal alkali-burning	
6	Dorzolamide HCl	i. Increased intensive treatment of glaucoma ii. More therapeutic efficacy iii. Better patient compliance	
7	Delta-8-tetrahydrocannabinol	i. Intense and long-lasting intraocular pressure depressant effect ii. No irritation	
8	Dexamethasone	Improved pharmacokinetic parameters	
9	Timolol	Bioavailability of timolol was increased	
10	Levobunolol	Higher apparent lipophilicity	
11	Chloramphenicol	Stability in the microemulsion formulations was increased	

6.5.6 Application of Nanoemulsions in Nasal Drug Delivery

The nasal cavity is easily accessible and rich in vascular plexus which permits topically administered drug uptake for rapid achievement of effective blood levels while avoiding intravenous catheters. It is most effectively accompanied by the distribution of drug solutions as a mist rather than as large drops which may aggregate and run off instead of being absorbed. Since the vascular bed in the nose is easily accessed, administration of drugs via the nasal route is emerging as an advantageous method of delivering medications directly into the blood stream. The same can eliminate the need for intravenous catheters and may achieve rapid, effective blood levels of the administered medication.

The advantages of applying medications via the nasal mucosa include the following:

1. The rich vascular plexus provides a straight route directly to the blood stream for treatments which can easily cross mucous membranes.
2. It avoids GI and hepatic destruction of drugs by acid and enzymes which allows drug cost efficiency, quick absorption, and probably more bioavailability than when administered orally.
3. In many intranasal medications, the absorption rates and plasma concentrations are comparable to intravenous administration and are typically better than intramuscular or subcutaneous routes.
4. Ease, convenience, and safety of administration.
5. Since nasal mucosa is very near to the brain, drug concentrations of cerebrospinal fluid (CSF) can exceed plasma concentrations. Therefore, they may rapidly reach therapeutic drug concentrations in the brain and spinal cord.

6.5.6.1 Nose–Brain Pathway

If the medication administered nasally comes in contact with the olfactory mucosa, there may be good evidence to suggest drug moiety transport can occur directly via the tissue and into the CSF (Banks et al., 2004; Henry, 1998; Sakane, 1991). The upper nasal cavity contains olfactory mucosa immediately behind the cribriform plate of the skull. It contains olfactory cells that traverse the cribriform plate and lengthen up to the cranial cavity. When medicated molecules come in contact with this particular mucosa they get rapidly crossed directly into the brain, avoiding the blood–brain barrier, and very quickly reaches CSF levels (faster than intravenously). The concept of molecule transfer from the nose to the brain is called the nose–brain pathway, which is important when centrally acting agents like sedatives, anti-seizure drugs, and opiates are administered nasally. Different authors reported that the nose–brain pathway leads to the immediate drug release of some nasal medicament to the CSF by avoiding the blood–brain barrier (Westin et al., 2006). Figure 6.8 represents the nose-to-brain delivery of medicaments.

6.5.6.2 Why Intranasal Is Better than Buccal or Oral

The most common reason for selecting the application of buccal or nasal routes is due to the slow onset of action via the oral route and relatively poor drug availability in the blood plasma. Due to the failure of the application of the nose–brain path, in oral medications, it results in much slower drug delivery to the brain. Surprisingly, oral medicaments are also generally refused by 30% of the pediatric group of patients, which makes them purely ineffective in this situation (Khalil et al., 2003). Buccal medications also need a cooperative patient who will retain the medication in their buccal and sublingual mucosa and will not swallow or spit it out. In general, even though buccal medications are given to volunteers in research settings, only approximately 56% of medicament remains in the buccal cavity for absorption (Anttila, 2003). Nasal drug delivery does have some problems based on the delivery method; one study reported that in 5.3% of pediatric patients they were not able to dispense drops into the nose because of resistance to this delivery technique (Yuen et al., 2008). However, nasal delivery is possible in the majority of cases since a number of devices have been developed to overcome delivery issues.

Kumar et al. (2008a) carried out an investigation for the preparation and characterization of olanzapine nanoemulsion (ONE) and olanzapine mucoadhesive nanoemulsion (OMNE) for nasal delivery and tried to evaluate their performance in animal models. It was hypothesized that the nanoemulsion/ mucoadhesive nanoemulsion-based alternative drug delivery route would lead to faster nose-to-brain transport and greater transport as well as distribution within the brain. This could reduce the adverse effects, reduce the dose and repeated administration requirements, and possibly reduce the cost of the therapy. They concluded that the study aptly demonstrated a rapid and larger extent of selective olanzapine nose-to-brain transport when compared with olanzapine solution in rats. This may be helpful in lowering the dose and dosing frequency and possibly enhances the therapeutic indices. But the clinical risk–benefit ratio of the formulations developed in this area will decide its future in the clinical practice for the treatment of schizophrenia.

Another study was carried out to formulate and characterize risperidone nanoemulsion (RNE) and mucoadhesive nanoemulsion (MRNE) and to evaluate their performance in an animal model. It was hypothesized that a nanoemulsion/MRNE-based drug delivery system would result in fast/immediate nose-to-brain transport of the risperidone and higher transport and distribution into and within the brain. This could lower the adverse effects, dose and time of administration, and probably the cost of the therapy. They made the conclusion that a significant quantity of risperidone could be quickly and effectively

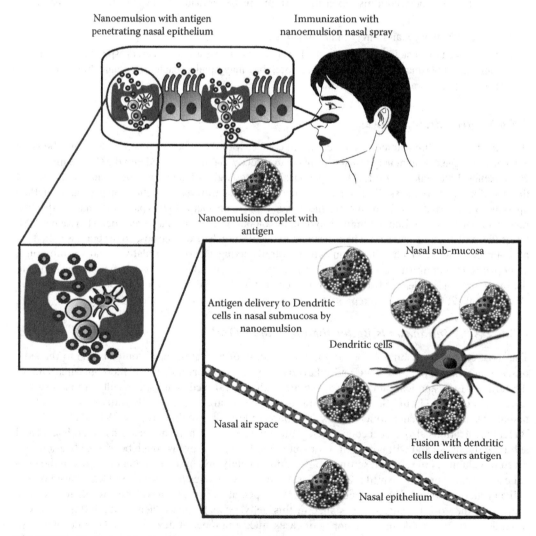

FIGURE 6.8 Mechanism of action of nanoemulsion vaccine.

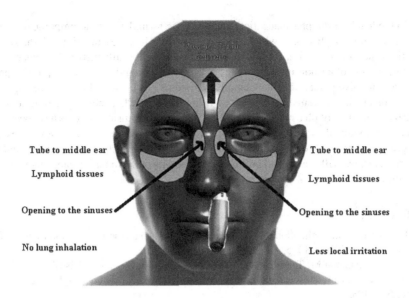

FIGURE 6.9 Nose to brain drug delivery system.

delivered to the brain via MRNE. Though animal studies showed the effectiveness of intranasal delivery of risperidone as an antipsychotic agent, clinical data must to be generated for further progress (Kumar et al. 2008b).

Central nervous system (CNS) diseases including schizophrenia, migraine, meningitis, Parkinson's disease, and Alzheimer's disease need drug delivery into the brain for treatment. However, such transport is challenging and problematic, especially for water-soluble drugs and drugs with a large molecular weight due to the impervious nature of the endothelial membrane partitioning the systemic circulation system and central interstitial fluid, the blood–brain barrier (Pardridge, 1999). Therefore, several medicaments may have been abandoned because sufficient drug concentration in the brain could not be achieved. "Biologics" are too big and hydrophilic to penetrate the blood–brain barrier and will be rapidly disrupted by GI enzymes/liver cytochromes if they are orally administered. A non-invasive therapy will be needed for patients requiring chronic dosing, as in dementia therapy. Animal and human investigations have shown that exogenous material transport directly from the nose-to-brain is a potential route for avoiding the blood–brain barrier (Illum, 2000). The route consists of the olfactory or trigeminal nerve systems that initialize in the brain and are terminated in the nasal cavity at the olfactory neuroepithelium or respiratory epithelium respectively. These are indeed the only externally exposed portions of the CNS and hence represent the straightest method of non-invasive penetration into the brain. However, drug solutions administered nasally that have been demonstrated to be transported directly from the nose-to-brain are too low in quantity, generally not more than 0.1%, and therefore are not suitable for use as therapeutics and no product is licensed specifically via this route (Illum, 2004). A nanoemulsion vaccine may be useful in that case. Figure 6.9 shows the mechanism of action of nanoemulsion vaccines.

The strategy of administering medicaments which are encapsulated in particulate vesicles (such as synthetic nanoparticles) into the olfactory epithelium can potentially improve the direct CNS delivery of drugs like biologics. If drugs are able to reach the CNS in a desirable quantity via this route, it will to generate interest in previously abandoned drug candidates and will enable an entirely novel prospect of CNS drug delivery.

6.6 Conclusions

Nanoemulsions are unique nanostructures, recently in focus for the delivery of drugs and biopharmaceuticals. By virtue of their typical composition and ultrastructure, these systems place an enabling technology in the hands of the pharmaceutical and cosmetic formulators to attempt the delivery of such drugs and cosmetics through routes which were not possible in an earlier time. Perhaps the biggest potential of the systems is in the improvement of the physiological availability of the bioactives through parenteral, enteral, and other non-invasive routes of delivery. Through this chapter, it is hoped that the spectrum of non-invasive potential of nanoemulsion systems has been highlighted. In the field of cosmetics, the nanoemulsion systems are helping to bring to the market more appealing and functional cosmeceuticals. A plethora of pharmaceutical and cosmetic nanoemulsion systems have been reported in research literature, though the translation to the clinical market is not as vivid. However, it is hoped that with the international research community intensifying its efforts to study, understand, and formulate these nanostructures, we will see several promising nanoemulsion formulations appearing in the market in the near future.

REFERENCES

Abdul BM. Development and validation of glibenclamide in nanoemulsion formulation by using RP-HPLC. J Pharm Biomed Sci. 2011;8:1–5.

Acheampong AA, Shackleton M, Tang-Liu DDS, Ding S, Stern ME, Decker R. Distribution of Cyclosporin A in ocular tissues after topical administration to albino rabbits and beagle dogs. Cur Eye Res. 1999;18(2):91–103.

Acosta E. Bioavailability of nanoparticles in nutrient and nutraceutical delivery. Curr Opin Colloids Interf Sci. 2009;14(1):3–15.

Agrawal AG, Kumar A, Gide PS. Formulation of solid self-nanoemulsifying drug delivery systems using N-methyl pyrrolidone as cosolvent. Drug Dev Ind Pharm. 2015;41(4):594–604.

Akhter S, Talegaonkar S, Khan ZI, Jain GK, Khar RK, Ahmad FJ. Assessment of ocular pharmacokinetics and safety of Ganciclovir loaded nanoformulation. Biomed Nanotechnol. 2011;7:144–5.

Ali M. Topical nanoemulsion of turmeric oil for psoriasis: Characterization, ex vivo and in vivo assessment. Int J Drug Del. 2012;4:184–97.

Ali MS. Formulation, characterization and in-vivo assessment of topical nanoemulsion of betamethasone valerate for psoriasis and dermatose. Int J Pharm. 2013;3:186–99.

Ali MS. Formulation, characterization and in-vivo study of nanoemulsion topical gel of beclomethasone dipropionate for psoriasis. World J Pharmacy Pharm Sci. 2012;1:839–57.

Alvarez-Figueroa MJ, Blanco-Mendez J. Transdermal delivery of methotrexate: Iontophoretic delivery from hydrogels and passive delivery from Microemulsions. Int J Pharm. 2001;215(1–2):57–65.

Amani A, Peter Y, Henry C, Clark BJ. Evaluation of a nanoemulsion-based formulation for respiratory delivery of budesonide by nebulizers. AAPS PharmSciTech. 2010;11(3):1147–51.

Amselem S, Friedman D. Solid fat nanoemulsions as drug delivery vehicles. US Patent 5576016. 1996.

André LS, Francisco AJ, Lourena MV, Lucymara FAL, Lucila CME, Anselmo GO, Eryvaldo STE. Physical factors affecting plasmid DNA compaction in stearylamine-containing nanoemulsions intended for gene delivery. Pharmaceuticals. 2012;5(6):643–54.

Anton N, Benoit JP, Patrick S. Design and production of nanoparticles formulated from nano-emulsion templates – A review. J Control Release. 2008;128(3):185–99.

Anttila M. Bioavailability of dexmedetomidine after extra vascular doses in healthy subjects. Br J Clin Pharmacol. 2003;56:691–3.

Araujo FA, Kelmann RG, Araujo BV, Finatto RB, Teixeira HF, Koester LS. Development and characterization of parenteral nanoemulsions containing thalidomide. Eur J Pharm Sci. 2011;42(5):238–45.

Ajazuddin A, Saraf S. Applications of novel drug delivery system for herbal formulations. Fitoterapia. 2010;81(7):680–89.

Baboota S, Shakeel F, Ahuja A, Ali J, Shafiq S. Design, development and evaluation of novel nanoemulsion formulations for transdermal potential of celecoxib. Acta Pharm. 2007;57:315–32.

Bali V, Ali M, Ali J. Study of surfactant combinations and development of a novel nanoemulsion for minimising variations in bioavailability of ezetimibe. Colloids Surf B Biointerfaces. 2010;76(2):410–20.

Banks WA, During MJ, Niehoff ML. Brain uptake of the glucagon-like peptide-1 antagonist extending after intranasal administration. J Pharmacol Exp Ther. 2004;309(2):469–75.

Barakat N. Formulation design of indomethacin-loaded nanoemulsion for transdermal delivery. Pharm Anal Acta. 2011;S2:1–8.

Behzad SM. Optimization of ibuprofen delivery through rat skin from traditional and novel nanoemulsion formulations. Iran J Pharm Res. 2012;11:47–58.

Bellot JL, Alió JL, Ruiz Moreno JM, Artola A. Corneal concentration and systemic absorption of cyclosporin-a following its topical application in the rabbit eye. Ophthalmic Res. 1992;24:351–6.

Bernardi DS. Formation and stability of oil-in-water nanoemulsions containing rice bran oil: In vitro and in vivo assessments. J Nanobiotech. 2011;9:1–9. DOI:10.1186/1477-3155-9-44.

Bhanushali RS, Gatne MM, Gaikwad RV, Bajaj AN, Morde MA. Nanoemulsion based intranasal delivery of anti-migraine drugs for nose to brain targeting. Indian J Pharm Sci. 2009;71(6):707–9.

Bivas-Benita M, Oudshoorn M, Romeijn S. Cationic submicron emulsions for pulmonary DNA immunization. J Control Rel. 2004;100(1):145–55.

Breitenbach J. Melt extrusion: From process to drug delivery technology. Eur J Pharm Biopharm. 2002;54:107–17.

Brussel F, Manoela L, Luisa BW, Michelle F, Koester LS, Teixeira HF. Nanoemulsions as parenteral drug delivery systems: A review. Chem News. 2012;35(9):34–9.

Burapapadh K, Takeuchi H, Sriamornsak P. Novel pectin-based nanoparticles prepared from nanoemulsion templates for improving in vitro dissolution and in vivo absorption of poorly water-soluble drug. Eur J Pharm Biopharm. 2012;82(2):250–61.

Calderilla-Fajardo SB, Cazares-Delgadillo J, Villalobos-Garcia R, Quintanar-Guerrero D, Ganem-Quintanar A, Robles R. Influence of sucrose esters on the in vivo percutaneous penetration of octyl methoxycinnamate formulated in nanocapsules, nanoemulsion, and emulsion. Drug Dev Ind Pharm. 2006;32:107–13.

Calvo P, Alonso MJ, Vila Jato JL, Robinson JR. Improved ocular bioavailability of indomethacin by novel ocular drug carriers. J Pharm Pharmacol. 1996;48:1147–52.

Calvo P, Remunan-Lopez C, Vila-Jato CL, Alonso MJ. Novel hydrophilic chitosan- polyethylene oxide nanoparticles as protein carriers. J App Poly Sci. 1997;63:125–32.

Caraglia M, Rosa GD, Abbruzzese A, Leonetti C. Nanotechnologies: New opportunities for old drugs. The case of aminobisphosphonates. J Nanomedic Biotherapeu Discover. 2011;1(1):103–104. DOI:10.4172/2155-983X.1000103e.

Charles L, Attama AA. Current state of nanoemulsions in drug delivery. J Biomat Nanobiotech. 2011;2:626–39.

Chavhan SS, Petkar KC, Sawant KK. Simvastatin nanoemulsion for improved oral delivery: Design, characterisation, in vitro and in vivo studies. J Microencapsul. 2013;30(8):771–9.

Chein YW, Cabana BE, Mares SE. Ocular controlled release drug administration. In: Chein YW, editor. Novel drug delivery systems: Fundamentals, development concepts, biomedical assessments (Drugs and the pharmaceutical sciences). New York: Marcel Dekker; 1982. p. 13–55.

Chen H, Khemtong C, Yang X, Chang X, Gao J. Nanonization strategies for poorly water-soluble drugs. Drug Discov Today. 2011;16(7–8):354–60.

Chhabra G, Chuttani K, Mishra AK, Pathak K. Design and development of nanoemulsion drug delivery system of amlodipine besilate for improvement of oral bioavailability. Drug Dev Ind Pharm. 2011;37(8):907–16.

Clark AM, Jepson MA, Hirst BH. Exploiting M cells for drug and vaccine delivery. Adv Drug Deliv Rev. 2001;50(1–2):81–106.

Courrier H, Krafft MP, Nakamura S, Shibata O, Vandamme T. Water-in-fluorocarbon reverse emulsion as a pulmonary drug delivery system: Effect on the lung as modelled by a phospholipid monolayer. STP Pharma Pratiques. 2003;13:22–6.

Csaba N, Garcia Fuentes M, Alonsa MJ. Nanoparticles for nasal vaccination. Adv Drug Deliv Rev. 2009;61(2):140–57.

De Campos AM, Sanchez A, Gref R, Calvo P, Alonso MJ. The effect of a PEG versus a chitosan coating on the interaction of drug colloidal carriers with the ocular mucosa. Eur J Pharm Sci. 2003;20:73–81.

De Campos AM, Diebold Y, Carvaiho ELS, Sanchez A, Alonso MJ. Chitosan nanoparticles as new ocular drug delivery system: In vitro stability, in vivo fate, and cellular toxicity. Pharm Res. 2004;21:803–10.

Delgado Charro MB, Iglesias Vilas G, Blanco Mendez J, Lopez Quintela MJ, Marty MA, Guy JP. Delivery of a hydrophilic solute through the skin from novel microemulsion systems. Eur J Pharm Biopharm. 1997;43(1):37–42.

Doh HJ, Jung Y, Balakrishnan P, Cho HJ, Kim DD. A novel lipid nanoemulsion system for improved permeation of granisetron. Colloids Surf B Biointerfaces. 2013;101:475–80.

Dreher F, Walde P, Walter P, Wehrli E. Interaction of a lecithin microemulsion gel with human stratum corneum and its effect on transdermal transport. J Control Rel. 1997;45(2):131–40.

Dua HS, Jindal VK, Gomes JAP, Amoaku WA, Donoso LA, Laibson PR, Mahlberg K. The effect of topical cyclosporin on conjunctiva-associated lymphoid tissue (CALT). Eye. 1996;10:433–8.

Effendy I. Therapeutics strategy of onychomycosis. J Eur Acad Derm Venereol. 1995;4:S10–3.

El-Aasser MS, Lack CD, Vanderhoff JW, Fowkes FM. Mini emulsification process different form of spontaneous emulsification. Colloids Surf. 1986;29:103–18.

Ethirajan A, Schoeller K, Musyanovych A, Ziener U, Landfester K. Synthesis and optimization of gelatin nanoparticles using the miniemulsion process. Biomacromolecules. 2008;9(9):2383–9.

Faiyaz S. Comparative pharmacokinetic profile of aceclofenac from oral and transdermal application. J Bioequiv Bioavail. 2009;1:13–7.

Felt O, Furrer P, Mayer JM, Plazonnet B, Buri P, Gurny R. Topical use of chitosan in ophthalmology: Tolerance assessment and evaluation of pre-corneal retention. Int J Pharm. 1999;180:185–93.

Feynman, Richard P, *There's Plenty of Room at the Bottom*. Engineering and Science.1960; 23:22–36. Available from: http://resolver.caltech.edu/CaltechES:23.5.1960Bottom [cited on 14/12/2018].

Foets B, Missotten L, Vanderveeren P, Goossens W. Prolonged survival of allogeneic corneal grafts in rabbits treated with topically applied cyclosporine: A systemic absorption and local immunosuppressive effect. Br J Ophthalmol. 1985;69:600–3.

Ganta S, Amiji M. Co administration of paclitaxel and curcumin in nanoemulsion formulation to overcome multidrug resistance in tumor cells. Mol Pharm. 2009;6(3):928–39.

Ganta S, Deshpande D, Korde A, Amiji M. A review of multifunctional nanoemulsion systems to overcome oral and CNS drug delivery barriers. Mol Membr Biol. 2010;27(7):260–73.

Ganta S, Singh A, Patel NR, Cacaccio J, Rawal YH, Davis BJ, Amiji MM, Coleman TP. Development of EGFR-targeted nanoemulsion for imaging and novel platinum therapy of ovarian cancer. Pharm Res. 2014a;31(9):2490–2.

Ganta S, Talekar M, Singh A, Coleman TP, Amiji MM. Nanoemulsions in translational research-opportunities and challenges in targeted cancer therapy. AAPS PharmSciTech. 2014b;15(3):694–708.

Gasco MR, Gallarate M, Pattarino F. In vitro permeation of azelaic acid from viscosized microemulsions. Int J Pharm. 1991;69:193–6.

Gaur PK, Mishra S, Purohit S, Dave K. Trans-dermal drug delivery system: A review. Asian J Pharm Clin Res. 2009;2(1):14–20.

Genta I, Conti B, Perugini P, Pavaneto F, Spadaro A, Puglisi G. Bioadhesive microspheres for ophthalmic administration of acyclovir. J Pharm Pharmacol. 1997;49:737–42.

Gidwani SK, Singnurkar SP. A process for preparing a nanoemulsion for delivery of dithranol. Indian Patent 208817 518/MUM/2003. 2003.

Goyal A, Kumar S, Nagpal M, Singh I, Arora S. Potential of novel drug delivery systems for herbal drugs. Indian J Pharm Educ Res. 2011;45:225–35.

Guglielmini G. Nanostructured novel carrier for topical application. Clin Dermatol. 2008;26:341–6.

Gunduz K, Ozdemir O. Topical cyclosporin treatment of keratoconjunctivitis sicca in secondary Sjogren's syndrome. Acta Ophthalmol. 1994;72:438–42.

Gupta AK, Albreski D, Rossp QJ, Konnikov N. The use of new oral antifungal agents itrconazole, terbinafine and flucanozole to treat onychomycosis and other dermatomycoses. Cur Prob Dermatol. 2001;13(4):213–46.

Haake HM, Lagrene H, Brands A, Eisfeld W, Melchior D. Determination of the substantivity of emollients to human hair. J Cosmet Sci. 2007;58:443–50.

Hamouda T, Flack M, Baker JR. Development of a novel antifungal drug (NB-002) for topical application in humans. Paper presented at: American Academy of Dermatology Annual Meeting; 2008; San Antonio, TX.

Heidi, M.M., Yun, Seok, R., Xiao, W., 2009. Nanomedicine in pulmonary delivery. Int. J.Nanomed. 4:299–319.

Henry RJ. A pharmacokinetic study of imidazolam in dogs: Nasal drop vs. atomizer administration. Paed Dent. 1998;20(5):321–6.

Hirano S, Seino H, Akiyama I, Nonaka I. Chitosan: A biocompatible material for oral and intravenous administration. In: Gebelein CG, Dunn RL, editors. Progress in biomedical polymers. New York: Plenum Press; 1990. p. 283–9.

Huafeng Z. Preparation and characterization of a lecithin nanoemulsion as a topical delivery system. Nanoscale Res Lett. 2010;5:224–30.

Hughes PM, Mitra AK. Overview of ocular drug delivery and iatrogenic ocular cytopathologies. In: Mitra AK, editor. Ophthalmic drug delivery systems. New York: Marcel Dekker; 1993. p. 1–27.

Huglin D, Roding JF, Supersaxo AW, Weder HG. Use of nanodispersions in cosmetic end formulations. US Patent 2005/0191330. 2005.

Hunter RJ. Foundations of colloid science: Volume II. Oxford: Oxford Science Publications; 1989.

Hussein OA, Salama HA, Ghorab M, Mahmoud AA. Nanoemulsion as a potential ophthalmic delivery system for dorzolamide hydrochloride. AAPS PharmSciTech. 2009;10(3):808–19.

Illum L. Is nose-to-brain transport of drugs in man a reality? J Pharm Pharmacol. 2004;56:3–17.

Illum L. Transport of drugs from the nasal cavity to the central nervous system. Eur J Pharm Sci. 2000;11:1–18.

Israelachvili J. Intermolecular and surface force. 2nd ed. London: Academic Press; 1992.

James R, Baker Jr, Tarek H, Amy S. Method of preventing and treating microbial infection. US Patent number: 6,506,803. 2003.

James R, Baker Jr, Tarek H, Amy S. Non-toxic antimicrobial compositions and methods of use. US Patent number: US 6,559,189 B2. 2003.

James R, Baker Jr, Tarek H, Amy S. Non-toxic antimicrobial compositions and methods of use. US Patent number: 6,635,676 B2. 2003.

Jean TS, Odile S, Sylvie L. Nanoemulsion based on oxyethylenated or non-oxyethylenated sorbitan fatty esters, and its uses in the cosmetics, dermatological, and/or ophthalmological fields. US Patent number: 6,335,022 B1. 2002.

Jignesh D, Modi JN, Patel K. Nanoemulsion-based gel formulation of aceclofenac for topical delivery. Int J Pharmacy Pharmaceut Sci Res. 2011;1:6–12.

Kakumanu S, Kambalapally S, Nicolosi RJ. A self assembling nanoemulsion of lovastatin (SANEL) decreases cholesterol accumulation and Apob-100 secretion greater than lovastatin alone a Hepg2 cell line. J Nanomed Nanotechol. 2012;3(8):1–4. DOI:10.4172/2157-7439.1000151.

Kaswan RL, Salisbury MA, Ward DA. Spontaneous canine keratoconjunctivitis sicca: A useful model for human keratoconjunctivitis sicca: Treatment with cyclosporine eye drops. Arch Ophthalmol. 1989;107(8):1210–16.

Kayser O, Lemke A, Hernandez-Trejo N. The impact of nanobiotechnology on the development of new drug delivery systems. Curr Pharm Biotechnol. 2005;6:3–5.

Kelmann RG, Kuminek G, Teixeira H, Koester LS. Carbamazepine parenteral nanoemulsions prepared by spontaneous emulsification process. Int J Pharm. 2007;342(1–2):231–9.

Kemken JA, Ziegler A, Muller BW. Influence of supersaturation on the pharmacodynamic effect of bupranolol after dermal administration using microemulsions as vehicle. Pharm Res. 1992;9(4):554–8.

Khalil S, Vije H, Kee S. A paediatric trial comparing midazolam/syrpalta mixture with premixed midazolam syrup (Roche). Paed Anaesth. 2003;13:205–9.

Kim BS, Won M, Lee KM, Kim CS. In vitro permeation studies of nanoemulsions containing ketoprofen as a model drug. Drug Del. 2008;15(7):465–9.

Knapczyk J, Kro'wczynski L, Krzck J, Brzeski M, Nirnberg E, Schenk D, Struszcyk H Requirements of chitosan for pharmaceutical and biomedical applications. In: Skak-Braek G, Anthonsen T, Sandford P, editors. Chitin and chitosan: Sources, chemistry, biochemistry, physical properties and applications. London: Elsevier; 1989. p. 657–63.

Kobayashi Y, Miyamoto M, Sugibayashi K, Morimoto Y. Drug permeation through three layers of human nail plate. J Pharm Pharmacol. 1999;51(3):271–8.

Kong M, Chen XG, Kweon DK, Park HJ. Investigations on skin permeation of hyaluronic acid based nano-emulsion as transdermal carrier. Carbohydr Polym. 2011;86(2):837–43.

Kreilgaard M, Pedersen EJ, Jaroszewski JW. NMR characterization and transdermal drug delivery potentials of microemulsion systems. J Control Rel. 2000;69(3):421–33.

Kreilgaard M. Dermal pharmacokinetics of microemulsion formulations determined by in-vitro microdialysis. Pharm Res. 2001;18(3):367–73.

Kreilgaard M, Kemme MJB, Burggraaf J, Schoemaker RC, Cohen AF. Influence of a microemulsion vehicle on cutaneous bioequivalence of a lipophilic model drug assessed by microdialysis and pharmacodynamics. Pharm Res. 2001a;18(5):593–9.

Kreilgaard M, Kemme MJB, Burggraaf J, Schoemaker RC, Cohen AF. Influence of a microemulsion vehicle on cutaneous bioequivalence of a lipophilic model drug assessed by microdialysis and pharmacodynamics. Pharm Res. 2001b;18(5):593–9.

Kriwet K, Muller Goymann CC. Diclofenac release from phospholipid drug systems and permeation through excised human stratum corneum. Int J Pharm. 1995;125(2):231–42.

Ktistis G, Niopas I. A study on the in-vitro percutaneous absorption of propranolol from disperse systems. J Pharm Pharmacol. 1998;50(4):413–19.

Kumar, M., Misra, A., Babbar, A. K., Misra A. K., Mishra, P. Intranasal nanoemulsion based brain targeting drug delivery system of risperidone. Int J Pharm. 1998;358(1–2):285–91.

Kumar M, Misra A, Babbar AK, Mishra AK, Mishra P, Pathak K. Intranasal nanoemulsion based brain targeting drug delivery system of risperidone. Int J Pharm. 2008a;358(1–2):285–91.

Kumar M, Misra A, Mishra AK, Mishra P, Pathak K. Mucoadhesive nanoemulsion-based intranasal drug delivery system of olanzapine for brain targeting. J Drug Target. 2008b;16(10):806–14.

Kumar S, Talegaonkar S, Lalit MN, Zeenat IK. Design and development of ciclopirox topical nanoemulsion gel for the treatment of subungual onychomycosis. Ind J Pharm Edu Res. 2012;46(4):303–11.

Kuo F, Subramanian B, Kotyla T, Wilson TA, Yoganathan S, Nicolosi RJ. Nanoemulsions of an anti-oxidant synergy formulation containing gamma tocopherol have enhanced bioavailability and anti-inflammatory properties. Int J Pharm. 2008;363(1–2):206–13.

Lala RR, Awari NG. Nanoemulsion-based gel formulations of COX-2 inhibitors for enhanced efficacy in inflammatory conditions. Appl Nanosci. 2014;4(2):143–51.

Lallemand FO, Felt-Baeyens K, Besseghir F, Behar-Cohen GR. Cyclosporine A delivery to the eye: A pharmaceutical challenge. Eur J Pharm Biopharm. 2003;56:307–18.

L'alloret F, Aubrun-Sonneville O, Simon Net JT. Nanoemulsion containing nonionic polymers, and its uses. US Patent 6998426. 2006.

Lawrence MJ, Rees GD. Microemulsion-based media as novel drug delivery systems. Adv Drug Deliv Rev. 2000;45(1):89–121.

Lee PJ, Langer R, Shastri VP. Novel microemulsion enhancer formulation for simultaneous transdermal delivery of hydrophilic and hydrophobic drugs. Pharm Res. 2003;20(2):264–9.

Lifei Z, Miao L, Junxing D, Yiguang J. Dimethyl silicone dry nanoemulsions inhalations: Formulation study and anti lung injury effect. Int J Pharm. 2015;491:292–8.

Liu CH, Yu SY. Cationic nanoemulsions as non-viral vectors for plasmid DNA delivery. Colloids Surf B Biointer. 2010;79(2):509–15.

Liu F, Yang J, Huang L, Liu D. Effect of non-ionic surfactants on the formation of DNA/emulsion complexes and emulsion-mediated gene transfer. Pharm Res. 1996;13(11):1642–6.

Losa C, Marchal-Heussler L, Orallo F, Vila Jato JL, Alonso MJ. Design of new formulations for topical ocular administration: Polymeric nanocapsules containing metipranolol. Pharm Res. 1993;10:80–7.

Luo Y, Chen D, Ren L, Zhao X, Qin J. Solid lipid nanoparticles for enhancing vinpocetine's oral bioavailability. J Control Release. 2006;114(1):53–9.

Ma Y, Li H, Guan S. Enhancement of the oral bioavailability of breviscapine by nanoemulsions drug delivery system. Drug Dev Ind Pharm. 2015;41(2):177–82.

Manchun S, Dass CR, Sriamornsak P. Designing nanoemulsion templates for fabrication of dextrin nanoparticles via emulsion cross-linking technique. Carbohydr Polym. 2014;101:650–55.

Mansour HM, Rhee YS, Wu X. Nanomedicine in pulmonary delivery. Int J Nanomed. 2009;4:299–319.

McClements DJ. Food emulsions: Principles, practice, and techniques. Boca Raton, FL: CRC Press; 2005.

McClements DJ. Emulsion design to improve the delivery of functional lipophilic components. Annu Rev Food Sci Technol. 2010;1(1):241–69.

McClements DJ. Nanoemulsions versus microemulsions: Terminology, differences, and similarities. Soft Matter. 2012;8(6):1719–29.

Mehnert W, Mader K. Solid lipid nanoparticles: Production, characterization and application. Adv Drug Deliv Rev. 2001;47(2–3):165–96.

Middleton DL, Leung SS, Robinson JR. Ocular bioadhesive delivery systems. In: Lenaerts V, Gurny R, editors. Bioadhesive drug delivery systems. Boca Raton, FL: CRC Press; 1990. p. 179–202.

Mistry A, Stolnik S, Illum L. Nanoparticles for direct nose-to-brain delivery of drugs. Int J Pharm. 2009;379(1):146–57.

Monteiro LM, Lione VF, do Carmo FA, do Amaral LH, da Silva JH, Nasciutti LE, et al. Development and characterization of a new oral dapsone nanoemulsion system: Permeability and in silico bioavailability studies. Int J Nanomed. 2012;7:5175–82.

Muller Goymann CC. Physicochemical characterization of colloidal drug delivery systems such as reverse micelles, vesicles, liquid crystals and nanoparticles for topical administration. Eur J Pharm Biopharm. 2004;58(2):343–56.

Murdan S. Drug delivery to nail following topical application. Int J Pharm. 2002;236(1–2):1–26.

Nagariya K. Nanoemulsion: Safe, stable and effective formulation system for ophthalmology. Am J Pharm Tech Res. 2013;3(3):252–67.

Nasr M, Nawaz S, Elhissi A. Amphotericin B lipid nanoemulsion aerosols for targeting peripheral respiratory airways via nebulization. Int J Pharm. 2012;436(1–2):611–6.

Nesamony J, Kalra A, Majrad MS, Boddu SH, Jung R, Williams FE, et al. Development and characterization of nanostructured mists with potential for actively targeting poorly water-soluble compounds into the lungs. Pharm Res. 2013;30(10):2625–39.

Nesamony J, Shah IS, Kalra A, Jung R. Nebulized oil-in-water nanoemulsion mists for pulmonary delivery: Development, physico-chemical characterization and in vitro evaluation. Drug Dev Ind Pharm. 2014;40(9):1253–63.

Niven RW, Schreier H. Nebulization of liposomes I: Effects of lipid composition. Pharm Res. 1990;7(11):1127–33.

Osborne DW, Ward AJ, Neil KJ. Microemulsions as topical delivery vehicles: In-vitro transdermal studies of a model hydrophilic drug. J Pharm Pharmacol. 1991;43(6):450–54.

Overbeek JTHG. The first rideal lecture. Microemulsions, a field at the border between lyophobic and lyophilic colloids. Faraday Discuss. Chem Soc. 1978;65:7–19.

Pardridge WM. Non-invasive drug delivery to human brain using endogenous blood brain barrier transport system. Pharm Sci Tech Today. 1999;2(2):49–59.

Patel R, Patel KP. Advances in novel parenteral drug delivery systems. Asian J Pharm. 2010;4(3):193–9.

Patel N, Marlow M, Lawrence MJ. Formation of nonionic fluorinated surfactant micro emulsions in hydro fluorocarbon 134a (HFC 134a). J Colloid Interface Sci. 2003;258(2):345–53.

Patton TF, Robinson JR. Quantitative precorneal disposition of topically applied pilocarpine nitrate in rabbit eyes. J Pharm Sci. 1976;65:1295–301.

Pleyer U, Lutz S, Jusko WJ, Nguyen KD, Narawane M, Rückert D, Mondino BJ, Lee VH, Nguyen K. Ocular absorption of topically applied FK506 from liposomal and oil formulations in the rabbit eye. Invest Ophthalmol Vis Sci. 1993;34(9):2737–42.

Poluri K. Formulation and preparation of felodipine nanoemulsions. Asian J Pharm Clin Res. 2011;4:116–17.

Pouton CW. Formulation of self emulsifying drug delivery systems. Adv Drug Deliv Rev. 1997;25:47–58.

Power WJ, Mullaney P, Farrell M, Collum LM. Effect of topical cyclosporin A on conjunctival T cells in patients with secondary Sjögren's syndrome. Cornea. 1993;12:507–11.

Price MO, Price FW Jr. Efficacy of topical cyclosporine 0.05% for prevention of cornea transplant rejection episodes. Am J Ophthalmol. 2006;113(10):1785–90.

Primo FL, Michieloto L, Rodrigues MAM, Macaroff PP, Morais PC, Lacava ZGM, Bently MVLB, Tedesco AC. Magnetic nanoemulsions as drug delivery system for Foscan: Skin permeation and retention in vitro assays for topical application in photodynamic therapy (PDT) of skin cancer. J Magnet Magnet Mat. 2007;311(1):354–57.

Quemin E. Translucent nanoemulsion, production method, and uses thereof in the cosmetic, dermatological and/or ophthalmological fields. US Patent 6902737. 2005.

Rachmawati H, Haryadi BM. The influence of polymer structure on the physical characteristic of intraoral film containing BSA loaded nanoemulsion. J Nanomed Nanotechnol. 2014;5:2–6.

Ravi TPU, Padma T. Nanoemulsions for drug delivery through different routes. Res Biotechnol. 2011;2(3):1–13.

Rhee YS, Choi JG, Park ES, Chi SC. Transdermal delivery of ketoprofen using microemulsions. Int J Pharm. 2001;228(1):161–70.

Ribier DA, Simonnet JT, Legret S. Transparent nanoemulsion less than 100 NM based on fluid non-ionic amphiphilic lipids and use in cosmetic or in dermopharmaceuticals. US Patent 5753241. 1998.

Rocha-Filho PA, Camargo MFP, Ferrari M, Maruno M. Influence of lavender essential oil addition on passion fruit oil nanoemulsions: Stability and in vivo study. J Nanomed Nanotechnol. 2014;5(2):1–11. DOI:10.4172/2157-7439.1000198.

Rodriguez L, Passerini N, Cavallari C, Cini M, Sancin P, Fini A. Description and preliminary evaluation of a new ultrasonic atomizer for spray-congealing process. Int J Pharm. 1999;183(2):133–43.

Rubio MC, Ariz IR, Gil J, Benito J, Rezusta A. Potential fungicidal effect of voriconazole against Candida spp. Int J Antimicrob Agents. 2005;25(3):264–7.

Rutvij JP, Gunjan JP, Bharadia PD, Pandya VM, Modi DA. Nanoemulsion: An advanced concept of dosage form. Int J Pharm Cosmetol. 2011;1(5):122–33.

Sahoo SK, Labhasetwar V. Nanotech approaches to drug delivery and imaging. Drug Discov Today. 2003;8(1):1112–20.

Sakane T. Transport of cephalexin to the cerebrospinal fluid directly from the nasal cavity. J Pharm Pharmacol. 1991;43(6):449–51.

Salim N. Phase behaviour, formation and characterization of palm-based esters nanoemulsion formulation containing ibuprofen. J Nanomedic Nanotechnol. 2011;2(4):1–5. DOI:10.4172/2157-7439.1000113.

Sandig AG, Campmany ACC, Campos FF, Villena MJM, Naveros BC. Transdermal delivery of imipramine and doxepin from newly oil-in-water nanoemulsions for an analgesic and anti-allodynic activity: Development, characterization and in vivo evaluation. Colloids Surf B Biointerfaces. 2013;103:558–65.

Sarker DK. Engineering of nanoemulsions for drug delivery. Curr Drug Deliv. 2005;2:297–310.

Schmalfus. O., 2001. Nanoemulsion a base de copolymeres blocs d'oxyde d'ethylene et d'oxyde de propylene, et ses utilisations dans les domaines cosmetique, dermatologique et/ou ophtalmologique. Available at: "https://patents.google.com/patent/FR2788007B1/fr" FR2788007B1.

Schmalfus U, Neubart R, Wohlrab W. Modification of drug penetration into human skin using microemulsions. J Control Rel. 1997;46(3):279–85.

Shah N. Formulation, design and characterization of microemulsion based system for topical delivery of antipsoriatic drug. World J Pharmacy Pharm Sci. 2013;3:1464–80.

Shah RB, Zidam AS, Funck T, Tawakkul MA, Nguyenpho A, Khan MA. Quality by design: Characterization of self-nanoemulsified drug delivery systems (SNEDDSs) using ultrasonic resonator technology. Int J Pharm. 2007;341(1–2):189–94.

Shakeel F, Baboota S, Ahuja A, Ali J, Mohammed A, Shafiq S. Nanoemulsions as vehicles for transdermal delivery of aceclofenac. AAPS PharmSciTech. 2007;8(4):191–9.

Shakeel F, Ramadan W. Transdermal delivery of anticancer drug caffeine from water-in-oil nanoemulsions. Colloids Surf B Biointer. 2010;75(1):356–62.

Sheela AY, Sushil KP, Singh DK, Shaikh A. A review: Nanoemulsion as vehicle for transdermal permeation of antihypertensive drug. Int J Pharm Pharm Sci. 2012;4(1):41–4.

Shen Q, Wang Y, Zhang Y. Improvement of colchicine oral bioavailability by incorporating eugenol in the nanoemulsion as an oil excipient and enhancer. Int J Nanomedicine. 2011;6:1237–43.

Shi AM, Li D, Wang LJ, Li BZ, Adhikari B. Preparation of starch-based nanoparticles through high-pressure homogenization and miniemulsion cross-linking: Influence of various process parameters on particle size and stability. Carbohydr Polym. 2011;83(4):1604–10.

Sieg JW, Robinson JR. Vehicle effects on ocular drug bioavailability II: Evaluation of pilocarpine. J Pharm Sci. 1977;66(9):1222–8.

Silva HR. Surfactant-based transdermal system for fluconazole skin delivery. J Nanomed Nanotechnol. 2014;5(5):1–10. DOI:10.4172/2157-7439.1000231.

Silva AP, Nunes BR, de Oliveira MC, Koester LS, Mayorga P, Bassani VL, Teixeira HF. Development of topical nanoemulsions containing the isoflavone genistein. Pharmazie. 2009;64(1):32–5.

Simonnet JT, Sonneville O, Legret S. Nanoemulsion based on phosphoric acid fatty acid esters and its uses in the cosmetics, dermatological, pharmaceutical, and/or ophthalmological fields. US Patent 6274150. 2001.

Singh BP. Development and characterization of a nanoemulsion gel formulation for transdermal delivery of carvedilol. Int J Drug Dev Res. 2012;4(1):151–61.

Singh G, Pai RS. In vitro and in vivo performance of supersaturable self-nanoemulsifying system of trans-resveratrol. Artif Cells Nanomed Biotechnol. 2014 [cited Epub ahead of print]. DOI:10.3109/21691401.2014.966192.

Singh KK, Vingkar S. Formulation, antimalarial activity and biodistribution of oral lipid nanoemulsion of primaquine. Int J Pharm. 2008;347(1–2):136–43.

Sivakumar M, Tang SY, Tan KW. Cavitation technology – A greener processing technique for the generation of pharmaceutical nanoemulsions. Ultrason Sonochem. 2014;21(6):206920–83.

Sole I, Maestro A, Gonzalez C, Solans C, Gutierrez JM. Optimization of nano-emulsion preparation by low energy methods in an ionic surfactant system. Langmuir. 2006;22(20):8326–32.

Sommerville ML, Hickey AJ. Aerosol generation by metered dose inhalers containing dimethyl ether/propane inverse microemulsions. AAPS PharmSciTech. 2003;4:455–61.

Sommerville ML, Johnson CS, Cain JB, Rypacek F, Hickey AJ. Lecithin microemulsions in dimethyl ether and propane for the generation of pharmaceutical aerosols containing polar solutes. Pharm Dev Technol. 2002;7(3):273–88.

Stuti V. Formulation of a novel nanoemulsion system for enhanced solubility of a sparingly water soluble antibiotic, clarithromycin. J Nanosci. 2014:1–7. DOI:10.1155/2014/268293.

Subhashis D, Satayanarayana J, Gampa VK. Nanoemulsion- a method to improve the solubility of lipophilic drugs. PHARMANEST – Int J Adv Pharm Sci. 2011;2(2–3):72–83.

Surbhi S, Kumkum S. Nanoemulsions for cosmetics. Int J Adv Res Pharma Bio Sci. 2012;2(3):408–15.

Tadros TF, Izquierdo P, Esquena J, Solans C. Formation and stability of nanoemulsions. Adv Colloid Interface Sci. 2004;108–109:303–18.

Tagne JB, Kakumanu S, Ortiz D, Shea T, Nicolosi RJ. A nanoemulsions formulation of tamoxifen increases its efficacy in a breast cancer cell line. Mol Pharm. 2008;5(2):280–6.

Talekar M, Ganta S, Singh A, Amiji M, Kendall J, Denny WA, Garg S. Phosphatidylinositol 3-kinase inhibitor (PIK75) containing surface functionalized nanoemulsion for enhanced drug delivery, cytotoxicity and pro-apoptotic activity in ovarian cancer cells. Pharm Res. 2012;29(10):2874–86.

Tamilvanan S. Submicron emulsions as a carrier for topical (ocular and percutaneous) and nasal drug delivery. Indian J Pharm Edu. 2004;38(2):73–8.

Tenjarla S. Microemulsions: An overview and pharmaceutical applications. Crit Rev Ther Drug Carrier Syst. 1999;16(5):461–521.

Trotta M. Influence of phase transformation on indomethacin release from microemulsions. J Control Release. 1999;60(2):399–405.

Trotta M, Pattarino F, Gasco MR. Influence of counter ions on the skin permeation of methotrexate from water-oil microemulsions. Pharm Acta Helv. 1996;71(2):135–40.

Ugwoke MI, Agu RU, Verbeke N, Kinget R. Nasal mucoadhesive drug delivery: Background, applications, trends, and future perspectives. Adv Drug Del Rev. 2005;57(11):1640–65.

Ugwoke MI, Vereyken IJ, Luessen H. Microparticles and liposomes as pulmonary drug delivery systems: What are the recent trends? In: Bechtold K, Luessen H, editors. Pulmonary drug delivery: Basics, applications and opportunities for small molecules and biopharmaceutics. Aulendorf: ECV Editio-Cantor-Verlag; 2007. p. 308–77.

Uhumwangho M, Okor R. Current trends in the production and biomedical applications of liposomes: A review. J Med Biomed Res. 2005;4:9–21.

Venkatesan N, Yoshimitsu J, Ito Y, Shibata N, Takada K. Liquid filled nanoparticles as a drug delivery tool for protein therapeutics. Biomaterials. 2005;26(34):7154–63.

Vyas TK, Shahiwala A, Amiji MM. Improved oral bioavailability and brain transport of Saquinavir upon administration in novel nanoemulsions formulations. Int J Pharm. 2008;347(1–2):93–101.

Walters KA, Flynn GL. Permeability characteristics of human nail plate. Int J Cosmet Sci. 1983;5(6):231–46.

Wang X, Jiang Y, Wang YW, Huang MT, Ho CT, Huang Q. Enhancing antiinflammation activity of cur cumin through o/w nanoemulsions. Food Chem. 2008;108(2):419–24.

Westin UE, Boström E, Gråsjö J, Hammarlund-Udenaes M, Björk E. Direct nose-to-brain transfer of morphine after nasal administration to rats. Pharm Res. 2006;23(3):565–72.

Yamamoto T, Poon D, Weil PA, Horikoshi M. Molecular genetics elucidation of the tripartite structure of the Schizosaccharomyces pombe 72 kDa TFIID subunit which contains a WD40 structural motif. Genes Cells. 1997;2(4):245–54.

Yashpal S, Tanuj H, Harsh K. Nanoemulsions: A pharmaceutical review. Int J Pharma Prof Res. 2013;4(2):928–35.

Yi SW, Yune TY, Kim TW, Chung H. A cationic lipid emulsion/DNA complex as a physically stable and serum-resistant gene delivery system. Pharm Res. 2000;17(3):314–20.

Yilmaz E, Borchert HH. Effect of lipid-containing, positively charged nanoemulsions on skin hydration, elasticity and erythema – An in vivo study. Int J Pharm. 2006;307(2):232–8.

Yoo BH, Kang BY, Yeom MH, Sung DS, Han SH, Kim HK, Ju HK. Nanoemulsion comprising metabolites of ginseng saponin as an active component and a method for preparing the same, and a skin-care composition for anti-aging containing the same. US Patent 2006/0216261. 2006.

Youenang Piemi MP, Korner D, Benita S, Marty JP. Positively and negatively charged submicron emulsions for enhanced topical delivery of antifungal drugs. J Control Rel. 1999;58(2):177–87.

Yu H, Huang Q. Improving the oral bioavailability of curcumin using novel organogel-based nanoemulsions. J Agric Food Chem. 2012;60(21):5373–9.

Yue, Z., 2014. Nanoemulsion for Solubilization, Stabilization, and In Vitro Release of Pterostilbene for Oral Delivery, AAPS PharmSciTech. 15, 1000-8.

Yuen VM, Hui TW, Irwin MG, Yuen MK. A comparison of intranasal dexmedetomidine and oral midazolam for premedication in pediatric anesthesia: A double-blinded randomized controlled trial. Anesth Analg. 2008;106(6):1715–21.

Zhang Y, Shang Z, Gao C, Du M, Xu S, Song H, Liu T. Nanoemulsion for solubilization, stabilization, and in vitro release of pterostilbene for oral delivery. AAPS PharmSciTech. 2014;15(4):1000–8.

Zhou WY, Wang M, Cheung WL, Guo BC, Jia DM. Synthesis of carbonated hydroxyapatite nanospheres through nanoemulsions. J Mater Sci Mater Med. 2008;19(1):103–10.

Zhou Y. Ultrasound-mediated drug/gene delivery in solid tumor treatment. J Healthc Eng. 2013;4(2):223–54.

7

Nanostructures for Improving the Oral Bioavailability of Herbal Medicines

Lalduhsanga Pachuau

CONTENTS

ABSTRACT: The science and promise of nanotechnology is spreading fast, becoming the major thrust area of research in all scientific disciplines, including drug delivery. The integration of nanotechnology with medicine has resulted in the development of safer and more effective drug delivery systems for several challenging drugs. Nanomedicine formulations have also been demonstrated, in recent times, to overcome the inherent bioavailability problems of different herbal drugs. Curcumin, quercetin, paclitaxel, and several other plant extracts have been successfully delivered through formulation into different types of nanoparticulate systems. Manipulating structures at the nanometer scale has enabled the enhancement of the stability, solubility, and membrane permeability of these herbal drugs, which are the major barriers that restricted their therapeutic applications. This chapter will provide a review on various nanostructure-based drug delivery systems used to enhance the bioavailability of herbal medicines.

KEY WORDS: *Nanotechnology, herbal drugs, bioavailability, drug delivery systems, oral absorption*

7.1 Introduction

Nanoparticles are smaller than can be seen with the naked eye. The prospect of manipulating structures at such a tiny scale has gained significant momentum in recent years. The engineering and manipulation of entities at the 1–100 nm size range is called nanotechnology. There is no scientific or engineering field that has not embraced this tiny technology to develop novel functional materials. Unconstrained by the size limitations, nanotechnology has been defined as the design, characterization, production, and application of structures, devices, and systems by controlled manipulation of size and shape at the nanometer scale (atomic, molecular, and macromolecular scale) that produces structures, devices, and systems with at least one novel or superior characteristic or property (Bawa, 2007). The impact of this technology extends from the medical and environmental applications to the engineering and material industries. At the nanometer scale, materials exhibit physicochemical properties that are different from their bulk solids as a result of the enhanced surface area available for interactions with other materials around them. A stronger, lighter, more durable, more reactive, better electrical conducting material, along with many other traits, can be effectively produced through this technology and already more than 800 everyday, commercial products rely on nanoscale materials and processes (NNI, 2017).

  Nanomedicine is the application of nanotechnology in developing more effective drug delivery systems for prophylactic, diagnostic, and therapeutic use, which has rapidly expanded in recent years. Nanotherapeutics have already been approved by the Food and Drug Administration (FDA) and are currently available for clinical use in the treatment of cancer, high cholesterol, autoimmune disease, fungal infections, macular degeneration, hepatitis, and many other conditions (Ventola, 2012). A preparation such as Doxil®, a polyethylene glycosylated (PEGylated) liposomal doxorubicin, was approved by the FDA as early as 1995, becoming the first FDA-approved nanodrug. Apparently, the prospect of improving the bioavailability of herbal drugs through the application of nanotechnology has also been realized over the years. An injectable, albumin-stabilized nanoparticle formulation of paclitaxel, the active constituent isolated from the yew tree *Taxus brevifolia*, is already in the market for treating different forms of cancer. The incorporation of nanoconstruct-based formulation to phytomedicine has offered several advantages including the improvement of solubility and bioavailability, safeguards from toxicity, enhancement of pharmacological activity, improvement of stability, increase in tissue macrophage distribution, sustained delivery, and protection from physical and chemical degradation (Liu & Feng, 2015). As most of the bioactive constituents of herbal drugs are flavonoids, which are poorly absorbed due to their large molecular size and poor miscibility with oils and other lipids, several novel formulation techniques have been attempted to overcome this limitation (Kidd & Head, 2005). These nanoconstruct techniques include Phytosomes, self-microemulsifying preparations, solid lipid nanoparticles and nanostructured lipid carriers, microemulsions, and liposomes. These nano-sized, novel drug delivery systems for herbal drugs have great potential in enhancing the activity of and overcoming problems associated with plant medicines.

  The pharmaceutical application of nanotechnology mainly evolved from the observed increase in solubility when the size of material was reduced to less than 100 nm, where quantum physics takes over and the materials begin to show entirely new properties such as enhanced solubility, resulting in the increased oral bioavailability of poorly water-soluble drugs (Bawarski et al., 2008).

  The uptake of nanoparticulate systems through the gastrointestinal (GI) tract is today a well-known and accepted phenomenon (Florence, 1997, 2004). Significant research has been dedicated towards the oral delivery of nanoparticles, as the oral route is the most preferred mode of drug administration in recent years. The binding of drugs to nanoparticles was also reported to significantly enhance and prolong the pharmacological activity of insulin and hydrocortisone (Kreuter, 1991). This chapter will concentrate particularly on the application of nanoconstruct formulations in enhancing the oral bioavailability of different herbal drugs.

7.2 Oral Drug Absorption

7.2.1 The Gastrointestinal Tract and Oral Drug Absorption

The oral route is the most convenient and preferred means of drug administration because it leads to better patient compliance. However, the barriers to drug absorption offered by the GI tract are immense when compared to other routes of administration. As a result, several compounds do not display the characteristics required for oral administration or exhibit insufficient systemic availability as required for therapeutic and other beneficial actions. To reach general circulation following oral administration, the drug molecules must first cross the intestinal epithelium, a typical cell membrane composed primarily of a phospholipid bilayer. The permeability of the drug from the absorption site into the systemic circulation depends significantly upon the properties of this cell membrane. Highly polar drugs that are insoluble in this phospholipid membrane are unable to penetrate this cellular membrane, whereas lipid-soluble drugs tend to dissolve in this membrane and penetrate it more easily. Therefore, efficient absorption of drugs from the GI tract is influenced by a combination of several factors, including the physicochemical properties of the drug substance, the dosage form in which it is prepared, and the anatomy and physiology of the GI tract. Factors such as surface area of the GI tract, stomach-emptying rate, GI mobility, and blood flow to the site of absorption are all important determinants to affect the rate and extent of drug absorption (Shargel et al., 2005). Many drugs such as erythropoietin, somatotropin, and insulin are not administered orally due to their potential for degradation in the stomach or the intestine. Also, since drugs are recognized as xenobiotics by the human body, they may undergo extensive metabolism, known as first-pass metabolism, before reaching the general circulation, terminating its effectiveness or reducing its bioavailability. Therefore, a good understanding of GI tract physiology, the properties of the drug, as well the methods of absorption will provide a sound foundation for the design of nanostructure-based drug delivery systems.

Irrespective of whether the cell membrane is a single layer or several layers thick, drugs are thought to penetrate these biological membranes in two general ways: by passive diffusion and through specialized transport mechanisms (Allen et al., 2011). In passive diffusion, drug molecules move from a region of higher concentration to a region of lower concentration, therefore the process is driven by a concentration gradient. Most drugs are absorbed through the biological membrane by passive diffusion. A specialized transport system, such as active transport, and facilitated diffusion exists for several drugs which are mainly carrier-mediated. Active transport is characterized by the transport of drug molecules against the concentration gradient and requires a specialized carrier system. A few lipid-insoluble drugs, such as 5-fluorouracil, that resemble natural physiologic metabolites are absorbed from the GI tract by this process (Shargel et al., 2005). The drug binds with the carrier to form a carrier-drug complex that shoots the drug across the membrane and then dissociates on the other side of the membrane to release the free drug molecules. In facilitated diffusion, the transport of the drug takes place along the concentration gradient, however, with the help of a carrier system. Other transport mechanisms such as the vesicular transport system, pore transport, and ion-pair transport play a minor role in overall drug absorption. However, the vesicular transport system, including pinocytosis and phagocytosis, which is the process of engulfing particles or dissolved materials by the cell, is involved in the oral absorption of the Sabin polio vaccine, various large proteins, and nanoparticles.

7.2.2 Anatomy and Physiology of the GI Tract

The GI tract is a highly specialized region of the body that has undergone advanced development through thousands of years of evolution. About 90% of drug products are administered through the oral route, having to travel through different regions of the GI tract before being available in the systemic circulation system, as demonstrated in Figure 7.1 (Abuhelwa et al., 2017). Therefore, the dosage form must

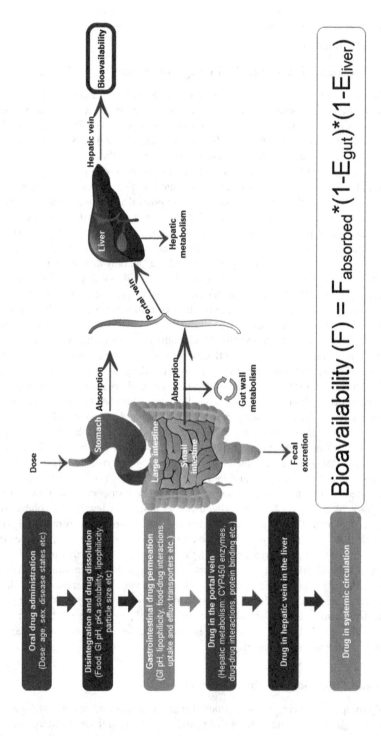

FIGURE 7.1 Model of oral drug absorption. (Reproduced with permission from Abuhelwa et al. (2017). Copyright©Elsevier Science BV 2016.)

be designed to withstand the extreme pH changes along the GI tract, the presence of food, degrading enzymes, varying drug permeability in the different regions of the intestine, and GI motility (Shargel et al., 2005). The alimentary system consists of the mouth to the anus canal that can be divided into the conduit region and the digestive and absorptive region. The whole alimentary canal is lined by a mucous membrane that consist of an epithelium, lamina propria, and muscularis mucosae, which is a thin layer of smooth muscle. The conduit region of the GI tract, such as the gums and hard palate in the mouth, do not offer much as an absorption site due to their partially keratinized, multilayer structure. The linings on the stomach, small intestine, and the colon are simple and are more involved in the absorption and secretion of several substances (Kokate et al., 2006).

7.2.2.1 Oral Cavity

The ingestion, grinding, and mixing of food with saliva takes place in the oral cavity. Saliva is secreted, which has a pH of about 7, and the digestion of carbohydrates is initiated. The secretion of mucin, a glycoprotein to lubricate the food, may interact with the drug.

7.2.2.2 Esophagus

The esophagus links the oral cavity with the stomach and is composed of a thick muscular layer about 250 mm long and 20 mm in diameter. The esophagus contains a well-differentiated squamous epithelium of non-proliferative cells and the pH is usually between 5 and 6. The transit of the dosage form through the esophagus is usually rapid, about 10–14 seconds, therefore very little dissolution of the drug occurs in this region.

7.2.2.3 Stomach

Following oral administration, the stomach is the first digestive and absorption site the drug product encounters in the alimentary canal. Apart from acting as a temporary storage organ, reduction of the ingested solids and mixing to slurry takes place at the stomach with the help of acid and enzyme secretions which are controlled by complex neural, muscular, and hormonal processes. The pH of the stomach is about 1.3 (pH 1.1–1.6) during fasting state, which may be increased to about 4.9 (pH 3.9–5.5) in the presence of food (Russell et al., 1993). Hydrochloric acid is secreted by the gastric secretory cells, known as the parietal cells, along the surface of the gastric epithelial layers, due to which basic drugs are solubilized rapidly in the stomach. The surface of the gastric mucosa is lined by an epithelial layer of column cells that secrete mucous, which is required to protect the epithelial surface of the stomach from acid, enzymes, and pathogens. A thick layer of about 1.0–1.5 mm of mucous covers this epithelial cell surface. These epithelial cells of the gastric mucosa are one of the most rapidly proliferating epithelial tissues, as they are shed by the normal stomach at a rate of about a half-million cells per minute and the surface of the epithelial layer is therefore renewed every one to three days (Mayersohn, 2002). The absorption of drugs in the stomach is very little and is limited due to the small absorptive surface area and short residence time for the dosage form in the stomach.

7.2.2.4 Small Intestine

The small intestine represents the longest and the most important part of the GI tract for digestion and absorption of drugs and nutrients. It consists of the duodenum, the jejunum, and the ileum. Due to the presence of bicarbonate that neutralizes the acids emptied from the stomach, the duodenum has a pH of about 6–6.5. An enzyme containing pancreatic juice is secreted into the duodenum from the bile duct, which results in the hydrolysis of proteins, fats, and other foods. The complex fluid at the duodenum helps in the solubilization of many drugs with limited aqueous solubility, however, the presence of proteolytic enzymes results in the instability and inadequate absorption of protein drugs. The region between the duodenum and the ileum is the jejunum, which has broader and thicker walls than the ileum. The ileum is the terminal part of the small intestine, having a pH of about 7 with the distal part as high as 8, and the presence of bicarbonate secretion helps in the dissolution of fats and hydrophobic drugs (Shargel et al., 2005). Compared to the jejunum, the folds in the ileum are lower and sparser, however, the ileum has distinctive lymphoid tissue patches, called Peyer's patches, which are the most important structural units of the gut associated with lymphoid tissue (GALT).

The small intestine is well perfused with a rich network of blood and lymphatic vessels on its walls. As the blood leaving the small intestine flows into the hepatic portal vein that carries it through the liver into systemic circulation, an absorbed drug may be metabolized in the liver before it reaches general circulation. This pre-systemic metabolism is called first-pass metabolism. Some drugs are so extensively pre-metabolized that their bioavailability from the oral route may eventually become insignificant.

On the structural aspects, the small intestine has a remarkable means of increasing its effective luminal surface area for the absorption of drugs and nutrients. The surface area of the small intestine is increased to about 600 times its corresponding simple cylinder size. This is achieved through the folds and projections on the epithelial surface. There is a submucosal fold known as the folds of Kerckring, which may be several millimeters in depth, then the finger-like projections called villi, ranging in length from 0.5 to 1.5 mm, which line the entire epithelial surface. Then, again, there is another fine structure called the microvilli, with an average length of 1 mm, that project from the surface of the villi. The presence of the villi and microvilli on the surface result in an extremely large absorptive surface area, making the small intestine the most conducive area for absorption.

7.2.2.5 Large Intestine

The colon, or the large intestine, is the final part of the GI tract. Its primary function is to absorb water and electrolytes and to store and compact feces. The colon is devoid of the villi and microvilli structure of the small intestine, however, the irregularly-folded mucosae result in a 10–15 times increase in surface area to that of a simple cylinder. A few drugs are absorbed well in this region of the GI tract and those drugs are also good candidates for formulation into oral sustained-release dosage forms. The pH of the colon may range from 6 to 6.5 in the cecum and from 7 to 7.5 in the distal part of the colon (Ashford, 2013). Due to the presence of a huge number of aerobic and anaerobic microorganisms that colonize the colon, several metabolic reactions are taking place, which may provide a good target for converting inactive conjugated drugs, such as prodrugs, into their active form. Several controlled release formulations have been developed over the years that target the metabolizing enzymes of the colonic microorganisms to affect drug release.

7.2.3 Factors Influencing GI Drug Absorption

Several factors may influence the absorption of the drug from the GI tract. For most drugs, the best site for absorption is the upper part of the small intestine or the duodenum region. Therefore, any factor that reduces or enhances the residence time of the drug in this region is going to have an influence on the rate and extent of absorption.

7.2.3.1 GI Motility

Following oral administration, the GI motility moves the drug along the alimentary canal, from the oral cavity to the large intestine. Several factors influence the transit time of the drug through the alimentary canal, including the properties of the drug, the dosage form, and the physiology. The presence of food or other drugs may also affect the GI motility.

7.2.3.2 Gastric Emptying Rate

This is the rate at which the stomach empties its contents into the small intestine. Since the duodenum is the site where the absorption of drugs is most favorable, anything that delays the stomach in emptying its contents into the duodenum will have a negative influence on the rate and extent of drug absorption.

Since many drugs are also extremely unstable at the acidic pH of the stomach, reducing the gastric emptying rate will result in decomposition of acid-labile drugs.

7.2.3.3 Intestinal Motility

It is essential that for complete and efficient absorption, the drug must reside in the absorption site (small intestine) for a sufficient period of time. The average small intestinal transit time is about 3–4 hours, which may increase during the fed state. Certain conditions such as diarrhea, where there is very high intestinal motility, may mean the drug does not stay long enough in the small intestine to be properly absorbed.

7.2.3.4 Perfusion of the GI Tract

Since most drugs are absorbed from the GI tract by passive diffusion, blood flow and proper perfusion of the absorption site is very important. Proper blood flow helps in maintaining the concentration gradient and sink condition that is required for the continuous absorption of the drug by passive diffusion.

7.2.3.5 Effect of Food

Oral drug absorption, and hence the bioavailability, can be affected by the presence of food in the GI tract. The extent and nature of influence may, however, be dependent on the nature of the food or the type of drugs. In general, the presence of food may delay the gastric emptying rate, stimulate the flow of bile secretions, change the pH of the GI tract, increase the blood flow, or physically or chemically interact with the drug product leading to formation of large complexes or degradation (Shargel et al., 2005).

7.2.4 Oral Absorption of Nanoparticles

The oral delivery of nanomedicines is not a new concept in drug delivery science. Today, the uptake of nanoparticulate systems through the GI tract is a well-known and accepted phenomenon that provides an additional route for drug administration. As early as the beginning of the twentieth century, the observation of intact starch particles in blood and urine was published in literature (Florence, 1997; Kreuter, 1991). Thereafter, more intensive investigations on the particulate uptake of different materials that were a few thousandths of a nanometer following oral administration were completed. These materials included corn starch, rice starch, potato starch, diatoms, polyvinyl chloride (PVC) particles, cellulose particles, and plant cells and all of these were regularly detected in intact particulate form in blood and urine (Volkheimer, 1977; Volkheimer & Schulz, 1968). However, due to the lack of sophisticated analysis equipment and technical knowledge about the properties of nanomaterials in the past, nanotechnology has only begun to blossom in the last 20 years. In recent years, a significant amount of research work has been dedicated to the development of nanoparticles for oral drug delivery. Several studies have shown that nanoparticulate formulations such as nanocapsules, nanostructured lipid carriers, microemulsions, and others resulted in superior bioavailability, residence time, and biodistribution of drugs when compared to free drug delivery (Liu et al., 2014; Pandey et al., 2005). However, oral delivery of therapeutic nanoparticles is not without difficulties. Several barriers exist and the uptake of nanoparticles from the GI tract is influenced by the particle size, the physicochemical properties of the particles, such as surface charge or zeta potential, and the attachment of cellular uptake promoters such as antibodies and lectins (Bhavsar & Amiji, 2007; Florence & Hussain, 2001). The basic physical and biological factors affecting the uptake of nanoparticulates from the GI tract are listed in Table 7.1.

The translocation or uptake of nanoparticles (Figure 7.2) from the walls of the GI tract is suggested to take place through three different pathways: (1) an uptake by the paracellular pathway, (2) intracellular uptake and transport through the epithelial cells lining the intestinal mucosa, and (3) a lymphatic uptake through the M-cells and Payer's patches (Kreuter, 1991). The paracellular translocation involves the uptake of nanoparticles between the cells of the intestinal walls. As evidenced from *in vivo* studies,

TABLE 7.1

Some of the Basic Physical and Biological Factors
Affecting Nanoparticle Uptake (Florence, 2004)

S. No.	Factors
1	Particle diameter
2	Surface characteristics
3	Physical stability of the carriers in the gut
4	Chemical stability of carrier and drug
5	Transit times in the GI tract
6	Residence times in regions of particle uptake
7	Interaction with gut contents
8	Transport through mucus
9	Adhesion to epithelial surfaces
10	Stimulus for cellular uptake

the appearance of the particles could be very rapid following oral administration. Particles could be detected in the blood within 10 minutes of oral dosing and this rapid appearance is believed to be due to the paracellular uptake or translocation (Volkheimer, 1977). The endocytotic uptake mechanism of drugs has been attributed to the reason behind the intracellular uptake of nanoparticles (Kreuter, 1991). The intracellular uptake of nanoparticles by the cells lining the intestinal mucosa was also supported by electron microscopic autoradiographic investigations. The uptake by the M-cells of the Peyer's patches is also hugely significant in the oral absorption of nanoparticles. The Peyer's patches are the most important structural units of the GALT, which are characterized by specialized cells called M-cells. These M-cells overlie the lymphoid tissue and are specialized in endocytosis and the transport of drugs into the intraepithelial spaces and adjacent lymphoid tissue. The nanoparticles bind to the apical membrane of these M-cells, after which they are rapidly internalized and then transferred to the lymphocytes (Gasco, 2007). This lymphatic absorption of a drug through the GALT has an advantage over the usual portal blood route as it avoids the pre-systemic, or first-pass, metabolism by the liver. The results from histologic investigations suggest the translocation of beads of 1 μm and smaller from the Peyer's patches to the mesenteric lymph nodes by the lymphatic system (Kreuter, 1991). Therefore, more than one absorption method is involved in the uptake of nanoparticles from the GI tract and these pathways may also occur

The *in vivo* situation

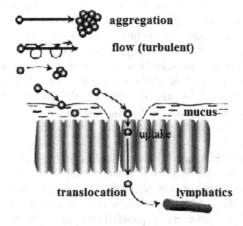

FIGURE 7.2   Some factors affecting particle uptake in the gut. (Reproduced with permission from Florence and Hussain (2001), Copyright©Elsevier Science BV 2001.)

simultaneously in the absorption of the same nanoparticle. The major pathways for the uptake of the nanoparticles may also differ along the different regions of the GI tract.

7.3 Oral Absorption and Bioavailability of Herbal Medicines

Herbal medicines or phytomedicines, as in Traditional Chinese Medicines (TCM) and the Indian System of Medicine, have been used widely for thousands of years around the world. Through the years, numerous dietary compounds obtained from fruits, vegetables, and spices have been isolated and evaluated for their therapeutic potential and other health benefits. However, the major challenge in herbal drug technology is to translate experimental findings into clinical situations. For instance, the (-)epigallocatechin-3-gallate (EGCG) was shown to prevent the growth of melanoma tumor cell lines *in vitro* with an half maximal inhibitory concentration (IC_{50}) value from 11 to 89 μm, but, in humans, the peak plasma concentration (C_{max}) of EGCG were only 0.237–0.328 μg/mL (51.7–71.6 nm) after taking 1 g of green tea extract or 250 g fresh grape plus EGCG enriched-nutrient mixture (Gao & Hu, 2010).

Similarly, trans-resveratrol is a compound which is reported to be active against all three major stages of initiation, promotion, and progression of carcinogenesis at a concentration greater than 1,000 ng/mL *in vitro* (de la Lastra & Villegas, 2005). The C_{max} that can be achieved with a single dose administration of trans-resveratrol (200 mg) in human volunteers ranged from 21.6 ± 9.7 to 28.0 ± 22.0 ng/mL or from 27.1 ± 14.4 to 34.5 ± 32.1 ng/mL in a multiple dose study of 200 mg trans-resveratrol thrice daily (Nunes et al., 2009). This indicates that many herbal drugs and extracts, despite their extraordinary potential *in vitro*, were far below the plasma level required *in vivo* for therapeutic activity following oral administration. Compensating for such poor oral bioavailability with higher doses could lead to higher plasma concentration of the active constituents, but this high dose size may be unacceptable to the patient or result in toxicity (Cheng et al., 2001).

The absorption and hence the bioavailability of the bioactive constituents of phytomedicines, such as polyphenols, largely depends upon the physicochemical properties of the phytochemicals. Even though the bioavailability of a compound cannot be accurately predicted, the analysis by Lipinski's "rule of five" provides some insight: in general, a compound will have better bioavailability when it contains not more than five hydrogen-bond donors, not more than 10 hydrogen-bond acceptors, has a molecular mass not greater than 500 daltons, a partition co-efficient log P-value of not greater than five, and contains less than 10 rotatable bonds (Lipinski et al., 1997, 2001, 2012). Polyphenols, such as curcumin and green tea polyphenols, do not fall within these "rule of five" specifications, and they exhibit low bioavailability (Lipinski et al., 2001). Meanwhile, compounds such as genistein and biochanin A, which, as per the rule, should be very well absorbed, showed limited bioavailability as they are excreted by an efflux mechanism into the gut at a high rate (Gao & Hu, 2010). Therefore, several other factors could also play a role in limiting bioavailability of phytochemicals and these factors may include the solubility of the compound, stability due to gastric and colonic pH, metabolism by gut microflora, absorption across the intestinal wall, active efflux mechanisms, and first-pass metabolic effects (Aqil et al., 2013).

Much research is still required to completely understand the mechanism by which the phytochemicals are absorbed from the GI tract (Rice-Evans et al., 2000). As with any other drug products, to get absorbed into systemic circulation, the phytochemicals must dissolve in the fluids of the GI tract, survive the harsh pH environment, and not get degraded or metabolized by intestinal enzymes such as the glycosidases, esterases, oxidases, and hydrolases. After dissolution in the GI tract fluid, the dissolved drug has to be partitioned through the phospholipid-based biological membrane into systemic circulation. Many plant bioactives, such as most polyphenols, are probably too hydrophilic, exhibiting poor partitioning properties, to penetrate the gut wall by passive diffusion, however, the membrane carriers that could be involved in polyphenol absorption have not been identified (Manach et al., 2004). In contrast, paclitaxel, a diterpenoid pseudoalkaloid extracted from the bark of *Taxus brevifolia* exhibits a poor aqueous solubility (<1 μg/ml) resulting in low oral bioavailability (<2%) (Yang et al., 2015).This low oral bioavailability of paclitaxel has been attributed to its poor aqueous solubility, low permeability restricted by P-glycoprotein (P-gp), and metabolism by P_{450} enzymes like CYP3A4 (Iqbal et al., 2011; Yang et al.,

2015). In most of the cases, all flavonoids, except flavanols, are found in glycosylated forms in foods, and this glycosylation influences absorption. Only aglycones and some glucoside forms of the drug can be absorbed in the small intestine, whereas polyphenols linked to a rhamnose sugar must reach the colon and be hydrolyzed by rhamnosidases of the microflora before absorption (Manach et al., 2004). Hollman et al. (1995) suggested that glucosides could be transported into enterocytes by the sodium-dependent glucose transporter SGLT1. Another pathway involves the lactase phloridzine hydrolase, a glucosidase of the brush border membrane of the small intestine that catalyzes extracellular hydrolysis of some gluco-sides, which is followed by diffusion of the aglycone across the brush border. Another reason for the poor bioavailability of many phytochemicals is their rapid conjugation, especially by glucuronidation in the intestine and liver, which is mediated by uridine diphosphate-glucuronosyl transferases (UDP-UGTs), and together with reactions of cytochrome P_{450} enzymes they represent more than 80% of the pathways by which compounds are metabolized (Aqil et al., 2013).

7.4 Nanostructural Approaches to Improve Oral Bioavailability of Herbal Medicines

Bioavailability is the rate and extent to which the administered drug reaches systemic circulation. Whether it is a herbal or non-herbal drug, there are three steps that an orally administered drug must undergo to become available in systemic circulation (Yu & Huang, 2013). The first step is solubilization in the GI fluid, or the intestinal lumen, which is mainly an aqueous base. Many phytomolecules, such as curcumin, paclitaxel, or luteolin, show low oral bioavailability because they fail in this first step. The second step is the absorption or partitioning across the biological membrane at the absorption site along the GI tract. In this respect, many polyphenols, such as flavonoids, are too hydrophilic and exhibit poor miscibility with oils and other lipids, which severely limits their ability to get absorbed and cross the biological membrane (Kidd & Head, 2005; Manach et al., 2004). Third, the drug must escape rapid pre-systemic and systemic metabolism, reach the site of action, and circulate long enough for therapeutic action. Resveratrol, a dietary antioxidant polyphenol found in grapes, red wine, and peanuts, has been strongly indicated to have protective effects against several diseases. The absorption of a dietary-relevant 25 mg oral dose was at least 70%, however only trace amounts of unchanged resveratrol (<5 ng/mL) could be detected in plasma. This is due to the rapid systemic metabolism of resveratrol through sulfate and glucuronic acid conjugation of the phenolic groups and hydrogenation of the aliphatic double bond. The extremely rapid sulfate conjugation by the intestine/liver appears to be the rate-limiting step in the bioavailability of resveratrol (Walle et al., 2004).

In the past, several attempts have been focused on approaches towards improving the oral bioavail-ability and pharmacokinetic properties of herbal drugs through the development of new drug deliv-ery systems. The oral bioavailability improvement schemes for these herbal drugs necessarily focuses on the above three bioavailability requirement steps of improving solubility in the GI fluid, enhancing absorption or lipid partitioning, and reducing metabolism. The application of novel drug delivery sys-tems resulted in several notable advantages including enhanced solubility and membrane permeabil-ity, improved pharmacokinetic properties through the sustained-release system, and reduced toxicity (Mukherjee et al., 2015).

The application of nanotechnology has become one of the more promising approaches to overcome low bioavailability and systemic toxicity of herbal drugs. These drug-loaded, nano-sized carriers offer to enhance systemic absorption, modulate the pharmacokinetic properties, increase the con-centration at the site of action, and enhance cellular uptake while reducing toxicity (Aqil et al., 2013; Mukherjee et al., 2015). The overall objectives of these nanostructural approaches to the formulation of herbal drugs include improving the solubility of the phytoconstituents, helping to enhance the per-meation of biological membranes, improving biodistribution, and sustaining the release of the active constituents to achieve better and prolonged therapeutic activity. This chapter will cover several dif-ferent types of nanotechnology-based formulations that are applied to enhance the bioavailability of phytochemicals.

7.4.1 Phyto-Phospholipid Complexation

To have good oral bioavailability, natural products must possess a good balance between hydrophilicity for dissolution in the GI tract and lipophilicity for crossing the biological membranes (Selmaty et al., 2010). Several phytochemicals, such as certain polyphenols, are highly polar and are readily soluble in water. However, in spite of their good water solubility, they are poorly absorbed along the GI tract due to their large molecular size for simple diffusion, and poor miscibility with oils and other lipids to penetrate the gut wall (Kidd, 2009; Manach et al., 2004). In contrast, other polyphenols, such as curcumin, exhibit poor oral absorption due to their extremely low aqueous solubility or extensive pre-systemic metabolism (Maiti et al., 2007). As a result, these phyto-constituents exhibit a low or sub-therapeutic bioavailability when given orally.

In recent times, the phyto-phospholipid complexation technique has emerged as one of the leading methods of improving the oral bioavailability of phytochemicals that are poorly miscible with lipids and do not cross the biological membranes (Khan et al., 2013). In this technique, a complex is formed between dietary phospholipids and those plant actives that exhibit potent *in vitro* pharmacological activities but failed to demonstrate similar *in vivo* responses, forming new amphiphatic cellular structures. The size of such resultant phyto-phospholipid complex molecules usually varies from nanometer to micrometer scale, depending upon the method of preparation (Bhattacharyya et al., 2013; Li et al., 2014). Since phospholipids are the constituents of all cell membranes present in food from plant and animal sources, they are also considered an excellent source of choline, an essential nutrient for nutritional supplements (Ramadan, 2012). In nature, such as in lecithin, phospholipids may occur as a mixture composed of phosphatidylcholine, phosphatidylethanolamine, and phosphatidylinositol. In this complex, the phytochemical is entrapped within the phospholipid such that the properties of the phospholipid are substantially retained (Li et al., 2014). Thus, the amphiphilic character of the complex facilitates its passive transport from the relatively aqueous environment of the GI tract to the predominantly lipid-rich biological membrane to improve its bioavailability.

A novel curcumin-phospholipid (a hydrogenated soy phosphatidyl choline) complex formulation was developed to overcome the limitation of oral absorption of curcumin and the protective effect of the complex on carbon tetrachloride-induced acute liver damage in rats was also investigated (Maiti et al., 2007). It was found that the complex provides better protection to the rats' livers than free curcumin at similar doses, the serum concentration of curcumin obtained from the complex was higher than pure curcumin, and the C_{max} achieved with the complex (1.2 µg/ml) was more than twice that of the C_{max} obtained with pure curcumin (0.5 µg/ml). The complex also maintained effective concentrations of curcumin for a longer period of time in rat serum. A silybin-phospholipid complex was also prepared to increase the oral bioavailability of the silybin, obtained from the seed extract of the milk thistle, *Silibum marianum* Gaertn (Yan-yu et al., 2006). The complex resulted in nanoscale structures and there was a remarkable increase in the bioavailability of silybin in rats following the oral administration of the silybin-phospholipid complex compared to silybin-*N*-methylglucamine, which can be attributed to an impressive improvement of the lipophilic property of the silybin-phospholipid complex along with an improvement in the biological effect of the silybin. The preparation of a bergenin-phospholipid complex also resulted in a complex with spherical particles of 169.2 ± 20.11 nm average size and with a zeta potential of −21.6 ± 2.4 mV. The solubility of the bergenin-phospholipid complex in water and n-octanol was effectively enhanced and the relative bioavailability (RB) was also significantly increased to 439% over the pure bergenin (Qin et al., 2010). The phospholipid complex of naringenin also produced better antioxidant activity than the free compound with a prolonged duration of action, and helps to reduce fast elimination of the molecule from the body (Maiti et al., 2006).

A patented technology called Phytosome® utilizes the phospholipid-herbal extract complex to enhance the oral bioavailability of the active phytochemicals. Phytosome® technology emerged in the year 1989 in Italy after it was observed that certain polyphenols exhibit strong bonding affinity for phospholipids in their intact plant tissue (Kidd, 2009). The term "phyto" means "plants" and "somes" means "cell-like". By complexing a polyphenol with a phospholipid, mainly phosphatidylcholine (PC), an intermolecular bonding between the individual polyphenol molecules and molecules of the PC was created. Since the PC is an amphipathic molecule, having a positively charged head group and two neutrally charged tail groups, it is miscible in both water and lipid environments. As a result, the polyphenol-PC complex Phytosome makes the polyphenol more suitable to cross the GI tract lumen (Kidd, 2009). Moreover, PC is also the main phospholipid that constitutes the major part of the lipid-bilayer biological membrane. The Phytosome shuttles the enwrapped polyphenols across the membrane into the cells, enhancing the blood levels of polyphenol constituents by factors of at least 2–6 times (Ajazuddin & Saraf, 2010; Kidd, 2009). A Phytosome differs from a liposome in that a liposome has hundreds of phospholipid molecules aggregated to form a spherule, within which other molecules are compartmentalized without specific bonding, whereas in the Phytosome the phosphatidyl molecules effectively enwrap a polyphenol molecule (Figure 7.3). Also, while the liposome concept remains unproven as an oral delivery vehicle, the Phytosome is known to dramatically enhance the bioavailability following oral delivery (Kidd, 2009).

Various preparation methods have been followed in the complexation of a polyphenolic phytoconstituent or mixture with a phospholipid to get the Phytosome. Depending on the product, the active phytoconstituent or mixture has been complexed with the phospholipid at different molar ratios ranging from 0.5:1 to 3:1. Recently, most of the research works on the preparation of Phytosomes preferred a stoichiometric drug to phospholipid ratio of 1:1 in the formulation of the phospholipid complex (Khan et al., 2013; Giacomelli et al., 2002; Selmaty et al., 2010). In the preparation of a bergenin-phospholipid complex, Qin et al. (2010) concluded the drug to phospholipid ratio of 0.9 to be the optimum condition. In another study, naringenin (Maiti et al., 2006) and gallic acid (Bhattacharyya et al., 2013) complexes were prepared taking a 1:1 molar ratio with hydrogenated soy phosphatidyl choline, while in the case of the alkaloid oxymatrine-phospholipid complex a drug to phospholipid ratio of 3:1 at 140°F for three hours produced a resultant complex with the highest yield (Yue et al., 2010). The preparation of phospholipid complex both in an aprotic solvent (Luo et al., 2006) as well as a protic solvent (Franceschi and

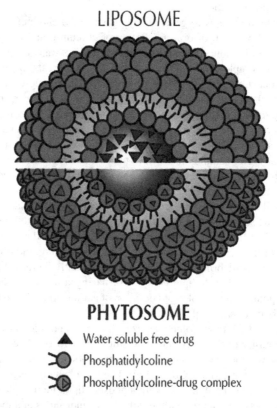

LIPOSOME

PHYTOSOME

▲ Water soluble free drug

🔾 Phosphatidylcoline

🔾 Phosphatidylcoline-drug complex

FIGURE 7.3 The difference between liposome and Phytosome®. (Reproduced with permission from Selmaty et al. (2010), Copyright © Elsevier Science BV 2009.)

Giori, 2010) have also been reported, which showed remarkable differences in their properties (Selmaty et al., 2010).

Several Phytosome products are already in the market and the technology has been applied to many popular herbal extracts, such as milk thistle, ginkgo biloba, green tea, grapeseed, hawthorn, curcumin, and ginseng to improve their oral bioavailability (Table 7.2). A study in young healthy subjects (16–26 years) has shown that oral administration of 360 mg doses of silybin in Siliphos® Phytosome® form and conventional form showed remarkable difference in bioavailability (Barzaghi et al., 1990). The plasma level of silybin achieved after eight hours with the Phytosome was almost three times that of the conventional non-complexed silybin and the total area under the curve (AUC) was also 4.6 times better in Phytosomes than in conventional form. Clinical efficacy in the treatment of obesity with highly bioavailable green tea extract, GreenSelect® Phytosome® has also been proven on obese subjects of both genders (Di Pierro et al., 2009). The oral consumption of 150 mg MonCam (coated tablet containing GreenSelect® Phytosome®) twice daily along with a hypocaloric diet has resulted in significant weight loss after 90 days (14 kg) compared to the diet-only group (5 kg). In recent years, nano-sized soybean phosphatidyl choline-drug complex Phytosome surface-functionalized with folate-PEG has already been evaluated as a carrier for mitomycin C, to achieve reduced toxicity and a superior therapeutic effect (Li et al., 2014). Thus, the Phytosome technology has potential application beyond the established antioxidant capabilities of phytoconstituents to interact with cell membrane-signaling molecular complexes, diffusible transcription factors inside cells, and intranuclear receptor complexes to provide the most clinical benefit (Kidd, 2009).

TABLE 7.2

Some Phytosomes® Available in the Market from Indena S.p.A, Italy

Brand	Botanical Origin	Biological Activity
Bosexil® Frankincense Phytosome®	*Boswellia serrata* Roxb. ex Colebr. – Resin	Soothing, lenitive, anti-photoageing
Casperome® Boswellia Phytosome®	*Boswellia serrata* Roxb. ex Colebr. – Resin	Healthy inflammatory response
Centella Asiatica Selected Triterpenes Phytosome®	*Centella asiatica* (L.) Urban – Leaf	Collagen restructurant, antiwrinkles agent
Escin β-sitosterol Phytosome®	*Aesculus hippocastanum* L. – Seed	Capillarotropic
GinkgoSelect® Phytosome® Ginkgo Biloba Phytosome®	*Ginkgo biloba* L. – Leaf	Cognition and circulation improver, antioxidant activity, vasokinetic
Virtiva® – Ginkgo Biloba Phosphatidylserine Phytosome®	*Ginkgo biloba* L. – Leaf	Cognitive enhancer
Ginkgo Biloba Dimeric Flavonoids Phytosome®	*Ginkgo biloba* L. – Leaf	Lipolytic, vasokinetic, phosphodiesterase inhibitor
GinSelect® Phytosome® Ginseng IDB Phytosome®	*Panax ginseng* C.A. Meyer – Root	Adaptogen, tonic, skin elasticity improver
GreenSelect® Phytosome® Green Tea Phytosome®	*Camellia sinensis* (L.) O. Kuntze – Young leaf	Antioxidant activity, weight loss agent
LeucoSelect® Phytosome® Grape Seed Phytosome®	*Vitis vinifera* L. – Seed	Healthy cardiovascular function, UV protectant, antioxidant activity
Proanthocyanidin A2 Phytosome®	*Aesculus hippocastanum* L. – Bark	UV protectant, firming and oval reshaping agent
Rexatrol® Resveratrol Phytosome®	*Polygonum cuspidatum* Sieb. e Zucc. – Rhizome	Antioxidant activity, anti-ageing, SIRT1 modulator
Siliphos® Silybin Phytosome®	*Silybum marianum* (L.) Gaertn. – Fruit	Healthy liver
Meriva® Turmeric Phytosome®	*Curcuma longa* L. – Rhizome	Joint health, healthy inflammatory response, soothing

Source: www.indena.com (last accessed on 13-12-2018)

7.4.2 Polymeric Nanoparticles

Numerous polymeric nanoparticles have been fabricated and tested for their potential as useful drug carriers for oral and parenteral administrations. The low systemic bioavailability of several herbal drugs following their oral administration is quite common, which is mainly attributable to poor solubility and poor absorption. Therefore, designing and manufacturing polymeric nanoparticles to improve the oral bioavailability of phytochemicals has become a promising strategy to enhance their efficacy *in vivo* (Yao et al., 2015). Apart from enhancing the bioavailability due to efficient permeability across the bio-membrane, encapsulation of an active constituent with a polymeric carrier resulted in enhanced stability of labile drugs, reduced toxicity, and controlled drug release (Pandey et al., 2005).

Several biodegradable, biocompatible, and approved polymers are available for the purpose of drug encapsulation. However, the most commonly employed polymers are poly(lactic acid) (PLA), poly(glycolic acid) (PGA), and their copolymer poly(lactide-coglycolide) (PLGA), as they also offer versatile degradation kinetics apart from being biocompatible and biodegradable (Liu & Feng 2015). An oral nanoparticulate delivery system for paclitaxel, one of the most important drugs widely used for the treatment of different kinds of cancers, was developed and investigated *in vitro* and *iv vivo* (Iqbal et al., 2011). The poor oral bioavailability of paclitaxel, the active constituent extracted from the pacific yew tree *Taxus brevifolia*, has been attributed to its poor aqueous solubility as well as P-gp-mediated efflux and metabolism by cytochrome P_{450} metabolic enzymes (Singla et al., 2002). The delivery system containing 0.5% poly(acrylic acid)-cysteine (PAA-cysteine) and 0.5% reduced glutathione (GSH) was found to improve the *in vitro* transport of paclitaxel across caco-2 monolayers by 6.7–7.4 folds when compared to paclitaxel alone. When tested *in vivo* on Sprague-Dawley (SD) rats, there was an improved paclitaxel plasma concentration and bioavailability following the oral administration of formulations containing paclitaxel, GSH, and PAA-cysteine. The plasma concentration–time AUC of paclitaxel was improved by 4.7–5.7 times, while the C_{max} was also enhanced by 6.3–7.3 times in comparison to the oral formulation containing paclitaxel alone.

Silymarin was also encapsulated in PLGA-based nanoparticles by the single emulsion solvent evaporation technique to improve its bioavailability and, hence, its therapeutic efficacy towards prostate cancer (Snima et al., 2014). Polymeric nanoparticles with an average size of less than 300 nm were successfully prepared, achieving 60% encapsulation efficiency and 13% loading efficiency. From an *in vitro* dissolution study, a slow and sustained release of the silymarin was obtained under physiological conditions. While the X-ray diffraction spectroscopy (XRD) and thermogravimetric analysis (TGA) studies showed that the preparation was amorphous in nature, the *in vitro* cell viability study showed that silymarin nanoparticles were biocompatible and exhibit preferential toxicity towards prostate cancer cells. The oral administration of 2 mg/day of curcumin encapsulated in PLGA nanoparticles was also reported to delay the progression of diabetic cataracts in a streptozocin-induced diabetic cataract model in rats (Grama et al., 2013).

Since most of the orally administered particles are not retained and undergo direct transit through the GI tract, mucoadhesive formulations have often been employed to improve the residence time of the particles in the GI tract (Ensign et al., 2012). Mucoadhesive formulations provide a prolonged and intimate contact of the dosage form with biological surfaces, leading to better absorption and bioavailability. A polycaprolactone-based nanoparticle loaded with curcumin was coated with mucoadhesive polymers of chitosan for buccal delivery (Mazzarino et al., 2012). Uniform particle size ranging between 114 and 125 nm was obtained and the encapsulation efficiency was remarkably high at 99%. The chitosan-coated nanoparticles exhibited a positive charge, showing their capacity to provide adhesion to the negatively-charged mucosal surface. A superior anticancer effect in a colorectal cancer model was also reported for the mucoadhesive, curcumin-containing chitosan nanoparticles (Chuah et al., 2014). An *ex vivo* study on an everted SD rat intestinal sac model was performed, which showed that curcumin-containing chitosan nanoparticles resulted in improved mucoadhesion and were also taken up to a greater extent by colorectal cancer cells when compared to free curcumin. The prolonged contact resulting from the mucoadhesion of curcumin-containing chitosan nanoparticles onto the cells leads to a greater reduction in percentage cell viability and a lower IC_{50}, which indicates its potential for an improved treatment outcome.

7.4.3 Liposomes

Nano-sized liposomes are artificial spherical vesicles that can be produced from natural phospholipids and cholesterol. They have a lipid bilayer membrane structure composed of phospholipids with hydrophilic heads and hydrophobic fatty acid tails. Liposomes may greatly vary in size from nanometers to tens of micrometers, however liposomes for drug delivery applications are usually unilamellar with diameters of about 50–150 nm (Aqil et al., 2013; Maurer et al., 2001). The uniqueness and, hence, the advantage of liposomes is that they can accommodate and encapsulate materials of varying properties such as polarity, charge, and size. Some liposome-based drugs have already got approval from the FDA and are available for treating different diseases (Wang et al., 2014). Both hydrophilic and lipophilic drugs are seamlessly entrapped in the liposomes due to the availability of the aqueous core as well as the phospholipid bilayer. Liposomes also have unique advantages as carriers for drug delivery, such as low toxicity, improved solubility and stability of the active ingredients, ability to alter tissue distribution and targeting to specific sites, and enhanced therapeutic effects (Liu & Feng, 2015). However, despite these several advantages, liposomes do suffer from certain drawbacks, including their instability in the plasma as the liposomes in the circulatory system are recognized as foreign particles by the reticuloendothelial system (RES) and are rapidly cleared, a process known as opsonization (Wang et al., 2014). Also, the electrostatic, hydrophobic, and van der Waals forces can disintegrate the liposomes, thus, steric stabilization is required, which can be achieved through coating with polymers such as PEG or poloxamers. Avoiding the RES and opsonization results in a prolonged circulation time for the liposomes and may result in a significant accumulation in highly vascularized, permeable tissues such as tumors, especially in cases involving active neoangiogenesis (Aqil et al., 2013).

Concern over the oral delivery of liposomes has always abounded, as there are questions about its ability to remain stable along the harsh GI tract conditions (Park et al., 2011). However, a liposome containing *Curcuma longa* (Ukon) extract (LUE) was found to enhance the bioactivity in providing liver protection as compared to the uncapsulated extract (Takahashi et al., 2008). Liposomes were prepared by the homogenization-microfluidization method, taking equal volumes of lecithin and ukon extract, which produced a small, unilamellar vesicle of approximately 100 nm in diameter under optimal conditions. The LUE solutions containing 5 wt % was found to remain well dispersed for at least 14 days and showed a 2-fold higher residual rate of curcumin than the uncapsulated extract under simulated gastric and intestinal fluids. The bioactivity of LUE was further examined for its suppressive effect on carbon tetrachloride-induced liver injury by using mice. Following the oral administration of LUE at 10 mg/kg dose (as the extract), there was a much higher suppressive effect on the serum aspartate aminotransferase (AST) and alanine aminotransferase (ALT) levels, compared to the uncapsulated extract at a dose of 33 mg/kg. A liposome-encapsulated curcumin (LEC) was also prepared from the commercially available lecithin by the same group to enhance the absorption of curcumin following oral administration (Takahashi et al., 2009). The LEC prepared at 5 wt % of soy bean lecithin (SLP-PC70) and 2.5 wt % of curcumin showed a good dispersibility, with 68% encapsulation efficiency, and the vesicles were unilamellar with diameters of approximately 263 nm. Following the oral absorption in SD rats, a high bioavailability of curcumin was obtained in the LEC, characterized by faster and better absorption when compared to uncapsulated curcumin and unformulated curcumin-lecithin mixture. In terms of pharmacokinetic parameters, the LEC gave higher C_{max} and shorter T_{max} (time to reach C_{max}) values, as well as a higher value for the blood concentration–time AUC at all time points, which indicated that liposome encapsulation enhanced the oral GI absorption of curcumin. The plasma antioxidant activity of the orally administered LEC was also significantly higher than that of the other treatments.

To improve the poor bioavailability of silymarin following oral administration, silymarin hybrid liposomes for buccal delivery were developed. Optimum conditions for the preparation of the liposomes, such as molar ratio for lecithin and cholesterol, and the effects of surfactants were studied by two-level, full factorial designs with three factors (2^3 designs) (El-Samaligy et al., 2006a). Results showed that the positively charged liposomes exhibit superior entrapment efficiency over neutral and negatively charged liposomes. The hybrid liposome formulation containing lecithin, cholesterol, stearyl amine, and Tween 20 in 9:1:1:0.5 molar ratios, respectively, gave the best drug absorption and permeation, showing steady state permeation through the chicken cheek pouch for six hours. Thus, this optimized liposomal preparation is expected to improve the bioavailability of silymarin following the buccal delivery system. The same group followed up the above investigation by studying the physical stability and *in vivo* performance of the optimized hybrid liposomes encapsulating silymarin in albino rats (El-Samaligy et al., 2006b). The optimized LEC was administered buccally to the albino rats and its hepatoprotective activity against carbon tetrachloride-induced oxidative stress was measured through biochemical parameters such as serum glutamic oxalacetate transaminase (SGOT) and serum glutamic pyruvate transaminase (SGPT). It was found that the silymarin hybrid liposomes produced a significant decrease in both transaminase levels when challenged with carbon tetrachloride (intraperitonially) in comparison with orally administered silymarin suspension which was also confirmed by histopathological study.

A proliposome system was also prepared by the film deposition method to enhance the stability and increase the oral bioavailability of silymarin. Proliposomes are dry, free-flowing particles that immediately form a liposomal suspension when they are in contact with water, and this solid property of the proliposomes enhances their stability without influencing their intrinsic properties (Yan-yu et al., 2006). A very stable proliposome containing silymarin was prepared with an encapsulation efficiency of more than 90% and an average particle size of about 196.4 nm. After oral administration of the formulation to beagle dogs, a remarkable improvement in the pharmacokinetic parameters was observed. The T_{max} for both the proliposome silymarin and silymarin only was 30 minutes, while the C_{max} was 472.62 and 89.78 ng/mL, respectively, and the AUC was 2606.21 and 697 ng/mL/h, respectively. Therefore, a highly bioavailable silymarin could be obtained by the orally administered proliposome as a result of the enhanced GI absorption of the silymarin from the proliposome.

7.4.4 Solid Lipid Nanoparticles and Nanostructured Lipid Carriers

Over the years, solid lipid nanoparticles (SLNs) and nanostructured lipid carriers (NLC) have received increased interest as efficient and non-toxic drug delivery carriers (Gainza et al., 2014). SLNs are colloidal particles containing highly purified lipids that are solid at room temperature, while NLCs use liquid lipids (oil) for their preparations. Both these lipid carriers offer several advantages as drug delivery systems, including controlled release of the encapsulated drug due to the solid lipid core, targeting to specific sites, protecting drugs against degradation, biocompatibility, low toxicity, the ability to carry both lipophilic and hydrophilic drugs, high encapsulation efficiency, absence of organic solvents, and adaptability to large scale production and sterilization (Liu & Feng 2015).

Li et al. (2009) prepared quercetin-loaded (QT)-SLNs by the emulsification-low temperature solidification method and evaluated it as an oral delivery carrier for the poorly water-soluble flavonoids. QT-SLNs of an average diameter 155.3 nm with entrapment efficiency, drug loading, and zeta potential of 91.1%, 13.2%, and −32.2 mV respectively was obtained. The absorption of the QT-SLN along the GI tract was studied by the *in situ* perfusion method in rats and found that the absorption of quercetin in the stomach was minimal, at only 6.2% in the first two hours. There was a striking enhancement in the bioavailability of quercetin when it was given in the form of QT-SLN, as the RB of QT-SLN to quercetin in sodium carboxymethyl cellulose (Na-CMC) suspension was 571.4%. The C_{max} obtained with the QT-SLN and quercetin suspension was 12.22 ± 2.15 µg/mL and 5.90 ± 1.24 µg/mL respectively, while the AUC of the QT-SLN was 5.71 times greater than the suspension. There was also a slight delay in the T_{max} from 5.00 ± 1.18 h in suspension to 8.00 ± 1.38 h in QT-SLN formulation.

Vinpocetine (VIN), an alkaloid extracted from the periwinkle plant, *Vinca minor*, possesses several pharmacological properties, however its bioavailability has always been hindered by its low aqueous solubility and extensive first-pass metabolism (Medina, 2010). A SLN formulation was prepared to enhance the oral bioavailability of VIN (Luo et al., 2006). The VIN-loaded SLN (VIN-SLN) was prepared by the ultrasonic solvent emulsification technique, which produced a particle size ranging from 70 to 200 nm with an entrapment efficiency of higher than 97% in all the formulations. A pharmacokinetic study in male rats following oral administration of the VIN-SLN at a dose equivalent to 10 mg/kg VIN showed that there was a significant increase in the bioavailability of VIN from the VIN-SLN preparation when compared to the VIN solution. The formulation of VIN into SLNs resulted in enhanced absorption and the AUC values from the VIN-SLNs were 4.16 to 4.17 times greater that obtained with the VIN solution.

In another study, an NLC formulation was compared against microemulsion in enhancing the oral bioavailability of the poorly water-soluble drug luteolin (LT), a flavone compound isolated from *Reseda odorata* (Liu et al., 2014). The luteolin-NLC (LT-NLC) was prepared by the hot high-pressure homogenization technique while the LT microemulsion (LT-ME) was prepared by dissolving LT in a mixture of Labrasol and Cremophor ELP followed by gentle stirring. Different pharmacokinetic parameters were compared to assess the ability of the LT-NLC and LT-ME formulations in improving the oral bioavailability of LT after they were administered to rats by oral gavage. The RB of LT-NLC to LT-suspension was calculated as 515.06% while the RB of LT-ME to LT-suspension was 885.46%. This increase in the AUC for LT-NLC and LT-ME to LT-suspension during the 48 hours drug absorption study in experimental rats indicates that both the formulations significantly enhanced the oral absorption of the LT.

A binary, lipid-based NLC prepared by high-pressure homogenization with glycerol distearates, oleic acids, lecithin, and Tween 80 was also reported to enhance the oral bioavailability of silymarin by 2.54- and 3.10-fold to that of Legalon® and silymarin solid dispersion pellets, respectively (Shangguan et al., 2014).

7.4.5 Self-Microemulsifying (SME) and Self-Nanoemulsifying (SNE) Drug Delivery Systems

The self-microemulsifying drug delivery system (SMEDDS) and self-nanoemulsifying drug delivery system (SNEDDS) are emerging approaches to improve the solubility, dissolution, and oral absorption for poorly water-soluble drugs. They are an isotropic mixture of lipid/oil, surfactant, co-surfactant, and drug substance, which rapidly forms a microemulsion under conditions of GI fluid and GI motility after oral administration, providing a large surface area for increased absorption (Cui et al., 2009; Dwivedi et al., 2014). The microemulsion thus formed has a particle/globlet size of less than 100 nm and the increasing solubility of the hydrophobic drug can enhance its absorption in the GI tract. Moreover, some of the SMEDDS formulations contain surfactants such as Cremophors or Polysorbate 80, which can inhibit drug efflux by P-gp (Liu & Feng 2015).

The SMEDDS formulation has been reported to enhance the dissolution of curcumin in buffer solutions of both acidic and neutral pH. While more than 95% of curcumin in SMEDDS was dissolved in 20 minutes at the buffer solutions, less than 2% of crude curcumin was dissolved in 60 minutes time (Cui et al., 2009). The oral absorption of the SMEDDS was found to follow the passive diffusion mechanism and the percentage of curcumin absorbed from the SMEDDS in 24 hours was 3.86 times that of the curcumin suspension.

A typical sustained release of silymarin was obtained from a SMEDDS formulation containing silymarin, Tween 80, ethyl alcohol, and ethyl linoleate (Wu et al., 2006). The RB of silymarin was dramatically enhanced from the SMEDDS at 1.88- and 48.82-fold to that of silymarin PEG 400 solution and suspension, respectively, confirming the increase in oral bioavailability of silymarin by the SMEDDS. The oral bioavailability of silymarin was also 2.2 times improved by the SMEDDS preparation when compared to the commercially available silymarin hard capsule formulation, Legalon® (Li et al., 2010).

A SNEDDS formulation has also been reported to enhance the oral bioavailability of several phytoconstituents. The pharmacokinetic and pharmacodynamic property of arteether was significantly enhanced by the SNEDDS formulations when compared to the arteether in ground nut oil preparation (Dwivedi et al., 2014). The oral RB of arteether was enhanced 257% by the SNEDDS formulations and a lower dose of the drug at 12.5 mg/kg for five days was also highly effective against the *Plasmodium yoelii nigeriensis* infected mice. Therefore, the SNEDDS formulation not only enhances the oral bioavailability, it also reduces the required dose, thereby eliminating the high dose exposure of the drug which will reduce the possible toxicity associated with it. An improvement in aqueous solubility and oral bioavailability of poorly water-soluble and less bioavailable nisoldipine was also rendered by the SNEDDS formulation (Krishnamoorthy et al., 2015). The bioavailability was enhanced by 2.22–2.5 times that of the drug in suspension.

7.4.6 Niosomes

Niosomes are non-ionic, surfactant-based vesicles that are formed from the self-assembly of non-ionic amphiphiles in aqueous media resulting in closed bilayer structures (Uchegbu & Vyas, 1998). In such an arrangement, the hydrophobic part of the molecules is shielded from the aqueous environment and the hydrophilic head group is exposed to the aqueous medium. They are analogues to the phospholipid vesicles (liposomes) with the ability to carry hydrophilic drugs by encapsulation and hydrophobic drugs by partitioning into the hydrophobic domain (Hu & Rhodes, 1999). Niosomes can be prepared by the hydration technique where the hydration of the surfactant results in the formation of colloidal dispersion that entraps the desired compounds (Pando et al., 2013). Niosomes have been attracting major attention as an alternative to colloidal lipid carriers such as liposomes, as they offer certain advantages such as higher chemical stability, lower cost, and the availability of several surfactants to choose from (Song et al., 2015). They have been applied as a carrier for various kinds of drugs delivered through different routes, including transdermal and oral routes. Multilamellar niosomes containing black tea extract were prepared for delivery through the dermal route (Yeh et al., 2013). The penetration rates for caffeine and gallic acid, the constituents of the black tea extract in the niosome formulation, were found to be higher than their dispersion in aqueous solution. Resveratrol-containing niosomes were prepared by using both Span 80 and Span 60 surfactants following a two-stage technique of mechanical agitation and sonication (Pando et al., 2013). When the two niosomes were compared, it was found that the niosomes prepared with Span 80 were very stable but the Span 60 niosomes provided better entrapment efficient, however the resveratrol release was slower in niosomes prepared with Span 60 than in those with Span 80.

The niosomal formulation can also be a promising delivery system for improving the oral bioavailability and prolonged release delivery for several drugs (Jadon et al., 2009). A niosomal oral delivery system for paclitaxel was designed and prepared using different kinds of surfactants, including Tweens, Spans, and Brijs, with an encapsulation efficiency ranging from 12.1 to 96.6% (Bayindir & Yuksel, 2010). The paclitaxel niosomes were highly stable and drug release was found to be diffusion controlled. A good GI stability of paclitaxel was obtained with the Span 40 niosomes and the slow release of the drug observed was considered to be beneficial for reducing the toxicity associated with paclitaxel.

In order to avoid the possible instability problems of niosomes, such as aggregation, fusion, and leakage of the entrapped drugs, dry and free-flowing proniosomes have been proposed, which immediately form niosomes on contact with hot water followed by agitation for a short period of time. Such proniosomes have the potential to enhance the oral bioavailability of poorly water-soluble drugs such as alkaloid VIN (Song et al., 2015). The proniosome prepared with Span 60, cholesterol, sorbitol, and VIN formed a niosome immediately on contact with water and the oral bioavailability of the drug was remarkably enhanced by about 4–4.9 fold.

7.4.7 Nanodispersion, Nanosuspension, and Nanocrystals

The nanosolid dispersion technique is one approach which combines the sustained release of the drug as well as improving the solubility of poorly water-soluble phytochemicals. The technique has been especially beneficial in improving the oral bioavailability of those drugs that belong to the Biopharmaceutical Classification System (BCS) Class II drugs, which are poorly water-soluble but highly permeable, such as curcumin. By using this technique, a molecular dispersion of the drug is attained in the polymeric carrier and the drug release from the dispersion is then modulated by the polymeric carrier. The oral bioavailability of curcumin from its cellulose acetate solid dispersion was enhanced remarkably as the AUC was increased from 156.36 ng/mL/h in curcumin suspension to 1192.34 ng/mL/h in curcumin-loaded solid dispersion, while the overall pharmacokinetic property was also strikingly improved (Wan et al., 2012). To overcome the drawbacks encountered in producing efficacious formulations of curcumin, nanocrystal solid dispersions (CSD-Cur), amorphous solid dispersions (ASD-Cur), and nanoemulsions (NE-Cur) were prepared to improve curcumin's physicochemical and pharmacokinetic properties (Onoue et al., 2010). A significant improvement in pharmacokinetic behavior was observed in the new formulations as there was a 12- (ASD-Cur), 16- (CSD-Cur), and 9-fold (NE-Cur) increase of oral bioavailability when compared to the unformulated curcumin.

Nanodispersion of taxifolin dihydrate, a naturally occurring flavonoid, prepared by co-precipitation-lyophilization resulted in the aggregation of free spherical particles with a mean size of about 150 nm and the prepared dispersion significantly enhanced the dissolution of taxifolin as a 90% taxifolin release was observed within the first 30 minutes (Shikov et al., 2009). The oral bioavailability of naringenin, a bioactive flavonoid, was also reported to be significantly enhanced after formulation into nanocrystals by the supercritical antisolvent method (Zhang et al., 2013). Compared to the coarse naringenin powder, the nanocrystal formulation exhibited a significantly decreased T_{max}, with a 3.6-fold higher C_{max} and 3.4-fold higher AUC.

7.4.8 Micelles

Micelles are nano-sized, spherical colloidal dispersions formed by the spontaneous aggregates of amphiphiles such as surfactants. Since the center of the micelle is hydrophobic, they can sequester hydrophobic drugs at the core until they are released by some drug delivery mechanism. Certain polymers with blocks or segments of alternating hydrophilic and hydrophobic character also form micelles in water, resulting in the formation of polymeric micelles (Husseini & Pitt, 2008). Micelles are known for their solubilization property and are also the perfect carrier for drugs, being used more than any other nanoparticulate platforms.

Polymeric micelles have often been applied to enhance the solubility and, hence, the bioavailability of orally administered drugs, including phytochemicals. The polymeric micelles composed of pluronic P123 and solutol HS15 were prepared by the thin film dispersion method and loaded with naringenin, a poorly soluble drug (Zhai et al., 2013). Sustained release of naringenin for 36 hours was achieved and the cytotoxicity of the drug-loaded polymeric micelle against cancer cell lines was also significantly enhanced when compared to the free naringenin. The bioavailability of paclitaxel entrapped in glycyrrhizic acid micelles following oral administration was also reported to be six times higher than that of taxol (Yang et al., 2015). There was also a report of 6-fold increments in the antiproliferative activity of curcumin against colorectal cancer cells when compared to those cells treated with a solution containing an equivalent concentration of free curcumin (Wang et al., 2012).

7.5 Conclusion

Reports from various researches over the last few years have demonstrated the potential of therapeutic nanostructures in improving the solubility and oral bioavailability of phytochemicals, which usually exhibit very low bioavailability following oral administration. Various nanostructural therapeutic carriers for these phytochemicals have been reviewed, including Phytosomes®, liposomes, polymeric nanoparticles, lipid nanoparticles, self-emulsifying systems, niosomes, nanodispersions, and micelles. The major objectives in the formulation of therapeutic phytochemicals into these nanostructures are mainly to improve their aqueous solubility, improve their absorption across the biological membrane, and achieve sustained release of the therapeutics. The results from these studies indicate that all these nanostructures possess a remarkable ability to enhance the therapeutics' solubility as well as improve their bioavailability following oral administration. Therefore, therapeutic nanostructures have great potential and are highly beneficial in improving the oral bioavailability of herbal drug substances.

REFERENCES

Abuhelwa AY, Williams DB, Upton RN, Foster DJR. Food, gastrointestinal pH, and models of oral drug absorption. Eur J Pharm Biopharm. 2017;112:234–48.

Ajazuddin, Saraf S. Applications of novel drug delivery system for herbal formulations. Fitoterapia. 2010;81:680–9.

Allen Jr. LV, Popovich NG, Ansel HC. Ansel's pharmaceutical dosage forms and drug delivery systems. 9th South Asian ed. Philadelphia, PA: Lippincott Williams & Wilkins; 2011.

Aqil F, Munagala R, Jeyabalan J, Vadhanam MV. Bioavailability of phytochemicals and its enhancement by drug delivery systems. Cancer Lett. 2013;334:133–41.

Ashford M. Gastrointestinal tract – physiology and drug absorption. In: Aulton ME, Taylor K, editors. Aulton's pharmaceutics: The design and manufacture of medicines. International ed. China: Churchill Livingstone Elsevier; 2013. p. 296–313.

Aulton ME, Taylor K, editors. Aulton's pharmaceutics: The design and manufacture of medicines. International ed. London (Printed in China): Churchill Livingstone Elsevier; 2013. p. 296–313.

Barzaghi N, Crema F, Gatti G, Pifferi G, Perucca E. Pharmacokinetic studies on IdB 1016, a silybin phosphatidylcholine complex, in healthy human subjects. Eur J Drug Metab Pharmacokinet. 1990;15:333–8.

Bawa R. Patents and nanomedicine. Nanomedicine. 2007;2:351–74.

Bawarski WE, Chidlowsky E, Bharali DJ, Mousa SA. Emerging nanopharmaceuticals. Nanomed Nanotechnol Biol Med. 2008;4:273–82.

Bayindir ZS, Yuksel N. Characterization of niosomes prepared with various nonionic surfactants for paclitaxel oral delivery. J Pharm Sci. 2010;99:2049–60.

Bhattacharyya S, Ahammed SK, Saha BP, Mukherjee PK. The gallic acid–phospholipid complex improved the antioxidant potential of gallic acid by enhancing its bioavailability. AAPS PharmSciTech. 2013;14:1025–33.

Bhavsar MD, Amiji MM. Polymeric nano- and microparticle technologies for oral gene delivery. Exp Opin Drug Delivery. 2007;4:197–213.

Cheng AL, Hsu CH, Lin JK, Hsu MM, Ho YF, Shen TS, et al. Phase I clinical trial of curcumin, a chemopreventive agent, in patients with high-risk or pre-malignant lesions. Anticancer Res. 2001;21:2895–900.

Chuah LH, Roberts CJ, Billa N, Abdullah S, Rosli R. Cellular uptake and anticancer effects of mucoadhesive curcumin-containing chitosan nanoparticles. Colloids Surf B Biointerfaces. 2014;116:228–36.

Cui J, Yu B, Zhao Y, Zhu W, Li H, Lou H, Zhai G. Enhancement of oral absorption of curcumin by self-microemulsifying drug delivery systems. Int J Pharm. 2009;371:148–55.

De la Lastra CA, Villegas I. Resveratrol as an anti-inflammatory and anti-aging agent: Mechanisms and clinical implications. Mol Nutr Food Res. 2005;49:405–30.

Di Pierro F, Menghi AB, Barreca A, Lucarelli M, Calandrelli A. GreenSelect® phytosome as an adjunct to a low-calorie diet for treatment of obesity: A clinical trial. Altern Med Rev. 2009;14:154–60.

Dwivedi P, Khatik R, Khandelwal K, Srivastava R, Taneja I, Rama Raju KS, et al. Self-nanoemulsifying drug delivery system (SNEDDS) for oral delivery of arteether: Pharmacokinetics, toxicity and antimalarial activity in mice. RSC Advs. 2014;4:64905–18.

El-Samaligy MS, Afifi NN, Mahmoud EA. Increasing bioavailability of silymarin using a buccal liposomal delivery system: Preparation and experimental design investigation. Int J Pharm. 2006a;308:140–8.

El-Samaligy MS, Afifi NN, Mahmoud EA. Evaluation of hybrid liposomes-encapsulated silymarin regarding physical stability and in vivo performance. Int J Pharm. 2006b;319:121–9.

Ensign LM, Cone R, Hanes J. Oral drug delivery with polymeric nanoparticles: The gastrointestinal mucus barriers. Adv Drug Delivery Rev. 2012;64:557–70.

Florence AT. The oral absorption of micro- and nanoparticulate: Neither exceptional nor unusual. Pharm Res. 1997;14:259–66.

Florence AT. Issues in oral nanoparticle drug carrier uptake and targeting. J Drug Targeting. 2004;12:65–70.

Florence AT, Hussain N. Transcytosis of nanoparticles and dendrimer delivery systems: Evolving vistas. Adv Drug Delivery Rev. 2001;50:S69–89.

Franceschi F, Giori A. Phospholipid complexes of olive fruits or leaves extracts having improved bioavailability. US Patent US20100068316 A1. 2010.

Gainza G, Pastor M, Aguirre JJ, Villullas S, Pedraz JL, Hernandez RM, Igartua M. A novel strategy for the treatment of chronic wounds based on the topical administration of rhEGF-loaded lipid nanoparticles: In vitro bioactivity and in vivo effectiveness in healing-impaired db/db mice. J Controlled Release. 2014;185:51–61.

Gasco MR. Gastrointestinal applications of nanoparticulate drug-delivery systems. In: Thassu D, Deleers M, Pathak Y, editors. Nanoparticulate drug delivery systems. New York: Informa Healthcare; 2007. p. 305–16.

Gao S, Hu M. Bioavailability challenges associated with development of anti-cancer phenolics. Mini Rev Med Chem. 2010;10:550–67.

Giacomelli S, Gallo D, Apollonio P, Ferlini C, Distefano M, Morazzoni P, Scambia G. Silybin and its bioavailable phospholipid complex (IdB 1016) potentiate *in vitro* and *in vivo* the activity of cisplatin. Life Sci. 2002;70:1447–59.

Grama CN, Suryanarayana P, Patil MA, Raghu G, Balakrishna N, Kumar MN, Reddy GB. Efficacy of biodegradable curcumin nanoparticles in delaying cataract in diabetic rat model. PLoS ONE. 2013;8:e78217.

Hollman PCH, de Fries JHM, van Leeuwen SD, Mengelers MJB, Katan MB. Absorption of dietary quercetin glycosides and quercetin in healthy ileostomy volunteers. Am J Clin Nutr. 1995;62:1276–82.

Hu C, Rhodes DG. Proniosomes: A novel drug carrier preparation. Int J Pharm. 1999;185:23–35.

Husseini GA, Pitt WG. Micelles and nanoparticles for ultrasonic drug and gene delivery. Adv Drug Delivery Rev. 2008;60:1137–52.

Iqbal J, Sarti F, Perera G, Bernkop-Schnurch A. Development and *in vivo* evaluation of an oral drug delivery system for paclitaxel. Biomaterials. 2011;32:170–5.

Jadon PS, Gajbhiye V, Jadon RS, Gajbhiye KR, Ganesh N. Enhanced oral bioavailability of griseofulvin via niosomes. AAPS PharmSciTech. 2009;10:1186–92.

Khan J, Alexander A, Ajazuddin, Saraf S, Saraf S. Recent advances and future prospects of phyto-phospholipid complexation technique for improving pharmacokinetic profile of plant actives. J Controlled Release. 2013;168:50–60.

Kidd P, Head K. A review of the bioavailability and clinical efficacy of milk thistle phytosome: A silybin-phosphatidylcholine complex (Siliphos®). Altern Med Rev. 2005;10:193–203.

Kidd PM. Bioavailability and activity of phytosome complexes from botanical polyphenols: The silymarin, curcumin, green tea, and grape seed extracts. Altern Med Rev. 2009;14:226–46.

Kokate A, Marasanapalle VP, Jasti BR, Li X. Physiological and biochemical barriers to drug delivery. In: Li X, Jasti BR, editors. Design of controlled release drug delivery systems. New York: McGraw Hill; 2006. p. 41–74.

Kreuter J. Peroral administration of nanoparticles. Adv Drug Delivery Rev. 1991;7:71–86.

Krishnamoorthy B, Habibur Rahman SM, Tamil Selvan N, Hari Prasad R, Rajkumar M, Siva Selvakumar M, et al. Design, formulation, *in vitro*, *in vivo*, and pharmacokinetic evaluation of nisoldipine-loaded self-nanoemulsifying drug delivery system. J Nanopart Res. 2015;17:34.

Li H, Zhao X, Ma Y, Zhai G, Li L, Lou H. Enhancement of gastrointestinal absorption of quercetin by solid lipid nanoparticles. J Controlled Release. 2009;133:238–44.

Li X, Yuan Q, Huang Y, Zhou Y, Liu Y. Development of silymarin self-microemulsifying drug delivery system with enhanced oral bioavailability. AAPS PharmSciTech. 2010;11:672–8.

Li Y, Wu H, Jia M, Cui F, Lin J, Yang X, et al. Therapeutic effect of folate-targeted and PEGylated phytosomes loaded with a mitomycin C–soybean phosphatidyhlcholine complex. Mol Pharm. 2014;11:3017–26.

Lipinski CA, Lombardo F, Dominy BW, Feeney PJ. Experimental and computational approaches to estimate solubility and permeability in drug discovery and development settings. Adv Drug Delivery Rev. 1997;23:3–25.

Lipinski CA, Lombardo F, Dominy BW, Feeney PJ. Experimental and computational approaches to estimate solubility and permeability in drug discovery and development settings. Adv Drug Delivery Rev. 2001;46:3–26.

Lipinski CA, Lombardo F, Dominy BW, Feeney PJ. Experimental and computational approaches to estimate solubility and permeability in drug discovery and development settings. Adv Drug Delivery Rev. 2012;64:4–17.

Liu A, Lou H, Zhao L, Fan P. Validated LC/MS/MS assay for curcumin and tetrahydrocurcumin in rat plasma and application to pharmacokinetic study of phospholipid complex of curcumin. J Pharm Biomed Anal. 2006;40:720–727.

Liu Y, Feng N. Nanocarriers for the delivery of active ingredients and fractions extracted from natural products used in traditional Chinese medicine (TCM). Adv Colloid Interface Sci. 2015;221:60–76.

Liu Y, Wang L, Zhao Y, He M, Zhang X, Niu M, Feng N. Nanostructured lipid carriers versus microemulsions for delivery of the poorly water-soluble drug luteolin. Int J Pharm. 2014;476:169–77.

Luo Y, Chen D, Ren L, Zhao X, Qin J. Solid lipid nanoparticles for enhancing vinpocetine's oral bioavailability. J Controlled Release. 2006;114:53–9.

Manach C, Scalbert A, Morand C, Remesy C, Jimenez L. Polyphenols: Food sources and bioavailability. Am J Clin Nutr. 2004;79:727–47.

Maurer N, Fenske DB, Cullis PR. Developments in liposomal drug delivery systems. Exp Opin Biol Ther. 2001;1:923–47.

Maiti K, Mukherjee K, Gantait A, Saha BP, Mukherjee PK. Enhanced therapeutic potential of naringenin–phospholipid complex in rats. J Pharm Pharmacol. 2006;58:1227–33.

Maiti K, Mukherjee K, Gantait A, Saha BP, Mukherjee PK. Curcumin–phospholipid complex: Preparation, therapeutic evaluation and pharmacokinetic study in rats. Int J Pharm. 2007;330:155–63.

Mayersohn M. Principles of drug absorption. In: Banker GS, Rhodes CT, editors. Modern pharmaceutics. New York: Marcel Dekker; 2002. p. 29–93.

Mazzarino L, Travelet C, Murillo SO, Otsuka I, Pignot-Paintrand I, Lemos-Senna E, Borsali R. Elaboration of chitosan-coated nanoparticles loaded with curcumin for mucoadhesive applications. J Colloid Interface Sci. 2012;370:58–66.

Medina AE. Vinpocetine as a potent anti-inflammatory agent. Proc Natl Acad Sci. 2010;107:9921–2.

Mukherjee PK, Harwansh RK, Bhattacharyya S. Bioavailability of herbal products: Approach towards improved pharmacokinetics. In: Mukherkee PK, editor. Evidence-based validation of herbal medicines. Amsterdam: Elsevier; 2015. p. 217–46.

National Nanotechnology Initiative (NNI). What is nanotechnology? NNI; 2017 [cited 6 November 2017]. Available from: http://www.nano.gov/nanotech-101/what/definition.

Nunes T, Almeida L, Rocha JF, Falcao A, Fernandes-Lopes C, Loureiro AI, et al. Pharmacokinetics of *trans*-resveratrol following repeated administration in healthy elderly and young subjects. J Clin Pharmacol. 2009;49:1477–82.

Onoue S, Takahashi H, Kawabata Y, Seto Y, Hatanaka J, Timmermann B, Yamada S. Formulation design and photochemical studies on nanocrystal solid dispersion of curcumin with improved oral bioavailability. J Pharm Sci. 2010;99:1871–81.

Pandey R, Ahmad Z, Sharma S, Khuller GK. Nano-encapsulation of azole antifungals: Potential applications to improve oral drug delivery. Int J Pharm. 2005;301:268–76.

Pando D, Gutierrez D, Coca J, Pazos C. Preparation and characterization of niosomes containing resveratrol. J Food Eng. 2013;117:227–34.

Park K, Kwon IC, Park K. Oral protein delivery: Current status and future prospect. React Funct Polym. 2011;71:280–7.

Qin X, Yang Y, Fan T, Gong T, Zhang X, Huang Y. Preparation, characterization and *in vivo* evaluation of bergenin-phospholipid complex. Acta Pharmacol Sinica. 2010;31:127–36.

Ramadan MF. Antioxidant characteristics of phenolipids (quercetin-enriched lecithin) in lipid matrices. Ind Crops Prod. 2012;36:363–9.

Rice-Evans C, Spencer JPE, Schroeter H, Rechner AR. Bioavailability of flavonoids. Drug Metab Drug Interact. 2000;17:291–310.

Russell TL, Berardi RR, Barnett JL, Dermentzoglou LC, Jarvenpaa KM, Schamltz SP, Dressman JB. Upper gastrointestinal pH in seventy-nine healthy, elderly, North American men and women. Pharm Res. 1993;10:187–96.

Selmaty A, Selmaty M, Rawat MSM, Francheschi F. Supramolecular phospholipids–polyphenolics interactions: The PHYTOSOME® strategy to improve the bioavailability of phytochemicals. Fitoterapia. 2010;81:306–14.

Shangguan M, Lu Y, Qi J, Tian Z, Xie Y, Hu F, Yuan H, Wu W. Binary lipids-based nanostructured lipid carriers for improved oral bioavailability of silymarin. J Biomater Appl. 2014;28:887–96.

Shargel L, Wu-Pong S, Yu ABC. Applied biopharmaceutics and pharmacokinetics. Singapore: McGraw Hill Education (Asia); 2005.

Shikov AN, Pozharitskaya ON, Miroshnyk I, Mirza S, Urakova IN, Hirsjarvi S, et al. Nanodispersions of taxifolin: Impact of solid-state properties on dissolution behavior. Int J Pharm. 2009;377:148–52.

Singla AK, Garg A, Aggarwal D. Paclitaxel and its formulations. Int J Pharm. 2002;235:179–92.

Snima KS, Arunkumar P, Jayakumar R, Lakshmanan VK. Silymarin encapsulated poly(D, L-lactic-co-glycolic acid) nanoparticles: A prospective candidate for prostate cancer therapy. J Biomed Nanotechnol. 2014;10:559–70.

Song S, Tian B, Chen F, Zhang W, Pan Y, Zhang Q, Yang X, Pan W. Potentials of proniosomes for improving the oral bioavailability of poorly water-soluble drugs. Drug Dev Ind Pharm. 2015;41:51–62.

Takahashi M, Kitamoto D, Imura T, Oku H, Takara K, Wada K. Characterization and bioavailability of liposomes containing a ukon extract. Biosci Biotechnol Biochem. 2008;72:1199–205.

Takahashi M, Uechi S, Takara K, Asikin Y, Wada K. Evaluation of an oral carrier system in rats: Bioavailability and antioxidant properties of liposome-encapsulated curcumin. J Agric Food Chem. 2009;57:9141–6.

Uchegbu IF, Vyas SP. Non-ionic surfactant based vesicles (niosomes) in drug delivery. Int J Pharm. 1998;172:33–70.

Ventola CL. The nanomedicine revolution. Part 2. Current and future clinical applications. Pharm Ther. 2012;37:582–91.

Volkheimer G. Persorption of particles: Physiology and pharmacology. Adv Pharmacol Chemother. 1977;14:163–87.

Volkheimer G, Schulz SH. The phenomenon of persorption. Digestion. 1968;1:213–8.

Walle T, Hsieh F, DeLegge MH, Oatis JE, Walle UK. High absorption but very low bioavailability of oral resveratrol in humans. Drug Metab Dispos. 2004;32:1377–82.

Wan S, Sun Y, Qi X, Tan F. Improved bioavailability of poorly water-soluble drug curcumin in cellulose acetate solid dispersion. AAPS PharmSciTech. 2012;13:159–66.

Wang H, Liu M, Du S. Optimization of madecassoside liposomes using response surface methodology and evaluation of its stability. Int J Pharm. 2014;473:280–5.

Wang K, Zhang T, Liu L, Wang X, Wu P, Chen Z, et al. Novel micelle formulation of curcumin for enhancing antitumor activity and inhibiting colorectal cancer stem cells. Int J Nanomed. 2012;7:4487–97.

Wu W, Wang Y, Que L. Enhanced bioavailability of silymarin by self-microemulsifying drug delivery system. Eur J Pharm Biopharm. 2006;63:288–94.

Yang FH, Zhang Q, Liang QY, Wang SQ, Zhao BX, Wang YT, Cai Y, Li GF. Bioavailability enhancement of paclitaxel via a novel oral drug delivery system: Paclitaxel-loaded glycyrrhizic acid micelles. Molecules. 2015;20:4337–56.

Yan-yu X, Yunmei S, Zhipeng C, Qineng P. Preparation of silymarin proliposome: A new way to increase oral bioavailability of silymarin in beagle dogs. Int J Pharm. 2006;319:162–8.

Yao M, McClements DJ, Xiao H. Improving oral bioavailability of nutraceuticals by engineered nanoparticle-based delivery systems. Curr Opin Food Sci. 2015;2:14–9.

Yeh MI, Huang HC, Liaw JH, Huang MC, Huang KF, Hsu FL. Dermal delivery by niosomes of black tea extract as a sunscreen agent. Int J Dermatol. 2013;52:239–45.

Yu H, Huang Q. Bioavailability and delivery of nutraceuticals and functional foods using nanotechnology. In: Bagchi D, Bagchi M, Moriyama H, Shahidi F, editors. Bio-nanotechnology: A revolution in food, biomedical and health sciences. West Sussex, UK: John Wiley & Sons; 2013. p. 595–604.

Yue PF, Yuan HL, Li XY, Yang M, Zhu WF. Process optimization, characterization and evaluation in vivo of oxymatrine–phospholipid complex. Int J Pharm. 2010;387:139–46.

Zhai Y, Guo S, Liu C, Yang C, Dou J, Li L, Zhai G. Preparation and *in vitro* evaluation of apigenin-loaded polymeric micelles. Colloids Surf A Physicochem Eng Aspects. 2013;429:24–30.

Zhang J, Huang Y, Liu D, Gao Y, Qian S. Preparation of apigenin nanocrystals using supercritical antisolvent process for dissolution and bioavailability enhancement. Eur J Pharm Sci. 2013;48:740–7.

8

Nanotechnology for Tissue Regeneration

Kumud Joshi, Pronobesh Chattopadhyay, and Bhaskar Mazumdar

CONTENTS

ABSTRACT: Loss of organ functionality and tissue damage due to injury, aging, or disease is one of the leading causes of morbidity, mortality, and disability. Currently, only some major organs like the kidney, liver, heart, and a few others can be transplanted. Moreover, these transplants are associated with immune graft rejections. Patients are compelled to take immunosuppressive therapy, which has its own share of demerits. Tissue engineering will bring a paradigm shift in the current transplantation practices. It intends to provide implantable scaffolds which provide a suitable environment for *in vivo* tissue regeneration from patients' own stem cells. These scaffolds are also suitable for the *ex vivo* development of whole tissue constructs suitable for implantation. This crossover of material science with biological science helps to achieve the desired goals of medical science. Most cellular structures have the basic building blocks of nano-dimensions. In order to effectively imitate the cellular structures, it is necessary to build them bottom-up from the nano-range. Nanotechnology utilizes materials of dimensions less than 1,000 nm and within that range material behavior becomes significantly different from the conventional behavior. Tissues develop from cells and the nano-dimensional architecture provides them with efficient blood flow, cell adhesion, nutrient uptake, and biochemical efficiency. Hence to achieve effective tissue regeneration, the utilization of nanotechnology becomes important. Nanoscaffolds are one of the most frequently applied platforms for bio-regeneration. In this chapter, we will provide insight into the various advancements of regenerative medicine, with a focus on nanoscience and nanotechnology, and discuss the future prospects in this area.

KEY WORDS: *Tissue regeneration, nanoscaffolds, nanomaterials, nanofibers, regenerative medicine, cell delivery, tissue engineering*

8.1 Introduction of Regenerative Medicine and Tissue Regeneration

All organisms possess the capability to regenerate damaged tissues, albeit to a different extent. Some species, like salamanders, have remarkable regeneration capability and can regenerate entire limbs. As for humans, this is limited to only certain organs with specialized tissues, like the liver (Brockes and Gates, 2014). Tissue regeneration was first described by de Réaumur, who observed it in crayfish claws in 1712 (de Réaumur, 1712). One of the most ambitious human endeavors in modern medical science is to develop artificial organs and tissue regeneration capability, which can possibly treat a plethora of diseases, including ones with high mortality, like cancer and organ failures, and many others with high morbidity, like diabetes and Alzheimer's. To achieve this ambitious goal, advancements in different spheres of sciences like nanotechnology, physical science, materials science, cell science, immunology, and others are brought together. This fusion is the genesis of an interdisciplinary field involving translational research called tissue engineering and regenerative medicine (Khademhosseini and Langer. 2016). It seeks to develop therapies which can recreate lost tissues and restore or improve tissue activity. This can treat patients afflicted with a variety of age and disease-related disorders/dysfunctions (Shafiee and Atala, 2017). For example, if successful regeneration of β-cells of islets of Langerhans is achieved, it might, in the future, eliminate the need for regular medication in diabetes and eliminate the lifelong need for insulin (Subash-Babu et al., 2009). The application of tissue engineering in achieving functional tissues and organs can be termed as regenerative medicine. It holds huge potential for the treatment of diseases with high mortality, like cancer, and severe tissue damage in serious injuries, like spinal cord injuries or damage to vital organs like cardiac and kidney damage (Figure 8.1).

8.1.1 Principle of Tissue Graft Development and Tissue Regeneration

The basic approach is to create bioengineered tissues or organs by combining the patient's own cells, generally pluripotent stem cells, with a natural and/or synthetic biomaterial scaffold under suitable culture conditions, providing a variety of growth factors, like fibroblast growth factor (FGF), vascular endothelial growth factor (VEGF), and bone morphogenic proteins (BMPs), and other conditions, like suitable micromechanical stress for stimulating cell division and differentiation in a suitable manner needed for the development of functional tissue (Hu et al., 2018). The cells are seeded with a biomaterial/scaffold in a bioreactor. Bioreactors provide and maintain the environmental conditions desired for cellular proliferation and differentiation. It controls different environmental parameters, like scaffold stretch, nutrient flow, etc., and imitates the natural environment for the cells. This process is performed in either static culture conditions or dynamic bioreactor systems (Kasper et al., 2009; Radisic et al., 2004). This allows cell division and tissue formation which after proper development can be transplanted *in vivo* (Freed et al., 2006; Goldstein and Christ, 2009; Zhao et al., 2016a). The cells that are cultured on a scaffold should have similar behavior to what they would have in the body without a synthetic scaffold.

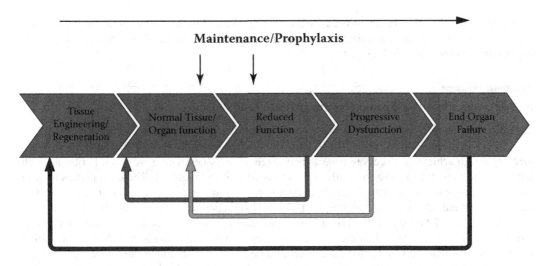

FIGURE 8.1 Scope of tissue regeneration.

8.2 Tissue Scaffold in Tissue Engineering

Normal biological tissues possess a highly specialized, composite structure at the molecular level. There are different macromolecules involved, providing a myriad of biochemical functions to cells. The biomolecules, such as structural proteins, including collagen and elastin, and polysaccharides, like glycosaminoglycans (GAGs), form a matrix in which cells are embedded. The matrix gives mechanical strength to the tissue to physically withstand normal wear and tear. Specialized proteins, such as fibrillin, fibronectin, and laminin, and proteoglycans can act as channels of cellular transport and modulators of communication for cell–cell interactions and specific functions like cell adhesion. The working dimensions of cellular structures lie in the nano-range and most of the cells allow transport of very small molecules of a mass range of 1,000 daltons. Scaffolds must mimic native extracellular matrices (ECM), providing a conducive environment for the cell division and differentiation needed for new tissue development or for restoring the existing tissues to their original healthy state. An ideal tissue scaffold should have the following characteristics (Freed et al., 2006; Gomes et al., 2002; Mohamed and Xing, 2012).

- It should allow systematic cell division and differentiation as needed for the development of normal tissue
- It should be biodegradable and biocompatible
- It should have appropriate permeability and porosity for the transfer of nutrients to the dividing cells and for taking the wastes away from them
- It should support efficient cell adhesion and spreading during tissue development
- It should be strong enough to withstand the mechanical stress during different times of tissue formation.

There are different techniques that have been developed for producing smart biomaterials, and these techniques provide scaffolds that can enhance target functions such as adhesion, proliferation, migration, and tissue differentiation. These techniques are classified into two classes: (1) internal modulation techniques and (2) external modulation techniques (Pérez et al., 2013). Strategies for internal modulation include surface tailoring with bio-mimetic molecules, such as peptides and adhesive proteins, to mimic the native ECM (Pérez et al., 2013), while external modulation uses smart biomaterials with drug delivering potential to deliver a suitable site and trigger the instructive cues necessary for tissue regeneration at the site of action. Typically, bio-factors such as proteins, growth factors, or chemical modulators are incorporated in many ways within the biomaterials engineered as scaffolds, as these agents need to permeate the membrane receptors or need intracellular uptake within the cytosol or nucleus. Therefore, the biomaterials are modulated to deliver bio-factors of interest at the targeted site for successful tissue regeneration (Pérez et al., 2013) (Figure 8.2).

8.3 Biomaterials for Tissue Regeneration

We know biomaterials play a very important role in the design of scaffolds for tissue regeneration. They act as carriers for cells, providing them with the necessary mechanical strength and environmental cues for achieving success in generating the requisite tissues. There are several criteria which a biomaterial must fulfill to successfully serve as tissue scaffold material. These properties include the desired mechanical strength, degradation rate, porosity, surface properties, and moldability. The choice of biomaterial for three-dimensional (3D) scaffolds also depends on the type of cells being seeded, the desired biology of the target tissue, the growth factors involved, and any biocompatibility issues (Li et al., 2017; Pérez et al., 2013). There are different kinds of polymers, including natural and synthetic polymers, that have been established as suitable materials for tissue regeneration. Natural polymers include collagen, alginate, chitosan (CHI), hyaluronic acid, silk, and others. Synthetic polymers include poly(L-lactic) acid (PLLA), polyglycolic acid (PGA), poly(ethylene oxide) (PEO), and others (Allo et al., 2012; Geiger et al., 2003; Guo and Ma, 2014; Parenteau-Bareil et al., 2010). Different polymers have different attributes. While natural polymers are biocompatible and biodegradable, allowing scaffolds to be built with excellent cell–matrix interaction, achieving tailored degradation is difficult with them (Asghari et al., 2017; Li et al., 2015; Parenteau-Bareil et al., 2010). In contrast to this, synthetic materials can be designed to obtain predictable and tunable properties, but usually have poor bioactivity and low mechanical properties (Allo et al., 2012; Pérez et al., 2013). Apart from polymers, other materials such as calcium phosphate (hydroxyapatite (HA), β-tricalcium phosphate) are also incorporated in tissue grafts, especially in bones for providing strength and rigidity. Bioglass has osteoconductivity and rigidity but it is fragile (Liu et al., 2013; Rahaman et al., 2011). Carbon nanomaterials, like graphene and carbon nanotubes, also serve as good biomaterials, especially when good strength and good conductivity are desired (Ghassemi et al., 2017; Kumar et al., 2016). Since there is often no ideal biomaterial, to overcome the limitations of a particular biomaterial we often combine their advantages. Polymers, bioceramics, carbon nanomaterials, and other suitable biomaterials can be combined to achieve tailored, desired properties.

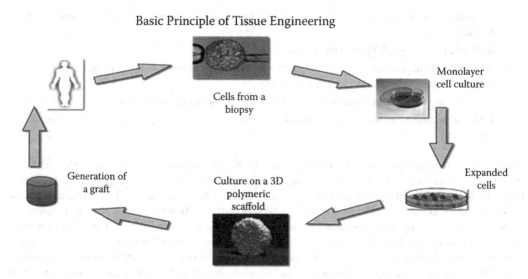

FIGURE 8.2   Basic principle of tissue regeneration.

8.4 Current Status of Regenerative Medicine

So far, few products for regenerative medicine are available on the market. Taking into consideration products for treating skin burns and injuries, tissue engineering has expanded its reach with graft applications to damages caused by various disease conditions, including bone grafts, connective tissue growth, and vascular grafts for heart surgeries, which have all been developed and approved. Developments in different fields, including stem cell research, material research, nanotechnology, and regenerative medicine, are creating newer paths of therapeutics which hold the promise of treatment for many clinical diseases that were previously hard to tackle, providing better health for mankind (Hunziker, 2010; Mishra et al., 2016) (Table 8.1).

8.5 Role of Nanotechnology in Tissue Regeneration

The major structural proteins present in native ECM are collagen and elastin, which, in fact, are nano-dimensional fibers. Any matrix that is meant to be used for tissue engineering should mimic the features of natural ECM to elicit favorable cellular behavior, including a high surface area for efficient absorption of different cellular proteins for healing. The high active surface area of nanoscaffolds provides better cell adhesion, proliferation, and differentiation in comparison to micro-structured scaffolds (Gupta et al., 2014). Cellular dimensions as well as fiber dimensions in every tissue, such as bone, skin, tendon, ligament, and so forth, lies within the size range of 50 to 1,000 nm. Thus, any carrier used for tissue regeneration must have similar nano-dimensional topography (Li et al., 2017; Singh et al., 2014). Tissue regeneration is a complex and specialized process targeting the delivery of essential molecules to cells, preferably in a controlled fashion. The biomaterial used must be functionalized by attaching ligands/epitomes for this targeted action to occur. It becomes imperative to mimic the nano-architecture of normal cells. In order to achieve better functionalization and control of material properties, nanotechnology is an ideal tool, as the bottom-up approach used in nanofabrication allows a higher control of the extent and degree of functionalization. Thus, to achieve the needed complex structural organization of tissue scaffolds, nanotechnology is an indispensable tool (Christ et al., 2013; Nel et al., 2009). Nanotechnology thus forms a very integral part of tissue engineering. The advancements in nanotechnology and material science have provided us with the capability to design smart composites of biomaterials with certain desired properties and functions, enabling us to produce biomaterial similar to native ECM organization and structure for achieving optimal tissue regeneration (Pérez et al., 2013).

8.6 Different Nanostructured Novel Therapeutic Platforms of Tissue Regeneration

The different types of nanoformulations that have been developed for use in tissue engineering include the following:

- Nanohydrogels
- Nanofibers
- Nanocell sheets
- Layer by layer structures.

8.6.1 Nanostructured Hydrogel-Based Materials

Injectable nanohydrogels are a promising class of materials for tissue engineering and regenerative medicine applications. Nanohydrogel has a high content of water and can form a swollen polymeric structure which resembles natural connective tissue 3D hydrogels. 3D gels are capable of entrapping cells inside

TABLE 8.1

Approved Cell Therapy Products by Different Regulatory Bodies Including US FDA, EU EMEA, and Other Countries (Boyce and Lalley, 2018; Hellman, 2008; Mishra et al., 2016)

S.No.	Product Name	Product Company	Product Details	Intended Therapeutic Use
Skin Tissue Regeneration Products				
1.	Apligraf® Skin	Organogenesis, Inc.	Viable allogeneic fibroblasts/ keratinocytes of type-1 bovine collagen	Standard therapeutic compression for treatment of non-infected partial and full-thickness skin ulcers
2.	Dermagraft®	Advanced Tissue Sciences, Inc.	Cryopreserved dermal substitute, allogeneic fibroblasts, bioabsorbable scaffold	Treatment of full-thickness diabetic foot ulcers
Musculoskeletal Tissue Regeneration				
3.	Carticel®	Genzyme Corporation	Autologous cultured chondrocytes	Repair of femoral condyle caused by acute or repetitive fracture
4.	OP-1 Implant	Stryker Biotech	Recombinant human osteogenic protein (rh OP-1), type-1 bovine bone collagen matrix	Alternative to autograft in recalcitrant long bone nonunions
5.	Infuse® bone graft	Medtronic, Inc.	Solution containing rhBMP-2 and an ACS used to fill the LT-CAGE® lumbar tapered fusion device	Expedite healing of fractures in the clavicle, tibia, femur, humerus, radius, and ulna
6.	GEM 21STM	Bio-mimetic Pharmaceuticals, Inc.	Growth factor enhanced matrix	Treatment for periodontal-related defects: intra-bony defects, furcation defects, gingival recession associated with periodontal defects
7.	Cartistem®	Medipost		Treatment for traumatic and degenerative osteoarthritis
Nerve Tissue Regeneration Products				
8.	Avance® Nerve Graft	AxoGen, Inc.	Processed nerve allografts are harvested from cadaveric nerve sources	Functional repair of nerve gaps
Cardiac Tissue Regeneration Products				
9.	Hearticellgram®-AMI	FCB-Pharmicell	Autologous bone marrow-derived mesenchymal stem cells	Heart repair post myocardial infarction

the swollen polymeric matrix with a high permeability for nutrients, oxygen, and metabolites (Billiet et al., 2012). They are prepared by the gelation method using swelling polymers (natural or synthetic) to form a gel, which is crosslinked by photo-crosslinking or chemical crosslinking, involving the formation of the covalent or non-covalent bond. For tissue engineering, reversible, non-covalent bond-based hydrogels are suitable (Wan and Ying, 2010). Nanohydrogel conjugated with biological peptides (Hartgerink et al., 2001; Yanlian et al., 2009), proteins, such as porcine decellularized nerve matrix (Lin et al., 2018a), or inorganics (Kim et al., 2018; Radhakrishnan et al., 2018) can form the basis of targeting hydrogels to specific locations in the body to enhance cell adhesion (Harrington et al., 2006), therapeutic cell delivery (Webber et al., 2015), or for targeting angiogenic growth regulators to the site of tissue damage (Stendahl et al., 2008). The application of nanohydrogels in conjugation with nanoparticles for the delivery of nanoparticles, like gold nanoparticles (AuNPs) and gold nanorods (AuNRs), for tissue regeneration is very interesting (Dvir et al., 2011).

8.6.2 Nanofibers

Nanofibers are one of the most widely used carriers for tissue regeneration and consist of three essential components: polymer matrix, cells to be cultured, and growth regulators for providing the environmental cues necessary for tissue growth. Nanofibers have two important characteristics, viz. pore size and fiber diameter, which influence cell growth and tissue development in a scaffold (Vasita and Katti, 2006). As collagen fibers are the main component of natural ECM, especially in muscular tissue, the fiber diameter ranges from 50 to 500 nm. The fiber must have a dimension in this range to provide a suitable environment enabling effective tissue growth. There are a variety of methods used for obtaining suitable nanofibers, including solvent casting, salt leaching, phase-separation, self-assembly, gas foaming, and electrospinning (Dahlin et al., 2011; Radisic et al., 2004). These nanofibers can be implanted into the tissue defect site. Nanofibrous scaffolds utilize a wide range of biodegradable polymers and copolymers (Jeong et al., 2010; Nasri-Nasrabadi et al., 2014). These scaffolds mimic the physical architecture of natural collagen, with fiber diameters of 50–500 nm as mentioned above (Woo et al., 2003; Yang et al., 2004). Nanofibrous scaffolds have also been developed with carbon nanotubes (CNTs). CNTs are cylindrical carbon tubes with diameters in the ~1 nm range and lengths that are typically hundreds or thousands of times greater. Moreover, these nanofibers can be functionalized to obtain desired characteristics (Yoo et al., 2009). They find applications in regeneration of various types of tissues, like nerve (Prabhakaran et al., 2013), bone (Balagangadharan et al., 2016), skin (Sundaramurthi et al., 2014), heart (Orlova et al., 2011), and others.

8.6.3 Scaffold-Free Single Cell Sheet

The scaffold-free single cell sheet is a unique technology for scaffold-free tissue regeneration. For successful tissue reconstruction, various biomaterials are used. However, whether natural or synthetic, these polymers have possible problems in immunological responses and biocompatibility (Roach et al., 2007). Even though the biocompatibility of polymers can be enhanced by different methods, it is desirable to obtain absolutely non-immunogenic scaffolds. Recently a unique technology called cell sheet technology (CST) has been developed for achieving this goal (Hannachi et al., 2009). This method involves the collection of cell sheets without breaking them and without using any proteolytic enzymes or instruments; this keeps their natural cell structures and networking preserved. The obtained cell sheets are spontaneously detached and harvested on a special thermoresponsive polymer, poly(N-isopropylacrylamide) (PIPAAm). PIPAAm is a non-ionic polymer which, on heating above a critical temperature, undergoes phase separation in water; this temperature is called the lower critical solution temperature (LCST). The PIPAAm shows a change in hydrophilicity at a temperature which is higher than the LCST of PIAPPAm. At 89.6°F, surfaces are hydrophobic and therefore suitable for cell adhesion. At temperatures lower than the LCST, the surface becomes hydrophilic, inhibiting cell adhesion and making the cell sheets detachable (Okano et al., 2006). Therefore, the cells adhered to the PIPAAm sheets above the LCST get detached upon temperature reduction. Cell sheets so obtained are collected and transplanted to a defective tissue *in vivo*, i.e., to the defective tissue of a patient (Haraguchi et al., 2012). Many other responsive systems for single cell sheet technology have been developed like photoresponsive, thermoresponsive, and pH-responsive systems (Patel and Zhang, 2013). The single cell sheets can also give us nanofiber scaffolding by electrospinning (Ahn et al., 2015). These single cell sheet structures find a variety of applications in tissue regeneration (Ahn et al., 2015; Cerqueira et al., 2014; Lin et al., 2013; Sekiya et al., 2006; Sekiya et al., 2013).

8.6.4 Layer by Layer Structures

Layer by layer structures are composed of polyelectrolyte molecules stacked over one another in the form of alternating anionic and cationic layers, producing multilayered film structures. The layers are held together by different molecular forces like hydrogen bonding, covalent bonds, electrostatic forces, or hydrophobic bonds (Shukla and Almeida, 2014). The layer by layer structure allows for the development of multicellular, heteromorphic, 3D arrangements, which allows the fabrication of complex tissues using differently charged species, including biological molecules like polypeptides, nucleic acid, and proteins.

This offers an attractive approach for the development of tissue scaffolds of macro- and nano-dimensions (Poon et al., 2011). The thickness of the layer can be controlled by modulating different parameters like coating time, ion concentration, pH. and polarity of the media. Layer by layer self-assembly finds an array of applications in tissue engineering.

8.7 Different Types of Technologies for the Development of Nanotissue Scaffolds

Different techniques have been developed for fabricating nanocomposites which can mimic complex structural nano-topographical attributes, which in turn provides the desired functional behavior to nanoscaffolds for achieving tissue regeneration. There are basically two approaches for scaffold development: the top-down approach or the bottom-up approach. Conventionally used methods employ the top-down approach, where cells are seeded onto a biocompatible and biodegradable polymeric scaffold, the seeded cells proliferate, and they replace the scaffold with ECM (Lu et al., 2013). Meanwhile, in the bottom-up approach the scaffold is developed by the aggregation of nano-sized units to form a well-organized cellular composite. This approach has the advantage that requisite cellular behavior can be elicited by controlling biochemical and mechanical characteristics like nanotopography in a well-defined manner (Cunha et al., 2011; Kang et al., 2013). Bio-mimetic scaffolds consist of a biodegradable polymer, fabricated in the form of ECM, for the growth of embedded self, and have been optimized for properties like porosity and nanotopography to achieve suitable functionality. These scaffolds are fabricated by several methods. Here, we focus on these fabrication methods for nanofibrous scaffolds and their successful applications in tissue engineering. The various methods used include electrospinning, self-Assembly, solvent casting, freeze-drying, gas foaming, bioprinting and lithography.

8.7.1 Electrospinning

Electrospinning provides us with ultrafine nanofibers which are similar to native ECM and are often used for tissue regeneration. These ultrathin fibers have a high surface area and porosity, providing favorable conditions for tissue regeneration. Electrospinning produces continuous fibers with diameters in the micro- to nanometer range by using a jet of polymer under the influence of an applied electric field without any physical contact (Li and Xia, 2004). In the electrospinning process, fibers ranging from 50 nm to 1,000 nm or greater can be produced (Bosworth and Downes, 2011), and different structures like beads, porous core shells, ribbons, and fibers have been produced by this technique (Ramakrishna et al., 2006). The fiber diameter and properties depend on polymer viscosity, flow rate, and applied electrical potential. Electrospinning is an efficient technique for producing two-dimensional (2D) mat-structured scaffolds. Tissue engineering scaffolds widely utilize electrospinning for producing scaffolds for application in soft tissue regeneration as well as for hard tissue like bone. Recently, 3D scaffolds have also been fabricated successfully by Lin et al. (2018b). Wu et al. (2015) successfully electrospun scaffold from poly-(l-lactide-co-caprolactone) (P(LLA-CL)), collagen, and chitosan. The scaffold had a multilayered symmetrical structure, and showed good proliferation for human coronary artery smooth muscle cells (hSMCs) growth (Wu et al., 2015). Electrospun nerve conduit fibers using poly(d,l-lactide)-co-poly(ethylene glycol) and polypyrrole (PELA-PPY) (20%, 30%, and 50%) were prepared by Zhou et al. (2016) and seeded with the rat pheochromocytoma PC-12 cell. The developed scaffold successfully overcame the nerve sciatic injury in the rats in 12 weeks (Zhou et al., 2016). Electrospinning has established itself as an effective technique for regeneration of a variety of other tissues, including bone (Li et al., 2006), cartilage (Xue et al., 2013), and tendon (Maghdouri-White et al., 2018), etc.

8.7.2 Self-Assembly

In this technique, the molecules assemble to spontaneously form a well-defined structure. It is a common phenomenon that molecules tend to take certain shapes and structures to achieve the most stable state. Applying the principle of self-assembly of nano-sized fibers to tissue engineering for achieving the desired 3D organization of cells and nanofibers in tissue scaffolds is an attractive approach (Koutsopoulos, 2016). The 3D scaffolds developed are generally developed using molecules like amphiphilic oligonucleotides with complimentary positive and negative charged residues, which easily self-assemble to produce macromolecules and nanofibers by utilizing different molecular forces, including Van der Waals forces, ionic bonds, and hydrophilic and hydrophobic interactions (Loo et al., 2015). Peptides under suitable conditions self-assemble to form hydrogels. Carbohydrates such as oligosaccharides, glycopolypeptides, and peptidoglycans also self-assemble to form complex structural matrices. Similarly, nucleic acids also self-assemble to form bio-motifs (Guven et al., 2015).

Self-assembly is particularly useful for *in vivo* use; as it is carried out in physiological media and not in organic solvents, the concern of toxicity gets resolved. Injectable self-assembling peptides with amino acids and bioactive motifs can be grafted onto scaffolds to achieve specificity of targeting, which enhances proliferation and attachment. Argenine-glycine-aspartic acid (RGD) peptides incorporated in nanofibers with mesenchymal stem cells (MSCs) showed good oestrogenic activity (Hosseinkhani et al., 2006). Self-assembly can find further application in wound healing (Schneider et al., 2008; Semino, 2008), bone tissue regeneration, cardiac tissue regeneration, neural regeneration, and so on (Koutsopoulos, 2017).

8.7.3 3D Biofabrication

With advancements in regenerative medicine, it is being realized that 3D topologies and geometries, and the orientation of cells, play a critical role in tissue functionality. High porosity, tunable pore sizes, and mechanical strength are desirable for tissue engineering applications. High porosity provides channels for easy diffusion of molecules like nutrients, cellular waste products, and gases within a 3D scaffold. It also facilitates cell migration, spreading, and proliferation. 3D scaffolds can be prepared by using both synthetic and naturally occurring biomaterials (Annabi et al., 2010; Mandrycky et al., 2016). Different techniques for the development of 3D scaffolds are as follows.

8.7.3.1 Solvent Casting or Particle Leaching

Solvent casting is a technology based on the principle of salting out. In this technique, a polymer solution is dissolved in a solvent and salt with a uniform size is added to it. The pore size of the scaffold depends on the diameter of the salt particles (Mikos et al., 1994). After the polymer has been dissolved completely the solvent is allowed to evaporate, allowing the polymer to aggregate out and producing a composite with uniformly distributed salt particles. This polymer matrix is then immersed in water to allow leaching of the salt particles. Finally, we obtain a highly porous 3D matrix of porosity ranging up to 90% and pore diameter of up to 500 µm. These scaffolds can be seeded with different types of cells (Mehrabanian and Nasr-Esfahani, 2011).

8.7.3.2 Freeze-Drying

Freeze-drying or lyophilization involves cooling down a polymer scaffold below the freezing point of the solvent used in a vacuum environment (Madihally and Matthew, 1999). The solvent molecules sublime under vacuum conditions, producing a highly porous, polymeric matrix suitable for cell seeding. The pore size of the scaffold is determined by the concentration of the polymer and the freezing regime (Kang et al., 1999a, 1999b).

8.7.3.3 Gas Foaming

Gas foaming is a technique where inert foaming gas such as carbon dioxide, nitrogen, or water is purged through the polymer scaffold during its formative stages and over time, as the scaffold solidifies, the pressure of the gas is reduced which leads to a decrease in solubility of the gas in the polymer (Mooney et al., 1996). This process gives us a highly porous structure with a size ranging from 100 to 500 µm, which can be seeded with cells and used for tissue regeneration purposes (Lee et al., 2010).

8.7.3.4 Bioprinting

Bioprinting has garnered a lot of interest for its possible application in the 3D printing of complex, 3D, functional living tissues from cells and biocompatible materials. This can address the need of regenerative medicine for tissues and organs for transplantation. Many multilayered complex tissues have been developed successfully using 3D bioprinting. This technique is a significant cornerstone for future tissue and organ printing (Murphy and Atala, 2014). It uses lithography, a group of techniques used for the fabrication of specific patterns and dimensions of materials using computer-aided techniques (CAD), which can be used for developing patterned cell structures to achieve specific cell responses. Lithography employs two important techniques for producing the nanocomposites: photolithography and stereolithography. Photolithography techniques use light, or photons, or laser beam guided systems for architecting specific geometric shapes onto a mask, which is generated on a light sensitive surface. For biomedical applications, photolithography has been successfully utilized for achieving uniform cell encapsulation in the 3D network of a biopolymer (Lee et al., 2008; Sun et al., 2009b). This technique offers specific advantages including uniform cell seeding in scaffolds. Also, as this technique produces a very low amount of heat, this prevents any damage to cells during seeding (Williams et al., 2005). So far, multitudes of cells have been successfully fabricated into scaffolds using this technique, including 3D scaffolds for culturing multiple cell types, such as hepatocytes (Liu Tsang et al., 2007), fibroblasts, C2C12 myoblasts, endothelial cells, cardiac stem cells (Aubin et al., 2010), HT1080 fibrosarcoma cells (Hahn et al., 2006), and mouse embryonic stem cells (mESCs) (Bajaj et al., 2013). Stereolithography is the second technique, which uses a mask-less photopatterning technique which involves the use of CAD to guide the ultraviolet (UV) laser to develop cell structures layer by layer, finally producing a 3D structure. It offers a unique advantage in that it does not need to develop a physical mask for the template. Also, being an automated technique, it allows for precise control of the thickness of the scaffold and help in the rapid fabrication of the scaffold. As with time, greater strides are being made in the advancement of fabrication techniques in tissue engineering. Still, the complexity of the existing technology is very high for utilizing them on a mass scale and there is the need to refine and simplify these complicated technologies to fabricate scaffolds which are readily available and more biocompatible with low potential for toxicity.

8.8 Application of Nanotechnology in Different Areas of Tissue Regeneration

8.8.1 Application of Nanotechnology in Bone Tissue Engineering

Hip and joint replacements are the most common advanced age ailment-associated surgeries. Also, various bone fractures, osteoarthritis, osteoporosis, and/or bone cancers represent common and significant clinical problems. In this scenario, bone tissue regeneration capability can play a vital role in bone and cartilage substitutes to regenerate bone and cartilage tissue at defect sites that will last the lifetime of the patient. As we know, bone consists of a nanostructured architect of different proteins like collagen and non-collagenous proteins (laminin, fibronectin, and vitronectin) which form a soft hydrogel matrix. The hard, inorganic components (hydroxyapatite (HA) ($Ca_{10}(PO4)_6(OH)_2$)) are typically of a dimension of 20–80 nm long and 2–5 nm thick (Zhang and Webster, 2009). The design of scaffolds for bone tissue engineering should match the properties of normal bone tissue in terms of mechanical strength, pore size, porosity, hardness, and overall 3D architecture (Wang et al., 2016). There are several factors that are at work while oestrogenic activity occurs in bone formation, and this includes different physical, biological, and chemical cues which control the microenvironment around the cells. These factors include growth factors, chemical mediators, and mechanical stress. To ensure proper cellular growth, artificial mechanical environments have to be created and maintained (Stegen et al., 2015). Different studies indicate nanotopography is an important factor influencing cell protein specific interactions and stimulating tissue regeneration (Degasne et al., 1999; Salmasi et al., 2015; Webster et al., 2000). Nanoceramics help better adhesion. Osteoblasts, along with nano-sized (67 nm grain size) HA, have been found to have efficiently inhibited fibroblast adhesion compared to conventional (179 nm grain size) HA within four hours of inoculation (Webster et al., 2000). The application of nanocomposites in bone and cartilage substitutes has achieved significant development in recent reports of osteogenic differentiation, and the self-renewal of MSCs are mediated by nanopatterned substrates that have been established (Dalby et al., 2007; McMurray et al., 2011; Wang et al., 2016; Xie et al., 2016). Several tissue scaffolds utilizing different types of biomaterials have been developed. Natural polymers including collagen, fibrin, gelatin, starch, hyaluronic acid, and chitosan, which principally provide good biocompatibility. Anionic collagen matrices have been found to be osteoconductive and have successfully healed bone defects in rats, therefore demonstrating bone formation (Maghdouri-White et al., 2018; Rocha et al., 2002). Other natural polymers like chitosan have been successfully used in combination with other polymers for producing suitable biocomposites with certain desired mechanical properties and degradation rates (Sachlos and Czernuszka, 2003; Xie et al., 2016). Synthetic polymers are also used, including poly-(α-hydroxy acids) such as poly-glycolic acid (PGA) (Toosi et al., 2016), poly-lactic acid (PLA) (Gregor et al., 2017), and poly-(ϵ-caprolactone) (Dong et al., 2017) and poly-hydroxyethyl methacrylate (poly-HEMA) (Çetin et al., 2011). Nanoscale ceramics, including nano-HA (a native component of bone), find a wide application in bone regeneration in combination with various polymers and carbon materials as well (Sachlos and Czernuszka, 2003; Xie et al., 2015). Nanoceramics, including alumina and zinc oxide, are also suitable to promote bone growth. In fact, nano-sized zinc oxide (23 nm) ceramics enhanced osteoblast adhesion by 146% and 200% (Colon et al., 2006). The excellent mechanical properties of metals also make them ideal candidates for orthopedic tissue engineering applications. Metals are morphologically and mechanically similar to trabecular bone scaffolding. Titanium and its alloys have high biocompatibility, strength, lightness, and high resistance to corrosion, and as such are the metal materials more commonly used for biomedical applications (Disegi, 2000; Niinomi, 2008). Using nano-dimensional titania has enhanced oestrogenic effects (Bose et al., 2018). Nano bio-glasses are promising new materials for bone tissue engineering, which they are fabricated by using an SiO_2-CaO-P_2O_5-Na_2O system, and have been found to have the suitable mechanical strength needed for bone development. They provide a surface which is suitable for osteocyte adhesion and proliferation, and for the attachment of hydroxycarbonate apatite or HA, providing a suitable interface for bonding of soft tissues with strong bone. As all the constituents of bioglass are biodegradable, after healing the glass will be replaced by bone and the glass will dissolve (Bakhtiyari et al., 2016).

Graphene and CNTs have attracted a lot of attention in recent years because of their superior properties, like high tensile strength, electric conductivity, and cytocompatibility. These properties make them suitable scaffold material for bone tissue and nerve tissue engineering applications. CNTs of a size range around 60 nm in diameter produce excellent osteoblast adhesion (Price et al., 2003). Ultra-short, single-walled (SW)CNT polymer nanocomposites have been found to enhance osteogenesis by 2–3 times in comparison with control polymers (Sitharaman et al., 2008). Thus, carbon nanomaterials are also promising materials for bone tissue regeneration (Silva et al., 2017),

8.8.2 Application of Nanotechnology in Nerve Tissue Regeneration

Peripheral nerve (PN) injury after traumas like accidents and burns are common; 300,000 cases are reported annually in Europe alone (Ciardelli and Chiono, 2006). Thus, effective regeneration capabilities can do a great service. The most commonly employed technique for nerve tissue regeneration is autografting, which involves transplanting the patient's own nerve graft from one donor site to the site of damage for regeneration (Hood et al., 2009). Allografting is a similar technique, which uses nerve conduit of non-self origin, either from human or primate source. It provides a temporary template for the development of new nerves in the host (Moore et al., 2011). Allografting provides a suitable substrate for the regeneration of damaged nerve tissue. There are a plethora of biomaterials which can be used for allograft nerve conduits, but these grafts carry the risk of transmitting animal diseases to humans (Yang et al., 2012). Nerve tissue engineering can provide a solution to many of the problems associated with the treatment of nerve damage. A successful design for suitable nerve conduit must incorporate features similar to the natural ECM. In the first place, it should be biocompatible, non-immunogenic, non-cytotoxic, and biodegradable. Additionally, it should mimic the functional environment of natural ECM, providing suitable topographical, electrical, and chemical cues for adhesion and proliferation of neural cells and enhanced neurite outgrowth. A scaffold should be flexible and should not increase tension at the lesion site for efficient axon regeneration. It is also necessary that scar formation should be inhibited after injury due to phagocytosis and to ensure adult neuronal viability for initiating axonal extension (Ellis-Behnke et al., 2006; Subramanian et al., 2009). Nanomaterials, due to their superior properties, are being explored for developing the ideal scaffold. Nanostructured prosthesis, like nanofibers and self-assembly peptide gels, have various advantages with their increased surface area and optimized scaffold geometry. The advantages include better cell seeding of different types of cells, like Schwann cells, and better delivery of growth factors, providing efficient nerve regeneration. Nanomaterials inhibit astrocyte activity and support axonal growth, restoring synaptic connections (Fraczek-Szczypta, 2014; Peran et al., 2013). Self-assembly peptides functionalized with neuro-selective epitomes, like isoleucine-lysine-valine-alanine-valine (IKVAV), have been found to promote the regeneration of nanofibers for both descending and ascending neurons within 11 weeks of treatment when injected in mouse spinal cord injury (Tysseling-Mattiace et al., 2008). CNTs have emerged as an efficient material for tissue regeneration. CNTs obtained by functionalizing with groups like NH_2, -SH, and -COOH have a positive effect on nerve growth, including enhanced neurite length and branching (Fraczek-Szczypta, 2014). Carbon-based nanomaterials such as graphene and CNTs have excellent electrical conductivity and enhance neural growth (Fraczek-Szczypta, 2014; Qian et al., 2018). It is also found that positively charged CNTs promote better tissue regeneration than negatively charged ones.

Growth factors like nerve growth factor (NGF) are special factors which enhance neural regeneration (Sofroniew et al., 2001), however achieving tissue targeting and the desired cellular concentration presents a significant challenge. Recently, Sun et al. (2009a) achieved the delivery of collagen-binding-domain nerve growth factor β (CBD-NGF β) to nerve ECM collagen to restore the peripheral nerve function in rat sciatic nerves. The integration of more self-sustained biological components, like stem cells, or biomolecules, such as arginine–glycine–aspartic acid (RGD)-peptides (de Mel et al., 2008; de Mel et al., 2009), can be examples of such improvements in which they affect cellular attachment and neurite outgrowth. Neural tissue engineering will provide efficient and effective recovery of nerve damage and will help to treat many complicated neural injuries which currently need complicated surgery, like neural grafting. However, challenges still remain to be tackled, including the availability of suitable cell sources, due to ethical concerns, and the availability of cells. In this regard, induced pluripotent stem cells (iPSC) (cells obtained from patients own blood) hold great promises.

8.8.3 Application of Nano Tissue Engineering in Wound Healing and Skin Regeneration

Human skin consists of a weaved mesh network of several cell types entrenched in an ECM composed of different types of proteins, like collagen, fibronectin, and keratin, showing anisotropic, mechanical properties due to the alignment of ECM fibers. Skin tissue plays an important role in protecting internal organs from damage from environmental factors. It is regularly exposed to harsh conditions and injuries, and skin damage is common. In some situations such as burns and diabetic disease, damage to the skin is extensive and this may lead to a greater risk of mortality or higher health costs (Shevchenko et al., 2010). Tissue-engineered replacements play an extremely important role in the treatment of chronic skin wounds. Tissue substitutes for skin tissue replacements should be adherent, providing good physical strength and elasticity, and should be non-antigenic (Liu and Cao, 2007). Any viable tissue-engineered material for skin regeneration should have low immunogenicity with good safety for the patient through high clinical efficacy and convenience of use. Tissue grafts developed from materials of animal origin (such as bovine collagen and murine feeder cells) and used for skin replacement carries a high risk of transmitting zoonotic viral or bacterial infections to humans.

Currently, there are several marketed products for the regenerative therapy of wounds. These are suitable for hastening the healing of different types of wounds, including burns. They include Epicel, Cell Spray, MySkin, Recell, AlloDerm®, and Dermagraft®, to name a few. The nanostructure of the scaffold plays an important role in skin grafting. A nano-porous structure in the scaffolding improves the adhesion of engineered tissues onto the matrix of underlying tissue, and allows efficient transport of low molecular weight solutes and nutrients into the cells (Peran et al., 2013). Hydrogel-based, *in situ*, cross-linkable polymers show good potential for use as tissue scaffolds. Collagen- or gelatin-based polymers are commonly used for developing hydrogel with tunable properties for tissue engineering (Zhao et al., 2016a). 3D bioactive hydrogels have been successfully prepared by two-photon laser scanning photolithography for guided 3D cell proliferation (Lee et al., 2008). Incorporating nanoparticles of drugs and proteins in hydrogels enhances skin healing by seeding cells to help form neo-skin tissue. Nano-porous microspheres have also been a very attractive approach for skin regeneration for larger wounds, as they can be used for easy cultivation and transport of autologous keratinocytes (Kim et al., 2005).

Fibroblasts play an important role in skin and are suitable for wound healing tissue engineering applications. Poly(L-lactic acid)-copoly ε-caprolactone and gelatin-based scaffolds have been found to promote fibroblast proliferation (Peran et al., 2013). In addition to the epithelial components (cultured epithelial autografts) of certain constructs, the dermal component improves critical communication pathways (i.e., between the epidermis and the dermis). Adult stem cells and MSCs obtained from bone marrow, umbilical cord, or adult mammalian cells exhibit high proliferation and differentiation and have low immunoreactivity. Their use in wound healing promotes quicker healing. Jin et al. cultivated an epidermal cell lineage on electrospun callogen/poly(L-lactic acid)-co-poly(3-caprolactone) scaffolds which showed great potential for wound healing. Fetal stem cells have the capability to produce scar-less wound healing, as fetal cells follow regenerative mechanisms in spite of repair mechanisms. Ramelet et al. (2009) hypothesized that the use of fetal stem cells, if engineered effectively, holds the greatest potential for treating severe wounds like burns without producing scars. The use of fetal biological bandages gave encouraging results, showing effective healing without secondary effects (Ramelet et al., 2009). Acellular constructs are also a very attractive avenue. Acellular dermal substitutes with permanent skin substitutes containing autologous cells have been shown to provide definitive wound closure in burns involving greater than 90% of the total body surface area (Boyce and Lalley, 2018). Skin tissue engineering holds great potential for better therapy of serious wounds like burns. Future research is focused on developing dermal and epidermal substitutes which are similar to natural skin and have low immunogenic potential. Advancements in nanotechnology and material sciences will help in this quest.

8.8.4 Application of Nanotechnology in Cardiovascular Regeneration

Cardiovascular diseases are the leading cause of death worldwide; in fact, death due to cardiac disease (26.6%) outnumbers the total deaths caused by cancer (22.8%) annually. Cardiovascular diseases are responsible for almost one-third of deaths worldwide (Kung et al., 2008). Congenital heart defects affect nearly 1% of live births. Although some of these defects can be mediated by medications only, others may require complex surgeries involving valve replacements and grafts for patient survival. Regenerative scaffolds can be an effective substitute to these surgeries. Currently, peripheral and coronary bypass surgery are the most common ways of treating serious cardiovascular problems, but none of the current therapies can fully restore the myocardium functionality. Functionalized, tissue-engineered cardiovascular scaffolds are an attractive approach, providing solutions to these problems. Stem cell therapy utilizing iPSCs (Nelson et al., 2009b), ESCs (Kraehenbuehl et al., 2011), and MSCs (Amado et al., 2005) have shown to improve the function of damaged cardiomyocytes. The current methods of injecting stem cells have certain drawbacks, like insufficient cell retention, survival, and engraftment, and differentiation rates. Providing a suitable carrier for tissue regeneration can improve the efficiency of the current methods. A cardiac tissue construct should mimic myocardial cells and possess similar anisotropic mechanical properties, like flexibility and stiffness, integrate with host tissues without causing any immunogenic reaction, and generate electrical and mechanical signals in synchrony with the myocardium (Xu and Guan, 2016). As many of these features are functions of topographical features of the cells, the nanotopography and organization of the scaffold should be highly ordered and similar to the nanoscale structure of collagen fibers. Nanofibers are one of the most suitable tissue substitutes with the desired nanotopography for cardiac regeneration. Aydogdu et al. (2018) successfully developed bio-mimetic, small-diameter blood vessels using polycaprolactone (PCL), ethyl cellulose (EC), and collagen type-1. Such vascular scaffolding can provide a therapeutic solution for cardiovascular diseases. Nanohydrogels also provide a suitable nanoformulation for cardiac tissue regeneration. Different natural and synthetic 3D hydrogel scaffolds have been developed which produce effective proliferation and differentiation of stem cells to cardiac cells and also support native cardiomyocytes (Tijore et al., 2018;

Zimmermann et al., 2002). Many re-absorbable biomaterials have been proposed for use in cardiac tissue matrix development, including polysaccharides (Silva et al., 2015), PLGA (Chi et al., 2013), and collagen (Serpooshan et al., 2013), and others among all these injectable bioabsorbable scaffolds, including PCL and PGS, have achieved clinical outcomes (Lee et al., 2013). IK-5001 is an alginate-based formulation marketed by Bellerophon BCM, LLC. It consists of an injectable hydrogel recommended for myocardial infarction of tissue. The hydrogel produces a temporary, bioabsorbable cardiac scaffold. IK-5001 treatment prevents ventricular remodeling in patients with myocardial infarction (Frey et al., 2014). CNTs, owing to their superior physical properties and electrical conduction properties, are also suitable for tissue regeneration. Their higher electrical conductivity appreciably enhances the propagation of cardiac electrical impulses. Pok et al. (2014) developed chitosan-gelatin hydrogel-based scaffolds incorporating SWCNTs. They found CNTs improved the conduction velocity of the action potential in primary rat cardiomyocytes (Pok et al., 2014). CNTs have been successfully utilized alone or in combination with other polymers for cardiac tissue regeneration (Dozois et al., 2017). In future cardiac tissue engineering, using programmable nanoscaffolds for remodeling and regeneration of cardiac tissue will strive for these smart, inductive scaffolds that will closely mimic cardiac ECM and be able to attain self-guided cell proliferation and differentiation, providing tissue regeneration in a selective and controlled fashion.

8.8.5 Application of Nanotechnology in Muscle Tissue Regeneration

Muscle tissue has higher regenerative potential than most tissues. Damage due to injury often recovers by the regenerative capacity of the body, but the loss of more than 20% of mass in a single muscle can cause serious damage to the functionality of the muscle (Valentin et al., 2010). Tissue engineering of skeletal muscle is still in its preliminary stages of development and it will be some time before its clinical applications will see the light of day. The first breakthrough in muscle tissue cultivation came when Vandenburgh et al. (1988) successfully cultivated myoblasts in collagen gels in 1988, which were able to achieve the contractility of muscles. Over last two decades, there has been an increase in understanding of the functioning of muscle tissue and the ways of designing suitable engineered tissue substrates for use in muscle regeneration, which can mimic muscle tissue in physiology as well as functionality. Skeletal muscle tissue is made up of a highly-organized, 3D matrix with a parallel orientation of muscle fibers that can generate longitudinal force on contraction of neural stimulation of the muscle (Hinds et al., 2011). The ECM provides a suitable environment for the differentiation and alignment of myoblasts (Cassell et al., 2001). Ideally, the ECM provides the framework for cell adhesion and tissue growth, which supports cell differentiation and proliferation. The parallel alignment of the natural ECM in skeletal muscle tissue is an important characteristic that needs to be incorporated into an artificial matrix. Recent developments in the field of nanotechnology have brought up new possibilities. There are different types of cells used for seeding muscle tissues, including satellite cells and iPSCs produced from adult fibroblasts. Nanowhiskers using bacterial cellulose have been found to have nanotopography which helps in the development of highly-oriented myoblasts and promotes myogenesis. This holds potential for the treatment of severe skeletal muscle diseases like Duchenne's muscular dystrophy (Dugan et al., 2012). Ideally, nanofiber matrices are a suitable substrate that can be used for cultivating myoblasts. They can produce parallel alignment of cells suitable for muscle tissue engineering (Huang et al., 2006). Furthermore, nanotopography plays important role in the self-assembly of ECM which mediates cell attachment and orientation and modulates physiological characteristics like protein absorption and integrin binding. Nano-grooving also regulates focal adhesion and fibrinogenesis (Kim et al., 2013). There are also some techniques which are used for generating suitable 3D matrices for muscle tissue engineering; freeze-drying of hydrogels is one such, which produces highly porous scaffolds produced by the sublimation of ice crystals (Madaghiele et al., 2008). This method is suitable for achieving high porosity. Such freeze-dried hydrogels are produced by biological materials like collagen (Mandal and Kundu, 2009) and fibroin. Other methods suitable for producing smart scaffolds with controllable drug release include nanoparticles and nanofibers. Drug delivering scaffolds-using nanopolymers are under phase I clinical trials (Nelson et al., 2009a). Nanofibers can successfully deliver important cellular growth factors which help in tissue regeneration (Sahoo et al., 2010). 3D constructs made of nanofibers and their composites can further lead to enhanced growth, increased contraction forces, and proper alignment of the muscle fibers (Cai et al., 2017).

8.8.6 Applications of Nanotechnology in Other Tissue Regeneration

Apart from the above stated major organ systems, tissue engineering nanotechnology also finds myriad applications in the regeneration of other tissues, including the liver, pancreas, kidney, urinary tract, etc. Lee et al. (2009) designed liver tissue spheroids with inverted colloidal crystal scaffolds. Dvir-Ginzberg et al. (2003) successfully engineered and studied alginate-based liver tissue scaffold.

Kidney failures and renal disease are the commonest serious health problems associated with advanced age. Tissue engineering research for kidney transplantation and restorative purposes is well underway and scaffolds have been successfully developed and tested on animal models. Different groups have been involved in kidney and lower tissue scaffold engineering. Ross et al. (2009) demonstrated the capability of embryonic stem cells to differentiate into kidney cells in suitable ECM. Song et al. (2013) successfully engineered ectopic kidney tissue using a decellularized matrix; these grafts produced rudimentary urine *in vitro* when perfused through their intrinsic vascular bed. When transplanted in an orthotropic position in the rat, the grafts were perfused by the recipient's circulation and produced urine through the urethral conduit *in vivo*, successfully demonstrating the potential of the development of a functional kidney. PCL-based, electrospun scaffold for the proliferation of human kidney epithelial cells has been investigated and the diameter of the nanofibers is known to play a crucial role in the differentiation of the cells (Burton et al., 2018). Similarly, advancements in the field of bladder tissue grafts are also underway and 3D nanografts utilizing urothelial cells, smooth muscle cells, and vascular growth factors have shown promising results in repairing bladder damage (Ling et al., 2017).

These efforts currently underway will fructify in future and promise to change the current scenario of regenerative medicine for the better. Challenges do exist in the path of clinical application of tissue scaffolds, but with the increasing understanding of biology coupled with advances in technology these hurdles will surely be crossed in the times to come.

8.9 Challenges and Future Prospects of Use of Nanotechnology in Tissue Regeneration

With advances in nanotechnology, there are simultaneous advances in biomaterials which are establishing new paradigms for tissue regeneration. However, there are significant challenges in the development of the ideal tissue scaffold. Nanomaterials, due to their high surface area and enhanced porosity, have a distinctive advantage over the micro or larger-sized materials, in that they provide better cell adhesion, proliferation, and differentiation, and provide the desired biochemical or mechanical cues for achieving the desired cell behavior and tissue development.

Nanomaterials have their own drawbacks which need to be tackled, like issues related to the toxicity and biocompatibility of nanostructures which must be addressed in designing the scaffold. Achieving the ideal nano-patterning and topography is also crucial. So far, 2D, functional cellular monolayers have been developed successfully, but achieving 3D, cellular structures is yet to be perfected. Thus, it can be concluded that if issues like toxicity and pharmacokinetic distribution of nanomaterials can be overcome, nanostructured scaffolds can achieve efficient and effective tissue regeneration. On the technological front, preparing suitable ECM synthetically is a big challenge. 3D, bio-mimetic scaffolds have been designed using different methods, including electrospinning, phase-separation, freeze-drying, self-assembly, and 3D bioprinting. 3D printing has opened vast avenues for designing scaffolds down to the cellular level. Electrospinning is one of most extensively used techniques to produce well-aligned, nano-dimensional fibers suitable for various tissue engineering applications like muscle, heart, and neural regeneration. Similarly, self-assembly techniques are also providing us with newer ways of developing injectable *in situ* scaffolds for tissue regeneration. In spite of many great achievements in the field of scaffold design, achieving functionalized, 3D, complex, multicellular tissue architecture for organs like the liver and the kidney is still a challenge. Future research will focus on designing smart, bioactive scaffolds which are not just carriers to cells, but that also provide different bioactive cues for integration of the scaffold with the native ECM and to guide cellular adhesion, differentiation, and proliferation in a controlled and selective manner for achieving specific tissue regeneration.

REFERENCES

Ahn H, Ju YM, Takahashi H, Williams DF, Yoo JJ, Lee SJ, Okano T, Atala A. Engineered small diameter vascular grafts by combining cell sheet engineering and electrospinning technology. Acta Biomater. 2015;16(1):14–22.

Allo BA, Costa DO, Dixon SJ, Mequanint K, Rizkalla AS. Bioactive and biodegradable nanocomposites and hybrid biomaterials for bone regeneration. J Funct Biomater. 2012;3(4):432–63.

Amado LC, Saliaris AP, Schuleri KH, St. John M, Xie JS, Cattaneo S, Durand DJ, et al. Cardiac repair with intramyocardial injection of allogeneic mesenchymal stem cells after myocardial infarction. Proc Natl Acad Sci. 2005;102(32):11474–9.

Annabi N, Nichol JW, Zhong X, Ji C, Koshy S, Khademhosseini A, Dehghani F. Controlling the porosity and microarchitecture of hydrogels for tissue engineering. Tissue Eng Part B Rev. 2010;16(4):371–83.

Asghari F, Samiei M, Adibkia K, Akbarzadeh A, Davaran S. Biodegradable and biocompatible polymers for tissue engineering application: A review. Artif Cells Nanomed Biotechnol. 2017;45(2):185–92.

Aubin H, Nichol JW, Hutson CB, Bae H, Sieminski AL, Cropek DM, Akhyari P, Khademhosseini A. Directed 3D cell alignment and elongation in microengineered hydrogels. Biomaterials. 2010;31(27):6941–51.

Aydogdu MO, Chou J, Altun E, Ekren N, Cakmak S, Eroglu M, Osman AA, et al. Production of the biomimetic small diameter blood vessels for cardiovascular tissue engineering. Int J Polym Mater Polym Biomater. 2018:1–13.

Bajaj P, Marchwiany D, Duarte C, Bashir R. Patterned three-dimensional encapsulation of embryonic stem cells using dielectrophoresis and stereolithography. Adv Healthcare Mater. 2013;2(3):450–8.

Bakhtiyari SSE, Karbasi S, Monshi A, Montazeri M. Evaluation of the effects of nano-TiO$_2$ on bioactivity and mechanical properties of nano bioglass-P3HB composite scaffold for bone tissue engineering. J Mater Sci Mater Med. 2016;27(1):2.

Balagangadharan K, Dhivya S, Selvamurugan N. Chitosan based nanofibers in bone tissue engineering. Int J Biol Macromol. 2016;104:1372–82.

Billiet T, Vandenhaute M, Schelfhout J, Van Vlierberghe S, Dubruel P. A review of trends and limitations in hydrogel-rapid prototyping for tissue engineering. Biomaterials. 2012;33(26):6020–41.

Bose S, Banerjee D, Shivaram A, Tarafder. Calcium phosphate coated 3D printed porous titanium with nanoscale surface modification for orthopedic and dental applications. Mater Des. 2018;151:102–12.

Bosworth LA, Downes S. Electrospinning for tissue regeneration. Elsevier: Swatson, UK; 2011.

Boyce ST, Lalley AL. Tissue engineering of skin and regenerative medicine for wound care. Burns Trauma. 2018;6(1):4.

Brockes JP, Gates PB. Mechanisms underlying vertebrate limb regeneration: Lessons from the salamander. Biochem Soc Trans. 2014;42:625–30.

Burton TP, Corcoran A, Callanan A. The effect of electrospun polycaprolactone scaffold morphology on human kidney epithelial cells. Biomed Mater (Bristol). 2018;13(1):015006.

Cai A, Horch RE, Beier JP. Nanofiber composites in skeletal muscle tissue engineering. Nanofiber Compos Biomed Appl. 2017:369–94.

Cassell OC, Morrison WA, Messina A, Penington AJ, Thompson EW, Stevens GW, Perera JM, et al. The influence of extracellular matrix on the generation of vascularized, engineered, transplantable tissue. Ann NY Acad Sci. 2001;944:429–42.

Cerqueira MT, Pirraco RP, Martins AR, Santos TC, Reis RL, Marques AP. Cell sheet technology-driven re-epithelialization and neovascularization of skin wounds. Acta Biomater. 2014;10(7):3145–55.

Çetin D, Kahraman AS, Gümüsderelioglu M. Novel scaffolds based on poly(2-hydroxyethyl methacrylate) superporous hydrogels for bone tissue engineering. J Biomater Sci Polym Ed. 2011;22(9):1157–78.

Chi NH, Yang MC, Chung TW, Chou NK, Wang SS. Cardiac repair using chitosan-hyaluronan/silk fibroin patches in a rat heart model with myocardial infarction. Carbohydr Polym. 2013;92(1):591–7.

Christ GJ, Saul JM, Furth ME, Andersson KE. The pharmacology of regenerative medicine. Pharmacol Rev. 2013;65(3):1091–133.

Ciardelli G, Chiono V. Materials for peripheral nerve regeneration. Macromol Biosci. 2006;6(1):13–26.

Colon G, Ward BC, Webster TJ. Increased osteoblast and decreased staphylococcus epidermidis functions on nanophase ZnO and TiO2. J Biomed Mater Res Part A. 2006;78(3):595–604. DOI:10.1002/jbm.a.30789.

Cunha C, Panseri S, Antonini S. Emerging nanotechnology approaches in tissue engineering for peripheral nerve regeneration. Nanomed Nanotechnol Biol Med. 2011;7(1):50–9.

Dahlin RL, Kasper FK, Mikos AG. Polymeric nanofibers in tissue engineering. Tissue Eng Part B Rev. 2011;17(5):349–64.

Dalby MJ, Gadegaard N, Curtis G, Adam S, Oreffo C, Richard O. Nanotopographical control of human osteo-progenitor differentiation. Curr Stem Cell Res Ther. 2007;2(2):129–38.

de Mel A, Jell G, Stevens MM, Seifalian AM. Biofunctionalization of biomaterials for accelerated in situ endothelialization: A review. Biomacromolecules. 2008;9(11):2969–79.

de Mel A, Punshon G, Ramesh B, Sarkar S, Darbyshire A, Hamilton G, Seifalian AM. In situ endothelialisation potential of a biofunctionalised nanocomposite biomaterial-based small diameter bypass graft. Bio-Med Mater Eng. 2009;19(4–5):317–31.

Degasne I, Baslé MF, Demais V, Huré G, Lesourd M, Grolleau B, Mercier L, Chappard D. Effects of roughness, fibronectin and vitronectin on attachment, spreading, and proliferation of human osteoblast-like cells (Saos-2) on titanium surfaces. Calcif Tissue Int. 1999;64(6):499–507.

Disegi JA. Titanium alloys for fracture fixation implants. Injury. 2000;31:D14–7.

Dong L, Wang SJ, Zhao XR, Zhu YF, Yu JK. 3D-printed poly(ε-caprolactone) scaffold integrated with cell-laden chitosan hydrogels for bone tissue engineering. Sci Rep. 2017;7(1):13412.

Dozois MD, Bahlmann LC, Zilberman Y, Tang XS. Carbon nanomaterial-enhanced scaffolds for the creation of cardiac tissue constructs: A new frontier in cardiac tissue engineering. Carbon. 2017;120: 338–49.

Dugan JM, Cartmell SH, Eichhorn SJ, Gough JE. Cellulose nanowhiskers and nechanotransduction as strategies for tissue engineering skeletal muscle from human mesenchymal stem cells. J Tissue Eng Regen Med. 2012;6:118.

Dvir T, Timko BP, Brigham MD, Naik SR, Karajanagi SS, Levy O, Jin H, Parker KK, Langer R, Kohane DS. Nanowired three-dimensional cardiac patches. Nat Nanotechnol. 2011;6(11):720–5.

Dvir-Ginzberg M, Gamlieli-Bonshtein I, Agbaria R, Cohen S. Liver tissue engineering within alginate scaffolds: Effects of cell-seeding density on hepatocyte viability, morphology, and function. Tissue Eng. 2003;9(4):757–66.

Ellis-Behnke RG, Liang YX, You SW, Tay DKC, Zhang S, So KF, Schneider GE. Nano neuro knitting: Peptide nanofiber scaffold for brain repair and axon regeneration with functional return of vision. Proc Natl Acad Sci USA. 2006;103(13):5054–9.

Fraczek-Szczypta A. Carbon nanomaterials for nerve tissue stimulation and regeneration. Mater Sci Eng C. 2014;34:35–49.

Freed LE, Guilak F, Guo XE, Gray ML, Tranquillo R, Holmes JW, Radisic M, Sefton MV, Kaplan D, Vunjak-Novakovic G. Advanced tools for tissue engineering: Scaffolds, bioreactors, and signaling. Tissue Eng. 2006;12(12):3285–305.

Frey N, Linke A, Süselbeck T, Müller-Ehmsen J, Vermeersch P, Schoors D, Rosenberg M, Bea F, Tuvia S, Leor J. Intracoronary delivery of injectable bioabsorbable scaffold (IK-5001) to treat left ventricular remodeling after ST-elevation myocardial infarction: A first-in-man study. Circ Cardiovasc Interv. 2014;7(6):806–12.

Geiger M, Li RH, Friess W. Collagen sponges for bone regeneration with rhBMP-2. Adv Drug Deliv Rev. 2003;55(12):1613–29.

Ghassemi T, Saghatolslami N, Matin MM, Gheshlaghi R, Moradi A. CNT-decellularized cartilage hybrids for tissue engineering applications. Biomed Mater (Bristol). 2017;12(6):065008.

Goldstein AS, Christ G. Functional tissue engineering requires bioreactor strategies. Tissue Eng Part A. 2009;15(4):739–40.

Gomes ME, Godinho JS, Tchalamov D, Cunha AM, Reis RL. Alternative tissue engineering scaffolds based on starch: Processing methodologies, morphology, degradation and mechanical properties. Mater Sci Eng C. 2002;20(1–2):19–26.

Gregor A, Filová E, Novák M, Kronek J, Chlup H, Buzgo M, Blahnová V, et al. Designing of PLA scaffolds for bone tissue replacement fabricated by ordinary commercial 3D printer. J Biol Eng. 2017;11(1), 31.

Guo B, Ma PX. Synthetic biodegradable functional polymers for tissue engineering: A brief review. Sci China Chem. 2014;57(4):490–500.

Gupta KC, Haider A, Choi YR, Kang IK. Nanofibrous scaffolds in biomedical applications. Biomater Res. 2014;18(1):5.

Guven S, Chen P, Inci F, Tasoglu S, Erkmen B, Demirci U. Multiscale assembly for tissue engineering and regenerative medicine. Trends Biotechnol. 2015;33(5):269–79.

Hahn MS, Miller JS, West JL. Three-dimensional biochemical and biomechanical patterning of hydrogels for guiding cell behavior. Adv Mater. 2006;18(20):2679–84.

Hannachi IE, Yamato M, Okano T. Cell sheet technology and cell patterning for biofabrication. Biofabrication. 2009;1(2):022002.

Haraguchi Y, Shimizu T, Yamato M, Okano T. Scaffold-free tissue engineering using cell sheet technology. RSC Adv. 2012;2(6):2184–2190.

Harrington DA, Cheng EY, Guler MO, Lee LK, Donovan JL, Claussen RC, Stupp SI. Branched peptide-amphiphiles as self-assembling coatings for tissue engineering scaffolds. J Biomed Mater Res Part A. 2006;78(1):157–67.

Hartgerink JD, Beniash E, Stupp SI. Self-assembly and mineralization of peptide-amphiphile nanofibers. Science. 2001;294(5547):1684–8.

Hellman KB. Tissue engineering: Translating science to product. Top Tissue Eng. 2008;4:1–28.

Hinds S, Bian W, Dennis RG, Bursac N. The role of extracellular matrix composition in structure and function of bioengineered skeletal muscle. Biomaterials. 2011;32(14):3575–83.

Hood B, Levene HB, Levi AD. Transplantation of autologous schwann cells for the repair of segmental peripheral nerve defects. Neurosurg Focus. 2009;26(2):E4.

Hosseinkhani H, Hosseinkhani M, Tian F, Kobayashi H, Tabata Y. Osteogenic differentiation of mesenchymal stem cells in self-assembled peptide-amphiphile nanofibers. Biomaterials. 2006;27(22):4079–86.

Hu S, Ogle BM, Cheng K. Body builder: From synthetic cells to engineered tissues. Curr Opin Cell Biol. 2018;54:37–42.

Huang NF, Patel S, Thakar RG, Wu J, Hsiao BS, Chu B, Lee RJ, Li S. Myotube assembly on nanofibrous and micropatterned polymers. Nano Lett. 2006;6(3):537–42.

Hunziker R. Regenerative medicine – Fact sheet. National Institutes of Health: Bethesda, Maryland; October 2010. p. 1–2.

Jeong SI, Krebs MD, Bonino CA, Khan SA, Alsberg E. Electrospun alginate nanofibers with controlled cell adhesion for tissue engineering. Macromol Biosci. 2010;10(8):934–43.

Jin G, Prabhakaran MP, Ramakrishna S. Stem cell differentiation to epidermal lineages on electrospun nanofibrous substrates for skin tissue engineering. Acta biomaterialia. 2011;7(8):3113–22.

Kang HW, Tabata Y, Ikada Y. Fabrication of porous gelatin scaffolds for tissue engineering. Biomaterials. 1999a;20(14):1339–44.

Kang HW, Tabata Y, Ikada Y. Effect of porous structure on the degradation of freeze-dried gelatin hydrogels. J Bioact Compat Polym. 1999b;14(4):331–43.

Kang Y, Jabbari E, Yang Y. Integrating top-down and bottom-up scaffolding tissue engineering approach for bone regeneration. In: Micro and nanotechnologies in engineering stem cells and tissues. IEEE Press: Piscataway, New Jersey. 2013. p. 142–58.

Kasper C, Van Griensven M, Pörtner R. Bioreactor systems for tissue engineering. Adv Biochem Eng Biotechnol. 2009:112.

Khademhosseini A, Langer R. A decade of progress in tissue engineering. Nat Protoc. 2016;11(10):1775–81.

Kim HN, Jiao A, Hwang NS, Kim MS, Kang DH, Kim DH, Suh KY. Nanotopography-guided tissue engineering and regenerative medicine. Adv Drug Deliv Rev. 2013;65(4):536–58.

Kim MH, Kim BS, Park H, Lee J, Park WH. Injectable methylcellulose hydrogel containing calcium phosphate nanoparticles for bone regeneration. Int J Biol Macromol. 2018;109:57–64.

Kim SS, Gwak SJ, Choi CY, Kim BS. Skin regeneration using keratinocytes and dermal fibroblasts cultured on biodegradable microspherical polymer scaffolds. J Biomed Mater Res Part B Appl Biomater. 2005;75(2):369–77.

Koutsopoulos S. Self-assembling peptide nanofiber hydrogels in tissue engineering and regenerative medicine: Progress, design guidelines, and applications. J Biomed Mater Res Part A. 2016;104(4):1002–16.

Koutsopoulos S. Self-assembling peptides in biomedicine and bioengineering: Tissue engineering, regenerative medicine, drug delivery, and biotechnology. In: Peptide applications in biomedicine, biotechnology and bioengineering. Woodhead Publishing: Duxford, UK; 2017. p. 387–408.

Kraehenbuehl TP, Ferreira LS, Hayward AM, Nahrendorf M, van der Vlies AJ, Vasile E, Weissleder R, Langer R, Hubbell JA. Human embryonic stem cell-derived microvascular grafts for cardiac tissue preservation after myocardial infarction. Biomaterials. 2011;32(4):1102–9.

Kumar S, Raj S, Sarkar K, Chatterjee K. Engineering a multi-biofunctional composite using poly(ethylenimine) decorated graphene oxide for bone tissue regeneration. Nanoscale. 2016;8(12):6820–36.

Kung HC, Hoyert DL, Xu J, Murphy SL. Deaths: Final Data for 2005. Natl Vital Stat Rep. 2008;56(10):1–120.

Lee JW, Kim JY, Cho DW. Solid free-form fabrication technology and its application to bone tissue engineering. Int J Stem Cells. 2010;3(2):85–95.

Lee J, Cuddihy MJ, Cater GM, Kotov NA. Engineering liver tissue spheroids with inverted colloidal crystal scaffolds. Biomaterials. 2009;30(27):4687–94.

Lee LC, Wall ST, Klepach D, Ge L, Zhang Z, Lee RJ, Hinson A, Gorman JH, Gorman RC, Guccione JM. Algisyl-LVR with coronary artery bypass grafting reduces left ventricular wall stress and improves function in the failing human heart. Int J Cardiol. 2013;168(3):2022–8.

Lee SH, Moon JJ, West JL. Three-dimensional micropatterning of bioactive hydrogels via two-photon laser scanning photolithography for guided 3D cell migration. Biomaterials. 2008;29(20):2962–8.

Li C, Vepari C, Jin HJ, Kim HJ, Kaplan DL. Electrospun silk-BMP-2 scaffolds for bone tissue engineering. Biomaterials. 2006;27(16):3115–24.

Li D, Xia Y. Electrospinning of nanofibers: Reinventing the wheel? Adv Mater. 2004;16(14):1151–70.

Li Q, Ma L, Gao C. Biomaterials for in situ tissue regeneration: Development and perspectives. J Mater Chem B. 2015;3(46):8921–38.

Li Y, Xiao Y, Liu C. The horizon of materiobiology: A perspective on material-guided cell behaviors and tissue engineering. Chem Rev. 2017;117(5):4376–421.

Lin T, Liu S, Chen S, Qiu S, Rao Z, Liu J, Zhu S, et al. Hydrogel derived from porcine decellularized nerve tissue as a promising biomaterial for repairing peripheral nerve defects. Acta Biomater. 2018a:1–13.

Lin SJ, Xue YP, Chang G, Han QL, Chen LF, Jia YB, Zheng YG. Layer-by-layer 3-dimensional nanofiber tissue scaffold with controlled gap by electrospinning. Mater Res Express. 2018b;5(2) 025401.

Lin YC, Grahovac T, Oh SJ, Ieraci M, Rubin JP, Marra KG. Evaluation of a multi-layer adipose-derived stem cell sheet in a full-thickness wound healing model. Acta Biomater. 2013;9(2):5243–50.

Ling Q, Wang T, Yu X, Wang SG, Ye ZQ, Liu JH, Yang SW, Zhu XB, Yu J. UC-VEGF-SMC three dimensional (3D) nano scaffolds exhibits good repair function in bladder damage. J Biomed Nanotechnol. 2017;13(3):313–23.

Liu Tsang V, Chen AA, Cho LM, Jadin KD, Sah RL, DeLong S, West JL, Bhatia SN. Fabrication of 3D hepatic tissues by additive photopatterning of cellular hydrogels. FASEB J. 2007;21(3):790–801.

Liu W, Cao Y. Application of scaffold materials in tissue reconstruction in immunocompetent mammals: Our experience and future requirements. Biomaterials. 2007;28(34):5078–86.

Liu X, Rahaman MN, Fu Q. Bone regeneration in strong porous bioactive glass (13-93) scaffolds with an oriented microstructure implanted in rat calvarial defects. Acta Biomater. 2013;9(1):4889–98.

Loo Y, Goktas M, Tekinay AB, Guler MO, Hauser CAE, Mitraki A. Self-assembled proteins and peptides as scaffolds for tissue regeneration. Adv Healthcare Mater. 2015;4(16):2557–86.

Lu T, Li Y, Chen T. Techniques for fabrication and construction of three-dimensional scaffolds for tissue engineering. Int J Nanomed. 2013;8:337 50.

Madaghiele M, Sannino A, Yannas IV, Spector M. Collagen-based matrices with axially oriented pores. J Biomed Mater Res Part A. 2008;85(3):757–67.

Madihally SV, Matthew HWT. Porous chitosan scaffolds for tissue engineering. Biomaterials. 1999;20(12):1133–42.

Maghdouri-White Y, Petrova S, Sori N, Polk S, Wriggers H, Ogle R, Ogle R, Francis M. Electrospun silk-collagen scaffolds and BMP-13 for ligament and tendon repair and regeneration. Biomed Phys Eng Express. 2018;4(2):025013.

Mandal BB, Kundu SC. Cell proliferation and migration in silk fibroin 3D scaffolds. Biomaterials. 2009;30(15):2956–65.

Mandrycky C, Wang Z, Kim K, Kim DH. 3D bioprinting for engineering complex tissues. Biotechnol Adv. 2016;34(4):422–34.

McMurray RJ, Gadegaard N, Tsimbouri PM, Burgess KV, McNamara LE, Tare R, Murawski K, Kingham E, Oreffo RO, Dalby MJ. Nanoscale surfaces for the long-term maintenance of mesenchymal stem cell phenotype and multipotency. Nat Mater. 2011;10(8):637–44.

Mehrabanian M, Nasr-Esfahani M. HA/nylon 6,6 porous scaffolds fabricated by salt-leaching/solvent casting technique: Effect of nano-sized filler content on scaffold properties. Int J Nanomed. 2011;6:1651–9.

Mikos AG, Thorsen AJ, Czerwonka LA, Bao Y, Langer R, Winslow DN, Vacanti JP. Preparation and characterization of poly(l-lactic acid) foams. Polymer. 1994;35(5):1068–77.

Mishra R, Bishop T, Valerio IL, Fisher JP, Dean D. The potential impact of bone tissue engineering in the clinic. Regener Med. 2016;11(6):571–87.

Mohamed A, Xing MM. Nanomaterials and nanotechnology for skin tissue engineering. Int J Burns Trauma. 2012;2(1):29–41.

Mooney DJ, Baldwin DF, Suh NP, Vacanti JP, Langer R. Novel approach to fabricate porous sponges of poly(D,L-lactic-co-glycolic acid) without the use of organic solvents. Biomaterials. 1996;17(14):1417–22.

Moore AM, MacEwan M, Santosa KB, Chenard KE, Ray WZ, Hunter DA, Mackinnon SE, Johnson PJ. Acellular nerve allografts in peripheral nerve regeneration: A comparative study. Muscle Nerve. 2011;44(2):221–34.

Murphy SV, Atala A. 3D bioprinting of tissues and organs. Nat Biotechnol. 2014;32(8):773–85.

Nasri-Nasrabadi B, Mehrasa M, Rafienia M, Bonakdar S, Behzad T, Gavanji S. Porous starch/cellulose nanofibers composite prepared by salt leaching technique for tissue engineering. Carbohydr Polym. 2014;108(1):232–8.

Nel AE, Mädler L, Velegol D, Xia T, Hoek EMV, Somasundaran P, Klaessig F, Castranova V, Thompson M. Understanding biophysicochemical interactions at the nano-bio interface. Nat Mater. 2009;8(7):543–57.

Nelson SF, Crosbie RH, Miceli MC, Spencer MJ. Emerging genetic therapies to treat duchenne muscular dystrophy. Curr Opin Neurol. 2009a;22(5):532–8.

Nelson TJ, Martinez-Fernandez A, Yamada S, Perez-Terzic C, Ikeda Y, Terzic A. Repair of acute myocardial infarction by human stemness factors induced pluripotent stem cells. Circulation. 2009b;120(5):408–16.

Niinomi M. Mechanical biocompatibilities of titanium alloys for biomedical applications. J Mech Behav Biomed Mater. 2008;1(1):30–42.

Okano T, Yamada N, Okuhara M, Sakai H, Sakurai Y. Mechanism of cell detachment from temperature-modulated, hydrophilic-hydrophobic polymer surfaces. In: The biomaterials: Silver jubilee compendium. 1995;16:297–303.

Orlova Y, Magome N, Liu L, Chen Y, Agladze K. Electrospun nanofibers as a tool for architecture control in engineered cardiac tissue. Biomaterials. 2011;32(24):5615–24.

Parenteau-Bareil R, Gauvin R, Berthod F. Collagen-based biomaterials for tissue engineering applications. Materials. 2010;3(3):1863–87.

Patel NG, Zhang G. Responsive systems for cell sheet detachment. Organogenesis. 2013;9(2):93–100.

Peran M, Garcia M, Lopez-Ruiz E, Jiménez G, Marchal JA. How can nanotechnology help to repair the body? Advances in cardiac, skin, bone, cartilage and nerve tissue regeneration. Materials. 2013;6(4):1333–59.

Pérez RA, Won JE, Knowles JC, Kim HW. Naturally and synthetic smart composite biomaterials for tissue regeneration. Adv Drug Deliv Rev. 2013;65(4):471–96.

Pok S, Vitale F, Eichmann SL, Benavides OM, Pasquali M, Jacot JG. Biocompatible carbon nanotube-chitosan scaffold matching the electrical conductivity of the heart. ACS Nano. 2014;8(10):9822–32.

Poon Z, Chang D, Zhao X, Hammond PT. Layer-by-layer nanoparticles with a pH-sheddable layer for in vivo targeting of tumor hypoxia. ACS Nano. 2011;5(6):4284–92.

Prabhakaran MP, Vatankhah E, Ramakrishna S. Electrospun aligned PHBV/collagen nanofibers as substrates for nerve tissue engineering. Biotechnol Bioeng. 2013;110(10):2775–84.

Price RL, Waid MC, Haberstroh KM, Webster TJ. Selective bone cell adhesion on formulations containing carbon nanofibers. Biomaterials. 2003;24(11):1877–87.

Qian Y, Zhao X, Han Q, Chen W, Li H, Yuan W. An integrated multi-layer 3D-fabrication of PDA/RGD coated graphene loaded PCL nanoscaffold for peripheral nerve restoration. Nat Commun. 2018;9(1):323.

Radhakrishnan J, Manigandan A, Chinnaswamy P, Subramanian A, Sethuraman S. Gradient nano-engineered in situ forming composite hydrogel for osteochondral regeneration. Biomaterials. 2018;162:82–98.

Radisic M, Park H, Shing H, Consi T, Schoen FJ, Langer R, Freed LE, Vunjak-Novakovic G. Functional assembly of engineered myocardium by electrical stimulation of cardiac myocytes cultured on scaffolds. Proc Natl Acad Sci USA. 2004;101(52):18129–34.

Rahaman MN, Day DE, Bal BS, Fu Q, Jung SB, Bonewald LF, Tomsia AP. Bioactive glass in tissue engineering. Acta Biomater. 2011;7(6):2355–73.

Ramakrishna S, Fujihara K, Teo WE, Yong T, Ma Z, Ramaseshan R. Electrospun nanofibers: Solving global issues. Mater Today. 2006;9(3):40–50.

Ramelet AA, Hirt-Burri N, Raffoul W, Scaletta C, Pioletti DP, Offord E, Mansourian R, Applegate LA. Chronic wound healing by fetal cell therapy may be explained by differential gene profiling observed in fetal versus old skin cells. Exp Gerontol. 2009;44(3):208–18.

de Réaumur RAF. Sur les diverses reproductions qui se font dans les ecrevisse, les omars, les crabes, etc. et entr'autres sur celles de leurs jambes et de leurs ecailles. Mem Acad Roy Sci. 1712:223–42.

Roach P, Eglin D, Rohde K, Perry CC. Modern biomaterials: A review – Bulk properties and implications of surface modifications. J Mater Sci Mater Med. 2007;18(7):1263–77.

Rocha LB, Goissis G, Rossi MA. Biocompatibility of anionic collagen matrix as scaffold for bone healing. Biomaterials. 2002;23(2):449–56.

Ross EA, Williams MJ, Hamazaki T, Terada N, Clapp WL, Adin C, Ellison GW, Jorgensen M, Batich CD. Embryonic stem cells proliferate and differentiate when seeded into kidney scaffolds. J Am Soc Nephrol. 2009;20(11):2338–47.

Sachlos E, Czernuszka JT. Making tissue engineering scaffolds work. Review: The application of solid freeform fabrication technology to the production of tissue engineering scaffolds. Eur Cells Mater. 2003;18(7):1263–77.

Sahoo S, Ang LT, Goh JC, Toh SL. Growth factor delivery through electrospun nanofibers in scaffolds for tissue engineering applications. J Biomed Mater Res Part A. 2010;93(4):1539–50.

Salmasi S, Kalaskar DM, Yoon WW, Blunn GW, Seifalian AM. Role of nanotopography in the development of tissue engineered 3D organs and tissues using mesenchymal stem cells. World J Stem Cells. 2015;7(2):266–80.

Schneider A, Garlick JA, Egles C. Self-assembling peptide nanofiber scaffolds accelerate wound healing. PLoS ONE. 2008;3(1):e1410.

Sekiya S, Shimizu T, Yamato M, Okano T. Hormone supplying renal cell sheet in vivo produced by tissue engineering technology. BioRes Open Access. 2013;2(1):12–9.

Sekiya S, Shimizu T, Yamato M, Kikuchi A, Okano T. Bioengineered cardiac cell sheet grafts have intrinsic angiogenic potential. Biochem Biophys Res Commun. 2006;341(2):573–82.

Semino CE. Self-assembling peptides: From bio-inspired materials to bone regeneration. J Dent Res. 2008;87(7):606–16.

Serpooshan V, Zhao M, Metzler SA, Wei K, Shah PB, Wang A, Mahmoudi M, et al. The effect of bioengineered acellular collagen patch on cardiac remodeling and ventricular function post myocardial infarction. Biomaterials. 2013;34(36):9048–55.

Shafiee A, Atala A. Tissue engineering: Toward a new era of medicine. Ann Rev Med. 2017;68(1):29–40.

Shevchenko RV, James SL, James SE. A review of tissue-engineered skin bioconstructs available for skin reconstruction. J R Soc Interface. 2010;7(43):229–58.

Shukla A, Almeida B. Advances in cellular and tissue engineering using layer-by-layer assembly. Wiley Interdiscip Rev Nanomed Nanobiotechnol. 2014;6(5):411–21.

Silva AKA, Juenet M, Meddahi-Pellé A, Letourneur D. Polysaccharide-based strategies for heart tissue engineering. Carbohydr Polym. 2015;116:267–77.

Silva E, de Vasconcellos LMR, Rodrigues BVM, dos Santos DM, Campana-Filho SP, Marciano FR, Webster TJ, Lobo AO. PDLLA honeycomb-like scaffolds with a high loading of superhydrophilic graphene/multi-walled carbon nanotubes promote osteoblast in vitro functions and guided in vivo bone regeneration. Mater Sci Eng C. 2017;73:31–9.

Singh D, Singh D, Zo S, Han SS. Nano-biomimetics for nano/micro tissue regeneration. J Biomed Nanotechnol. 2014;10(10):3141–61.

Sitharaman B, Shi X, Walboomers XF, Liao H, Cuijpers V, Wilson LJ, Mikos AG, Jansen JA. In vivo biocompatibility of ultra-short single-walled carbon nanotube/biodegradable polymer nanocomposites for bone tissue engineering. Bone. 2008;43(2):362–70.

Sofroniew MV, Howe CL, Mobley WC. Nerve growth factor signaling, neuroprotection, and neural repair. Ann Rev Neurosci. 2001;24:1217–81.

Song JJ, Guyette JP, Gilpin SE, Gonzalez G, Vacanti JP, Ott HC. Regeneration and experimental orthotopic transplantation of a bioengineered kidney. Nat med. 2013;19(5):646.

Stegen S, van Gastel N, Carmeliet G. Bringing new life to damaged bone: The importance of angiogenesis in bone repair and regeneration. Bone. 2015;70:19–27.

Stendahl JC, Wang LJ, Chow LW, Kaufman DB, Stupp SI. Growth factor delivery from self-assembling nanofibers to facilitate islet transplantation. Transplantation. 2008;86(3):478–81.

Subash-Babu P, Ignacimuthu S, Agastian P, Varghese B. Partial regeneration of β-cells in the islets of Langerhans by Nymphayol a sterol isolated from *Nymphaea stellata* (Willd.) flowers. Bioorg Med Chem. 2009;17(7):2864–70.

Subramanian A, Krishnan UM, Sethuraman S. Development of biomaterial scaffold for nerve tissue engineering: Biomaterial mediated neural regeneration. J Biomed Sci. 2009;16:108.

Sun JG, Tang J, Ding JD. Cell orientation on a stripe-micropatterned surface. Chin Sci Bull. 2009b;54(18):3154–9.

Sun W, Sun C, Lin H, Zhao H, Wang J, Ma H, Chen B, Xiao Z, Dai J. The effect of collagen-binding NGF-Beta on the promotion of sciatic nerve regeneration in a rat sciatic nerve crush injury model. Biomaterials. 2009a;30(27):4649–56.

Sundaramurthi D, Krishnan UM, Sethuraman S. Electrospun nanofibers as scaffolds for skin tissue engineering. Polym Rev. 2014;54(2):348–76.

Tijore A, Irvine SA, Sarig U, Mhaisalkar P, Baisane V, Venkatraman S. Contact guidance for cardiac tissue engineering using 3D bioprinted gelatin patterned hydrogel. Biofabrication. 2018;10(2):025003.

Toosi S, Naderi-Meshkin H, Kalalinia F, Peivandi MT, HosseinKhani H, Bahrami AR, Heirani-Tabasi A, Mirahmadi M, Behravan J. PGA-incorporated collagen: Toward a biodegradable composite scaffold for bone-tissue engineering. J Biomed Mater Res Part A. 2016;104(8):2020–8.

Tysseling-Mattiace VM, Sahni V, Niece KL, Birch D, Czeisler C, Fehlings MG, Stupp SI, Kessler JA. Self-assembling nanofibers inhibit glial scar formation and promote axon elongation after spinal cord injury. J Neurosci. 2008;28(14):3814–23.

Valentin JE, Turner NJ, Gilbert TW, Badylak SF. Functional skeletal muscle formation with a biologic scaffold. Biomaterials. 2010;31(29):7475–84.

Vandenburgh HH, Karlisch P, Farr L. Maintenance of highly contractile tissue-cultured avian skeletal myotubes in collagen gel. In Vitro Cell Dev Biol. 1988;24(3):166–74.

Vasita R, Katti DS. Nanofibers and their applications in tissue engineering. Int J Nanomed. 2006;1(1):15–30.

Wan ACA, Ying JY. Nanomaterials for in situ cell delivery and tissue regeneration. Adv Drug Deliv Rev. 2010;62(7–8):731–40.

Wang Q, Yan J, Yang J, Li B. Nanomaterials promise better bone repair. Mater Today. 2016;19(8):451–63.

Webber MJ, Tongers J, Renault MA, Roncalli JG, Losordo DW, Stupp SI. Reprint of: Development of bioactive peptide amphiphiles for therapeutic cell delivery. Acta Biomater. 2015;23(S):S42–51.

Webster TJ, Ergun C, Doremus RH, Siegel RW, Bizios R. Specific proteins mediate enhanced osteoblast adhesion on nanophase ceramics. J Biomed Mater Res. 2000;51(3):475–83.

Williams CG, Malik AN, Kim TK, Manson PN, Elisseeff JH. Variable cytocompatibility of six cell lines with photoinitiators used for polymerizing hydrogels and cell encapsulation. Biomaterials. 2005;26(11):1211–8.

Woo KM, Chen VJ, Ma PX. Nano-fibrous scaffolding architecture selectively enhances protein adsorption contributing to cell attachment. J Biomed Mater Res Part A. 2003;67:531–7.

Wu T, Huang C, Li D, Yin A, Liu W, Wang J, Chen J, Hany EH, Al-Deyab SS, Mo X. A multi-layered vascular scaffold with symmetrical structure by bi-directional gradient electrospinning. Colloids Surf B Biointerfaces. 2015;133:179–88.

Xie J, Peng C, Zhao Q, Wang X, Yuan H, Yang L, Li K, Lou X, Zhang Y. Osteogenic differentiation and bone regeneration of iPSC-MSCs supported by a biomimetic nanofibrous scaffold. Acta Biomater. 2016;29:365–79.

Xie X, Hu K, Fang D, Shang L, Tran SD, Cerruti M. Graphene and hydroxyapatite self-assemble into homogeneous, free standing nanocomposite hydrogels for bone tissue engineering. Nanoscale. 2015;7(17):7992–8002.

Xu Y, Guan J. Biomaterial property-controlled stem cell fates for cardiac regeneration. Bioact Mater. 2016;1(1):18–28.

Xue J, Feng B, Zheng R, Lu Y, Zhou G, Liu W, Cao Y, Zhang Y, Zhang WJ. Engineering ear-shaped cartilage using electrospun fibrous membranes of gelatin/polycaprolactone. Biomaterials. 2013;34(11):2624–31.

Yang F, Xu CY, Kotaki M, Wang S, Ramakrishna S. Characterization of neural stem cells on electrospun poly(L-lactic acid) nanofibrous scaffold. J Biomater Sci Polym Ed. 2004;15(12):1483–97.

Yang RG, Zhong HB, Zhu JL, Zuo TT, Wu KJ, Hou SX. Clinical safety about repairing the peripheral nerve defects with chemically extracted acellular nerve allograft. Zhonghua Wai Ke Za Zhi [Chin J Surg]. 2012;50:74–6.

Yanlian Y, Ulung K, Xiumei W, Horii A, Yokoi H, Shuguang Z. Designer self-assembling peptide nanomaterials. Nano Today. 2009;4(2):193–210.

Yoo HS, Kim TG, Park TG. Surface-functionalized electrospun nanofibers for tissue engineering and drug delivery. Adv Drug Deliv Rev. 2009;61(12):1033–42.

Zhang L, Webster TJ. Nanotechnology and nanomaterials: Promises for improved tissue regeneration. Nano Today. 2009;4(1):66–80.

Zhao J, Griffin M, Cai J, Li S, Bulter PEM, Kalaskar DM. Bioreactors for tissue engineering: An update. Biochem Eng J. 2016a;109:268–81.

Zhao X, Lang Q, Yildirimer L, Lin ZY, Cui W, Annabi N, Ng KW, Dokmeci MR, Ghaemmaghami AM, Khademhosseini A. Photocrosslinkable gelatin hydrogel for epidermal tissue engineering. Adv Healthcare Mater. 2016b;5(1):108–18.

Zhou ZF, Zhang F, Wang JG, Chen QC, Yang WZ, He N, Jiang YY, Chen F, Liu JJ. Electrospinning of PELA/PPY fibrous conduits: Promoting peripheral nerve regeneration in rats by self-originated electrical stimulation. ACS Biomater Sci Eng. 2016;2(9):1572–81.

Zimmermann WH, Schneiderbanger K, Schubert P, Didié M, Münzel F, Heubach JF, Kostin S, Neuhuber WL, Eschenhagen T. Tissue engineering of a differentiated cardiac muscle construct. Circ Res. 2002;90(2):223–30.

9

Nanotechnology in Preventive and Emergency Healthcare

Nilutpal Sharma Bora, Bhaskar Mazumder, Manash Pratim Pathak, Kumud Joshi, and Pronobesh Chattopadhyay

CONTENTS

ABSTRACT: Globally there is growing demand for the need to understand the impact of medical nanotechnology, especially in the field of emergency and preventive healthcare. Nanomedicine is defined as a science which highlights the use of molecular tools and knowledge of the human body for preventing diseases and traumatic injury, diagnosing, treating, preserving, and improving human health. The utilization of nanomedicine in the field of emergency trauma mitigation and preventive healthcare may provide medical professionals with the means to diagnose and overcome life threatening conditions and also paralyze the propagation of infectious diseases. Nano-based drug delivery systems, nano-chip-based diagnostics, nano-surgical interventions, protective nanoformulations, and nanodevices are a few of the novel medical tools that have currently surfaced in the field of healthcare.

KEY WORDS: *Nanotechnology, nanomedicine, emergency, preventive, diagnostics*

9.1 Introduction

Nanotechnology was first idealized by physicist Richard P. Feynman, when his talk "There's plenty of room at the bottom", delivered at the yearly convention of the American Physical Society during the late 1950s, raised serious interest and debate over maneuvering and managing things on a very small scale (Feynman, 1960). Since then, these manipulations of matter have evolved tremendously and now involve particles as small as 1 to 100 nanometers (nm) for the upliftment of multidisciplinary innovation. Numerous applications of nanomaterials are observed nowadays, ranging from incorporating light weight nanoparticles into automobiles and sports equipment and fabricating nanoscale diagnostic and drug delivery systems to nanoscale energy systems that aid in broad spectrum scientific and technological advancement (Kang and Ceder, 2009; Kayser et al., 2005; National Nanotechnology Initiative, 2011). Nanotechnology has been considered a frontier where the field of medicine and emergency healthcare is concerned. Nanotechnology, along with background knowledge of human anatomy and physiology, has been providing healthcare professionals with the luxury of working at the molecular and intracelluler levels (Duncan, 2004). The culpable reputation of nanomaterials, in the territory of medicine and biotechnology, is attributed to its shape, size, solubility, and surface properties, etc., which facilitate optimum biological effects due to their larger surface area and higher permeability (Biswal and Yusoff, 2017). The application of nanotechnology in medicine is especially important because products for targeted drug delivery, molecular imaging, and disease diagnosis are already being developed or undergoing trials (Sahoo et al., 2007).

Therefore, the most imperative purpose for nanotechnology may be related to the field of pharmaceutical research, due to the problems associated with stability, therapeutic absorption, frequency of drug delivery, and drug bioavailability, etc. (Kayser et al., 2005; Li et al., 2004; Pawar et al., 2004). Although there has been quite a lot of evolution in the field of nanotechnology, it is still a new field of science and hence its potential has not been fully explored. Drug delivery devices are vital tools for pharmaceutical industries to improve their drug market. Therefore, an estimated 13% of the current pharmaceutical market focuses on the sale of medicines which include a drug delivery system (Mazzola, 2003). Nanotechnology is beneficial in these aspects as it facilitates the tissue- or organ-specific targeting of drugs, along with multifunctional nanosystems that include the combined benefits of targeting, diagnostic, and therapeutic actions (Sahoo et al., 2007). Nanotechnological systems can also be designed to achieve drug targeting, controlled release of drugs, and recovery of drugs possessing low bioavailability (Prego et al., 2006). The delivery of vaccines is also an important aspect of medicine that has been associated with nanotechnology. Micro- and nano-particles synthesized using newer techniques of encapsulation have been verified to be proficient in enhancing the process of immunization in animal models (Cui and Mumper, 2003; Diwan et al., 2003; Köping-Höggård et al., 2005). Targeted drug delivery, gene delivery, and the advancement of molecular medicines are the milestones of nanotechnology, which are evident from the numerous advantages that are enjoyed by nano-drug delivery systems (Sahoo et al., 2007). The US Food and Drug Administration (US FDA) have approved the use of a number of nanotechnological products and many are still under clinical trials and preclinical development. Among these, the first-generation nanotechnology products like liposomes and drug–polymer conjugates are very common. However, the prime aspect of novel nanotechnology is to fabricate multifunctional products that are more effective, safe, and capable of effective drug targeting, synergistic combinational drug delivery, and sustained drug delivery (Bhatia and Ingber, 2016).

In the next ten years the world is predicted to observe a boom in the field of nanotechnology products as a number of US patents related to nanotechnology expired in 2008, which translated to a revenue loss of 70–80 billion US dollars in 2011. Nanomaterials and nanodevices are the two main types of tools in pharmaceutical nanotechnology. Nanomaterials can be further classified into nanocrystalline and nanostructured materials which can be used as biomaterials in dental implants, tissue-engineering scaffolds, or can be surface modified for increasing biocompatibility. Nanodevices, on the other hand, are minuscule devices which include microarray, nano- or microelectrical circuits, and microfluidics, and are used for the detection of airborne pathogens, disease status, or biological hazards (Bhatia and Ingber, 2016). It is evident from the above discussion that both nanomaterials and nanodevices possess the functionalities that enable us to use them in preventive and emergency healthcare medicine. This chapter focuses on the applications of biomedical nanotechnology products that are effectively used for emergency treatments and preventive healthcare regimens. The recent innovations in this field are discussed along with the potential toxicological implications of the same. Both *in vivo* and *in vitro* nanotechnology products are discussed here, which have the prospective to emerge as serious contenders in the realm of emergency and preventive healthcare.

9.2 Novel Applications of Nanotechnology in Medical Science

9.2.1 Drug Delivery Devices and Tools

Conventional drugs can be pharmacologically improved by using drug delivery systems (DDS), which include lipid or polymeric carriers, along with their pharmacokinetic modifications and modulations of their biodistribution patterns. These DDS are known to possess various advantages over usual dosage forms, for example, sustained release and controlled release. The involvement of nanotechnology in drug delivery devices and tools can address the different non-ideal properties of drugs, like lower solubility, tissue injury or extravasation, rapid drug degradation *in vivo*, adverse pharmacokinetics or reduced biodistribution, and deprivation of selectivity for target tissues (Allen, 2002; Allen and Cullis, 2004; Duncan, 2003; Lavan et al., 2003; Liu et al., 2000).

Nano-DDS and tools are a result of the amalgamation of the cellular and molecular components and engineered materials, which possess novel electronic, optical, and structural properties that are unavailable in bulk solids or individual molecules. Nanoscale devices are constituted of various nanoconstructs and self-assembled, biodegradable nanoparticles, for use in organ-targeted drug delivery of antitumor drugs or as agents for contrast imaging which are customizable as per individual needs. These vehicles are competent to deliver hefty doses of chemotherapeutic agents or genetic therapy into malignant cells while excluding healthy cells. Nanoparticle drug delivery acts via two main mechanisms: passive and active targeting. Passive targeting is accomplished by three forms: leaky vasculature (enhancing permeability and retention phenomenon), tumor microenvironment (conjugating the drug to a tumor-specific molecule), and direct local application. Active targeting is attained by conjugating the nanoparticle with a targeting moiety and directs them to the cell surface carbohydrates, receptors, and antigens, henceforth classified as carbohydrate-targeted, receptor-targeted, and antigen-targeted (Guo and Szoka, 2003; Sinha et al., 2006; Yamazaki et al., 2000).

Information technology has also converged with biological applications to pave the way for nano-electromechanical systems (NEMS)-based drug delivery devices, which can completely control drug release, meeting requirements for on-demand, repeated, and customized continuous dose administration for sustained periods. NEMS technologies have undergone significant development in recent years, serving the multifaceted requirements of data management, controlling devices, and device communication. Examples of such systems are glucose sensors for continuous glucose monitoring and porfimer sodium in combination with therapeutic photodynamic balloons equipped with a fiber optic diffuser for ablation of high-grade dysplasia in patients afflicted with Barrett's esophagus. The drug–device and biological–device combinations are intended to react to specific signals or circumstances and provide more receptiveness towards adjustments in the dosing of drugs (Staples et al., 2006; Yamazaki et al., 2000).

Nanoscale drug delivery technologies can be categorized into polymeric, biologic, carbon-based, silicon-based, or metallic. The forms of nanostructures available for drug delivery can be summarized as vesicles, nanotubes, rings, nanoparticles, micelles, dendrimers, porous nanoparticles, nanoneedles, fullerenes, and nanoshells. These DDS can be utilized as gene delivery vectors, included in pulmonary therapies, and to stabilize drug molecules with a shorter shelf-life. Reduced toxicity and more proficient drug distribution are the other valuable advantages available when nanoscale DDS are used (Courrier et al., 2002; Desai et al., 1996; Hughes, 2005; Senior, 1998). Physiological barriers can be overcome and targeted organs can be reached by using nanoscale DDS. The utilization of nanoscale DDS have facilitated improved pulmonary drug delivery, increased gastrointestinal and transcutaneous absorption. Drug molecules can now be targeted directly into cells to achieve increased antitumor activity and genetic alterations have become possible by using nanoscale delivery systems as carriers for deoxyribonucleic acid (DNA) and/or ribonucleic acid (RNA) (Drummond et al., 1999; Hussain et al., 2001; Martin and Kohli, 2003).

These advantages make nanoscale systems a very promising option for use as devices and tools for the delivery of drugs molecules and biologicals into targeted organs or tissues. This section deals with the different devices and tools that are available within the nanoscale range.

9.2.1.1 Brain Targeting Nanoparticles

Drug targeting to the brain has been a challenge as crossing the blood–brain barrier (BBB) is one of the greatest challenges faced by a formulator. Numerous attempts have been taken up for the development of novel drug delivery devices capable of crossing the BBB, which possess the ability to penetrate the tight endothelial cell junctions of the capillaries within the brain and the central nervous system (CNS). Some regulators of the brain's functions, like peptides, are gifted with the capacity to pass through the BBB from the blood vessels into the brain, such as transferrin, cytokines, endorphins, and enkephalins, etc. On the other hand, other entities, like some amino acids, penetrate poorly through the BBB. It is this poor penetration of the BBB that makes drug targeting to the brain a strenuous and complicated process (Raeissi and Audus, 1989; Schröder and Sabel, 1996; Schroeder et al., 1998).

Polysorbate-coated nanoparticles are one of the few methods to facilitate the penetration of drugs through the BBB. The mechanism involved in enhancing the transport of drugs from the coated nanoparticles through BBB is (1) by passive diffusion of the drugs from the blood stream to the brain, which is facilitated by the binding of the nanoparticles to the inner endothelial lining of the brain capillaries, thereby providing a large concentration gradient, and (2) by uptake of the drugs from the brain endothelial lining by phagocytosis (Alyautdin et al., 1995). Borchard et al. (1994) demonstrated that polysorbate-80-coated poly(methyl methacrylate) nanoparticles exhibited an increased uptake by the microvessel endothelial cell monolayers in bovines. Polysorbate-coated (alkyl cyanoacrylate) nanoparticles were found to be superior over the uncoated nanoparticles to transport drugs across the BBB (Steiniger et al., 1999). Tröster et al. (1990) showed that the accumulation of radioactivity caused by intravenously administered polysorbate-80-coated ^{14}C-polymethylmethacrylate (^{14}C-PMMA) nanoparticles in the brain was nine times higher when compared to non-coated controls. Polybutylcyanoacrylate (PBCA) nanoparticles and non-biodegradable polystyrene (PS) nanoparticles have also been fabricated to study the transfer of dalargin (analgesic peptide) to the brain (Olivier et al., 1999). However, the therapeutic potential of polysorbate-80-coated PBCA nanoparticles is limited because a very high systemic nanoparticle concentration is necessary for the delivery of drugs to the CNS (Soppimath et al., 2001).

Chitosan nanoparticles have also been used as tools to deliver medication and genetic equipment into the CNS, predominantly by endocytosis and accumulation around the cell nucleus. The grafting of chitosan with polyethylenimine (PEI) has yielded a higher solubility and transfection rate and a lower toxicity in cultured cells. Biocompatible chitosan-based films and microspheres made of chitosan can effectively execute growth factor delivery to the CNS and also promote regeneration of the CNS after injury by serving as a structural support for transplanted cells (Demeule et al., 2008; Malatesta et al., 2012; Peng et al., 2014; Skop et al., 2013). *In vivo* studies involving murine models have demonstrated that Tween 80®-coated chitosan nanoparticles have improved the targeting efficiency of gallic acid to the brain and also enhanced the acceptance levels of the drug when compared with intraperitoneal or oral administration (Nagpal et al., 2012).

Due to their superior protection from physiologically hostile conditions and antigenecity, collagen-based nanoparticles have been used extensively for delivering drugs, growth factors, and genetic materials. Glial-derived growth factor can be delivered to the CNS by stereotatic injection into the rat striatum using *in situ* gelling type I collagen hydrogel (Posadas et al., 2016).

Lipids and cationic lipids are one of the most studied and clinically recognized drug nanocarriers due to their successful history, low toxicity, and ability to function as a carrier for both lipophilic and hydrophilic drugs. Liposomes are composed of one central, aqueous compartment surrounded by a single bilayer or multiple bilayers of lipids. Comparitively huge amounts of drugs can be included into the aqueous or lipid compartments (Ishii et al., 2013). Liposomal formulations have been used to deliver various drugs into the CNS, including anti-retrovirals, chemotherapeutic compounds, drugs used to treat Alzheimer's disease, and anti-ischemia drugs. Liposomes have been increasingly used for the treatment of brain tumors, wherein ultrasound-mediated BBB disruption is combined to augment the efficiency of anti-neoplastic drugs via intravenous administrations (Wong et al., 2010). Surface modifications for targeted delivery can further increase the potential of liposomes for CNS delivery. For example, OX26 monoclonal antibody (mAb) immunoliposomes favor receptor-mediated transcytosis across the BBB *in vivo* without altering the BBB integrity. Similary, sodium borocaptate was successfully delivered using liposomes conjugated with anti-epidermal growth factor receptor (EGFR) antibodies into EGFR-overexpressing glioma cells *in vitro* and mice bearing brain tumors. Solid state tumor results in the disruption of the BBB and hence it enables the immunoliposomes to cross the BBB. The same can be achieved with peptides and transferrins, however the presence of transferrin transporters in the liver and bone marrow limits the selectivity of these compounds (De Boer et al., 2003).

Solid lipid nanoparticles (SLNs) are also widely used nanocarriers for drug delivery into the CNS, mainly in cases of brain tumor, human immunodeficiency virus (HIV), and Alzheimer's disease. Physiological lipids, such as mono-, di-, and triglycerides, fatty acids, waxes, and mixtures of glycerine which remain solid at body and room temperatures, are used to construct matrices which form the primary component of SLNs. These require surfactants for stabilization and to facilitate the sustained and controlled release of medications (Gaur et al., 2014; Yusuf et al., 2013). Cholesterol-based SLNs fabricated with esterquat and stearylamine increased the permeability of saquinavir into the BBB tested in single-layered, human brain-microvascular endothelial cells *in vitro* (Kuo and Wang, 2014).

Lipoplexes are another novel complex which have been increasingly used for the delivery of genetic substances into the CNS via the BBB. Lipoplexes are cationic lipid–DNA/RNA complexes that are formed by cationic lipids binding and compacting DNA or RNA through electrostatic interactions. After successful systemic administration, the lipoplexes are absorbed by endocytosis and the lipid component fuses with the endosomal vesicle and facilitates the quick release of the nucleic acid into the cytoplasm (Sahay et al., 2013). However, the reduced organ-targeting efficiency, short degradation period, and low efficiency of expression have led to the need for local injections in order to obtain significant transfection in the CNS. "Trojan Horse Liposome" (THL) technology is a significant breakthrough for overcoming the genetic material delivery obstacle. This technology uses the transcytosis pathway to enter the brain, avoiding the BBB disruption. The *in vivo* stability of THLs can be accredited to the immunoliposomes carrying genetic materials which are engineered with known lipids containing polyethylene glycol (PEG). The required tissue-target specificity of THLs is achieved when they are conjugated to peptidomimetic mAb, which conjugates with specific endogenous receptors of transferrin or insulin located within the BBB and the cellular membranes of the brain. Other applications of THLs include tissue-specific gene expression with tissue-specific promoters and the expression of therapeutic genes within the brain in Parkinson's disease animal models (Boado and Pardridge, 2011).

Pyridinium cationic lipids are also a new string of cationic lipids which are recognized to increase the efficiency of transfection via liposomal delivery. They consist of a dopamine backbone connected to the hydrophobic alkyl tails through an ether linkage and have been successfully used to transfect the dopaminergic neurons of the nucleus accumbens into cortical neurons located in the frontal cortex and within the primary coculture of neurons and glial cells (Savarala et al., 2013).

Synthetic, polymeric nanoparticles, increasingly used nowadays, are compact, colloidal systems made of polymers, wherein the therapeutic agent is either coated upon the particle surface by conjugation or adsorption or is entrapped within the colloid matrix. Examples of such synthetic polymers include poly(glycolic acid) (PGA), poly-acrylate, PEI, poly-L-lysine (PLL), poly(D,L-lactide) (PLA), and poly(D,L-lactide-co-glycolide) (PLGA). PEI can bind to various nucleic acids electrostatically since it contains numerous amino groups within its backbone, which can be protonated to get a high cationic-charge density potential. But a major drawback is that PEI can be toxic to certain cell types, including neurons, depending on the molecular weight (MW), structure, and concentration of the molecule used. High MW PEIs cause necrotic-like changes due to alterations in the plasma membrane, followed by activation of a mitochondrial-mediated apoptosis in several cell types (Lee et al., 2014; Zhong et al., 2013). Low MW PEIs are much less toxic but have negligible transfection efficiency. A substantial amount of labor is being given to perk up the biocompatibility of high MW PEIs by varying their chemical homogeneity, biodegradability, and solubility. PLL is a cationic polymer that contains positively charged hydrophilic amino groups consisting of amino acid lysine as a repeat unit, which is biodegradable and rapidly released from endosomes. But PLLs show a low level of transfection efficiency and suffer from immunogenicity and toxicity caused by their amino acid backbones (Zhang et al., 2010).

Poly(acrylic acid), and poly(methacrylic acid) and poly(butyl cyanoacrylate) (PACA) nanoparticles are other synthetic polymers which are increasingly used nowadays. These polymers come under the class of poly-acrylate polymers, which are polymers consisting of a vinyl group and a carboxylic acid terminus (based on acrylic acid). PACA nanoparticles have the ability to absorb or encapsulate an extensive range of drugs and the successful development of nanoparticles for intravenous administration has been possible (Vauthier et al., 2003). But the synthetic polymer still suffers the limitation of a deficient biocompatibility and drug release rate. Moreover, poly-acrylate polymers suffer the disadvantage of being neurotoxic and not hydrolytically degradable. In the dearth of surface modification with target molecules, polymeric nanoparticles have a partial capacity to cross the BBB, which translates into the fact that more work is still needed to achieve efficient polymer-based transfection vectors via intravenous administration (Ulery et al., 2011).

Dendrimers are a group of nanoparticles which possess a significant upperhand when compared to others, as their advantages include versatility in the composition and structurally controlled nanoscale building, easy functionalization, and well-defined reactive groups on the particle surface. Dendrimers are technically repeatedly branched polymer molecules containing a cascade of branches grown from one or several cores, containing three essential, fundamental structural domains: (1) the core, to which the branches are joined, (2) the shell, which consists of the branches surrounding the core, and (3) the multivalent surface formed by the branches' termini (Albertazzi et al., 2012; Tomalia et al., 1985). To date, more than 200 kinds of dendrimers and dendrons exist that can be classified into different families, including the poly(propylene imine) (PPI), poly(amidoamine) (PAMAM), phosphodendrimer, poly(triazene), polyamide ether, poly(thiophosphate), and poly(bis phenylpropyl) families and a dendritic family based on the 2,2-bismethylolpropionic acid (bis-MPA) monomer (Tomalia et al., 2012).

Endocytosis via a clathrin- and caveolaemediated cellular internalization is the main mechanism by which dendrimers transport molecules across cell membranes or biological barriers. The secondary and tertiary amines on their surface enable them to act as a proton attracter once they enter the cell, thus escaping from the endosomes and lysosomes, finally releasing their cargo molecules to the cytosol of the cells (Pérez-Martínez et al., 2011). PAMAM dendrimers are the most studied and utilized type of dendrimer, which suffers the major drawback of reduced CNS penetration in the absence of a meningeal inflammation, even after an intraventricular or subarachnoid administration. Most of them remain deposited in the ventricle lining within the ependymal layer (Lesniak et al., 2013; Mishra et al., 2014). This reduced penetration of the dendrimers has been addressed by coupling specific BBB-targeting ligands onto the surface of dendrimers like transferrin, rabies virus glycoprotein (RVG29), epidermal growth factor (EGF), and angiopep2 peptides (Gao et al., 2013). The use of dendrimers for brain drug delivery, imaging, and diagnosis is also limited by the toxicity that they exhibit due to the positive charge present on the facade of the cationic dendrimers. For example, PAMAM dendrimers are known to cause extensive DNA damage and display increased reactive oxygen species (ROS) synthesis, along with the combined woe of neuronal toxicity (Nyitrai et al., 2013; Wang et al., 2014). Appropriate modifications of the surface of the dendrimers have resulted in a significant reduction in their toxicity, with the transformation of PAMAM amine groups into pyrrolidone derivatives translating to minimum toxicity *in vitro* (Ciolkowski et al., 2013). Although the dendrimers have shown promising results in terms of transfection efficiency and as carriers of gene cargo in numerous cell lines *in vitro*, their use as CNS DDS is currently under investigation as crossing the BBB is still a major hurdle *in vivo* (Pérez-Carrión and Ceña, 2013).

Carbon nanotechnology is becoming increasingly popular day-by-day, and is extensively used in more and more medical applications. Fullerenes, carbon nanotubes (CNTs), and carbon nanohorns (CNHs) are the primary materials which come under carbon nanotechnology (Kostarelos et al., 2009). CNTs are structurally single- or multiwalled tubes made up of graphite sheets with a diameter between 0.5–20 nm and a length of 1.5–5,000 nm (Upadhyay, 2014). The inherent properties of these nanomaterials enable them to be used in different fields of nanomedicine, such as the thermal treatment of tumors due to their thermal conductivity, and also in drugs, genes, and vaccines (Xu et al., 2013). CNTs also constitute another major form of drug delivery device known as "nanoneedles", which are discussed later in this chapter.

Quantum dots (QDs) are essentially colloidal sheets which consist of a shield enclosing a metalloid crystalline core, rolled into single or multiwalled tubes (2–10 nm), and are a form of nanoscale product that is extensively used nowadays. These QDs are semiconductor nanocompositions which are formed by a combination of the elements of the groups II, III, and V of the periodic table (Masserini, 2013). Long-term photostability and bright fluorescence over narrow and size-tunable emission bands are a few of the advantages enjoyed by QDs which allow for multiplexing with multiple emission wavelengths from a single excitation source. These are very unique properties, which are indigenious to QDs only. Surface functionalization with PEG, lipids, and dextrans render them soluble in biologically relevant solvents, as they are otherwise insoluble due to their hydrophobicity. Decorating the surface of QDs with various functional groups, like carboxylic acids, primary amines, and thiols, enables them to be used for conjugation with biomolecules, such as antibodies, nucleic acids, or peptides, by either covalent or non-covalent interactions. Cadmium Selenide/Zinc-Sulfide (CdSe/ZnS), Cadmium Selenide/Zinc-Cadmium-Sulfide (CdSe/ZnCdS), Cadmium-Telluride/Cadmium-Selenide (CdTe/CdSe), and Indium Phosphide/Zinc Sulfide (InP/ZnS) are the most popular core/shell used in various medicinal and biological purposes and research. However, their toxicities, like oxidative stress and cytotoxicity, have limited their use for *in vivo* purposes (Dubertret et al., 2002).

9.2.1.2 Nanoneedles

Nanoneedles are a form of CNTs which are used to ferry the administered drug to the target site, while minimizing non-target accumulation and achieving penetration of various biological membranes. The core idea of a nanoneedle "mechanism" is to passively cross different membranes by following a translocation mechanism. These CNTs are perceived as intensely multifunctional and innovative carriers, which feature the covalent and non-covalent introduction of several pharmaceutically active ingredients and also feature the advantage of being highly functionalizable, with different functional groups for the simultaneous transport of several moieties for imaging, therapy, and targeting (Prato et al., 2007).

It has been established that needle-shaped particles can cross the plasma membrane depending upon their size and shape. For example, genetic material can be delivered into mammalian cells using asbestos whiskers. However, their use has been limited due to their toxicity. Biodegradable PLGA needles have been designed which can revert to spherical shape on demand to reduce the toxicity while maintaining the efficiency of drug delivery. These nanoneedles can be used to deliver genetic material into cells, which has been successfully demonstrated by their ability to deliver small interfering (si)RNA *in vitro* using green florescent protein (GFP)-expressing endothelial cells (Guth, 1981; Jang et al., 2010; Kolhar et al., 2011). The needle-like form of these nanoparticles facilitates greater penetrability and permeability along the classical phagocytotic pathway. Nanoneedles attached to an atomic force microscopy tip can also be used for the delivery of specific genetic material or macromolecules to an individual targeted cell (Obataya et al., 2005).

CNTs covalently attached with small MW molecules seem to penetrate plasma membranes via an energy-independent mechanism to a substantial extent, and further passively traversing the membrane by behaving like tiny needles (Kostarelos et al., 2007; Pantarotto et al., 2004). Examples of functionalized CNTs include P-gp-antibody-*f* CNTs loaded with doxorubicin, which demonstrated a four-fold increase in the cytotoxicity towards multidrug resistant (MDR) leukemic K562 cells than that of free doxorubicin (Li et al., 2010).

Folic acid-*f* CNTs conjugated with Platinum (Pt) (IV) has demonstrated their efficiency to suitably deliver platinum-based anticancer agents with enhanced target specificity towards folate receptors, which are found extensively within tumor cells (Dhar et al., 2008). Sandvik Bioline, RK 91™ needles are an example of nanoneedles being used in dentistry for surgical interventions (Kumar and Vijayalakshmi, 2006).

9.2.1.3 Vector-Based Gene Therapy

Gene therapy is increasingly used to tackle genetic mutations, cellular proliferation, angiogenesis, metastasis, and tumor immunogenicity, etc. It is one of the most commonly used therapies in cases of human carcinoma and malignancies. Gene therapy is also utilized in cases of genetic abnormalities and deficiencies such as autoimmune disorders, acquired immune deficiency syndrome (AIDS), and cardiovascular diseases (Jin et al., 2014). Gene therapy utilizes the fundamental procedure of gene transfection, which is used to study gene function and protein expression or manipulate cells in molecular biology.

The techniques available for gene transfection include: (1) the viral (transduction) method, (2) the physical (direct micro injection) method, and (3) chemical methods. On its own, the viral method is considered to be the most superior due to its specificity and efficiency in penetrating cells. It uses the theory of functional gene insertion into a nonspecific site within the viral genome, which in turn uses the host cell's biosynthetic apparatus to expresses the genes *in vivo* (Mulligan, 1993). However, there is a limit to the length of the genes that can be synthesized and there are numerous difficulties that are faced with respect to its handling and large-scale production. Moreover, disadvantages like immunogenicity also limit the use of viral methods extensively (Ledford, 2007). Physical methods are used to directly penetrate the gene into the cells by utilizing fine needle punctures, high pressure gas, or simulations of electrical impulses. The primary physical methods available for gene therapy include biolistics, electroporation, ultrasound, hydrodynamic injection, and jet injection (Gao and Huang, 1995; Naldini et al., 1996; Orio et al., 2012; Panje et al., 2012). Though they have much lesser side effects when compared with viral methods, the physical methods still suffer major disadvantages related to cellular damage, low transfection, necessity of expensive instrumentation, and difficult and labor intensive protocols and manipulations (Robinson and Pertmer, 2001).

Vector-based gene therapy is largely based on the "transformation principle" first reported by British bacteriologist Frederick Griffith in 1928. In this report, the transformation of a non-virulent pneumococcal type into a virulent type was accounted (Griffith, 1928). A few years later, the findings of Frederick Griffith were further established by Dawson and Sia, and a method for achieving this transformation *in vitro* was also developed (Dawson and Sia, 1931). Thereafter, young scientist James L. Alloway took this study one step further and was able to achieve transformation by adding the filtered intracellular extract of the "S" form of pneumococcus to a growing culture of the "R" form of pneumococcus. He concluded that an unknown entity present within the cell free extract was accountable for the transformation. This "something" is called as "transforming principle" (Alloway, 1932). In 1941, Avery and McCarthy were able to isolate this transforming principle and in 1944 they established that transformation is caused by DNA (Avery et al., 1944). This extraordinary development in the field of molecular biology changed the perspective of the understanding of the molecular basis of life. Thereafter, DNA became a subject of passionate research exploration. The chief attractions of the development of gene therapy medicines are depicted in Table 9.1.

Figure 9.1(A) depicts the present state of affairs wherein cancer is the most common disorder treated by gene therapy, followed by monogenetic and cardiovascular diseases (Figure 9.1). Over 1,800 successful clinical trials on gene therapy have been conducted and more are still ongoing, with adenoviral vectors, retroviral vectors, and naked plasmid being the most commonly used gene transfer vectors. Figure 9.1(B) shows the different vectors used in genetic therapy and their percentage in terms of popularity. Gendicine™ (developed by SiBiono Gene Tech Co. in China in 2003) is the brand name of the first gene therapy product to be approved for clinical use. It is prescribed for the treatment of squamous cell carcinoma of the head and neck, and is essentially a non-replicative adenoviral vector wherein the E1 gene is replaced using the human p53 complementary (c)DNA (Peng, 2005; Wilson, 2005). Oncorine™ (developed by Sunway Biotech Co. Ltd. in China in 2005), a genetic therapy product with approval for marketing, is an adenovirus which possesses a conditionally replicative ability. It is used in combination with chemotherapy for the treatment of late stage refractory nasopharyngeal cancer (Heise et al., 1996). Cerepro® (developed by Ark Therapeutics Group plc in 2004) is another gene therapy product which received the first good manufacturing practice

TABLE 9.1

Important Highlights and Milestones in the Development of Genetic Therapy Products and Medicines (Gonçalves, 2008)

Year	Milestone
1928	"Transforming principle" described by Frederick Griffith.
1944	Avery, MacLeod, and McCarthy describe that genetic material is transported in the form of DNA.
1952	Joshua Lederberg introduces transduction as a mechanism of genetic transfer.
1953	The double-helical structure of DNA is described by Watson and Crick.
1961	Howard Temin discovered that virus infection could result in genetic mutation.
1962	First documented genetic transfer in mammalian cell lines performed by Waclaw Szybalski.
1968	Rogers and Pfeuderer demonstrated proof of virus-mediated gene transfer.
1989	First officially approved genetic transfer in humans performed by Steven A. Rosenburg.
1990	First therapeutic gene transfer in adenosine deaminase deficiency (ADA) patients.
1999	Death of Jese Gelsinger (first person publicly identified as having died in a clinical trial for gene therapy).
2003	China approves the first genetic therapy product for clinical use.
2009	First Phase III clinical trial for a genetic therapy-based product conducted in the EU.
2012	For the first time, EMA recommended a genetic therapy product for approval in the EU.

(GMP) certification in the European Union (EU) for the treatment of brain malignant tumors. Cerepro® is an adenoviral vector containing the herpes simplex virus thimidine kinase (HSV-tk) and acts as a pro-drug activating enzyme to change the nucleotide analogue ganciclovir (GCV) to GCV-monophosphate. This GCV-monophosphate is further converted to GCV-diphosphate by the cells' endogenous kinases and eventually manifests into its toxic metabolite GCV-triphosphate. GCV-triphosphate, being cytotoxic, successfully inhibits cancer proliferation by promoting 339 inhibition of the DNA polymerase, thereby preventing

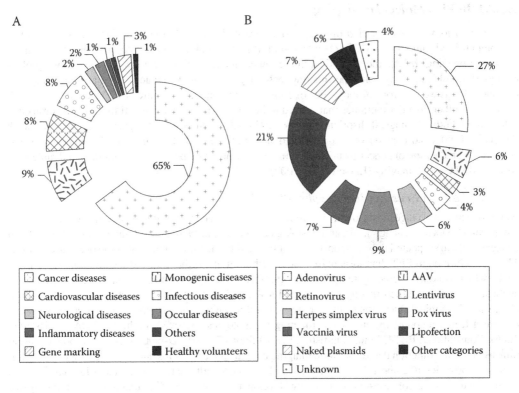

FIGURE 9.1 (A) Different diseases managed and treated using genetic therapy. (B) Different types of vectors utilized for genetic therapy along with their popularity.

DNA replication. In 2008, Cerepro® was the only gene therapy drug that had successfully completed Phase III clinical trials, after successful completion of Phase IIa and IIb clinical trials (Immonen et al., 2004; Moolten and Wells, 1990; Sandmair et al., 2002; Wirth et al., 2006).

Glybera, the first recommended gene therapy product of the European Medicines Agency (EMA) in 2012, is an adenoviral vector proposed for the management of severe deficiencies of lipoprotein lipase. Conceptualised by Amsterdam Molecular Therapeutics, Glybera is now marketed by UniQure, and represents the state of gene therapy drugs and the challenges faced by the same. The Committee on Human Medicinal Products (CHMP), which gives the final recommendations for marketing authorization in the EU, had rejected Glybera three times before passing a verdict in its favor on the fourth time (Ylä-Herttuala, 2012; Wirth et al., 2013).

9.2.2 Molecular Imaging-Based Diagnostics

Molecular imaging technique is a relatively old technique used to diagnose disease progression. Radioactive, as well as recently added fluorescent labeled, contrast agent binds to specific proteins and genes to visualize, quantify, and characterize complex biological progressions at the molecular level in a non-invasive manner. Nuclear imaging was the first of a kind molecular imaging technique to study spatially localized neuro-receptor systems in the human brain with the help of single photon emission computed tomography (SPECT) and positron emission tomography (PET) (Garnett et al., 1983; Wagner et al., 1984). With the advent of time, from being a technique to study neuro-receptor location imaging techniques evolved as begetters of human genome sequencing that caters to the diverse pool of genes and specific proteins which directly or indirectly related to human diseases. With the aid of combinatorial chemistry and mass screening techniques that produced scores of ligands, imaging techniques reached a new height in the area of biological target interaction (Cherry, 2004).

9.2.2.1 In Vivo *Molecular Imaging*

There are various types of *in vivo* imaging modalities which include the like of (1) SPECT, (2) PET, (3) computerised tomography (CT), (4) ultrasound, (5) optical bioluminescence imaging, (6) magnetic resonance imaging (MRI), and (7) optical fluorescence imaging, to name a few. Nanotechnology plays an important role in aiding molecular imaging technology for a quick and accurate diagnosis. Magnetic nanoparticles as well as iron carbonyl (FeCo)/graphitic-shell nanocrystals are new players in cancer cell therapy and other disorder therapies that proved to be a multifunctional, clinical tool for cancerous cell detection and other biological disorientation by MRI contrast enhancement (Hadjipanayis et al., 2010; Seo et al., 2006). Gold-coated iron oxide hybrid nanoparticles are another nano-based technology visualized for the eradication of cancer by utilization in combination with photothermal therapy and molecular specific MRI/optical imaging (Larson et al., 2007).

9.2.2.1.1 Positron Emission Tomography (PET)

Any biochemical change in an organ is followed by a change in physiological activity. PET detects any normal/abnormal change in organ physiological activity, *in vivo*, and non-invasively. Biochemical changes that are suspected to be responsible for altering the functional integrity of the organ are studied. The image formed in PET diagnosis marks the distribution of a radionucleotide within the organ under study, which is administered systematically in the form of a selected, radiologically labeled, pharmaceutical entity prior to the imaging procedure (Ter-Pogossian, 1983). PET is fast emerging as a tool of choice for the clinical diagnosis of cancer staging and restaging for various malignant anomalies and is a chart leader amongst the molecular imaging technologies approved by the US FDA (Weissleder, 2006). Conventional PET/CT imaging using 2-[18F] fluoro-2-deoxy-D-glucose (FDG) produces three-dimensional (3D) functional and anatomic imaging data in cancer patients but is unable to specify malignant transformation (Gambhir, 2002; Petersen et al., 2011). So, alternate specific tumor biomarkers such as lipid nanoparticles or liposomes are being developed and evaluated as diagnostic imaging agents (Goins, 2008). These agents possess immense potential as scintigraphic radiotracers for the rapid detection of metastases and tumors, generate high resolution images, and permit the direct quantification of tissue biodistribution and excretion through blood. Petersen et al. (2011) developed a procedure by which a copper-radionuclide (^{64}Cu) could be remotely loaded and transported across the membranes of preformulated liposomes by using a new ionophore called 2-hydroxyquinoline. This liposomal system can be transported to an encapsulated copper-chelator and can be used as a facilitator for performing PET in healthy and tumor-bearing mice. Though a high concentration of ^{64}Cu helps in generating high resolution images, the liposome coating prohibited the exchange of ^{64}Cu with the biological environment due to the defensive barrier of the liposomes. This ionophore offered exceptionally competent loading and retention stability, that made the ^{64}Cu liposomes extremely appropriate to be used as PET imaging agents. Nanotechnology-based imaging technology has been used not only in oncology, but has also been utilized to diagnose in other disorders, including atherosclerosis.

Nahrendorf et al. (2008) developed a PET detectable imaging agent which was a Diethylenetriamine Pentaacetic Acid (DTPA)-modified, dextranated magnetofluorescent nanoparticle of 20 nm diameter. The nanoparticles in this PET detection system were labeled with the PET tracer ^{64}Cu (1 millicurie per 0.1 mg of nanoparticles). The peak PET activity displayed the uptake of this agent into the arethomata mice deficient in apolipoprotein E after 24 hours of intravenous injection. The arethomata of these mice were found to be proficient with macrophage accumulation, which were detected by Oil Red O staining followed by fluorescence and flow cytometry of the cells. This study also established the aptitude of a nano-agent to correlate PET signals to identify macrophages in atherosclerosis using a conventional biomarker like CD68.

9.2.2.1.2 Single Photon Emission Computed Tomography (SPECT)

SPECT is a type of diagnostic imaging technique in which tomographs of a radionuclide distribution are generated from gamma photons detected at numerous positions about the distribution. SPECT, as routinely performed in nuclear medicine clinics, uses for photon detection/data acquisition an imaging system composed of one or more rotating gamma cameras. In a process known as image reconstruction, tomographs are computed from the data using software that inverts a mathematical model of the data acquisition process. Nanoparticles present significant benefits over conventional molecular imaging using antibodies. The number of modifications that an antibody can accommodate is far lower compared to that which a nanoparticle of a similar size can carry. This is primarily due to the loss of antibody activity with the increase in the number of modifications; a direct result of conformation changes in the protein structure and the consequent loss in the antigen identification sites (Debbage and Jaschke, 2008; Minchin and Martin, 2010; Patel et al., 2012). SPECT is analogous to PET and provides extensive whole-body and quantitative molecular imaging by employing contrast agents, such as ^{177}Lu, ^{99m}Tc, ^{111}In, and ^{123}I. PET and SPECT tracers are also useful beyond the realm of contrast agents and have been utilized for organ and blood vessel characterization, functional imaging, pharmacokinetics, and treatment response (Amen et al., 2008; Brom et al., 2010; Dijkgraaf and Boerman et al., 2010; Ji and Travin, 2010; Valotassiou et al., 2010).

Liposomes, with their unique ability to easily penetrate any cellular barrier, their size, and amphiphilic nature, are highly maneuverable for the targeting of specific ligands (Torchilin, 2007). ^{111}In-DTPA-labelled PEGylated liposomes have been proven to be successful in targeting tumors of the brain, head and neck, breast, and lungs by using SPECT imaging (Harrington et al., 2001). The pharmacokinetic characteristics and biodistribution of dendrimers can be altered by modifying the peripheral groups of the dendrimers, therby paving a path for extensive active as well as passive targeting (Wijagkanalan et al., 2011). Imaging techniques like PET, optical imaging, MRI, CT, PET, and SPECT have used dendrimers as contrast agents (Barrett et al., 2009). A combination of SPECT/CT imaging and human sodium/ iodide symporter (hNIS) ectopic expression in endothelial cells (ECs) have been used to establish the practicability of a highly sensitive, non-invasive, and functional assessment of engineered endothelium in expanded polytetrafluoroethylene (ePTFE) tubes. This methodology may also prove to be useful in conditions where non-invasive, live cell tracking is desired, by manifesting itself as a novel bio-imaging tool for regenerative medicine and additional cell-based therapies (Jiang et al., 2017).

9.2.2.1.3 Optical Bioluminescence Imaging

The phenomenon of emission of visible light from living organisms is termed "bioluminescence". As a tool for the molecular imaging of laboratory animals, the bioluminescence imaging technique has undergone extensive development over the last few decades, facilitating researchers to study different biological processes.

Bioluminescence imaging is executed by introducing and expressing a gene construct to fabricate the protein "luciferase" within the body. This "luciferase" is an enzyme that facilitates the translation of D-luciferin to oxyluciferin with the simultaneous emission of visible light which grants the imaging contrast (Lim et al., 2009; Zhang et al., 2001, 2009). Luciferases obtained from the firefly are routinely used for bioluminescence imaging in animal models. Recently, a study was performed by encapsulating luciferin within a lipid nanocarrier system (Nano-Luc) for prolonged delivery of this substrate within the animal. Nano-Luc was able to achieve significant bioluminescent radiance for over 24 hours by expressing luciferase when administered to tumor-bearing mice either by intravenous, subcutaneous, or intraperitoneal routes. Nanoparticles of D-luciferin-1,1'-Dioctadecyl-3,3,3',3'-tetramethylindotricarbocyanine iodide (Nano-Luc-DiR), therefore, was able to provide new avenues for tumor multimodality imaging and may also serve as a system for the characterization of *in vivo* activities in preclinical studies (Patel et al., 2014).

9.2.2.1.4 Optical Fluorescence Imaging

Optical fluorescence imaging technique utilizes a highly sensitive camera for the detection of fluorescent emissions from fluorophores within the bodies of live organisms. Fluorophores with long emission at the near-infrared (NIR) region, like indocarbocyanine dyes, are generally preferred in these cases to surmount the phenomenon of photon attenuation in living tissue (Rao et al., 2007). Exogenous fluorescent probes are injected into the body which emit light when excited with a suitable wavelength, and can be modified in such a way that it targets and marks specific biological organs for visualization with high contrast imaging via the utilization of suitable optical filters to minimize the auto-fluorescence involvement of the neighboring tissues (Mérian et al., 2012). Usually, fluorescent dyes with a molecular $\leq 1,500$ Da for NIR fluorophores are preferred due to their properly defined chemical structure, along with the added benefit of the ability to customize them to obtain desired chemical and optical properties. The most common organic fluorophore families include rhodamines, BODIPY, indocyanines, porphyrines, and phthalocyanines, but this list is far from comprehensive (Gonçalves, 2008). Small, organic, fluorescent dyes present drawbacks that arise especially when it comes to *in vivo* applications. First, due to their aromatic structure, these molecules are generally poorly soluble in aqueous medium, and are thus poorly bioavailable. This inadequate water solubility issue is chiefly important in the case of NIR dyes, as the addition of pi-conjugated bonds further increases the hydrophobicity of the molecule. Nanotechnologies can benefit in three ways from the design of efficient fluorescent probes dedicated to *in vivo* imaging: (1) by taking benefit of the nanometer-size governed biodistribution of the probe, (2) by designing, in an easy and versatile manner, complex and modular tracers associating different functionalities, and (3) by obtaining brighter fluorescent tracers (Mérian et al., 2012). The intensity of fluorescent dyes can diminish drastically when numerous fluorescent dyes are conjugated to one molecule, for example antibodies, due to dye–dye quenching. On the contrary, the utilization of a viral capsid as a scaffold for labeling enables the loading of more than 40 Cy5 dyes onto a single virus particle by precise chemical coupling. This technique also provides the added advantage of negligible fluorescence quenching due to vast intermolecular distances, thereby synthesizing fluorescent viral nanoparticles which are structurally well defined and possess a diameter of 30 nm (Rao et al., 2007; Soto et al., 2006).

9.2.2.1.5 Magnetic Resonance Imaging (MRI)

MRI has manifested itself as the gold standard in bio-imaging techniques. The protons present in the living tissues that contain proteins, water, lipids, and other macromolecules are imaged in the MRI technique. The application of Larmor, or resonance, frequency causes the protons to excite by the absorption of energy and subsequently "relax" back into their original position, which is determined by T1 and T2 relaxation times. Tumors and other tissues with a large amount of freely mobile water usually appear dark in T1-weighed images but bright in proton-density-weighed and T2-weighed images, thereby providing a thorough anatomical image onto film (Edelman and Warach, 1993). MRI specially finds its use in the visualization of soft tissues, nerves, and discs. Plasmonic layer-coated nanoparticles or gold-coated iron oxide hybrid nanoparticles are the new type of entrants in the realm of MRI/optical imaging and in the photothermal therapy of tumor cells (Larson et al., 2007). The use of iron oxide nanoparticles (IONPs) results in a much larger magnetic moment surrounding each particle, which increases the MRI contrast compared to the use of conventional gadolinium-based contrast agents (Bowen et al., 2002). Due to which IONPs may be coated with many materials with desirable properties without losing their T2 enhancement properties. In addition to being biodegradable and less toxic, the added advantage of using IONPs is that they can evade the immune system and target cancer cells for destruction at the same time as providing MRI contrast (Moore et al., 1997). Though nanotechnology-based imaging is in its nascent stage, thorough research may result in promising outputs.

9.2.2.2 In Vitro *Molecular Imaging*

Nanoparticles are an emerging class of functional materials in the field of medical imaging defined by size-dependent properties. Various methodologies have been developed for the imaging of molecular functions within the cells. One method employs the dye rhodamine 123, which can be used as an indicator of mitochondrial membrane potential, while the other method employs staining with adenosine triphosphate (ATP)-specific bioluminescence dye, which can allow researchers to histochemically observe mitochondrial oxidative phosphorylation (Bose et al., 1998; Hagen et al., 1998; Hata et al., 1998; Holmbom et al., 1993; Kogure and Alonso, 1978; Liszczak et al., 1984). The collective command of live-cell imaging technologies and genetically encoded fluorescent proteins (FPs) allows researchers and clinicians to study the assembly of various protein complexes within cells. These tools can offer significant insight into protein conformational changes, protein–protein interactions, and the behavior of pathway signaling molecules within the cellular structure. Confocal microscopy is one such tool that can visualize and quantify the location of various specific entities in the cell. More extensive techniques like fluorescence lifetime imaging microscopy (FLIM), fluorescence resonance energy transfer (FRET), and fluorescence recovery after photobleaching (FRAP) have also come into being to analyze the ever evolving occurrences within the organelles, sub-organelle components, and within the cell as a whole (Day and Schaufele, 2005; Swift and Trinkle-Mulcahy, 2004).

9.2.2.2.1 *Fluorescence Recovery after Photobleaching (FRAP)*

The FRAP technique analyzes the dynamics of molecular moieties within the cell. The advantage of FRAP over traditional fluorescence microscopy lies in the fact that the former illuminates the dynamics of proteins of interest whereas the later only indicates the localization of molecules in the cells in a quantitative manner. In the FRAP technique, the protein under consideration for investigation must be suitably conjugated to a fluorophore.

To follow the properties of the complex biological membranes, selective and swift molecular replacement over the barrier is very important. Vertically aligned carbon nanofiber (VACNF) forests are one such technique of membrane mimics that has shown tremendous promise due to their compatibility with microfabrication technologies, their stable mechanical properties, and the ability to tailor their morphology and surface properties. FRAP is combined with Monte Carlo simulations and finite differences, which enables it to quantify diffusive transport in VACNF containing microfluidic structures. Recently in a study, FRAP was utilized to verify the fluidity of the membrane surrounding the nanopillars of living cells where clathrin-mediated endocytosis was mediated in plasma membrane that led to inward budding and nanoscale bending by which cells control both membrane protein distribution and the entry of extracellular species (Zhao et al., 2017).

9.2.2.2.2 *Fluorescence Lifetime Imaging Microscopy (FLIM)*

FLIM, unlike the FRAP technique, relies on the behavior of the molecular environment of the fluorophore, which in turn depends on the fluorescent lifetime of the fluorophore. This phenomenon is not dependent on the concentration of the fluorophore, due to which effects at the molecular level in a biological sample can be studied independent of the fluorophore concentration (Becker, 2012). The excitatory lifetime (in nanoseconds) of a particular fluorophore and its spatial movement is charted by FLIM within microscopic images. The FLIM system has been applied both in the frequency as well as time domain. In the frequency domain, the lifetime mapping is carried out by using modulated detectors and excitation light modulated on the basis of sinusoidal intensity, whereas in the time domain, pulsed excitation sources are used along with time-gated or time-correlated detection (van Munster and Gadella, 2005). Even though QD nanoparticles have developed immensely owing to extensive research and development, emission intensity-based techniques have always had the upper hand when it comes to bio-imaging and sensing requirements. Mercaptopropionic acid-capped QD (MPA-QD) nanoparticles, when combined with FLIM methodologies, improved sensitivity more than any other fluorescent dye that is used for pH imaging, using the concept of photoluminescence (PL) lifetime (Orte et al., 2013). Intracellular doxorubicin release, along with theranostic nanoparticle uptake, was analysed utilizing FLIM in a study, which endowed IONPs with the capability to penetrate MCF-7 breast cancer cell lines and H1299 lung cancer cell lines. The availability of an aldehyde functionality offered the attachment of doxorubicin to the nanoparticle shells via imine bonds, thereby facilitating the controlled release of doxorubicin within the acidic environment (Basuki et al., 2013).

9.2.2.2.3 *Fluorescence Resonance Energy Transfer (FRET)*

FRET is an important phenomenon between two molecules where the understanding revolves around the biological system and also finds application in the development of thin film devices and optoelectronics (Holmbom et al., 1993; Liszczak et al., 1984). In optical microscopy, the technique of FRET is applied to determine the advancement between two molecules within several nanometers. Owing to its strong sensitivity in the determination of distance, FRET is utilized to measure interactions at the molecular level. Quantitative as well as qualitative improvements were recently made in the technique, especially in the field of sensitivity, the increment of the spatial resolution, and the distance range (Hussain, 2009). Applying the technology of FRET, several biosensors and novel chemical sensors are being developed out of FRET nanomaterials. In comparison with the small molecule-based traditional FRET system, the advancement of the surface chemistry of nanomaterials has led to the development of multiple probes based on linked recognition molecules such as small-molecule ligands, peptides, and nucleic acids, etc. (Chen at al., 2013). Most recently, a number of materials based on nanotechnology have began to be utilized in FRET assays, including gold nanoparticles (AuNPs), semiconductor QDs, graphene oxide (GO), graphene QDs (GQDs), and upconversion nanoparticles (UCNPs) (Shi et al., 2015). Recently, FRET technology was applied for the detection of vitamin B12, where thermally reduced carbon dots (t-CD) were used for the detection of concentration ranging from 1 to 12 mg/mL and the limit of detection (LOD) was 0.1 mg/mL (Wang et al., 2015). Detecting and imaging tumor-related messenger (m)RNA within living cells helps in the detection of early cancer. But most of the probes which were designed for imaging intracellular mRNA face a lot of intrinsic inferences that arise due to the complex nature of biological matrices that result in the production of false positive signals. Based on FRET technology, this problem is tackled by the development of DNA tetrahedron nanotweezer (DTNT), an intracellular DNA nanoprobe, which is capable of imaging tumor-related mRNA in living cells and works on an "off" to "on" signal read out mode (He et al., 2016).

9.2.3 Chip-Based Molecular Diagnostics

9.2.3.1 *Biochip Technology*

The key strategy to prevent disease progression is rapid diagnosis and thereby treatment (Nayak et al., 2016). Presently, diagnosis for disease biomarkers is performed in centralized laboratories using sophisticated, automated machines. The processing of samples consumes a lot of time as samples need to be transferred to the laboratory and trained staff are required to process the sample, meaning longer waiting times coupled with higher costs (Uludağ et al., 2016). Hence, the development of new diagnostic devices is the need of the hour, devices that would require minimal samples as well as reagent, are rapid, and provide user friendly operation at a low cost. "point-of-care" (POC) diagnostics has been introduced into healthcare as a tool within personalized medicine. POC devices presume that every test can be conducted at or near the site of patient care (McPartlin and O'Kennedy, 2014; Omidfar et al., 2012; Tang et al., 2017). In the clinical laboratory diagnostics setting, the introduction of POC-based, near-patient settings mark a new era in the traditional diagnostics test setup where physicians get timely diagnostic information that enables better diagnosis and quick treatment regimen implementation (Gomez, 2014). Nano-based technology leads to the miniaturization of devices in the area of sensor technology, electronics, and microfluidics application, which enables multiple testing of a variety of analytes. Therefore, biosensor technology has potential in the manufacturing of POC testing devices (Tothill, 2009). Recently, biochip technology has made a tremendous development utilizing various nano products such as CNTs, nanospheres, graphene, and metal nanoparticles, as well as their analogous nanocomposites (Kim et al., 2015; Shivakumar et al., 2014).

9.2.3.1.1 Lab-on-a-Chip

Innovative and portable lab-on-a-chip systems (LOCs) are of tremendous utility in customizing therapeutic algoroithms for POCs. LOCs can be designed to obtain measurements from processing small volumes of complex fluids with effciency and speed, and without the need for an expert operator. This unique set of capabilities addresses all main requirements for POC medical diagnostic systems Yager et al., 2006). Any research which involves miniaturization of the biological and chemical processes is widely termed as LOC technology, although it is not a well-defined scientific term. LOC deals with a broad spectrum of miniaturized systems, for example arrays, microfluidic chips, and sensors (Dittrich and Manz, 2006).

LOC-based microfluidic chips include optical biosensors, electrochemical biosensors, paramagnetic particle biosensors, mass-loading methodology-based biosensors, multiplexing biosensors (Alexander, 2015), fluorescence/chemiluminescence biosensors (Ye et al., 2004), microcantilever biosensors (Hwang et al., 2004; Lee et al., 2005; Wee et al., 2005), and quartz crystal microbalance (QCM) biosensors (Uludag and Tothill, 2010; Zhang et al., 2007). Although publications on the topic of "microfluidics" are rising per year, commercial products based on microfluidics are still very low (Mark et al., 2010) and there are few documented biomarkers validated for routine clinical practice (Poste, 2011). Though there are lots of LOCs in the market, most cease to exist within a few years of their launch. LOCs like Genie II HIV-1/HIV-2, Efoora HIV Rapid, Hema-Strip® HIV 1/2, DoubleCheckGold™ HIV 1/2 are presently not available in the market (HIV Assays, LOC 1). This trend defines the fact that even though a LOC is commercially launched in the market owing to research laboratory setting investigations, it may not survive due to the lack of proper research in real-world scenarios. Other causes, such as knowledge of specific biomarkers and their low concentration and the rigorous sample preparation process, make any disease more difficult for LOCs when compared to antibodies in the blood. Also, many microfluidic companies that have launched their products in the market refrain from publishing their evaluation data in open literature.

9.2.3.1.2 Organ-on-a-Chip

The mere presence of blood flow marks the availability of a dynamic environment along with the added benefit of mechanical stimuli, which cannot be found in usual cell culture-based *in vitro* systems. The cellular microcnvironment in living cells is often heterogeneous, whereas cell culture-based screening systems are usually constructed on homogeneous substrates, which restricts the showcasing of the efficacy of a drug candidate in the human body, where interactions between multiple organs exist (Lee and Sung, 2013). The failure of two-dimensional (2D) cell culture in reconstituting the cellular microenvironment lead to the development of 3D cell culture models, which allow cells to grow within extracellular matrix (ECM) gels (Huh et al., 2011). But 3D cell culture models also failed in some respects, such as the highly variable sizes and shapes of organs and, within the structures, the inability to maintain the positions of cells consistently. The lack of fluid flow in the model has put a question mark on the interaction of cultured cells with the circulating blood cells as well as the immune cells. Organ-on-a-chip systems are micrometer-sized chambers, which are perfused with miniature microfluidic devices for culturing living cells, that mimic the physiological functions of organs and tissues. The aim of this procedure is to develop the minimum number of functional units that can copy tissue systems or an organ as a whole. A prime example of the simplest unit of an organ-on-a-chip model is a perfused, single microfluidic assembly containing a single kind of cultured cell (Bhatia, 2014).

Initially silicon was used for the purpose of microsystems, however it was too costly, involved micro-machining and microfabrication techniques, and was only able to be accessed by specialized engineers. In due course, cost-effective materials such as poly(dimethylsiloxane) (PDMS) (Whitesides et al., 2001) and silicon rubbers were developed, which aid in exploring new opportunities in the field of cellular biology. For survival and other cellular reactions, free-flow oxygen supply is a must. Oxygenators were commonly attached to devices made of glass, silicon, and plastic, but the development of PDMS has eliminated the need for oxygenators due to its high gas permeability, which guarantees sufficient, continuous supply of oxygen to cells in the microchannels. The porosity of PDMS has proven to be advantageous for cell cultures with high metabolic demands, like liver epithelial cells, that are critical for studies in toxicology (Leclerc et al., 2004).

PDMS microfluidic systems also enabled researchers to fabricate feasible and practical human tissues comprised of kidney epithelial cells (Baudoin et al., 2007), epidermal keratinocytes (O'Neill et al., 2008), osteoblasts (Jang et al., 2008; Leclerc et al., 2006), and chondrocytes (Chao et al., 2005) for drug screening and mechanotransduction studies. Microfluidic organ-on-a-chip, although being a superior technology, is not liberated of technical challenges. Fabrication requires specialized micro-engineering capabilities. One of the issues that needs to be tackled is the removal of bubbles, which hinder the fabrication of chips as well as their control and may lead to tissue injury. Although high levels of long-term surviving cells supported by uninterrupted perfusion proved to be beneficial, ECM coatings or gels may pose a threat to matrix deterioration or contraction over time. Apart from the above mentioned issues, some other issues, such as preventing contamination by microbes, attaining consistent and robust seeding of cells in the microfluidic system, and managing cell–cell or cell–ECM interactions, need to be worked out to maintain a precise tissue structure–function relationship (Bhatia, 2014).

Organ-on-a-chip technologies, being relatively new, are in a dearth of further validation and maturation, which will facilitate more efficient drug development via the prospective of predicting clinical responses in humans. This can translate into the production of more profound effects on drug innovation, environmental toxicology testing, and the advancement of artificial organs.

9.2.3.1.3 Human-on-a-Chip

Prior to phase II clinical trials, preclinical screening is done on groups of several hundreds of animals for months or years to investigate the pharmacological profile of a drug candidate for a particular disease. Scientists are exploring translational alternatives to these pharmacological screening models carried out on animals, which may diminish the phylogenetic gap between human beings and laboratory animals and therefore close the gap between human beings and human organs-on-a-chip in terms of bio-similarity. The main objective of an organ-on-a-chip is to mimic normal human physiology and pathology *in vitro* within a controlled system that possesses measurement accessibility of a high caliber. At this point, microfluidics technology is quite developed and mature and has unmatched control over culture conditions that provide stimuli such as time-dependent biochemical stimulations, chemical gradients, mechanical properties, and spatial homogeneity. Due to the miniature size, physical inaccessibility, and lower number of cells, microfluidic cells and tissue culture systems possess analytical challenges. As per the principle, any process of interest or analyte may be detected by any detection technology as long as the technique can be accomplished within the engineering limitations of the cell culture device.

9.2.3.2 Microarray Analysis Techniques

Microarrays are miniaturized devices, very specific and sensitive, that are used to detect human diseases through selected proteins, DNA sequences, and mutated genes (Pedroso and Guillen, 2006). The potential objective of microarrays is to target the selective binding of complementary, single-stranded, nucleic acid sequences. All microarrays share a common principle of arranging the immobilized DNA molecule, possessing a known sequence of an unknown sample although their production is hitherto unknown, hybridized to an order of array. These arrays constitute scores of DNA probe sequences, which are different from each other in that they are arranged in a uniform matrix supported by silicon or glass. With the aid of microarray technology, gene analysis can be done on a large scale as, unlike the normal nucleic acid hybridization, this technique has the ability to simultaneously identify thousands of genes at the same time (Gershon, 2002). Gene microarray technology has provided the scope to integrate thousands of DNA sequences onto a tiny surface, usually a slide made of glass which is often referred to as a "chip". These DNA fragments are arranged in an orderly manner in specific columns and rows in such a way that the fragments can be identified on the array by their location. Tissue microarray (TMA) and gene expression microarray are the two types of microarrays available. The reverse transcriptase-polymerase chain reaction (RT-PCR) and the northern blot can test only a number of genes per experiment, but the microarray has the ability to test thousands of different genes. Apart from the much higher order of magnitude in examining genes than was previously possible, the microarray is also known as the "global expression profile", providing the advantage of selecting the genes of interest independent of the pre-selection influence of genes (Govindarajan et al., 2012). Fluorescence imaging is another commonly used molecular labeling method, but it proved to be very expensive due to its requirements for sophisticated instruments to hybridize, design platforms, as well as interpret images from studies as compared to microarray-based studies. Nanotechnology proved to be a cost-effective technology, sharing similar specificity and similarity to other contemporary techniques involving gene analysis. DNA and the protein functional nanoparticles (FNPs) are new entrants in the field of gene analysis that may be used as hybridization probes in the screening of single nucleotide polymorphisms (SNP), as well as for the detection of biological markers for cardiovascular diseases, cancer, and infection (Pedroso and Guillen, 2006). Superparamagnetic iron oxide nanoparticles (SPIONs) are highly cytotoxic and target cancer cell selective nanoparticles whose molecular mechanisms are elucidated using bioinformatics and DNA microarray techniques (He et al., 2016). Local hyperthermia applied to cancerous cells may alter the permeability of the cell membrane and receptors, which results in the alteration of the cell structure and enzyme activity and may lead to apoptosis of the cells. The impact of heat mostly influences the synthesis (S) phase by inactivating protein replication and damaging the chromatin structure. So to confirm the mechanism of local hyperthermia on cancer cells, a group of researchers, with the help of bioinformatic tools and DNA microarrays, investigated the gene networks and gene expression pattern in oral squamous cell carcinoma (OSCC) HSC-3 cells. Microarray analysis unveiled some 14 genes that were involved in the ongoing biological functioning, such as cellular movement and cell death, including DUSP1, JUN, and ATF3, and a further 13 genes, such as HSPA1B, BAG3, and DNAJB1, were found to be associated with cellular maintenance, assembly, organization, and function (Tabuchi et al., 2012). Another recent work involved the development and characterization of DNA and protein-coated Cyanine5 silica nanoparticles that use DNA microarrays in detection of targets with no nonspecific binding. These newly developed nanoparticles were found to be quite sensitive, having a lower LOD than compared to assays involving free-dye labels. Due to the requirement for very small reagent concentration in these microarray assays, they have been found to be handy in pathogen detection in medical diagnostics involving POC devices (Flynn et al., 2016). Circulating micro (mi)RNAs are emerging as potential blood-based biomarkers for cancer, cardiac disease, and other critical diseases. A microarray platform, developed using the combination of carboxyl-PEG as a functional layer and aminated hairpin nucleic acid molecules as target-specific capture probes (CPs), decreased the detection levels of miRNA to as low as 10 fentometers by inducing an anti-fouling effect, which otherwise targets a relatively huge volume of human serum in an untreated stage. Compared to the commonly used miRNA microarrays, this newly developed technology demands less total assay as well as sample preparation time as it is freed from the process of RNA extraction, target amplification, and labeling (Roy et al., 2016).

To measure the immunoglobulin-E (IgE)-mediated hypersensitivity, testing of the IgE levels in the blood is one of the best parameters in the field of diagnostics. Numerous nanotechnology-based testing tools are devised in the form of nanoparticles either to enhance the magnitude of signal or to employ in capturing analytes. ImmunoCAP Immuno Solid-phase Allergen Chip (ISAC) (developed by ThermoFisher Scientific) is one such microarray technology based on nanotechnology, which is composed of 112 allergens which were collected from various venoms, weed pollens, grasses, trees, food, and protein groups such as nonspecific lipid transfer protein (nsLTP), PR-10, and polcacin, etc. ImmunoCAP ISAC, with approximately 4 hours of assay time, is performed mainly in laboratories for research activities, particularly in studying epidemiology that is highlighted by the yielding, plentiful data attained from the Medall Chip, a modified version of ImmunoCAP ISAC (Märki and Rebeaud, 2017). The development of microarray technology holds good in the field of cancer, especially prostate cancer. The label-free aptasensor is a new microarray-based innovation capable of directly distinguishing prostate specific antigens (PSA), reported to consist of (1) a grafting matrix that immobilizes the short aptamer strands, and (2) a redox transducer using a quinone-containing, conducting copolymer. The detection of PSA is possible owing to its generation of decreasing current above a limit of quantification in the range of ng/mL, which is measured by square wave voltammetry. The current change is utilized in the determination of the PSA–aptamer dissociation constant, KD. To further verify this concept, a competitive exchange, heterogeneous in nature with a cDNA strand, is studied that is responsible for the PSA–aptamer interactions. This double check further cemented the condition of ideally outlined specific recognition (Souada et al., 2015). One of the major public health problems that is emerging nowadays is food borne diseases, which are spreading across the globe both in developed and developing countries. Water and food that are contaminated are the major cause of diarrhea, a symptom of food borne diseases. Utilizing the nanoparticle labeling and multiplex PCR, a recent development was made in the field of gene-based diagnostics where four major food borne pathogens, viz. *Campylobacter jejuni*, *Salmonella enteric*, *Vibrio cholera*, and *Escherichia coli* O157:H7, can be detected simultaneously. For detection of pathogens, the whole procedure was carried out in an oligonucleotide array where the hybridization of PCR products containing 5′biotinylated single strands was done and visualized with the help of streptavidin-coated magnetic nanoparticles (SA-MNPs). The oligonucleotide microarray, utilizing the SA-MNPs as visible labels, was able to differentiate all four pathogens present with a detection sensitivity up to 316 CFU/mL (Li et al., 2013).

9.2.4 Biosensors and Biodevices

9.2.4.1 *Implantable Devices and Bionics*

Implants are usually defined as devices that reinstate any specific body part or act as a part or whole biological structure. Various fields of cardiovascular science, orthopaedics, pacemakers, defibrillators, prosthetics, and DDS use different types of implants (Khan et al., 2014; Regar et al., 2001). However, at times, many induce an immune compromised situation due to their interaction with neighboring tissues, fluids, and proteins, which leads to the development of non-conductive glial tissues (Webster et al., 2003). Selective formation of desired tissues, by employing approaches like the alteration of material coating and the utilization of biodegradable polymers and composites of both synthetic and natural materials, are under constant development to fulfill the aim of obtaining improved next generation biomaterials. Nanomaterials specifically enjoy the benefit of being of a suitable material of selection for bioimplants, due to their similarity with biological organs like bone and the nervous system (Ayad et al., 1998).

Recent advances in the realm of nanotechnology have paved the way for new fields of research related to brain-specific diseases and therapies, which translates into the field of nano-neuroscience, a combined specialty territory of nanotechnology and neuroscience (Kafa et al., 2016). Nano-neuroscience employs the application of novel techniques for the elucidation of cellular and molecular aspects of neural disorders and provides a superior understanding of cellular and molecular cognition and behavior (Fernanda et al., 2016). Nanotechnology can provide answers to the problems pertaining to the inferior success rate and longer development times of CNS drugs, by facilitating the selective penetration of drugs into the BBB, blood–cerebro-spinal fluid barrier (BCSFB), and systemic distribution coupled with clearance. Nanoparticles like albumin nanoparticles, nanoemulsion, SLNs, liposomes, and polymeric nanoparticles are known to have the ability to cross the BBB due to their high plasma protein content and high efflux rate. Additional modifications with PEG, immunoliposomes, or multifunctional nanoparticulates resulting from chimeric proteins have been successful to provide DDS which facilitate longer circulation times, targeting specific receptors of the brain and multifunctional utilities respectively (Shah et al., 2013).

9.2.4.2 *Neurology*

9.2.4.2.1 *Nanoformulation in Neurology*

Liposomes, polymeric micelles, nanoshells, nanotubes, gold nanoparticles, and iron oxides are some of the nanoformuations that are being concurrently used in CNS disorders. Liposomes are miniature, spherical vesicles, fabricated from cholesterol and nontoxic phospholipids, with variable sizes ranging from 0.025 to 2.500 μm and have proven themselves to be very gifted systems for the delivery of drugs and genetic materials (Akbarzadeh et al., 2013). Liposome-based nanoparticles have evolved as a novel plan for the delivery of active ingredients for the treatment of various neuro diseases. A suitable example in this case are nano sterically stabilized liposomes (NSSLs), which are composed of two parts. The first part consists of a water-soluble, amphipathic, weak acid, like a glucocorticosteroid pro-drug, methyl-prednisolone hemisuccinate (MPS), and an amphipathic, weak, base nitroxide called tempamine (TMN) that has been proven to be useful for the treatment for autoimmune encephalomyelitis (EAE) in experimental mice (Turjeman et al., 2015). Similarly, the endothelin B receptor agonist, IRL-1620, encapsulated in PEGylated liposomes has been established to be able to annul neuronal cell death pathways and stimulate neuroprotection by the promotion of vascular and neuronal escalation in murine models. These types of liposomal formulations can be helpful in the treatment of most neurodegenerative diseases, including Alzheimer's disease, wherein neuronal cell death is observed due to the generation of ROS, apoptosis, and oxidative stress. Serum-deprived differentiated PC-12 cells were found to propagate and grow when treated with liposomal IRL-1620, with the increase in expression of the anti-apoptotic marker BCL-2 and the subsequent decrease in that of the pro-apoptotic marker BAX (Joshi et al., 2016).

Polymeric micelles, with sizes ranging from 10 to 100 nm, have also materialized as potential nanocarriers for delivering genetic material, drugs, and imaging agents to tumor cells. Polymeric micelles, due to their small size, can selectively penetrate from the circulatory system into the tumor sites, owing to their superior permeability and retention effect, thereby emerging as the carriers of hydrophobic drugs, leading to improved therapeutic efficacy and circulatory half-life (Mohamed et al., 2014). Polyplex nano-micelles, which are based on the self-assembly of PEG-polyamino acid block copolymers, have been found to be suitable for the administration of brain derived neurotrophic factors (BDNF) expressing mRNA via the intranasal route in mice, thereby promoting normal olfactory epithelial architecture and an extraordinary recovery of functions of the olfactory senses (Baba et al., 2015). Sterically stabilized (PEGylated) phospholipid nanomicelles (SSM) have been found to inhibit the aggregation of β-Amyloid (Aβ), which is a primary principle for the pathogenesis of disease. These PEGylated phospholipids nullify the conversion of Aβ-42 to the amyloidogenic β-sheeted form, thereby eliciting neuroprotection *in vitro* (Pai et al., 2006).

The accidental discovery of CNTs by Sumio Iijima in 1991 marked a new era in the field of nanotechnology. With sizes of around 1 nm, two main types of these miniature miracles of nanotechnology exist, both with superior structural flawlessness (Iijima et al., 1993). On the one hand there are single-walled nanotubes (SWNTs), which are comprised of one graphite sheet wrapped faultlessly into a cylindrical tubular structure, and on the other hand there are multiwalled nanotubes (MWNTs), which consist of an arrangement of such nanotubes laid out in concentric circles (Baughman et al., 2002). All these structures have found their utility in the treatment and management of neural degeneration, cellular transductional imperfections, stroke damage, brain trauma, and tumors (Hajra et al., 2016; Vidu et al., 2014). For example, SWNTs suitably eliminated autophagic substrates from CRND8 mice-derived primary glial cells that superexpress mutant amyloid precursor protein (APP) by the reversal of the anomalous activation of mammalian target of rapamycin (mTOR) signaling and deficits in lysosomal proteolysis (Xue et al., 2014). Studies have proven that MWCNT-anandamide complex has more therapeutic efficiency than that of anandamide (AEA) alone in PC12 cell lines assaulted by prolonged exposure periods to oxygen-glucose deprivation (OGD). AEA alone suffers from major disadvantages like inferior solubility properties and short half-life (Hassanzadeh et al., 2016).

9.2.4.2.2 Nanodevices in Neurology

The most complex entity in the universe is undoubtedly the human brain and, up to this point in time, no functioning theory of the brain has been universally accepted. In order to replicate the human brain, understanding it is the key to success, but due to a lack of proper technology and machinery, a complete human brain map is still under development. Non-invasive and invasive are the only two methods by which brain mapping may be accomplished. The problem with the non-invasive method is that it does not allow scientists to carry out proper mapping of the human brain in live condition. Invasive mapping technology is built around two techniques, brain implants and nanorobots. Neurosurgery is required to place the brain implant, which stimulates and records the function of neurons via telemetry signals, but the safety profile of the whole process is low in both the long or short term. In the year 2014, IBM announced the most advanced brain-like chip to be developed to date called TrueNorth. The efficiency of this brain-like chip is 768 times more than any other contemporary chip ever built (Flores, 2016; Salinas, 2015).

Although nanorobots, or little robots, travel throughout the circulatory system, this technology is still in the nascent stage and needs further research and development. Fabricated nanorobots are injected into the systemic circulation of the patient for therapeutic or diagnostic purposes at the nanoscale level. Recently, a team of researchers developed nanorobots that were employed to deliver drugs in the form of MNPs into the CNS, which disrupted the BBB hypothermically by introducing the MNPs within an air-conditioned field in an experimental mouse model (Aguilar, 2012; Sheikh, 2014; Tabatabaei et al., 2012). In another study, a prototype of a nanorobot was developed, embedded with biosensors to give telemetric control, to transport data for cerebral aneurysm by detecting the overexpression of nitric oxide synthase (NOS). This nanorobot acts through detecting changes in the chemical gradient in the bloodstream and this information is relayed back to detect its position inside the blood vessel and ultimately detect the intracranial aneurysm (Cavalcanti et al., 2009).

A recent technology in the field of nano-neurotechnology is the "Brain-pacing Sticker", which is believed to have revolutionized device miniaturization and interface at the nano-level. This brain-pacing technology, that will be applied for correcting mental disorders, will be armed with a sticker device that carries a feedback-loop of nano-size. This feedback-loop will calculate and correct brain optimization by "feeding back" corrective signals remotely. The main advantage of this technology is that it is non-invasive and the nanoparticles will be delivered to the exact position by intranasal administration (Peled, 2016).

9.2.4.3 Cardiology

According to the World Health Organization (WHO) (2017), 17.7 million people die each year from cardiovascular diseases (CVD), out of which >75% of the deaths occur in low-income and middle-income countries. The same report states that 80% of all CVD deaths are due to heart attacks and strokes, and these account for 31% of all global deaths (WHO, 2017). Until now, various treatment strategies have been employed to tackle CVD using various interdisciplinary scientific backgrounds. The death of cardiomyocytes leads to myocardial infarction, which results in the restriction of blood supply to the surrounding myocardium. Apoptosis of cardiomyocytes activates the myofibroblast and macrophase migration towards the area for cardiac repair. These migrating cells create scar tissue that ultimately reduces the contractile ability of the heart and leads to heart failure (Amezcua et al., 2016). Another major cause of CVD is atherosclerotic plaque that causes heart attacks and strokes, which involves the progressive accumulation of lipid, inflammatory, and cellular material within the artery wall. Although current treatments, which are based on the mechanical removal of the vessel-blockage, have achieved great clinical success, several post-treatment complications such as restenosis and in-stent restenosis, late-stage clotting, etc., are frequently encountered. Additionally, the high treatment cost of current techniques, the sedentary lifestyle, and the ageing population have led researchers to investigate other alternatives for effective, intelligent, and comparatively inexpensive treatments and/or diagnostics (Wang et al., 2016). Most recently, the employment of nanotechnology in mitigating CVD has given many breakthroughs, especially in the fields of polymer-based nano-carriers, nano-scaffolding in tissue engineering and regenerative medicine, nano-patterning, and nano-fabrication, etc.

Cardiomyocytes, which are terminally differentiated, have a very low proliferative potential and may defect or damage one of the four heart valves, leading to heart failure. Correction measures involve replacement surgery by implanting valves from mechanical or biological backgrounds (xenografts or homografts). But engineering the valve is a real challenge for the researcher because it should match the natural one and mimic the contractile and elastic properties. Moreover, the engineered pump should be guaranteed against arrhythmias and dysfunction, which may further prevent the natural motion of blood to entire body parts. The advancement of nanotechnology propelled researchers to design and fabricate compatible scaffolds that mimic the native structure and normal physiology of the heart (Perán et al., 2013). Nanofiber-based scaffolding is the current technology and a challenging task for such tissue engineering processes where scaffolds are synthesized at the nanoscale level. Currently three processes are being employed for the fabrication of nanofibers, viz. phase separation, electrospinning, and self-assembly. Among these three processes, electrospinning has been shown to produce promising results due to the abundance of varieties of synthetic and natural biomaterials (Namdari et al., 2016). Nanofibers prepared by the electrospinning technique can mimic the anisotropic alignment of the native cardiac tissue as well as their nano- or microenvironment, and exceptional mechanical properties provide a better environment for the maturation of cardiacmyocytes (Figure 9.2) (Orlova et al., 2011).

Researchers face a great challenge in mimicking the anisotropic cardiac structure, and therefore designing a 3D cellular orientation of the fabricated scaffolds plays a significant role in cardiac tissue regeneration. The nanofibers yarns network (NFYs-NET), a 3D hybrid scaffold based on the principles of aligned conductive NFYs-NET comprised of CNTs, silk fibroin, and polycaprolactone encased within a hydrogel shell, has proven to be very successful in mimicking the structure of native cardiac tissue. These 3D, hybrid scaffolds formulated by the electrospinning technique have demonstrated ulterior possibilities in controlling cardiac cellular course, augmenting the maturation of cardiomyocytes, along with the ability to promote selective elongation and arrangement of cardiomyocytes as different levels as well as gain access to the individual control of each cellular layers in a 3D setup (Wu et al., 2017).

Engineered nanofiber scaffolds come with some disadvantages, such as intraluminal thrombosis and intimal hyperplasia, which can be tackled by loading these scaffolds with specific factors so that a circulation of endothelial progenitor cells (EPCs) can be achieved, which in turn can promote *in situ* endothelialization. So in view of the critical role played by EPCs, a team of researchers synthesized, fabricated, and characterized a polyurethane (PU) scaffold containing PEG-coated cerium oxide nanoparticles (CNPs-PEG) loaded with antioxidant properties via utilization of the electrospinning technique. The sustained release of CNPs-PEG/vascular endothelial growth factor (VEGF), along with its superior stability and biocompatibility, effectively suppresses EPC apoptosis induced by hydrogen peroxide (H_2O_2). Electrospun PU scaffolds, comprised of a combination of CNPs-PEG and VEGF, assist in *in vitro* endothelialization, and thus pave new avenues in the synthesis of artificial blood vessels of minute diameter (Dai et al., 2017). Currently, researchers are employing nanofibers for coating bare metal stents for eluting drugs. Poly(ε-caprolactone) (PCL) and PU- blended nanofibers coupled with paclitaxel (PTX) in both the shell and the core, fabricated via electrospinning, allowed the controlled release of PTX, which superiorly suppressed the *in vitro* proliferation of L6 cells. This drug-eluting PCL/PU coaxial nanofiber is also expected to drastically reduce restenosis, which is the declarative side effect of surgeries involving the implantation of stents (Son et al., 2017).

In conclusion, mimicking the structure of native tissues is the most challenging task for researchers. Fabricating cardiomyocytes at nano- or micro-scale levels is the most important target in the field of tissue engineering and regenerative medicine. Although some simple nanofiber scaffolds for cardiac graft have been fabricated and applied, multidisciplinary research is needed for more complex cardiovascular tissue application. Uniform vascularization inside the fabricated tissue is the single most important target

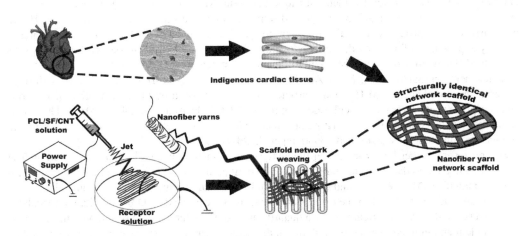

FIGURE 9.2 Schematic representation of the fabrication process of 3D, hybrid, nanofiber yarn, network-based scaffolds using the electrospinning technique.

to be achieved. Scientists need device technology to manipulate cells behavior using new methods so as to minimize graft rejection and post-operative side effects.

9.2.4.4 Orthopedics

Bones, apart from supporting and protecting the internal organs, generate movement by coordinating the muscle and stored blood cells (Hing, 2004). But various natural and accidental conditions, such as osteoporosis and accidental fracture, disrupt the normal architecture of the bone. Numerous treatment modalities are utilized by medical professionals to mitigate such conditions, as well as several materials and drugs that have been developed to treat fractures and other bone disorders. One such technique is orthopedic implants, which are widely used for treating bone related disorders. An orthopedic implant includes partial and total joint replacement, spinal devices, plates, nails, and anchors, etc. Although implants have been proven to be a successful treatment modality, they have some drawbacks too, namely the increasing incidence of failed implant–tissue interaction, aseptic loosening accounts due to inflammatory response to abrasion particles from the articulation, and long-lasting osseointegration or stress-shielding (Streicher et al., 2007). The orthopedic implant industry has grown rapidly over the past decades. An estimated 152,000 hip replacements were performed in 2000, and by the year 2030 a projected quantity of 272,000 per year is estimated. But in 1997, 12.8% of total hip arthroplasties were revision surgeries due to failed hip replacements (Park and Webster, 2005). Another statistic reflects the number of joint implants in the United States to be 600,000 per annum and growing steadily (Garimella and Eltorai, 2017). To counter failed implant–tissue interaction, scientists are continuously investigating novel orthopedic materials which can bring desirable responses from the surrounding cells for better integration and tissue acceptance. Proteins get adsorbed from the surrounding bone marrow, blood. and other tissues before osteoblasts adhere to the implanted surfaces. In this way, proteins control the cell adhesion by getting adsorbed into the surface of the implants. So, nanotechnology best fits to fill-up the gap. A pictorial description of the scopes of nanotechnological products for orthopedic utilization is depicted in Figure 9.3.

Human bone is developed from nano-sized organic and mineral phases, like calcium phosphate crystallites, which are similar to hydroxyapatite (HA) and type I collagen fibers to emulate natural bone structures (Sato and Webster, 2004). HA, a ceramic mineral present in bone, can be synthesized and used for long periods owing to their ionic bonding mechanism that is compatible with osteoblast (or bone-forming cells) function (Parkand Webster, 2005). A team of researchers created nanometer crystalline HA and amorphous calcium phosphate compacts functionalized with the arginine-glycine-aspartic acid (RGD) peptide sequence. This study showed that a decrease in particle size to the nanometer level may promote good osteoblast adhesion (Balasundaram et al., 2006). Most recently, another group of researchers designed HA nanoparticles loaded with folic acid (HapNP-FA) as an effective bone regenerative system, to induce osteoblast differentiation and improve bone regeneration. The fabrication of HA to nanoparticles uses citric acid as a tailoring agent for particle morphology that leads to the grouping of carboxylic pendants followed by the immobilization of folic acid (FA) to form HapNP-FA. The results show the differentiation of osteoblasts, which confirms the potential applicability of HapNP-FA in local bone regeneration (Santos et al., 2017). Another group of researchers utilized green chemistry through mineral-substituted apatite (M-HA) nanoparticles that were prepared by the precipitation of minerals and phosphate reactants in choline chloride-thiourea (ChCl-TU) deep eutectic solvent (DES). The results show excellent biocompatibility, consisting of cell co-cultivation and cell adhesion, *in vivo* according to surgical implantation in Wistar rats (Govindaraj et al., 2017).

Several recent studies have focused on the analysis of nanophase metals in bone regeneration. The metals investigated to date include titanium, alpha-beta titanium alloy Ti6Al4V, and cobalt-chromium-molybdenum (CoCrMo) (Webster and Ejiofor, 2004). Porous titanium implants provide good bone fixation and are widely employed in the orthopedics field. In a recent study, 3D porous titanium implants were investigated for surface treatments via quantification of the interaction of serum proteins and cells with single titanium microbeads (300–500 μm in diameter) where the samples were nanostructured by anodization. 3D porous titanium implants in the form of discs of 2 mm thickness showed better integration and increased hydrophilicity, protein adsorption. and roughness (Rieger et al., 2015).

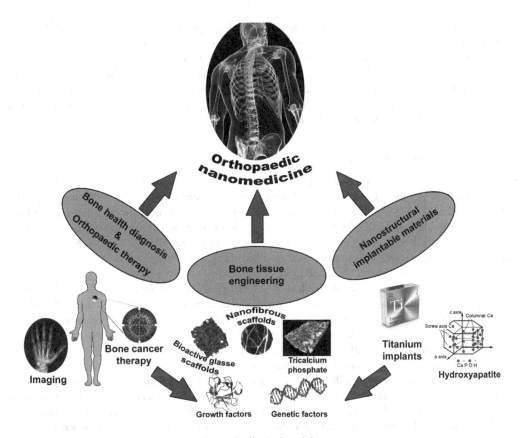

FIGURE 9.3 Scope of nanotechnology for orthopaedic diagnosis and therapy.

Another finding showed, for the first time, the potential of elemental selenium nanocluster, chemopreventive, titanium orthopedic material coating. Elemental selenium is an established chemopreventive agent and when selenite salts were glutathione-reduced in the presence of immersed titanium substrate, elemental selenium was nucleated and formed hemispherical nanoclusters that became an adherent, nanostructured, composite surface. This nanostructured elemental selenium composite surface-inhibited cancerous bone cell proliferation in mouse osteosarcoma cells and showed good adhesion, proliferation, alkaline phosphatase activity, and calcium deposition compared to conventional, untreated titanium (Tran et al., 2010).

Titania nanotubes (TNTs) co-engineered with minimally invasive, drug-releasing, titanium wire implants loaded with specific drugs can be inserted into bones while a simultaneous maintenance of bone viability is carried out using a ZetOSTM bioreactor system. The TNT structure proved to be stable, with adherence to bone cells and acceptable histological architecture observed (Rahman et al., 2016).

It is evident from the findings and studies that nanotechnology will change healthcare and thus human life in the future. But nanotechnology in the field of orthopedics is still in the nascent stage and requires much more research before being able to utilize their full potential. Before promotion commercially, nanostructured orthopedic implants need be cleared of certain drawbacks such as the estheticity of composite resin to ceramic materials, osseointegration with surrounding osteoblasts and proteins, and other regulatory safety concerns.

9.2.5 Nanotechnology in Therapeutics and Surgical Interventions

Nanotechnology finds its way into medicine and surgical interventions via the concepts of nanomedicine and nanosurgery. While the former involves the use of nanoscale DDS and vector-based genetic therapy drugs, the later has opened up new avenues in surgery through the refinement of microsurgical procedures. Nano-biotechnology has been pivotal in the development of biosensor-based nanotechnology that can detect emerging diseases in the body at a stage when they may be curable. Exemplary utility can be mentioned, as in the case of nano-radiotransmitters which are small enough to be put into a cell and nanoacoustic devices which can monitor and record heart beats. However, it is necessary to note that although the clinical application of nanotechnology is rare, there are still promising avenues which are worth mentioning. The applications of nanotherapeutics and nanosurgery in cases of respiratory disease, oncology, wound care, and surgery will be discussed in detail (Omlor et al., 2015).

9.2.5.1 Respiratory Medicine

Nanoparticles are considered a serious candidate when it comes to drug delivery and interventions for treating respiratory disorders. Diseases like lung cancer, tuberculosis, and pulmonary fibrosis can be targeted specifically by the use of nanomedicine. Vector-based gene therapy for the treatment of cystic fibrosis and lung diagnosis with CT or MRI are some of the greatest uses of nanomedicine in the case of respiratory diseases. However, nanoscale respiratory medicine has numerous disadvantages in terms of nanotoxicity, like agglomeration, oxidative stress, genotoxicity, and absorption, the symptoms of which are vastly different from classical toxicity (Omlor et al., 2015).

The applications of nanomedicine in the field of respiratory diseases include drug delivery, hyperthermia, diagnostics, and gene therapy. Due to their size, nanomaterials have the capacity to reach almost any tissue, and the advantage of nanoparticle binding with drug molecules via linker compounds or encapsulation makes them a clear winner when it comes to the treatment of diseases involving the pulmonary system. For example, in the case of pulmonary tumors, the leaky vasculature of the tumor tissues can serve as an advantage to achieve passive targeting of chemotherapeutic-loaded nanoparticles (Hobbs et al., 1998). Pulmonary nanomedicine mostly utilizes this advantage of leaky vasculature, which is also known as the enhanced permeability and retention (EPR) effect, to exert its action. Genoxol-PM, a first-generation type respiratory nanomedicine comprised of a polymeric paclitaxel-loaded poly(lactic acid)-block PEG micelle-formulation, has been tested through phase II trials in patients with advanced non-small cell lung cancer (NSCLC). The response rate was found to be 46.5%, with a lower incidence of emetogenicity. However, adverse events like neutropenia and pneumonia were observed (Ahn et al., 2014; Sanna et al., 2014). Nano-DDS possessing ligands that target specific sites form the second-generation type of respiratory nanomedicine. The ligands can be aptamers, antibodies, proteins, or small molecules. Tumor specific mAb with active targeting capacity are already widely used in cancer therapy. Polyglycolic acid nanoparticles conjugated with cetuximab antibodies for targeting and loaded with the drug paclitaxel palmitate significantly increased the survival rate of mice with A549-luc-C8 lung tumors when compared with the control group (Karra et al., 2013). Aptamers (synthetic oligonucleotides) are another approach by which specific drug targeting can be achieved. Their advantages include specificity, small size, and lack of immunogenicity. Most tumors express a high density of folate and protein receptors. Since folate is a small molecule, it can be effectively used as a ligand in order to achieve drug targeting. Also, the overexpression of protein receptors can be exploited to achieve specific drug targeting by the use of transferrin and similar proteins as ligands in nanomedicine (Libutti et al., 2010). Numerous nanomedicines are already being used clinically for the treatment of lung cancer, like nanocarriers with Aurimmune Cyt-6091 and Bind-014. The former, Aurimmune Cyt-6091, has been used against adenocarcinoma of the lung in a phase I clinical trial (with future plans for phase II trials), and is a DDS based on AuNPs functionalized with PEG and tumor necrosis factor alpha (TNF-α), with TNF-α acting as both a targeting and therapeutic agent. The later, Bind-014, is composed of a PLA core, within which docetaxel is physically entrapped, and is under phase II trials for the treatment of NSCLC. It has been observed that Bind-014 is well tolerated, active, and exerts fewer side effects when compared to solvent-based docetaxel. It has also demonstrated marked effects on patients with KRAS mutations, where ordinary antitumor drugs fail to exert any effects (Biosciences, 2013; Hrkach et al., 2012; Libutti et al., 2010; Natale et al., 2014; Omlor et al., 2015).

Nanomaterials such as FA-modified dendrimer entrapped AuNPs and gadolinium-tetraxetan (also known as DOTA) have surfaced as a boon in the field of pulmonary X-ray diagnosis of lung tumor and MRI diagnosis of lung tissue, respectively. The former have been found to have the ability to gather in the lysosomes of FA receptor-expressing lung adenocarcinoma cells (SPC-A1), serving as probes for lung specific CT imaging. Superior biocompatibility, the absence of adverse effects related to apoptosis, morphology modification, and the disruption of cell cycles are a few of the advantages of these nanomaterials. In murine models, the intratracheal administration of gadolinium-DOTA nanoparticles has been shown to improve the identification of various disorders in organs via the utilization of ultrashort-echo-time-proton-MRI (Bianchi et al., 2013; Wang et al., 2013).

9.2.5.2 Cancer Therapy

Nanomaterials used for cancer therapy should possess a few favorable characteristics, like stability, slow degradation, target specificity, and low chemotherapy-induced toxicity, when compared to conventional anticancer drugs. They should also possess the ability to survive any immunological attack by the normal cells of the patient and should not exert any cytotoxic effects on normal cells. A nanomaterial used for the treatment of cancer may use active or passive targeting of the drug to the tumor cells and will avoid any contact with the normal cells of the body.

Passive targeting, as discussed earlier, utilizes the leaky vasculature of the tumor tissue for increasing the permeability and retention of the drugs at the target site. Leaky and defective lymphatics and blood vasculature always assist in the accumulation and retention of the drugs carried by nanomaterials, which have the highest penetrating power due to their size (Wang and Thanou, 2010). Active tumor targeting, on the other hand, uses mAb against tumor-specific antigens attached to nanoparticles, which facilitate the direct transfer of these nanoparticles into the tumor site. QDs, with their excellent optical properties of high brightness, tunable wavelength, and resistance to photobleaching, are excellent candidates for tumor imaging. QDs are also useful in the early detection of cancer and conjugated QDs are applicable in tumor targeted therapy (Zhang et al., 2008). SWCNT are also used in cases of cancer treatment, as their surface can be modified and conjugated with functional groups that can modulate functional and biological properties. The photochemical destruction of cancer cells is possible using SWCNTs because their large surface area allows for extensive manipulation of the physical and surface dimensions. Their surface area can be engineered to enhance drug efficacy, compatibility, specificity, and dispersibility (Rastogi et al., 2014).

SPIONs are nanoparticles synthesized from iron oxide (Fe_3O_4, magnetite) measuring <20 nm, which exhibit characteristic supermagnetic properties. They find their utility in cancer hyperthermia and thermal surgery. Due to the oscillating magnetic fields of the SPION, heat is generated and thus thermal treatment was found to be possible in animals with prostate cancer implants in peritoneum. The heat did not affect the normal cells of the body. SPION-enhanced MRI imaging can be useful in distinguishing between tumor and normal tissue. Ultra-small supermagnetic iron oxide (USPIO) nanoparticles are known to be taken up by the reactive phagocytic cells at the tumor margin, which could enable them to be used in cases of intracranial tumors. They can also be used as contrast material in MRI and high-resolution ultrasonography (USG) imaging as they show better delineation of tumor margin as they are retained longer in the parenchyma due to their decreased diffusion and phagocytosis (Varallyay et al., 2013). Polymer nanoparticle devices composed of natural or synthetic, biodegradable polymers can be useful in the treatment of cancer. The anticancer drug could be dissolved, adsorbed, encapsulated, or entrapped covalently by linkage to the polymer backbone through simple ester amide bonds that can be hydrolyzed *in vivo* through a change in pH. Natural polymers used for nanomedicine include chitosan, alginate, and gelatin, but synthetic polymers like PGA, PEG, PLA, and their co-polymers have also been used extensively due to their bioavailability. After synthesis of the nanoparticles, the drug of interest can be adsorbed, encapsulated, or conjugated to the structures to achieve specific site targeting, improving kinetic properties or imaging capabilities (Bajpai et al., 2008). Dendrimers are also used in cancer therapy as discussed earlier. The branching generations of the active groups in a dendrimer nanoparticle can provide a large surface area for the attachment of cytotoxic drugs or ligands using an adsorption or conjugation process. Formulators also have the option to store the anticancer drug in the hydrophobic core through hydrophobic interaction, hydrogen bonding, or chemical interaction. Peptide dendrimer-based nanoparticles have a vast applicability as vehicles for drug delivery in cancer treatment or as carriers for moieties of imaging tumor tissues. Dendritic polymers are suitable for a variety of applications, like the early diagnosis of cancer or the delivery of drugs directly into the cancer cells, with PAMAM dendrimers being the most commonly used due to their biodegradability, solubility, and compatibility properties.

They are also greatly modifiable and can be uniformly produced with a diameter of only 5 nm. The G5 PAMAM dendrimer is one such nanodevice dendrimer which has been used to deliver methotreaxate into animal models bearing tumors. The surface charge of the medicated dendrimer was reduced by modifying peripheral amines with acetyl groups. It was then conjugated with methotrexate as a cytotoxic drug and folate as a targeting molecule, which observed good accumulation on FA-expressing experimental tumors (Kukowska-Latallo et al., 2005).

Micelles are biodegradable nanocarriers, formed by the self-assembly of block polymers with hydrophilic and hydrophobic ends, with sizes varying from 10 to 200 nm in diameter. They have hydrophobic inner cores and hydrophilic corona-like shells. The inner core minimizes exposure to the environment and the hydrophilic shell ensures stability by maintaining direct contact with water. The core is ideal for carrying lipophilic drugs, like ether and cremophor EL, with reduced toxicity. The outer hydrophilic shell maintains its stability when the material is in blood and also provides outer functional groups for further modification of the nanomolecule. Micelles can also be altered in such a manner so as to provide timely release of the drugs through biodegradation and pH changes, or to directly deliver drugs to the cancer cells via the conjugation of antibody or peptide ligands, thereby reducing toxicity to the adjoining cells (Oerlemans et al., 2010). Liposomes are another class of drug that are used to deliver anticancer drugs into the body. The major advantages enjoyed by liposomes include ease of preparation, biodegradability, and lower toxicity. Liposomal nanomaterials can be divided into three generations. The first generation (or naked liposome) has an unmodified phospholipid surface, the second generation (or stealth liposome) has a hydrophilic carbohydrate, and the third generation incorporates surface ligands to improve the therapeutic index (Qiu et al., 2008; Torchilin, 2005).

As stated earlier, nanotechnology has been used extensively in tumor imaging for tumour localization as it is the most important step before planning any antitumor therapy. Nano-sized ferromagnetic materials have been used for the accurate localization of tumors. Ferumoxtran-10 is the most extensively used to detect subcentimeter lymph node metastasis (Saksena et al., 2006). CNT-based X-ray imaging is another avenue where nanomaterials are used. Each CNT in CNT-based X-ray imaging serves as an individual electron emitter to yield X-ray images of higher resolution (Cao et al., 2009). Gold nanoparticles (GNPs) have been used in PSAs in 18 prostate cancer patients who were undetectable in conventional laboratory tests. Levels of PSA as low as 330 fg/mL were detected in this novel bio-barcode assay (Thaxton et al., 2009). Real-time monitoring of tumor cells using nanotechnology as a technique for the detection of circulating microvesicles secreted from glioblastoma multiforme tumor cells has already been developed (Hughes et al., 2012). Nanotechnology can also be used as an improved radiosensitizer when compared to conventional agents. An example of such an agent is the hafnium oxide nanoparticle (NBTXR3), which measures just 50 nm in size and has shown promising results as a radiosensitizer in animal tumour xenograft models with two sarcoma and colorectal cancer cell lines. The phase I clinical trial of NBTXR3 is underway for use in the treatment of extremity soft tissue sarcoma and hypopharyngeal cancers. Gold is another material which can be a superb radiosensitizing nanomaterial, due to its inertness and biocompatibility (Maggiorella et al., 2012; Wang and Tepper, 2014). Liposomal doxorubicin is a nanomedicine which is clinical trials for its concurrent use with chemoradiotherapy in the treatment of NSCLC and head and neck cancers, exploiting the fact that nanomaterial can penetrate and accumulate inside the tumor cells due to their leaky vasculature. Liposomal cisplatinum has also been used in cases of head and neck cancers, however skin and mucosal toxicity was not observed to diminish in spite of promising hypothetical claims (Koukourakis et al., 1999; Rosenthal et al., 2002). Nanobrachytherapy is another field which utilizes the scope of nanomaterials to the fullest to directly deliver radiation doses to tumors. Nanoscale radionuclides can be designed to specifically target tumor tissues. A single intratumoral dose of a simple, fabricated, poly radioactive gold (^{198}Au) dendrimer composite nanodevice (CND) was able to deliver a dose of 74 microCi, which resulted in a 45% decrease in tumor volume with no incidence of toxicity (Chatterjee et al., 2013; Khan et al., 2008).

9.2.5.3 Wound Care

The innovative and novel applications of nanomedicine also have an impact on the field of wound care and tissue regeneration. Nanotechnology opens up a plethora of avenues in wound healing due to the advantages of the ability to enter into the cytoplasmic space across cellular barriers, modulate drugs to achieve better bioavailability and biocompatibility, activate repair mechanisms, and increase safety profiles. Nanotechnology also offers the advantage of customizing therapeutic modules to suit the patient's profile. Thus, nanotechnology can be regarded as the answer to improving the therapeutic regimen of impaired wound healing, which is considered a serious health care issue (Medintz et al., 2008; Sandhiya et al., 2009; Suri et al., 2007; Tocco et al., 2012). Nano-biotechnology combined with the knowledge of cellular and subcellular tissue regenerative dysfunctional events can prove to be very beneficial when treating chronic wounds like pressure ulcers and diabetic ulcers, which affect a cumulative estimate of 1.3 to 3 million US individuals and 10 to 15% of 20 million diabetic individuals respectively (Kuehn, 2007; Tocco et al., 2012).

Nanoparticle-bearing endogenous molecules like soluble active proteins (growth factors and cytokines) find their use in the treatment of chronic dehiscent wounds. However, recombinant human platelet-derived growth factor-beta (PDGF-BB) is the only recombinant growth hormone that has been approved by the US FDA, and only for use in diabetic foot ulcers. The clinical use of endogenous factors as pure compounds is limited due to various factors like low plasma half-life caused by the breakdown by the proteolytic enzymes that enrich the wound site. Therefore, the development of nanosystems to increase the efficiency of endogenous compounds like nitric oxide, thrombin, opioids, growth factors, and protease inhibitors, have gained importance in recent times. The conjugation of thrombin with iron oxide nanoparticles (γ-Fe$_2$O$_3$ conjugation) has yielded promising results when tested for wound response in animal models as compared with free-thrombin-treated wounds and untreated wounds (Lenz and Mansson, 1991; Murphy and Nagase, 2009; Senet et al., 2011; Singer and Clark, 1999; Tocco et al., 2012; Werner and Grose, 2003).

Nitric oxide (NO) is a small, free radical that is synthesized by the inflammatory cells in the proliferative phase of wound healing from the amino acid, L-arginine. It has been found to regulate cell proliferation, collagen formation, and wound contraction, and also plays the role of an antimicrobial agent by interfering directly with DNA replication and cell respiration through the inactivation of zinc metalloproteins (Schäffer et al., 1997; Weller et al., 2001). However, NO therapies have been found to have little application in the treatment of cutaneous infection in acute and chronic wounds. Silane hydrogel-based nanotechnology is used to engineer NO to promote the storage and delivery of small, gaseous, short-lived NO free radicals to exert their antimicrobial activity (Englander and Friedman, 2010; Friedman et al., 2008). NO-releasing nano delivery systems have been shown to exhibit strong antimicrobial action against methicillin resistant *Staphylococcus aureus* (MRSA), *Pseudomonas aeruginosa*, and *Acinetobacter baumannii*. NO knockout in experimental models have been found to decrease the rate and effectiveness of wound healing, while the restoration of NO production via the supplementation of arginine or by genetic transfection has been found to reverse the process (Barraud et al., 2009; Han et al., 2010; Martinez et al., 2009; Miller et al., 2009). Therefore, NO-releasing nanoparticles may enhance the wound healing process not only by eliminating bacteria, but also by playing the vital role of angiogenesis promoter and tissue remodeler (Peleg et al., 2008; Yamasaki et al., 1998).

The use of growth factors for wound healing has become a recent topic of interest. Wound healing can be significantly accelerated by the use of various growth factors, like Transforming growth factor-beta (TGF-β), fibroblast growth factor (FGF), PDGF, VEGF, etc., by promoting cell migration, the proliferation of fibroblasts and epithelial cells, the initiation of new blood vessel formation, and, finally, remodelling the scars (Weller et al., 2009). Different types of DDS have been designed from various degradable and non-degradable, natural and synthetic polymers like PLA and PLGA to enhance the *in vivo* efficacy of growth factors, proteolytic degradation, protein structure stability, and prolonged delivery time. Examples include particulate delivery systems like TGF-β-embedded gelatine microparticles, EGF embedded in PLA microspheres, FGF embedded in gelatin microspheres, and PDGF-embedded PLGA nanospheres (Tocco et al., 2012). Newer and more advanced methods for the fabrication of such delivery systems have been proposed which possess the advantages of being environmentally safe and inexpensive. Hyaluronan-based (HYAFF11) porous nanoparticles embedded with PDGF is one such DDS used for the *in vivo* treatment of skin ulcers, designed by the application of such new novel methods of synthesis. This system has the ability to absorb growth factors and release them in a controlled event-based manner along with the timed, localized release of PDGF, which promotes wound healing and tissue regeneration (Knighton et al., 1986; Lepidi et al., 2006; Zavan et al., 2009). Opioids have also been found to play a major role in tissue regeneration via assisting in the migration of keratinocytes, along with the traditional pain-relieving action. Nanoparticulate carriers have been found to increase the skin penetration and delay the release of opioids. Opioids have been found to increase the cell migration and closure of experimental wounds on human keratinocyte-derived cell line HaCaT (Bigliardi et al., 2009; Wolf et al., 2009). Morphine-loaded nanoparticles have been found to increase the re-epithelialization process in human-based, 3D-wound healing models (Küchler et al., 2010).

Proper wound care demands a strict regimen to control infections or sepsis at the wound site, which can be achieved with the use of antibiotics. Nanoparticles carrying antibiotics have been designed for inhibiting the invasion of the wound site by pathological microbes. Vancomycin is one such antibiotic which has been incorporated into nanoparticulate systems for the treatment of *Staphylococcal* infections. Vancomycin-conjugated nanoparticles have also been proposed for the treatment of endophthalmitis via occular instillation (Hachicha et al., 2006). N-methylthiolated β-lactams have been formulated into a poly-acrylate nanoemulsion, where the drug monomer was incorporated into the polymeric matrix, for treatment against *Staphylococcus* bacteria, including MRSA. *In vitro* screenings have declared the product to be nontoxic to human dermal fibroblasts and stable in blood serum for up to 24 hours (Turos et al., 2007). Silver nanoparticles have been found to exert antimicrobial activity since time immemorial and this is still a topic of interest when it comes to silver-based nanotechnology. Silver nanoparticles of 7–20 nm demonstrated the highest amount of antibacterial and antifungal activity when used in combination with traditional antibiotics (Jain et al., 2009). Silver nanoparticles have also been found to reduce cytokine-modulated inflammation via the induction of apoptotic neutrophils and the reduction of matrix metalloproteinases (MMPs) and TGF-β activity to induce the promotion of wound healing, along with the combined effect of the reduction of hypertrophic scarring. Silver nanoparticles are a gold standard in the field of promotion of keratinocyte/fibroblast proliferation, tissue differentiation, and migration, which translates into superior wound healing activities (Widgerow, 2010).

Nanofibrous scaffolds, which are known to possess dramatic similarities to the morphology and structural integrity of the ECM of native tissues, are used by health professionals on a regular basis at present. The scaffolds are known to facilitate cell proliferation, adhesion, and differentiation, thereby promoting wound healing by mimicking the fibrous architecture of the ECM (Tran et al., 2009). Nanofibrous scaffolds of poly(L-lactic acid)-co-poly(ε-caprolactone) (PLACL) treated with plasma (PLACL/plasma) and PLACL/gelatin complexes have been shown to increase fibroblast proliferation and secretion of collagen when treated with plasma as compared to gelatin. Tegaderm-nanofiber (TG-NF) constructs, consisting of nanofibrous scaffolds directly electrospun onto PU dressings (Tegaderm, 3M Medical), have been shown to establish themselves as a suitable candidate for autogenous fibroblast populations by promoting superior cell adhesion, growth, and proliferation when tested in a cell culture environment (Chong et al., 2007). Enhanced regenerative effects can be achieved by designing the scaffolds in such a way that they will act as a delivery system for drugs, cytokines, or growth factors (Yoo et al., 2009). Specifically engineered PLGA microspheres in nanofibrous scaffolds have already proven to be effective in controlling the release of PDGF *in vivo*. This approach has shown promising results in the parameters of cell penetration, vasculogenesis, and tissue neogenesis, thereby providing a technology to accurately control growth factor release to promote soft tissue regeneration *in vivo* (Jin et al., 2008).

Polymeric gene delivery systems are also used in the wound treatment realm where necessary. Genetic material loaded within polymeric systems was observed to have advantages like protection from degradation by nuclease and prolonged action due to controlled release. pFlt23k, an anti-angiogenic DNA, was incorporated into biocompatible and biodegradable PLGA polymers and was shown to reduce the secretion of VEGF by epithelial cells, which is typically useful in the treatment of wound disorders in which VEGF is elevated (Kim et al., 2005; Mayo et al., 2010). Chitosan is another such polymer which has been found to be useful when it comes to the long-term release of drugs loaded on to polymers. Chitosan has been utilized in this regard due to its biochemical properties and compatibility with several new methods of nanoparticle synthesis (Masotti et al., 2008). A highly useful method reported by Masotti and Ortaggi (2009) is based on the osmotic nanofabrication of small DNA-containing chitosan nanospheres (38 ± 4 nm) for various biomedical functions, utilizing synthetic or natural polymers for the fulfillment of their fabrication purpose.

During the last decade, the use of nanomaterials in the isolation, maintenance, and regulation of stem cells has exhibited a dramatic increase. Nano 3D architecture has already been developed, and found to be useful in controlling stem cell proliferation, differentiation, and maturation. Nowadays, angiogenesis promotion can be achieved by the use of VEGF high-expressing, transiently modified stem cells to achieve tissue regeneration. Human VEGF gene incorporated into non-viral, biodegradable polymeric nanoparticles can be transported into human mesenchymal stem cells (hMSCs) and human embryonic stem cell-derived cells (hESdCs) to surmount the tribulations of low cell viability and inadequate expression of angiogenic factors. Nanoscaffolds containing VEGF-expressing stem cells (hMSCs and hESdCs) have been shown to promote an increase in vessel density and enhance angiogenesis and limb restoration, while reducing fibrosis and muscle degeneration (Yang et al., 2010). The use of stem cells in the treatment of wounds is still a new avenue which needs to be explored, albeit it has been proven that stem cells engineered with biodegradable polymer nanoparticles may be useful for attaining higher vascularization in regenerative medicine and the treatment of ischemic diseases.

9.2.5.4 Surgery

Surgeries during the twenty-first century have evolved in such a way so as to minimize trauma both during and post surgery. The evolution of microsurgery using tubular devices (laproscopy) and catheters via small incisions and the use of robotics and implants have marked this process of miniaturization in surgery (Sandhiya et al., 2009).

We are all aware that surgeries performed by robots have already become popular in the field of medicine due to their high precision and functionality. Taking this process further, nanorobots have been developed which can be implanted into the human body with prior programming to act as an independent, on-site surgeon inside the human body. Concepts like *in vivo* corrections of disorders via manipulation of these nanorobots at the cellular level have become possible, which was once considered science-fiction. Ingestible and precisely positionable, endoscopic microcapsules are being developed by scientists at the NanoRobotics Laboratory in the Carnegie-Mellon University (Pittsburgh, PA, US). These microcapsules will have the capability to attach to the gastric lumen and enable physicians to diagnose any illness inside the body via less invasive methods. Prospects are observed in cases where these nanobots can be modified to carry a treatment mechanism so that it can have the double benefit of a diagnostic and treatment agent. The development of nanorobots have also been considered for different parts of the body, as well as for the prophylaxis of cancer. The early detection and personalized treatment of cancer would be possible with such devices, which will facilitate the continuous, accurate analysis of biomarkers *in vivo* (Jain, 2005).

Nanoscale laser surgery is another concept which is becoming increasingly popular due to its ability to target specific structures at the cellular level within the body. Femtosecond (one millionth of a billionth of a second) laser pulses, which are already being used in ocular surgery, are observed as a serious contender for improving the realm of nanosurgery. With the ability to deliver 100 quadrillion W/m^2 of photons in a single pulse of 1 μm width for only 10–15 femtoseconds (fs), these laser surgery methods can perform various precise tasks like disrupting electrons on target, carving channels less than 1 μm within a cell, and selectively targeting strands within genetic material (Sacconi et al., 2005).

9.2.6 Nanotechnology in Disease Control and Occupational Hazards

9.2.6.1 Infectious Disease Control

The re-emergence of virulent infectious diseases is a matter of great concern for the medical fraternity. The history of the pre-antibiotic era was marred with epidemics of infectious agents which claimed the lives of hundreds of thousands of people in various parts of the world. Understanding the microbial causes of these infections was one of the most important milestones of modern medicine for humanity. The discovery of antibiotics drastically brought down the mortality rates associated with infectious diseases. In recent decades, the overuse and irrational, indiscriminate use of antibiotics has detrimentally affected the efficacy of antibiotics and is responsible for the rise of drug resistant strains of microbes and for blunting the effects of antibiotics against microbes (Andersson and Hughes, 2010). With rising antibiotic resistance, the increasing severity of infections, and the paucity of antibiotics to fight against resistance, innovation is the need of the hour to devise novel approaches and newer methods of combating antibiotic resistance. Nanotechnology, owing to its distinct advantages in improving the efficacy of existing drugs, is being seen as an important tool to develop newer means of fighting drug resistance and infections (Burke, 2003). The scope of nanotechnology is also extended to help mankind to fight the infectious diseases as well as control and manage them.

9.2.6.1.1 Metal Nanoparticles

Metal nanoparticles are amongst the earliest and most widely utilized nanoformulations for their antimicrobial action. There are different metals which have been reduced to nano-size and used for antimicrobial action, most notably silver, gold, copper, and iron.

9.2.6.1.1.1 Silver Nanoparticles

Silver is amongst the most potent antimicrobials known since ancient times (Jain et al., 2009). With the advent of nanotechnology, the potency of silver has increased as silver nanoparticles (AgNPs) are amongst the most widely studied nanoparticles. Reviews by Durán et al. (2016), and El Khoury et al. (2015) on AgNP synthesis suggest that the synthesis of AgNPs is based on the chemical reduction process, where a reducing agent such as sodium citrate reduces the silver metal to colloidal silver. Recently, a newer approach of "green" synthesis is gaining popularity as it is considered eco-friendly, cost-effective, and superior to the traditional chemical synthesis methods (Ahmad et al., 2015).

AgNPs are very potent agents against microbes and can effectively fight common infectious diseases. They are widely used topically for the treatment of burns and wounds to prevent microbial contamination. Moreover, recent studies found that some green nanoparticles were effective against extensively drug-resistant (XDR) strains as well (Mapara et al., 2015). Green AgNPs are also proving to be potent against mycobacterium tuberculosis (Singh et al., 2015b).

A recent novel way to effectively utilize the antimicrobial action of AgNPs is by combining them with other antimicrobials, like antibiotics, to provide synergistic antimicrobial action. Several novel molecules are also combined with AgNPs to control their size, shape, cytotoxicity, biodegradability, or biocompatibility and to enhance activity; such agents are capping agents, like polyvinyl alcohol (PVA), or polymers, like chitosan, and are found to increase activity against MDR strains (Cavassin et al., 2015). Thus, AgNPs are in fact most widely used and an effective nano-agent against microbes.

9.2.6.1.1.2 Gold Nanoparticles

Gold is an inert metal and is nontoxic as well. AuNPs can be synthesized either by chemical reduction or by utilizing green synthesis (Suganya et al., 2015). Modified AuNPs, for instance by bifunctionalization with mesoporous silica (MSN) (MSN-AuNPs), have shown to have peroxidase and oxidase-like generating ROS (Tao et al., 2015). The MSN-AuNPs were effective against both Gram-negative (*E. coli*) and Gram-positive (*S. aureus*) bacterial stains. They were also found to inhibit biofilm formation of *Bacillus subtilis* and even managed to breakdown existing biofilm (Vinoj et al., 2015).

Nanoparticles were found to have synergistic effects when combined with other antimicrobial agents. For example, coating AuNPs with N-acylated homoserine lactonase proteins (AiiA-AuNPs) enhanced their antimicrobial effectiveness (Vinoj et al., 2015). Boda et al. (2015) demonstrated an interesting insight into intrinsic antibacterial activity of a particular kind of ultra-small AuNPs with core diameters of 0.8 and 1.4 nm, stabilized with a triphenylphosphine-monosulfonate shell; these nanoparticles have higher bactericidal as well as bacteriostatic activity in very low doses than the conventional nanoparticles, suggesting their high clinical utility.

9.2.6.1.2 Metal Oxides

Metal oxide nanoparticles consist of nanoparticles derived from metal oxides and they too possess high antimicrobial activity. Their antimicrobial action is dependent upon the generation of ROS.

9.2.6.1.2.1 Zinc Oxide (ZnO) Nanoparticles

Zinc oxide (ZnO) nanoparticles today represent one of the major nanoparticles. Like metal nanoparticles, ZnO nanoparticles are effective against bacteria, and also inhibit biofilm formation (Hsueh et al., 2015). Like AgNPs, ZnO nanoparticles cause toxicity via several mechanisms through the release of Zn^{+2} ions, and ROS generation by ZnO nanoparticles producing bacterial toxicity (Arakha et al., 2015). ZnO nanoparticles produced significant antimicrobial action against *V. cholera* and *S. aureus* (Lakshmi Prasanna and Vijayaraghavan, 2015; Sarwar et al., 2016).

9.2.6.1.2.2 Copper (Cu) Nanoparticles

Copper has antimicrobial activity and copper nanoparticles (CuNPs) are found to be effective against animal and plant pathogens. They have activity against microbial biofilm as well as including those of multidrug resistant bacteria (LewisOscar et al., 2015), such as *P. aeruginosa* and MRSA (Kruk et al., 2015; Zhang et al., 2015). In fact, these can be a cheaper alternative to AgNPs and AuNPs for antimicrobial action. CuNPs also inhibit the growth of *Staphylococcal* biofilms (Singh et al., 2015a). The mechanism of activity of CuNPs is lipid peroxidation of cells and chromosomal degradation by ROS generation (Chakraborty et al., 2015).

9.2.6.1.3 Nanocombinations

Combining nanoparticles with other organic and inorganic materials is also an attractive approach to enhance activity, reduce toxicity, and provide specific properties to nanoparticles. For example, combining AgNPs with magnetic Fe_3O_4 is found to enhance their capability to penetrate biofilms. Similarly, combining AgNPs with graphene is found to enhance antibacterial activity and stability (Ghaseminezhad and Shojaosadati, 2016; Ma et al., 2015; Zhang et al., 2016a).

9.2.6.1.4 Nanoparticles as Carriers of Antimicrobial Agents

Apart from metal nanoparticles, which themselves possess antimicrobial activity and are used for the same purpose, many other organic and inorganic nanoparticles are also used as delivery platforms for antimicrobial agents. Nanoparticles can serve as vehicles to deliver antibacterial agents such as antibiotics, and they enhance the activity of antimicrobial agents by protecting them from degradation and by enhancing their cellular penetration and activity. Nanoformulations also provide sustained release of the drug, increasing their potency.

Nanoparticles based on organic polymers, like those of chitosan or lecithin, are biocompatible and nontoxic. They can be loaded with antimicrobial agents. Nanoparticles using chitosan provided deep cellular penetration in treating deep tissue infections of *S. aureus* (Dhanalakshmi et al., 2016). Similarly, coating nanoparticles with lecithin enhanced skin penetration as well as stability in alkaline pH (Sonvico et al., 2006). With rising drug resistance, treating severe bacterial infections, especially biofilm forming bacteria, is becoming challenging. Nanotechnology can effectively tackle this challenge.

SLNs can enable sustained drug delivery over time (Mosallaei et al., 2013). Rifampin-loaded SLNs effectively eradicated *Staphylococcus epidermidis* biofilm with a higher efficacy than free rifampin (Bazzaz et al., 2016). Using conjugation technology and attaching lytic enzymes like deoxyribonuclease (DNase) I, means SLNs can be created which are capable of degrading extracellular DNA, a key factor involved in the development of biofilm (Baelo et al., 2015). Functionalized nanoparticles developed by the conjugation of PLGA with ciprofloxacin and DNase I prevented the formation of *P. aeruginosa* biofilm and also eradicated the established biofilm very effectively (Alipour et al., 2009). Similar results were also obtained with DNase I conjugated with aminoglycoside nanoparticles. Such a smart antibiotic delivery system, capable of controlling antibiotic release and damaging biofilms, has great promise for treating chronic infections, like those involved in cystic fibrosis (Baelo et al., 2015). One of the major applications of nanoparticles is in the delivery of poorly soluble drugs. The effectiveness of many antibiotics is decreased due to the lack of solubility. Delivering such drugs by nanoparticles is an effective way to achieve their therapeutic potency. For example, roxithromycin can be delivered via nanoparticles (Masood et al., 2016), increasing the antibiotic intracellular concentration which may otherwise be subtherapeutic and promote antibiotic resistance (Abed et al., 2015). Achieving higher therapeutic concentrations helps in tackling antibiotic resistance. This premise has been proven by treatment of different drug resistant strains, including MRSA and MDR *E. coli* (i.e., enteroaggregative *E. coli*, enteropathogenic *E. coli*, and extraintestinal pathogenic *E. coli*) (Masood et al., 2016).

9.2.6.2 Radiation Exposure Protection

Radiation is energy which is produced when unstable atoms undergo fragmentation or decay and release energy waves. Radiation, like light, travels through space, and depending on its energy they can either be ionizing radiation or non-ionizing radiation. Ionizing radiation is a high energy radiation capable of knocking out electrons from the atoms they strike. This process of ionization produces ionic particles. They pose a serious risk and can cause severe damage to the DNA of living cells and so damage genes. Generally, ionizing radiation is released by radioactive elements, cosmic particles from outer space, and X-ray machines.

Non-ionizing radiation is also a form of radiation, albeit carrying a lower energy. It has the capability of producing vibrational energy in the molecules, with or without generating heat, but it cannot push electrons out of atoms. Examples of this kind of radiation are radio waves, visible light, and microwaves (United Nations, 2009). Such radiation does not pose serious health risks.

Radiation exposure damage ranges from less severe damage by ultraviolet (UV) radiation from the sun, cosmic radiation or medical radiations such as X-rays, to severe damage ensuing from exposure to nuclear radiation, which can cause genetic damage due to its ionizing effect. As DNA is the storehouse of genetic information needed for the synthesis of biomolecules, like proteins, and for other normal cellular processes, any damage to DNA causes irreversible damage to a cell, which may be to some extent be recovered or can lead to cell death. Further, any damage to DNA of germline cells gets transmitted to the next generation of the species. On exposure to ionizing radiation, some cells may die or become abnormal, either temporarily or permanently, while damage to the genetic material (DNA) can result in cancer. Though our bodies do have an efficient system for repairing cell damage, high exposure to radiation leads to severe irreparable damage. Major ionizing radiation types which can cause cellular damage are discussed below.

There are several types of radiation and different types of radiation have different effects on human health, which are summarized in Table 9.2 (United Nations Scientific Committee on the Effects of Atomic Radiation, 2000; United States Nuclear Regulatory Commission, 2017).

TABLE 9.2

Types of Radiation and Its Health Effects

S. No.	Radiation	Nature and Penetration	Effect on Health
1	X-rays and gamma rays	Like light, they represent energy transmitted in a wave without the movement of material and have a high penetrability.	Decreased white blood cell (WBC) count
2	Alpha particles	They consist of two protons and two neutrons, in the form of atomic nuclei. They thus have a positive electrical charge and are emitted from naturally occurring heavy elements such as uranium and radium. They have low penetration through the intact skin but can enter through cuts and body cavities.	Decreased WBC count, myelosuppression, bone marrow damage, intestinal damage
3	Beta particles	They are fast-moving electrons ejected from the nuclei of many kinds of radioactive atoms. These particles are much smaller than alpha particles and can penetrate up to 1 to 2 cm of water or human flesh.	Decreased WBC count, myelosuppression, bone marrow damage, intestinal damage
4	Cosmic radiation	Consists of very energetic particles, mostly protons, which bombard the Earth from outer space. It is more intense at higher altitudes than at sea level.	Decreased WBC count, myelosuppression, bone marrow damage, intestinal damage
5	Neutrons	They are particles which are also very penetrating. Neutrons can easily penetrate through the human skin and protection can only be achieved using water as they cannot completely pass though water. On Earth they mostly come from the splitting, or fissioning, of certain atoms inside a nuclear reactor.	Decreased WBC count, myelosuppression, bone marrow damage, intestinal damage

In the modern world, there are several sources of radiation exposure. The global average of radiation exposure of a modern human from natural sources is 2.4 mSv per year. Radiation like radio waves from cell phones and microwaves are examples of non-ionizing radiation. There is little evidence of damage from this type of radiation. Occasional exposure to ionizing radiation, such as from medical devices like X-rays, also has a very low propensity of causing any severe health damage. The average diagnostic radiation exposure is 0.4 mSv, which still is well below the safety limits of radiation exposure (US EPA, 2017). Table 9.3 presents different sources of exposure to radiation.

Severe acute exposure, as in the case of nuclear explosion or accidental exposure to highly reactive industrial materials, may lead to single or multiple exposure of very high doses of radiation through external exposure or after entering into the body through wounds, oral intake, or respiratory exposure. Exposure in excess of 75 rad (unit of absorbed dose) in a short amount of time can cause acute health effects like radiation sickness. Radiation sickness is divided into four stages, namely the prodromal stage (N-V-D stage), the latent stage, the manifest illness stage, and recovery or death. While acute exposure to high doses of radiation can produce symptoms of radiation sickness, as discussed above, chronic exposure to low doses of radiation can have a negative effect on health, including damage to the thyroid gland and an increased risk of cancer. Long-term risks associated with exposure to radiation include lung fibrosis, skin burns, hypothyroidism, thyroiditis, cataracts, suppression of ovulation, suppression of sperm count, increased risk of cancers, as well as heritable genetic damage (Centers for Disease Control and Prevention, 2017a; US EPA, 2017).

There are many conventional treatments for acute radiation exposure. Potassium iodide (KI) is one of the therapeutic drugs that blocks the absorption of radioactive iodine and thus can be administered post-radiation exposure to inhibit damage to the thyroid gland, which is the main target of radiation. Prussian blue is another ideal first-line drug for radiation therapy due to its capacity to entrap radioactive molecules, like cesium and thallium, in the intestines. Prussian blue forms complexes with cesium and thallium in intestinal cavities, prevents further reabsorption, and eliminates the complexes through feces, thereby reducing the half-life of cesium and thallium in the body and preventing the damages associated with them. Diethylenetriamine pentaacetate (TPA) is a drug developed specifically for the removal of radioactive plutonium, americium, and curium. TPA can form complexes with these radioactive elements that are eventually excreted out of the body, reducing the residence time and half-life of the specified radioactive elements in the body. Neupogen® (Filgrastim) is a drug indicated in cancer therapy to treat damage associated with chemotherapy, like granulocytopenia and neutropenia. It is capable of boosting the production of white blood cells (granulocytes or neutrophils) and has been successfully used in cancer. Owing to this property, Neupogen is suitable to use in bone marrow damage associated with exposure to high doses of radiation, which causes a fall in blood cell count (Centers for Disease Control and Prevention, 2017a, 2017b).

Several radiation protective agents have been investigated for the protection of normal cells from radiation damage during chemotherapy. Radiation damage during therapy produces high amounts of

TABLE 9.3

Average Levels of Different Radiation Hazards Measured Worldwide

Source	Source Subtype	Worldwide Average (mSv/year)
Natural	Cosmic rays	0.4
	Terrestrial gamma rays	0.5
	Internal exposure	
	Inhalation (mostly via radon)	1.2
	Ingestion	0.3
	Total natural	**2.4**
Artificial	Diagnostic medical	0.4
	Atmospheric nuclear testing	0.005
	Chernobyl accident	0.002
	Nuclear power production	0.0002

Source: US EPA, 2017.

free radicals. Cellular damage from ionizing radiation produces several free radicals that are capable of destroying normal tissues. Once formed, these free radicals can damage DNA and RNA, causing detrimental molecular changes which ultimately causes cell death. Though our body has elaborate defense mechanisms for preventing tissue damage, including enzymes like superoxide dismutase (SOD), glutathione, and metallothionine which scavenge the free radicals or repair damaged cells, sometimes these inherent measures are insufficient. Nanotechnology has modified the paradigms of medicine and drug delivery. It has been proven to reduce drug toxicity, enhance bioavailability, and help in achieving higher cellular concentrations of drugs, providing better therapeutic outcomes. Recently, there have been attempts to utilize the above stated advantages of nanotechnology in the field of radiation protective drugs.

9.2.6.2.1 Nanoparticles of N-(2-mercaptoethyl)1,3-diaminopropane (WR-1065)

Radiation causes severe oxidative stress, and to overcome this several free-radical scavengers were developed for the treatment of radiation-associated damage to prevent mutagenesis. Amifostine is amongst the few that gained clinical approval for radioprotection in different cancer therapies. The active metabolite of amifostine, N-(2-mercaptoethyl)1,3-diaminopropane (WR-1065) suffers from a short half-life. Intravenous (IV) and subcutaneous (SC) administration of amifostine is associated with various drawbacks such as hypotension and nausea. To overcome these problems, Pamujula et al. (2008) developed nanoparticles of WR-1065 by the spray drying technique, using PLGA as the polymer matrix. They determined the radioprotection by measuring reductions in radiation-induced toxicities, survival in 30 days, and other indicators of radiation toxicities including bone marrow suppression and intestinal injury following 9 Gray (Gy) units of whole-body gamma irradiation in mice. The result of their studies indicated that when treatments of nanoparticles of WR-1065 were given 1 hour pre-irradiation and were tested at the dose of 500 mg/kg, there was significant survival of the treated group in comparison to non-treated control mice, and significant radioprotection was demonstrated in the treated group in terms of bone marrow suppression and intestinal injury as well. These findings clearly demonstrate the feasibility of developing an effective oral formulation of WR-1065 as a radioprotective agent (Pamujula et al., 2008).

9.2.6.2.2 Melanin Nanoparticles for Protection of Bone Marrow during Radiation Therapy for Cancer

Melanin is the natural pigment responsible for skin and hair color, which is generally produced on exposure to UV sunlight and has UV-protective action. It possesses free-radical scavenging activity which eliminates free radical generation by radiation exposure. Dr. Dadachova and colleagues showed that melanin mitigates the radiation damage due to its free-radical scavenging activity. Their experiment involved proving prophylactic use of melanin nanoparticles in preventing radiation damage in 24 hour-duration gamma radiation exposed mice. While one group of mice was injected with nanoparticles 3 hours prior to exposure, the other group was left untreated. Post-radiation exposure, both groups were observed for signs of radiation damage, including that of bone marrow damage such as decreased numbers of white blood cells and platelets, for a period of 30 days. While severe symptoms of radiation sickness appeared in the untreated group, with platelet count falling up to 60% below normal, the melanin-treated grouped suffered a fall in platelet count of only 10%. Furthermore, levels of white blood cells and platelets returned to normal much more quickly than in the control mice (Dadachova and Casadevall, 2016; Schweitzer et al., 2010).

9.2.6.2.3 Palladium Lipoic Acid Nanocomplex

Lipoic acid is an excellent antioxidant, which can fight the adverse effects of radiation exposure. Nanoparticles of palladium lipoic acid were developed by El-Marakby et al. (2013) for radio protection. In their experiment, they exposed adult male rats to 6 Gy single dose from a Cs-137 source, followed by measuring the electrical properties of erythrocytes suspension in the frequency range of 40 kHz to 5 MHz, as well as monitoring other haematologically important parameters also. Prophylactic doses of 2 mL/kg of poly-MVA were administered three times a day starting from two weeks prior to exposure and continuing to two weeks post exposure. The results suggest that poly-MVA mitigated and reversed radiation-induced damage, thus proving its utility as an effective radioprophylatic and therapeutic agent (El-Marakby et al., 2013).

9.2.6.2.4 *Chitosan-based, Poly(Sulfobetaine Methacrylate) (PSBMA)-Ferulic Acid Long-circulating Nanocarrier for Radioprotection*

Ferulic acid is an excellent antioxidant, and encapsulating it in chitosan nanoparticles increases its half-life and circulation time in the body. Its use in radioprotection has been attempted by Zhou et al. (2015) and they found them to be effective in providing protection from radiation-induced damage.

9.2.6.2.5 *Radiation Protection with CeO₂ Nanoparticles*

The radioprotective efficacy of cerium oxide (CeO_2) nanoparticles was determined by Tarnuzzer et al. in 2005. As CeO_2 is an excellent antioxidant and free-radical scavenger, its nanoparticles can provide protection from oxidative damage induced by radiation, and thus salvage cells and tissues from severe oxidation and cell death. Their study involved treating human cells with nano-ceria followed by radiation treatment, where a survival rate of up to 99% was achieved in normal cells (Baker, 2011; Tarnuzzer et al., 2005).

9.2.6.2.6 *Cysteine-Protected Molybdenum Sulfate Nanodots with Renal Clearance for Radiation Protection*

Cysteine-protected molybdenum sulfate (MoS_2) dots as radioprotectants were evaluated by Zhang et al. (2010). They synthesized nanodots of sizes less than 5 nm, having a high antioxidant efficacy and radical scavenging activity. They evaluated their radioprotective potential on mice, and a survival rate of up to 79% was achieved in the treated group in comparison to the untreated group, owing to their DNA protective antioxidant effects. Further, the nanoparticles have a good clearance of up to 80% within 24 hours of administration, with no other noticeable toxicity, thereby proving their utility as a safe and effective radioprotectant (Zhang et al., 2016b).

9.2.6.2.7 *NanoDTPA™ Capsules to Treat Radiation Exposure*

NanoDTPA (developed by Talton, Nanotherapeutics, Inc., 2015) is the first nano-product approved by the US FDA for the use in treatment of radiation damage. It is indicated in cases of suspected contamination with plutonium, americium, or curium, and has been granted the status of "orphan drug" by the US FDA, and is used as an alternative to US FDA-approved injectable Ca-DTPA and Zn-DTPA products. The NanoDTPA™ capsule allows the DTPA to be absorbed from the gastrointestinal tract, offering a less invasive route of drug administration (Reddy et al. 2012).

9.3 Future Scope and Prospects

The current chapter deals with the amazing and remarkable realm of nanotechnology in the field of preventive and emergency medicine. Nanotechnology is the fusion of medical science and engineering marvels that leads to unprecedented advances in preventing diseases, managing acute trauma, and envisaging any critical emergency conditions related to human health. Many novel nanotechnological advances are already listed and their benefits have been explained in this chapter previously. To name a few more, we can include abraxane (for the treatment of metastatic breast cancer), doxil (as an anti-neoplastic agent), and emend (as an anti-emetic agent). Despite pathbreaking advances in nanotechnology, few products have made it to the market via approval from the US FDA. The regulatory and toxicological aspects of nanocarriers and nanomedicines make them a matter of concern, hence these products have not been particularly approved for the market. Clinical transparency in the testing of nanoparticles for market release is the need of the hour. Since nanotechnology bears the capability of providing efficient and controlled DDS, it has become essential to introduce cost-effective and safe biomaterials which are fit for clinical applications (Mukherjee et al., 2014).

In comparison to nanomaterials, like CNTs and AuNPs, nanoliposomes are the most researched nanomaterial and possess the highest possibility of entering the market. Genexol-PM, which is an amphiphilic diblock copolymer (PEG-D, L-Lactic acid) conjugated with paclitaxel, is available in the market for the treatment of breast cancer, ovarian cancer, and lung cancer. Genexol-PM is designed to deliver paclitaxel at a controlled rate and is patented by a Korean company by the name of Samyang Biopharmaceuticals Corporation (Samyang Biopharmaceuticals Corporation, 2017). Similarly, amphotericin B containing nanoliposomes are also in the market for the treatment of fungal infections associated with acute leukemia and for central line fungal infections, doxorubicin-loaded nanoliposomes are stealth marketed for hepatocellular carcinoma (Perinelli et al., 2017), and doxorubicin hydrochloride (HCl)-loaded nanoliposome has recently been released in the market for the treatment of ovarian neoplasms (Cancer Research UK, 2017). Nanomechanical devices in the form of nanorobots have become the new approach through which various diseases and disorders can be detected and diagnosed. The means of detection of many diseases, which were a matter of fiction a few years ago, have become possible due to the advent of nanotechnology and nanomedicine. Nanotechnology is expected to play a major role in the near future and provide solutions to medical problems which will comprehensively improve the quality of life. The only aspect in which nanotechnology is lacking is that of regulatory affairs. The toxicity and safety issues that surface in the realm of nanotechnological products represent a major hurdle in their market capitalization via regulatory clearance. Efforts should be made to sensitize academic organizations, government agencies, and industrial establishments to develop and establish credible testing procedures for medical nanodevices and DDS. Several organizations have been setup to execute this tremendous task, which include the likes of the International Council on Nanotechnology (ICON) and the International Organization for Standardization (Geneva, Switzerland). In 1996, the National Nanotechnology Initiative (NNI) was established by the United States, which facilitates the coordination of tasks related to the development of nanoscience and technology within multiple government agencies like the US FDA, the Department of Labor through the Occupational Safety and Health Administration (OSHA), the National Institute for Occupational Safety and Health (NIOSH), and the Environmental Protection Agency (EPA).

In conclusion, it is essential to state that nanotechnology has brought a revolution in the field of medicine, particularly in the area of emergency and preventive healthcare, which has the potential to propel drug therapy strategies to tremendous heights in the near future. Short term and long term toxicological screening of nanodevices and nanomedicines would help to uplift this position and will also provide a path for the large scale market capitalization and utilization of nanotechnology to endow mankind with a utilitarian fortification against multiple diseases and disorders.

REFERENCES

Abed N, Saïd-Hassane F, Zouhiri F, Mougin J, Nicolas V, Desmaële D, Gref R, Couvreur P. An efficient system for intracellular delivery of beta-lactam antibiotics to overcome bacterial resistance. Sci Rep. 2015;5:1–14.

Aguilar ZP. Nanomaterials for medical applications. Newnes, Elsevier Press: Cambridge, Massachusetts; 2012.

Ahmad N, Bhatnagar S, Ali SS, Dutta R. Phytofabrication of bioinduced silver nanoparticles for biomedical applications. Int J Nanomed. 2015;10:7019.

Ahn HK, Jung M, Sym SJ, Shin DB, Kang SM, Kyung SY, Park JW, Jeong SH, Cho EK. A phase II trial of Cremorphor EL-free paclitaxel (Genexol-PM) and gemcitabine in patients with advanced non-small cell lung cancer. Cancer Chemother Pharmacol. 2014;74(2):277–82.

Akbarzadeh A, Rezaei-Sadabady R, Davaran S, Joo SW, Zarghami N, Hanifehpour Y, Samiei M, Kouhi M, Nejati-Koshki K. Liposome: Classification, preparation, and applications. Nanoscale Res Lett. 2013;8(1):102.

Albertazzi L, Gherardini L, Brondi M, Sulis Sato S, Bifone A, Pizzorusso T, Ratto GM, Bardi G. *In vivo* distribution and toxicity of PAMAM dendrimers in the central nervous system depend on their surface chemistry. Mol Pharm. 2012;10(1):249–60.

Alipour M, Suntres ZE, Omri A. Importance of DNase and alginate lyase for enhancing free and liposome encapsulated aminoglycoside activity against *Pseudomonas aeruginosa*. J Antimicrob Chemother. 2009;64(2):317–25.

Allen TM. Ligand-targeted therapeutics in anticancer therapy. Nat Rev Cancer. 2002;2(10):750.

Allen TM, Cullis PR. Drug delivery systems: Entering the mainstream. Science. 2004;303(5665):1818–22.

Alloway JL. The transformation *in vitro* of R pneumococci into S forms of different specific types by the use of filtered pneumococcus extracts. J Exp Med. 1932;55(1):91–9.

Alyautdin R, Gothier D, Petrov V, Kharkevich D, Kreuter J. Analgesic activity of the hexapeptide dalargin adsorbed on the surface of polysorbate 80-coated poly (butyl cyanoacrylate) nanoparticles. Eur J Pharm Biopharm. 1995;41(1):44–8.

Amen DG, Hanks C, Prunella J. Predicting positive and negative treatment responses to stimulants with brain SPECT imaging. J Psychoactive Drugs. 2008;40(2):131–8.

Amezcua R, Shirolkar A, Fraze C, Stout DA. Nanomaterials for cardiac myocyte tissue engineering. Nanomaterials. 2016;6(7):133.

Andersson DI, Hughes D. Antibiotic resistance and its cost: Is it possible to reverse resistance? Nat Rev Microbiol. 2010;8(4):260–71.

Arakha M, Saleem M, Mallick BC, Jha S. The effects of interfacial potential on antimicrobial propensity of ZnO nanoparticle. Sci Rep. 2015;5:1–10.

Avery OT, MacLeod CM, McCarty M. Studies on the chemical nature of the substance inducing transformation of Pneumococcal types. J Exp Med. 1944;79(2):137–58.

Ayad S, Boot-Handford R, Humphries M, Kadler K, Shuttleworth A. The extracellular matrix factsbook. Academic Press: London, UK; 1998.

Baba M, Itaka K, Kondo K, Yamasoba T, Kataoka K. Treatment of neurological disorders by introducing mRNA *in vivo* using polyplex nanomicelles. J Control Release. 2015;201:41–8.

Baelo A, Levato R, Julián E, Crespo A, Astola J, Gavaldà J, Engel E, Mateos-Timoneda MA, Torrents E. Disassembling bacterial extracellular matrix with DNase-coated nanoparticles to enhance antibiotic delivery in biofilm infections. J Control Release. 2015;209:150–8.

Bajpai AK, Shukla SK, Bhanu S, Kankane S. Responsive polymers in controlled drug delivery. Prog Polym Sci. 2008;33(11):1088–118.

Baker CH. Radiation protection with nanoparticles. In: Nanomedicine in health and disease, JH Ross, and RP Victor (eds.), Cambridge University Press: Cambridge, Massachusetts; 2011:268–92.

Balasundaram G, Sato M, Webster TJ. Using hydroxyapatite nanoparticles and decreased crystallinity to promote osteoblast adhesion similar to functionalizing with RGD. Biomaterials. 2006;27(14):2798–805.

Barraud N, Schleheck D, Klebensberger J, Webb JS, Hassett DJ, Rice SA, Kjelleberg S. Nitric oxide signaling in *Pseudomonas aeruginosa* biofilms mediates phosphodiesterase activity, decreased cyclic di-GMP levels, and enhanced dispersal. J Bacteriol. 2009;191(23):7333–42.

Barrett T, Ravizzini G, Choyke PL, Kobayashi H. Dendrimers in medical nanotechnology. IEEE Eng Med Biol Mag. 2009;28(1):12–22.

Basuki JS, Duong HT, Macmillan A, Erlich RB, Esser L, Akerfeldt MC, Whan RM, Kavallaris M, Boyer C, Davis TP. Using fluorescence lifetime imaging microscopy to monitor theranostic nanoparticle uptake and intracellular doxorubicin release. ACS nano. 2013 Oct 16;7(11):10175–89.

Baudoin R, Griscom L, Monge M, Legallais C, Leclerc E. Development of a renal microchip for *in vitro* distal tubule models. Biotechnol Prog. 2007;23(5).1245–53.

Baughman RH, Zakhidov AA, De Heer WA. Carbon nanotubes – The route toward applications. Science. 2002;297(5582):787–92.

Bazzaz BSF, Khameneh B, Zarei H, Golmohammadzadeh S. Antibacterial efficacy of rifampin loaded solid lipid nanoparticles against *Staphylococcus epidermidis* biofilm. Microb Pathog. 2016;93:137–44.

Becker W. Fluorescence lifetime imaging–techniques and applications. Journal of microscopy. 2012 Aug;247(2):119–36.

Bhatia S. Nanoparticles types, classification, characterization, fabrication methods and drug delivery applications. In: Natural polymer drug delivery systems, B. Saurabh (ed.). Springer International Publishing; Basel, Switzerland; 2016. p. 33–93.

Bhatia SN, Ingber DE. Microfluidic organs-on-chips. Nature Biotechnol. 2014;32(8):760–72.

Bianchi A, Lux F, Tillement O, Crémillieux Y. Contrast enhanced lung MRI in mice using ultra-short echo time radial imaging and intratracheally administered Gd-DOTA-based nanoparticles. Magn Reson Med. 2013;70(5):1419–26.

Bigliardi PL, Tobin DJ, Gaveriaux-Ruff C, Bigliardi-Qi M. Opioids and the skin – Where do we stand? Exp Dermatol. 2009;18(5):424–30.

Biosciences BIND. A phase 2 study to determine the safety and efficacy of BIND-014 (docetaxel nanoparticles for injectable suspension) as second-line therapy to patients with non-small cell lung cancer [Internet]. Bethesda, MD: National Library of Medicine (US); 2013 [cited July 2013]. Available from: ClinicalTrials. gov.

Biswal BM, Yusoff Z. Application of nanotechnology in cancer treatment. In: Engineering applications of nanotechnology, S. K. Vishwanatha, and B. H. N. Hisham (eds.). Springer International Publishing: Basel, Switzerland; 2017. p. 269–311.

Boado RJ, Pardridge WM. The Trojan horse liposome technology for nonviral gene transfer across the blood–brain barrier. J Drug Deliv. 2011;2011:1–12.

Boda SK, Broda J, Schiefer F, Weber-Heynemann J, Hoss M, Simon U, Basu B, Jahnen-Dechent W. Cytotoxicity of ultrasmall gold nanoparticles on planktonic and biofilm encapsulated gram-positive *Staphylococci*. Small. 2015;11(26):3183–93.

Borchard G, Audus KL, Shi F, Kreuter J.. Uptake of surfactant-coated poly (methyl methacrylate)-nanoparticles by bovine brain microvessel endothelial cell monolayers. Int J Pharm. 1994;110(1):29–35.

Bose B, Osterholm JL, Berry R. A reproducible experimental model of focal cerebral ischemia in the cat. Brain research. 1984 Oct 8;311(2):385–91.

Bowen CV, Zhang X, Saab G, Gareau PJ, Rutt BK. Application of the static dephasing regime theory to super-paramagnetic iron-oxide loaded cells. Magn Reson Med. 2002;48(1):52–61.

Brom M, Andraojc K, Oyen WJ, Boerman OC, Gotthardt M. Development of radiotracers for the determination of the beta-cell mass *in vivo*. Curr Pharm Des. 2010;16(14):1561–7.

Burke JP. Infection control – A problem for patient safety. N Engl J Med. 2003;348(7):651.

Cancer Research UK. London, UK. Liposomal doxorubicin [Internet]. 2017 [cited 16 November 2017]. Available from: http://www.cancerresearchuk.org/about-cancer/cancer-in-general/treatment/cancer-drugs/drugs/liposomal-doxorubicin.

Cao G, Lee YZ, Peng R, Liu Z, Rajaram R, Calderon-Colon X, An L, Wang P, Phan T, Sultana S, Lalush DS. A dynamic micro-CT scanner based on a carbon nanotube field emission x-ray source. Phys Med Biol. 2009;54(8):2323.

Cavalcanti A, Shirinzadeh B, Fukuda T, Ikeda S. Nanorobot for brain aneurysm. Int J Robotics Res. 2009;28(4):558–70.

Cavassin ED, de Figueiredo LFP, Otoch JP, Seckler MM, de Oliveira RA, Franco FF, Marangoni VS, Zucolotto V, Levin ASS, Costa SF. Comparison of methods to detect the *in vitro* activity of silver nanoparticles (AgNP) against multidrug resistant bacteria. J Nanobiotechnol. 2015;13(1):64.

Centers for Disease Control and Prevention. Acute radiation syndrome: A fact sheet for clinicians. Center for Preparedness and Response (CPR): Georiga, 2017a [cited 16 November 2017]. Available from: https://emergency.cdc.gov/radiation/arsphysicianfactsheet.asp.

Centers for Disease Control and Prevention. Neupogen® (Filgrastim): General information for clinicians Center for Preparedness and Response (CPR): Georiga, 2017b [cited 16 November 2017]. Available from: https://emergency.cdc.gov/radiation/physicianneupogenfacts.asp.

Chakraborty R, Sarkar RK, Chatterjee AK, Manju U, Chattopadhyay AP, Basu T. A simple, fast and cost-effective method of synthesis of cupric oxide nanoparticle with promising antibacterial potency: Unraveling the biological and chemical modes of action. Biochim Biophys Acta. 2015;1850(4):845–56.

Chao PG, Tang Z, Angelini E, West AC, Costa KD, Hung CT. Dynamic osmotic loading of chondrocytes using a novel microfluidic device. J Biomech. 2005;38(6):1273–81.

Chatterjee DK, Wolfe T, Lee J, Brown AP, Singh PK, Bhattarai SR, Diagaradjane P, Krishnan S. Convergence of nanotechnology with radiation therapy – Insights and implications for clinical translation. Transl Cancer Res. 2013;2(4):256.

Chen G, Song F, Xiong X, Peng X. Fluorescent nanosensors based on fluorescence resonance energy transfer (FRET). Industrial & Engineering Chemistry Research. 2013 Feb 8;52(33):11228–45.

Cherry SR. *In vivo* molecular and genomic imaging: New challenges for imaging physics. Phys Med Biol. 2004;49(3):R13.

Chong EJ, Phan TT, Lim IJ, Zhang YZ, Bay BH, Ramakrishna S, Lim CT. Evaluation of electrospun PCL/gelatin nanofibrous scaffold for wound healing and layered dermal reconstitution. Acta Biomaterialia. 2007;3(3):321–30.

Ciolkowski M, Petersen JF, Ficker M, Janaszewska A, Christensen JB, Klajnert B, Bryszewska M. Surface modification of PAMAM dendrimer improves its biocompatibility. NBM. 2012;8(6):815–7.

Courrier HM, Butz N, Vandamme TF. Pulmonary drug delivery systems: Recent developments and prospects. Crit Rev Ther Drug Carrier Syst. 2002;19(4–5):1–64.

Cui Z, Mumper RJ. Microparticles and nanoparticles as delivery systems for DNA vaccines. Crit Rev Ther Drug Carrier Syst. 2003;20:103–37.

Dadachova E, Casadevall A. Oral administration of melanin for protection against radiation. US Patent Application 15/230,617. 2016.

Dai WW, Guo HF, Qian DH, Qin ZX, Lei Y, Hou XY, Wen C. Improving endothelialization by the combined application of polyethylene glycol coated cerium oxide nanoparticles and VEGF in electrospun polyurethane scaffolds. J Mater Chem B. 2017;5(5):1053–61.

Dawson MH, Sia RH. *In vitro* transformation of pneumococcal types: I: A technique for inducing transformation of pneumococcal types *in vitro*. J Exp Med. 1931;54(5):681.

Day RN, Schaufele F. Imaging molecular interactions in living cells. Molecular endocrinology. 2005 Jul 1;19(7):1675–86.

De Boer AG, Van Der Sandt ICJ, Gaillard PJ. The role of drug transporters at the blood–brain barrier. Annu Rev Pharmacol Toxicol. 2003;43(1):629–56.

Debbage P, Jaschke W. Molecular imaging with nanoparticles: Giant roles for dwarf actors. Histochem Cell Biol. 2008;130(5):845–75.

Demeule M, Currie JC, Bertrand Y, Che C, Nguyen T, Regina A, Gabathuler R, Castaigne JP, Beliveau R. Involvement of the low-density lipoprotein receptor-related protein in the transcytosis of the brain delivery vector Angiopep-2. J Neurochem. 2008;106(4):1534–44.

Desai MP, Labhasetwar V, Amidon GL, Levy RJ. Gastrointestinal uptake of biodegradable microparticles: Effect of particle size. Pharm Res. 1996;13(12):1838–45.

Dhanalakshmi V, Nimal TR, Sabitha M, Biswas R, Jayakumar R. Skin and muscle permeating antibacterial nanoparticles for treating *Staphylococcus aureus* infected wounds. J Biomed Mater Res B. 2016;104(4):797–807.

Dhar S, Liu Z, Thomale J, Dai H, Lippard SJ. Targeted single-wall carbon nanotube-mediated Pt (IV) prodrug delivery using folate as a homing device. J Am Chem Soc. 2008;130(34):11467–76.

Dijkgraaf I, Boerman OC. Molecular imaging of angiogenesis with SPECT. Eur J Nucl Med Mol Imaging. 2010;37(1):104–13.

Dittrich PS, Manz A. Lab-on-a-chip: Microfluidics in drug discovery. Nat Rev Drug Discov. 2006;5(3):210.

Diwan M, Elamanchili P, Lane H, Gainer A, Samuel J. Biodegradable nanoparticle mediated antigen delivery to human cord blood derived dendritic cells for induction of primary T cell responses. J Drug Target. 2003;11(8–10):495–507.

Drummond DC, Meyer O, Hong K, Kirpotin DB, Papahadjopoulos D. Optimizing liposomes for delivery of chemotherapeutic agents to solid tumors. Pharmacol Rev. 1999;51(4):691–744.

Dubertret B, Skourides P, Norris DJ, Noireaux V, Brivanlou AH, Libchaber A. *In vivo* imaging of quantum dots encapsulated in phospholipid micelles. Science. 2002;298(5599):1759–62.

Duncan R. The dawning era of polymer therapeutics. Nat Rev Drug Discov. 2003;2(5):347.

Duncan R. Nanomedicines in action. Pharm J. 2004;273:485–8.

Durán N, Durán M, de Jesus MB, Seabra AB, Fávaro WJ, Nakazato G. Silver nanoparticles: A new view on mechanistic aspects on antimicrobial activity. NBM. 2016;12(3):789–99.

Edelman RR, Warach S. Magnetic resonance imaging. N Engl J Med. 1993;328(11):785–91.

El Khoury E, Abiad M, Kassaify ZG, Patra D. Green synthesis of curcumin conjugated nanosilver for the applications in nucleic acid sensing and anti-bacterial activity. Colloids Surf B. 2015;127:274–80.

El-Marakby SM, Selim NS, Desouky OS, Ashry HA, Sallam AM. Prophylaxis and mitigation of radiation-induced changes in the electrical properties of erythrocytes by palladium lipoic acid nano-complex (Poly-MVA). Rom J Biophys. 2013;23:171–90.

Englander L, Friedman A. Nitric oxide nanoparticle technology: A novel antimicrobial agent in the context of current treatment of skin and soft tissue infection. J Clin Aesthetic Dermatol. 2010;3(6):45.

Fernanda Veloz-Castillo M, West RM, Cordero-Arreola J, Arias-Carrión O, Méndez-Rojas MA. Nanomaterials for neurology: State-of-the-art. CNS Neurol Disord Drug Targets. 2016;15(10):1306–24.

Feynman RP. There's plenty of room at the bottom. Eng Sci. 1960;23(5):22–36.

Flores DS. The secret of neuroscience boom: Are there secret human experiments in Latin América? Egypt J Intern Med. 2016;28(1):1.

Flynn SP, Kelleher SM, Acorn JN, Kurzbuch D, Daniels S, McDonagh C, Clancy E, Smith TJ, Nooney R. Ultrasensitive microarray bioassays using cyanine5 dye-doped silica nanoparticles. Nanotechnology. 2016;27(46):465501.

Friedman AJ, Han G, Navati MS, Chacko M, Gunther L, Alfieri A, Friedman JM. Sustained release nitric oxide releasing nanoparticles: Characterization of a novel delivery platform based on nitrite containing hydrogel/glass composites. Nitric Oxide. 2008;19(1):12–20.

Gambhir SS. Molecular imaging of cancer with positron emission tomography. Nat Rev Cancer. 2002;2(9):683.

Gao XZ, Huang LB. Cationic liposome-mediated gene transfer. Gene Ther. 1995;2(10):710–22.

Gao X, Qian J, Zheng S, Xiong Y, Man J, Cao B, Wang L, Ju S, Li C. Up-regulating blood–brain barrier permeability of nanoparticles via multivalent effect. Pharm Res. 2013;30(10):2538–48.

Garimella R, Eltorai AE. Nanotechnology in orthopedics. J Orthop. 2017;14(1):30–3.

Garnett ES, Firnau G, Nahmias C. Dopamine visualized in the basal ganglia of living man. Nature. 1983;305(5930):137–8.

Gaur PK, Mishra S, Bajpai M, Mishra A. Enhanced oral bioavailability of efavirenz by solid lipid nanoparticles: *In vitro* drug release and pharmacokinetics studies. Biomed Res Int. 2014;2014:1–9.

Gershon D. Microarray technology: An array of opportunities. Nature. 2002;416(6883):885–91.

Ghaseminezhad SM, Shojaosadati SA. Evaluation of the antibacterial activity of Ag/Fe3O4 nanocomposites synthesized using starch. Carbohydr Polym. 2016;144:454–63.

Goins BA. Radiolabeled lipid nanoparticles for diagnostic imaging. Expert Opin Med Diagn. 2008;2(7):853–73.

Gomez FA. Paper microfluidics in bioanalysis. Bioanalysis. 2014 Nov;6(21):2911–4.

Gonçalves MST. Fluorescent labeling of biomolecules with organic probes. Chem Rev. 2008;109(1):190–212.

Govindaraj D, Rajan M, Munusamy MA, Alarfaj AA, Sadasivuni KK, Kumar SS. The synthesis, characterization and *in vivo* study of mineral substituted hydroxyapatite for prospective bone tissue rejuvenation applications. NBM. 2017;13(8):2661–2669.

Govindarajan R, Duraiyan J, Kaliyappan K, Palanisamy M. Microarray and its applications. J Pharm Bioallied Sci. 2012;4(Suppl 2):S310.

Griffith F. The significance of pneumococcal types. Epidemiol Infect. 1928;27(2):113–59.

Guo X, Szoka FC. Chemical approaches to triggerable lipid vesicles for drug and gene delivery. Acc Chem Res. 2003;36(5):335–41.

Guth JH. The Mechanism of Asbestos-Induced Carcinogenesis: Calcium and Plasma Membrane Integrity. Paper presented at: The Virginia Academy of Sciences 59th Annual Meeting; 1981; Norfolk, VA.

Hachicha W, Kodjikian L, Fessi H. Preparation of vancomycin microparticles: Importance of preparation parameters. Int J Pharm. 2006;324(2):176–84.

Hadjipanayis CG, Machaidze R, Kaluzova M, Wang L, Schuette AJ, Chen H, Wu X, Mao H. EGFRvIII antibody-conjugated iron oxide nanoparticles for magnetic resonance imaging-guided convection-enhanced delivery and targeted therapy of glioblastoma. Cancer Res. 2010;70(15):6303–12.

Hagen TM, Ingersoll RT, Wehr CM, Lykkesfeldt J, Vinarsky V, Bartholomew JC, Song MH, Ames BN. Acetyl-L-carnitine fed to old rats partially restores mitochondrial function and ambulatory activity. Proceedings of the National Academy of Sciences. 1998 Aug 4;95(16):9562–6.

Hajra A, Bandyopadhyay D, Hajra SK. Future in neuromedicine: Nanotechnology. JNRP. 2016;7(4):613.

Han G, Martinez LR, Mihu MR, Friedman AJ, Friedman JM, Nosanchuk JD. Nitric oxide releasing nanoparticles are therapeutic for *Staphylococcus aureus* abscesses in a murine model of infection. PloS One. 2009;4(11):e7804, 1–7.

Harrington KJ, Mohammadtaghi S, Uster PS, Glass D, Peters AM, Vile RG, Stewart JSW. Effective targeting of solid tumors in patients with locally advanced cancers by radiolabeled pegylated liposomes. Clin Cancer Res. 2001;7(2):243–54.

Hassanzadeh P, Arbabi E, Atyabi F, Dinarvand R. Carbon nanotube-anandamide complex exhibits sustained protective effects in an *in vitro* model of stroke. Physiol Pharmacol. 2016;20(1):12–23.

Hata R, Mies G, Wiessner C, Fritze K, Hesselbarth D, Brinker G, Hossmann KA. A reproducible model of middle cerebral artery occlusion in mice: hemodynamic, biochemical, and magnetic resonance imaging. Journal of Cerebral Blood Flow & Metabolism. 1998 Apr;18(4):367–75.

He C, Jiang S, Jin H, Chen S, Lin G, Yao H, Wang X, Mi P, Ji Z, Lin Y, Lin Z. Mitochondrial electron transport chain identified as a novel molecular target of SPIO nanoparticles mediated cancer-specific cytotoxicity. Biomaterials. 2016;83:102–14.

Heise C, Horn S, Muna M, Ng L, Nye JA. An adenovirus mutant that replicates selectively in p53-deficient human tumor cells. Science. 1996;274:373.

Hing KA. Bone repair in the twenty-first century: Biology, chemistry or engineering? Philos Trans Royal Soc A. 2004;362(1825):2821–50.

Hobbs SK, Monsky WL, Yuan F, Roberts WG, Griffith L, Torchilin VP, Jain RK. Regulation of transport pathways in tumor vessels: Role of tumor type and microenvironment. Proc Natl Acad Sci USA. 1998;95(8):4607–12.

Holmbom B, Näslund U, Eriksson A, Virtanen I, Thornell LE. Comparison of triphenyltetrazolium chloride (TTC) staining versus detection of fibronectin in experimental myocardial infarction. Histochemistry. 1993 Apr 1;99(4):265–75.

Hrkach J, Von Hoff D, Ali MM, Andrianova E, Auer J, Campbell T, De Witt D, Figa M, Figueiredo M, Horhota A, Low S. Preclinical development and clinical translation of a PSMA-targeted docetaxel nanoparticle with a differentiated pharmacological profile. Sci Transl Med. 2012;4(128):128–39.

Hsueh YH, Ke WJ, Hsieh CT, Lin KS, Tzou DY, Chiang CL. ZnO nanoparticles affect *Bacillus subtilis* cell growth and biofilm formation. PLoS One. 2015;10(6):e0128457.

Hughes GA. Nanostructure-mediated drug delivery. NBM. 2005;1(1):22–30.

Hughes AD, Mattison J, Powderly JD, Greene BT, King MR. Rapid isolation of viable circulating tumor cells from patient blood samples. J Vis Exp. 2012;(64):1–5.

Huh D, Hamilton GA, Ingber DE. From 3D cell culture to organs-on-chips. Trends Cell Biol. 2011;21(12):745–54.

Hussain N, Jaitley V, Florence AT. Recent advances in the understanding of uptake of microparticulates across the gastrointestinal lymphatics. Adv Drug Deliv Rev. 2001;50(1):107–42.

Hussain SA. An introduction to fluorescence resonance energy transfer (FRET). arXiv preprint arXiv:0908.1815. 2009 Aug 13.

Hwang KS, Lee JH, Park J, Yoon DS, Park JH, Kim TS. *In situ* quantitative analysis of a prostate-specific antigen (PSA) using a nanomechanical PZT cantilever. Lab Chip. 2004;4(6):547–52.

Iijima S, Ichihashi T. Single-shell carbon nanotubes of 1 nm diameter. Nature. 1993;363(6430):603–5.

Immonen A, Vapalahti M, Tyynelä K, Hurskainen H, Sandmair A, Vanninen R, Langford G, Murray N, Ylä-Herttuala S. AdvHSV-tk gene therapy with intravenous ganciclovir improves survival in human malignant glioma: A randomised, controlled study. Mol Ther. 2004;10(5):967–72.

Ishii T, Asai T, Oyama D, Agato Y, Yasuda N, Fukuta T, Shimizu K, Minamino T, Oku N. Treatment of cerebral ischemia-reperfusion injury with PEGylated liposomes encapsulating FK506. FASEB J. 2013;27(4):1362–70.

Jain KK. Role of nanobiotechnology in developing personalized medicine for cancer. Technol Cancer Res Treat. 2005;4(6):645–50.

Jain J, Arora S, Rajwade JM, Omray P, Khandelwal S, Paknikar KM. Silver nanoparticles in therapeutics: Development of an antimicrobial gel formulation for topical use. 2009;10;6(5):1388–1401.

Jang J, Lim DH, Choi IH. The impact of nanomaterials in immune system. Immune Netw. 2010;10(3):85–91.

Jang K, Sato K, Igawa K, Chung UI, Kitamori T. Development of an osteoblast-based 3D continuous-perfusion microfluidic system for drug screening. Anal Bioanal Chem. 2008;390(3):825–32.

Ji SY, Travin MI. Radionuclide imaging of cardiac autonomic innervation. J Nucl Cardiol. 2010;17(4):655–66.

Jiang B, Wu Y, Haney CR, Duan C, Ameer GA. Assessment of an engineered endothelium via single-photon emission computed tomography. Biotechnol Bioeng. 2017;114(10):2371–8.

Jin Q, Wei G, Lin Z, Sugai JV, Lynch SE, Ma PX, Giannobile WV. Nanofibrous scaffolds incorporating PDGF-BB microspheres induce chemokine expression and tissue neogenesis *in vivo*. PLoS One. 2008;3(3):1–9.

Jin L, Zeng X, Liu M, Deng Y, He N. Current progress in gene delivery technology based on chemical methods and nano-carriers. Theranostics. 2014;4(3):240.

Joshi MD, Oesterling BM, Wu C, Gwizdz N, Pais G, Briyal S, Gulati A. Evaluation of liposomal nanocarriers loaded with ET B receptor agonist, IRL-1620, using cell-based assays. Neuroscience. 2016;312:141–52.

Kafa H, Wang JTW, Jamal KTA. Current perspective of carbon nanotubes application in neurology. Int Rev Neurobiol. 2016;130:229–63.

Kang B, Ceder G. Battery materials for ultrafast charging and disaharging [J]. Nature. 2009;458:190.

Karra N, Nassar T, Ripin AN, Schwob O, Borlak J, Benita S. Antibody conjugated PLGA nanoparticles for targeted delivery of paclitaxel palmitate: Efficacy and biofate in a lung cancer mouse model. Small. 2013;9(24):4221–36.

Kayser O, Lemke A, Hernandez-Trejo N. The impact of nanobiotechnology on the development of new drug delivery systems. Curr Pharm Biotechnol. 2005;6(1):3–5.

Khan MK, Minc LD, Nigavekar SS, Kariapper MS, Nair BM, Schipper M, Cook AC, Lesniak WG, Balogh LP. Fabrication of {198 Au 0} radioactive composite nanodevices and their use for nanobrachytherapy. NBM. 2008;4(1):57–69.

Khan W, Muntimadugu E, Jaffe M, Domb AJ. Implantable medical devices. In: Focal controlled drug delivery. Springer US: Boston, Massachusetts; 2014. p. 33–59.

Kim IS, Lee SK, Park YM, Lee YB, Shin SC, Lee KC, Oh IJ. Physicochemical characterization of poly (L-lactic acid) and poly (D, L-lactide-co-glycolide) nanoparticles with polyethylenimine as gene delivery carrier. Int J Pharm. 2005;298(1):255–62.

Kim MI, Park TJ, Paskaleva EE, Sun F, Seo JW, Mehta KK. Nanotechnologies for biosensor and biochip. J Nanomater. 2015;2015:1–2.

Knighton DR, Ciresi KF, Fiegel VD, Austin LL, Butler EL. Classification and treatment of chronic nonhealing wounds: Successful treatment with autologous platelet-derived wound healing factors (PDWHF). Ann Surg. 1986;204(3):322.

Kolhar P, Doshi N, Mitragotri S. Polymer nanoneedle-mediated intracellular drug delivery. Small. 2011;7(14):2094–100.

Köping-Höggård M, Sánchez A, Alonso MJ. Nanoparticles as carriers for nasal vaccine delivery. Expert Rev Vaccines. 2005;4(2):185–96.

Kostarelos K, Bianco A, Prato M. Promises, facts and challenges for carbon nanotubes in imaging and therapeutics. Nat Nanotechnol. 2009;4(10):627–33.

Kostarelos K, Lacerda L, Pastorin G, Wu W, Wieckowski S, Luangsivilay J, Godefroy S, Pantarotto D, Briand JP, Muller S, Prato M. Cellular uptake of functionalized carbon nanotubes is independent of functional group and cell type. Nat Nanotechnol. 2007;2(2):108–13.

Koukourakis MI, Koukouraki S, Giatromanolaki A, Archimandritis SC, Skarlatos J, Beroukas K, Bizakis JG, Retalis G, Karkavitsas N, Helidonis ES. Liposomal doxorubicin and conventionally fractionated radiotherapy in the treatment of locally advanced non-small-cell lung cancer and head and neck cancer. J Clin Oncol. 1999;17(11):3512–21.

Kruk T, Szczepanowicz K, Stefańska J, Socha RP, Warszyński P. Synthesis and antimicrobial activity of monodisperse copper nanoparticles. Colloids Surf B. 2015;128:17–22.

Küchler S, Wolf NB, Heilmann S, Weindl G, Helfmann J, Yahya MM, Stein C, Schäfer-Korting M. 3D-wound healing model: Influence of morphine and solid lipid nanoparticles. J Biotechnol. 2010;148(1):24–30.

Kuehn BM. Chronic wound care guidelines issued. JAMA. 2007;297(9):938–9.

Kukowska-Latallo JF, Candido KA, Cao Z, Nigavekar SS, Majoros IJ, Thomas TP, Balogh LP, Khan MK, Baker JR. Nanoparticle targeting of anticancer drug improves therapeutic response in animal model of human epithelial cancer. Cancer Res. 2005;65(12):5317–24.

Kumar SR, Vijayalakshmi R. Nanotechnology in dentistry. Indian J Dent Res. 2006;17(2):62–5.

Kuo YC, Wang CC. Cationic solid lipid nanoparticles with cholesterol-mediated surface layer for transporting saquinavir to the brain. Biotechnol Prog. 2014;30(1):198–206.

Lakshmi Prasanna V, Vijayaraghavan R. Insight into the mechanism of antibacterial activity of ZnO: Surface defects mediated reactive oxygen species even in the dark. Langmuir. 2015;31(33):9155–62.

Larson TA, Bankson J, Aaron J, Sokolov K. Hybrid plasmonic magnetic nanoparticles as molecular specific agents for MRI/optical imaging and photothermal therapy of cancer cells. Nanotechnology. 2007;18(32):325101.

Lavan DA, McGuire T, Langer R. Small-scale systems for *in vivo* drug delivery. Nature Biotechnol. 2003;21(10):1184.

Leclerc E, Sakai Y, Fujii T. Microfluidic PDMS (polydimethylsiloxane) bioreactor for large-scale culture of hepatocytes. Biotechnol Prog. 2004;20(3):750–5.

Leclerc E, David B, Griscom L, Lepioufle B, Fujii T, Layrolle P, Legallaisa C. Study of osteoblastic cells in a microfluidic environment. Biomaterials. 2006;27(4):586–95.

Ledford H. HIV vaccine may raise risk. 2007;450(7168):325–325.

Lee JB, Sung JH. Organ-on-a-chip technology and microfluidic whole-body models for pharmacokinetic drug toxicity screening. Biotechnol J. 2013;8(11):1258–66.

Lee JH, Hwang KS, Park J, Yoon KH, Yoon DS, Kim TS. Immunoassay of prostate-specific antigen (PSA) using resonant frequency shift of piezoelectric nanomechanical microcantilever. Biosens Bioelectron. 2005;20(10):2157–62.

Lee CH, Kasala D, Na Y, Lee MS, Kim SW, Jeong JH, Yun CO. Enhanced therapeutic efficacy of an adenovirus-PEI-bile-acid complex in tumors with low coxsackie and adenovirus receptor expression. Biomaterials. 2014;35(21):5505–16.

Lenz GR, Mansson PE. Growth factors as pharmaceuticals. Pharm Technol. 1991;15(1).

Lepidi S, Abatangelo G, Vindigni V, Deriu GP, Zavan B, Tonello C, Cortivo R. *In vivo* regeneration of small-diameter (2 mm) arteries using a polymer scaffold. FASEB J. 2006;20(1):103–5.

Lesniak WG, Mishra MK, Jyoti A, Balakrishnan B, Zhang F, Nance E, Romero R, Kannan S, Kannan RM. Biodistribution of fluorescently labeled PAMAM dendrimers in neonatal rabbits: Effect of neuroinflammation. Mol Pharm. 2013;10(12):4560–71.

LewisOscar F, MubarakAli D, Nithya C, Priyanka R, Gopinath V, Alharbi NS, Thajuddin N. One pot synthesis and anti-biofilm potential of copper nanoparticles (CuNPs) against clinical strains of *Pseudomonas aeruginosa*. Biofouling. 2015;31(4):379–91.

Li S, Liu H, Deng Y, Lin L, He N. Development of a magnetic nanoparticle microarray for simultaneous and simple detection of foodborne pathogens. J Biomed Nanotechnol. 2013;9(7):1254–60.

Li KC, Pandit SD, Guccione S, Bednarski MD. Molecular imaging applications in nanomedicine. Biomed Microdevices. 2004;6(2):113–6.

Li R, Wu RA, Zhao L, Wu M, Yang L, Zou H. P-glycoprotein antibody functionalized carbon nanotube overcomes the multidrug resistance of human leukemia cells. ACS Nano. 2010;4(3):1399–408.

Libutti SK, Paciotti GF, Byrnes AA, Alexander HR, Gannon WE, Walker M, Seidel GD, Yuldasheva N, Tamarkin L. Phase I and pharmacokinetic studies of CYT-6091, a novel PEGylated colloidal gold-rhTNF nanomedicine. Clin Cancer Res. 2010.

Lim E, Modi KD, Kim J. *In vivo* bioluminescent imaging of mammary tumors using IVIS spectrum. J Vis Exp. 2009;(26):1–2.

Liu M, Kono K, Fréchet JM. Water-soluble dendritic unimolecular micelles: Their potential as drug delivery agents. J Control Release. 2000;65(1):121–31.

Luni C, Feldman HC, Pozzobon M, De Coppi P, Meinhart CD, Elvassore N. Microliter-bioreactor array with buoyancy-driven stirring for human hematopoietic stem cell culture. Biomicrofluidics. 2010;4(3):034105.

Ma S, Zhan S, Jia Y, Zhou Q. Highly efficient antibacterial and Pb (II) removal effects of Ag-CoFe$_2$O$_4$-GO nanocomposite. ACS Appl Mater Interfaces. 2015;7(19):10576–86.

Maggiorella L, Barouch G, Devaux C, Pottier A, Deutsch E, Bourhis J, Borghi E, Levy L. Nanoscale radiotherapy with hafnium oxide nanoparticles. Future Oncol. 2012;8(9):1167–81.

Malatesta M, Giagnacovo M, Costanzo M, Conti B, Genta I, Dorati R, Galimberti V, Biggiogera M, Zancanaro C. Diaminobenzidine photoconversion is a suitable tool for tracking the intracellular location of fluorescently labelled nanoparticles at transmission electron microscopy. Eur J Histochem. 2012;56(2):123–128.

Mapara N, Sharma M, Shriram V, Bharadwaj R, Mohite KC, Kumar V. Antimicrobial potentials of *Helicteres isora* silver nanoparticles against extensively drug-resistant (XDR) clinical isolates of *Pseudomonas aeruginosa*. Appl Microbiol Biotechnol. 2015;99(24):10655–67.

Mark D, Haeberle S, Roth G, Von Stetten F, Zengerle R. Microfluidic lab-on-a-chip platforms: Requirements, characteristics and applications. In: Microfluidics based microsystems, Kakac, S., Kosoy, B., Li, D., Pramuanjaroenkij, A (eds.). Dordrecht: Springer; 2010. p. 305–76.

Märki I, Rebeaud F. Nanotechnologies for *in vitro* IgE testing. Curr Allergy Asthma Rep. 2017;17(7):50.

Martin CR, Kohli P. The emerging field of nanotube biotechnology. Nat Rev Drug Discov. 2003;2(1):29.

Martinez LR, Han G, Chacko M, Mihu MR, Jacobson M, Gialanella P, Friedman AJ, Nosanchuk JD, Friedman JM. Antimicrobial and healing efficacy of sustained release nitric oxide nanoparticles against *Staphylococcus aureus* skin infection. J Invest Dermatol. 2009;129(10):2463–9.

Masood F, Yasin T, Bukhari H, Mujahid M. Characterization and application of roxithromycin loaded cyclodextrin based nanoparticles for treatment of multidrug resistant bacteria. Mater Sci Eng C. 2016;61:1–7.

Masotti A, Ortaggi G. Chitosan micro-and nanospheres: Fabrication and applications for drug and DNA delivery. Mini-Rev Med Chem. 2009;9(4):463–9.

Masotti A, Bordi F, Ortaggi G, Marino F, Palocci C. A novel method to obtain chitosan/DNA nanospheres and a study of their release properties. Nanotechnology. 2008;19(5):055302.

Masserini M. Nanoparticles for brain drug delivery. ISRN Biochem. 2013;2013:1–18.

Mayo AS, Ambati BK, Kompella UB. Gene delivery nanoparticles fabricated by supercritical fluid extraction of emulsions. Int J Pharm. 2010;387(1):278–85.

Mazzola L. Commercializing nanotechnology. Nature Biotechnol. 2003;21(10):1137–43.

McPartlin DA, O'Kennedy RJ. Point-of-care diagnostics, a major opportunity for change in traditional diagnostic approaches: Potential and limitations. Expert Rev Mol Diagn. 2014;14(8):979–98.

Medintz IL, Mattoussi H, Clapp AR. Potential clinical applications of quantum dots. Int J Nanomedicine. 2008;3(2):151.

Mérian J, Gravier J, Navarro F, Texier I. Fluorescent nanoprobes dedicated to *in vivo* imaging: From preclinical validations to clinical translation. Molecules. 2012;17(5):5564–91.

Miller C, McMullin B, Ghaffari A, Stenzler A, Pick N, Roscoe D, Ghahary A, Road J, Av-Gay Y. Gaseous nitric oxide bactericidal activity retained during intermittent high-dose short duration exposure. Nitric oxide. 2009;20(1):16–23.

Minchin RF, Martin DJ. Minireview: Nanoparticles for molecular imaging – An overview. Endocrinology. 2010;151(2):474–81.

Mishra MK, Beaty CA, Lesniak WG, Kambhampati SP, Zhang F, Wilson MA, Blue ME, Troncoso JC, Kannan S, Johnston MV, Baumgartner WA. Dendrimer brain uptake and targeted therapy for brain injury in a large animal model of hypothermic circulatory arrest. ACS Nano. 2014;8(3):2134–47.

Mohamed S, Parayath NN, Taurin S, Greish K. Polymeric nano-micelles: Versatile platform for targeted delivery in cancer. Ther Deliv. 2014;5(10):1101–1121.

Moolten FL, Wells JM. Curability of tumors bearing herpes thymidine kinase genes transfered by retroviral vectors. JNCI. 1990;82(4):297–300.

Moore A, Weissleder R, Bogdanov A. Uptake of dextran-coated monocrystalline iron oxides in tumor cells and macrophages. J Magn Reson Imaging. 1997;7(6):1140–5.

Mosallaei N, Jaafari MR, Hanafi-Bojd MY, Golmohammadzadeh S, Malaekeh-Nikouei B. Docetaxel-loaded solid lipid nanoparticles: Preparation, characterization, *in vitro*, and *in vivo* evaluations. J Pharm Sci. 2013;102(6):1994–2004.

Mukherjee B, Dey NS, Maji R, Bhowmik P, Das PJ, Paul P. Current status and future scope for nanomaterials in drug delivery. In: Application of nanotechnology in drug delivery. InTech; 2014.

Mulligan RC. The basic science of gene therapy. Science. 1993;260:926–32.

Murphy G, Nagase H. Progress in matrix metalloproteinase research. Mol Aspects Med. 2008;29(5):290–308.

Nagpal K, Singh SK, Mishra DN. Nanoparticle mediated brain targeted delivery of gallic acid: *In vivo* behavioral and biochemical studies for improved antioxidant and antidepressant-like activity. Drug Deliv. 2012;19(8):378–91.

Nahrendorf M, Zhang H, Hembrador S, Panizzi P, Sosnovik DE, Aikawa E, Libby P, Swirski FK, Weissleder R. Nanoparticle PET-CT imaging of macrophages in inflammatory atherosclerosis. Circulation. 2008;117(3):379–87.

Naldini L, Blomer U, Gallay P, Ory D, Mulligan R, Gage FH, Verma IM, Trono D. *In vivo* gene delivery and stable transduction of nondividing cells by a lentiviral vector. Science. 1996;272(5259):263–7.

Namdari M, Eatemadi A. Nanofibrous bioengineered heart valve – Application in paediatric medicine. Biomed Pharmacother. 2016;84:1179–88.

Natale R, Socinski M, Hart L, Lipatov O, Spigel D, Gershenhorn B, Weiss G, Kazmi S, Karaseva N, Gladkov O, Moiseyenko V. Clinical activity of BIND-014 (docetaxel nanoparticles for injectable suspension) as second-line therapy in patients (pts) with stage III/IV non-small cell lung cancer. Eur J Cancer. 2014;50:19.

Nayak S, Sridhara A, Melo R, Richer L, Chee NH, Kim J, Linder V, Steinmiller D, Sia SK, Gomes-Solecki M. Microfluidics-based point-of-care test for serodiagnosis of Lyme Disease. Sci Rep. 2016;6:35069.

Nyitrai G, Héja L, Jablonkai I, Pál I, Visy J, Kardos J. Polyamidoamine dendrimer impairs mitochondrial oxidation in brain tissue. J Nanobiotechnol. 2013;11(1):9.

Obataya I, Nakamura C, Han S, Nakamura N, Miyake J. Nanoscale operation of a living cell using an atomic force microscope with a nanoneedle. Nano Lett. 2005;5(1):27–30.

Oerlemans C, Bult W, Bos M, Storm G, Nijsen JFW, Hennink WE. Polymeric micelles in anticancer therapy: Targeting, imaging and triggered release. Pharm Res. 2010;27(12):2569–89.

Olivier JC, Fenart L, Chauvet R, Pariat C, Cecchelli R, Couet W. Indirect evidence that drug brain targeting using polysorbate 80-coated polybutylcyanoacrylate nanoparticles is related to toxicity. Pharm Res. 1999;16(12):1836–42.

Omidfar K, Khorsand B, Larijani B. Development of a new sensitive immunostrip assay based on mesoporous silica and colloidal Au nanoparticles. Mol Biol Rep. 2012;39(2):1253–9.

Omlor AJ, Nguyen J, Bals R, Dinh QT. Nanotechnology in respiratory medicine. Respir Res. 2015;16(1):64.

O'Neill AT, Monteiro-Riviere NA, Walker GM. Characterization of microfluidic human epidermal keratinocyte culture. Cytotechnology. 2008;56(3):197.

Orio J, Coustets M, Mauroy C, Teissie J. Electric field orientation for gene delivery using high-voltage and low-voltage pulses. J Membrane Biol. 2012;245(10):661–6.

Orlova Y, Magome N, Liu L, Chen Y, Agladze K. Electrospun nanofibers as a tool for architecture control in engineered cardiac tissue. Biomaterials. 2011;32(24):5615–24.

Orte A, Alvarez-Pez JM, Ruedas-Rama MJ. Fluorescence lifetime imaging microscopy for the detection of intracellular pH with quantum dot nanosensors. ACS nano. 2013 Jun 28;7(7):6387–95.

Pai AS, Rubinstein I, Önyüksel H. PEGylated phospholipid nanomicelles interact with β-amyloid (1–42) and mitigate its β-sheet formation, aggregation and neurotoxicity *in vitro*. Peptides. 2006;27(11):2858–66.

Pamujula S, Kishore V, Rider B, Agrawal KC, Mandal TK. Radioprotection in mice following oral administration of WR-1065/PLGA nanoparticles. Int J Radiat Biol. 2008;84(11):900–8.

Panje CM, Wang DS, Pysz MA, Paulmurugan R, Ren Y, Tranquart F, Tian L, Willmann JK. Ultrasound-mediated gene delivery with cationic versus neutral microbubbles: Effect of DNA and microbubble dose on *in vivo* transfection efficiency. Theranostics. 2012;2(11):1078.

Pantarotto D, Singh R, McCarthy D, Erhardt M, Briand JP, Prato M, Kostarelos K, Bianco A. Functionalized carbon nanotubes for plasmid DNA gene delivery. Angew Chem. 2004;116(39):5354–8.

Park GE, Webster TJ. A review of nanotechnology for the development of better orthopedic implants. J Biomed Nanotechnol. 2005;1(1):18–29.

Patel V, Papineni RV, Gupta S, Stoyanova R, Ahmed MM. A realistic utilization of nanotechnology in molecular imaging and targeted radiotherapy of solid tumors. Radiat Res. 2012;177(4):483–95.

Patel AR, Lim E, Francis KP, Singh M. Opening up the optical imaging window using nano-luciferin. Pharm Res. 2014;31(11):3073–84.

Pawar R, Ben-Ari A, Domb AJ. Protein and peptide parenteral controlled delivery. Expert Opin Biol Ther. 2004;4(8):1203–12.

Pedroso S, Guillen IA. Microarray and nanotechnology applications of functional nanoparticles. Comb Chem High Throughput Screen. 2006;9(5):389–97.

Peled A. Brain-pacing Sticker [BpS]. 2016.

Peleg AY, Seifert H, Paterson DL. *Acinetobacter baumannii*: Emergence of a successful pathogen. Clin Microbiol Rev. 2008;21(3):538–82.

Peng Z. Current status of gendicine in China: Recombinant human Ad-p53 agent for treatment of cancers. Hum Gene Ther. 2005;16(9):1016–27.

Peng YS, Lai PL, Peng S, Wu HC, Yu S, Tseng TY, Wang LF, Chu IM. Glial cell line-derived neurotrophic factor gene delivery via a polyethylene imine grafted chitosan carrier. Int J Nanomedicine. 2014;9:3163.

Perán M, García MA, Lopez-Ruiz E, Jiménez G, Marchal JA. How can nanotechnology help to repair the body? Advances in cardiac, skin, bone, cartilage and nerve tissue regeneration. Materials. 2013;6(4):1333–59.

Pérez-Carrión MD, Ceña V. Knocking down HMGB1 using dendrimer-delivered siRNA unveils its key role in NMDA-induced autophagy in rat cortical neurons. Pharm Res. 2013;30(10):2584–95.

Pérez-Martínez FC, Guerra J, Posadas I, Ceña V. Barriers to non-viral vector-mediated gene delivery in the nervous system. Pharm Res. 2011;28(8):1843–58.

Perinelli DR, Cespi M, Bonacucina G, Rendina F, Palmieri GF. Heating treatments affect the thermal behaviour of doxorubicin loaded in PEGylated liposomes. Int J Pharm. 2017;534(1–2):81–88.

Petersen AL, Binderup T, Rasmussen P, Henriksen JR, Elema DR, Kjær A, Andresen TL. 64 Cu loaded liposomes as positron emission tomography imaging agents. Biomaterials. 2011;32(9):2334–41.

Posadas I, Monteagudo S, Ceña V. Nanoparticles for brain-specific drug and genetic material delivery, imaging and diagnosis. Nanomedicine. 2016;11(7):833–49.

Poste G. Bring on the biomarkers. Nature. 2011;469(7329):156–7.

Prato M, Kostarelos K, Bianco A. Functionalized carbon nanotubes in drug design and discovery. Acc Chem Res. 2007;41(1):60–8.

Prego C, Torres D, Fernandez-Megia E, Novoa-Carballal R, Quiñoá E, Alonso MJ. Chitosan-PEG nanocapsules as new carriers for oral peptide delivery: Effect of chitosan pegylation degree. J Control Release. 2006;111(3):299–308.

Qiu L, Jing N, Jin Y. Preparation and *in vitro* evaluation of liposomal chloroquine diphosphate loaded by a transmembrane pH-gradient method. Int J Pharm. 2008;361(1):56–63.

Raeissi S, Audus KL. *In vitro* characterization of blood–brain barrier permeability to delta sleep-inducing peptide. J Pharm Pharmacol. 1989;41(12):848–52.

Rahman S, Gulati K, Kogawa M, Atkins GJ, Pivonka P, Findlay DM, Losic D. Drug diffusion, integration, and stability of nanoengineered drug releasing implants in bone *ex vivo*. J Biomed Mater Res A. 2016;104(3):714–25.

Rao J, Dragulescu-Andrasi A, Yao H. Fluorescence imaging *in vivo*: Recent advances. Curr Opin Biotechnol. 2007;18(1):17–25.

Rastogi V, Yadav P, Bhattacharya SS, Mishra AK, Verma N, Verma A, Pandit JK. Carbon nanotubes: An emerging drug carrier for targeting cancer cells. J Drug Deliv. 2014;2014. doi.org/10.1155/2014/670815.

Reddy JD, Cobb RR, Dungan NW, Matthews LL, Aiello KV, Ritter G, Eppler B, Kirk JF, Abernethy JA, Tomisaka DM, Talton JD. Preclinical toxicology, pharmacology, and efficacy of a novel orally administered diethylenetriaminepentaacetic acid (DTPA) formulation. Drug Development Research. 2012 Aug;73(5):232–42.

Regar E, Sianos G, Serruys PW. Stent development and local drug delivery. Br Med Bull. 2001;59(1):227–48.

Rieger E, Dupret-Bories A, Salou L, Metz-Boutigue MH, Layrolle P, Debry C, Lavalle P, Vrana NE. Controlled implant/soft tissue interaction by nanoscale surface modifications of 3D porous titanium implants. Nanoscale. 2015;7(21):9908–18.

Robinson HL, Pertmer TM. Nucleic acid immunizations. In: Current protocols in immunology; 2001. p. 2–14.

Rosenthal DI, Yom SS, Liu L, Machtay M, Algazy K, Weber RS, Weinstein GS, Chalian AA, Miller L, Rockwell K, Tonda M. A phase I study of SPI-077 (Stealth® liposomal cisplatin) concurrent with radiation therapy for locally advanced head and neck cancer. Invest New Drugs. 2002;20(3):343–9.

Roy S, Soh JH, Ying JY. A microarray platform for detecting disease-specific circulating miRNA in human serum. Biosens Bioelectron. 2016;75:238–46.

Sacconi L, Tolic IM, Antolini R, Pavone FS. Combined intracellular three-dimensional imaging and selective nanosurgery by a nonlinear microscope. J Biomed Opt. 2005;10(1):014002–25.

Sahay G, Querbes W, Alabi C, Eltoukhy A, Sarkar S, Zurenko C, Karagiannis E, Love K, Chen D, Zoncu R, Buganim Y. Efficiency of siRNA delivery by lipid nanoparticles is limited by endocytic recycling. Nat Biotechnol. 2013;31(7):653–8.

Sahoo SK, Parveen S, Panda JJ. The present and future of nanotechnology in human health care. NBM. 2007;3(1):20–31.

Saksena MA, Saokar A, Harisinghani MG. Lymphotropic nanoparticle enhanced MR imaging (LNMRI) technique for lymph node imaging. Eur J Radiol. 2006;58(3):367–74.

Salinas D. True North: IBM's brain chip are their secret human experimentations. Int J Elec Eng. 2015;8:221–4.

Samyang Biopharmaceuticals Corporation. Korea. 2013. Genexol PM Injection. https://www.samyangbio-pharm.com/eng/ProductIntroduce/injection01 (accessed November 16, 2017).

Sandhiya S, Dkhar SA, Surendiran A. Emerging trends of nanomedicine – An overview. Fundam Clin Pharmacol. 2009;23(3):263–9.

Sandmair AM, Loimas S, Puranen P, Immonen A, Kossila M, Puranen M, Hurskainen H, Tyynelä K, Turunen M, Vanninen R, Lehtolainen P. Thymidine kinase gene therapy for human malignant glioma, using replication-deficient retroviruses or adenoviruses. Hum Gene Ther. 2000;11(16):2197–205.

Sanna V, Pala N, Sechi M. Targeted therapy using nanotechnology: Focus on cancer. Int J Nanomedicine. 2014;9:467.

Santos C, Gomes P, Duarte JA, Almeida MM, Costa ME, Fernandes MH. Development of hydroxyapatite nanoparticles loaded with folic acid to induce osteoblastic differentiation. Int J Pharm. 2017;516(1): 185–95.

Sarwar S, Chakraborti S, Bera S, Sheikh IA, Hoque KM, Chakrabarti P. The antimicrobial activity of ZnO nanoparticles against *Vibrio cholerae*: Variation in response depends on biotype. NBM. 2016;12(6):1499–509.

Sato M, Webster TJ. Nanobiotechnology: Implications for the future of nanotechnology in orthopedic applications. Expert Rev Med Devices. 2004;1(1):105–14.

Savarala S, Brailoiu E, Wunder SL, Ilies MA. Tuning the self-assembling of pyridinium cationic lipids for efficient gene delivery into neuronal cells. Biomacromolecules. 2013;14(8):2750–64.

Schäffer MR, Efron PA, Thornton FJ, Klingel K, Gross SS, Barbul A. Nitric oxide, an autocrine regulator of wound fibroblast synthetic function. J Immunol. 1997;158(5):2375–81.

Schröder U, Sabel BA. Nanoparticles, a drug carrier system to pass the blood–brain barrier, permit central analgesic effects of IV dalargin injections. Brain Res. 1996;710(1):121–4.

Schroeder U, Sommerfeld P, Sabel BA. Efficacy of oral dalargin-loaded nanoparticle delivery across the blood–brain barrier. Peptides. 1998;19(4):777–80.

Schweitzer AD, Revskaya E, Chu P, Pazo V, Friedman M, Nosanchuk JD, Cahill S, Frases S, Casadevall A, Dadachova E. Melanin-covered nanoparticles for protection of bone marrow during radiation therapy of cancer. Int J Radiat Oncol Biol Phys. 2010;78(5):1494–502.

Senet P, Vicaut E, Beneton N, Debure C, Lok C, Chosidow O. Topical treatment of hypertensive leg ulcers with platelet-derived growth factor-BB: A randomized controlled trial. Arch Dermatol. 2011;147(8):926–30.

Senior K. Nano-dumpling with drug delivery potential. Mol Med Today. 1998;4(8):321.

Seo WS, Lee JH, Sun X, Suzuki Y, Mann D, Liu Z, Terashima M, Yang PC, McConnell MV, Nishimura DG, Dai H. FeCo/graphitic-shell nanocrystals as advanced magnetic-resonance-imaging and near-infrared agents. Nat Mater. 2006;5(12):971.

Shah L, Yadav S, Amiji M. Nanotechnology for CNS delivery of bio-therapeutic agents. Drug Deliv Transl Res. 2013;3(4):336–51.

Sheikh BY. Recent advances in nanotechnology: Potential prospects in neuromedicine and neurosurgery. Nanosci Technol. 2014;1(3):1–13.

Shi J, Tian F, Lyu J, Yang M. Nanoparticle based fluorescence resonance energy transfer (FRET) for biosensing applications. Journal of materials chemistry B. 2015;3(35):6989–7005.

Shivakumar N, Poornima S, Raghu S. The future and prospects of bio-chips. Innovare J Eng Technol. 2014:2(2):5–9.

Singer AJ, Clark RA. Cutaneous wound healing. N Engl J Med. 1999;341(10):738–46.

Singh A, Ahmed A, Prasad KN, Khanduja S, Singh SK, Srivastava JK, Gajbhiye NS. Antibiofilm and membrane-damaging potential of cuprous oxide nanoparticles against *Staphylococcus aureus* with reduced susceptibility to vancomycin. Antimicrob Agents Chemother. 2015a;59(11):6882–90.

Singh R, Nawale LU, Arkile M, Shedbalkar UU, Wadhwani SA, Sarkar D, Chopade BA. Chemical and biological metal nanoparticles as antimycobacterial agents: A comparative study. Int J Antimicrob Agents. 2015b;46(2):183–8.

Sinha R, Kim GJ, Nie S, Shin DM. Nanotechnology in cancer therapeutics: Bioconjugated nanoparticles for drug delivery. Mol Cancer Ther. 2006;5(8):1909–17.

Skop NB, Calderon F, Levison SW, Gandhi CD, Cho CH. Heparin crosslinked chitosan microspheres for the delivery of neural stem cells and growth factors for central nervous system repair. Acta Biomater. 2013;9(6):6834–43.

Son YJ, Kim HS, Choi DH, Yoo HS. Multilayered electrospun fibrous meshes for restenosis-suppressing metallic stents. J Biomed Mater Res B. 2017;105(3):628–35.

Sonvico F, Cagnani A, Rossi A, Motta S, Di Bari MT, Cavatorta F, Alonso MJ, Deriu A, Colombo P. Formation of self-organized nanoparticles by lecithin/chitosan ionic interaction. Int J Pharm. 2006;324(1):67–73.

Soppimath KS, Aminabhavi TM, Kulkarni AR, Rudzinski WE. Biodegradable polymeric nanoparticles as drug delivery devices. J Control Release. 2001;70(1):1–20.

Soto CM, Blum AS, Vora GJ, Lebedev N, Meador CE, Won AP, Chatterji A, Johnson JE, Ratna BR. Fluorescent signal amplification of carbocyanine dyes using engineered viral nanoparticles. J Am Chem Soc. 2006;128(15):5184–9.

Souada M, Piro B, Reisberg S, Anquetin G, Noël V, Pham MC. Label-free electrochemical detection of prostate-specific antigen based on nucleic acid aptamer. Biosens Bioelectron. 2015;68:49–54.

Staples M, Daniel K, Cima MJ, Langer R. Application of micro- and nano-electromechanical devices to drug delivery. Pharm Res. 2006;23(5):847–63.

Steiniger S, Zenker D, Briesen HV, Begley D, Kreuter J. The influence of polysorbate 80-coated nanoparticles on bovine brain capillary endothelial cells *in vitro*. Proc Int Symp Control Rel Bioact Mater. 1999;26:789–90.

Streicher RM, Schmidt M, Fiorito S. Nanosurfaces and nanostructures for artificial orthopedic implants. Nanomedicine. 2007;2(6):861–74.

Suganya KU, Govindaraju K, Kumar VG, Dhas TS, Karthick V, Singaravelu G, Elanchezhiyan M. Blue green alga mediated synthesis of gold nanoparticles and its antibacterial efficacy against Gram positive organisms. Mater Sci Eng C. 2015;47:351–6.

Suri SS, Fenniri H, Singh B. Nanotechnology-based drug delivery systems. J Occup Med Toxicol. 2007;2(1):16.

Swift SR, Trinkle-Mulcahy L. Basic principles of FRAP, FLIM and FRET. InProceedings of the Royal Microscopical Society 2004 (Vol. 39, No. 1, pp. 3–11). London: Royal Microscopical Society, [1966]-c2004.

Tabatabaei SN, Duchemin S, Girouard H, Martel S. Towards MR-navigable nanorobotic carriers for drug delivery into the brain. Paper presented at: 2012 IEEE International Conference on Robotics and automation (ICRA): St. Paul, Minnesota; May 2012. p. 727–32.

Tabuchi Y, Wada S, Furusawa Y, Ohtsuka K, Kondo T. Gene networks related to the cell death elicited by hyperthermia in human oral squamous cell carcinoma HSC-3 cells. Int J Mol Med. 2012;29(3):380–6.

Talton JD. Compositions and methods for oral delivery of encapsulated diethylenetriaminepentaacetate particles. US Patent 9,040,091. 2015.

Tang RH, Yang H, Choi JR, Gong Y, Feng SS, Pingguan-Murphy B, Huang QS, Shi JL, Mei QB, Xu F. Advances in paper-based sample pretreatment for point-of-care testing. Crit Rev Biotechnol. 2017;37(4):411–28.

Tao Y, Ju E, Ren J, Qu X. Bifunctionalized mesoporous silica-supported gold nanoparticles: Intrinsic oxidase and peroxidase catalytic activities for antibacterial applications. Adv Mater. 2015;27(6):1097–104.

Tarnuzzer RW, Colon J, Patil S, Seal S. Vacancy engineered ceria nanostructures for protection from radiation-induced cellular damage. Nano Lett. 2005;5(12):2573–7.

Ter-Pogossian MM. Positron emission tomography (PET). In: Diagnostic imaging in medicine. Dordrecht, RA. Robb (Ed.): Springer; 1983. p. 273–7.

Thaxton CS, Elghanian R, Thomas AD, Stoeva SI, Lee JS, Smith ND, Schaeffer AJ, Klocker H, Horninger W, Bartsch G, Mirkin CA. Nanoparticle-based bio-barcode assay redefines "undetectable" PSA and biochemical recurrence after radical prostatectomy. Proc Natl Acad Sci USA. 2009;106(44):18437–42.

Tocco I, Zavan B, Bassetto F, Vindigni V. Nanotechnology-based therapies for skin wound regeneration. J Nanomater. 2012;2012:4.

Tomalia DA, Baker H, Dewald J, Hall M, Kallos G, Martin S, Roeck J, Ryder J, Smith P. A new class of polymers: Starburst-dendritic macromolecules. Polym J. 1985;17(1):117–32.

Tomalia DA, Christensen JB, Boas U. Dendrimers, dendrons, and dendritic polymers: Discovery, applications, and the future. Cambridge University Press: New York, New York; 2012.

Torchilin VP. Recent advances with liposomes as pharmaceutical carriers. Nat Rev Drug Discov. 2005;4(2):145–60.

Torchilin VP. Targeted pharmaceutical nanocarriers for cancer therapy and imaging. AAPS J. 2007;9(2):E128–47.

Tothill IE. Biosensors for cancer markers diagnosis. In: Seminars in cell and developmental biology (Vol. 20, No. 1). Academic Press; 2009. p. 55–62.

Tran PA, Sarin L, Hurt RH, Webster TJ. Titanium surfaces with adherent selenium nanoclusters as a novel anticancer orthopedic material. J Biomed Mater Res A. 2010;93(4):1417–28.

Tran PA, Zhang L, Webster TJ. Carbon nanofibers and carbon nanotubes in regenerative medicine. Adv Drug Deliv Rev. 2009;61(12):1097–14.

Tröster SD, Müller U, Kreuter J. Modification of the body distribution of poly (methyl methacrylate) nanoparticles in rats by coating with surfactants. Int J Pharm. 1990;61(1–2):85–100.

Turjeman K, Bavli Y, Kizelsztein P, Schilt Y, Allon N, Katzir TB, Sasson E, Raviv U, Ovadia H, Barenholz Y. Nano-drugs based on nano sterically stabilized liposomes for the treatment of inflammatory neurodegenerative diseases. PLoS One. 2015;10(7):e0130442.

Turos E, Shim JY, Wang Y, Greenhalgh K, Reddy GSK, Dickey S, Lim DV. Antibiotic-conjugated polyacrylate nanoparticles: New opportunities for development of anti-MRSA agents. Bioorganic Med Chem Lett. 2007;17(1):53–6.

Ulery BD, Nair LS, Laurencin CT. Biomedical applications of biodegradable polymers. J Polym Sci B Polym Phys. 2011;49(12):832–64.

Uludağ Y, Tothill IE. Development of a sensitive detection method of cancer biomarkers in human serum (75%) using a quartz crystal microbalance sensor and nanoparticles amplification system. Talanta. 2010;82(1):277–82.

Uludag Y, Narter F, Sağlam E, Köktürk G, Gök MY, Akgün M, Barut S, Budak S. An integrated lab-on-a-chip-based electrochemical biosensor for rapid and sensitive detection of cancer biomarkers. Anal Bioanal Chem. 2016;408(27):7775–83.

United Nations Scientific Committee on the Effects of Atomic Radiation. Sources and effects of ionizing radiation: Sources (Vol. 1). United Nations Publications; 2000.

United Nations Scientific Committee on the Effects of Atomic Radiation. Effects of ionizing radiation: UNSCEAR 2006 Report to the General Assembly, with scientific annexes (Vol. 2). United Nations Publications; 2009.

United States National Nanotechnology Initiative. 2011. What Is It and How It Works. https://www.nano.gov/nanotech-101/what (accessed August 5, 2017).

United States Nuclear Regulatory Commission (NRC). Radiation Basics [Internet]. 2017 [cited 16 November 2017]. Available from: https://www.nrc.gov/about-nrc/radiation/health-effects/radiation-basics.html.

Upadhyay RK. Drug delivery systems, CNS protection, and the blood–brain barrier. BioMed Res Int. 2014;2014.

US EPA. Radiation health effects [Internet]. 2017 [cited 16 November 2017]. Available from: https://www.epa.gov/radiation/radiation-health-effects.

Valotassiou V, Archimandritis S, Sifakis N, Papatriantafyllou J, Georgoulias P. Alzheimer's disease: Spect and pet tracers for beta-amyloid imaging. Curr Alzheimer Res. 2010;7(6):477–86.

Varallyay CG, Nesbit E, Fu R, Gahramanov S, Moloney B, Earl E, Muldoon LL, Li X, Rooney WD, Neuwelt EA. High-resolution steady-state cerebral blood volume maps in patients with central nervous system neoplasms using ferumoxytol, a superparamagnetic iron oxide nanoparticle. J Cereb Blood Flow Metab. 2013;33(5):780–6.

Vauthier C, Dubernet C, Chauvierre C, Brigger I, Couvreur P. Drug delivery to resistant tumors: The potential of poly (alkyl cyanoacrylate) nanoparticles. J Control Release. 2003;93(2):151–60.

Vidu R, Rahman M, Mahmoudi M, Enachescu M, Poteca TD, Opris I. Nanostructures: A platform for brain repair and augmentation. Front Syst Neurosci. 2014;8.

Vinoj G, Pati R, Sonawane A, Vaseeharan B. *In vitro* cytotoxic effects of gold nanoparticles coated with functional acyl homoserine lactone lactonase protein from *Bacillus licheniformis* and their antibiofilm activity against *Proteus* species. Antimicrob Agents Chemother. 2015;59(2):763–71.

Wagner HN Jr, Burns HD, Dannals RF, Wong DF, Langstrom B, Duelfer T, Frost JJ, Ravert HT, Links JM, Rosen-Bloom SB, Lukas SE. Imaging dopamine receptors in the human brain by positron tomography. J Comput Assist Tomogr. 1984;8(1):190.

Wang J, Wei J, Su S, Qiu J. Novel fluorescence resonance energy transfer optical sensors for vitamin B 12 detection using thermally reduced carbon dots. New Journal of Chemistry. 2015;39(1):501–7.

Wang YJ, Larsson M, Huang WT, Chiou SH, Nicholls SJ, Chao JI, Liu DM. The use of polymer-based nanoparticles and nanostructured materials in treatment and diagnosis of cardiovascular diseases: Recent advances and emerging designs. Prog Polym Sci. 2016;57:153–78.

Wang S, Li Y, Fan J, Wang Z, Zeng X, Sun Y, Song P, Ju D. The role of autophagy in the neurotoxicity of cationic PAMAM dendrimers. Biomaterials. 2014;35(26):7588–97.

Wang AZ, Tepper JE. Nanotechnology in radiation oncology. J Clin Oncol. 2014;32(26):2879–85.

Wang M, Thanou M. Targeting nanoparticles to cancer. Pharmacol Res. 2010;62(2):90–9.

Wang H, Zheng L, Peng C, Shen M, Shi X, Zhang G. Folic acid-modified dendrimer-entrapped gold nanoparticles as nanoprobes for targeted CT imaging of human lung adencarcinoma. Biomaterials. 2013;34(2):470–80.

Webster TJ, Ejiofor JU. Increased osteoblast adhesion on nanophase metals: Ti, Ti6Al4V, and CoCrMo. Biomaterials. 2004;25(19):4731–39.

Webster TJ, Waid MC, McKenzie JL, Price RL, Ejiofor JU. Nano-biotechnology: Carbon nanofibres as improved neural and orthopaedic implants. Nanotechnology. 2003;15(1):48.

Wee KW, Kang GY, Park J, Kang JY, Yoon DS, Park JH, Kim TS. Novel electrical detection of label-free disease marker proteins using piezoresistive self-sensing micro-cantilevers. Biosens Bioelectron. 2005;20(10):1932–8.

Weibel DB, Whitesides GM. Applications of microfluidics in chemical biology. Curr Opin Chem Biol. 2006;10(6):584–91.

Weissleder R. Molecular imaging in cancer. Science. 2006;312(5777):1168–71.

Weller R, Price RJ, Ormerod AD, Benjamin N, Leifert C. Antimicrobial effect of acidified nitrite on dermatophyte fungi, Candida and bacterial skin pathogens. J Appl Microbiol. 2001;90(4):648–52.

Werner S, Grose R. Regulation of wound healing by growth factors and cytokines. Physiol Rev. 2003;83(3):835–70.

Whitesides GM, Ostuni E, Takayama S, Jiang X, Ingber DE. Soft lithography in biology and biochemistry. Annu Rev Biomed Eng. 2001;3(1):335–73.

Widgerow AD. Nanocrystalline silver, gelatinases and the clinical implications. Burns. 2010;36(7):965–74.

Wijagkanalan W, Kawakami S, Hashida M. Designing dendrimers for drug delivery and imaging: Pharmacokinetic considerations. Pharm Res. 2011;28(7):1500–19.

Wilson JM. Gendicine: The first commercial gene therapy product; Chinese translation of editorial. Hum Gene Ther. 2005;16(9):1014–5.

Wirth T, Hedman M, Makinen K, Manninen H, Immonen A, Vapalahti M, Yla-Herttuala S. Safety profile of plasmid/liposomes and virus vectors in clinical gene therapy. Curr Drug Saf. 2006;1(3):253–7.

Wirth T, Parker N, Ylä-Herttuala S. History of gene therapy. Gene. 2013;525(2):162–9.

Wolf NB, Küchler S, Radowski MR, Blaschke T, Kramer KD, Weindl G, Kleuser B, Haag R, Schäfer-Korting M. Influences of opioids and nanoparticles on *in vitro* wound healing models. Eur J Pharm Biopharm. 2009;73(1):34–42.

Wong HL, Chattopadhyay N, Wu XY, Bendayan R. Nanotechnology applications for improved delivery of antiretroviral drugs to the brain. Adv Drug Deliv Rev. 2010;62(4):503–17.

World Health Organization (WHO). HIV assays: Operational characteristics, HIV rapid diagnostic tests (detection of HIV-1/2 antibodies) [Internet]. Geneva: World Health Organization; 2013 [cited 20 December 2016]. Report No.: 17.

World Health Organization (WHO). Cardiovascular diseases (CVDs) [Internet]. Geneva: World Health Organization; 2017 [cited 12 November 2017]. Available from: http://www.who.int/cardiovascular_diseases/en/.

Wu Y, Wang L, Guo B, Ma PX. Interwoven aligned conductive nanofiber yarn/hydrogel composite scaffolds for engineered 3D cardiac anisotropy. ACS Nano. 2017.

Xu G, Mahajan S, Roy I, Yong KT. Theranostic quantum dots for crossing blood–brain barrier *in vitro* and providing therapy of HIV-associated encephalopathy. Front Pharmacol. 2013;4.

Xue X, Wang LR, Sato Y, Jiang Y, Berg M, Yang DS, Nixon RA, Liang XJ. Single-walled carbon nanotubes alleviate autophagic/lysosomal defects in primary glia from a mouse model of Alzheimer's disease. Nano Lett. 2014;14(9):5110–7.

Yager P, Edwards T, Fu E, Helton K, Nelson K, Tam MR, Weigl BH. Microfluidic diagnostic technologies for global public health. Nature. 2006;442(7101):412.

Yamasaki K, Edington HD, McClosky C, Tzeng E, Lizonova A, Kovesdi I, Steed DL, Billiar TR. Reversal of impaired wound repair in iNOS-deficient mice by topical adenoviral-mediated iNOS gene transfer. J Clin Invest. 1998;101(5):967.

Yamazaki N, Kojima S, Bovin NV, Andre S, Gabius S, Gabius HJ. Endogenous lectins as targets for drug delivery. Adv Drug Deliv Rev. 2000;43(2):225–44.

Yang X, Shah JD, Wang H. Nanofiber enabled layer-by-layer approach toward three-dimensional tissue formation. Tissue Eng *Part* A. 2008;15(4):945–56.

Yang F, Cho SW, Son SM, Bogatyrev SR, Singh D, Green JJ, Mei Y, Park S, Bhang SH, Kim BS, Langer R. Genetic engineering of human stem cells for enhanced angiogenesis using biodegradable polymeric nanoparticles. Proc Natl Acad Sci USA. 2010;107(8):3317–22.

Ye Z, Tan M, Wang G, Yuan J. Preparation, characterization, and time-resolved fluorometric application of silica-coated terbium (III) fluorescent nanoparticles. Anal Chem. 2004;76(3):513–8.

Ylä-Herttuala S. Endgame: Glybera finally recommended for approval as the first gene therapy drug in the European Union. Mol Ther. 2012;20(10):1831–2.

Yoo HS, Kim TG, Park TG. Surface-functionalized electrospun nanofibers for tissue engineering and drug delivery. Adv Drug Deliv Rev. 2009;61(12):1033–42.

Yusuf M, Khan M, Khan RA, Ahmed B. Preparation, characterization, *in vivo* and biochemical evaluation of brain targeted Piperine solid lipid nanoparticles in an experimentally induced Alzheimer's disease model. J Drug Target. 2013;21(3):300–11.

Zavan B, Vindigni V, Vezzù K, Zorzato G, Luni C, Abatangelo G, Elvassore N, Cortivo R. Hyaluronan based porous nano-particles enriched with growth factors for the treatment of ulcers: A placebo-controlled study. J Mater Sci Mater Med. 2009;20(1):235–47.

Zhang W, Feng JQ, Harris SE, Contag PR, Stevenson DK, Contag CH. *Rapid in vivo* functional analysis of transgenes in mice using whole body imaging of luciferase expression. Transgenic Res. 2001;10(5):423–34.

Zhang B, Zhang X, Yan HH, Xu SJ, Tang DH, Fu WL. A novel multi-array immunoassay device for tumor markers based on insert-plug model of piezoelectric immunosensor. Biosens Bioelectron. 2007;23(1):19–25.

Zhang H, Yee D, Wang C. Quantum dots for cancer diagnosis and therapy: Biological and clinical perspectives. Nanomedicine. 2008;3(1):83–91.

Zhang N, Lyons S, Lim E, Lassota P. A spontaneous acinar cell carcinoma model for monitoring progression of pancreatic lesions and response to treatment through noninvasive bioluminescence imaging. Clin Cancer Res. 2009;15(15):4915–24.

Zhang X, Oulad-Abdelghani M, Zelkin AN, Wang Y, Haîkel Y, Mainard D, Voegel JC, Caruso F, Benkirane-Jessel N. Poly (L-lysine) nanostructured particles for gene delivery and hormone stimulation. Biomaterials. 2010;31(7):1699–706.

Zhang Y, Zhu P, Li G, Wang W, Chen L, Lu DD, Sun R, Zhou F, Wong C. Highly stable and re-dispersible nano Cu hydrosols with sensitively size-dependent catalytic and antibacterial activities. Nanoscale. 2015;7(32):13775–83.

Zhang HZ, Zhang C, Zeng GM, Gong JL, Ou XM, Huan SY. Easily separated silver nanoparticle-decorated magnetic graphene oxide: Synthesis and high antibacterial activity. J Colloid Interface Sci. 2016a;471:94–102.

Zhang XD, Zhang J, Wang J, Yang J, Chen J, Shen X, Deng J, Deng D, Long W, Sun YM, Liu C. Highly catalytic nanodots with renal clearance for radiation protection. ACS Nano. 2016b;10(4):4511–9.

Zhao W, Hanson L, Lou HY, Akamatsu M, Chowdary PD, Santoro F, Marks JR, Grassart A, Drubin DG, Cui Y, Cui B. Nanoscale manipulation of membrane curvature for probing endocytosis in live cells. Nature nanotechnology. 2017 Aug;12(8):750.

Zhong D, Jiao Y, Zhang Y, Zhang W, Li N, Zuo Q, Wang Q, Xue W, Liu Z. Effects of the gene carrier polyethyleneimines on structure and function of blood components. Biomaterials. 2013;34(1):294–305.

Zhou Y, Hua S, Yu J, Dong P, Liu F, Hua D. A strategy for effective radioprotection by chitosan-based long-circulating nanocarriers. J Mater Chem B. 2015;3(15):2931–4.

10

Prospects of Nanotechnology in Brain Targeting Drug Delivery

Srijita Chakrabarti, Probin Kr Roy, Pronobesh Chattopadhyay, and Bhaskar Mazumder

CONTENTS

ABSTRACT: Drug delivery to the brain is one of the most challenging areas of research in the pharmaceutical field not only due to the blood–brain barrier (BBB) but also for the blood–cerebro-spinal fluid barrier. Being a delicate organ, the brain has a complex physiological structure and also a complicated mechanism of action. Tight junctions between the cerebral endothelial cells at the blood–brain interface restrict the entry of 98% of small molecule drugs and 100% of large molecule drugs, specially hydrophilic substances to maintain the homeostasis of the brain. This drug delivery challenge is an obstacle in the treatment of central nervous system (CNS) disorders such as epilepsy, Alzheimer's disease, Parkinson's disease, and brain tumors, etc. Therefore, to deliver drugs to the brain we must overcome these problems. Researchers have found some solutions such as drug modification, and new delivery system strategies like nanocarriers are used to deliver drugs to the brain. These nanocarriers are advantageous over conventional delivery systems because they not only overcome the limitation properties of the drug or the delivery system to cross the BBB but also provide some extra opportunities, such as sustained release and dose reduction of the drug as well as a reduction of toxicity, and protection of the drug from chemical or enzymatic degradation. In light of these many advantages, nanotechnology is a main focus in every field of research now-a-days. Nanotechnology-based drug delivery systems are a more effective and more promising strategy because of their controlled release or site specific action, particularly to deliver drugs to the brain. So, this chapter enlightens the nanotechnology-based delivery approach to the brain, which may help researchers, academia, and industrialists.

KEY WORDS: *Blood–brain barrier, blood–cerebro-spinal fluid barrier, brain targeting, drug delivery to the CNS, nanocarriers*

10.1 Introduction

The design and development of brain targeted drug delivery systems have gained significant interest among researchers around the globe due to the diversified nature of different disorders of the brain. Brain targeted drug delivery systems are challenging work for researchers because central nervous system (CNS) disorders are very complicated and need serious care. Despite numerous research studies done on this topic, there are still a large number of patients living with several CNS diseases such as meningitis, encephalitis, epilepsy, depression, Parkinson's disease, Alzheimer's disease, and brain tumors, etc. (Misra et al., 2003). Moreover, the strategic location of the blood–brain barrier (BBB) and blood–cerebro-spinal fluid barrier (BCSFB) have intensified the problem of development. Enough potent and effective drugs are available, however the delivery of these therapeutic entities in specific sites of the brain in a therapeutically effective concentration imposes a serious threat to scientists. Now-a-days, researchers are interested in site specific drug delivery due to its reduction of doses as well as side effects and its increase of treatment efficiency. But there is a limitation in the ability of these delivery systems to cross the BBB and the BCSFB, which makes them especially challenging. The BBB is an unique system that keeps the brain apart from the systemic circulation system and maintains the chemical composition of the neuronal microenvironment for proper neuronal functions. It also protects the brain from potentially harmful substances present in the blood stream and supplements the brain with nutrients.

However, besides these advantages, the BBB has a major drawback in that it restricts the transportation of some drugs into the brain because of their larger size or their hydrophilic nature. Approximately 98% of the small molecule drugs and nearly 100% of the large molecule drugs (such as peptides, proteins, and nucleic acids) cannot substantially cross this barrier (Pardridge et al., 2005).

There are several methods to deliver drugs across the barrier, but nanotechnology-based delivery systems are preferred over the other conventional delivery systems. In the last decade, nanotechnologies applied to brain drug delivery assumed great significance and immense potential because of their several advantages (Craparo et al., 2011). These types of formulations have a subcellular size, the possibility to use biocompatible and biodegradable materials, and the ability to minimize the dose of the drug and thus side effects are minimized and dose frequency can be minimized because of the sustained release, therefore patient compliance increases. Some approaches that are used to deliver drugs to the brain include liposomes, polymeric micelles, polymeric nanoparticles, solid lipid nanoparticles, inorganic nanoparticles, nanogels, nanoemulsions, polymerosomes, exosomes, dendrimers, and quantum dots.

10.2 Barriers for the Delivery of Drugs to the CNS

In spite of having a substantial blood flow, the brain is amongst the organs which are most challenging for a drug to access. The brain is uniquely protected by several barriers such as BBB, BCSFB, blood–tumor barrier, and blood–arachnoid layer. BBB and BCSFB are the main two physiological barriers which separate the brain from the free transport of molecules via the blood system. The surface area of the human BBB is approximately 5,000 times that of the BCSFB, rendering the BBB the main hurdle for the uptake of molecules (such as drugs) into the brain and the target of delivering drugs to the brain (Allon & Gavish, 2017). The two main barriers, the BBB and the BCSFB, are briefly described here.

10.2.1 Blood–Brain Barrier (BBB)

The BBB accounts for the control of the internal environment of the brain and the maintenance of a stable environment (homeostasis), ensuring proper CNS functionality (Nagpal et al., 2013). The BBB is defined by microvasculature of the brain, which composed a monolayer of polarized brain capillary endothelial cells that are merged together by tight junctions (TJs). TJs almost completely close the capillary wall so that nothing can be exchanged across the wall except the smaller molecules and lipid-soluble molecules (Tiwari and Amiji, 2006). Because of the TJs, the transendothelial electrical resistance (TEER) of the BBB can be as high as 8,000 Ωcm^2. The structure of the BBB has been well characterized. The cerebral microvessels in the human brain are almost 650 km long, and the total surface area of the human brain microvessels is approximately 12 m^3. In a gram of brain, there is approximately 8 µL of capillary volume (Yang et al., 1999). TJs are composed of dot-like fusion sites between the outer leaflets of the plasma membranes. Occludin, claudins, and junctional adhesion molecules (JAM) play a vital role in the functioning of TJs. TJs perform two major functions, namely as the gate function and the fence function. Prevention of the paracellular diffusion of macromolecules is called the gate function and when apical and the basolateral fraction of the plasma membrane gets separated it is called the fence function (Nagpal et al., 2013).

The blood capillaries of the CNS are structurally different from other tissue capillaries and there is a permeability barrier between the blood within the brain capillaries and the extracellular tissue. Vertebrate brain capillaries lack the small pores which allow rapid movement of solutes from circulation into other organs. This permeability barrier, comprising the brain capillary endothelium, is known as the BBB. The BBB prevents most of the CNS drugs from entering into the brain, except for highly lipid-soluble molecules under a threshold of 400–600 Da (Yang et al., 1999). The brain parenchyma is made up of neurons and neuroglia cells. Neuroglia cells provide structural support to the neighbouring neurons (Araque et al., 1999; Compston et al., 1997). They were classified as neuroglia or "nerve glue" owing to their spindle-like form and their "soft, medullary, fragile nature". However, the neuroglia cells are now known not only for their structural roles, but also for having multiple functions in regulating an optimal interstitial environment (Lee et al., 2001). There are two primary types of neuroglia cells that comprise the brain parenchyma: the macroglia and microglia.

The macroglia is comprised of oligodendrocytes and astrocytes, which, like neurons, possess ecto-dermal origins and proliferate throughout life, particularly triggered by an injury (Peters et al., 1991). Astrocytes have a star-shaped morphology and are comprised of numerous cytoplasmic fibrils of which the glial acidic fibrillary protein is the chief component (Walz, 2000). Astrocytes form the structural framework for the neurons and control their biochemical environment. Astrocytes are nonexcitable cells with a large membrane potential, which is sensitive to extracellular potassium ion (K^+) concentration changes (Kuffler et al., 1966). Astrocytes play an active role in the homeostatic maintenance of the CNS by locally removing excess K^+ which has been released from active neurons. Astrocytes are also involved in the initiation and regulation of immune and inflammatory events during injury and infection. They can inactivate neurotransmitters and regulate and produce growth factors and cytokines. Many of these astrocytes are involved in the production of apolipoprotein E (ApoE) (Lee et al., 2001). The function of the BBB is dynamically regulated by various cells, including astrocytes, neurons, and pericytes.

Spanish neuroanatomist del Rio-Hortega (1932) first described microglia, which are blood-derived mononuclear macrophages and represent 5 to 20% of the total glial population within the CNS (Lawson et al., 1990; Raivich et al., 1999). Microglia are smaller than macroglias. Moreover, they are able to pro-liferate in response to injury and are considered to be the resident immune cells of the brain.

Microglia are also capable of proliferating in response to injury. However, the origin of microglia, whether mesodermal or neuroectodermal, still remains under debate and it is a controversial issue due to the lack of unique cell markers (Lee et al., 2001).

Oligodendrocytes are responsible for the formation and maintenance of the myelin sheath, which sur-rounds the axons and is essential for the fast transmission of action potentials by salutatory conduction. The BBB is additionally protected by a high concentration of P-glycoprotein (P-gp), which is an active drug-efflux-transporter protein residing in the luminal membranes of the cerebral capillary endothelium. Drug molecules are removed by P-gp from the endothelial cell cytoplasm before they pass through the brain parenchyma. ApoE is a glycoprotein which interacts with the low-density lipoprotein receptors (LDLR), and thus significantly modulates plasma lipoprotein and cholesterol levels (Hauser et al., 2011). ApoE is synthesized primarily by astrocytes in adult brain tissue, however sometimes it is synthesized by microglia and neurons under certain physiological and pathological conditions. During nerve regenera-tion, ApoE acts as a scavenger of lipophilic molecules. In the peripheral circulation in the brain, the role of ApoE involves cholesterol transport and intercellular exchange of metabolites between neurons and glial cells (Nagpal et al., 2013).

10.2.2 Blood Cerebro-Spinal Fluid Barrier (BCSFB)

This is another barrier that systemically administered drugs encounter before entering the CNS. Since the cerebro-spinal fluid (CSF) can replace molecules with the interstitial fluid of the brain parenchyma, the passage of blood-borne molecules into the CSF is regulated by the BCSFB. Physiologically, the barrier is in the epithelium of the choroid plexus (CP), which is arranged in a manner that confines the passage of molecules and cells into the CSF. The CP and the arachnoid membrane act together as the barriers between the blood and the CSF. The BCSFB plays a vital role in the selectivity and permeability of the CP membrane to various nutrients and xenobiotics. The CP are a network of capillaries, covered by ependymal cells and a highly vascular organ that protrudes into the ventricles. The BCSFB permits certain substances to enter and excludes other harmful sub-stances. Substances which enter into the CSF from the CP cannot leak between the ependymal cells as these cells are joined by TJs. The BCSFB regulates the passage of blood-borne molecules into the CSF because the CSF can exchange molecules with the interstitial fluid of the brain. This barrier is accredited to the intractable diffusion distances required for equilibration between the CSF and the brain interstitial fluid. Therefore, entry into the CSF does not guarantee the penetration of drugs into the brain. Once the CSF is formed, it is rapidly moved by bulk flow over the cerebral convexities and reabsorbed into general circulation at the upper regions of the brain through the arachnoid villi

(Sherwood, 2015). The adult human brain has approximately 100–140 mL of CSF and the rate of CSF production is comparable to the rate of CSF absorption into the peripheral blood stream, which is about 20 mL/hr. Therefore, the entire CSF volume in the human brain is cleared every 5 hours. The relative surface area of the CP epithelium, which forms the BCSFB compared to the tight BBB is 1:1,000 (Pardridge, 1992).

10.3 Mechanism of Drug Transport in the Brain

The drug transport mechanism across the BBB depends on some physicochemical properties of the drug, such as lipophilicity, molecular weight, and ionic state (Spector, 1990) (Figure 10.1). Generally, small, non-ionic, lipid-soluble molecules penetrate more easily across the BBB than larger, water-soluble, and/ or ionic molecules. Protein and polypeptide drugs have a limited ability to enter the CNS by means of passive diffusion due to their polar nature, but recently this view has changed because many peptide analogs are able to cross the BBB because of their lipophilicity, hydrogen bonding, and conformation (Lee et al., 2001). It is difficult to explain the rate of entry and distribution of some drugs in the CNS by passive processes that depend on the physicochemical characteristics stated above. Many drug transporters that have been well characterized in peripheral tissues and are known to be involved in the influx and efflux of drugs, such as the organic cation, organic anion, nucleoside, P-gp, and multiple receptor protein (MRP) transporters, have been identified in the brain. These drug transporters may influence many pharmacokinetic characteristics of drugs in the processes of absorption, distribution, and elimination (Lee et al., 2001).

When a substance is transported into the brain, the rate of transport is mediated by a transport system at the BBB. This transport system is one of two types depending on the movement of the substance. If the solutes move from the plasma to the brain, it is defined as an influx transport system, whereas the efflux transport system is defined as the transfer of substances from the brain to the systemic circulation system at the BBB. The net flux is the difference between the two unidirectional rates, and is wholly dependent on the nature of the BBB. The net BBB flux is a main factor determining the brain concentrations of protein and polypeptide (P/P) drugs (Neuwelt, 2004).

The passage of substances across the BBB can be briefed as follows.

10.3.1 Paracellular (Aqueous) Diffusion

This diffusion mechanism is defined as the diffusion of small, water-soluble molecules through the BBB by apparently passing through the TJs (Brasnjevic et al., 2009).

10.3.2 Transcellular (Lipophilic) Diffusion

This diffusion mechanism is lipophilic in nature. In this pathway, lipid-soluble substances such as oxygen (O_2), carbon dioxide (CO_2), alcohol, and steroid hormones dissolve in their lipid plasma membrane and therefore can easily cross endothelial cells (Brasnjevic et al., 2009; Tsuji, 2000).

10.3.3 Saturable (Carrier-Mediated) Transport

Some substances such as nutrients and endogenous substances belong to this group. They are actively transported by highly selective membrane bound-carrier systems. The expression of these carriers is often polarized (co-localized on both the luminal and abluminal membranes of the brain microvessel endothelia) to optimize the transport of substrates such as nutrients and endogenous substances into the brain. Several carrier systems have been described in brain capillaries, including those specific for organic anions and cations, small-molecule peptides, hexoses, amino acids, monocarboxylic acids, nucleosides, and neurotransmitters. Although the exact mechanisms of carrier-mediated influx of many substrates is unidentified, this process probably involves the creation of transient narrow pores induced

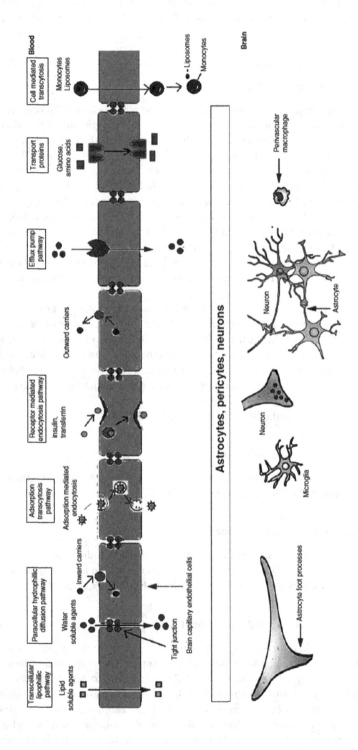

FIGURE 10.1 Mechanisms of transport from blood to brain. Reproduced from Nagpal et al. (2013) with the permission of Taylor & Francis.

by the binding of the respective substrate to the carrier, allowing only the passage of the specific substrate molecule (Brasnjevic et al., 2009).

10.3.4 Receptor-Mediated Endocytosis

A mechanism for the selective uptake of specific macromolecules is known as receptor-mediated endocytosis. Several numbers of proteins and peptides such as insulin, insulin-like growth factors (IGF-I, IGF-II), angiotensin II, transferrin, interleukin (IL)-1, and atrial and brain natriuretic peptide transport across cellular barriers. However, the receptor-mediated endocytosis mechanism is involved with the transportation of a few proteins and polypeptides across the BBB. They are insulin, transferrin, certain cytokines, and leptin (Brasnjevic et al., 2009).

The major steps of the receptor-mediated endocytosis mechanism are summarized below (Brasnjevic et al., 2009; Broadwell et al. 1998):

- Receptor-mediated endocytosis begins with the binding of a ligand (protein and peptide) to its specific membrane receptor localized in the luminal membrane of the brain microvessel endothelia.
- Then the receptor–ligand complex induces an endocytic event in the luminal membrane that probably involves the aggregation of receptor–ligand complexes within the pits.
- The coated pits further trigger the formation of endocytic vesicles of about 100 nm diameters (endocytosis). These endocytic vesicles can then enter a pathway, which carries them across the endothelial cell during which they lose their clathrin coat and are fused with an endosome.
- Then the ligand detaches from the receptor, and the
- ligand-containing vesicles pinch into two parts: the free ligand and the empty receptor.
- Some ligand-containing vesicles can be packed into export vesicles directed to exocytosis on the abluminal face of the endothelial cell, resulting in the transport of peptides/proteins across the BBB, and the others may enter into a pathway that causes them to fuse with a lysosome, forming a secondary lysosome, which then constitutes a dead-end and may result in degradation of the contained ligand.
- Finally the ligands fuse with lysosomes and free receptors are recycled to the cell surface.

10.3.5 Adsorptive Endocytosis

The mechanism by which peptides and proteins with a basic isoelectric point can be taken up by the brain is called adsorptive endocytosis (Brasnjevic et al., 2009). This mechanism may not be involved in specific plasma membrane receptors. The initial binding to the luminal endothelia membrane is governed by electrostatic interactions between cationic moieties of the peptide/protein and the anionic region of the plasma membrane surface region, which induces endocytosis (Gonatas et al., 1984). The main difference between adsorptive endocytosis and receptor-mediated endocytosis lies in the affinity and capacity of these two pathways. Some molecules which penetrate the BBB via adsorptive endocytosis are cationic proteins such as protamine, histone, avidin, cationized albumin, and cationized polyclonal bovine immunoglobulin (Brasnjevic et al., 2009).

10.3.6 Efflux Transport Systems

The BBB and the BCSFB have provided a protective barrier function by removing substances from the brain or the CSF and transferring them to the systemic circulation system, respectively. They are limiting factors in determining the brain concentration of P/P drugs (Brasnjevic et al., 2009). To date, several transporters have been identified in the efflux of P/P drugs from the brain, like organic cation transporters, organic anion transporters, multidrug resistance (MDR) transporters, and monocarboxylate transporters (MCT). Efflux transport is important for drug development as it improves brain penetration and avoids drug–drug interactions by involving these transporters and their subsequent side effects (Kusuhara and Sugiyama, 2001).

10.3.7 Organic Cation Transport System

The transport of organic cations has been reported in various epithelia such as the kidney, liver, intestine, CP, and placenta (Zhang et al., 1998). The functional characteristics of organic cation transport in the various epithelia have been determined by a variety of experimental techniques, such as perfused or non-perfused tissue, isolated apical and basolateral membrane vesicles, primary cell culture, and continuous cell lines. Though the tissues used in characterizing organic cation transport mechanisms have multiple carriers with overlapping driving forces and substrate selectivities, the exact mechanisms of the transport of organic cations by particular transporters are still mostly unknown (Meijer et al., 1990; Zhang et al., 1998). Moreover, it was reported that to understand the organic cation transport mechanisms, the isolation of transporters and their expression in heterologous expression systems is essential. However, the approach of isolating the transport proteins by affinity chromatography and photoaffinity labeling (Holohan et al. 1992; Kimura et al., 1995) has been hindered by a number of intrinsic problems, many of which are related to the development of specific ligands to the native transport proteins. In addition, the low abundance of transport proteins poses a major problem in their isolation and purification.

Many important endogenous metabolites (e.g., choline), therapeutic drugs (e.g., amiloride), and agricultural agents (e.g., paraquat) are organic cations. Accumulation of these compounds within the body may alter a wide variety of cellular functions. However, rapid elimination of organic cations from plasma by the liver and kidney protects the body from their potentially toxic effects (Rennick et al., 1981; Zhang et al., 1998). Although organic cation transport mechanisms in the kidney and liver have been well characterized, the major obstacle for the experimental assessment of the mechanisms used in other tissues, such as the CP, is the small size and anatomic inaccessibility of the transporting epithelia. Some fundamental knowledge about CP organic cation transport has been obtained using *in situ* ventriculocisternal perfusion and preparations of isolated CP and apical membrane vesicles. Yet, these methods do not offer direct access to both faces of the intact CP epithelium and hence limit the experimental ability to analyze the mechanisms of organic cation transport across the BCSFB (Villalobos et al., 1997).

10.3.8 Organic Anion Transport System

Organic anion transport systems have two main families, namely the organic anion transporter (OAT) polypeptide, which includes different isoforms such as oatp1, oatp2, oatp3, OAT-K1, OAT-K2, OATP-A, prostaglandin transporter (PGT), and liver-specific transporter (LST)-1, and the organic anion transporter families, such as OAT1, OAT2, OAT3, and OAT4. Oatp1, a bidirectional organic anion cloned from rat brain, was expressed in the liver, the apical membrane of the CP, and the kidney. It has broad substrate specificity and can mediate the transport of steroid hormones, bile salts, leukotriene C4, and a variety of bulky organic cations (Angeletti et al., 1997; Li et al., 1998; Sekine et al., 1997). Oatp2 was also cloned from rat brain and expressed in the kidney, liver, basolateral membrane of the CP, and brain capillaries. It helps in the uptake of cholate, taurocholate, ouabain, digoxin, [D-Pen(2), D-Pen(5)]enkephalin (DPDPE), and estrogen conjugates (Asaba et al., 2000; Gao et al., 2001; Noe et al., 1997). Oatp3 was isolated from rat retina and expressed in the kidney and retina. It transports thyroxine, triiodothyronine, and taurocholate (Abe et al., 1998). OAT-K1 and OAT-K2 are both localized to the luminal membrane of the renal proximal tubule. OAT-K1 is involved in the transport of methotrexate and folate, whereas OAT-K2 transports methotrexate, folate, taurocholate, and prostaglandin E2 (Masuda et al., 1999). OATP-A is the cloned human liver organic anion carrier that transports cholate, taurocholate, taurochenodeoxycholate, glycocholate, bromosulfophthalein, and tauroursodeoxycholate and opioid peptides such as deltorphin II and DPDPE (Kakyo et al., 1999a; Kullak-Ublick et al., 1995). It is expressed in the human lung, liver, kidney, brain endothelial cells, and testes (Gao et al., 1996). There are another two oatp isoforms known as PGT and LST-1, in which the tissue expression and substrate specificity of PGT are unknown and LST-1 is expressed in the liver and mediates the uptake of taurocholate (Kakyo et al., 1999b; Kanai et al., 1995). The OAT family has four isoforms, OAT1, OAT2, OAT3, and OAT4, and these are responsible for the elimination of organic anions. OATs are categorized into three groups depending on their energy requirements, namely active OATs, which require adenosine triphosphate (ATP), sodium-dependent OATs, and sodium-independent facilitators or exchangers. The active and sodium-independent OATs possess broad substrate specificity and are primarily involved in the secretion of organic anions in both the kidney and liver. On the other hand, the sodium-dependent OATs have a narrow substrate specificity and they have an important role in the reabsorption of essential anionic substances into the proximal tubules of the kidney (Sekine et al., 2000). OAT1, cloned from rats and humans, is expressed in the kidney and brain and it possesses broad specificity, such as p-aminohippurate, dicarboxylates, cyclic nucleotides, prostaglandin E, urate, µ lactum antibiotics, nonsteroidal anti-inflammatory drugs, and diuretics (Sekine et al., 1997). OAT2 is also a liver and kidney specific organic anion transporter, cloned from rats only, and it accepts salicylate, acetylsalicylate, p-aminohippurate, prostaglandin E, and dicarboxylates as substrates (Sekine et al., 1998). OAT3 was isolated from rat brain by reverse transcription-polymerase chain reaction (RT-PCR) cloning methods (Kusuhara et al., 1999) and the messenger ribonucleic acid (mRNA) of OAT3 is expressed in the liver, brain, kidney, and eye. It mediates the transport of p-aminohippurate, estrone sulphate, and cimetidine. Acidic metabolites of neurotransmitters such as epinephrine, norepinephrine, dopamine, and serotonin inhibited the uptake of estrone sulfate by OAT3, suggesting its role in the excretion/detoxification of endogenous anionic substrates from the brain (Kusuhara et al., 1999). OAT4, expressed in the placenta and kidney, is a novel member of the multispecific OAT family exhibiting approximately 38 to 44% amino acid sequence homology to the other members of the OAT family (Cha et al., 2000). It mediates the transport of estrone sulfate, dehydroepiandrosterone sulfate, and a variety of anionic compounds (i.e., bile salts, sulfobromophthalein, diuretics) in a sodium-independent manner.

10.3.9 Multidrug Resistance (MDR) Transporters

MDR proteins (MRPs) are also energy dependent pumps, like P-gp. The proteins belong to the ATP-binding cassette (ABC) gene family of transporters. They can confer MDR and may participate in BBB efflux transport. P-gp and MRPs share some functional similarities, like overlapping substrate specificities, even though they possess only a 15% amino acid homology (Löscher and Potschka, 2005a). The mammalian MRP family consists of seven proteins, which are range from 1,325 to 1,545 amino acids (Borst et al., 2000; König et al., 1999a). All MRPs contain two transmembrane domains of six α-helices and each of them is connected to a cytoplasmic linker. In addition, MRP1, MRP2, MRP3, and MRP6 contain up to six additional membrane-spanning helices at the amidogen (NH_2) terminal domain (Borst et al., 2000; Klein et al., 1999; Lautier et al. 1996), which is not required for drug transport, but the linker region is important to maintain the protein transport properties (Bakos et al., 1998; Gao et al., 1996). MRPs are found in different organs. MRP2, MRP3, and MRP6 are found mainly in the liver and kidney, MRP4 is found in high concentrations in the prostate, MRP1 and MRP5 appear to be ubiquitous and both proteins are expressed in the brain (Klein et al., 1999). MRP1 is a 190-kDa plasma membrane-bound protein and the most characterized of the MRP family members. It has been implicated in MDR (e.g., anticancer drugs) including anthracyclines, epipodophyllotoxins, and several vinca alkaloids. Physiologically, MRP1 plays a major role in the transport of several glucuronide, glutathion, and sulfate conjugates, including conjugated leukotrienes, steroid glucuronides, and glutathion disulfide (Lautier et al., 1996; Leier et al., 1996; Loe et al., 1996a,1996b). Thus, MRP1 may also play a role in the regulation of the intracellular redox potential and flux of ions, inflammatory mediation, and elimination of potentially toxic endo- and xenobiotics (Lee et al., 2001). Furthermore, MRP1 can transport heavy metal oxyanions such as sodium arsenate and antimony potassium tartrate, hence it suggests that MRP1 may play a protective role against environmental toxins. MRP2 has an important role in the biliary excretion of various glutathion and glucuronide conjugates, including bilirubin glucuronide (König et al., 1999a). MRP3 is located in the basolateral membrane of hepatocytes and may be involved in bile acid uptake from the gut (Borst et al., 2000; König et al., 1999b). The physiological functions of the remaining MRP family members are unknown, however some literature demonstrates that the ability of the transport nucleotide analogs of MRP4 and MRP5 suggests these two MRP homologs may be involved in nucleotide/nucleoside transport *in vivo* (Jedlitschky et al., 2000; Schuetz et al., 1999; Wijnholds et al., 2000).

10.3.10 Nucleoside Transport System

It is well known that purine and pyrimidine nucleosides and their metabolic products are the nucleic acids such as deoxyribonucleic acid (DNA) and RNA precursors and participate in different biological brain processes. In general, nucleosides are synthesized endogenously via de novo synthetic pathways (Carver, 1999). However, a number of tissues, including the brain, are deficient in de novo nucleotide synthetic pathways and rely on the salvage of exogenous nucleosides to maintain nucleoside pools and to meet their metabolic demands (Fox and Kelley, 1978). Therefore, the brain is dependent on a continuous and balanced supply of purine and pyrimidine nucleoside constituents from both synthesis *in situ* and the blood. Nucleoside transport systems have only been identified in selected mammalian cells: macrophages, CP, microglia, leukemia cells, splenocytes, intestinal cells, and renal brush border membrane vesicles. Several nucleoside transport systems such as concentrative nucleoside transporters and equilibrative nucleoside transporters have been identified in the BBB and BCSFB. Early *in vivo* studies revealed that at the BBB, purine ribonucleosides cross from the blood to the brain by facilitated diffusion, whereas pyrimidine deoxyribonucleosides did not show appreciable uptake into the brain (Lee et al., 2001).

10.3.11 P-Glycoprotein (P-gp)

The best known 170 kDa plasma membrane, ATP-dependent efflux transporter is P-gp. It restricts harmful exposure to drugs, toxins, and xenobiotics in the body by extruding/expelling them out of cells. It can moderate the pharmacokinetic properties of therapeutic agents and also modulate drug transporters through induction or inhibition by various drugs or herbs. Therefore, drug–drug and drug–herb interactions happen (Nagpal et al., 2013). Moreover, the genetic polymorphism of P-gp may affect the drug disposition, produce variable drug effects, and may alter susceptibility to disease risk (Tandon et al., 2006). P-gp was discovered in Chinese hamster ovary cells selected for colchicine resistance. These cells exhibited broad cross-resistance to a number of naturally occurring, structurally diverse anti-neoplastic agents such as anthracyclines, vinca alkaloids, and taxanes (Lee et al., 2001). Consequently, this phenomenon was termed MDR (Biedler and Riehm, 1970; Kessel and Bosmann, 1970). P-gp is a product of the MDR gene. In humans, two MDR genes, MDR1 and MDR2 or MDR3, have been cloned and sequenced. P-gp of MDR1 genes in humans have broad tissue specificity. They help in the transport of organic cation and lipophilic compounds into different organs such as the intestine, brain, heart, and kidney. Although the MDR1 protein is involved in the MDR phenotype, the protein encoded by the human MDR2 gene functions as a phosphatidyl translocase in the liver (Lee et al., 2001). In addition to the human P-gp homologs, hamster pgp1, 2, and 3 (Gerlach et al., 1986) and mouse mdr1a, 1b, and 2 (Gros et al., 1986) also exist. Similar to their human counterparts, murine P-gp encoded by mdr1a/b confers the MDR phenotype (Gros et al., 1986; Ueda et al., 1997), whereas the mdr2 gene product serves in the transport of hepatic phospholipid into bile (Smit et al., 1993).

10.4 Factors Affecting Drug Penetration through the BBB

Lipinski's rule of five is also known as Pfizer's rule of five or simply the "rule of five" (RO5). This RO5 has been used to assess drug likeness or to determine if a chemical compound with a certain pharmacological or biological activity has characteristics that would make it a likely orally active drug in humans. The rule was given by Christopher A. Lipinski in 1997, finding that most medication drugs are relatively small and lipophilic (Lipinski et al., 1997).

The following features are considered as thumb rule of the RO5:

1. Five H-bond donors
2. Ten H-bond acceptors
3. The molecular weight (MWT) is greater than 500 Da
4. The calculated log P (clogP) is greater than 5 (or MlogP > 4.15).

However, there are some exceptions to the rule, as some drugs do not obey this rule. These therapeutic classes lie outside the RO5: vitamins, antibiotics, antifungals, and cardiac glycosides. Although this rule generated some interest for researchers, especially in medicinal chemistry, the Lipinski's RO5 is of limited significance now-a-days (Lipinski et al., 1997; Pavan et al., 2008).

10.4.1 Molecular Weight (MWT)

The free diffusion of molecules across the BBB depends on the MWT of the molecules. Some lipophilic drugs with pharmacological significance and with a MWT more than 400–500 Da do not cross the BBB (Fischer et al., 1998). This threshold of MWT is probably due to the temporary formation of pores within the phospholipid bilayer, which are formed as the free fatty acyl side-chain links at the time of normal molecular transportation. These pores are of finite size and physically restrict the motion of small molecules with a spherical volume in excess of the pore volume (Marrink et al., 1996; Nagpal et al., 2013).

10.4.2 Hydrogen-Bonding

According to the weak acid model, the proton transport rate is high because of weakly acidic contaminants, which act as protonophores, which are a possible transport mechanism for protons. According to the transient hydrogen-bonded chain (tHBC) model, proposed by Nagle, a strand of water molecules can connect through the membrane to the opposing water layer because of thermal fluctuations, thus forming a water pore (Nichols and Deamer, 1980, Nichols et al. 1980) and protons can be transported very fast by a combined mechanism (Marrink et al., 1996). The permeability of a drug across the BBB decreases by one log of magnitude for each pair of hydrogen (H)-bonds, which are present in the form of polar functional group(s) and added to the molecule (Pardridge et al., 1983). If the number of H-bonds does not obey Lipinski's RO5, the probability of crossing the drug through the BBB via the lipid-mediated free diffusion process and with pharmacologically significant amounts is much lower. The permeation of the molecule is more likely when there are up to five H-bond donors (expressed as the sum of OHs and NHs) (Diamond and Wright, 1969; Lipinski et al., 1997).

10.4.3 Plasma Area under the Curve (pAUC)

Plasma area under the curve (pAUC) comes under pharmacokinetics and this parameter is very useful in determining the availability of a drug in the brain. If the CNS is considered a separate pharmacokinetics compartment, the rate and extent of drug access to the brain will depend on several factors, such as:

1. The plasma concentration, defined by the drug absorption, distribution, metabolism, and excretion characteristics.
2. The degree of plasma–protein binding, as only the unbound fraction diffuses across the barrier.
3. The effective permeability across the BBB, which depends on the combination of the passive permeability and the contribution of efflux and influx carrier-mediated transport (CMT).
4. The metabolic modification by barrier enzymes and the "sink effect" of the continual drainage of CSF.
5. The non-specific binding to brain tissue (Alavijeh et al., 2005; Feng et al., 2002).
6. The concentration of drug transported to the brain is directly proportional to both its pAUC and BBB permeability coefficient (Pe).
7. Lipidization may increase Pe and its biodistribution in the body and may alter its plasma clearance (Greig et al., 1990).

10.4.4 Active Efflux Transporters (AET)

Different transport routes exist, through which solutes move across the BBB, such as transcellular transport, adsorptive transcytosis, cell-mediated transcytosis, CMT, and active efflux transport (Nagpal et al., 2013). Some lipophilic molecules which are small in size dissolve in the lipid membrane and follow the transcellular mechanism, but small hydrophilic molecules diffuse through TJs to a small extent. They have some specific properties such as a molecular weight less than 500 Da, unionized, with a log P value of approximately 2, not more than ten cumulative hydrogen bonds, and they exhibit transcellular transport by passive diffusion. Adsorptive transport is required for some substances like glucose, amino acids, and proteins/carriers to cross the BBB. In the case of CMT, the protein transporter upon getting bound to the solute (such as a glucose/amino acid) to one side of the membrane, changes its conformation, resulting in the transport of the substance along the concentration gradient. Active efflux transporter (AET) substrates cannot effectively pass through the brain, but they are responsible for secreting multiple drugs from the CNS into the bloodstream. The AET play a pivotal role in extruding xenobiotics from the brain tissue back into the circulation system, and can significantly limit the invasion of the substrate into the brain parenchyma and the accumulation of many therapeutic agents in the brain. The best known and most extensively characterized ATP-dependent efflux transporter is P-gp. It is expressed on the luminal membrane of the BBB, MRP, and breast cancer resistance protein (BCRP). It is also responsible for limiting the transport of cationic and lipophilic compounds such as antibiotics, hormones, anticancer drugs, and human immunodeficiency virus (HIV) protease inhibitors into the brain and modulation of pharmacokinetic properties of many therapeutic agents (Begley, 2004; Löscher and Potschka, 2005b). The influx transporters, like members of the organic ion transporting polypeptide family (OATPs) and OATs can facilitate not only substrate delivery into the brain, but also their efflux (Chen and Liu, 2012; Nagpal et al., 2013).

10.5 Nanocarrier-Based Delivery in the CNS

Nanoscale drug carriers consist of particles in the size range of 10 to 1,000 nm. The ideal properties of nanocarriers for drug delivery across the BBB are listed below (Bhaskar et al., 2010; Chen and Liu, 2012; Koo et al., 2006):

- Nontoxic, biodegradable, and biocompatible
- Particle size less than 100 nm (except if transport is via monocytes or macrophages)
- Physically stable in blood (no aggregation or dissociation)
- Avoidance of the mononuclear phagocytic system (MPS) (no opsonization) Prolonged blood circulation time
- Non-immunogenic
- BBB-targeted moiety (receptor or adsorptive mediated mechanism, or uptake by monocytes or macrophages)
- Well maintained parent drug stability
- Possible modulation of drug release profiles
- Able to carry small molecules, proteins, peptides, or nucleic acids
- Minimal nanoparticle excipient-induced drug alteration (e.g., chemical degradation/alteration, protein denaturation) (Olivier, 2005).

Here, some of the nanocarriers such as nanoliposomes, nanoparticles, nanogels, nanoemulsions, polymeric micelles, polymerosomes, dendrimers, and quantum dots are described briefly. In addition, some of the preclinical and clinical applications of nano-based delivery systems have been enumerated in Table 10.1.

10.5.1 Nanoliposomes

Liposomes were the first novel drug delivery system (Micheli et al., 2012). Liposomes are vesicular, self-assembling, colloidal structures composed of one core aqueous compartment, in which hydrophilic drugs are encapsulated, and a lipid bilayer, in which lipophilic drugs are incorporated. Liposomes are small vesicles enclosed by a lipid bilayer membrane. They are useful as efficient delivery vehicles for drugs, plasmids, peptides, proteins, viruses, and bacteria as they have a high solubility and bioavailability, and have low toxicity and are non-immunogenic (Guo et al., 2017). Liposome-based carriers enhance drug and/or substance delivery across the BBB. The liposome helps in the transfer of any desired modified or unmodified molecules, such as therapeutic molecules, labelling molecules, or markers, through the BBB without the unwanted disposition effect of the drugs/substances into other, non-target organs or

TABLE 10.1

Some of the Preclinical and Clinical Applications of Nano-Based Delivery Systems

S. No.	Therapeutic Agents	Polymer Used	Delivery System	Route of Administration	Clinical Indications	References
1.	Tanshinone IIA	PEG	Conjugated nanoparticles	Intravenous	Cerebral ischaemia	Liu et al. (2013)
2.	Dopamine	PLGA	Nanoparticles	Intravenous	Increased levels of dopamine and its metabolites	Pahuja et al. (2015)
3.	Human glial cell line-derived neurotrophic factor gene	PAMAM	Nanoparticles	Intravenous	Improved locomotor activity, reduced dopaminergic neuronal loss and enhanced monoamine neurotransmitter levels	Huang et al. (2009)
4.	Urocortin	PEG-PLGA	Nanoparticles	Intravenous	Parkinson's disease	Hu et al. (2011)
5.	Nerve growth factor	Poly(butyl cyanoacrylate)	Nanoparticles	Intravenous	Reversed scopolamine-induced amnesia and improved recognition and memory, significantly reduced the basic symptoms of Parkinsonism such as oligokinesia, rigidity, and tremor	Kurakhmaeva et al. (2009)
6.	Indomethacin	Poly(e-caprolactone)	Nanocapsules	Organotypic hippocampal slice cultures medium	Neuroprotective formulation for cerebral ischemia	Bernardi et al. (2010)
7.	JM-20	–	–	Organotypic hippocampal slice cultures medium	Cerebral ischaemia	Ramírez-Sánchez J et al. (2015)
8.	C21 compound	–	–	Intraperitoneal (i.p.) injections	Cerebral ischaemia	Mateos et al. (2016)
9.	Doxorubicin	Glutathione-PEG	Liposomes	Human brain endothelial cell line hCMEC/D3	Brain tumors	Gaillard et al. (2014)
10.	Basic fibroblast growth factor and inhibitors of caspases (e.g., benzyloxyc arbonyl-Asp(OMe)-Glu(OMe)-Val-Asp(OMe)-fluoromethylketone (z-DEVD-FMK))	Chitosan	Nanoparticles	Intracerebroventricular administration	Neuroprotective in stroke, reduced the infarct volume, provided 3-hour therapeutic window	Yemisci et al. (2015)

(Continued)

TABLE 10.1 (CONTINUED)

Some of the Preclinical and Clinical Applications of Nano-Based Delivery Systems

S. No.	Therapeutic Agents	Polymer Used	Delivery System	Route of Administration	Clinical Indications	References
11.	Squalenoyl adenosine	–	Nanoassemblies	Intravenous infusion	Decreased the infarct volume in a dose dependent manner, decreased the cell death processes in the ischaemic areas, treatment of traumatic injury of the spinal cord	Gaudin et al. (2014)
12.	Apolipoprotein E4	–	Lipid nanoparticles	Intravenous administration	Apolipoprotein E4-polysorbate 80-coated nanoparticles increased nanoparticle translocation into brain parenchyma, improved brain nanoparticle accumulation by 3-fold compared to undecorated particles	Dal Magro et al. (2018)
13.	Tetraethyl orthosilicate	PEG methyl ether amine	Silica nanoparticles	Intraperitoneal administration	Efficient permeable delivery vehicle which can cross the BBB and reach the brain tissues via receptor-mediated endocytosis by the endothelial cells of the brain capillary followed by transcytosis and ApoE, LDL, and GLUT transporters support	Tamba et al. (2018)
14.	Pramipexole dihydrochloride	Chitosan	Nanoparticles	Intranasal route	Treatment of Parkinson's disease via non-invasive nasal route	Raj et al. (2018)
15.	g7-glicopeptide	Poly(D,L–lactide-co-glycolide)	Hybrid nanoparticles	Nanoparticles were incubated with rat hippocampal culture medium for 6 hours	Nanoparticles are efficiently taken up by neurons, able to escape lysosomes, and release cholesterol into the cells resulting in an efficient modification in expression of synaptic receptors that could be beneficial in Huntington's disease	Daniela et al. (2018)

(Continued)

TABLE 10.1 (CONTINUED)

Some of the Preclinical and Clinical Applications of Nano-Based Delivery Systems

S. No.	Therapeutic Agents	Polymer Used	Delivery System	Route of Administration	Clinical Indications	References
16.	Angiopep-2 MS2 bacteriophage conjugation	–	Nanoparticle carriers	Intracerebroventricular administration	Diagnosis and treatment of intractable brain conditions, such as tinnitus	Apawu et al. (2018)
17.	L-pGlu-(1-benzyl)–L-His–L-ProNH2 (NP-355) and L-pGlu-(2-propyl)–L-His–L-ProNH2 (NP-647)	Poly-lactide-co-glycolide, chitosan	Polymeric nanoparticles	Intranasal administration	Antiepileptic potential	Kaur et al. (2018)
18.	Tarenflurbil	Poly-lactide-co-glycolide	Polymeric nanoparticles	Intranasal administration	Alzheimer's disease	Muntimadugu et al. (2016)
19.	Carbamazepine	Carboxymethyl chitosan	Nanoparticles	Intranasal administration	Epilepsy	Liu et al. (2018)
20.	Gold nanoparticles, plasmid DNA, nerve growth factor	Chitosan	Nanocomposites	Gene delivery	Parkinson's disease	Hu et al. (2018)
21.	MicroRNA-124	Poly-lactide-co-glycolide, poly-vinyl alcohol	Biocompatible and traceable polymeric nanoparticles	Intracerebral administration	Ameliorate motor symptoms of 6-hydroxydopamine mice thus effective in Parkinson's disease	Saraiva et al. (2016)
22.	Anti-green fluorescent protein (GFP) and small interfering RNA (siRNA)	Chitosan	Manganese nanoparticles	Both *in vivo* (intranasal instillation in mice) and *in vitro* (mouse fibroblasts (NIH3T3) cell line)	This gene delivery approach has a significant impact on the disease-modifying therapeutics of neurodegenerative diseases	Sanchez-Ramos et al. (2018)
23.	Docetaxel	Maleimide-poly-(ethylene glycol)-poly-(lactic-co-glycolic acid)	Nanoparticles	Intravenous injection via tail vein	Glioma therapy	Hua et al. (2018)

(Continued)

TABLE 10.1 (CONTINUED)

Some of the Preclinical and Clinical Applications of Nano-Based Delivery Systems

S. No.	Therapeutic Agents	Polymer Used	Delivery System	Route of Administration	Clinical Indications	References
24.	Amyloid-β peptide 1-42 (Aβ1-42)	Poly[(hexadecyl cyanoacrylate-co-rhodamine B cyanoacrylate-co-methoxypoly(ethylene glycol cyanoacrylate)], poly[methoxypoly (ethylene glycol) cyanoacrylate-co-Biotin-poly (ethylene glycol) cyanoacrylate-co-hexadecyl cyanoacrylate]	Nanoparticles	Intravenous injection	Memory recovery in Alzheimer's disease	Carradori et al. (2018)
25.	ZnSO$_4$	Poly(D,L-lactide-co-glycolide), Gly-l -Phe-D-Thr-Gly-1-Phe -l-Leu-LSer(O-β-D-Gl ucose)-CONH2 (g7)	Polymeric nanoparticles	Chronic injection of Zn-loaded nanoparticles in wild type and APP23 mice	Alzheimer's disease	Vilella et al. (2017)
26.	Temoporfin	–	Silica-based nanoparticles	Both *in vitro* (murine 4T1 mammary carcinoma and human breast carcinoma MDA-MB-231) and *in vivo* (injected subcutaneously into hind flanks of the Nu/ Nu mice) studies	Anticancer photodynamic therapy	Ingrid et al. (2017)

(*Continued*)

TABLE 10.1 (CONTINUED)

Some of the Preclinical and Clinical Applications of Nano-Based Delivery Systems

S. No.	Therapeutic Agents	Polymer Used	Delivery System	Route of Administration	Clinical Indications	References
27.	Docetaxel	Dodecylamine covalently linked with terpolymer of poly(methacrylic acid)	Polymer-lipid nanoparticles	Intravenous injection via tail vein	Treatment of brain metastasis of triple negative breast cancer	He et al. (2017)
28.	Iron oxide	PEG, PEI	Superparamagnetic iron oxide nanoparticles	Injected unilaterally into the left substantia nigra of rat brains	Efficient drug delivery system for deep brain stimulation of neurons in a magnetic field	Su et al. (2017)
29.	Andrographolide	Compritol 888 ATO (a mixture of mono-, di- and triglycerides of behenic acid)	Solid lipid nanoparticles	Intravenous injection	Deliver andrographolide into the brain	Graverini et al. (2018)
30.	Peptide iAβ$_5$	Poly(lactic-co-glycolic acid)	Nanoparticles	*In vitro* porcine brain capillary endothelial cell line	Alzheimer's disease	Loureiro et al. (2016)

tissues and without meddling with the integrity of the endothelial tissue which supports and protects the brain from molecules carried by the vascular system. Liposomes can ensure the crossing of the drug over the BBB in a safe, controlled, and effective manner without disrupting the integrity and/or the selectivity of the BBB (Allon & Gavish, 2017). They can be of different size ranges (up to 500 nm in diameter). Liposomes have unilamellar or multilamellar lipid bilayers. According to their size and preparation method, liposomes can be classified as follows (Samad et al., 2007; Sihorkar and Vyas, 2001):

1. Small unilamellar vesicles (SUV) (diameter between 20 and 50 nm)
2. Large unilamellar vesicles (LUV) (100 nm)
3. Reverse phase evaporation vesicles (REV) (0.5 µm)
4. Multilamellar large vesicles (MLV) (diameter between 2 and 10 µm).

Relatively large amounts of the drug or agent can be incorporated into these structures, providing the possibility for significant delivery to the CNS. The surface of the liposome can be modified and groups can be attached so that the construct can be targeted to the CNS via specific BBB mechanisms.

It was demonstrated that the delivery of anticancer drug 5-fluorouracil (5-FU) was improved with transferrin surface-conjugated liposomes after intravenous (i.v.) administration of the drug targeted to the animal's brain (Pasha and Gupta, 2010). It was also reported that topotecan (TPT) liposomes modified with tamoxifen (TA) and wheat germ agglutinin (WGA) showed dual targeting properties. It significantly improves drug transport across the BBB and targets brain tumors to reduce the spheroid volume change ratio in murine C6 glioma spheroids from days 6 to 10 in the range of 65.2–45.9% after applying a dose of 10 µm at day 5, in comparison with free TPT and TPT liposomes. In immunoliposome-treated mice, the amount of 10B in the tumor reached 28.36 mg/g after 24 hours and remained high until their dropped to 21.38 mg/g at the 48 hour mark, compared with the 10B content of tumors at 24 hours and 48 hours being 3.45 and 2.97 mg/g, respectively, in liposome-treated animals after i.v. injection at a dose of 35 mg 10B/kg in a brain tumor model of U87 EGFR cells in mice (Feng et al., 2009). Craparo et al. (2011) demonstrated that immunoliposome utilization in brain drug delivery also has applicability in gene therapy. In this regard, surface-PEGylated liposomes coated with transferrin receptor (TfR) monoclonal antibodies (mAbs) or human insulin receptor (HIR) mAbs have been constructed for the delivery of DNA across the BBB (Pardridg, 2007; Schlachetzki et al., 2004; Zhang et al., 2004; Zhu et al., 2004). Gupta et al. (2007) investigated the promise of cell penetrating peptide trans-activating transcriptional activator (TAT)-modified liposomes to enhance the delivery of a model gene (i.e., the plasmid encodes the green fluorescent protein (pEGFP-N1)) to human brain tumor U87MG cells in an intracranial model in nude mice. This nanocarrier can be used as a potential protective agent for the treatment of neurodegenerative disorders such as Parkinson's, Alzheimer's, and stroke. A study reported that significant therapeutic effects of cyclosporine A (CsA), a neuroprotective agent against cerebral reperfusion, have been observed in high systemic doses or by manipulating the BBB, resulting in systemic side effects and toxicity (Partoazar et al., 2017). Therefore, liposome nanocarriers have been developed to overcome these problems and for the efficient delivery of proteins and peptides. In addition, the liposomal CsA could improve cerebral (I/R) injuries. This formulation at a dose of 2.5 mg/kg was prepared in this study to assess the brain injury outcomes in a 90 minute middle cerebral artery occlusion stroke model followed by 48 hours of reperfusion in treating rats (Partoazar et al., 2017). Another study reported that docetaxel does not cross through the BBB due to the physicochemical and pharmacological characteristics of the drug, therefore docetaxel-loaded nanoliposome has been delivered to rat brains for the treatment of brain tumors (Shaw et al., 2017). In addition, a previous work has reported that the nasal delivery of nanoliposome-encapsulated ferric ammonium citrate can increase the iron content of a rat brain and this intranasal administration is a promising route for delivering drugs effectively to the brain across the BBB. It was reported that iron-encapsulated liposomes can be absorbed effectively by intragastric administration, but nanoliposomes were more efficient to transport drugs across the BBB due to their optimized size, shape, and surface coating. Additionally, as conventional iron supplements have a limited ability to cross the BBB, which decreases their efficacy, a drug delivery system with a higher absorption efficiency and patient compliance could be widely implemented (Guo et al., 2017). Another study reported that nanoliposomes can enhance the delivery of anti-schizophrenic agents to the brain through the nasal route. It was also reported that nasal delivery is easily accessible, convenient, and a dependable system, with a porous endothelial membrane and a highly vascularized epithelium that aids in the quick assimilation of the compound into the systemic circulation system, circumventing the hepatic first pass disposal. In addition to that, intranasal drug delivery allows reduction of the dose, rapid therapeutic level attainment of the drug into the blood, faster onset of pharmacological activity, and less side effects (Parmar et al., 2011; Upadhyay et al., 2017).

10.5.2 Nanoparticles

Over the last few decades, nanoparticles have been the main focus in drug delivery systems, as they overcome some vital challenges which are related to systemic drug delivery. Drugs that are generally poorly distributed to target tissues may be easily loaded into a nanocarrier, which can produce higher therapeutic concentrations of the drug in the brain than would previously have been achieved, via interactions with the endothelial capillaries of the BBB. Nanoparticles are able to easily pass through the BBB capillaries due to their small size and diffuse into target cells, releasing the loaded drug directly into the desired site of action (Sahoo and Labhasetwar, 2003; Stockwell et al., 2014). Nanocarriers are colloidal systems in which the drug is either coated on the particle surface via conjugation or absorption, or entrapped within the colloid matrix of the nanoparticle (Wong et al., 2011). Assessment of the loaded nanoparticle's stability, pH, zeta potential, and drug release kinetics shows that nanoparticles are generally safe and nontoxic, while being able to easily off-load the therapeutic agent once it arrives at the target site (Cruz et al., 2006). The mechanism of barrier opening by nanoparticles is not exactly known, but the delivered nanoparticles enter into the brain by crossing the BBB by various endocytotic mechanisms. Numerous mechanisms are reported by which nanoparticles achieve maximum drug concentration in the brain, including the increased retention of drugs in BBB capillaries combined with an adsorption to capillary walls as a higher concentration gradient increase transport, increasing the fluidization of the BBB membrane and opening the TJs between endothelial cells, etc. Various types of nanoparticles have been tested for diagnostic and therapeutic purposes, such as polymeric, lipid-based, and magnetic nanoparticles (Stockwell et al., 2014).

10.5.2.1 Polymeric Nanoparticles

Polymeric nanoparticles have significant potential for drug delivery to the CNS. Polymeric nanoparticles are composed of a core polymer matrix in which drugs can be embedded (Bhavna et al., 2014; Madan et al., 2013; Zhang et al., 2013), with sizes usually between 60 and 200 nm. A range of materials have been employed for the delivery of drugs. The polymeric nanoparticles can deliver traditional, small molecule drugs as well as nucleic acids, proteins, and diagnostic agents; these features can be attributed to the array of polymers and surface modification techniques currently available (Stockwell et al., 2014). Moreover, they also exhibit a good potential for surface modification and functionalization with different ligands, provide excellent pharmacokinetic control, and are suitable to encapsulate and deliver a plethora of therapeutic agents (Parveen et al., 2012). Polymeric nanoparticles are some of the most extensively studied nanoparticles. In recent years, some polymers have been designed primarily for medical applications and have entered the grounds of controlled release of bioactive agents. They are commonly made of acrylics, such as poly butyl cyanoacrylate (PBCA), or polyesters, and are also made of polylactides (PLA), polyglycolides (PGA), poly(lactide-co-glycolides) (PLGA), polyanhydrides, polycyanoacrylates, and polycaprolactone. Acrylic polymers exhibit rapid degradation, preventing polymer accumulation in the CNS and consequently preventing the adverse effects associated with excess tissue accumulation. Polyester nanoparticles degrade into water and carbon dioxide, making them a safer choice compared with acrylic polymers (Patel et al., 2011). One of the limitations to delivery of these drugs to the CNS is the P-gp efflux system, which actively exports molecules from the CNS. However, with polymeric nanoparticles this phenomenon is circumvented (Wong et al., 2007). Portioli et al. (2017) reported that the use of polymeric nanoparticles to enhance brain penetration could provide alternative strategies to CMT or the generation of prodrugs, which may induce side effects for chemical changes or drug inactivation due to early enzymatic cleavage. Koziara et al. (2004) reported that entrapment of paclitaxel in nanoparticles considerably increases the drug brain uptake and its toxicity toward P-gp-expressing tumor cells. It was postulated that paclitaxel nanoparticles had the ability to mask paclitaxel characteristics and hence limit its binding to P-gp, which in turn would give rise to the brain and tumor cell uptake of the otherwise effluxed drug. It is reported that polymeric nanoparticles are used to deliver drugs to the brain and have the potential for surface modifications, such as apolipoproteins absorbed on the surface of the nanoparticles, or PEGylation of the polyester. This helps improve the nanoparticle selectivity and increases their concentration at the target site in the CNS (Barbu et al., 2009; Olivier, 2005). Another study reported a novel functionalization strategy, based on the LDLR binding domain, to increase the brain targeting efficacy of PLGA nanoparticles. Custom-made PLGA nanoparticles were functionalized with an ApoE modified peptide (pep-ApoE) responsible for LDLR binding, or with lipocalin-type prostaglandin-D-synthase (L-PGDS). It was identified that both ApoE and L-PGDS could bind to LDLR, and these ligands are effective for brain targeting and the delivery of medication in the absence of unwanted side effects on immune cells. It was also mentioned that this strategy provides an alternative system for the use of PLGA nanoparticles for drug delivery in pathological conditions (Portioli et al., 2017). Additionally, curcumin-loaded PLGA nanoparticles modified with g7 ligand was reported as a promising brain delivery system in the treatment of Alzheimer's disease. It was reported that there is no apparent toxicity of the formulated nanoparticles, and in response to the curcumin-loaded nanoparticles the aggregation of β-amyloid, a hallmark of Alzheimer's disease, was decreased significantly (Barbara et al., 2017).

10.5.2.2 Solid Lipid Nanoparticles (SLNs)

SLNs are prepared with lipids, which maintain a solid state at body temperature. They are commonly made of fatty acids or mono-, di-, and triglycerides. They can easily cross the BBB because of their lipophilic nature, even with surface modifications. These lipid-based nanocarriers have solid hydrophobic lipid cores in which the drug can be dissolved or dispersed. They are generally of small size (around 40–200 nm), allowing them to cross the tight endothelial cells of the BBB and escape from the reticuloendothelial system (RES) (Pardeshi et al., 2012). The melted lipid is mixed with the drug during their fabrication and it is generally dispersed in an aqueous surfactant by high-pressure homogenization or microemulsification. The benefits of SLN are their biocompatibility, drug entrapment efficiency, which is relatively higher than that of other nanoparticles, and ability to ensure a continuous release of the medicament for many weeks (Mishra et al., 2010). Additionally, the composition of SLNs can be controlled by modifying their surface properties to target molecules to the brain and to limit RES uptake (Blasi et al., 2007). Surface modifications of SLNs are produced by coating them with hydrophilic molecules, such as poly(ethylene glycol) (PEG) derivatives, to improve plasma stability and biodistribution, and the subsequent bioavailability of the drugs entrapped (Masserini, 2013). Several reports are available describing an enhanced drug delivery to the brain mediated by the SLNs. They have been used for delivering chemotherapeutic cancer drugs as they facilitate increased brain accumulation. For instance, SLN carries a calcium channel blocker drug, which, after i.v. administration into rodents, showed that the drug was taken up to a greater extent by the brain and maintained at high drug levels for a longer time compared to the free drug suspension. A study has reported that SLNs and PEGylated SLNs have also been used for increasing the concentration of antitumor drugs, including camptothecin, doxorubicin, and paclitaxel in the CNS (Blasi et al., 2007; Pasha and Gupta, 2010; Wong et al., 2007). Oral administration of camptothecin-loaded SLNs at a dose of 1.3 camptothecin/kg body weight (wt) of mice showed enhanced and sustained release of the same at 38.5 and 70.3 ng/g of brain tissue after 15 minutes and 4.5 hours, respectively, in comparison with the drug concentration of 49.7 ng/g from camptothecin control solution after 15 minutes but nil after 4.5 hours. These results revealed that the drug was protected in the camptothecin-loaded SLNs, which was not favored in the camptothecin control solution (Yang et al., 1999). The intracerebral bioavailability of doxorubicin and paclitaxel carried by the SLN was evaluated compared with the free solution after systemic treatment at an equivalent therapeutic dosage of 5 and 1.5 mg/kg, respectively. Doxorubicin vehiculated by SLNs achieved intratumoral concentrations ranging from 12- to 50-fold higher compared to free solutions in a rat brain glioma model after 30 minutes and 24 hours, respectively. Paclitaxel incorporated in SLNs produced 10-fold higher drug concentrations in brain tissue than paclitaxel free solutions when administered by i.v. route to normal rabbits (Zara et al., 2002).

SLNs present some rewarding properties, such as low intrinsic cytotoxicity, physical stability, protection of labile drugs from degradation, controlled release, easy preparation, and biodegradability of the lipids used in their preparation, which makes them very useful candidates for brain delivery and particularly for the treatment of brain tumors. Another study reported the synthesis of 3',5-dioctanoyl-5-fluoro-2-deoxyuridine (DO-FUdR) to solve the problem of limited access to the drug FUdR and its incorporation into SLNs. The results showed that DOFUdR-SLN had brain targeting efficiency *in vivo* of about 2 times that of free FUdR. These authors say that SLNs can enhance the ability of the drug to penetrate the BBB and is a promising drug targeting system to treat CNS disorders (Martins et al., 2012; Wang et al., 2002). Moreover, a study reported olanzapine SLNs (OLZ-SLNs) using a minimal dose of OLZ to enhance the brain efficacy as well as to reduce the side effects associated with OLZ. When OLZ-SLNs were administrated by the i.v. route, the relative bioavailability of OLZ in the brain was increased up to 23-fold and clearance was decreased (Natarajan et al., 2017). As OLZ, an antipsychotic drug, is lipophilic in nature, it was quite easy to prepare OLZ-loaded SLNs. So SLNs are a promising drug delivery for OLZ. Additionally, a study reported the development, characterization, and potential use of embelin-loaded nanolipid carriers (NLCs) for brain targeting. The embelin-loaded NLCs were developed as a beneficial carrier to achieve sustained release and brain targeting through the nasal route (Sharma et al., 2017). In addition, resveratrol and grape extract-loaded SLNs were reported for the treatment of Alzheimer's disease. In this study, it was reported that SLNs were functionalized with an antibody, the anti-TfR monoclonal antibody (OX26 mAb), which was used as a possible carrier to transport the extract to the brain. Experiments on human brain-like endothelial cells showed that the cellular uptake of the OX26 SLNs is substantially more efficient than that of normal SLNs. Therefore, the transcytosis ability of these different SLNs is higher when functioned with OX26 (Loureiro et al., 2017). Another study reported surface-modified, PEGylated SLNs with anti-contactin-2 or anti-neurofascin for brain targeted delivery for the treatment of multiple sclerosis, a chronic CNS inflammation. It was reported that the uptake of surface-modified SLNs was higher in the brain tissue compared with the PEGylated SLNs. The report of this study may help scientists design more efficient nanocarriers for the treatment of multiple sclerosis (Gandomi et al., 2017). Another study reported SLNs functionalization with ApoE. Since ApoE receptors are predominantly expressed in the brain, ApoE has been attached in albumin nanoparticles for brain delivery. ApoE binds to the ApoE receptors with high affinity. This functionalization is promising as it shows specific recognition between ApoE molecules on the surface of SLNs and the LDLR that are overexpressed on the BBB. Although this is a highly regulated and energy-dependent process, it can allow the whole nanocarrier and loaded drug to transport across the BBB (Neves et al., 2017). Another study reported that SLNs conjugated with TA and lactoferrin (Lf) were applied to carry anticancer carmustine across the BBB for enhanced antiproliferation against glioblastoma multiforme (GBM). The capacity of TA and Lf on SLNs was investigated to transport carmustine across the BBB and to further inhibit the growth of GBM (Kuo and Cheng, 2016).

10.5.2.3 Inorganic Nanoparticles

Inorganic nanoparticles have potential opportunities in the field of neuroscience. They can be used as a potential drug delivery system to treat CNS-related diseases and may help with the promotion of neuronal differentiation and regeneration. Inorganic nanoparticles can possess potentially broad applicability on the imaging and therapeutics of CNS compared to organic nanomaterials due to their excellent multi-modal imaging capabilities, such as magnetic resonance imaging (MRI), optical imaging, and positron emission tomography (PET) and therapeutic multivalencies for tagging drugs. Therefore, the development of BBB-permeable inorganic nanoparticles is important (Yim et al., 2012). Inorganic materials such as silica, carbon, metallic nanoparticles, and magnetic nanoparticles have been widely used in brain targeting drug delivery systems. Silica nanoparticles can be utilized to carry various drugs and other functional agents due to their unique properties, such as high surface area, large pore volume, good biocompatibility, and ease of functionalization (Li et al., 2017). A study reported the preparation and penetration of PEGylated fluorescein-doped magnetic silica nanoparticles (FMSNs) through the BBB. FMSNs have been covalently conjugated with the second generation (G2) polyamidoamine (PAMAM) dendrimers through 3-(triethoxysilyl) propyl isocyanate (ICP) to yield poly (P)FMSNs, followed by the reaction with tresylated methoxy PEG-5000 (MPEG) to yield PEGylated PFMSNs. PEGylated PFMSNs could penetrate the BBB through the transcytosis of vascular endothelial cells, subsequently diffusing into the cerebral parenchyma and being distributed in the neurons. This may offer great opportunities to develop a multifunction platform of drug delivery and imaging for brain diseases (Ku et al., 2010).

Carbon nanotubes have the ability to penetrate the BBB like a needle due to their unique structure. But they have not been severely explored due to some of their characteristics, such as low reproducibility, insolubility in most of the solvents, and some toxicity issues (Meng et al., 2017). A study reported targeting ligand angiopep-2 modified PEGylated, oxidized, multi-walled carbon nanotubes (O-MWNTs) dual-targeting drug delivery system to deliver anticancer drugs to treat brain glioma (Ren et al., 2012). O-MWNTs and angiopep-2 were conjugated by 1,2 distearoyl sn glycero 3 phosphoethanolamine N [methoxy(poly ethylene glycol) 2000] (DSPE-PEG2000), the complex was loaded with doxorubicin, and then the obtained complex was labelled with fluorescein isothiocyanate (FITC) for investigating their intracellular distribution. This dual-targeting drug delivery system was reported as a potential targeted delivery system for brain glioma.

Magnetic nanoparticles (MNPs) are a powerful and versatile diagnostic tool in the field of medicine. They are able to target specific areas and reduce systemic distribution of the cytotoxic compounds. They enhance drug uptake at the target site, resulting in effective treatment at lower doses. Magnetic iron oxide (Fe_3O_4) particles without any surface coatings have hydrophobic surfaces with a large surface area to volume ratio, which leads to particle agglomeration and formation of large clusters, resulting in increased particle size. This inherent aggregation behavior of magnetic nanoparticles is a crucial limiting factor that reduces the intrinsic superparamagnetic properties and triggers the opsonization process. Therefore, their surfaces are modified to minimize the aggregation. This limitation can be easily dealt with using a surface modification of the MNPs, making these nanoparticles more stable and minimizing remnant magnetization once the external magnetic field is removed. Surface modification also makes them more biostable, biodegradable, and nontoxic (Parveen et al., 2012). A limitation of MNPs is that their function is size dependent, meaning that the nanoparticles with a diameter more than 200 nm are sequestered by the spleen and that those smaller than 10 nm are rapidly removed by renal clearance (Silva, 2008). This may greatly compromise the action of these particles at their target site. Delivery systems containing magnetic properties such as iron oxide nanoparticles also showed a great deal of promise in concentrating the particles at a desired magnetic field and delivering encapsulated drugs into the CNS. Another study reported ferric-cobalt electro-magnetic nanomaterial which bound to small interfering (si)RNA targeting beclin1 to cross the BBB and attenuate the neurotoxic effects of HIV-1 infection in the CNS. It has been reported that this nanoformulation decreases beclin1 expression and attenuates HIV-1 replication and viral-induced cytokine and chemokine release through the transmigration of an *in vitro* BBB without illicit toxicity to brain cells (Rodriguez et al., 2017).

Metallic nanoparticles, such as gold nanoparticles, have several characteristics that make them a promising carrier for drug transport across the BBB. The characteristics include their small size, large surface area, minimal or no tissue reactivity, low toxicity, and the neutral surface charge of modified gold nanoparticles (Meng et al., 2017). A study reported the preparation of gold nanospheres via cross-linking with dithiol-PEG as a targeted drug delivery platform for enhanced brain tumor treatment and increased nanoparticle retention in the tumor tissue via the passive targeting effect. As this nanoassembly can overcome the rapid clearance of small particles and the slow clearance of stable nanostructures, it may have the promising potential as a favorable platform for constructing excellent drug delivery systems for brain tumor therapy (Feng et al., 2017). Moreover, it was reported that covalently coated gold glyconanoparticles with glucose, OEG-amine/galactose, or poly(ethylene glycol) (OEG)-amine/galactose/insulin showed higher uptake efficiency into both human brain endothelial cells (hCMEC/D3) and primary astrocytes. A cargo molecule of DNA oligonucleotide was attached to the galactose-coated gold glyconanoparticles to form single-stranded (ss)DNA/galactose nanoparticles. DNA oligonucleotide cargo can cross the BBB and enter into the brain cells, therefore < 5 nm gold glyconanoparticles may be a useful carrier of oligonucleotides into the brain (Gromnicova, 2017). Additionally, a brain delivery probe based on the PEG-coated Fe_3O_4 nanoparticles was reported to be used as an MRI contrast agent. This probe was prepared by covalently conjugating lactoferrin (Lf) to a PEG-coated Fe_3O_4 nanoparticle in order to facilitate the transport of the nanoparticles across the BBB by receptor-mediated transcytosis via the Lf receptor present on cerebral endothelial cells. PEG-coated Fe_3O_4 nanoparticles were adopted to couple with Lf for constructing a receptor-mediated transcytosis probe because PEG has several advantages, such as it can reduce protein adsorption and limit immune recognition, thus effectively increasing the blood circulation time of the underlying particles, and it can also increase the endothelial permeability of the nanoprobes and thus facilitate their BBB passage (Qiao et al., 2012).

10.5.3 Polymeric Micelles

In recent years, polymeric micelles have gained scientific attention and been introduced as a promising platform for different pharmaceutical applications. Polymeric micelles are nanostructures with a hydrophilic shell and a hydrophobic core, formed by amphiphilic block copolymers whose aggregation in aqueous media leads to spheroidal structures. They can be divided into two types: hydrophobically assembled micelles, which consist of amphiphilic copolymers with a hydrophobic block and hydrophilic block, and polyion-complex micelles. Polymeric micelles are suitable for drug delivery purposes for their inherent and modifiable properties. One of the most utilized polymers is the Pluronic type, block copolymer based on ethylene oxide and propylene oxide. The use of these nanoparticles for brain targeted drug delivery has already been reported. Chitosan-conjugated Pluronic nanocarriers with a specific target peptide for the brain (rabies virus glycoprotein (RVG29)) following i.v. injection in mice showed *in vivo* brain accumulation either of a quantum dot fluorophore conjugated to the nanocarrier, or of a protein loaded into the carrier (Kim et al., 2013; Masserini, 2013; Zhang et al., 2012). Polymeric micellar systems modified with TAT show potential drug delivery vehicular action across the BBB, and polymeric micelles self-assembled from the TAT-PEG-block-cholesterol (TAT-PEG-b-Chol) come under this category (Liu et al., 2008a, 2008b). The distribution of FITC in hippocampus brain sections of rats was observed at 2 hours after i.v. injection of FITC and FITC-loaded micelles (1 mL, 0.0512 mg/mL of FITC). FITC molecules were unable to cross the BBB; by contrast, FITC-loaded TAT-PEG-b-Chol micelles crossed the BBB. The micelles with TAT may afford a probable vehicle for delivering ciprofloxacin or other antibiotics across the BBB to treat brain infections. It is not necessary that all polymeric micelle formulations show effective delivery of antibiotics to the CNS. A study demonstrated that the macrolide antibiotic rapamycin incorporated in PEG-block-poly(a-caprolactone) (PEG-b-PCL) micelles formulated with and without the addition of a-tocopherol observed reduced brain uptake compared with control rapamycin dissolved in Tween 80/PEG 400/N,N dimethyl acetamide. Rapamycin in PEG-b-PCL exposed a major decline in concentration in the brain by 75%. Rapamycin in PEG-b-PCL with a-tocopherol resulted in a significant reduction in concentration in the brain by 75%, whereas the concentration increased significantly by 69% in the urinary bladder. In conclusion, polymeric micelles with Pluronic and those with TAT-PEG-b-Chol offer excellent opportunities for the management of various brain pathologies (Yáñezet al., 2008).

10.5.4 Nanogels

Gels are physically or chemically crosslinked, 3-dimensional (3D) polymeric networks. Hydrogels exhibit a thermodynamic compatibility with water which allows them to swell in aqueous media (Peppas et al., 2000) and this swollen network can retain a large amount of absorbed liquids (Li et al., 2017). Polymeric gels have two types of supporting media, aqueous and organic liquids, which lead to the formation of hydrogels and organic gels respectively. They have good biocompatibility, high water content, and the possibility of pharmaceutical molecules to diffuse into (drug loading) and out of (drug release) the swollen polymeric networks (Li et al., 2017). Hydrogels are used for numerous applications in the medical and pharmaceutical fields, for example as membranes for biosensors, materials for contact lenses or artificial skin, and linings for artificial hearts. Moreover, they are used for 3D cell culture and as drug delivery devices (Bastiancich et al., 2016). Hydrogels also emerged as excellent candidates for controlled release, bioadhesive, and/or targeted drug delivery as they are able to encapsulate biomacromolecules including proteins and DNA, as well as hydrophilic or hydrophobic drugs. Mauri et al. (2017) reported double conjugated nanogels to carry hydrophilic molecules for selective intracellular drug delivery without losing the drug content in body fluids. These double conjugated nanogels offer a double strategy for the traceability of nanogels and the linkage between drug molecules and polymeric chains. Therefore, this nanogel is a promising tool for the selective administration of pharmacological compounds in microglia cells (Mauri et al., 2017). In addition, an ionizing radiation-engineered, carboxyl-functionalized poly(N-vinyl pyrrolidone) nanogel system was reported as an insulin nanocarrier for the development of a new approach for the treatment of Alzheimer's disease. This nanogel system may be a suitable nanocarrier as it can overcome the limits related to the administration of "free" insulin by using it as a vehicle. Moreover, this nanogel was reported as a very effective delivery system due to its high biocompatibility, absence of toxicity, absence of proliferative, immunogenic, and thrombogenic responses, insulin protection against protease degradation, greater insulin activation signalling in comparison with free insulin, and major efficiency in the transport of insulin across the BBB model (Picone et al., 2016). Additionally, another study reported that hydrogels have been used to treat GBM as mimicking platforms in 3D, *in vitro* tumor microenvironment models to study the tumor cell biology, motility, migration, and angiogenesis behavior. This delivery system is used as a tool for preclinical screening to grow *ex vivo* cultures of GBM and assess their sensitivity to radiation and drugs, and it is also useful as an anticancer drug delivery system for the treatment of GBM (Bastiancich et al., 2016).

10.5.5 Nanoemulsions

Hyaluronic, acid-based lipidic nanoemulsion was reported, which was encapsulated with two polyphenols, resveratrol, and curcumin, for the transnasal treatment of neurodegenerative diseases. This nanoemulsion system was proven to be a promising carrier because of its lipidic nature, small particle size surfactant content, and mucoadhesiveness (Nasr, 2016). Another study reported the evaluation of the therapeutic efficacy of cationic nanoemulsion encapsulated tumor necrosis factor-alpha (TNF-α) siRNA in the prevention of experimental neuroinflammation for intranasal brain delivery. siRNA nanoemulsions showed site specific downregulation of TNFα cytokines when tested in a lipopolysaccharide (LPS)-induced model of neuroinflammation. Cationic lipid-based nanoemulsions were reported as a safe and effective drug delivery strategy to enhance the delivery of biologics, such as TNFα siRNA, through the nasal route to target the brain (Yadav et al., 2016). A study reported thymoquinone mucoadhesive nanoemulsion targeted to the brain via non-invasive nasal route administration to enhance the bioavailability of drugs. Although thymoquinone, an antioxidant, has a potential role in the amelioration of cerebral ischemia, its low solubility and poor absorption means it exhibits low serum and tissue levels. Therefore, this nanoemulsion may be a promising and safe brain targeted drug delivery system for thymoquinone because of its novelty and effectiveness for the treatment of cerebral ischemia (Ahmad et al., 2016). Moreover, another study reported the effect of safranal-loaded, mucoadhesive nanoemulsion on oxidative stress markers in cerebral ischemia. Safranal, an antioxidant, has a potential role in the amelioration of cerebral ischemia, however its low solubility and poor absorption properties also result in low serum and tissue bioavailability. This nanoemulsion enhances the bioavailability of drugs and hence evaluates the drug targeting in the brain via non-invasive nasal route administration and may be a potential delivery system for the treatment of cerebral ischemia (Ahmad et al., 2016). Another study reported the effects of a quercetin-loaded nanoemulsion with the free form of the drug in a collagenase-induced intracerebral haemorrhage rat model. Quercetin is highly lipophilic and has an antioxidant effect *in vitro* but low bioavailability *in vivo*. As nanoemulsion is a potential reservoir for lipophilic drugs and it can also facilitate the transport of these drugs through semi-permeable membranes, this quercetin-loaded nanoemulsion may be a promising delivery system for the treatment of intracerebral haemorrhage (Galho et al., 2016).

10.5.6 Polymerosomes

Polymersomes are versatile nanosystems with tremendous potential due to their increased colloidal stability, tunable membrane properties, chemical versatility, good pharmacological properties, good biodegradability and biocompatibility, and the ability to accommodate a broad range of drugs and biomolecules (proteins and genes). Moreover, their tailorable loading, the releasing mechanisms, and the targeting strategies are very promising to increase the cellular specificity and chemical versatility that allows them to be designed to overcome cellular barriers. Polymersomes have distinctive bio-similar properties that offer the possibility to mimic the biological cells, cellular compartments, and their biological processes. Therefore, polymersomes offer vast opportunities in nanotechnology for developing novel and effective nanocarrier systems (Balasubramanian et al., 2016). In addition, the design and delivery of tumor-targeted quantum dot (QD) and doxorubicin (DOX)-encapsulated PEG-PLGA nanopolymersomes has been reported. QD- and DOX-encapsulated nanopolymersomes were conjugated with folate for folate-binding protein receptor guided delivery for the imaging and chemotherapy of breast cancer (Alibolandi et al., 2016b). Another study reported that robust, multifunctional, pH-responsive, and photo-crosslinked polymersomes can be used as carrier systems and also for applications in microsystem devices and nanotechnology. Polymersomes were decorated with adamantane and azide groups and were used for the subsequent postconjugation of polymersomes, started with a covalent approach and finalized by a noncovalent approach for host–guest interactions of adamantane units with β cyclodextrin molecules, and due to these properties polymersomes can be a smart and useful nanocarrier for different applications (Iyisan et al., 2016). Additionally, a study reported docetaxel encapsulated folate-antennae decorated dextran (DEX)-PLGA polymersomes for breast cancer chemotherapy. The targeted polmersomes demonstrated an increased therapeutic potency, over both non-targeted and free drugs, and the targeted drug delivery system was able to control the docetaxel release. This study may open a new horizon for the preparation of folate receptor targeted polymersomes based on DEX-PLGA copolymers for translational purposes (Alibolandi et al., 2016a).

10.5.7 Exosomes

Exosomes were initially thought to be a mechanism for removing unneeded membrane proteins from reticulocytes (Johnstone, 2006). Exosomes were first described in the mid-eighties by P. Stahl and R. Johnstone while studying red blood cell maturation from reticulocytes, which are considered to be disease markers for various human ailments, into red blood corpuscles, which can be monitored by the release of the TfR from the plasma membrane to the extracellular medium (Giau et al., 2016). Molecular constituents, such as exosomal proteins and micro (mi)RNAs, were reported as novel and promising biomarkers for clinical diagnosis. Exosomal miRNAs within biological fluids were also reported as good disease-related markers, and a powerful tool for solving many difficulties in both the diagnosis and treatment of Alzheimer's disease patients. Exosomes as well as miRNAs play an important role in cell-to-cell communication, and influence both physiological and pathological processes (Giau et al., 2016). Exosomes play a major role in communication in the brain and are prominent mediators of neurodegenerative diseases, such as prion disease, Alzheimer's disease, and Parkinson's disease. Exosomes contain neurodegenerative disease-associated proteins such as the prion protein, β-amyloid, and α-synuclein. These novel approaches pave the way to clinical trials aimed to deliver siRNA or next generation drugs by means of exosomes to delay Alzheimer's disease and Parkinson's disease. Exosomes' function as Trojan horses was also reported, which facilitates the accumulation and spread of neurodegenerative disease-associated protein in the brain, with *in vivo* data to support this hypothesis in Alzheimer's disease and Parkinson's disease forthcoming. By understanding the physical nature of exosomes, it may be possible to manipulate their contents to deliver therapeutic factors to delay the onset of neurodegeneration (Vella et al., 2016). Another study reported that siRNA delivery in the brains of transgenic mice by means of exosomes can reduce α-synuclein aggregates, the main component of Lewy bodies, which are the characteristic pathological feature in Parkinson's disease. Modified exosomes have been injected peripherally to achieve widespread delivery of siRNAs to the brain expressing Ravies virus glycoprotein loaded with siRNA (Cooper et al., 2014). This technology opens a new arena for the treatment of a wide variety of neurological diseases, including Alzheimer's disease, Huntington's disease, and prion disease.

10.5.8 Quantum Dots (QDs)

Quantum dots (QDs) are semiconductor nanocrystals with a 2 to 10 nm size range. In comparison with conventional dyes, QDs have exhibited many unique optical advantages such as narrow emission spectra, wide excitation wavelengths, high quantum yield, superior photostability, low photo bleaching, excellent fluorescence intensity, and strong resistance to degradation (Huang et al., 2017). A study reported the efficacy of asparagine-glycine-arginine peptide-modified, PEGylated cadmium selenide (CdSe)/zinc sulfide (ZnS) QDs to cross the BBB for targeted fluorescence imaging of glioma and tumor vasculature. Although QDs are toxic to cells, due to their cadmium cation (Cd^{2+}) release from the CdSe core, the addition of a ZnS shell and incorporation with PEG at the CdSe/ZnS core/shell QD surface reduced toxicity and made sure of its biocompatibility. Moreover, is was also reported that QDs provided an unprecedented opportunity to transport across the BBB and reach brain parenchyma based on their nanoscale size (Huang et al., 2017). Another study reported cadmium sulphoselenide (CdSSe)/ZnS loaded PEI-conjugated QD-based nanocarriers for *in vitro* gene delivery to CNS tumor cells. This QD nanocarrier has been reported as a safe and efficient gene nanocarrier for siRNA delivery and a new and effective way for combating glioblastoma gene therapy (Lin et al., 2017).

10.5.9 Dendrimers

Vögtle et al. (1978), first reported dendrimers at the end of 1970. Dendrimers emerged from the "cascade molecules" (Leiro et al., 2017; Vögtle et al., 1978). The word dendrimer comes from the Greek "dendron" (tree or branch) and "meros" (part), and refers to the characteristic organization of their branched building units (Leiro et al., 2017). Dendrimers are well-defined, 3D, highly branched, monodispersive, symmetric polymeric macromolecules with 1–10 nm diameter range (Dwivedi et al., 2016). They are widely used in various biomedical applications. They are typically composed of three major components: (1) the core, (2) the interior layers, or the generations (G), made of repeated units, and (3) the shell, or the exterior, which may terminate in a variety of functional groups (such as the cationic amines, anionic carboxyls, or neutral hydroxyls) under the typical cellular pH of 7 (Srinageshwar et al., 2017). The presence of numerous surface groups allows for high drug payload and/or multifunctionality. Combined with their nanometer size range, dendrimers are attractive carrier vehicles for drug delivery (Li et al., 2017). Due to their easy tuning regarding composition, structure, and size, dendrimers are multifaceted systems that can serve as vectors for many biomedical applications such as brain delivery and diagnosis. They have been reported as prospective carriers of chemical drugs, therapeutic nucleic acids (NAs), proteins and peptides, macromolecular contrast agents, and biosensor platforms for CNS therapies, imaging, and diagnosis (Leiro et al., 2017). In spite of the wide variety of dendrimers available, a commonly known and commercially available class of dendrimers is PAMAM dendrimers, which have an amide backbone similar to proteins with various cores and surface groups. The major advantages of PAMAM dendrimers with specific surfaces are biocompatibility, non-immunogenicity, and water solubility, which are highly suitable for drug or gene delivery. It was also reported that PAMAM dendrimers cross the BBB when administered through the carotid artery in C57BL/6J mice. These dendrimers offer a safer alternative in terms of cell toxicity and can be useful in the delivery of drugs or small RNA molecules, *in vivo*, to the brain via the intra-carotid injection method (Srinageshwar et al., 2017). Li et al. (2012) have synthesized doxorubicin-loaded fourth generation PAMAM dendrimer-based, pH-sensitive, dual-targeting carriers conjugated with Tf and TA for treating brain gliomas. Tf and TA help to strengthen the BBB transporting ability and in concentrating the drugs in the glioma cells. This nanocarrier could transport across the BBB and could effectively inhibit the growth of glioma spheroids, thus offering a promising delivery system for treating brain tumors (Li et al., 2012). Furthermore, it was demonstrated that G6 PAMAM dendrimers with a hydroxyl group-functionalized surface are better vehicles for targeting injured CNS cells in a 30 kg canine brain injury model than G4 dendrimers. As it has been reported that G6 dendrimers have greater brain accumulation and penetrating efficiency into the CSF than many clinically used drugs, G6 dendrimer may be a powerful vehicle to deliver potent drugs to the CNS (Zhang et al., 2017). Dendrimers have various advantages, such as they deliver active therapeutic molecules as well as assess brain function and diagnose CNS diseases, and they can even perform a combination of these. This significant capability bolsters their remarkable potential as a precise theranostic multifunctional agent (Leiro et al., 2017).

10.6 Parameters to Study Brain Permeation

There are a number of parameters to study brain permeation. Among them the two main parameters for studying the rate of brain penetration are the unidirectional influx constant (K_{in}) and the permeability–surface area (PS) product.

10.6.1 Influx Clearance into Brain

K_{in} is the unidirectional influx constant from the blood to the brain. K_{in} is expressed in mL/min/g brain. K_{in} is mostly determined after i.v. injection or after *in situ* perfusion of the molecule. It can be determined in a specific brain region. The i.v. injection technique is considered the gold standard for brain uptake experiments, as it involves complete physiological conditions (Bickel, 2005; Smith, 2003). With this technique, a radiolabeled compound is injected intravenously. Blood is collected at various time points. A single brain tissue sample can be obtained at the terminal time point. After the i.v. administration of drugs, K_{in} is calculated using the equation:

$$K_{in} = Q_{br} / AUC_{(0 \to T)}$$

Here, Q_{br} is the quantity of drug in the brain, without intravascular content (mass/g brain) (Bickel, 2005; Van and An, 2016) and $AUC_{(0 \to T)}$ is the integral of plasma concentration from $t=0$ to $t=T$.

The *in situ* perfusion method complements the i.v. injection method. It was originally developed for rats, but it has been expanded for mice, guinea pigs, and rabbits (Hervonen and Wall, 1984; Murakami et al., 2000; Zlokovic et al., 1986). Takasato et al. (1984) developed this method in the rat, which is anesthetized and, after a series of artery ligations, the perfusion fluid is infused up the common carotid artery. Perfusion can be stopped at a predetermined time point. Similar to the i.v. injection method, the perfusion fluid remaining in the brain can be flushed out, or a vascular marker can be included in the fluid. The brain is taken out for analysis of the compound. K_{in} can be obtained using the following equation:

$$K_{in} \frac{1}{4} Q_{br} = C_{pf} T$$

Where Q_{br} is the quantity of compound in the brain, without intravascular content (mass/g), C_{pf} is the concentration of the compound in the perfusion fluid (mass/ml), and T is the perfusion time.

10.6.2 Permeability–Surface Area (PS) Product

The PS product is a measure of unidirectional clearance from the blood to the brain (Pardridge, 2004). It denotes the volume of plasma, which gives up its contents at a particular solute to interstitial fluid per unit time rate (Pinter et al., 1974). The unit of PS is the same as K_{in} and it is also determined after i.v. injection or after *in situ* perfusion of the compound and can be measured in a particular brain region.

A study reported the influence of various large, neutral amino acid carriers and acivicin derivatives on the parietal PS product of tracer concentration (0.0015 mm) of [^{14}C] acivicin. This was determined by the *in situ* rat brain perfusion technique. Using K_{in} values, the PS product was determined using the Crone–Renkin model of capillary transfer:

$$PS = -Fl_n \left(1 - K_{in} / F\right)$$

Here, F is the regional perfusion fluid flow (mL/s/g) and K_{in} is the unidirectional blood-to-brain transfer constant (Chikhale et al., 1995).

10.6.3 Brain/Plasma Ratio

Commonly used in the pharmaceutical industry is the brain/plasma ratio. It indicates the distribution coefficient of total drug between brain and plasma and is extensively employed in brain permeation studies (Reichel, 2009). The ratio is a measure of the extent of brain penetration, not of the rate of brain penetration. The test drug is administered to the animal by the desired route. At a predetermined time point, the blood is sampled and the brain is taken out. The brain is homogenized and the drug concentration is determined in both brain and plasma. If multiple animals were used for multiple time points, the AUC of both the brain and plasma can be obtained. The brain concentration is then divided by the plasma concentration. This can be the ratio of one time point or the ratio of the AUCs (Doran et al., 2005). Generally the residual drug in the brain vasculature is not taken into account (van Rooy et al., 2011).

10.6.4 Brain Uptake Index (BUI)

The brain uptake index (BUI) represents the relative uptake of a drug compared to a reference substance (Bonate, 1995; Oldendorf, 1970). The reference is freely diffusible across the BBB, such as ^{14}C-butanol. The test compound is also radiolabeled, for example with ^{3}H. A small aliquot of buffer having the test compound as well as the reference is quickly injected into the common carotid artery of anesthetized animals (e.g,. in the rat, 0.2 mL in less than 0.5 s). The bolus passes through the brain in less than 2 seconds after injection. After 5–15 seconds, the brain is isolated, and the radioactivity in the brain tissue and injected buffer is determined. The BUI can be calculated using the following equation:

$$BUI = \frac{3H\,brain\,/\,14C\,brain}{3H\,injected\,/\,14C\,injected}$$

The BUI can be expressed as a percentage by multiplying it by 100. The BUI represents the net uptake of the drug normalized by the net uptake of the reference compound. It is therefore a direct function of the single pass extraction (E) (Riant et al., 1988). If the extraction of the reference is known (e.g., 100% for butanol (Pardridge and Mietus, 1979)), the extraction of the drug can be calculated:

$$E_{drug} = E_{reference} \times BUI$$

The BUI can be related to the PS product using the Renkin-Crone equation:

$$E = 1 - e^{-PS/F}$$

The main advantage of the BUI technique is that it is fast, while its main disadvantage is its low sensitivity. Additionally, drugs that are taken up slowly cannot be studied with this method (Bickel, 2005). Examples of BUI values obtained *in vivo* are 1.4% of sucrose and 90% of caffeine (Lundquist et al., 2002).

10.6.5 Distribution Coefficient of Drugs between Brain and Plasma on the Logarithmic Scale (log BB)

To study the brain permeability of drugs, another physiochemical parameter to be considered is a blood–brain partitioning. Previous literature was based on small molecules and the linear relationships between the BBB permeability and lipid-to-water partition coefficient in different solvent systems. However, during the past decade, progress in computational tools along with the development of efficient modelling algorithms led to the routine development of more complicated models. But now it has become feasible to calculate the permeability of large molecules. The selection of parameters (electronic, steric, or structural) which adequately describe the CNS distribution of a drug is a challenging task. In the *in silico* studies, log BB is the distribution coefficient of the total drug between the brain and plasma on the logarithmic scale (Boriss, 2010) and is calculated by equation:

$$Log\,BB = \log \frac{c_{brain}}{c_{plasma}}$$

$$Log\,BB = \log \frac{Brain}{Plasma}\,ratio$$

$$= \log K_p$$

K_p is the most commonly used parameter to evaluate brain penetration and has been used as the key parameter to optimize brain drug delivery in CNS drug discovery. Although, log BB is a measure of the extent of partitioning into the brain similar to K_p. However, some experts believe that log BB or K_p does not actually describe the BBB permeability. It may not reflect the relevant drug exposure in the brain due to the fact that it is highly affected by the relative binding affinity of compounds to plasma proteins and brain tissue. In fact, total brain–blood partitioning reflects nothing but the process of an inert partitioning of the drug into lipid material (Van and An, 2016). This can be attributed to these parameters being supremely influenced by the relative binding affinity of compounds for protein and lipid contents of both sides (Hammarlund-Udenaes et al., 2008).

10.6.6 Total Brain to Total Plasma Concentration Ratio ($K_{p, free}$)

$K_{p, free}$ is the distribution coefficient of free drug between brain and plasma and is determined using the following equation (Boriss, 2010):

$$Kp, free = \frac{fu\, plasma \times total\, concentration\, of\, drug\, in\, brain}{fu\, brain \times total\, concentration\, of\, drug\, in\, plasma}$$

$$fu\, plasma = \frac{Cu, plasma}{C\, plasma}$$

$$fu\, brain = \frac{Cu, brain}{C\, brain}$$

$$Kp, free = \frac{Cu, brain}{Cu, plasma}$$

The relation between K_p and $K_{p,free}$ is:

$$Kp = \frac{fu, plasma}{fu, brain} \times Kp, free$$

Where, C_{plasma} and $C_{u, plasma}$ are the total and unbound plasma concentration respectively, C_{brain} and $C_{u,brain}$ are the total and unbound brain concentration respectively, and $f_{u,plasma}$ and $f_{u,brain}$ are the plasma and brain unbound fraction respectively. A low K_p can be due to the high, non-specific binding in plasma, a low binding in the brain, or low $K_{p, free}$.

10.6.7 Apparent Permeability

P_{app} is the rate of permeability of the drug across a cell monolayer or organic solvent. It is calculated by the following equation (Artursson, 1990):

$$P = \frac{dQ}{dt} \frac{1}{Ac060}\, cm/s$$

Here, dQ/dt is the amount of test compound transported per minute (ng/min), A is the surface area of the filter (e.g., an *in vitro* model comprising brain capillary endothelial cells (BCEC) and astrocytes co-cultured on semi-permeable filter inserts) (cm), and C is the initial concentration of the test compound (ng/mL). The apparent BBB permeability coefficient (P_{app}) of a drug accounts for the permeability characteristics of a pharmaceutical ingredient; it is independent of the experimental design, as it is corrected for the transport surface area, the duration of the experimentation, and the applied concentration (Artursson, 1990; Gaillard and de Boer 2000). Therefore, P_{app} is an excellent parameter for comparing permeability values of drugs between experiments.

10.6.8 Brain Unbound Concentration

Brain unbound concentration ($C_{u,brain}$) is an essential parameter in understanding the pharmacokinetics and pharmacodynamics of CNS targeted drug candidates, as only the unbound drug is available to elicit a pharmacological effect. Its unit is ng/mL. The unbound concentration in the brain acts as surrogate for brain interstitial fluid (ISF) levels. It is directly related to the target affinity (Reichel, 2009). According to the brain free drug hypothesis, the *in vivo* efficacy of a drug is independent of total brain drug concentration but depends on the unbound brain drug concentration. The unbound brain concentration represents the drug concentration at the biophase and should be used to evaluate brain penetration. Furthermore, a large total brain/total plasma ratio does not necessarily indicate a large unbound brain/plasma ratio (Liu et al., 2009).

10.6.9 Brain Unbound Volume of Distribution ($V_{u,brain}$)

$V_{u,brain}$ is another pharmacologically relevant parameter which indicates whether a compound is distributed in brain ISF ($V_{u,brain}$ around 0.2 mL/g brain), throughout the brain water space, that is, into both ISF and intracellular fluid (ICF) ($V_{u,brain}$ around 0.8 mL/g brain), or has a tendency to non-specifically bind to brain tissue ($V_{u,brain} > 0.8$ mL/g brain) (Reichel, 2009).

10.6.10 Brain Unbound Fraction

The brain free fraction is also known as a brain unbound fraction (Reichel, 2009). A brain tissue binding experiment using equilibrium dialysis (with rat brain homogenates) is a widely accepted method for determining the free fraction of drug candidates. It is believed that parameters that were calculated based on the unbound fraction of drugs in CNS should be used (Mehdipour and Hamidi, 2009). Some authors have suggested using the BBB PS product for the prediction of the level of free drug in the brain instead of log BB (Martin, 2004; Pardridge, 2004). Nevertheless, there is a second opinion that the PS product by itself cannot envisage the unbound fraction of drug in the CNS because PS product is an estimate of net influx clearance and is affected by the possible association of pharmaceutical ingredients with active influx or efflux (Mehdipour and Hamidi, 2009). The unbound concentration gradient is not dependent on blood–protein binding or protein binding to the tissue component of the brain, which is one of the main drawbacks of using log BB (Mehdipour and Hamidi, 2009). The fraction of unbound drug that had been used to estimate drug binding to various tissues and a robust and interpretable quantitative structure–activity relationship (QSAR) for fu prediction has been proposed for the assessment of CNS drug delivery (Wan et al., 2007). It may be concluded that just one factor cannot explain all aspects of drug permeation into the brain. Therefore, it is always advisable to use at least three factors: permeability clearance, unbound drug in the brain, and intra-brain distribution of the drug for assessment of the rate and extent of drug delivery to the brain (Hammarlund-Udenaes et al., 2008).

10.7 Conclusions and Future Perspectives

Some brain targeting drugs have the ability to treat CNS diseases, but they are not able to do so because of their inability to cross the BBB. A molecule should have specific physiochemical properties to cross the BBB. Therefore, to overcome the drawbacks, some systems, such as novel drug delivery systems, have been developed for the treatment of such diseases. Nanotechnology represents an innovative and promising approach. Some nanoparticles are available with different physiochemical properties and applications, which facilitate the delivery of neuroactive molecules such as drugs, growth factors, genes, and cells to the brain. Nanotechnology solves some major problems by offering some clinical advantages for drug delivery, which are mentioned below:

- Decreased drug dose, reduced side effects, and better patient compliance
- Increased drug half-life

- The possibility to enhance drug transport across the BBB
- Biodegradable and able to be eliminated from the brain, providing the brain targeted drug delivery systems with biological safety
- Uniform preparation methods developed to make the nanoparticles more homogenous and predictable
- Factors that influence *in vivo* behaviour of nanoparticles evaluated and well elaborated (Gao et al., 1996).
- However, considerable improvement should be made before their clinical application.

The BBB permeation of drugs or formulations can be assessed by their therapeutic effect, which needs to be maximized within the safe range of the therapeutic window. It might be possible for some drugs to cross the BBB in an appreciable amount which is not enough as a therapeutic concentration, so the therapeutic efficacy of drugs may be absent. On the other hand, with other drugs, the higher permeability of the drug may lead to unwanted toxic effects. Thus, to overcome such situations, pharmacodynamic tests are important criteria to determine the therapeutic activity of the drug. In this way, pharmacodynamic tests are complimentary to pharmacokinetic studies. Some cell lines which may act as biological surrogates for BBB modelling can be used to improve the permeation of some drugs across the BBB. Some cell lines, such as HMEC-1, HCMEC/D3, TY08, caco-2 cells, ARPE-19, and HRPE, etc., can mimic the BBB with suitable modifications in a better way as they can incorporate the role of transporters and rheological disturbances in drug permeation through the BBB (Naik and Cucullo, 2012). An ideal *in vitro* BBB model is still awaited and may be made possible by the combined effort of biologists, biotechnologists, and pharmaceutical researchers (Nagpal et al., 2013). Although nanotechnology overcomes the problem of transportation of some drugs across the BBB, the enhancement of brain delivery obtained with drug-loaded nanoparticles are still quantitatively limited in comparison with free drugs. Consequently, nanoparticle-related toxicity is another concern for pharmaceutical researchers. Overall, remarkable development has occurred in brain targeting drug delivery systems during past two decades. Although they are still in development, they will play an increasingly important role in treating CNS disorders.

REFERENCES

Abe T, Kakyo M, Sakagami H, Tokui T, Nishio T, Tanemoto M, et al. Molecular characterization and tissue distribution of a new organic anion transporter subtype (oatp3) that transports thyroid hormones and taurocholate and comparison with oatp2. J Biol Chem. 1998;273:22395–401.

Ahmad N, Ahmad R, Alam MA, Samim M, Iqbal Z, Ahmad FJ. Quantification and evaluation of thymoquinone loaded mucoadhesive nanoemulsion for treatment of cerebral ischemia. Int J Biol Macromol. 2016;88:320–32.

Alavijeh MS, Chishty M, Qaiser MZ, Palmer AM. Drug metabolism and pharmacokinetics, the blood–brain barrier, and central nervous system drug discovery. NeuroRx. 2005;2:554–71.

Alibolandi M, Abnous K, Hadizadeh F, Taghdisi SM, Alabdollah F, Mohammadi, M, et al. Dextran-poly lactide-co-glycolide polymersomes decorated with folate-antennae for targeted delivery of docetaxel to breast adenocarcinima in vitro and in vivo. J Controlled Release. 2016a;241:45–56.

Alibolandi M, Abnous K, Sadeghi F, Hosseinkhani H, Ramezani M, Hadizadeh F. Folate receptor-targeted multimodal polymersomes for delivery of quantum dots and doxorubicin to breast adenocarcinoma: *In vitro* and *in vivo* evaluation. Int J Pharm. 2016b;500:162–78.

Allon N, Gavish M. Liposomes for in-vivo delivery. US Patent 9655848 B2. 2017. p. 201.

Angeletti RH, Novikoff PM, Juvvadi SR, Fritschy JM, Meier PJ, Wolkoff AW. The choroid plexus epithelium is the site of the organic anion transport protein in the brain. Proc Natl Acad Sci. 1997;94:283–6.

Apawu AK, Curley SM, Dixon AR, Hali M, Sinan M, Braun RD, Castracane J, Cacace AT, Bergkvist M, Holt AG. MRI compatible MS2 nanoparticles designed to cross the blood–brain barrier: Providing a path towards tinnitus treatment. Nanomed Nanotechnol Biol Med. 2018. DOI:10.1016/j.nano.2018.04.003.

Araque A, Sanzgiri RP, Parpura V, Haydon PG. Astrocyte-induced modulation of synaptic transmission. Can J Physiol Pharmacol. 1999;77:699–706.

Artursson P. Epithelial transport of drugs in cell culture, I: A model for studying the passive diffusion of drugs over intestinal absorptive (Caco-2) cells. J Pharm Sci. 1990;79:476–82.

Asaba H, Hosoya KI, Takanaga H, Ohtsuki S, Tamura E, Takizawa T, Terasaki T. Blood–brain barrier is involved in the efflux transport of a neuroactive steroid, dehydroepiandrosterone sulfate, via organic anion transporting polypeptide 2. J Neurochem. 2000;75:1907–16.

Balasubramanian V, Herranz-Blanco B, Almeida PV, Hirvonen J, Santos HA. Multifaceted polymersome platforms: Spanning from self-assembly to drug delivery and protocells. Prog Polym Sci. 2016;60:51–85.

Barbara R, Belletti D, Pederzoli F, Masoni M, Keller J, Ballestrazzi A, Vandelli MA, Tosi G, Grabrucker AM. Novel curcumin loaded nanoparticles engineered for blood–brain barrier crossing and able to disrupt abeta aggregates. Int J Pharm. 2017;526:413–24.

Barbu E, Molnàr É, Tsibouklis J, Górecki DC. The potential for nanoparticle-based drug delivery to the brain: Overcoming the blood–brain barrier. Exp Opin Drug Deliv. 2009;6:553–65.

Bastiancich C, Danhier P, Préat V, Danhier F. Anticancer drug-loaded hydrogels as drug delivery systems for the local treatment of glioblastoma. J Controlled Release. 2016;243:29–42.

Begley DJ. Efflux mechanisms in the central nervous system: A powerful influence on drug distribution within the brain. In: Blood-spinal cord and brain barriers in health and disease. 2004. p. 83–97.

Bernardi A, Frozza RL, Horn AP, Campos MM, Calixto JB, Salbego C, Pohlmann AR, Guterres SS, Battastini AMO. Protective effects of indomethacin-loaded nanocapsules against oxygen-glucose deprivation in organotypic hippocampal slice cultures: Involvement of neuroinflammation. Neurochem Int. 2010;57:629–36.

Bhaskar S, Tian F, Stoeger T, Kreyling W, de la Fuente JM, Grazu V, et al. Multifunctional nanocarriers for diagnostics, drug delivery and targeted treatment across blood–brain barrier: Perspectives on tracking and neuroimaging. Part Fibre Toxicol. 2010;7:3.

Bhavna MDS, Ali M, Baboota S, Sahni JK, Bhatnagar A, Ali J. Preparation, characterization, in vivo biodistribution and pharmacokinetic studies of donepezil-loaded PLGA nanoparticles for brain targeting. Drug Dev Ind Pharm. 2014;40:278–87.

Bickel U. How to measure drug transport across the blood–brain barrier. NeuroRx. 2005;2:15–26.

Biedler JL, Riehm H. Cellular resistance to actinomycin D in Chinese hamster cells in vitro: Cross-resistance, radioautographic, and cytogenetic studies. Cancer Res. 1970;30:1174–84.

Blasi P, Giovagnoli S, Schoubben A, Ricci M, Rossi C. Solid lipid nanoparticles for targeted brain drug delivery. Adv Drug Deliv Rev. 2007;59:454–77.

Bonate PL. Animal models for studying transport across the blood–brain barrier. J Neurosci Methods. 1995;56:1–5.

Boriss H. Brain availability is the key parameter for optimising the permeability of central nervous system drugs. Drug Discov. 2010;7:58–62.

Borst P, Evers R, Kool M, Wijnholds J. A family of drug transporters: The multidrug resistance-associated proteins. J Natl Cancer Inst. 2000;92:1295–302.

Brasnjevic I, Steinbusch HW, Schmitz C, Martinez-Martinez P. European NanoBioPharmaceutics Research Initiative: Delivery of peptide and protein drugs over the blood–brain barrier. Prog Neurobiol. 2009;87:212–51.

Broadwell RD, Balin BJ, Salcman M. Transcytotic pathway for blood-borne protein through the blood–brain barrier. Proc. Natl. Acad. Sci. U.S.A. 1988;85:632–636.

Carradori D, Balducci C, Re F, Brambilla D, Le Droumaguet B, Flores O, Gaudin A, Mura S, Forloni G, Ordoñez-Gutierrez L, Wandosell F. Antibody-functionalized polymer nanoparticle leading to memory recovery in Alzheimer's disease-like transgenic mouse model. Nanomed Nanotechnol Biol Med. 2018;14:609–18.

Carver JD. Dietary nucleotides: Effects on the immune and gastrointestinal systems. Acta Paediatr. 1999;88:83–8.

Cha SH, Sekine T, Kusuhara H, Yu E, Kim JY, Kim DK, et al. Molecular cloning and characterization of multispecific organic anion transporter 4 expressed in the placenta. J Biol Chem. 2000;275:4507–12.

Chen Y, Liu L. Modern methods for delivery of drugs across the blood–brain barrier. Adv Drug Deliv Rev. 2012;64:640–65.

Chikhale EG, Chikhale PJ, Borchardt RT. Carrier-mediated transport of the antitumor agent acivicin across the blood-brain barrier. Biochem Pharmacol. 1995;49:941–5.

Compston A, Zajicek J, Sussman J, Webb A, Hall G, Muir D, et al. Glial lineages and myelination in the central nervous system. J Anat. 1997;190:161–200.

Cooper JM, Wiklander PB, Nordin JZ, Al-Shawi R, Wood MJ, Vithlani M, et al. Systemic exosomal siRNA delivery reduced alpha-synuclein aggregates in brains of transgenic mice. Mov Disord. 2014;29:1476–85.

Craparo EF, Bondì ML, Pitarresi G, Cavallaro G. Nanoparticulate systems for drug delivery and targeting to the central nervous system. CNS Neurosci Ther. 2011;17:670–7.

Cruz L, Soares LU, Dalla Costa T, Mezzalira G, da Silveira NP, Guterres SS, Pohlmann AR. Diffusion and mathematical modeling of release profiles from nanocarriers. Int J Pharm. 2006;313:198–205.

Dal Magro R, Albertini B, Beretta S, Rigolio R, Donzelli E, Chiorazzi A, Ricci M, Blasi P, Sancini G. Artificial apolipoprotein corona enables nanoparticle brain targeting. Nanomed Nanotechnol Biol Med. 2018;14:429–38.

Daniela B, Andreas G, Francesca P, Isabel M, Angela VM, Giovanni T, Jason DT, Flavio F, Barbara R. Hybrid nanoparticles as a new technological approach to enhance the delivery of cholesterol into the brain. Int J Pharm. 2018;543:300–10.

Diamond JM, Wright EM. Molecular forces governing non-electrolyte permeation through cell membranes. Proc R Soc London B Biol Sci. 1969;172:273–316.

Doran A, Obach RS, Smith BJ, Hosea NA, Becker S, Callegari E, et al. The impact of P-glycoprotein on the disposition of drugs targeted for indications of the central nervous system: Evaluation using the MDR1A/1B knockout mouse model. Drug Metab Dispos. 2005;33:165–74.

Dwivedi N, Shah J, Mishra V, Iqbal MdC, Amin Md, Iyer AK, Kesharwani P. Dendrimer mediated approaches for the treatment of brain tumor. J Biomater Sci Polym Ed. 2016;27:557–80.

Feng M. Assessment of blood–brain barrier penetration: *In silico, in vitro* and *in vivo*. Curr Drug Metab. 2002;3:647–57.

Feng B, Tomizawa K, Michiue H, Miyatake SI, Han XJ, Fujimura A. Delivery of sodium borocaptate to glioma cells using immunoliposome conjugated with anti-EGFR antibodies by ZZ-His. Biomaterials. 2009;30:1746–55.

Feng Q, Shen Y, Fu Y, Muroski ME, Zhang P, Wang Q, et al. Self-assembly of gold nanoparticles shows microenvironment-mediated dynamic switching and enhanced brain tumor targeting. Theranostics. 2017;7:1875–89.

Fischer H, Gottschlich R, Seelig A. Blood–brain barrier permeation: Molecular parameters governing passive diffusion. J Membr Biol. 1998;165:201–11.

Fox IH, Kelley WN. The role of adenosine and 2'-deoxyadenosine in mammalian cells. Annu Rev Biochem. 1978;47:655–86.

Gaillard PJ, de Boer AG. Relationship between permeability status of the blood–brain barrier and *in vitro* permeability coefficient of a drug. Eur J Pharm Sci. 2000;2:95–102.

Gaillard PJ, Appeldoorn CC, Dorland R, van Kregten J, Manca F, Vugts DJ, Windhorst B, van Dongen GA, de Vries HE, Maussang D, van Tellingen O. Pharmacokinetics, brain delivery, and efficacy in brain tumor-bearing mice of glutathione pegylated liposomal doxorubicin (2B3-101). PloS one. 2014;9:e82331.

Galho AR, Cordeiro MF, Ribeiro SA, Marques MS, Antunes MF, Luz DC, et al. Protective role of free and quercetin-loaded nanoemulsion against damage induced by intracerebral haemorrhage in rats. Nanotechnology. 2016;27:175101.

Gandomi N, Varshochian R, Atyabi F, Ghahremani MH, Sharifzadeh M, Amini M, Dinarvand R. Solid lipid nanoparticles surface modified with anti-Contactin-2 or anti-Neurofascin for brain-targeted delivery of medicines. Pharm Dev Technol. 2017;22:426–35.

Gao B, Meier PJ. Organic anion transport across the choroid plexus. Microsc Res Tech. 2001;52:60–4.

Gao M, Loe DW, Grant CE, Cole SP, Deeley RG. Reconstitution of ATP-dependent leukotriene C4 transport by co-expression of both half-molecules of human multidrug resistance protein in insect cells. J Biol Chem. 1996;271:27782–7.

Gerlach JH, Endicott JA, Juranka PF, Henderson G, Sarangi F, Deuchars KL, and Ling V. Homology between P-glycoprotein and a bacterial haemolysin transport protein suggests a model for multidrug resistance. Nature (Lond). 1986;324:485–489.

Gaudin A, Yemisci M, Eroglu H, Lepetre-Mouelhi S, Turkoglu OF, Dönmez-Demir B, Caban S, Sargon MF, Garcia-Argote S, Pieters G, Loreau O. Squalenoyl adenosine nanoparticles provide neuroprotection after stroke and spinal cord injury. Nat Nanotechnol. 2014;9:1054.

Gonatas NK, Stieber A, Hickey WF, Herbert SH, Gonatas JO. Endosomes and Golgi vesicles in adsorptive and fluid phase endocytosis. J Cell Biol. 1984;99:1379–90.

Graverini G, Piazzini V, Landucci E, Pantano D, Nardiello P, Casamenti F, Pellegrini-Giampietro DE, Bilia AR, Bergonzi MC. Solid lipid nanoparticles for delivery of andrographolide across the blood-brain barrier: *In vitro* and *in vivo* evaluation. Colloids Surf B Biointerfaces. 2018;161:302–13.

Greig NH, Daly EM, Sweeney DJ, Rapoport SI. Pharmacokinetics of chlorambucil-tertiary butyl ester, a lipophilic chrambucil derivative that achieves and maintains high concentrations in brain. Cancer Chemother Pharmacol. 1990;25:320–5.

Gromnicova R. Gold nanoparticles as a delivery system of oligonucleotides into the brain [dissertation]. The Open University, Faculty of Science, Technology, Engineering and Mathematics; Department of Life, Health and Chemical Sciences; Walton Hall, Milton Keynes, UK; 2017. p.1–219. https://core.ac.uk/download/pdf/82984335.pdf.

Gros P, Croop J, and Housman D. Mammalian multidrug resistance gene: complete cDNA sequence indicates strong homology to bacterial transport proteins. Cell. 1986;47:371–380.

Guo X, Zheng H, Guo Y, Wang Y, Anderson GJ, Ci Y et al. Nasal delivery of nanoliposome-encapsulated ferric ammonium citrate can increase the iron content of rat brain. J Nanobiotechnol. 2017;15:42.

Gupta B, Levchenko TS, Torchilin VP. TAT peptide-modified liposomes provide enhanced gene delivery intracranial human brain tumor xenografts in nude mice. Oncol Res. 2007;16:351–9.

Hammarlund-Udenaes M, Fridén M, Syvänen S, Gupta A. On the rate and extent of drug delivery to the brain. Pharm Res. 2008;25:1737–50.

Hauser PS, Narayanaswami V, Ryan RO. Apolipoprotein E: From lipid transport to neurobiology. Prog Lipid Res. 2011;50:62–74.

He C, Cai P, Li J, Zhang T, Lin L, Abbasi AZ, Henderson JT, Rauth AM, Wu XY. Blood–brain barrier-penetrating amphiphilic polymer nanoparticles deliver docetaxel for the treatment of brain metastases of triple negative breast cancer. J Controlled Release. 2017;246:98–109.

Hervonen H, Wall OS. Endothelial surface sulfhydryl-groups in blood–brain barrier transport of nutrients. Acta Physiol. 1984;121:343–51.

Holohan PD, White KE, Sokol PP, Rebbeor J. Photoaffinity labeling of the organic cation/H+ exchanger in renal brush border membrane vesicles. J Biol Chem. 1992;267:13513–9.

Hu K, Shi Y, Jiang W, Han J, Huang S, Jiang X. Lactoferrin conjugated PEG-PLGA nanoparticles for brain delivery: Preparation, characterization and efficacy in Parkinson's disease. Int J Pharm. 2011;415:273–83.

Hu K, Chen X, Chen W, Zhang L, Li J, Ye J, Zhang Y, Zhang L, Li CH, Yin L, Guan YQ. Neuroprotective effect of gold nanoparticles composites in Parkinson's disease model. Nanomed Nanotechnol Biol Med. 2018;14:1123–36.

Hua H, Zhang X, Mu H, Meng Q, Jiang Y, Wang Y, Lu X, Wang A, Liu S, Zhang Y, Wan Z. RVG29-modified docetaxel-loaded nanoparticles for brain-targeted glioma therapy. Int J Pharm. 2018;543:179–89.

Huang R, Han L, Li J, Ren F, Ke W, Jiang C, Pei Y. Neuroprotection in a 6-hydroxydopamine-lesioned Parkinson model using lactoferrin-modified nanoparticles. J Gene Med. 2009;11:754–63.

Huang N, Chengb S, Zhanga X, Tiana Q, Pi J, Tanga J, et al. Efficacy of NGR peptide-modified PEGylated quantum dots for crossing the blood–brain barrier and targeted fluorescence imaging of glioma and tumor vasculature. Nanomed Nanotechnol Biol Med. 2017;13:83–93.

Ingrid B, Kamil Z, Jarmila K, Alla S, Hana A, Pavel U, Pavla P, Martin H, Petr Š, Vladimír K. Silica-based nanoparticles are efficient delivery systems for temoporfin. Photodiagn Photodyn Ther. 2017. DOI:10.1016/j.pdpdt.2017.12.014.

Iyisan B, Kluge J, Formanek P, Voit B, Appelhans D. Multifunctional and dual-responsive polymersomes as robust nanocontainers: Design, formation by sequential post-conjugations, and pH-controlled drug release. Chem Mater. 2016;28:1513–25.

Jedlitschky G, Burchell B, Keppler D. The multidrug resistance protein 5 functions as an ATP-dependent export pump for cyclic nucleotides. J Biol Chem. 2000;275:30069–74.

Johnstone RM. Exosomes biological significance: A concise review. Blood Cells Mol Dis. 2006;36:315–21.

Kakyo M, Sakagami H, Nishio T, Nakai D, Nakagomi R, Tokui T, Naitoh T, Matsuno S, Abe T, Yawo H. Immunohistochemical distribution and functional characterization of an organic anion transporting polypeptide 2 (oatp2). Febs Lett. 1999a;445:343–6.

Kakyo M, Unno M, Tokui T, Nakagomi R, Nishio T, Iwasashi H, Nakai D, Seki M, Suzuki M, Naitoh T, Matsuno S. Molecular characterization and functional regulation of a novel rat liver-specific organic anion transporter rlst-1. Gastroenterology. 1999b;117:770–5.

Kanai N, Lu R, Satriano JA, Bao Y, Wolkoff AW, Schuster VL. Identification and characterization of a prostaglandin transporter. Science (Wash DC). 1995;268:866–869.

Kaur S, Manhas P, Swami A, Bhandari R, Sharma KK, Jain R, Kumar R, Pandey SK, Kuhad A, Sharma RK, Wangoo N. Bioengineered PLGA-chitosan nanoparticles for brain targeted intranasal delivery of antiepileptic TRH analogues. Chem Eng J. 2018, 346:630–9. DOI:10.1016/j.cej.2018.03.176.

Kessel D, Bosmann HB. On the characteristics of actinomycin D resistance in L5178Y cells. Cancer Res. 1970;30:2695–701.

Kim JY, Choi WI, Kim YH, Tae G. Brain-targeted delivery of protein using chitosan-and RVG peptide-conjugated, pluronic-based nano-carrier. Biomaterials. 2013;34(4):1170–8.

Kimura M, Nabekura T, Katsura T, Takano M, Hori R. Identification of organic cation transporter in rat renal brush-border membrane by photoaffinity labeling. Biol Pharm Bull. 1995;18:388–95.

Klein I, Sarkadi B, Váradi A. An inventory of the human ABC proteins. Biochim Biophys Acta-Biomembr. 1999;1461:237–62.

König J, Nies AT, Cui Y, Leier I, Keppler D. Conjugate export pumps of the multidrug resistance protein (MRP) family: Localization, substrate specificity, and MRP2-mediated drug resistance. Biochim Biophys Acta-Biomembr. 1999a;1461:377–94.

König J, Rost D, Cui Y, Keppler D. Characterization of the human multidrug resistance protein isoform MRP3 localized to the basolateral hepatocyte membrane. Hepatology. 1999b;29:1156–63.

Koo YE, Reddy GR, Bhojani M, Schneider R, Philbert MA, Rehemtulla A, et al. Brain cancer diagnosis and therapy with nanoplatforms. Adv Drug Deliv Rev. 2006;58:1556–77.

Koziara JM, Lockman PR, Allen DD, Mumper RJ. Paclitaxel nanoparticles for the potential treatment of brain tumors. J Controlled Release. 2004;99:259–69.

Ku S, Yan F, Wang Y, Sun Y, Yang N, Ye L. The blood–brain barrier penetration and distribution of PEGylated fluorescein-doped magnetic silica nanoparticles in rat brain. Biochem Biophys Res Commun. 2010;394:871–6.

Kuffler SW, Nicholls JG, Orkand RK. Physiological properties of glial cells in the central nervous system of amphibia. J Neurophysiol. 1966;29:768–87.

Kullak-Ublick GA, Hagenbuch B, Stieger B, Schteingart CD, Hofmann AF, Wolkoff AW, Meier PJ. Molecular and functional characterization of an organic anion transporting polypeptide cloned from human liver. Gastroenterology. 1995;109:1274–82.

Kuo YC, Cheng SJ. Brain targeted delivery of carmustine using solid lipid nanoparticles modified with tamoxifen and lactoferrin for antitumor proliferation. Int J Pharm. 2016;499:0–19.

Kurakhmaeva KB, Djindjikhashvili IA, Petrov VE, Balabanyan VU, Voronina TA, Trofimov SS, Kreuter J, Gelperina S, Begley D, Alyautdin RN. Brain targeting of nerve growth factor using poly(butyl cyanoacrylate) nanoparticles. J Drug Target. 2009;17:564–74.

Kusuhara H, Sekine T, Utsunomiya Tate N, Tsuda M, Kojima R, Cha SH, et al. Molecular cloning and characterization of a new multispecific organic anion transporter from rat brain. J Biol Chem. 1999;274:13675–80.

Kusuhara H, Sugiyama Y. Efflux transport systems for drugs at the blood–brain barrier and blood–cerebrospinal fluid barrier (Part 1). Drug Discov Today. 2001;6:150–6.

Lautier D, Canitrot Y, Deeley RG, Cole SP. Multidrug resistance mediated by the multidrug resistance protein (MRP) gene. Biochem Pharmacol. 1996;52:967–77.

Lawson LJ, Perry VH, Dri P, Gordon S. Heterogeneity in the distribution and morphology of microglia in the normal adult mouse brain. Neuroscience. 1990;39:151–70.

Lee G, Dallas S, Hong M, Bendayan R. Drug transporters in the central nervous system: Brain barriers and brain parenchyma considerations. Pharmacol Rev. 2001;53:569–96.

Leier I, Jedlitschky G, Buchholz U, Center M, Susan PC, Deeley RG, Keppler D. ATP-dependent glutathione disulphide transport mediated by the MRP gene-encoded conjugate export pump. Biochem J. 1996;314:433–7.

Leiro V, Santos SD, Lopes CDF, Pêgo AP. Dendrimers as powerful building blocks in central nervous system disease: Headed for successful nanomedicine. Adv Funct Mater. 2017;28(12):1700313.

Li L, Lee TK, Meier PJ, Ballatori N. Identification of glutathione as a driving force and leukotriene C4 as a substrate for oatp1, the hepatic sinusoidal organic solute transporter. J Biol Chem. 1998;273:16184–91.

Lipinski CA, Lombardo F, Dominy BW, Feeney PJ. Experimental and computational approaches to estimate solubility and permeability in drug discovery and development settings. Advanced drug delivery reviews. 1997;23:3–25.

Li X, Tsibouklis J, Weng T, Zhang B, Yin G, Feng G, Cui Y, Savina IN, Mikhalovska LI, Sandeman SR, Howel CA. Nano carriers for drug transport across the blood–brain barrier. Journal of drug targeting. 2017;25:17–28.

Li Y, He H, Jia X, Lu W, Lou J, Wei Y. A dual-targeting nanocarrier based on poly(amidoamine) dendrimers conjugated with transferrin and tamoxifen for treating brain gliomas. Biomaterials. 2012;33:3899–908.

Lin G, Chen T, Zou J, Wang Y, Wang X, Li J, Huang Q, Fu Z, Zhao Y, Lin MC, Xu G. Quantum dots-siRNA nanoplexes for gene silencing in central nervous system tumor cells. Front Pharmacol. 2017;8.

Liu L, Guo K, Lu J, Venkatraman SS, Luo D, Ng KC, et al. Biologically active core/shell nanoparticles self-assembled from cholesterol-terminated PEG-TAT for drug delivery across the blood–brain barrier. Biomaterials. 2008a;29:1509–17.

Liu L, Venkatraman SS, Yang YY, Guo K, Lu J, He B, Moochhala S, Kan L. Polymeric micelles anchored with TAT for delivery of antibiotics across the blood–brain barrier. Pept Sci. 2008b;90:617–23.

Liu X, Vilenski O, Kwan J, et al. Unbound brain concentration determines receptor occupancy: A correlation of drug concentration and brain serotonin and dopamine reuptake transporter occupancy for eighteen compounds in rats. Drug Metab Dispos. 2009;37:1548–56.

Liu X, Ye M, An C, Pan L, Ji L. The effect of cationic albumin-conjugated PEGylated tanshinone IIA nanoparticles on neuronal signal pathways and neuroprotection in cerebral ischemia. Biomaterials. 2013;34:6893–905.

Liu S, Yang S, Ho PC. Intranasal administration of carbamazepine-loaded carboxymethyl chitosan nanoparticles for drug delivery to the brain. Asian J Pharm Sci. 2018;13:72–81.

Loe DW, Almquist KC, Cole SP, Deeley RG. ATP-dependent 17-estradiol 17-(-D-glucuronide) transport by multidrug resistance protein (MRP) inhibition by cholestatic steroids. J Biol Chem. 1996a;271:9683–9.

Loe DW, Almquist KC, Deeley RG, Cole SP. Multidrug resistance protein (MRP)-mediated transport of leukotriene C and chemotherapeutic agents in membrane vesicles demonstration of glutathione-dependent vincristine transport. J Biol Chem. 1996b;271:9675–82.

Löscher W, Potschka H. Drug resistance in brain diseases and the role of drug efflux transporters. Nature Rev Neurosci. 2005a;6:591.

Löscher W, Potschka H. Role of drug efflux transporters in the brain for drug disposition and treatment of brain diseases. Prog Neurobiol. 2005b;76:22–76.

Loureiro JA, Gomes B, Fricker G, Coelho MA, Rocha S, Pereira MC. Cellular uptake of PLGA nanoparticles targeted with anti-amyloid and anti-transferrin receptor antibodies for Alzheimer's disease treatment. Colloids Surf B Biointerfaces. 2016;145:8–13.

Loureiro JA, Andrade S, Duarte A, Neves AR, Queiroz JF, Nunes C, et al. Resveratrol and grape extract-loaded solid lipid nanoparticles for the treatment of Alzheimer's disease. Molecules. 2017;22:277.

Lundquist S, Renftel M, Brillault J, Fenart L, Cecchelli R, Dehouck MP. Prediction of drug transport through the blood–brain barrier *in vivo*: A comparison between two *in vitro* cell models. Pharm Res. 2002;19(0724–8741; 7):976–81.

Madan J, Pandey RS, Jain V, Katare OP, Chandra R, Katyal A. Poly (ethylene)-glycol conjugated solid lipid nanoparticles of noscapine improve biological half-life, brain delivery and efficacy in glioblastoma cells. Nanomed Nanotechnol Biol Med. 2013;9:492–503.

Marrink SJ, Jahnig F, Berendsen HJC. Proton transport across transient single-file water pores in a lipid membrane studied by molecular dynamics simulations. Biophys J. 1996;71:632–47.

Martin I. Prediction of blood–brain barrier penetration: Are we missing the point? Drug Discov Today. 2004;9:161–2.

Martins S, Tho I, Reimold I, Fricker G, Souto E, Ferreira D, Brandl M. Brain delivery of camptothecin by means of solid lipid nanoparticles: Formulation design, *in vitro* and *in vivo* studies. Int J Pharm. 2012;439:49–62.

Masserini M. Nanoparticles for brain drug delivery. ISRN Biochem. 2013;2013:1–18.

Masuda S, Ibaramoto K, Takeuchi A, Saito H, Hashimoto Y, Inui K. Cloning and functional characterization of a new multispecific organic anion transporter, OAT-K2, in rat kidney. Mol Pharmacol. 1999;55:743–52.

Mateos L, Perez-Alvarez MJ, Wandosell F. Angiotensin II type-2 receptor stimulation induces neuronal VEGF synthesis after cerebral ischemia. Biochim Biophys Acta-Mol Basis Dis. 2016;1862:1297–308.

Mauri E, Veglianese P, Papa S, Mariani A, De Paola M, Rigamonti R, Chincarini GM, Vismara I, Rimondo S, Sacchetti A, Rossi F. Double conjugated nanogels for selective intracellular drug delivery. RSC Adv. 2017;7:30345–56.

Mehdipour AR, Hamidi M. Brain drug targeting: A computational approach for overcoming the blood–brain barrier. Drug Discov Today. 2009;14:1030–6.

Meijer DK, Mol WE, Muller M, Kurz G. Carrier-mediated transport in the hepatic distribution and elimination of drugs, with special reference to the category of organic cations. J Pharmacokinet Biopharm. 1990;18:35–70.

Meng J, Agrahari V, Youm I. Advances in targeted drug delivery approaches for the central nervous system tumors: The inspiration of nanobiotechnology. J Neuroimmune Pharmacol. 2017;12:84–98.

Micheli MR, Bova R, Magini A, Polidoro M, Emiliani C. Lipid-based nanocarriers for CNS-targeted drug delivery. Recent Pat CNS Drug Discov. 2012;7:71–86.

Mishra BB, Patel BB, Tiwari S. Colloidal nanocarriers: A review on formulation technology, types and applications toward targeted drug delivery. Nanomed Nanotechnol Biol Med. 2010;6:9–24.

Misra A, Ganesh S, Shahiwala A, Shah SP. Drug delivery to the central nervous system: A review. J Pharm Pharm Sci. 2003;6:252–73.

Muntimadugu E, Dhommati R, Jain A, Challa VGS, Shaheen M, Khan W. Intranasal delivery of nanoparticle encapsulated tarenflurbil: A potential brain targeting strategy for Alzheimer's disease. Eur J Pharm Sci. 2016;92:224–34.

Murakami H, Takanaga H, Matsuo H, Ohtani H, Sawada Y. Comparison of blood–brain barrier permeability in mice and rats using in situ brain perfusion technique. Am J Physiol-Heart Circ Physiol. 2000;279:H1022–8.

Nagpal K, Singh SK, Mishra DN. Drug targeting to brain: A systematic approach to study the factors, parameters and approaches for prediction of permeability of drugs across BBB. Exp Opin Drug Deliv. 2013;10:927–55.

Naik P, Cucullo L. *In vitro* blood–brain barrier models: Current and perspective technologies. J Pharm Sci. 2012;101:1337–54.

Nasr M. Development of an optimized hyaluronic acid-based lipidic nanoemulsion co-encapsulating two polyphenols for nose to brain delivery. Drug Deliv. 2016;23:1444–52.

Natarajan J, Baskaran M, Humtsoe LC, Vadivelan R, Justin A. Enhanced brain targeting efficacy of Olanzapine through solid lipid nanoparticles. Artif Cells Nanomed Biotechnol. 2017;45:364–71.

Neuwelt EA. Mechanisms of disease: the blood–brain barrier. Neurosurgery. 2004;54:131–140 discussion. 141–132.

Neves AR, Queiroz JF, Lima SAC, Reis S. Apo E-functionalization of solid lipid nanoparticles enhances brain drug delivery: Uptake mechanism and transport pathways. Bioconjugate Chem. 2017;28:995–1004.

Nichols JW, Deamer DW. Net proton-hydroxyl permeability of large unilamellar liposomes measured by an acid-base titration technique. Proc Natl Acad Sci. 1980;77(4):2038–42.

Nichols JW, Hill MW, Bangham AD, Deamer DW. Measurement of net proton-hydroxyl permeation of large unilamellar liposomes with the fluorescent pH probe 9-aminoacridine. Biochim Biophys Acta-Biomembr. 1980;596:393–403.

Noe B, Hagenbuch B, Stieger B, Meier PJ. Isolation of a multispecific organic anion and cardiac glycoside transporter from rat brain. Proc Natl Acad Sci. 1997;94:10346–50.

Oldendorf WH. Measurement of brain uptake of radiolabeled substances using a tritiated water internal standard. Brain Res. 1970;24:372–6.

Olivier JC. Drug transport to brain with targeted nanoparticles. NeuroRx. 2005;2:108–19.

Pahuja R, Seth K, Shukla A, Shukla RK, Bhatnagar P, Chauhan LKS, Saxena PN, Arun J, Chaudhari BP, Patel DK, Singh SP. Trans-blood–brain barrier delivery of dopamine-loaded nanoparticles reverses functional deficits in parkinsonian rats. ACS Nano. 2015;9:4850–71.

Pardeshi C, Rajput P, Belgamwar V, Tekade A, Patil G, Chaudhary K, Sonje A. Solid lipid based nanocarriers: An overview/Nanonosači na bazi čvrstih lipida: Pregled. Acta Pharm. 2012;62:433–72.

Pardridge WM. Recent developments in peptide drug delivery to the brain. Basic Clin Pharmacol Toxicol. 1992;71:3–10.

Pardridge WM. Log(BB), PS products and *in silico* models of drug brain penetration. Drug Discov Today. 2004;9:392–3.

Pardridge WM. The blood–brain barrier and neurotherapeutics. Neurotherapeutics. 2005;2:1–2.

Pardridge WM. shRNA and siRNA delivery to the brain. Adv Drug Deliv Rev. 2007;59:141–52.

Pardridge WM, Mietus LJ. Transport of steroid hormones through the rat blood–brain barrier: Primary role of albumin-bound hormone. J Clin Invest. 1979;64:145–54.

Pardridge WM, Sakiyama R, Fierer G. Transport of propranolol and lidocaine through the rat blood–brain barrier: Primary role of globulin-bound drug. J Clin Invest. 1983;71(4):900–8.

Parmar H, Bhandari A, Shah D. Recent techniques in nasal drug delivery: A review. Int J Drug Dev Res. 2011;3:99–106.

Partoazar A, Nasoohi S, Rezayat SM, Gilani K, Mehr SE, Amani A, Rahimi N, Dehpour AR. Nanoliposome containing cyclosporine A reduced neuroinflammation responses and improved neurological activities in cerebral ischemia/reperfusion in rat. Fundam Clin Pharmacol. 2017;31:185–93.

Parveen S, Misra R, Sahoo SK. Nanoparticles: A boon to drug delivery, therapeutics, diagnostics and imaging. Nanomed Nanotechnol Biol Med. 2012;8:147–66.

Pasha S, Gupta K. Various drug delivery approaches to the central nervous system. Exp Opin Drug Deliv. 2010;7:113–35.

Patel T, Zhou J, Piepmeier JM, Saltzman WM. Polymeric nanoparticles for drug delivery to the central nervous system. Adv Drug Deliv Rev. 2011;64:701–5.

Pavan B, Dalpiaz A, Ciliberti N, Biondi C, Manfredini S, Vertuani S. Progress in drug delivery to the central nervous system by the prodrug approach. Molecules. 2008;13:1035–65.

Peppas NA, Bures P, Leobandung W, Ichikawa H. Hydrogels in pharmaceutical formulations. Eur J Pharm Biopharm. 2000;50:27–46.

Peters A, Palay SL, Webster HF. The neuroglia cells. In: Peters A, Palay SL, Webster HF editors. The fine structure of the nervous system. Neurons and their supporting cells; 3rd edition. Oxford university press, Inc., New York, New York, 10016. 1991. p. 273–311.

Picone P, Ditta LA, Sabatino MA, Militello V, Biagio PLS, Giacinto MLD, et al. Ionizing radiation-engineered nanogels as insulin nanocarriers for the development of a new strategy for the treatment of Alzheimer's disease. Biomaterials. 2016;80:179–94.

Pinter GG, Atkins JL, Bell DR. Albumin permeability times surface area (PS) product of peritubular capillaries in kidney. Experientia. 1974;30:1045.

Portioli C, Bovi M, Benati D, Donini M, Perduca M, Romeo A, Dusi S, Monaco HL, Bentivoglio M. Novel functionalization strategies of polymeric nanoparticles as carriers for brain medications. J Biomed Mater Res Part A. 2017;105:847–58.

Qiao R, Jia Q, Hüwel S, Xia R, Liu T, Gao F, Galla HJ, Gao M. Receptor-mediated delivery of magnetic nanoparticles across the blood–brain barrier. ACS Nano. 2012;6:3304–10.

Raivich G, Bohatschek M, Kloss CU, Werner A, Jones LL, Kreutzberg GW. Neuroglial activation repertoire in the injured brain: Graded response, molecular mechanisms and cues to physiological function. Brain Res Rev. 1999;30:77–105.

Raj R, Wairkar S, Sridhar V, Gaud R. Pramipexole dihydrochloride loaded chitosan nanoparticles for nose to brain delivery: Development, characterization and *in vivo* anti-Parkinson activity. Int J Biol Macromol. 2018;109:27–35.

Ramírez-Sánchez J, Pires ENS, Nuñez-Figueredo Y, Pardo-Andreu GL, Fonseca-Fonseca LA, Ruiz-Reyes A, Ochoa-Rodríguez E, Verdecia-Reyes Y, Delgado-Hernández R, Souza DO, Salbego C. Neuroprotection by JM-20 against oxygen-glucose deprivation in rat hippocampal slices: Involvement of the Akt/GSK-3β pathway. Neurochem Int. 2015;90:215–23.

Reichel A. Addressing central nervous system (CNS) penetration in drug discovery: Basics and implications of the evolving new concept. Chem Biodivers. 2009;6:2030–49.

Rennick BR. Renal tubule transport of organic cations. Am J Physiol. 1981;240:F83–F89.

Ren J, Shen S, Wang D, Xi Z, Guo L, Pang Z, Qian Y, Sun X, Jiang X. The targeted delivery of anticancer drugs to brain glioma by PEGylated oxidized multi-walled carbon nanotubes modified with angiopep-2. Biomaterials. 2012;33:3324–33.

Riant P, Urien S, Albengres E, Renouard A, Tillement JP. Effects of the binding of imipramine to erythrocytes and plasma proteins on its transport through the rat blood–brain barrier. J Neurochem. 1988;51:421–5.

Rodriguez M, Kaushik A, Lapierre J, Dever S, El-Hage N, Nair M. Electro-magnetic nanoparticle bound Beclin1 siRNA crosses the blood–brain barrier to attenuate the inflammatory effects of HIV-1 infection *in vitro*. J Neuroimmune Pharmacol. 2017;12:120–32.

Sahoo SK, Labhasetwar V. Nanotech approaches to drug delivery and imaging. Drug Discov Today. 2003;8:1112–20.

Samad A, Sultana Y, Aqil M. Liposomal drug delivery systems: An update review. Curr Drug Deliv. 2007;4:297–305.

Sanchez-Ramos J, Song S, Kong X, Foroutan P, Martinez G, Dominguez-Viqueria W, Mohapatra S, Mohapatra S, Haraszti RA, Khvorova A, Aronin N. Chitosan-mangafodipir nanoparticles designed for intranasal delivery of siRNA and DNA to brain. J Drug Deliv Sci Technol. 2018;43:453–60.

Saraiva C, Paiva J, Santos T, Ferreira L, Bernardino L. MicroRNA-124 loaded nanoparticles enhance brain repair in Parkinson's disease. J Controlled Release. 2016;235:291–305.

Schlachetzki F, Zhang YF, Boado RJ, Pardridge WM. Gene therapy of the brain. Neurology. 2004;62:1275–81.

Schuetz JD, Connelly MC, Sun D, Paibir SG, Flynn PM, Srinivas RV, et al. MRP4: A previously unidentified factor in resistance to nucleoside based antiviral drugs. Nat Med. 1999;5:1048–51.

Sekine T, Watanabe N, Hosoyamada M, Kanai Y, Endou H. Expression cloning and characterization of a novel multispecific organic anion transporter. J Biol Chem. 1997;272:18526–9.

Sekine T, Cha SH, Tsuda M, Apiwattanakul N, Nakajima N, Kanai Y, Endou H. Identification of multispecific organic anion transporter 2 expressed predominantly in the liver. Febs Lett. 1998;429:179–82.

Sekine T, Cha SH, Endou H. The multispecific organic anion transporter (OAT) family. Pfluegers Arch. 2000;89:337–44.

Sharma N, Bhandari S, Deshmukh R, Yadav AK, Mishra N. Development and characterization of embelin-loaded nanolipid carriers for brain targeting. Artif Cells Nanomed Biotechnol. 2017;45:409–13.

Shaw TK, Mandal D, Dey G, Pal MM, Paul P, Chakraborty S, Ali KA, Mukherjee B, Bandyopadhyay AK, Mandal M. Successful delivery of docetaxel to rat brain using experimentally developed nanoliposome: A treatment strategy for brain tumor. Drug Deliv. 2017;24:346–57.

Sherwood L. Human physiology: From cells to systems. Cengage Learning; 2015.

Sihorkar V, Vyas SP. Polysaccharide coated liposomes for oral drug delivery: Development and characterization. In: Vyas SP, Dixit VK, editors. Advances in liposomal therapeutics. New Delhi: CBS Publishers; 2001.

Silva GA. Nanotechnology approaches to crossing the blood–brain barrier and drug delivery to the CNS. BMC Neurosci. 2008;9:S4.

Smit JJ, Schinkel AH, Oude Elferink RP, Groen AK, Wagenaar E, van Deemter L, et al. Homozygous disruption of the murine mdr2 P-glycoprotein expressed in a porcine kidney epithelial cell line LLCPK1. J Pharmacol Exp Ther. 1993;263:840–5.

Smith QR. A review of blood–brain barrier transport techniques. Methods Mol Med. 2003;89:193–208.

Spector R. Drug transport in the central nervous system: Role of carriers. Pharmacology. 1990;40:1–7.

Srinageshwar B, Peruzzaro S, Andrews M, Johnson K, Hietpas A, et al. PAMAM dendrimers cross the blood–brain barrier when administered through the carotid artery in C57BL/6J mice. Int J Mol Sci. 2017;18:628.

Stockwell J, Abdi N, Lu X, Maheshwari O, Taghibiglou C. Novel central nervous system drug delivery systems. Chem Biol Drug Des. 2014;83:507–20.

Su L, Zhang B, Huang Y, Zhang H, Xu Q, Tan J. Superparamagnetic iron oxide nanoparticles modified with dimyristoylphosphatidylcholine and their distribution in the brain after injection in the rat substantia nigra. Mater Sci Eng C. 2017;81:400–6.

Takasato Y, Rapoport SI, Smith QR. An *in situ* brain perfusion technique to study cerebrovascular transport in the rat. Am J Physiol. 1984;247:H484–93.

Tamba BI, Streinu V, Foltea G, Neagu AN, Dodi G, Zlei M, Tijani A, Stefanescu C. Tailored surface silica nanoparticles for blood–brain barrier penetration: Preparation and *in vivo* investigation. Arabian J Chem. 2018;11:981–990. DOI:10.1016/j.arabjc.2018.03.019.

Tandon VR, Kapoor B, Bano G, Gupta S, Gillani Z, Kour D. P-glycoprotein: Pharmacological relevance. Int J Pharmacol. 2006;38:13–24.

Tiwari SB, Amiji MM. A review of nanocarrier-based CNS delivery systems. Curr Drug Deliv. 2006;3:219–32.

Tsuji A. Specific mechanisms for transporting drugs into the brain. New York: Marcel Dekker; 2000. p. 121–44.

Ueda K, Taguchi Y, Morishima M. How does P-glycoprotein recognize its substrates? Cancer Biol. 1997;8:151–9.

Upadhyay P, Trivedi J, Pundarikakshudu K, Sheth N. Direct and enhanced delivery of nanoliposomes of anti schizophrenic agent to the brain through nasal route. Saudi Pharm J. 2017;25:346–58.

van Giau, V An SSA. Emergence of exosomal miRNAs as a diagnostic biomarker for Alzheimer's disease. J Neurol Sci. 2016;360:141–152.

van Rooy I, Cakir-Tascioglu S, Hennink WE, Storm G, Schiffelers RM, Mastrobattista E. *In vivo* methods to study uptake of nanoparticles into the brain. Pharm Res. 2011;28(3):456–71.

Van Giau V, An SS. Emergence of exosomal miRNAs as a diagnostic biomarker for Alzheimer's disease. J Neurol Sci. 2016;360:141–52.

Vella LJ, Hill AF, Cheng L. Focus on extracellular vesicles: Exosomes and their role in protein trafficking and biomarker potential in Alzheimer's and Parkinson's disease. Int J Mol Sci. 2016;17:173.

Vilella A, Belletti D, Sauer AK, Hagmeyer S, Sarowar T, Masoni M, Stasiak N, Mulvihill JJ, Ruozi B, Forni F, Vandelli MA. Reduced plaque size and inflammation in the APP23 mouse model for Alzheimer's disease after chronic application of polymeric nanoparticles for CNS targeted zinc delivery. J Trace Elem Med Biol. 2017;49:210–221. DOI:10.1016/j.jtemb.2017.12.006.

Villalobos AR, Parmelee JT, Pritchard JB. Functional characterization of choroid plexus epithelial cells in primary culture. J Pharmacol Exp Ther. 1997;282:1109–16.

Vogtle F, Buhleier EW, Wehner W. Cascade and Nonskid-Chain-Like Syntheses of Molecular Cavity Topologies. Synthesis. 1978;2:155–158.

Walz W. Controversy surrounding the existence of discrete functional classes of astrocytes in adult gray matter. Glia. 2000;31:95–103.

Wan H, Rehngren M, Giordanetto F, Bergström F, Tunek A. High-throughput screening of drug–brain tissue binding and *in silico* prediction for assessment of central nervous system drug delivery. J Med Chem. 2007;50:4606–15.

Wang JX, Sun X, Zhang ZR. Enhanced brain targeting by synthesis of 3', 5'-dioctanoyl-5-fluoro-2'-deoxyuridine and incorporation into solid lipid nanoparticles. Eur J Pharm Biopharm. 2002;54:285–90.

Wijnholds J, Mol CA, van Deemter L, de Haas M, Scheffer GL, Baas F, et al. Multidrug-resistance protein 5 is a multispecific organic anion transporter able to transport nucleotide analogs. Proc Natl Acad Sci. 2000;97:7476–81.

Wong HL, Bendayan R, Rauth AM, Li Y, Wu XY. Chemotherapy with anticancer drugs encapsulated in solid lipid nanoparticles. Adv Drug Deliv Rev. 2007;59:491–504.

Wong HL, Wu XY, Bendayan R. Nanotechnological advances for the delivery of CNS therapeutics. Adv Drug Deliv Rev. 2011;64:686–700.

Yadav S, Gandham SK, Panicucci R, Amiji MM. Intranasal brain delivery of cationic nanoemulsion-encapsulated TNFα siRNA in prevention of experimental neuroinflammation. Nanomed Nanotechnol Biol Med. 2016;12:987–1002.

Yáñez JA, Forrest ML, Ohgami Y, Kwon GS, Davies NM. Pharmacometrics and delivery of novel nanoformulated PEG-b-poly (ε-caprolactone) micelles of rapamycin. Cancer Chemother Pharmacol. 2008;61:133–44.

Yang S, Zhu J, Lu Y, Yang C. Body distribution of camptothecin solid lipid nanoparticles after oral administration. Pharm Res. 1999;16:751–7.

Yemisci M, Caban S, Gursoy-Ozdemir Y, Lule S, Novoa-Carballal R, Riguera R, Fernandez-Megia E, Andrieux K, Couvreur P, Capan Y, Dalkara T. Systemically administered brain-targeted nanoparticles transport peptides across the blood–brain barrier and provide neuroprotection. J Cereb Blood Flow Metab. 2015;35:469–75.

Yim YS, Choi JS, Kim GT, Kim CH, Shin TH, Kim DG, Cheon J. A facile approach for the delivery of inorganic nanoparticles into the brain by passing through the blood–brain barrier (BBB). Chem Commun. 2012;48:61–3.

Zara GP, Cavalli R, Bargoni A, Fundarò A, Vighetto D, Gasco MR. Intravenous administration to rabbits of non-stealth and stealth doxorubicin-loaded solid lipid nanoparticles at increasing concentrations of stealth agent: Pharmacokinetics and distribution of doxorubicin in brain and other tissues. J Drug Target. 2002;10:327–35.

Zhang L, Brett CM, Giacomini KM. Role of organic cation transporters in drug absorption and elimination. Annu Rev Pharmacol Toxicol. 1998;38:431–60.

Zhang Y, Zhang YF, Bryant J, Charles A, Boado RJ, Pardridge WM. Intravenous RNA interference gene therapy targeting the human epidermal growth factor receptor prolongs survival in intracranial brain cancer. Clin Cancer Res. 2004;10:3667–77.

Zhang P, Hu L, Yin Q, Feng L, Li Y. Transferrin-modified c[RGDfK]-paclitaxel loaded hybrid micelle for sequential blood–brain barrier penetration and glioma targeting therapy. Mol Pharmaceutics. 2012;9:1590–8.

Zhang X, Chen G, Wen L, Yang F, Shao AL, Li X, et al. Novel multiple agents loaded PLGA nanoparticles for brain delivery via inner ear administration: *In vitro* and *in vivo* evaluation. Eur J Pharm Sci. 2013;48:595–603.

Zhang F, Magruder JT, Lin YA, Crawford TC, Grimm JC, Sciortino CM, Wilson MA, et al. Generation-6 hydroxyl PAMAM dendrimers improve CNS penetration from intravenous administration in a large animal brain injury model. J Controlled Release. 2017;249:173–82.

Zhu C, Zhang Y, Zhang YF, Yi LJ, Boado RJ, Pardridge WM. Organ-specific expression of the lacZ gene controlled by the opsin promoter after intravenous gene administration in adult mice. J Gene Med. 2004;6:906–12.

Zlokovic BV, Begley DJ, Djuricic BM, Mitrovic DM. Measurement of solute transport across the blood–brain barrier in the perfused guinea pig brain: Method and application to N-methyl-alpha-aminoisobutyric acid. J Neurochem. 1986;46:1444–51.

11

Recent Advances and Potential Applications of Nanoemulsions in Food Stuffs: Industrial Perspectives

Bankim Chandra Nandy, Pranab Jyoti Das, and Sandipan Dasgupta

CONTENTS

ABSTRACT: Nanoemulsion-based delivery systems have a high demand in the food industry due to their unique properties. Nanoemulsion systems have gained more attention for the incorporation of functional compounds to prevent their degradation and enhance their bioavailability, solubility, and consumer acceptance. The present chapter illustrates and gives an overview of the nanoemulsion production system, ingredients used, applications in the food industry, and also the current analytical techniques used for the identification and characterization of nanoemulsions. Moreover, the physical properties of nanoemulsions and the possible factors that affect the design of functional nanoemulsions are briefly described. The chapter also throws some light on the toxicological concerns and regulatory aspects of food grade nanoemulsions.

KEY WORDS: *Nanoemulsion, functional food, characterization, food industry, toxicology, stability*

11.1 Introduction

In the field of food processing and packaging, rapid technological advances have been extended from micro to nano-sized (10^{-9} m) particles (Huang et al., 2010). The adaptation of nanotechnology for the food arena allows the modification of physical characteristics like texture, strength, and shelf-life of food, and it also improves the solubility, thermal stability, and bioavailability of functional compounds (Huang et al., 2010; McClements et al., 2007, 2009). Among the many types of nanometric systems, nanoemulsions are now being progressively more used for encapsulating bioactive compounds and acting as transport vehicles in probiotic health foods and other functional foods.

Conventional formulations like tablets and capsules of functional compounds are easily available to provide health benefits such as the prevention or treatment of diseases, but due to their low bioavailability these solutions may not sustain the health benefits of the functional compounds, in particular for lipophilic compounds (Chen et al., 2006; Chu et al., 2007; Spernath and Aserin, 2006). Gastrointestinal absorption is an important step pertaining to the bioavailability of the lipophilic functional compounds. Thus, enhancing the absorption of the lipophilic compounds in the gastrointestinal tract would address the problem of low bioavailability. The development of nanotechnology offers potential benefits for the improvement of water solubility and the bioavailability of lipophilic functional compounds (Chu et al., 2007).

Nanotechnology is the technology related to the research, development, and control of structures from 1 to 100 nm in size (Quintanilla-Carvajal et al., 2010). In addition, nanotech sensors in the packages are able to detect pathogens, toxins, and other chemicals, enabling consumers to avoid food spoilage.

Ongoing research areas include "interactive foods", like beverages which change flavors and colors, and foods that can adjust to a person's nutritional needs.

Among all the nanoformulations, nanoemulsions are one of the most amazing areas of application with immense immediate potential for industrial adoption. Several studies found that nanoemulsions can act as carriers or delivery systems for lipophilic compounds, such as nutraceuticals, drugs, flavors, antioxidants, and antimicrobial agents (Jiahui et al., 2004; Kesisoglou et al., 2007; McClements et al., 2007; Sanguansri and Augustin, 2006; Weiss et al., 2008; Wissing et al., 2004). Generally, nanoemulsions have better stability due to the domination of Brownian motion over gravitational force as well as smaller particle size as compared to the other conventional emulsion formulations (McClements, 2005). As the particle size decreases in the nanoemulsion, the range of attractive forces acting between the droplets decreases, leading to lesser droplet aggregation even as the range of steric repulsion is less (McClements, 2005; Tadros et al., 2004). A decrease in the particle size leads to an increase in the surface to volume ratio of nanoemulsion-based delivery systems, which helps to improve the bioavailability of the encapsulated components (Acosta, 2009). A range of methods such as emulsification– evaporation, emulsification–diffusion, solvent displacement, and precipitation have been developed to prepare nano-size globule diameters in lipophilic functional compounds, as mentioned in Table 11.1. The methods can be further classified as high or low energy approaches (Acosta, 2009; Chu et al., 2007; Leong et al., 2009; Tadros et al., 2004).

In this chapter, the industrial applications of nanoemulsions in food and food ingredients processing, preservation, and decontamination are presented. A description of the analytical techniques used for nanoemulsion characterization in this domain is also provided and their implications for the safety of foods. Various toxicological aspects and regulatory requirements for nanoemulsions are discussed.

11.1.1 Production Methods for Nanoemulsion

Nanoemulsion is a biphasic dosage form consisting of an aqueous continuous phase and an oil dispersal phase in the nanometric size range, in which oil droplets are covered with surfactant molecules (Acosta, 2009; McClements et al., 2007, 2009). Nanoemulsions are produced via various methods, like low energy and high energy methods, as given in Table 11.1.

In high energy methods, the devices used have high mechanical force, leading to more uniform size droplets. In this method, due to high shear force, some chemicals may be degraded during the production process (Date et al., 2010).

The preparation methods followed in the high energy process are

- High pressure homogenization
- High shear homogenization
- Ultrasonication.

On the other side, the low energy method is very simple, and generally depends upon low energy and the intrinsic physicochemical properties of surfactants and the oily phase for the production of nanoemulsions (Date et al., 2010).

The preparation methods followed in the low energy process are

- Membrane emulsification
- Spontaneous emulsification
- Solvent displacement
- Emulsion inversion point
- Phase inversion point.

The low energy process in particular needs the development of a comprehensive phase diagram for guidance in formulation development.

The materials used in nanoemulsion production are laid out as follows.

TABLE 11.1

An Overview of Techniques Used for the Preparation of Nanoemulsions

Techniques	Oil Phase/ Solvent	Bioactive Component	Surfactant/ Co-Surfactant	Particle Diameter	References
High pressure homogenization	Medium-chain triglycerides (MCT)	β-carotene	Tween 20, 40, 60, 80	132–148 nm	Yuan et al. (2008b)
	MCT	β-carotene	Tween 20, decaglycerol monolaurate, whey protein isolate, modified starch	115–178 nm	Mao et al. (2009)
	Oleoresin capsicum/ethyl acetate	Poly-caprolactone	Pluronic F68	320–460 nm	Choi et al. (2009)
	Olive oil	Poly-caprolactone	Pluronic F68, phospholipid	239–274 nm	Wulff-Perez et al. (2010)
	MCT	Curcumin	Tween 20	160 nm	Wang et al. (2008)
Microfluidizer	Peanut oil		Whey protein concentrate, whey protein hydrolysate, sodium caseinate, modified starch	150–600 nm	Jafari et al. (2007)
			Tween 80, sodium dodecyl sulfate (SDS)	120 nm	Wooster et al. (2008)
	Corn oil/ethyl acetate		Whey protein isolate	174 nm	Lee and McClements (2010)
	MCT	Coenzyme Q10	Non-ionic surfactant, soybean lecithin, glycerol	60 nm	Hatanaka et al. (2008)
	Corn oil/ octadecane		SDS, Tween 20, beta-lactoglobulin	60–150 nm	Qian and McClements (2011)
	MCT	α-Tocopherol	Glycerol, decaglyceryl monooleate, soybean lecithin	80–400 nm	Hatanaka et al. (2010)
	Hexadecane		Tween 80, SDS	80 nm	Wooster et al. (2008)
Ultrasonic homogenizer	MCT	2-(Butylam ino)-1-phenyl-1 -ethanethiosulf uric acid (BphEA)	Tween 80/ethanol	20–60 nm	Amani et al. (2010)De Araújo et al. (2007)
	Sunflower oil, canola oil		Tween 80, Span 80, SDS	40 nm	Leong et al. (2009)
	Flax seed oil		Tween 40	150 nm	Kentish et al. (2006)
	Olive oil, sesame oil, soybean oil		Pluronic F68	<500 nm	Wulff-Perez et al. (2009)
	soybean oil		Phosphatidylcholine/ sodium palmitate, sucrose palmitate	58 nm	Takegami et al. (2008)

(Continued)

TABLE 11.1 (CONTINUED)

An Overview of Techniques Used for the Preparation of Nanoemulsions

Techniques	Oil Phase/ Solvent	Bioactive Component	Surfactant/ Co-Surfactant	Particle Diameter	References
Spontaneous emulsification	MCT	α-Tocopherol/ acetone	Span 80, Span 85, Tween 20, Tween 80/ acetone, ethanol, ethyl acetate	100–600 nm	Bouchemal et al. (2004)
		Oligothymidylate	Egg lecithin/glycerol	225 nm	Martini et al. (2008)
		Quercetin or methylquercetin	Egg lecithin	300 nm	Fasolo et al. (2007)
		Castor oil	Soybean lecithin, polyoxyl 35 castor oil	150–200 nm	Kelmann et al. (2007)
	Palm oil	Vitamin E	Tween 80, Pluronic F68	94 nm	Teo et al. (2010)
Rotor/stator (ultra-turrax)	n-Hexane	β-carotene	Tween 20	150 nm	Silva et al. (2011)
	Chloroform	Itraconazole	Pectin	200–400 nm	Burapapadh et al. (2010)
Solvent displacement technique	MCTs and lipoid E-80 dissolved in ethanol		Pluronic F18	185–208 nm	Chaix et al. (2003)
Phase inversion temperature	Tetradecane		Brij 30	80–120 nm	Rao and McClements (2010)
	Water/ polyoxyethylene 4-lauryl ether ($C_{12}E_4$)/ hexadecane system		$C_{12}E_4$	50–130 nm	Izquierdo et al. (2004)
	Water/$C_{12}E_4$/ isohexadecane system		$C_{12}E_4$	29–80 nm	Izquierdo et al. (2004)

11.1.1.1 Oil Phase

In nanoemulsion preparation, the oil phase consists of nonpolar components like triglycerols, triacylglycerols, monoacylglycerol, free fatty acids, flavor oils, essential oils, mineral oils, fat substitutes, and waxes. In addition to that, various weighting agents, oil-soluble vitamins, and lipophilic nutraceuticals such as carotenoids, curcumin, phytosterols, and co-enzyme Q are also used. The physicochemical characteristics of the oil phase such as polarity, water-solubility, interfacial tension, refractive index, viscosity, density, phase behavior, and chemical stability helps in the formation, stability, and properties of nanoemulsions (Anton et al., 2007; McClements, 2005; Tadros et al., 2004; Wooster et al., 2008).

In the food industry, triglycerides are frequently used in the preparation of nanoemulsions due to their low cost, high availability, and functional or nutritional characteristics, e.g., corn, soybean, sunflower, safflower, olive, flaxseed, algae, or fish oils. The majority of these oils used in the food industry consist of long-chain triacylglycerols (LCT). Even though medium-chain triacylglycerols (MCT) and short-chain triacylglycerols (SCT) are also being used in food applications, their use is often challenging due to their relatively low polarity, high interfacial tension, and high viscosity. Triglycerides also have the advantage of being metabolized easily, thereby affording high tolerance and very low toxicity potential.

11.1.1.2 Aqueous Phase

In the preparation of a nanoemulsion, the continuous phase usually consists of water, but to improve the stability and appearance of the formulation's polar components, including co-solvents such as simple alcohols and polyols, carbohydrates, proteins, minerals, acids, and bases are also being used. The concentration of these components in the formulation is critical because they govern the different physiochemical characteristics, like polarity, interfacial tension, refractive index, rheology, density, phase behavior, pH, and ionic strength, of the aqueous phase. The optimization of nanoemulsion formulations is critical and it depends upon the aqueous phase composition. In the high pressure homogenization method, optimization of the dispersed to continuous phase viscosity ratio is supported by adding crucial, water-soluble co-solvents to the aqueous phase.

11.1.1.3 Stabilizers

Nanoemulsions are formulations which contain two immiscible components, that is oil and water, so just mixing the oil and water together by homogenization would not stabilize the formulation. The system would rapidly breakdown through a variety of mechanisms, including droplet flocculation, coalescence, Ostwald ripening, and gravitational separation. For this purpose, it is always necessary to stabilize the formulation by adding various kinds of stabilizing agents to improve long-term stability, thus affording a meaningful shelf-life of the product.

11.1.1.4 Emulsifiers/Co-Emulsifiers

Emulsifiers and the co-emulsifiers are the most important components for the preparation of nanoemulsions. Hence, the selection of an appropriate emulsifier (or combination of emulsifiers) is one of the most critical factors to consider for the proper design of a nanoemulsion. The appropriate combination of emulsifier and co-emulsifier can protect the droplet surface from the disruption of droplets and droplet aggregation (McClements, 2005). The high energy method favors the production of small droplets by lowering the interfacial tension in the presence of emulsifiers, which enables droplet disruption. In low energy methods, emulsifiers are able to produce stable nanoemulsions by lowering interfacial tension, which leads to the spontaneous formation of small droplets. Changes in environmental conditions like pH, ionic strength, heating, cooling, or long-term storage leads to the formation of unstable emulsions, so the stability of these formulations is primarily dependent on the kind of emulsifiers used. In the food industry, the most important types of emulsifiers are small molecule surfactants, phospholipids, proteins, and polysaccharides.

The main emulsifiers/surfactants used in nanoemulsion production and their respective types are given in Table 11.2.

11.1.2 Functional Compounds

Different types of lipophilic functional compounds such as bioactive lipids, flavoring agent, antimicrobial agent, antioxidant compound, nutrients, and drugs are encapsulated in nanoemulsion preparation (Chen et al., 2006; McClements et al., 2007; Shefer and Shefer, 2003a), leading to an escalation of their bioactivity, desirability, and palatability (McClements et al., 2007). There are four categories of major lipophilic functional compounds

- Fatty acids (e.g., omega three fatty acids)
- Carotenoids (e.g., β-carotene)
- Antioxidants (e.g., tocopherol)
- Phytosterols (e.g., stigmasterol).

Table 11.3 shows a list of lipophilic functional compounds that have been encapsulated into nanoemulsion systems, their expected benefits and their fields of application.

TABLE 11.2

Emulsifiers/Surfactants Used in Nanoemulsion Production and Their Respective Type

Emulsifier/Surfactant	Type	References
Polyethylene glycol (35) castor oil (Cremophor EL®)	Surfactant (non-ionic)	Usón et al. (2004)
Polyoxyethylene-660-12-hydroxy stearate	Surfactant (non-ionic)	Anton et al. (2007)
Polyoxyethylene 4-lauryl ether ($C_{12}E_4$)	Surfactant (non-ionic)	Izquierdo et al. (2004)
Polyoxyethylene 6-lauryl ether	Surfactant (non-ionic)	Izquierdo et al. (2005)
Span 20	Surfactant (non-ionic)	Porras et al. (2008)
Span 80	Surfactant (non-ionic)	Porras et al. (2008)
Tween 20	Surfactant (non-ionic)	Porras et al. (2008)
Tween 40	Surfactant (non-ionic)	Yuan et al. (2008a)
Tween 60	Surfactant (non-ionic)	Yuan et al. (2008a)
Tween 80	Surfactant (non-ionic)	Porras et al. (2008)
Sucrose fatty acid ester (L1695)	Surfactant (non-ionic)	Yin et al. (2009)
Decaglycerol monolaurate (ML750)	Surfactant (non-ionic)	Yin et al. (2009)
Polyglycerol esters of fatty acids (PGEs)	Surfactant (non-ionic)	Tan and Nakajima (2005b)
Alkanol – XC	Surfactant (anionic)	Howe and Pitt (2008)
Sodium dodecyl sulfate	Surfactant (anionic)	McClements (2000)
Dodecyl trimethyl ammonium bromide	Surfactant (cationic)	McClements (2000)
Gelatin	Protein (amphiphilic)	Ribeiro et al. (2008)
Whey protein isolate (WPI)	Protein (amphiphilic)	Lee and McClements (2010)
Whey protein concentrate (WPC)	Protein (amphiphilic)	Chu et al. (2007)
Whey protein hydrolysate (WPH)	Protein (amphiphilic)	Chu et al. (2007)
Sodium caseinate (SC)	Protein (amphiphilic)	Yin et al. (2009)
Modified starch	Hydrocolloid (cationic)	Jafari et al. (2007)
Pectin	Hydrocolloid (anionic)	Burapapadh et al. (2010)
Lipoid S 75-3®	Phospholipid (amphiphilic)	Anton et al. (2007)

11.1.3 Applications of Nanoemulsions

The applications of nanoemulsions according to the Food Safety Authority of Ireland (2008) are

- The delivery of active compounds to the body
- The stabilization of biologically active ingredients
- Prolonged shelf-life due to increased stability
- Maintaining the viscosity
- Preservative- and antimicrobial-encapsulated formulations kill pathogenic bacteria in food.

Research and development activities have shown that the controlled release of nutraceuticals and other bioactive components have been achieved in food stabilized by mono-dispersed oil-in-water (o/w) or water-in-oil (w/o) nanoemulsion systems (Weiss et al., 2008). Sophisticated processing technology combined with advanced technologies to improve new nanoencapsulated products allows for the controlled release of foods and bioactives in the gastrointestinal tract. A wide range of products like ready-to-drink or powdered formulations loaded with functional ingredients are available in the market.

In Table 11.1, the preparation methods like high pressure homogenization and micro-fluidization are described in detail. These techniques have been used to prepare stable, uniformly-dispersed o/w or w/o nanoemulsions. The physical properties like rheological and microstructural properties, phase separation behavior, and stability in the food products of emulsions prepared by conventional

TABLE 11.3

Examples of Bioactive/Functional Compounds Encapsulated into Nanoemulsion Systems, Their Benefits/
Advantages, and Principal Applications

Bioactive/Functional Compound	Benefits/Advantages	Application	References
α-Tocopherol	Antioxidant, enhanced permeability and diffusion into deeper dermis	Medicine, transdermal delivery vehicle for drugs and cosmetic applications	Cheong et al. (2008); Kong et al. (2011)
β-carotene	Antioxidant, vitamin A precursor, improved solubility of carotenoids into aqueous systems	Natural food color, functional food preparation	Yin et al. (2009); Yuan et al. (2008b)
Phytosterols	Cholesterol absorption inhibitor	Medicine, cosmetic, food additive	Garti et al. (2005)
Curcumin	Antioxidant, anti-inflammatory, anticarcinogenic, oral administration, increased cellular uptake and cytotoxicity	Natural food color, enhancement and targeting of anti-inflammation activity, pharmaceutical application against breast cancer	Huang et al. (2010); Wang et al. (2008); Mulik et al. (2010)
D-limonene	Increased antimicrobial activity	Solubilization and dispersion in foods	Donsi et al. (2010c)
Vitamin A	Increased skin permeation and release rate in comparison to nanoemulsions	Food supplement, transdermal delivery	Jenning et al. (2000)
Lycopene	Antioxidant, potential agent for the prevention of some types of cancers, particularly prostate cancer	Natural food color	Garti et al. (2005)
Lidocaine	Local anesthetic and antiarrhythmic drug	Medicine	Sadurní et al. (2005)
Itraconazole	Antifungal agent	Medicine	Burapapadh et al. (2010)
2-(Butylamino)-1-phenyl-1-ethanethiosulfuric acid (BphEA)	Schistosomicidal drug	Medicine	De Araújo et al. (2007)
Carbamazepine	Anticonvulsant, mood stabilizing	Medicine	Kelmann et al. (2007)
Quercetin	Anti-inflammatory, antioxidant	Medicine	Fasolo et al. (2007)
Cisplatin	Anticancer drug	Medicine	Velinova et al. (2004)
Thalidomide	Immunomodulatory agent	Medicine	Araujo et al. (2011)

techniques are considerably different from those of nanoemulsions. Plant-derived nutraceuticals, e.g., plant sterols, lycopene, limonenes, CoQ10, and protein-based bioactives incorporated successfully into the oil phase (o/w emulsions) or water phase (w/o emulsions) have already been successfully carried out using this technology. The functionality of nanoemulsions containing bioactive compounds are given in Table 11.4.

A nanoemulsion was produced by a company which is used to stimulate satiety signals to reduce food intake after passing the lower regions of the small intestine without releasing the drug in the stomach. As nanoemulsions have small particle sizes, they can be used to deliver various new drugs, such as low-fat products, due to the viscosity that is imparted at low oil droplet concentrations.

TABLE 11.4

Functionality of Nanoemulsions Containing Bioactive Compounds

Bioactive Compounds	Concentration	Lipid Carrier	Stabilizers	Functionality	References
β-carotene	5 mg/kg	Corn oil, medium-chain triacylglycerols (MCT), orange oil	Tween 20	β-carotene bioaccessibility decreased in the order: corn oil > MCT > orange oil	Qian et al. (2012)
β-carotene	3 mg/kg	MCT	Modified starch	β-carotene bioaccessibility improved significantly after encapsulation in nanoemulsions	Liang et al. (2013)
β-carotene	5 mg/kg	Corn oil and miglycol 812	Tween 20	In low fat nanoemulsions (1%), the bioaccessibility of β-carotene increases with increasing the fatty acids chain length	Salvia-Trujilo et al. (2013)
β-carotene	5 mg/kg	Corn oil	Tween 20	β-carotene bioaccessibility decreased as the initial droplet size increased	Salvia-Trujilo et al. (2013)
Vitamin E	8 mg/kg	MCT	Tween 80	Bioaccessibility of vitamin E encapsulated in nanoemulsions was higher compared with conventional emulsions	Mayer et al. (2013)
Coenzyme Q10 heptadecanoic acid	1 mg/kg	Corn oil, mineral oil	Tween 80	Bioavailability of coenzyme Q10 and heptadecanoic acid was the highest when they were encapsulated in droplets with the smallest size	Cho et al. (2014)
Curcumin	1.5 mg/kg	Long, medium, and short chain triacylglycerols	β-lactoglobuli n	The higher the chain length of fatty acids, the higher the bioaccessibility of curcumin	Ahmed et al. (2012)
Curcumin	1 mg/kg	Corn oil	Tween 20 or sodium dodecyl sulfate	Curcumin exhibited high bioavailability in the presence of Tween 20 in comparison with nanoemulsions of dodecyltrimethylammonium bromide	Pinheiro et al. (2013)

11.2 Physicochemical Properties of Nanoemulsion

11.2.1 Optical Properties

The overall appearance of a nanoemulsion is affected by droplet characteristics and it is an important factor to be considered when designing food products (Mason et al., 2006; McClements and Chanamai, 2002). Some food products are required to be clear or slightly hazy (e.g., clear fortified waters, soft drinks, juices, jellies, jams, desserts, and sauces) and so the incorporation of any adduct should not make them look cloudy or opaque. Other foodstuffs are optically opaque (e.g., cloudy or opaque dressings, mayonnaise, sauces, creams, and yogurts). The scattering of light by the oil droplets in a nanoemulsion system can make an important contribution to the overall appearance. In general, the optical properties of a colloidal dispersion are characterized in terms of their opacity and color, which can be quantified in terms of coordinate tristimulus color, such as in a L*a*b* system (McClements, 2005; McClements and Chanamai, 2002). Conventional emulsions show optical properties (usually very cloudy or opaque) which are generally characterized using reflectance measurements, while the opacity is expressed in terms of lightness (L*). Moreover, the optical properties of nanoemulsions (usually transparent or slightly turbid) are generally characterized by transmission measurements, and their opacity is expressed in terms of turbidity. In general, the overall optical properties of nanoemulsions are mainly obtained by the relative refractive index, the concentration of globule, and the globule size distribution.

11.2.2 Rheological Properties

The rheological properties of a nanoemulsion change due to the droplet characteristics, which is important in manufacturing food grade nanoemulsions. Some nanoemulsions demand low viscosity (such as beverages) and so the globules present should not appreciably escalate the overall viscosity. On the other hand, some nanoemulsions are highly viscous or gel-like (e.g., dressings, dips, deserts) and the globules usually play an important role by thickening the system or forming a gel-like structure. The rheological property also helps in determining the customer compliance of food grade nanoemulsion.

11.2.3 Stability

Various physicochemical mechanisms like gravitational separation, flocculation, coalescence, Ostwald ripening, and chemical degradation lead to instability of the nanoemulsion because of its metastable nature.

11.2.3.1 Gravitational Separation

Because of the difference in relative densities of dispersed globules and dispersion media, the stability of a nanoemulsion may change through creaming or sedimentation. Creaming shows upward shifting and sedimentation shows downward shifting of globules.

11.2.3.2 Droplet Aggregation

Nanoemulsions have improved stability to particle coalescence (flocculation) due to the influence of their small globule size in the colloidal interactions (Tadros et al., 2004). Such interactions depend on the bulk physical and chemical nature of the different phases (e.g., dielectric constant and refractive index), particle-core characteristics (such as diameter), shell characteristics (such as thickness, charge, packing arrangement, rheological property, and hydrophobicity), and the properties of the intervening fluid (such as pH, ionic strength, osmotic pressure, and temperature).

11.2.3.3 Ostwald Ripening

Ostwald ripening is defined as the process where the average size of the globules in a nanoemulsion increases over time due to the diffusion phenomenon of the dispersed phase from small to large globules through the intervening fluid.

11.2.3.4 Chemical Stability

The chemical degradation of bioactives due to oxidation or hydrolysis also leads to the instability of the nanoemulsion.

11.2.4 Molecular Distribution and Release Characteristics

The bioactive compounds may be solubilized in both the oil phase and aqueous phase of the nanoemulsion. The droplet size can also be varied to modify the release characteristics of the bioactive molecule.

11.3 Factors in Designing Functional Nanoemulsions

The important particle characteristics that can be controlled to obtained functional nanoemulsions by changing their functional properties, like optical, rheological, stability, digestibility, and release characteristics, are described here under.

11.3.1 Particle Composition

The composition of particles in the nanoemulsion system can be controlled by using the appropriate ingredients and processing operations. Various non-polar food ingredients are used to create the lipophilic core, including triacylglycerols, diacylglycerols, monoacylglycerols, free fatty acids, different oils, fat substitutes, waxes, wetting agents, oil-soluble vitamins, and nutraceuticals (such as carotenoids, curcumin, phytosterols, and co-enzyme Q). Different varieties of surface-active food ingredients can be used to create the shell surrounding the lipophilic core, including surfactants, phospholipids, proteins, polysaccharides, minerals, and solid particles. The choice of these excipients plays a crucial role in determining globule properties, such as physical stability, interactions, release characteristics, and digestibility. The dependence of globule composition on globule size will impact those properties (such as refractive index, density, and permeability) that alter the bulk physicochemical properties and stabilities of nanoemulsions (such as optical properties, creaming rate, and release characteristics).

11.3.2 Particle Concentration

The droplet concentration in a colloidal dispersion system is generally expressed in terms of the number of globules, mass of globules, or volume of globules per unit volume or mass of the total system. The globule concentration within a nanoemulsion is usually controlled during manufacturing by controlling the proportions of the two immiscible liquids (oil and water) used to prepare it. However, the concentration can also be varied after a nanoemulsion formation by using dilution (e.g., by adding more continuous phase) or concentration (e.g., by gravitational separation, filtration, centrifugation, or evaporation) methods. The effective size of the globules in a nanoemulsion may also be changed by using emulsifiers that form different adsorbed layer thicknesses, or by altering solution conditions that impact the range and magnitude of electrostatic interactions between charged particles (such as the pH or ionic strength). The dimension of particles also plays a key role within a nanoemulsion system as it influences the optical, rheological, stability, and release mechanisms, as well as potentially altering its biological fate.

11.3.3 Particle Charge

The surface charge of the globules in the nanoemulsion system is mainly because of the adsorption of ionized emulsifiers, mineral ions, or biopolymers to their surfaces. These surface charges play an important role in the performance and stability of the nanoemulsion, e.g., aggregation stability, interaction with other food components, and ability to adhere to non-biological (such as packaging materials) and biological surfaces (such as the tongue, mucous layer, or other locations within the gastrointestinal tract).

11.3.4 Interfacial Characteristics

The existence of interfacial layers will selectively allow only certain kinds of molecules to pass through, e.g., based on globule size, morphology, or surface charge. The consideration of interfacial permeability while designing a controlled release system may be important because it controls the release of functional agents from inside the globules or where two or more chemically-reactive substances are used. The rheological property can often be manipulated by controlling the type and interactions of the surfactant present. Some types of surfactant form relatively low-viscous interfacial layers (e.g., small molecule surfactants, flexible biopolymers), whereas others form more viscous or gel-like networks (e.g., surface- or thermally-denatured globular proteins, or crosslinked biopolymers). Interfacial rheology may be important in determining the stability of globules towards aggregation, or the bulk rheology in systems where there is extensive interfacial layer overlap. The interfacial characteristics of the globules in a nanoemulsion system can be controlled by the proper selection of an emulsifier, such as a surfactant, phospholipid, protein, or polysaccharide.

11.3.5 Particle Physical State

Nanoemulsion fabrication requires liquid oil, but it is also possible to prepare nanoemulsions from oil phases that are able to fully or partially crystallize at the final temperature of application (Müller et al., 2002). The crystallization temperature of an emulsified fat in a nanoemulsion may be below that of the same fat in bulk phase because of supercooling effects (Gibout et al., 2007). Moreover, the crystals formed in a nanoemulsion may be different from those formed by a bulk fat because of:

- Curvature affects
- The limited volume present for crystal growth in droplets
- The lack of secondary nucleation sites.

The concentration, nature, and location of the fat crystals within the lipid droplets in a nanoemulsion are generally controlled by the proper selection of oil (e.g., solid fat content versus temperature profile, polymorphic nature), thermal history (e.g., temperature versus time), emulsifier type (e.g., tail group characteristics), and droplet size.

11.4 Techniques Used in the Identification and Characterization of Nanoemulsions

It is quite essential to identify and characterize nanoformulation systems for a better understanding of the benefits as well as from a toxicity point of view. The various procedures that can be used for the identification and characterization of nanoemulsions are listed below.

11.4.1 Separation Techniques

It is very difficult to separate nanoemulsions from food matrices. Therefore, appropriate separation techniques are necessary to isolate the nanoemulsion from food prior to characterization. The various separation techniques for the isolation of the nanoemulsion system are below.

11.4.1.1 Chromatography

The most suitable type of liquid chromatography for the separation of nanoemulsion from the food matrix are size exclusion chromatography and ion exchange chromatography. In the case of size exclusion chromatography, compounds are separated on the basis of size, with larger molecules eluted faster as compared to smaller molecules. However, the shape of the compound also affects the separation process.

In the case of ion exchange chromatography, the separation of compounds depends on the charge of the compounds; low charged compounds elute faster than highly charged ones (Luykx et al., 2005, 2008).

High performance liquid chromatography (HPLC) is another form of liquid chromatography technique which is used in food analysis for measuring numerous compounds, e.g., carbohydrates, vitamins, additives, mycotoxins, amino acids, proteins, lipids, and pigments. This technique is robust and reproducible. Selective detectors are available in HPLC techniques and their selection depends on the type of compound to be analyzed. The most common detectors for the HPLC system are Ultraviolet-Visible (UV-Vis), fluorescence, and electrochemical detectors. Functional compounds encapsulated in a nanoemulsion system can also be quantified using the HPLC systems, viz. β-carotene (Tan and Nakajima, 2005a, 2005b) and α-tocopherol (Cheong et al., 2008).

11.4.1.2 Field Flow Fractionation (FFF)

Field flow fractionation (FFF) is a flow-assisted process for the separation of analytes from macromolecules, such as proteins in the nanometer range to micrometer-sized particles. FFF separates particles based on their Stokes diameter (Dulog and Schauer, 1996; Jores et al., 2004). In this method, smaller particles are moved faster and elute earlier for parabolic flow. For cross flow with particles of the same volume and different shape, the isometric particles will be eluted first, before the asymmetric particles (Jores et al., 2004). The advantages of the FFF method over other methods are the reduction of sample carry-over and simple sterility issues. The separation times usually range from a few minutes to 30 minutes. The FFF has the disadvantage of being easily overloaded by higher concentrations of the compounds. Dilutions should be made in order to overcome this problem.

11.5 Physical Characterization Techniques

These are the techniques that are used to characterize nanoemulsions from a physical perspective (e.g., the size, size distribution, zeta potential, and crystallinity of the nanoemulsions).

11.5.1 Photon Correlation Spectroscopy

Photon correlation spectroscopy or differential light scattering is a technique used for the determination of the size distribution of globules in a nanoemulsion system. Differential light scattering experiments measure the Brownian movement of globules and relates the size of the globules through the Stokes–Einstein equation. Through the illumination of the particles with a laser beam and observing the intensity fluctuations in the scattered light, differential light scattering methods allow the user to calculate the size of the globules in a nanoemulsion system. Differential light scattering gives rapid and reliable evaluation of globule size in nanoemulsions and is often used to evaluate the size distribution of nanoemulsions, as well as their size stability through storage (Silva et al., 2011).

11.5.2 Zeta Potential

The electrokinetic potential of a colloidal system is known as the zeta potential (Mills et al., 1993). In colloidal dispersion, the zeta potential is defined as the difference in potential between the dispersion medium and the stagnant layer of fluid attached to the dispersed globules. A value of 30 mV (positive or negative) can be taken as the arbitrary value that separates low-charged surfaces from highly-charged surfaces (Preetz et al., 2010). The zeta potential value attributes to the stability of colloidal dispersions. So, colloids with a high zeta potential (negative or positive) are electrically stabilized, while colloids with a low zeta potential tend to aggregate. Briefly, zeta potentials from 0 to ±30 mV indicate instability, while zeta potentials higher than ±30 mV indicate stability. There are several factors which influence the zeta potential of a colloidal dispersion, such as ionic strength, type of surfactant used, pH of the solution, and globule size, etc.

11.5.3 Differential Scanning Calorimetry

Differential scanning calorimetry (DSC) is a thermo-analytical technique used to detect the phase transition temperature, including the melting of crystalline regions, and to analyze the proportion of solid fat or the proportion of ice crystals in the nanoemulsion. The fat crystallization affects the emulsion stability depending on the type of emulsifier used. The thermal decomposition follows the melting of the functional compound encapsulated within the core. DSC can be used to determine the crystallization temperature of a mixture of surfactants used in a nanoemulsion system.

11.5.4 Fourier Transform Infrared Spectroscopy

Fourier transform infrared spectroscopy is an analytical technique used for the identification and comparative evaluation of the purity of compounds. It is based on the principle that when infrared radiation passes through a sample, a part of it is absorbed and a part of it is transmitted. The resulting spectrum represents the molecular absorption and transmission, creating a molecular fingerprint of the sample. Since each material is a unique combination of atoms, no two compounds produce the same infrared spectrum. Hence this technique can also be used for qualitative as well as quantitative determination, and also to interpret the compatibility of two compounds.

11.5.5 Nuclear Magnetic Resonance

The nuclear magnetic resonance (NMR) technique is used for structure elucidation. It is an additional supporting technique to optical spectroscopy and mass spectrometry, leading to precise information on the structural formula, stereochemistry, and preferred conformation of the molecules. Very few reports of application of the NMR technique in nanoemulsion characterization have been exploited to date. However, some researchers reported the incorporation of medium-chain triglyceride oil in a matrix of a solid, long-chain glyceride (glyceryl behenate) and the structure of the liquid lipids inside the matrix of solid lipid nanoparticles was characterized through ^{1}HNMR.

Other analytical techniques used are the X-ray diffraction technique, and small angle X-ray scattering, etc.

11.5.6 Imaging

Microscopy is the direct imaging technique used to gather information about the size, shape, and surface morphology of globules in a nanoemulsion system. Scanning electron microscopy, transmission electron microscopy, atomic force microscopy, field emission scanning electron microscopy, and cryo transmission electron microscopy are the various imaging tools used in the characterization of nanoemulsion systems.

11.6 Nanoemulsions in Food Technology

In general, nanoencapsulation involves the incorporation, absorption, or dispersion of bioactive compounds in small vesicles with nano (or submicron) diameters. The incorporated bioactive compounds may be protected against degradation and have improved stability and solubility (e.g., solubilizing a hydrophilic compound in hydrophobic matrices and vice versa), and therefore might increase the bioavailability and delivery to cells and tissues. Reducing the size of the encapsulates into the nanoscale offers opportunities related to the prolonged gastrointestinal retention time caused by bio-adhesive improvements in the mucus covering the intestinal epithelium (Bouwmeester et al., 2009).

Depending upon the nature of the bioactive compounds, two types of nanoemulsions can be prepared. For water insoluble hydrophobic compounds like certain vitamins, minerals, aroma volatiles, flavor components, antioxidants, carotenoids, and luteins, o/w emulsion is suitable. For water soluble bioactive compounds like the water-soluble vitamins, w/o/w emulsion is used. The compounds are entrapped in the aqueous core of such emulsions. In a $w_1/o/w_2$ emulsion system, upon consumption the external w_2 phase will be perceived but the internal w_1 phase will be shielded from the taste receptors during the time scales of eating. This helps to mask the bitter tastes and undesirable flavors of the bioactive compounds (Lakkis, 2007).

Various nonpolar components like lipids, flavors, antimicrobials, antioxidants, and vitamins are successfully encapsulated by nanoemulsions. Water insoluble or sparingly soluble molecules can be made water soluble and effectively used in food products by potentially changing their bioavailability once ingested. For example, a combination of coenzyme Q10 (CoQ10) and alfa lipioc acid was entrapped inside a single nanocarrier with an average diameter of 30 nm, which aimed to produce a synergistic effect to reduce body weight and is targeted at the weight management market. Some marketed formulations with their company name are given in Table 11.5.

Nanoemulsion has been designed to deliver vitamins or other nutrients in food and beverages without affecting the taste or appearance. These nanoparticles encapsulate the nutrients and carry them via the gastrointestinal tract into the bloodstream, increasing their bioavailability.

Recently, the application of nanoemulsions in the food industry have been fast increasing, as per the details given in Table 11.6. The encapsulation and release of functional compounds is one of the rising fields of nanotechnology as applied to the food industry.

Recently companies like NutraLease have been working to improve the bioavailability of functional compounds. Another company, Aquanova, developed their NovaSol beverage solution, established to improve the bioavailability of functional compounds. Encapsulated functional compounds like phytosterols, isoflavones, lycopene, lutein, β-carotene, omega-3, vitamins A, D3, and E, and coenzyme Q10 in beverages improves bioavailability in the human body (AquaNova, 2011; Halliday, 2007; NutraLease, 2011a). NutraLease nanoemulsion not only improves the bioavailability and encapsulation capacity, but also protects the flavor compounds from manufacturing conditions and throughout the beverages' shelf-life. In addition to this, NutraLease captures the flavor and protects it from temperature, oxidation, enzymatic reactions, and hydrolysis, and is thermodynamically stable at a wide range of pH values (NutraLease, 2011b). Aquanova also claims to improve the stability (both in terms of pH and temperature) of natural colorants (β-carotene, apocarotenal, chlorophyll, curcumin, lutein, and sweet pepper extract) (AquaNova, 2011).

In addition to that, antimicrobial nanoemulsions are used for the decontamination of food materials and food equipment, and also to improve the solubility of hydrophobic and hydrophilic compounds.

Other applications of nanoemulsion, like ice cream preparation with low fat content without compromising with the taste, were introduced by Unilever and Nestlé (Martins et al., 2007; Unilever, 2011).

TABLE 11.5

Reported Applications of Nanoemulsions in the Food Industry (2010)

Food Product Name	Company	Product Description
Canola Active Oil	Shemen, Israel	Nanoencapsulation of fortified phytosterols
Fortified Fruit Juice	High Vive, US	Nanoencapsulation of fortified vitamins, lycopene, theanine, and SunActive iron
NanoResveratrol™	Life Enhancement, US	Plant-based lipid, such as a solid triglyceride (fat), encased by a shell consisting of a natural phospholipid, such as phosphatidylcholine, delivery system
Nutri-Nano™ Co Q-10 3.1x Softgels	Solgar, US	30 nm o/w system
Spray for Life® Vitamin Supplements	Health Plus International® Inc., US	Nanoencapsulation of fortified vitamin beverage
"Daily Boost"	Jamba Juice, US	Nanoemulsion of fortified vitamin or bioactive components beverage
"Color emulsion"	Wild Flavors Inc., US	Beta-carotenal, apo-carotenal, or paprika nanoemulsions

TABLE 11.6

Nanotechnology Used in the Food Industry and Other Related Industries

Product	Details	References
Functional beverage	Fortified flavoured water and milk with vitamin, mineral and other functional ingredients using nanoemulsion technology for incorporation and controlled release of Bioactives.	Decker (2003); Hazen (2003)
Personalized beverages and foods	Nanoemulsions that releases different flavors through activation of heat, Ultrasonic frequency, pH or other triggers. Food that adjust its colour, flavor or nutrient content to accommodate a person's taste and health conditions.	Choi (2002); Gardner (2003)
Smart filters	Selective nano-filters that can distinguish molecules based on shape as well as size enabling the removal of toxins or adjustment of flavours.	Gardner (2003)
Smart sensors	Packaging with nanosensors that indicate when a product is compromised and not safe for consumption.	Gardner (2003)
Sanitizer	Nanoemulsion of vegetable oil, surfactants surrounded by trace amounts of alcohols suspended in water with properties to kill bacteria.	Newman (2002), Hamouda and Baker (2000)
Novel encapsulation system	Solid hydrophobic nanosphere composed of a blend of food-approved hydrophobic materials encapsulated in moisture sensitive or pH sensitive bio-adhesive microspheres to enable controlled release of active ingredients.	Shefer et al. (2003a, 2003b)
Drug delivery	Complex coacervates of DNA and chitosan used as delivery vehicle in gene therapy e.g. treatments of food allergies such as peanuts.	Roy et al. (1999)

The advantages of bioactive or functional compounds encapsulated into nanoemulsion systems and their principal applications are given in Table 11.3.

11.7 Nanoemulsions in the Food Industries

Nanoemulsions have been developed for use in the decontamination of food packaging equipment and in the packaging of food. A typical example is a nanomicelle-based product claimed to contain natural glycerin. It removes pesticide residues from fruits and vegetables, as well as the oil/dirt from cutlery. Nanoemulsions have in recent times received a lot of awareness from the food industry due to their high clarity. These allow the addition of nanoemulsified bioactives and flavors to an infusion without a change in product appearance. Nanoemulsions are effective against a variety of food pathogens, including Gram-negative bacteria. They can be used for the surface decontamination of food processing plants and for the reduction of surface contamination of chicken skin. The growth of *Salmonella typhimurium* colonies have been eliminated by treatment with nanoemulsion (Shah, 2014). Based on the physicochemical properties of microencapsulated fish oil, sugar beet pectin must be considered as an alternative to milk proteins and gum arabic for the encapsulation of functional food ingredients (Drusch, 2007; Gharsallaoui et al., 2007). The nanoemulsions showed great promise for use in beverages and other applications (Wooster, 2010). Various types of nanoemulsion, including single-layer, double-layer, and triple-layer nanoemulsions, could be produced, depending on the polyelectrolytes, such as alginate and chitosan (Choi et al., 2010). Solid lipid nanoparticles are formed by the controlled crystallization of food nanoemulsions and have been reported for use in the delivery of bioactives, such as lycopene and carotenoids (Weiss et al., 2006, 2008). The major advantages of solid lipid nanoparticles include large-scale production without the use of organic solvents, high concentration of functional compounds in the system, long-term stability, and the ability to be spray dried into powder form. Megestrol acetate oral suspension (MAOS) is an appetite stimulant indicated for cachexia in patients with AIDS. It is available in its original formulation, Megace® (MAOS), and as a nanocrystal dispersion, Megace® ES (MA-ES) (from Bristol-Myers Squibb, New York, US). The bioavailability and absorption were found to be greater for MA-ES than MAOS in fasting subjects, thereby suggesting MA-ES as the preferred formulation of megestrol acetate when managing cachectic patients whose caloric intake is reduced (Deschamps et al., 2009).

11.8 Food "Construction Up" during Nanotechnology

Nanotech companies are trying to fortify processed dairy and food products with nanoencapsulated nutrients, boost their appearance and taste through nano-developed colors, remove or disable their fat and sugar content by nano-modification, and improve mouth-feel. Food "defense" will be used to enhance nutritional claims, e.g., the addition of medically-valuable nanocapsules will soon enable chocolate chip cookies or hot chips to be marketed as health promoting or artery cleansing. Nanotechnology will also enable junk foods like ice cream and chocolate to be modified to reduce the amount of fats and sugars that the body can absorb. This is possible by using nanoparticles to prevent the body from digesting or absorbing these components of the food. In this way, the nano industry could market vitamins and fiber-fortified, fat, and sugar-blocked junk food as health promoting and weight reducing (Miller, 2008).

11.9 Nanostructures and Nanoparticles in Food

Most polysaccharides and lipids are linear polymers with thicknesses less than a nanometer, while food proteins are often globular structures of 1–10 nm in size. The functionality of many raw materials and the processing of foods arise from the presence, modification, and generation of forms of self-assembled nanostructures (Chen et al., 2006). The crystalline structures in starch and processed starch-based foods that determine gelatinization and influence the nutritional benefits during digestion, the fibrous structures that control the melting, setting, and texture of gels, and the two-dimensional (2D) nanostructure formed at oil-water and air-water interfaces are what control the stability of food/dairy foams and emulsions (Rudolph, 2004). For example, the creation of foams (e.g., the head on a glass of beer) or emulsions (e.g., sauces, creams, yoghurts, butter, and margarine) involves generating gas bubbles, or droplets of fat or oil, in a liquid medium. This requires the production of an air-water or oil-water interface and the molecules present at this interface determine its stability. These structures are one molecule thick and are examples of 2D nanostructures. A source of instability in most foods is the presence of mixtures of proteins and other small molecules such as surfactants (soap-like molecules or lipids) at the interface (Morris, 2005). Atomic force microscopy has allowed researchers to visualize and understand these interactions, and to improve the stability of the protein networks that can be widely and simultaneously applied in the dairy, baking, and brewing industries.

The knowledge gained in nanotechnology in the field of medicine, and electronics, etc., could be adapted in the field of food and dairy processing, more specifically in food safety (e.g., detecting pesticides and microorganisms), in environmental protection (e.g., water purification), and in the delivery of nutrients (Chau, 2007; Roco, 2003). The area that has led to the most debate on nanotechnology and food is the incidental or deliberate introduction of manufactured nanoparticles into food materials.

11.10 Processed Nanostructures in Food

A key area of application of nanotechnology in food processing involves the development of nanostructures (also termed nanotextures) in foodstuffs. The mechanisms commonly used for producing nanostructured food products include nanoemulsions, surfactant micelles, emulsion bilayers, double or multiple emulsions, and reverse micelles (Weiss et al., 2006). Examples of nanotextured foodstuffs include spreads, mayonnaise, cream, yoghurts, and ice creams, etc.

The nanotexturing of foodstuffs has been claimed to give new tastes, improved textures, and consistency and stability of emulsions when compared with equivalent conventionally processed products. A typical benefit of this technology could be in the form of a low-fat nanotextured food product that is as buttery as the full-fat alternative, and hence offers a strong purchase option to the consumer. Currently, there is no clear example of a proclaimed nanostructured food product that is available commercially, although some products are believed to be at the research and development stage, and some may be nearing the market. One such example is a mayonnaise, which is an o/w emulsion that contains nanodroplets of water inside the oil droplets. The mayonnaise may offer taste and texture attributes similar to the full-fat equivalent, but with a substantial reduction in fat intake by the consumer. Another area of application involves the use of nano-sized or nanoencapsulated food additives. This type of application is expected to exploit a much larger segment of the health food sector, and encompasses colors, preservatives, flavorings, and supplements. The main advantages claimed include better dispersion of water-insoluble additives in foodstuffs without the use of additional fat or surfactants, and enhanced tastes and flavors owing to the enlarged surface area of nano-sized additives when compared with conventional forms. A number of consumer products containing nano-sized additives are already available in some food sectors, including foods, health foods, supplements, and nutraceuticals. These include minerals, antimicrobials, vitamins, and antioxidants, etc. Virtually all of these products are claimed to have improved absorption and bioavailability in the body compared with their conventional equivalents. Another example is the increasing trend towards nano-milling of functional herbs and other plants, such as in the manufacture of green tea and ginseng.

11.11 Nanoemulsions: Manufacturing Perspectives and Relevance in the Market of Food

In current years, nanotechnology has enlarged into a huge scale business, becoming a multimillion dollar industry. It is also clear throughout a numeral of reports, investigations, and patent applications that the aim of nanotechnology previously had a large contact on diverse aspects of the food and food related industries (Chen et al., 2006; Neethirajan and Jayas, 2011). According to a report, the number of companies that presently compose the use of nanotechnology in the food area may be about 400 (Chau et al., 2007; Neethirajan and Jayas, 2011). Nanoemulsion fabrication used for encapsulation and delivery of purposeful compounds is one of the up-and-coming fields of nanotechnology to be applied in the food industry. Some examples of this application are given below.

NutraLease, a technology get-underway company launched by a scientific team, is working to recover the bioavailability of functional compounds. Infusions including encapsulated functional compounds, for instance lycopene, lutein, coenzyme Q10, β-carotene, omega-3, vitamins A, D3, and E, phytosterols, and isoflavones, are obtainable (NutraLease, 2011a). This technology is based on self-collected nano-emulsions wherever a better encapsulation rate over and above a recovered bioavailability in the human body can be realized (Halliday, 2007; NutraLease, 2011b). Nanoemulsions can defend flavor compounds from the manufacturing situation through to the infusions shelf-life period. It is declared that nanoemulsions can imprison the flavor and defend it from oxidation, enzymatic reactions, temperature, and hydrolysis, and are thermodynamically constant by a large range of pH values (NutraLease, 2011c).

A different example of nanotechnology in the industry of food is provided by AquaNova, with their NovaSol infusion solutions (AquaNova, 2011; Halliday, 2007). The NovaSol collection is divided into two classes: healthy functional compounds (vitamins A, D, D3, E, and K, omega-3 fatty acids, coenzyme Q10, and DL-α-tocopherol acetate) and natural colorants (apocarotenal, β-carotene, chlorophyll, curcumin, lutein, and extract of sweet pepper) (AquaNova, 2011). AquaNova claims better constancy (both in terms of temperature and pH) of encapsulated useful compounds and standardized preservative concentrations (AquaNova, 2011). Several huge companies in the food industry, for example Nestlé and Unilever, are also applying nanotechnology to their food products. Unilever has made ice cream better without compromising on taste throughout the appliance of nanoemulsions. The purpose is to manufacture ice cream with a lower content of fat, achieving a fat decrease from the actual 16% to 1%

(Martins et al., 2007; Unilever, 2011). Nestlé has exclusive rights in w/o emulsions (10–500 nm), and is aspiring to achieve simpler and faster melting during the addition of polysorbates and other micelle-forming substances; these are asserted to contribute to a regular melting of frozen foods in the microwave (Möller et al., 2009). An additional application of nanoemulsions containing antimicrobial nanoemulsions in the food industry for the sanitization of food apparatus, packaging, or food (Gruère et al., 2011). As well evaluated is the opportunity of delivering hydrophobic or hydrophilic compounds (useful compounds), in short to improve their bioavailability and solubility. Although nanotechnology has previously been applied to the industry of food, there is still a major gap in the rigid structure, and most countries are still relying on presenting legislation to control nanomaterials (Gruère et al., 2011). Civilizing the real legislation structure is a vital step to avoid consumers' propaganda regarding nanotechnology applied to foods. Misinformation was the focal dependent for the real suspicions that numerous consumers have towards heritably adapted organisms and their use in foods; this is still of great concern to consumers and has generated avoidance since the occurrence of nanotechnology in foodstuffs. Whereas legislation is still being modified, extra actions may be taken to get better consumer assurance for foods integrating nanotechnology-supported ingredients, e.g., the warranty of the use of food grade materials for nanosystems production. One more significant portion is the cost; scaling-up of high energy approaches is still very exclusive owing to the pressures required to develop high volumes. Such costs need to be condensed, by additional technical enhancements in the methods or by acceptance of easy methods of nanoemulsion production. Since a general suggestion, when functional to the food industry nanotechnology must be reasonable, easy to use, and with voluntarily apparent rewards in order to be an actual substitute to the regular methodologies. Table 11.7 shows the various applications of nanotechnology in agriculture and food.

11.12 Nanotechnology in Food and Dairy Processing

Cell membranes, hormones, and deoxyribonucleic acids (DNA), etc., that exist in nature are an example of nanostructures and the food molecules, proteins, fats, carbohydrates, etc., are not exceptional nor the result of nanoscale merges between sugars, fatty acids, and amino acids (Powell and Colin, 2008). "Nanofoods" from the Helmut Kaiser Consultancy (2009) estimate an increasing growth in the development of food and dairy related nanoproducts and patent applications. Nanotechnology can be applied to develop nanoscale materials, controlled delivery systems, contaminant detection, and to create nanodevices for molecular and cellular biology from how food is grown to how it is packaged. The application of nanotechnology with respect to the food and dairy industry will be covered under two major heads, viz. food additives (nano inside) and food and dairy packaging (nano outside).

11.13 Food Additives (Nano inside)

11.13.1 Nanodispersions and Nanocapsules

Purposeful ingredients (such as vitamins, drugs, flavorings, antioxidants, antimicrobials, preservatives, and colorants, etc.) come in physical forms and special molecular forms, e.g., nonpolar, polar, amphiphilic, low to high molecular weights, and physical states (liquid, solid, and gas). These ingredients are not often utilized straight in pure form; as an alternative, they are regularly included into a few forms of delivery systems.

Weiss et al. (2006) revealed that a system of delivery of necessity should achieve a numeral of diverse roles. First, it serves the same function as a vehicle for carrying the functional ingredient to the preferred action site. Second, it may have to shield the functional ingredient from biological and chemical deprivation (i.e., oxidation) for the period of storage, processing, and exploitation; this maintains the practical ingredient in its energetic state. Third, it may have to be able to release the controlling of the functional ingredient, for instance the liberate rate or the specific conditions of the environment that activate release

TABLE 11.7

Overview on Applications of Nanotechnology in Agriculture and Food

Food Processing	Food Packaging	Supplements	Agriculture
Nanocapsules to improve the bioavailability of nutraceuticals in standard ingredients such as cooking oils	Antibodies attached to fluorescent nanoparticles to detect chemicals or foodborne pathogens	Nano-size powders to increase the absorption of nutrients	Single molecule detection to determine enzyme/substrate interaction
Nanoencapsulated flavor enhancers	Biodegradable nanosensors for temperature, moisture, and time monitoring	Cellulose nanocrystal composites as drug carriers	Nanocapsules for the more efficient delivery of pesticides, fertilizers, and other agrichemicals
Nanotubes and nanoparticles as gelation and viscosifying agents	Nanoclays and nanofilms as barrier materials to prevent spoilage and prevent oxygen absorption	Nanoencapsulation of nutraceuticals for better absorption, better stability, or target delivery	Delivery of growth hormones in a controlled fashion
Nanocapsule infusion of plant-based steroids to replace meat cholesterol	Electrochemical nanosensors to detect ethylene	Nanocochleates (coiled nanoparticles) to deliver nutrients more efficiently to cells without affecting color or taste of food	Nanosensors for monitoring soil conditions and crop growth
Nanoparticles to selectively bind and remove chemicals or pathogens from food	Antimicrobial and antifungal surface coatings with nanoparticles (silver, magnesium, zinc)	Vitamin sprays dispersing active molecules into nanodroplets for better absorption	Nanochips for identity preservation and tracking
Nanoemulsions and particles for better availability and dispersion of nutrients	Lighter, stronger, and more heat-resistant films with silicate nanoparticles		Nanosensors for the detection of animal and plant pathogens
	Modified permeation behavior of foils		Nanocapsules to deliver vaccines
			Nanoparticles to deliver DNA to plants (targeted genetic engineering)

(e.g., ionic strength, pH, or temperature). Fourth, the system of delivery has to be companionable with the supplementary components in the system, in addition to being companionable with the physicochemical and qualitative attributes (texture, taste, appearance, and shelf-life) of the last product. In order to achieve the above said objectives, a number of potential delivery systems based on nanotechnology could be used, such as association colloids, biopolymeric nanoparticles, and nanoemulsion.

11.14 Nanotechnology and Food Safety

The meaning of food safety is that all food products must be protected from biological, physical, chemical, and radiation contamination during handling, processing, and distribution. Thus far, this chapter has focused on the purpose of nanotechnology in the food and dairy processing, as well as packaging. Nanotechnology has brought rebellion into the non-food sectors, but it is slowly gaining popularity in the food processing and dairy sector. Although consumers are thrilled at the exciting food and dairy products emerging through the application of nanotechnology, there is a serious question about safety that will require attention from the industry as well as the policy makers. It is important to note that nanomaterials (due to their increased contact surface area) might have toxic effects in the body that are not apparent in bulk materials (Dowling, 2004). Despite the lack of regulation and risk knowledge, a wide variety of food and nutrition products containing nanoscale additives are already in the market (e.g., iron in nutritional drink mixes, micelles that carry vitamins, minerals, and phytochemicals in oil, and zinc oxide in breakfast cereals, etc.) and nanoclays have been incorporated into plastic beer bottles.

The additives commonly accepted, like those classified as GRAS (generally regarded as safe), will have to be reexamined when used at the nanoscale level. Nanoparticles are extra reactive, additionally mobile, and likely to be more toxic. This toxicity is one of the important issues that must be addressed. There is a strong possibility that nanoparticles in the body can result in increased oxidative stress that, in turn, can generate free radicals, leading to DNA mutation, cancer, and possible fatality. It is also not completely understood whether increasing the bioavailability of certain food additives or nutrients might negatively influence human health (Moraru et al., 2003). The ingredients in these nanoparticles must undergo a full safety assessment by the relevant scientific advisory association before they are permitted to be used in dairy and food products, including packaging (U.K.RS/RAE, 2004).

11.15 Nano-Based Food Contact Materials and Antimicrobial Packaging

Distinct from trigger-dependent chemical release packaging, which is considered to release biocides in response to the humidity, growth of a microbial population, or other changing conditions, other packaging and food contact materials include antimicrobial nanomaterials that are designed not to be released, thus the packaging itself acts as an antimicrobial. These products generally use nanoparticles of silver, although some use nano zinc oxide or nano chlorine dioxide (AzoNano, 2007; LeGood and Clarke, 2006). Nano copper oxide, nano magnesium oxide, nano titanium dioxide, and carbon nanotubes are also predicted for future use in antimicrobial food packaging (ElAmin, 2007). A few examples are shown in Table 11.8.

11.16 Potential Applications of Nanoemulsions in the Food Industry

The pharmaceutical industry is the dominant field where most applications of nanoemulsions are proposed. Extensive research has been conducted on a variety of drug delivery systems with enhanced solubilization of poorly soluble drugs and improved bioavailability following incorporation into nanoemulsions (Shah et al., 2010; Shafiq et al., 2007). In the industry of food, many food-derived bioactive

TABLE 11.8

Nano-Based Antibacterial Food Packaging and Food Contact Materials (PEN 2007)

S. No.	Company/Institution	Application
1.	Com Song Sing Nano Technology Co., Ltd	Food cling wrap treated with nano zinc oxide
2.	Sharper Image	Food plastic storage bags treated with nano silver
3.	BlueMoonGoods, A-DO Global, Quan Zhou Hu Zheng Nano Technology Co., Ltd, and Sharper Image	Food storage containers treated with nano silver
4.	Daewoo, Samsung, and LG	Refrigerators treated with nano silver
5.	Baby Dream® Co., Ltd	Baby cup treated with nano silver
6.	A-DO Global	Chopping board treated with nano silver
7.	Com Song Sing Nano Technology Co., Ltd	Tea pot treated with nano silver
8.	Nano Care Technology, Ltd	Kitchenware treated with nano silver

Note: List based on the project on "Emerging Nanotechnology Consumer Products Inventory, 27 February 2008".

compounds show significant health benefits when consumed in relatively high concentrations (Flanagan and Singh, 2006; Garti et al., 2003; 0232095). Unfortunately, mainly of these compounds exhibit poor bioavailability and solubility in aqueous-based foods. In recent times, the development of nanoemulsions full of lipophilic food components has demonstrated the potential of nanoemulsions as a carrier to deliver lipophilic actives in food applications (Garti et al., 2005; Shefer and Shefer, 2003b).

Extensively, emulsions are used in infusion products as it is frequently necessary to incorporate water-insoluble flavors into an aqueous beverage. Yet, emulsion is frequently preferred to the infusion as it is clear in looks. The present solution for this particular requirement is to use a co-solvent, mostly alcohol, in this system (regularly introduced like the flavor solvent). On the other hand, flavors of alcohol-base have transportation issues caused by their low flash point. The alcoholic use also disqualifies the product from obtaining a Halal certificate, which is attractive in numerous regions of the globe. Nanoemulsions can solve these troubles while emulsions through mean droplet diameter (MDD) under 100 nm have the prospective to offer a clear look that can be included in infusions without losing that clearness (Skiff et al., 2007; Chanamai, 2007). Additional, non-weighted flavor emulsions are another exclusive application of nanoemulsions. For unadventurous flavor emulsions, weighing agents (e.g., brominated vegetable oil and ester gum) are normally used to evaluate the oil phase and decrease the density gap between oil and aqueous phases. Considerably, these weighing agents condense the creaming rapidity of droplets, although they have severe usage limits caused by their toxicity. Weighing agents can be removed in nanoemulsions because of their high constancy besides creaming, which provides a cleaner label.

11.17 Patented Nanoemulsion

A few examples of patented nanoemulsions shown in Table 11.9.

11.18 Toxicity Concerns of Food Grade Nanoemulsion

Nanoemulsions have many potential advantages for certain applications (optical clarity, improved bioavailability, good physical stability), and so they have a wide application in the food industry to

TABLE 11.9

Patent Nanoemulsion with Assignee and Patent Number

S.No.	Patent Name	Assignee	Patent Number
1.	Method of preventing and treating microbial infection	NanoBio Corporation (US)	6,506,803
2.	Nanoemulsion based on phosphoric acid fatty acid ester and its uses in the cosmetics, dermatological, pharmaceutical, and ophthalmological field	L'Oreal (Paris, FR)	6,274,50
3.	Nanoemulsion based on ethylene oxide block and propylene oxide block copolymer and its uses in the cosmetic, dermatological, and ophthalmological fields	L'Oreal (Paris, FR)	6,464,990
4.	Nanoemulsion based on phosphoric acid fatty acid esters and its uses in cosmetics, dermatological, pharmaceutical, and/or ophthalmological fields	L'Oreal (Paris, FR)	6,274,150
5.	Non-toxic antimicrobial compositions and methods of use	NanoBio Corporation (US)	6,559,189 and 6,635,676
6.	Nanoemulsion based on oxyethylenated or non-oxyethylenated sorbitan fatty esters, and its uses in the cosmetics, dermatological, and/or ophthalmological fields	L'Oreal (Paris, FR)	6,335,022
7.	Nanoemulsion based on glycerol fatty esters, and its uses in the cosmetics, dermatological, and/or ophthalmological fields	L'Oreal (Paris, FR)	6,541,018

encapsulate functional compounds. However, there is increasing concern that the presence of ultrafine particles in food grade nanoemulsions may have adverse effects on biological systems. Current knowledge of the impact of nanoformulation characteristics on the biological fate of encapsulated components within nanoemulsions is limited. However extensive research has been carried out to find information about the impact of specific particle characteristics of nanoemulsions on (1) the mechanisms of transportation of nanodroplets through the gastrointestinal tract, (2) the bioavailability pattern, and (3) the potential toxicity of encapsulated material. This knowledge is critical for the rational design and development of nanotechnology-based food strategies for the improvement of human health. Consequently, food grade beverage manufacturers can establish the most appropriate levels of functional materials to incorporate into suitable nanoemulsions so that they exhibit beneficial effects, without introducing unwanted adverse effects (e.g., potential toxicity). This information will in turn lead to the development of a highly integrated, multidisciplinary approach for testing food grade nanoemulsions, which may provide a useful model for establishing the safety of other types of engineered nanoformulations intended for food and beverage applications, such as those manufactured from proteins, polysaccharides, or inorganic materials. The toxicity study may include an *in vitro* cell line study and *in vivo* animal study. These studies were carried out using *in vitro* human CaCo-2 intestinal cell model and *in vivo* mice model. The other studies in toxicity include the lactate dehydrogenase leakage assay and the measurement of transepithelial electrical resistance (TEER). The schematic protocol for testing the potential toxicity of food grade nanoemulsions includes an *in vitro* hazard identification test, like nanoparticle fate (modification, digestion, absorption, uptake), cellular responses (morphology, cytotoxicity, mutagenicity, oxidative damage), and bioactive fate (chemical modification, physical location), etc. and then conformation of these *in vitro* findings using *in vivo* studies. Thus, there is an urgent need to have a precise and validated toxicological test method that can be used to screen for potential hazards, and also to develop new methodology for the measurement of nanoparticles in biological matrices, in order to assess human exposure. Nanoscale materials such as proteins, carbohydrates, and fats are unlikely to be a source of toxicity in their own right. However, little information is available about possible interactions of nanomaterials with other food components, the integrity of the nanomaterials following passage through the digestive system, and their pharmacokinetic behavior, that is how they are absorbed, distributed, and excreted (cleared) from the body.

11.19 Regulatory Aspects

Although advances in nanotechnology have been applied in the food industry, there is still a major gap in the regulatory framework, and most countries are still relying on existing legislation to regulate nanomaterials (Gruère et al., 2011). It is very important to improve and modify the actual legislation framework to prevent consumer misinformation regarding nanotechnology applied to the food industry. Such misinformation is mainly responsible for the actual suspicions that many consumers have towards genetically modified organisms and their use in foods; a great deal of care is needed to prevent that from happening with nanotechnology. While legislation is still being adapted, other actions may be taken to improve consumers' confidence in foods integrating nanotechnology-based ingredients, such as the guarantee of the use of food grade materials for nanosystem production.

Upon reflection on the deficiencies in the available information on the current applications of nanotechnology in the agriculture-food sector and also the information needed to carry out a meaningful risk assessment, there are a number of recommendations, including the following:

1. Nanograde food manufacturers should conduct risk assessments on all foods involving the introduction of new nanoparticles into foods and packaging or manipulation of the size distribution of natural food molecules in food matrices, in order to meet their legal obligations to produce safe food.
2. Legal provisions should be considered to ensure that food and feed should be reevaluated in terms of safety whenever the properties are changed/re-engineered to the nanoscale.

3. Urgent consideration should be given to whether additional controls are required on the disposal and/or recycling of nanoparticle-containing food contact and other materials.

4. Food surveillance programs should include investigation of the potential for nanoparticles, particularly inorganic molecules such as titanium dioxide and clay particles used in packaging, to migrate into foods and also to be recycled in the environment and enter the food chain indirectly.

In addition, in order to protect and inform consumers, it is necessary that food or food packaging in contact with food and incorporating nanoparticles should be labeled.

Moreover, it is also necessary to undertake research to increase the reliability of the assessment of possible risks of nanotechnology in food, including on issues such as the fate and behavior of nanoparticles within the human body and animals, their potential toxicity, methods for the safe and effective disposal of used or waste nanoparticles, the stability/lability of nanoparticles in various foods, and their potential interactions with other food components.

11.20 Nanotechnology in Food Preservation

Food preservation has been defined as the process of treating and handling food to stop or greatly slow down spoilage in order to prevent loss of quality, edibility, or nutritive value that is caused or accelerated by microorganisms, which aim to prevent the growth of microorganisms like bacteria or fungi. The modifications in food preservation are needed not only with the aim to improve the food products but also to meet the requirements of consumers. Some food grade antimicrobials have been encapsulated in nanoemulsions. Essential oils are GRAS food additives according to the United States Food and Drug Administration (FDA) and show activities against human pathogenic and food spoilage microorganisms, however water solubility problems, evaporation, and sensory properties have limited their application in food products. Therefore, the incorporation of essential oils in nanoemulsion systems represents an available and efficient approach to improve their physical stability and reduce the mass transfer resistance of the active molecules to the sites of action. Eugenol can be used in o/w nanoemulsion using sesame oil, Tween 80, and water prepared by the ultrasound cavitation method. Reports say that a nanoemulsion with 0.003% of eugenol shows stability for more than one month and exhibits antibacterial activity against *Staphylococcus aureus*. Nanoemulsions containing carvacrol and eugenol with triacylglyceride (Miglyol 812N) or Tween 80, prepared by high pressure homogenization and ultrasonication, show antimicrobial activity against *Escherichia coli*.

11.21 Conclusion

Nanoemulsion based-delivery systems are gaining popularity in many industrial applications. Due to the very fine size of the dispersed globules, light scattering does not occur and they are highly stable against creaming and coalescence. In addition, the small size of the droplets in nanoemulsions means that they are able to greatly increase the bioavailability of their entrapped lipophilic substances. Nanoemulsion is a promising delivery system for better solubility, bioavailability, and functionality of hydrophobic compounds. There are various methods of preparation of nanoemulsions available, but the most suitable methods are used for the preparation of optimized nanoemulsions on an industrial scale. The characterization of nanoemulsions is also important and is done by various sophisticated analytical techniques. In short, chromatographic techniques are used for the separation of functional compounds, photon correlation spectroscopy and differential light scattering techniques are used for the determination of globule size, zeta potential measurements indicate the stability of the nanoemulsion, and electron microscopic images confirm the hydrodynamic diameter of globules obtained from differential light scattering experiments and the morphological characteristics of nanoemulsion droplets.

Although there is a rapid escalation in the application of nanoscience and technology in the food industry, it still faces some obstacles that need to be addressed, both in terms of the manufacturing process, and especially its cost, and the characterization of both the resulting nanoemulsions and food systems to which they will be applied in terms of product safety and consumer acceptance. In general, when applied to the food industry, nanotechnology needs to be cost effective and simple to use, with readily perceived advantages, in order to be a real alternative to the standard methodologies. Moreover, there is a need for information regarding the interaction of the nanosystems with food matrices, the mechanisms of the release of nanoencapsulated functional components, and toxicity studies. To apply nanotechnology in the food industry, future research has to focus on the design of scalable methods and the identification of relatively low-cost excipients. Antimicrobial food nanocarriers may be suitable for controlling spoilage and the growth of pathogenic microorganisms in foodstuff. Thus, nanotechnology has play an important role in stimulating or accelerating worldwide economic growth and has been increasingly important in the food industry.

11.22 Future Prospects

There is a need for further studies, including a wide range of active compounds loaded in nanoemulsions, to elucidate the real benefits of nanoemulsions depending on the kind of lipid nanoparticle. Despite the fact that lipid nanoparticles show a similar digestibility pattern when compared to conventional emulsions, their toxicological safety cannot be assured. There is an increase in demand for the use of nanotechnology-based food products because of their better encapsulation efficiency, solubility, bioavailability, and functionality as compared to conventional delivery systems for certain applications like transparent foods and beverages, increased bioavailability, and improved physical stability. However, before the wide application of nanoemulsion is applied, there are numerous potential challenges that need to be overcome, including the following:

1. Suitable food grade ingredients need to be used for the production of food grade nanoemulsions
2. Suitable processing operations need to be identified for the economic and robust production of food grade nanoemulsions
3. Proper investigation of toxicological parameters needs to be standardized for the production of food grade nanoemulsions.

REFERENCES

Acosta E. Bioavailability of nanoparticles in nutrient and nutraceutical delivery. Curr Opin Colloid Interface Sci. 2009;14(1):3–15.

Ahmed K, Li Y, McClements DJ, Xiao H. Nanoemulsion-and emulsion-based delivery systems for curcumin: Encapsulation and release properties. Food Chem. 2012;132(2):799–807. DOI:10.1016/j.foodchem.2011.11.039.

Amani A, York P, Chrystyn H, Clark B. Factors affecting the stability of nanoemulsions: Use of artificial neural networks. Pharm Res. 2010;27(1):37–45.

Anton N, Gayet P, Benoit JP, Saulnier P. Nano-emulsions and nanocapsules by the PIT method: An investigation on the role of the temperature cycling on the emulsion phase inversion. Int J Pharm. 2007;344(1–2):44–52.

AquaNova; 2011 [Cited 13 April 2011]. Available from: http://www.aquanova.de/media/public/pdf_produ kteunkosher/NovaSOL_beverage.pdf.

Araújo FA, Kelmann RG, Araújo BV, Finatto RB, Teixeira HF, Koester LS. Development and characterization of parenteral nanoemulsions containing thalidomide. Eur J Pharm Sci. 2011;42(3):238–45.

Bouchemal K, Briançon S, Perrier E, Fessi H. Nano-emulsion formulation using spontaneous emulsification: Solvent, oil and surfactant optimization. Int J Pharm. 2004;280(1–2):241–51.

Bouwmeester H, Dekkers S, Noordam MY, Hagens WI, Bulder AS, De Heer C, Ten Voorde SE, Wijnhoven SW, Marvin HJ, Sips AJ. Review of health safety aspects of nanotechnologies in food production. Regul Toxicol Pharmacol. 2009;53(1):52–62.

Burapapadh K, Kumpugdee-Vollrath M, Chantasart D, Sriamornsak P. Fabrication of pectin-based nanoemulsions loaded with itraconazole for pharmaceutical application. Carbohydr Polym. 2010;82(2):384–93.

Chaix C, Pacard E, Elaïssari A, Hilaire JF, Pichot C. Surface functionalization of oil-in-water nanoemulsion with a reactive copolymer: Colloidal characterization and peptide immobilization. Colloids Surf B Biointerfaces. 2003;29(1):39–52.

Chen L, Remondetto GE, Subirade M. Food protein-based materials as nutraceutical delivery systems. Trends Food Sci Technol. 2006;17(5):272–83.

Cheong JN, Tan CP, Man YBC, Misran M. α-Tocopherol nanodispersions: Preparation, characterization and stability evaluation. J Food Eng. 2008;89(2):204–209.

Cho HT, Salvia-Trujillo L, Kim J, Park Y, Xiao H, McClements DJ. Droplet size and composition of nutraceutical nanoemulsions influences bioavailability of long chain fatty acids and coenzyme Q10. Food Chem. 2014;156:117–22.

Choi C. For personalized beverage just add microscopic liquid [Internet]. United Press International: Malaga, Spain. 2002 [cited 18 August 2003]. Available from: www.smalltimes.com.

Choi SJ, Decker EA, Henson L, Popplewell LM, McClements DJ. Stability of citral in oil-in-water emulsions prepared with medium chain triacylglycerols and triacetin. J Agric Food Chem. 2009;57(23):11349–53.

Chu BS, Ichikawa S, Kanafusa S, Nakajima M. Preparation of protein-stabilized β-carotene nanodispersions by emulsification–evaporation method. J Am Oil Chemists' Soc. 2007;84(11):1053–62.

Date AA, Desai N, Dixit R, Nagarsenker M. Self-nanoemulsifying drug delivery systems: Formulation insights, applications and advances. Nanomedicine. 2010;5(10):1595–616.

de Araújo SC, de Mattos ACA, Teixeira HF, Coelho PMZ, Nelson DL, de Oliveira MC. Improvement of in vitro efficacy of a novel schistosomicidal drug by incorporation into nanoemulsions. Int J Pharm. 2007;337(1–2):307–15.

Decker KJ. Wonder waters: Fortified and flavoured waters. Food Prod Des. 2003;13(5);57–74.

Donsi F, Sessa M, Ferrari G. Nanoencapsulation of essential oils to enhance their antimicrobial activity in foods. Journal of Biotechnology. 2010;150:S67.

Dulog L, Schauer T. Field flow fraction for particle size determination. Prog Org Coat. 1996;28(1):25–31.

Fasolo D, Schwingel L, Holzschuh M, Bassani V, Teixeira H. Validation of an isocratic LC method for determination of quercetin and methylquercetin in topical nanoemulsions. J Pharm Biomed Anal. 2007;44(5):1174–7.

Food Safety Authority of Ireland. The relevance for food safety of applications of nanotechnology in the food and feed industries. Cambridge, UK, Publisher: The Royal Society of Chemistry. 2008. p. 1–82.

Gardner E. Brainy food: Academia, industry sink their teeth into edible nano [Internet]. Small Time Correspondent CRC Press, London, UK; 2003 [cited 22 February 2005]. Available from: www.smalltimes.com.

Garti N, Spernath A, Aserin A, Lutz R. Nano-sized self assemblies of nonionic surfactants as solubilization reservoirs and micro reactors for food systems. Soft Matter. 2005;1(3):206–18.

Gruère G, Narrod C, Abbott L. Agricultural, food, and water nanotechnologies for the poor [Internet]. The International Food Policy Research Institute (IFPR): Washington, DC. 2011. Available from: http://www.ifpri.org/sites/default/files/publications/ifpridp01064.pdf.

Halliday J. EFSA opens the floor on nanotechnology [Internet]. Springer International Publishing: Basel, Switzerland. 2007 [cited 13 April 2011]. Available from: http://www.foodnavigator.com/Financial-Industry/EFSA-opens-the-floor-on-nanotechnology.

Hamouda T, Baker Jr. JR. Antimicrobial mechanism of action of surfactant lipid preparations in enteric Gram-negative bacilli. J Appl Microbiol. 2000;89:397–403.

Hatanaka J, Chikamori H, Sato H, Uchida S Debari, K, Onoue S, Yamada S. Physicochemical and pharmacological characterization of alpha-tocopherol-loaded nano-emulsion system. Int J Pharm. 2010;396(1–2):188–93.

Hatanaka J, Kimura Y, Lai-Fu Z, Onoue S, Yamada S. Physicochemical and pharmacokinetic characterization of water-soluble coenzyme Q(10) formulations. Int J Pharm. 2008;363(1–2):112–7.

Hazen C. Formulating function into beverages. Food Prod Des. 2003;12(10):36–70.

Howe AM, Pitt AR. Rheology and stability of oil-in water nanoemulsions stabilized by anionic surfactant and gelatin addition of homologous series of sugar-based co-surfactants. Adv Colloid Interface Sci. 2008;144(1–2):30–7.

Huang Q, Yu H, Ru Q. Bioavailability and delivery of nutraceuticals using nanotechnology. J Food Sci. 2010;75(1):R50–7.

Izquierdo P, Esquena J, Tadros TF, Dederen JC, Feng J, Garcia-Celma MJ, et al. Phase behavior and nanoemulsion formation by the phase inversion temperature method. Langmuir. 2004;20(16):6594–8.

Izquierdo P, Feng J, Esquena J, Tadros TF, Dederen JC, Garcia MJ, et al. The influence of surfactant mixing ratio on nano-emulsion formation by the pit method. J Colloid Interface Sci. 2005:285(1):388–94.

Jafari S, He Y, Bhandari B. Optimization of nano-emulsions production by micro fluidization. Eur Food Res Technol. 2007;225(5):733–41.

Jenning V, Gysler A, Schafer-Korting M, Gohla SH. Vitamin A loaded solid lipid nanoparticles for topical use: Occlusive properties and drug targeting to the upper skin. Eur J Pharm Biopharm. 2000;49:211–8.

Jiahui H, Johnston KP, Williams III RO. Nanoparticle engineering processes for enhancing the dissolution rates of poorly water soluble drugs. Drug Dev Ind Pharm. 2004;30(3):233–45.

Jores K, Mehnert W, Dreschler M, Bunjes H, Johann C, Mäder K. Investigations on the structure of solid lipid nanoparticles (SLN) and oil-loaded solid lipid nanoparticles by photon correlation spectroscopy, field-flow fractionation and transmission electron microscopy. J Control Release. 2004;95:217–27.

Kelmann R, Kuminek G, Teixeira H, Koester L. Carbamazepine parenteral nanoemulsions prepared by spontaneous emulsification process. Int J Pharm. 2007;342(1–2):231–9.

Kentish S, Wooster TJ, Ashokkumar A, Balachandran S, Mawson R, Simons L. The use of ultrasonics for nanoemulsion preparation. Paper presented at: The 3rd Innovative Foods Centre Conference on Food Innovation, Emerging Science, Technologies and Applications; 2006; Melbourne, Australia.

Kesisoglou F, Panmai S, Wu YH. Application of nanoparticles in oral delivery of immediate release formulations. Curr Nanosci. 2007;3:183–90.

Kong M, Chen XG, Kweon DK, Park HJ. Investigations on skin permeation of hyaluronic acid based nanoemulsion as transdermal carrier. Carbohydr Polym. 2011;86:837–43.

Lakkis JM. Encapsulation and controlled release technologies in food systems. Blackwell Publishing: Barcelona, Spain; 2007. p. 235–239. DOI:10.1002/9780470277881.

Lee SJ, McClements DJ. Fabrication of protein-stabilized nanoemulsions using a combined homogenization and amphiphilic solvent dissolution/evaporation approach. Food Hydrocolloids. 2010:24(6–7): 560–9.

Leong TSH, Wooster TJ, Kentish SE, Ashokkumar M. Minimising oil droplet size using ultrasonic emulsification. Ultrason Sonochem. 2009;16(6):721–7.

Liang R, Shoemaker CF, Yang X, Zhong F, Huang Q. Stability and bioaccessibility of β-carotene in nanoemulsions stabilized by modified starches. J Agric Food Chem. 2013;61:1249–57. DOI:10.1021/jf303967f.

Luykx DMAM, Goerdayal SS, Dingemanse PJ, Jiskoot W, Jongen PMJM. HPLC and tandem detection to monitor conformational properties of biopharmaceuticals. J Chromatogr B. 2005;821(1):45–52.

Luykx DMAM, Peters RJB, van Ruth SM, Bouwmeester H. A review of analytical methods for the identification and characterization of nano delivery systems in food. J Agric Food Chem. 2008;56(18):8231–47.

Mao LK, Xu D, Yang J, Yuan F, Gao YX, Zhao J. Effects of small and large molecule emulsifiers on the characteristics of beta-carotene nanoemulsions prepared by high pressure homogenization. Food Technol Biotechnol. 2009;47:336–42.

Martini E, Fattal E, de Oliveira M, Teixeira H. Effect of cationic lipid composition on properties of oligonucleotide/emulsion complexes: Physico-chemical and release studies. Int J Pharm. 2008;352(1–2):280–6.

Martins P, Dulley R, Ramos S, Barbosa M, Assumpção R, Junior S, Lacerda A. Nanotecnologias na indústria de alimentos [Internet]. 2007 [cited 14 April 2011]. Available from: http://www.pucsp.br/eitt/downloads/vi_ciclo_paulomartins_marisabarbosa_nano_puc.pdf.

Mason TG, Wilking JN, Meleson K, Chang CB, Graves SM. Nanoemulsions: Formation, structure, and physical properties. J Phys Condens Matter. 2006;18(41):R635.

Mayer S, Weiss J, McClements DJ. Behavior of vitamin acetate delivery systems under simulated gastrointestinal conditions: Lipid digestion and bioaccessibility of low-energy nanoemulsions. J Colloid Interface Sci. 2013;404:215–22. DOI:10.1016/j.jcis.2013.04.048.

McClements DJ. Isothermal titration calorimetry study of pectin–ionic surfactant interactions. J Agric Food Chem. 2000;48(11):5604–11.

McClements DJ. Food emulsions: Principles, practice, and techniques. 2nd ed. Boca Raton, FL: CRC Press; 2005.

McClements DJ, Chanamai RN. Physiochemical properties of monodispersed oil-in-water emulsions. J Disp Sci Tech. 2002;23:125–34.

McClements DJ, Decker EA, Park, Y, Weiss J. Structural design principles for delivery of bioactive components in nutraceuticals and functional foods. Crit Rev Food Sci Nutr. 2009;49(6):577–606.

McClements DJ, Decker EA, Weiss J. Emulsion-based delivery systems for lipophilic bioactive components. J Food Sci. 2007;72(8):R109–24.

Mills I, Cvitas T, Homann K, Kallay N, Kuchitsu K. IUPAC quantities, units and symbols in physical chemistry. 2nd ed. Oxford: Blackwell; 1993.

Mulik RS, Monkkonen J, Juvonen RO, Mahadik KR, Paradkar AR. Transferrin mediated solid lipid nanoparticles containing curcumin: Enhanced in vitro anticancer activity by induction of apoptosis. Int J Pharm. 2010;398:190–203.

Müller RH, Radtke M, Wissing SA. Solid lipid nanoparticles (SLN) and nanostructured lipid carriers (NLC) in cosmetic and dermatological preparations. Adv Drug Deliv Rev. 2002;54(1):131–55.

Newman H. New use for vegetable oil: It's like bleach, but edible [Internet]. Free Press; 2002 [cited 1 March 2005]. Available from: www.freep.com/tech.

NutraLease. 2011a. Available from: http://www.nutralease.com/Nutra/Templates/showpage.asp?.

NutraLease. 2011c. Available from: http://www.nutralease.com/Nutra/Templates/showpage.asp?.

Pinheiro AC, Lad M, Silva HD, Coimbra MA, Boland M, Vicente AA. Unravelling the behavior of curcumin nanoemulsions during in-vitro digestion: Effect of the surface charge. Soft Matter. 2013;9:3147–54. DOI:10.1039/c3sm27527b.

Porras M, Solans C, González C, Gutiérrez JM. Properties of water-in-oil (W/O) nano-emulsions prepared by a low-energy emulsification method. Colloids Surfaces A Physicochem Eng Aspects. 2008;324(1–3):181–8.

Powell M, Colin M. Nanotechnology and food safety: Potential benefits, possible risks. CAB Rev Perspect Agric Vet Sci Nutr Nat Res 2008;3:123–42.

Preetz C, Hauser A, Hause G, Kramer A, Mäder K. Application of atomic force microscopy and ultrasonic resonator technology on nanoscale: Distinction of nanoemulsions from nanocapsules. Eur J Pharm Sci. 2010;39(1–3):141–51.

Qian C, McClements D. Formation of nanoemulsions stabilized by model food grade emulsifiers using high pressure homogenization: Factors affecting particle size. Food Hydrocolloids. 2011;25:1000–8. DOI:10.1016/j.foodhyd.2010.09.017.

Quintanilla-Carvajal M, Camacho-Díaz B, Meraz-Torres L, Chanona-Pérez J, Alamilla-Beltrán L, Jimenéz-Aparicio A, et al. Nanoencapsulation: A new trend in food engineering processing. Food Eng Rev. 2010;2(1):39–50.

Rao J, McClements DJ. Stabilization of phase inversion temperature nanoemulsions by surfactant displacement. J Agric Food Chem. 2010;58(11):7059–66.

Ribeiro HS, Chu BS, Ichikawa S, Nakajima M. Preparation of nanodispersions containing β-carotene by solvent displacement method. Food Hydrocolloids. 2008;22(1):12–7.

Roy K, Mao HQ, Huang SK, Leong KW. Oral gene delivery with chitosan BDNA nanoparticles generates immunologic protection in a murine model of peanut allergy. Nature Med. 1999;5:387–391.

Sadurní N, Solans C, Azemar N, García-Celma MJ. Studies on the formation of O/W nano-emulsions, by low-energy emulsification methods, suitable for pharmaceutical applications. Eur J Pharm Sci. 2005;26(5):438–45.

Salvia-Trujillo L, Qian C, Martín-Belloso O, McClements DJ. Influence of particle size on lipid digestion and β-carotene bioaccessibility in emulsions and nanoemulsions. Food Chem. 2013a;141(2):1472–80. DOI:10.1016/j. foodchem.2013.03.050.

Salvia-Trujillo L, Qian C, Martín-Belloso O, McClements DJ. Modulating β-carotene bioaccessibility by controlling oil composition and concentration in edible nanoemulsions. Food Chem. 2013b;139(1–4):878–84. DOI:10.1016/j.foodchem.2013.02.024.

Sanguansri P, Augustin MA. Nanoscale materials development – A food industry perspective. Trends Food Sci Technol. 2006;17(10):547–56.

Shefer A, Shefer SD. Novel encapsulation system provided controlled release of ingredients. Food Technol. 2003a;57(11):40–42.

Shefer A, Shefer SD. Multicomponent biodegradable bioadhesive controlled release system for oral care products. United States Patent US6589562. 2003b.

Silva H, Cerqueira M, Souza B, Ribeiro C, Avides M, Quintas M, Coimbra J, Carneiro-da-Cunha M, Vicente A. Nanoemulsions of β-carotene using a high-energy emulsification-evaporation technique. J Food Eng. 2011;102(2):130–5.

Spernath A, Aserin A. Microemulsions as carriers for drugs and nutraceuticals. Adv Colloid Interface Sci. 2006;128–130:47–64.

Tadros T, Izquierdo P, Esquena J, Solans C. Formation and stability of nano-emulsions. Adv Colloid Interface Sci. 2004;108–109:303–18.

Takegami S, Kitamura K, Kawada H, Matsumoto Y, Kitade T, Ishida H, Nagata C. Preparation and characterization of a new lipid nano-emulsion containing two cosurfactants, sodium palmitate for droplet size reduction and sucrose palmitate for stability enhancement. Chem Pharm Bull. 2008:56(8):1097–102.

Tan CP, Nakajima M. Effect of polyglycerol esters of fatty acids on physicochemical properties and stability of β-carotene nano dispersions prepared by emulsification/evaporation method. J Sci Food Agric. 2005b;85(1):121–6.

Teo BS, Basri M, Zakaria MR, Salleh AB, Rahman RN, Rahman MB. A potential tocopherol acetate loaded palm oil esters-in-water nanoemulsions for nanocosmeceuticals. J Nanobiotechnol. 2010;8(4):1–12.

Unilever. 2011 [cited 13 April 2011]. Available from: http://www.unilever.com/innovation/ productinnovations/coolicecreaminnovations/?WT.LHNAV=Cool_ice_cream_innovations.

Usón N, Garcia MJ, Solans C. Formation of water-in oil(W/O) nano-emulsions in a water/mixed non-ionic surfactant/oil systems prepared by a low-energy emulsification method. Colloids Surf A Physicochem Eng Aspects. 2004;250(1–3):415–21.

Velinova MJ, Staffhorst RWHM, Mulder WJM, Dries AS, Jansen BAJ, de Kruijff B, et al. Preparation and stability of lipid-coated nanocapsules of cisplatin: Anionic phospholipid specificity. Biochimica et Biophysica Acta (BBA)-Biomembranes. 2004;1663(1–2):135–42.

Wang XY, Jiang Y, Wang YW, Huang MT, Ho CT, Huang QR. Enhancing anti-inflammation activity of curcumin through O/W nanoemulsions. Food Chem. 2008;108(2):419–24.

Weiss J, Decker E, McClements D, Kristbergsson K, Helgason T, Awad T. Solid lipid nanoparticles as delivery systems for bioactive food components. Food Biophys. 2008;3(2):146–54.

Wissing SA, Kayser O, Müller RH. Solid lipid nanoparticles for parenteral drug delivery. Adv Drug Deliv Rev. 2004;56(9);1257–72.

Wooster T, Golding M, Sanguansri P. Impact of oil type on nanoemulsion formation and Ostwald ripening stability. Langmuir. 2008;24(22):12758–65.

Wulff-Perez M, Galvez-Ruiz, MJ, de Vicente J, Martin-Rodriguez A. Delaying lipid digestion through steric surfactant Pluronic F68: A novel in vitro approach. Food Res Int. 2010;43(6):1629–33.

Wulff-Perez M, Torcello-Gomez A, Galvez-Ruiz MJ, Martin-Rodriguez A. Stability of emulsions for parenteral feeding: Preparation and characterization of O/W nanoemulsions with natural oils and pluronic F68 as surfactant. Food Hydrocolloids. 2009;23(4):1096–102.

Yin LJ, Chu BS, Kobayashi I, Nakajima M. Performance of selected emulsifiers and their combinations in the preparation of β-carotene nanodispersions. Food Hydrocolloids. 2009;23(6):1617–22.

Yuan Y, Gao YX, Mao L, Zhao J. Optimization of conditions for the preparation of beta-carotene nanoemulsions using response surface methodology. Food Chem. 2008a;107:1300–6.

Yuan Y, Gao YX, Zhao J, Mao L. Characterization and stability evaluation of β-carotene nanoemulsions prepared by high pressure homogenization under various emulsifying conditions. Food Res Int. 2008b;41(1):61–8.

12

Toxic Effects of Metal and Metal Oxide Nanoparticles on Humans and the Environment

Sanjay Dey, Bhaskar Mazumder, and Supriya Datta

CONTENTS

ABSTRACT: The use of nanoparticles (NPs) in commercial products and industrial applications has been prominently increased in recent years. However, the augmented use of NPs widens the likelihood of environmental exposure to NPs and poses a threat to specific NP-associated hazards. The anticipated toxicity of NPs has raised questions about the mechanisms through which they affect human life and their environment. The toxicity of NPs varies with their size, shape and other basic properties, therefore it is essential to study their basic properties as well as to assess their biological hazards. Among the existing NPs, there is a substantial interest in metal and metal oxide NPs for commercial development and the concerns surrounding their eco-toxicological effects and their effects on humans. The present chapter addresses the current understanding of the physico-chemical properties, possible mechanisms involved, and the toxicity of various metal and metal oxide NPs on humans and the environment.

KEY WORDS: *Eco-toxicity, metal, metal oxide, nanoparticles, reactive oxygen species*

12.1 Introduction

Nanoparticles (NPs) are engineered structures with dimensions between 1 to 100 nm. Now-a-days, the rapid advance of nanotechnology has resulted in an enormous array of NPs. The development of materials and products at the nano-scale has become a major investment area on a global level. According to "The Project on Emerging Nanotechnologies" (2005) inventory, the production of nanotechnology-based consumer products has increased 24% since 2010, reaching a total of 1,628 products on the market worldwide in March 2014 (http://www.nanotechproject.org/news/archive/9242/). In recent years, the use of NPs in commercial products and industrial applications has been increased greatly. NPs are currently being used for commercial purposes, such as fillers, opacifiers, catalysts, semiconductors, cosmetics, microelectronics, and drug carriers. However, the increased use of NPs increases the likelihood of environmental exposure to NPs and poses a huge question as to the specific NP-associated hazards. Despite the remarkable benefits of NPs, there are questions about how the NPs used in day-to-day life may affect human health and the ecological environment (Ray et al., 2009). The human exposure and environmental burden are perceptibly escalated as the application of NPs expands. One of the challenges in the field of nanotechnology is environmental health and safety (EHS), which is focussed on the consideration of the properties of NPs that could pose hazards to the environment and human beings (Thomas et al., 2001). The potential toxicity of NPs has received more and more attention because of the possible entering of NPs into the environment at any stage between production and disposal of consumer products (Behra and Krug, 2008). The possible eco-toxicological risks of NPs may arise during material fabrication, handling, usage, and waste disposal (Wiesner and Bottero, 2007; Wiesner et al., 2006). Due to the seriousness of the environmental risk posed by the release of NPs into the environment, it becomes necessary to set specific standards for the manufacture, use, and disposal of NPs (Federici et al., 2007).

This chapter aims to address the current understanding of the physico-chemical properties, possible mechanisms involved, and the toxicity of metal and metal oxide NPs on human beings and the environment.

12.2 Release of NPs in the Environment

Increased production and widespread use of NPs in various industries causes their frequent release into the environment. NPs enter into the environment either intentionally or unintentionally via atmospheric emissions and solid or liquid waste streams from production sites. The intentional release of NPs into the environment includes the use of NPs for the remediation of contaminated soil and water (Klaine et al., 2008). A proportion of NPs used as one of the components of paints, fabrics, personal health care products, cosmetics, etc., also find routes to enter into the environment. Several routine human activities, e.g., wear of car tyres and urban air pollution, also significantly release NPs into the environment.

NPs are present in the environment from the beginning of the earth's history, therefore the release of NPs in nature is becoming a concern to regulatory agencies. Natural NPs often disappear from the environment by dissolution or mostly through aggregation. But NPs may persist in the environment due to their stabilization by capping or fixing agents (Handy et al., 2008). Additionally, NPs may contain chemically toxic components that do not occur naturally.

12.3 Physico-Chemical Properties of NPs

The toxicity associated with NPs is mainly due to the generation of reactive oxygen species (ROS). The production of ROS by NPs may proceed through a variety of mechanisms. The formation of ROS from a NP is dependent on both the physical and chemical properties of the NPs (Shaligram and Campbell, 2013). The chemical and physical structural determinants of the NP toxicity include molecular size, shape, oxidation status, surface area, bonded surface species, surface coating, solubility, and degree of aggregation and agglomeration (Lu et al., 2010; Nel et al., 2006; Ray et al., 2009; Shaligram and Campbell, 2013; Xia et al., 2008). Through these intrinsic determinants, certain NPs can stimulate and generate inflammation that chemically or catalytically converts into less toxic oxidants, such as superoxide, hydrogen peroxide (H_2O_2), and more reactive free radicals, such as hydroxyl radicals. In addition, environmental factors, such as light, also influence the interaction of NPs with biological tissue, thereby affecting the generation of ROS and resulting in toxicity. Several critical determinants that profoundly determine NP-induced toxicity are discussed below.

12.3.1 Size and Shape

NP-induced toxicity is determined by the unique physical and chemical properties based on their size and surface-to-volume ratio. Because of their tiny size, NPs can easily penetrate cell membranes and other biological barriers of living organisms, resulting in cellular dysfunction (Nel et al., 2006; Xia et al., 2008). The amount of cellular uptake decreases with the increase of particle size (He et al., 2012; Sakai et al, 2011; Wang et al., 2010). The smaller the particle size, the greater will be the tendency to enter the subcellular organelles (Boyoglu et al., 2013). NPs, with their small size, can reach easily into the lungs and lead to several adverse effects (Oberdörster et al., 2005). Yoshida et al. (2012) demonstrated that particle size was a critical determinant of the intracellular distribution of amorphous silica and its induced ROS formation, leading to deoxyribonucleic acid (DNA) damage in human skin HaCaT cells. Moreover, bactericidal or toxic effects increase as the size of the NPs decreases (Auffan et al., 2009a). Li et al (2010) reported that alpha-manganese oxide (α-MnO_2) nanowire-induced cytotoxicity, DNA oxidative damage, and apoptosis in HeLa cells were mediated by ROS and oxidative stress. In cultured fibroblasts, long nanowires caused failed cell division, DNA damage, and increased ROS, while vertical nanowire arrays induced cell motility and proliferation (Persson et al., 2013). Nano-titanium dioxide (TiO_2) exposed in cultured WISH cells (a human amnion-derived cell line) induced cytotoxicity, morphological alterations, generation of intracellular ROS, and DNA damage (Saquib et al., 2012). Jiang et al. (2011) found that for a fixed total surface area, a S-shaped curve dependence for ROS generation per unit of surface area occurred as a function of the particle size (4–195 nm with the same crystal phase) (Braydich-Stolle et al., 2009).

The photocytotoxicity of nano-TiO_2 of four different sizes (<25 nm, 31 nm, <100 nm, and 325 nm) and two different crystal forms (anatase and rutile) in human skin keratinocytes was studied by Yin et al. (2012). All nano-TiO_2 particles induced photocytotoxicity and cell membrane damage in a light-dose and nano-TiO_2-dose-dependent manner upon ultraviolet A (UVA) irradiation. The nano-TiO_2 with a smaller particle size induced greater cell damage. The anatase form of nano-TiO_2 induced higher photocytotoxicity than the rutile form. All the induced photocytotoxic damage was mediated by ROS, generated during UVA irradiation (Yin et al., 2012).

Oh et al. (2010) determined that the 50-nm silica-titania hollow NPs exerted the greatest toxicity on macrophages among the different sizes (25 nm, 50 nm, 75 nm, 100 nm, and 125 nm) of silica-titania hollow NPs. On the other hand, the shape of the NPs is a critical determinant of nanomaterial-induced toxicity. Zinc oxide (ZnO) with snowflake particles seems to be the most active among all available morphologies (whisker, snowflake, and spherical shapes) (Przybyszewska and Zaborski, 2009). Hexagonal, plate-like ZnO nanocrystals were also reported to exhibit significantly higher activity than rod-shaped crystals (McLaren et al., 2009). In addition, the external morphology may directly influence uptake into cells in the following order: rods/spheres > cylinders > cubes (Chithrani et al., 2006; Qiu et al., 2010). Moreover, changes in the aspect ratio of nanorods tend to affect total cell uptake (Chithrani et al., 2006). The larger the contact area of the NPs, the higher the biocompatibility they reveal. The aforementioned properties may endow upon respective NPs huge differences in interactions with biological systems. For instance, a higher toxicity to zebrafish embryos was observed for dendritic nickel (Ni)-NPs compared to spherical Ni-NPs (Ispas et al., 2009). This indicates that the molecular size and shape of NPs are the specific determinants of intracellular response, degree of cytotoxicity, and potential mechanisms of toxicity. Moreover, these NPs induced cell specific responses, resulting in variable toxicity and subsequent cell damage.

12.3.2 Particle Surface, Surface Positive Charges, and Surface Containing Groups

Iron oxide magnetic NPs (nano-Fe_3O_4) have been used for biomedical applications. Nano-Fe_3O_4 coated with poly(ethylenimine) (PEI) and poly(ethylene glycol) (PEG) possess different surface positive charges. Hoskins et al. (2012) determined that nano-Fe_3O_4-PEI possesses a greater surface charge than nano-Fe_3O_4-PEI-PEG, and thereby exhibits greater cytotoxicity and ROS formation in human SH-SY5Y, MCF-7, and U937 cell lines.

Wang et al. (2012) demonstrated that hematite nano-Fe_2O_3 and maghemite nano-Fe_2O_3 have different surface structures and are responsible for the generation of hydroxyl radicals at different levels.

In the field of biomedical applications, especially in drug delivery systems including labelling and tissue engineering, mesoporous silica NPs are one of the most studied (Lee et al., 2011). Mesoporous silica and colloidal silica have different pore architecture, with a specific surface area and pore volume, and have different nanotoxicity during cellular uptake and immune responses.

Shi et al. (2012) demonstrated that copper NPs specially engineered with different surface ligands, e.g., 8-mercaptooctanoic acid, 12-mercaptododecanoic acid, and 16-mercaptohexadecanoic acid, exhibited different surface oxidation reactions and resulted in the generation of ROS at different levels.

Maccormack et al. (2012) determined that nano-silicon (Si), nano-gold (Au), and nano-cadmium selenide (CdSe) inhibited lactate dehydrogenase (LDH) activity and the inhibition was dependent on particle core and surface functional-group composition. These NPs were bound with abundant proteins non-specifically via a charge interaction (Maccormack et al., 2012). Oh et al. (2010) described cationic silica-titania hollow NPs as exhibiting the greatest toxicity and uptake efficiency on J774A.1 macrophages among the different sizes and surface functionalities of silica-titania hollow NPs. These results illustrate size-dependent and surface functionality-dependent nanotoxicity and uptake of silica-titania hollow NPs must be considered for the application of silica-titania hollow NPs as drug delivery and imaging probes (Oh et al., 2010). The effect of the composition of quantum dots on quantum dot-induced toxicity has been well studied (Hauck et al., 2010; Hoshino et al., 2011; Lewinski et al., 2008; Winnik and Maysinger, 2013). Winnik and Maysinger (2013) evaluated quantum dot-induced toxicity both *in vitro* and *in vivo* and they concluded that the quantum dot-induced toxicity is dependent on the composition, size, and properties of the surface-capping materials of quantum dots. Hauck et al. (2010) studied the toxicity of quantum dots and found that the composition lead sulphide (PbS) versus cadmium sulphide (CdS), size, shape (spheres vs. rods), and surface chemistry (amino vs. carboxylic groups) were critical determinants in quantum dot-induced toxicity and their biodistribution. Zhu et al. (2011) showed that both the chemical structure and particle size of quantum dots has a great influence on the stability of their monolayer surface coating.

NPs possess a large specific surface area to potentially absorb transition metals, such as iron (Fe), nickel (Ni), copper (Cu), and chromium (Cr), onto their surface. Fenton-like reactions and Habere–Weiss reactions generate hydroxyl radicals that directly attack DNA (Petersen and Nelson, 2010). Limbach et al. (2007) found that nano-silica doped with transition metals generates high levels of ROS in A549 cells.

Carbon black NPs absorb chemicals such as polycyclic aromatic hydrocarbons (PAHs) due to their large surface area. Upon biotransformation, PAHs oxidize to PAH quinines (redox-active) and reduce to PAH semiquinones and superoxide anion radicals. Superoxide anion radicals dismutate into H_2O_2 upon reacting with transition metals and produce hydroxyl radicals (Petersen and Nelson, 2010).

C_{60} is one of the fullerenes with a spherical shape. Water-soluble, monodisperse, or colloidal fullerenes aggregate and generate superoxide anions and induce lipid peroxidation, resulting in cytotoxicity (Nel et al., 2006). Different functional groups attached to the fullerenes determine the fullerene-induced toxicity (Nel et al., 2006). Thus, prooxidative (generation of free radicals) or antioxidative (scavenging free radicals) activities of fullerenes can be done by structural and surface modifications. Fullerenes can also be phototoxic to excite the fullerene surface upon visible or UV light irradiation. Fullerenes in their excited triplet-state transfer energy to molecular oxygen, thereby forming singlet oxygen, and also transfer an electron, which induces superoxide anion radicals, resulting in lipid peroxidation and thereby leading to cytotoxicity (Nel et al., 2006). Modifying the fullerene surface by attacking one or more malonyl groups yields derived fullerenes possessing antioxidant activity (Nel et al., 2006).

Several fullerenes, like fullerenols and endohedral metallo fullerenols, act as free-radical scavengers (Yin et al., 2009). Yin et al. (2009) studied the ROS-scavenging capability of $[Gd@C_{82}(OH)_{22}]_n$ NPs and compared them with other functionalized fullerenes, like functionalized C_{60}-fullerenols and

C_{60}-carboxyfullerenes. The *in vitro* electron spin resonance (ESR) measurements demonstrated that Gd@ $C_{82}(OH)_{22}$, $(C_{60}(OH)_{22})$ (a fullerenol), and $C_{60}(C(COOH)_2)_2$ (a carboxyfullerene) efficiently scavenged different types of free radicals, like superoxide anion radicals, hydroxyl radicals, and singlet oxygen (Yin et al., 2009). Fullerene derivatives have the capability of reducing H_2O_2-induced cytotoxicity, free radical formation, and mitochondrial damage in human lung adenocarcinoma A549 cells, as well as in rat brain capillary endothelial cells. The fullerene derivatives also have the capability to scavenge intracellular ROS. These observed, different, free-radical scavenging capabilities suggest that both chemical properties, such as surface chemistry (induced different electron affinity), and physical properties, such as degree of aggregation, influence the biological and biomedical activities of functionalized fullerenes (Yin et al., 2009).

According to Yang and Watts (2005), the particle surface plays an important role in the phytotoxicity of alumina NPs. In the year 2005, after studying the mechanism of nano-silver-induced toxicity, Yang and Watts (2005) came to the conclusion that both dissolved and surface coating of alumina NPs are important factors attributing to their toxicity.

12.3.3 Solubility and Particle Dissolution

Studer et al. (2010) revealed that intercellular solubility is a determinant, affecting NP-induced cytotoxicity. According to their findings, soluble copper metal is less toxic than copper oxide NPs. Shaligram and Campbell (2013) reported a similar kind of solubility-dependent toxic effect in NPs. The toxicity of copper salts was affected by its solubility profile and the cell type tested.

Shen et al. (2013) found that there is a good correlation among the nano-ZnO-induced cytotoxicity and the free intracellular zinc concentration in human immune cells. It was found that nano-ZnO dissolution and contact with the cells was required to elicit cytotoxicity. The level of nano-ZnO-induced ROS also correlated with the cytotoxicity and the level of intracellular free zinc. From the results, it was concluded that the intracellular dissolution of Zn-NPs is required to show cytotoxicity. Although, nano-ZnO-induced cytotoxicity was not affected by antioxidants that reduce ROS. Hence, it cannot be said that all nano-Zno-induced cytotoxicity is caused by ROS.

Mahto et al. (2010) reported that quantum dot core/shell (CdSe/ZnSe), when dispersed in aqueous media, induced ROS formation, and thereby released cadmium which led to toxicity. Nano-ZnO, when suspended in Dulbecco's modified Eagle's medium (DMEM), induced dose-dependent cytotoxicity resulting in cell membrane damage in human bronchoalveolar carcinoma A549 cells, human hepatocellular carcinoma HepG2 cells, human skin fibroblast cells, human skin HaCaT keratinocytes, and rat primary neuronal cells. The relative cytotoxic potential of nano-ZnO in these cell systems is in the order: new born rat forebrain primary cells > human skin fibroblasts > human skin HaCaT keratinocytes > human bronchoalveolar carcinoma A549 cells > human hepatocellular carcinoma HepG2 cells (Chiang et al., 2012).

A comparison was made on the cytotoxicity of nano-ZnO in serum-free DMEM and DMEM supplemented with 10% foetal bovine serum (FBS). It was found that nano-ZnO in DMEM supplemented with 10% FBS induced cytotoxicity at a similar level to that without the 10% FBS. Hence cytotoxicity is not only mediated by ROS. These results conclude that the mechanism of cytotoxicity is medium dependent, suggesting that cellular growth conditions play a significant role in the induction of cytotoxicity by nano-ZnO (Chiang et al., 2012). The results also conclude that not all nanomaterial-induced cell death is mediated by ROS.

12.3.4 Metal Ions Released from Metal and Metal Oxide NPs

Beer et al. (2012) studied the role of the silver (Ag) ion fraction of nano-Ag suspensions on nano-Ag-induced cytotoxicity in human A549 lung cells. They found that at high silver ion fractions the nano-Ag failed to induce measurable levels of toxicity. However, at 2.6% silver ion fraction, lower nano-Ag suspensions showed more toxic effects in comparison to their supernatant.

Faisal et al. (2013) observed nickel oxide NPs (nano-NiO) induced cellular ROS, oxidative stress, mitochondrial dysfunction, apoptosis/necrosis, and lipid peroxidation in tomato seedling roots. There is an enhancement in the activities of antioxidative enzymes, including catalase, glutathione, and SOD. They concluded that the induction of cell death by triggering the mitochondrion-dependent intrinsic pathway was due to the dissolution of Ni ions from nano-NiO.

Foldbjerg et al. (2011) studied the effects of well characterized polyvinyl pyrrolidone (PVP)-coated nano-Ag and silver ions (Ag⁺) in the human alveolar cell line A549. Both PVP-coated nano-Ag- and Ag⁺-induced cytotoxicity is dose-dependent. The cytotoxicity of both PVP-coated nano-Ag and Ag⁺ was inhibited by the antioxidant, N-acetylcysteine. There remains a strong correlation between the levels of ROS formation, mitochondrial damage, early apoptosis, and ROS-induced DNA damage. The level of bulky DNA-adduct formation was well correlated with the cellular ROS levels and was inhibited by pre-treatment with N-acetylcysteine. Hence, from the results, it can be seen that nano-Ag-induced geno-toxicity is mainly mediated by ROS (Foldbjerg et al., 2011).

When NPs are in a suspending medium or in a biological system where particle dissolution can take place, the ionic species generated by the nanomaterial dissolution also has the ability to show toxic-ity (Franklin et al., 2007). Franklin et al. (2007) observed that Zn^{+2} ions released from the dissolved nano-ZnO were highly toxic to aquatic organisms. Hence, the disruption of cellular zinc homeostasis, lysosomal and mitochondrial damage, and cell death are due to nano-ZnO dissociation. Subsequently, Xia et al. (2008) also determined that dissolution plays an important role in ZnO-induced cytotoxicity.

It has been seen that Ag⁺ can be released from the surface of nano-Ag by surface oxidation. Factors that influence the release of ions include the size of the nano-Ag particles, the sulfur concentrations contained in the nano-Ag, temperature, oxygen, pH, and light. The molecular oxygen reacts with the Ag⁺ ions to generate superoxide radicals and other ROS, resulting in apoptosis and the expression of stress response-related genes (Limbach et al., 2007; Lubick, 2008). Experimental results suggested that the Ag⁺ ion was the reactive species leading to the toxicity of the nano-Ag.

Zhu et al. (2013) studied the toxicity of nano-copper oxide (CuO), nano-cadmium oxide (CdO), and nano-TiO_2. CuO NPs were found to possess the most potent cytotoxicity and to cause the most damage to the DNA. Nano-CdO showed less cytotoxicity than nano-CuO. Nano-TiO_2 was found to show low cytotoxicity, without an increase in 8-oxo-2′-deoxyguanosine (8-OHdG) levels.

The redox properties of nano-metals and nano-metal oxides can enhance the formation of ROS by forming hydroxyl radicals through Fenton reactions, Fenton-like reactions, or the Habere–Weiss cycle reaction (Gonzalez et al., 2008; He et al., 2012; Wang et al., 2012). Hydroxyl radicals (·OH) from H_2O_2 and superoxide anion radicals (O_2^-) were generated due to the Habere–Weiss cycle reaction.

Fenton reaction:

$$Fe^{2+} + H_2O_2 \rightarrow Fe^{3+} + \bullet OH + : OH^-$$

Fenton-like reaction:

$$Cu^+ + H_2O_2 \rightarrow Cu^{2+} + \bullet OH + OH^-$$

$$Ag + H_2O_2 \rightarrow Ag^{2+} + \bullet OH + OH^-$$

Haber-Weiss cycle reaction:

$$Fe^{3+} + \bullet O_2^- \rightarrow F^{2+} + O_2$$

$$Fe^{2+} + H_2O_2 \rightarrow Fe^{3+} + \bullet OH + OH^-$$

12.3.5 Light Activation

The reactivity of NPs can be severally altered by exposure to UV and visible light. Light-induced changes in reactivity may affect the stability of a product containing NPs. Upon light irradiation, NPs can be excited and then reacted with molecular oxygen to generate ROS and produce cytotoxicity (i.e., photocytotoxicity). Nano-TiO_2 and nano-ZnO are two commercially important examples. The bonding in these materials gives rise to valence bands and conduction bands whose energy difference is similar to that in UV light (for nano-TiO_2 and nano-ZnO this energy is in the UV spectral region). Upon photo-excitation of the NPs, electrons are excited from the valence band to the conduction band, giving rise to excitons or electron hole pairs. These electron hole pairs can recombine and give rise to a chemical transformation. Alternatively, electrons can combine with surface-bound molecules (such as oxygen) to form a radical (such as a superoxide anion radical). Similarly, powerful oxidants such as the electron hole pairs can react with surface-bound molecules (such as water) and give rise to radicals (such as the hydroxyl radical). Additional intermediate radical reactions may result in the formation of other ROS-like singlet oxygen (Daimon and Nosaka, 2007; Ghosh et al., 2010; Hirakawa and Hirano, 2006; Kangwansupamonkon et al., 2009; Mura et al., 1999; Sycheva et al., 2011; Vevers and Jha, 2008).

Upon UV irradiation, nano-ZnO induced dose-dependent cytotoxicity, LDH release, lipid peroxidation, and DNA damage in a low dose. Nano-ZnO also induced the breakage of single-stranded DNA in supercoiled FX174 plasmid DNA. Under the illumination of visible light, nano-ZnO induced LDH leakage, hydroxyl radical generation, and 8-OHdG formation in a dose-dependent manner. Collectively, these results demonstrated the photocytotoxic and photogenotoxic effects of nano-ZnO on human skin keratinocytes *in vitro*. Upon receiving the UV irradiation, the resulting photo-excited nano-ZnO favourably extracted electrons from water or hydroxide ions to generate hydroxyl radicals. The overall result indicates that photoirradiation of nano-ZnO-generated ROS (mainly hydroxyl radicals) causes DNA cleavage, induces lipid peroxidation, and generates endogenous DNA adducts (8-OHdG) (Wang et al., 2013).

Zhao et al. (2013) studied the phototoxicity of bismuth oxychloride (BiOCl) single crystalline nanosheets with various surface structures under UV light irradiation. It was determined that the generation of ROS is dependent on the surface structure.

12.3.6 Aggregation and Mode of Interaction with Cells

Graphene oxide and graphene are two-dimensional, carbon-based NPs and are potential candidates for biomedical applications, including sensors, cell labelling, bacterial inhibition, and drug delivery (Liao et al., 2011). Liao et al. (2011) found that the toxicity of graphene and graphene oxide depended on their aggregation and on the mode of interaction with the cells (i.e., suspension vs. adherent cell types).

Ekstrand-Hammarstrom et al. (2012) determined that the level of ROS and cytotoxicity elicited by TiO_2 in turn depends on the crystal structure, surface area, and degree of aggregation and agglomeration. Kim and Ryu (2013) discussed the physiochemical characteristics that can determine nano-Ag-induced toxicity based on size, release of Ag^+ ions, and agglomeration/aggregation.

12.3.7 Crystalline State of Nanoparticles

McCarthy et al. (2012) found that an inflammatory effect in the lungs was caused due to the presence of amorphous nano-silicon dioxide (SiO_2). They also observed that the amorphous nano-SiO_2 induced inflammation in submucosal cells and generated ROS, leading to apoptosis and decreased cell survival.

12.3.8 pH of the System

Fe_3O_4 magnetic NPs exhibited an intrinsic peroxidase-like activity under acidic conditions and a catalase-like activity at neutral pH (Wang et al., 2012). The authors proposed that this pH-dependent dual enzyme activity can be utilized for cancer treatment through producing ROS in tumor tissues (Wang et al., 2012).

He et al. (2012) determined that in the presence of H_2O_2, nano-Ag can induce the formation of hydroxyl radicals through a Fenton-like mechanism, and the level of hydroxyl radical formation is greatly dependent on the experimental pH condition. The level of hydroxyl radical formation was in the following order: pH 1.2 > pH 3.6 > pH 4.6. At pH 7.4, no significant formation of hydroxyl radicals was observed. Wang et al. (2012) also determined that both hematite nano-Fe_2O_3 and maghemite nano-Fe_2O_3 induced hydroxyl radicals, with an intensity in the following order: pH 1.2 > pH 4.2 > pH 7.2. Apparently, all the above described results indicate that lower pH conditions facilitate Fenton and Fenton-like reactions to generate hydroxyl radicals.

12.4 Methodologies to Study the Toxicity of NPs

Both *in vitro* and *in vivo* methods provide valuable tools to investigate nanotoxicity. The evaluation of the toxicity of NPs by *in vitro* toxicity testing is cost-effective and there are too many NPs to evaluate each *in vivo*. Some advantages of *in vitro* studies using various cell lines are that they: (1) reveal the effects of target cells in the absence of secondary effects caused by inflammation, (2) permit the identification of primary mechanisms of toxicity in the absence of physiological and compensatory factors which confound the interpretation of whole animal studies, (3) are efficient, rapid, and cost-effective, and (4) can be used to improve the design of subsequent, expensive whole animal studies. For instance, human bronchoalveolar carcinoma-derived cells (A549), astrocytoma cells (U87), human monoblastoid cells (U937), mouse Leydig TM3 cells, human V79 and L929 fibroblasta, human SCCVII, B16F10, and FsaR tumour cells, and mouse peritoneal macrophage (RAW264.7) cell lines have been used to characterize the oxidative stress-related signalling pathway (Fahmy and Cormier, 2009; Ivankovic et al., 2006; Kang et al., 2007; Lai et al., 2008; Lin et al., 2006a, 2006b, 2008, 2009; Vamanu et al., 2008; Xia et al., 2008). A non-transformed human lung cell line (BEAS-2B) has been used for the expression of genes pertaining to oxidative stress and cell death pathways (Huang et al., 2010; Sarkar et al., 2007). Another valuable technique to manipulate environmental influences in *in vitro* studies is by co-culture of cell lines (Braydich-Stolle et al., 2010).

A limitation of *in vitro* testing is that cells in culture do not experience the range of pathogenic effects observed *in vivo*, partly related to issues of translocation, toxicokinetics, and coordinated tissue responses.

12.5 Mechanism of Toxicity of Metallic NPs (MNPs)

The toxicity of dissolved and micron-sized metallic NPs (MNPs) is dependent on the speciation of the dissolved metallic compounds, the chemical stability of the metallic particles, the oxidation state of the metals (Cr, arsenic (As), Fe, Mn, etc.), their association with specific ligands, and their concentration in solution (Leonard and Lauwerys, 1980; Thomas, 2011). Whether the MNPs are modified or not, the *in vitro* toxicity depends on the pH, ionic strength, thermodynamic feasibility, concentration of the species, kinetic facility of electron transfers, and also the redox conditions of the biological media (Pourbaix, 1974; Schoonen et al., 2006). The toxic action of MNPs can potentially involve three distinct mechanisms: (1) the release of metals in solutions (dissolution process), (2) the catalytic properties of MNPs, and (3) the redox evolution of the surface which can oxidize proteins, generate ROS, and induce oxidative stress (Adams et al., 2006; Brunner et al., 2006).

Oxidative stress is a state of redox disequilibrium in which ROS production (by the cell or by the NPs itself) overcomes the antioxidant defence capacity of the cell, thereby leading to adverse biological consequences (Chaudiére and Ferrari-Iliou, 1999; Gardes-Albert et al., 2003; Xia et al., 2006) and the damage of macromolecules, lipids, DNA, or proteins resulting in excess proliferation, apoptosis, lipid peroxidation, or mutagenesis (Kappus, 1987). The ROS formation by redox reactions may be crucial if the protective mechanisms of cells are overcome by a strong affinity with NPs. MNPs are known to induce oxidative stress by ROS generation during redox cycling by disruption of the electronic and ionic flux, perturbation of the permeability transition pores, and depletion of the cellular glutathione content (Limbach et al., 2007). The pathways of metabolism of NP-induced oxidative stress and toxicity are presented in Figure 12.1.

12.6 Toxicity of MNPs

The different kinds of MNPs are used in many applications, including electronics, optics, textiles, medical applications, cosmetics, food packaging, water-treatment technology, fuel cells, catalysts, biosensors, and agents for environmental remediation (Gerloff et al., 2012; Handy et al., 2008). The wide-spread

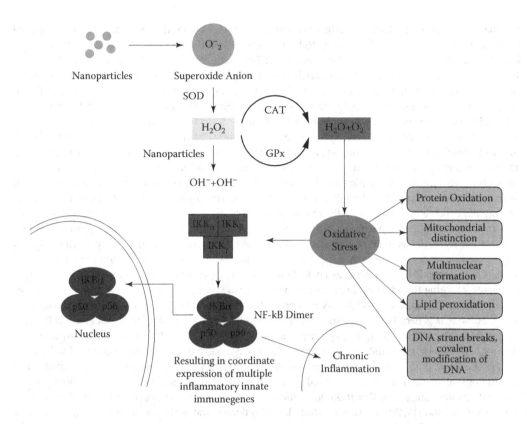

FIGURE 12.1 Pathways of metabolism of nanoparticle (NP)-induced oxidative stress and toxicity. ROS generation is capable of inducing oxidative DNA damage, strand breaks, protein denaturation, and lipid peroxidation, thereby demonstrating the mutagenic and carcinogenic characteristics associated with NPs. The simplified NF-κB signalling pathway is activated by oxidative stress and thereby chronic inflammation.

application of MNPs leads to exposure in the environment and in humans. Consequently, different metals in the form of MNPs have gained access to tissues, cells, and biological molecules within the human body (Mukherjee et al., 2012). The impact of MNPs on human health and the environment has not been fully assessed to date (Christen and Fent, 2012). A lot of research has been carried out to assess the toxic potential of MNPs, however they have presented some serious and far-reaching challenges. Therefore, there remains an urgent need for well-designed studies that will generate data so that the risk of MNPs can be assessed (Ngwa et al., 2011).

Now-a-days, AgNPs are one of the most widely commercially used NPs because of their unique physico-chemical and biological properties (Chen and Schluesener, 2008). For example, AgNPs are well known for their antimicrobial action, and are extensively used in the field of medicine and personal healthcare (Edward-Jones, 2009; Morones et al., 2005). These are also used in applications such as preservatives in cosmetics, textiles, water purification systems, coating in catheters, and wound dressings. In the near future, the anticipated widespread exposure to AgNPs will prompt governmental bodies and the public to raise questions about the safety of such applications (Christensen et al., 2010; Schrand et al., 2010; Wijnhoven et al., 2009). The increasing use of these commercial NPs is of great concern regarding their release into the aquatic environment and consequent negative impact on fish (Benn and Westerhoff, 2008; Farkas et al., 2011; Griffitt et al., 2009). In scientific literature, many studies reported the effects of various types of AgNPs in different test systems. A recent study in rats with AgNPs of three different sizes demonstrated that after 28 days of intravenous administration, particles were mostly distributed to organs containing high numbers of phagocytizing cells such as the liver, spleen, and lung (Lankveld et al., 2010). At the same time, a number of studies have shown that AgNPs may induce cytotoxicity in phagocytizing cells such as mouse peritoneal macrophages and human monocytes

(Foldbjerg et al., 2009; Park et al., 2010; Shavandi et al., 2011; Soto et al., 2007). It has been suggested that the cytotoxic effects were induced by ROS, resulting in cellular apoptosis at low concentration and short incubation times (Braydich-Stolle et al., 2005; Foldbjerg et al., 2009; Nishanth et al., 2011; Piao et al., 2011). Two previous studies have described the transport pattern of AgNPs into the chorion layers and inner mass of zebrafish eggs (Asharani et al., 2008; Lee et al., 2007). It was found that the NPs are transported into zebrafish embryos through chorion pore canals and are localized in the brain, heart, yolk, and blood. Neutron activation analysis and transmission electron microscopy (TEM) also confirm an efficient adsorption and uptake of AgNPs by zebrafish eggs (Bar-Ilan et al., 2009; Laban et al., 2010). Numerous developmental malformations are observed in these treated embryonic zebrafish, including fanfold abnormality, abnormal body axes, twisted notochord, slow blood flow, pericardial oedema, cardiac malformation, turbid chorionic fluid, hatching delays, and growth retardation. Similarly, two recent studies have demonstrated that AgNPs cause significant morphological malformations, delayed development, and oedema in embryonic fathead minnows (*Pimephales promelas*) and medaka fish (*Oryzias latipes*) (Laban et al., 2010; Wu et al., 2010). However, the mechanism of AgNP developmental toxicity to freshwater fish remains unclear. Lee et al. (2007) have characterized the transport pattern of AgNPs into the chorion layers of zebrafish eggs. The result suggests that the NPs dock into the chorion pore canals. These trapped NPs may serve as nucleation sites and aggregate with incoming NPs to form larger aggregates, clogging the chorion canals, affecting membrane transport, and destroying the structure of the egg chorion. The chorion, which acts as a protective barrier and maintains the osmotic balance for the fish embryo, is generally believed to prevent eggs from chemical insults (Boudreau et al., 2005; Hill et al., 2004). Taken all together, we hypothesized that AgNPs could produce impairments in chorionic integrity and osmotic imbalances in fish eggs, which may contribute to the developmental toxicity of AgNPs. Oxidative damage is known to be associated with certain pathological processes involved in the aetiology of many fish diseases caused by environmental pollutants (Livingstone, 2001). Earlier investigations have suggested that ROS generation is responsible for the observed bactericidal activity of AgNPs (Choi and Hu, 2008). Another study has also demonstrated the mechanism of AgNP toxicity to bacteria via oxidative damage (Hwang et al., 2008). AgNPs are also reported to increase ROS levels, resulting in the activation of antioxidant enzymes and the depletion of glutathione (GSH) content in rat liver cells (Hussain et al., 2005), mouse germline stem cells (Braydich-Stolle et al., 2005), and human cells (Asharani et al., 2008). Additional concern arises from the knowledge that ionic silver that can be released from AgNPs (Borm et al., 2006), which is known to bioaccumulate in estuarine/marine environments, where its toxicity may be of particular concern (Luoma et al., 1995). Silver is also highly toxic in freshwater environments (Hogstrand et al., 1996; Nebeker et al., 1983). The environment is hampered by exposure to AgNPs (Blaser et al., 2008) due to a lack of knowledge regarding its environmental fate and transport, as well as its original toxicity (Wiesner et al., 2006; Wiesner et al., 2009). AgNP toxicity has been reported in bacteria (Choi and Hu, 2008; Choi et al., 2008; Lok et al., 2006; Morones et al., 2005; Sondi and Salopek-Sondi, 2004; Yoon et al., 2007), cell culture systems (Ahamed et al., 2008; Hussain et al., 2005), *Drosophila melanogaster* (Ahamed et al., 2009), zebrafish (Asharani et al., 2008; Bar-Ilan et al., 2009; Griffitt et al., 2008), oyster embryos (Ringwood et al., 2009), and *Daphnia pulex* (Griffitt et al., 2008). However, it is unclear whether the toxicity of AgNPs is due to the dissolution of the silver or the potentially unique properties of silver NPs. Some studies support the hypothesis that AgNPs exert toxicity in a "nano specific" way, i.e., a size-dependent or crystalline state-dependent toxicity (Pal et al., 2007) that could not be predicted simply by the release of dissolved silver species (Kawata et al., 2009). Photosensitive NPs like AgNPs are shown to cleave double-strand DNA upon light exposure (Badireddy et al., 2007). Many *in vitro* studies indicate that AgNPs are toxic at higher concentrations to mammalian cells derived from the skin, liver, lung, brain, vascular system, and reproductive organs (Wijnhoven et al., 2009). AgNPs induce DNA damage, oxidative stress, damage of mitochondria, cell cycle arrest, apoptosis in human Jurkat T cells, lung fibroblast and glioblastoma cells (Eom and Choi, 2010), and oxidative changes in the hepatoma cell line HepG2 (Nowrouzi et al., 2010). It was shown that AgNPs or their aggregates are taken up by embryos of medaka (Kashiwada, 2006) and zebrafish (Yeo and Yoon, 2009), whereas SiNPs are not able to penetrate the chorion of zebrafish eggs (Fent et al., 2010). AgNPs lead to a concentration-dependent increase in mortality and delayed hatching in fish embryos (Asharani et al., 2008; Bar-Ilan et al., 2009), and alter the development and behaviour of zebrafish larvae depending on size, coating, and composition (Powers et al., 2011). Size-dependent uptake of AgNPs and the induction of oxidative stress in gills was also observed in brown trout (Scown et al., 2010). In zebrafish, increased levels of ROS and induced expression of pro-apoptotic genes is reported (Choi et al., 2010). In contrast, the effects of silica NPs on fish are less well known and silver-doped silica NPs have not yet been investigated for their potential effects.

Studies have shown the *in vivo* toxicity of gold NPs (AuNPs) when they were injected into experimental animals. The organ distribution of gold NPs was found to be related to their particle size. The gold content in different organs after 24 hours was determined upon intravenous injection of AuNPs with a diameter of 10, 50, 100, and 250 nm in the tail vein of rats (De Jong et al., 2008). AuNPs with a diameter of 10 nm were found to have an extensive tissue distribution in the blood, liver, spleen, kidney, testis, thymus, heart, lung, and brain, while the larger particles were only detectable in the blood, liver, and spleen. This shows that the tissue distribution of AuNPs is size-dependent, and the smaller NPs spread to more organs than the larger ones. The foetal exposure of NPs in murine pregnancy was influenced by both the gestational age and NP surface composition. As per the influence of gestational age, a critical time window existed in murine pregnancy and impacted on the extent of foetal exposure to maternally-administered AuNPs. The time window was found to be between ~9.5 and ~11.5 weeks when the transfers of AuNPs from the mother to the foetus were dramatically reduced during the maturation of the murine placental blood supply. In addition to gestational age, the cytotoxicity of AuNPs depended on the surface modification by ferritin, PEG, and citrate. This novel finding provided a biosafety implication for AuNP administration in pregnancy in humans (Yang et al., 2011).

From the toxicological viewpoint, the potential toxicity of AuNPs is due to their accumulation in different organs. After a single intravenous injection of 0.2 mL (15.1 μg/mL) AuNPs dispersion in rats, the AuNPs were found to be rapidly and consistently accumulated in the liver and spleen within two weeks. This accumulation significantly altered the *in vivo* expression of genes related to detoxification, lipid metabolism, cell cycle, defence response, and circadian rhythm (Balasubramanyam et al., 2009). AuNPs caused potential nephrotoxicity in rats after one week of NP exposure. They also induced an alteration in renal tubules marked with cloudy swelling and vacuolar degeneration in a particle size-dependent manner. These were possibly related with AuNP-induced oxidative stress (Abdelhalim and Jarrar, 2011). Similar ROS-induced histological alterations in the cardiac tissue were observed in rats with exposure to AuNPs (Abdelhalim, 2011). When male Wister rats were intraperitoneally injected with AuNPs (20 μg/kg body weight) for three days, oxidative stress-induced DNA damage and cell death were observed in the brain, accompanied with a decreased expression in antioxidant enzymes and the appearance of lipid peroxidation (Siddiqi et al., 2012). These toxic responses can be altered by the surface modification of AuNPs. Surface coating is considered a crucial factor in determining the biodistribution, clearance, and toxicity of AuNPs. For instance, BSA (bovine serum albumin)- and GSH-protected AuNPs differed in their pharmacokinetics. The GSH-protected AuNPs could be quickly removed by renal clearance and thus the toxicity was significantly decreased, however the BSA-protected AuNPs had low-efficient renal clearance and largely accumulated in the liver and spleen, causing an irreparable toxicity response (Zhang et al., 2012b).

12.7 Mechanism of Toxicity of Metal Oxide NPs

Among the existing NPs, there is substantial interest in metal oxide NPs for commercial development and, correspondingly, concerns surrounding their eco-toxicological effects (Auffan et al., 2009b). For example, nano-TiO$_2$ is often used as a pigment or additive for paints, paper, ceramics, plastics, foods, and other products. The toxic action of metal oxide NPs can potentially involve three distinct mechanisms (Brunner et al., 2006). First, the particle may release toxic substances into the exposure media, e.g., free Ag^{2+} ions from the silver particles. Second, surface interactions with the media may produce toxic substances, e.g., chemical radicals or ROS. Third, the particles or their surfaces may interact directly with and disrupt biological targets, e.g., carbon nanotube interaction with membranes or intercalation with DNA. Because nano-ZnO partially dissolves in water, exposures in aquatic systems are expected to involve both soluble and particulate species, suggesting that these three mechanisms of toxic action are tenable for ZnO. Solubilized Zn^{2+} from nano-ZnO has proven to contribute substantially to the cytotoxicity of these NPs (Brunner et al., 2006; Heinlaan et al., 2008).

An understanding of the mechanisms of interaction at the molecular level between NPs and biological systems is lacking (Maynard, 2006). In some of these products, such as skin creams and toothpastes, NPs are in direct contact with the user's body or can enter the environment on a continual basis from the removal (e.g., by washing) of such products (Daughton and Ternes, 1999), or, even worse, a fatal accident during the production of NPs could release a large quantity of NPs into the environment (Moore, 2006). Though the detailed mechanism of toxicity caused by NPs is not yet elucidated, a few mechanisms like the damage of membrane integrity, protein destabilization and oxidation, damage to nucleic acids, production of ROS, interruption of energy transduction, and the release of harmful and toxic components are likely involved in the damage caused by NPs (Klaine et al., 2008). The mechanism of toxicity of NPs is presented in Figure 12.2.

The cell membrane is a potential target of damage for NPs. NPs stimulate the production of ROS in organisms and cause damage in possibly every cell component. These ROS oxidize double bonds of fatty acids in the cell membrane, resulting in an increased permeability, rendering it more susceptible to osmotic stress. The mechanism underlying NP-induced cytotoxicity is only incompletely understood. It has been suggested to be related to the inflammatory response, oxidative stress, and p53 activation (Hussain et al., 2005; Xia et al., 2006). Because of their greater surface area per unit mass, NPs are more biologically active than larger particles of the same compounds, and, as a result, they have unique biological properties (Colvin, 2003; Papageorgiou et al., 2007). However, accurate toxicological information is required to ensure that NPs are manufactured and used safely. The ROS-mediated toxicities of metal oxide NPs are summarized in Table 12.1.

The production of ROS is regarded as the main mechanism underlying the toxicity of metal oxide NPs, including iron oxide NPs. For example, the γ-Fe_2O_3 NPs (20–40 nm in diameter) produced highly reactive hydroxyl radicals that mediated the toxicity of NPs (Voinov et al., 2011). L-glutamic acid-coated Fe_2O_3 NPs induced a disturbance on the cell redox status in Chinese hamster lung (CHL) cells. They caused ROS production, glutathione depletion, and the inactivation of some antioxidant enzymes, including glutathione reductase and superoxide dismutase, but not catalase (Zhang et al., 2012a). The Fe_2O_3 NPs could further induce oxidative stress-related apoptosis with the loss of mitochondrial membrane potential and apoptotic chromatin condensation in human umbilical endothelial (ECV304) cells (Zhu et al., 2010). Furthermore, the NPs that generated ROS also mediated microtubule remodelling and

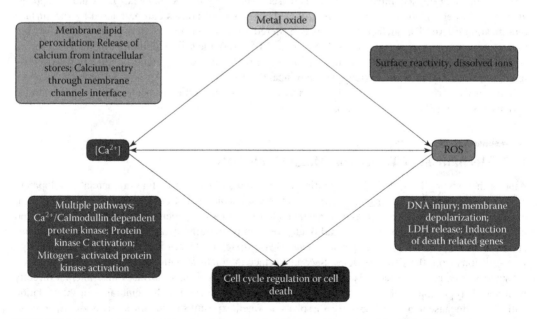

FIGURE 12.2 The possible mechanism of toxicity of metal and metal oxide nanoparticles in the human body.

TABLE 12.1

List of ROS-Dependent Effects of Metal Oxide Nanoparticles (Manake et al., 2013)

Nanoparticles	ROS-Dependent Effect
Iron oxide	Necrosis apoptosis in murine macrophage (J774) cells
	Human microvascular endothelial cell permeability
	Activation of NF-κB and AP-1, inflammation in human epidermal keratinocytes (HEK) and murine epidermal cells (JB6P(+))
Copper oxide	Genotoxicity in human lung epithelial cells
	Mitochondrial dysfunction, oxidative DNA damage, cell death in A549 cell line
	Cytotoxicity *in vitro* in Hep-2 cell line
	Nephrotoxicity and hepatotoxicity *in vivo*
Zinc oxide	Mitochondrial dysfunction, morphological modification, and apoptosis *in vitro* in human foetal lung fibroblasts
	Cellular oxidant injury, excitation of inflammation, and cell death in BEAS-2B and RAW 264.7 cells
	Mitochondrial damage, apoptosis, and IL-8 release *in vitro* in LoVo human colon carcinoma cell line
	Mitochondrial damage, genotoxic and apoptotic effects *in vitro* human liver cells
	Apoptosis in human alveolar adenocarcinoma cells via p53, surviving, and Bax/Bcl-2 pathways
Titanium dioxide	Apoptotic cell death through ROS-mediated Fas upregulation and Bax activation
	Cytotoxic and genotoxic effects *in vitro* in human amnion epithelial (WISH) cell line
	Cytotoxic and apoptotic cell death *in vitro* in HeLa cell line
Aluminum oxide	Mitochondria-mediated oxidative stress and cytotoxicity in human mesenchymal stem cells

increased cell permeability in human microvascular endothelial cells, while a pre-treatment of a ROS scavenger catalase significantly inhibited cell permeability (Apopa et al., 2009). The Fe_2O_3 NP-induced neurotoxicity was also attributed to the significant oxidative stress caused by the NPs, in which the altered activity of oxidative stress-associated enzymes was determined (Wang et al., 2009). Hence, it seems that ROS production and the sequent oxidative stress-induced cellular damages are responsible for the potential toxicity of Fe_2O_3 NPs *in vivo* and *in vitro*.

12.8 Toxicity of Metal Oxide NPs

12.8.1 Toxicity of Nano-TiO₂

TiO_2 NPs have been widely used as additives in cosmetics, pharmaceuticals, food colorants, paints, solar cells, photocatalytic water purification, sunscreens, and coatings for self-cleaning windows (Botta et al., 2011; Klaine et al., 2008). The annual production of nano-TiO_2 was estimated to be 5,000 metric tons in 2006–2010, 10,000 metric tons in 2011–2014, and around 2.5 million metric tons by 2025 (Menard et al., 2011). Therefore, nano-TiO_2 is inevitably present in the aquatic environment (Botta et al., 2011; Gottschalk and Nowack, 2011; Gottschalk et al., 2009). A large number researchers observed the toxicity of nano-TiO_2 (Menard et al., 2011; Sharma, 2009) on bacteria (Adams et al., 2006; Jiang et al., 2011), aquatic organisms (Hund-Rinke and Simon, 2006; Ji et al., 2011; Mohammed Sadiq et al., 2011), plants (Asli and Neumann, 2009), and mammalians (Warheit et al., 2007). Therefore, according to the US National Institute of Occupational Safety and Health, nano-TiO_2 is categorized as a potential human carcinogen (Maynard, 2011). However, the mechanism of the toxicity of nano-TiO_2 is still in dispute. The fate of NPs in the environment is depicted in Figure 12.3.

Different research groups applied different toxicity assays techniques and reached different conclusions on nanotoxicity and its underlying mechanism (Menard et al., 2011). Moreover, most of the reported studies used laboratory culture systems which were dramatically different from the natural environment. The natural environment contains complex constituents, especially natural organic matter (NOM), which would interact with the released NPs and consequently alter their physico-chemical properties (Gao et al., 2009). More studies are required to describe the nano-TiO_2 toxicity and its

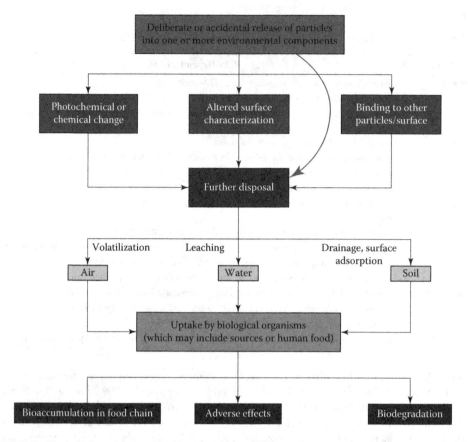

FIGURE 12.3   The fate of metal and metal oxide nanoparticles in the environment.

underlying mechanism, especially in the presence of NOM. Nano-TiO_2 may also react with a wide range of organic and biological molecules and then produce toxic effects in various cell lines with or without photoactivation (Chen et al., 2010; Jin et al., 2008; Wadhwa et al., 2011). Upon exposure to UV light, TiO_2 generate ROS due to its photocatalyst properties (Khus et al., 2006; Zhao et al., 2005). ROS production can lead to DNA strand breaks, crosslinking, and adducts of the bases or sugars (Cabiscol et al., 2000). In aqueous media, TiO_2 absorbs substantial UV radiation and yields hydroxyl species. These species may cause substantial damage to DNA (Dunford et al., 1997), resulting in environmental hazards. The aggregation of TiO_2 NPs influences their toxicity and leads to varying biological behaviour in organisms (Baveye and Laba, 2008). Exposure to nano-TiO_2 resulted in gill injury and oxidative stress in rainbow trout (Federici et al., 2007). Chronic exposure to nano-TiO_2 induced inhibition of growth, decrease of liver weight ratio, and histopathological changes to gills in zebrafish (Chen et al., 2011). A previous study has compared the *in vivo* toxicities of different NPs at relatively high concentrations on zebrafish embryos and larvae, however the toxicity comparison was not performed on NPs of identical nano-levels (Zhu et al., 2008). Therefore, it is necessary to perform the toxicity comparison of different NPs at environmentally relevant concentrations with identical nano-sizes. It has been reported that acute exposure to nano-TiO_2 at relatively high concentrations causes an increase in mortality, inhibition of growth, reduction of reproduction, decrease of movement, or alteration of gene expression patterns in *Caenorhabditis elegans* (Khare et al., 2011; Ma et al., 2009; Roh et al., 2010; Wang et al., 2009).

Nano-TiO_2 is one of the most commonly used photocatalysts. They are highly efficient in initiating light-induced redox activity with molecules adsorbed on their surfaces. TiO_2 exhibits size-dependent photocatalytic activity accompanied by size-dependent biological effects. TiO_2 crystallites are known to be highly sensitive to phase transformation as the size decreases. There are two main crystalline structures of TiO_2: anatase and rutile (Zhang and Banfield, 1998). These two types have different crystal faces (anatase: {011} and {110}, and rutile: {001} and {011}) and this change seems to have an implication on the oxidation and reduction sites (Ohno et al., 2002). Indeed, anatase is observed to be more biologically active than rutile in terms of cytotoxicity or oxidative DNA damage (Hirakawa et al., 2004; Sato and Taya, 2006). Jang et al. (2007) have shown that for nano-TiO_2 ranging between 15 and 30 nm, the bactericidal effects towards *Escherichia coli* and *Pseudomonas aeruginosa* decreases and the mass fraction of anatase increases.

12.8.2 Toxicity of Nano-ZnO

Nano-ZnO is widely used in the production of pigments, semiconductors, sunscreens, and food additives. Nano-ZnO has an inherently high risk of water contamination and can reach high concentrations in surface waters, posing a significant threat to aquatic ecosystems (Gottschalk et al., 2009). As far as metal oxide NPs are concerned, the release of metal ions is one of the most investigated pathways. The dissolved Zn^{2+} ions are considered the most relevant toxic agents of nano-ZnO in aqueous media (Aruoja et al., 2009; Franklin et al., 2007; Xia et al., 2008). Free Zn^{2+} appears toxic to many aquatic organisms, including marine phytoplankton (Miller et al., 2010), however some studies reveal that the adverse impact of nano-ZnO could not be completely explained by their metal ion release (Maness et al., 1999) and that metal NPs may be more toxic than either their ionic forms or their parent compounds (Farré et al., 2009; Navarro et al., 2008). At the current level of knowledge, the observed ZnO toxicity cannot be attributed univocally to the release of Zn ions or to their nano-size in the peculiar surface interactions of these NPs with target organisms and/or with the media (i.e., water, biological media, etc.). Therefore, the toxicity pathways are still to be established (Miao et al., 2010; Miller et al., 2010). Lin and Xing (2007) studied the effects of Zn and nano-ZnO in seed germination and root elongation tests. The results indicated that Zn and nano-ZnO caused a significant inhibition of seed germination and root growth. Acute exposure of nano-ZnO killed embryos, retarded the embryo hatching, reduced the body length of larvae, and caused tail malformation in zebrafish (Bai et al., 2010). ZnO NPs were more toxic towards algae than ZnO, but relatively less toxic towards crustaceans and fish (Wong et al., 2010). Nano-ZnO exposure increased mortality, negatively affected metamorphosis, and inhibited the growth of *Xenopus laevis* (Nations et al., 2011). ZnO is also a photocatalyst and promotes the generation of ROS under irradiation with energy at or above its band gap energy of 3.37 eV (equivalent to wavelength 368 nm) and produces significant toxicity to higher organisms (Ma et al., 2011).

12.8.3 Toxicity of Nano-SiO$_2$

SiO_2 NPs are widely used in various fields, particularly in chemical mechanical polishing as additives, and recently in medical applications. To date, their safety to human health and the environment has not been fully assessed. Nano-SiO_2 can enter human cells and cytotoxic effects are dose-, time-, and size-dependent (Napierska et al., 2010). Interactions of SiO_2 NPs with different cell types have been investigated, in particular on cells of the respiratory tract and immune cells (Napierska et al., 2010). Induction of oxidative stress, membrane damage, and disturbed calcium homeostasis were found to be the underlying mechanisms of cellular toxicity of nano-SiO_2 (Ariano et al., 2011). Survival/proliferation of neural cells was observed by 50 nm, but not 200 nm, sized silica NPs. Generally, smaller NPs appear to induce more marked effects than bigger ones (Lu et al., 2011; Napierska et al., 2010). SiO_2 NPs induced hepatocytic necrosis and increased serum aminotransferase and inflammatory cytokines in exposed mice. They were detected in mononuclear phagocytic cells in the liver and spleen (Liu et al., 2011) and clearance from the blood stream and tissues can be very slow (Cho et al., 2009). SiO_2 NPs can also be doped with varying amounts of Ag to convey antimicrobial activities. SiO_2-AgNPs can be used to make antimicrobial disposable bandages where Ag contributes to the antimicrobial activity (Tarimala et al., 2006). Such NPs may also find use in technical applications such as waste water technology. So far, the toxicity of Ag-doped NPs has not been investigated, in contrast to silica NPs or AgNPs.

12.8.4 Toxicity of Nano-Al₂O₃

While the toxicity of aluminum is relatively well known and has been discussed in many publications, the toxicity of aluminum oxide NPs (nano-Al_2O_3) remains almost unknown and is still an object of experimental work. The possibility of an impact of these NPs on the environment has been demonstrated by their bioaccumulation in various organisms, including *Tubifex tubifex* Muller, *Hyalella azteca* Saussure, *Lumbriculus variegates* Muller, and *Corbicula fluminea* Muller (Stanley et al., 2010). The result of a study by Coleman et al. (2010) indicated that nano-sized Al_2O_3 can accumulate and impact the reproduction and behaviour of *Eisenia fetida* Savigny, whereas at high levels they are unlikely to be found in the natural environment. This fact means that Al_2O_3 NPs may possibly enter the food chain and be responsible for toxicity in animals. The oral exposure of nano-Al_2O_3 in rats has been demonstrated as the potential cause of genotoxic damage (Balasubramanyam et al., 2009). Work by Chen et al. (2008) using the model of microvascular endothelial cells of the human brain indicates that nano-Al_2O_3 affects the cerebral vasculature in time- and concentration-dependent ways. An intraperitoneal injection of nano-sized Al_2O_3 in Sprague-Dawley rats showed an effect on the innate immune system of the brain (Li et al., 2009). The cytotoxic effect of these NPs on models of UMR 106 cells has been studied and their transport and accumulation in L929 and BJ cells (Radzium et al., 2011) and L5178Y and BEAS-2B cells (Kim et al., 2009) have been described. Possible mechanisms for the cytotoxicity of nano-sized Al_2O_3 particles are still being discussed, but oxidative stress (Prabhakar et al., 2012) and DNA damage (Kim et al., 2009) may be responsible for their cytotoxic effect. Some published research work focused on the possible cytotoxicity of nano-Al_2O_3 on cell models and animals, but their toxic effects on plants, e.g., phytotoxicity, remain almost entirely unknown. Possible interactions between these NPs and the cell walls of algae (*Scenedesmus* sp. and *Chlorella* sp.) have been described in the work of Sadiq et al. (2011). These authors demonstrated a decrease in the chlorophyll content upon exposure to Al_2O_3 NPs.

12.8.5 Toxicity of Nano-Fe₂O₃

The *in vivo* behaviour of ferric oxide NPs are crucial for the assessment of their potential health risk. Intratracheally installed Fe_2O_3 NPs (22 nm) in rats at a dose of 4 mg/rat were detected to penetrate the alveolar–capillary barrier and were transported into systemic circulation within 10 minutes. The NPs selectively and specifically distributed into the phagocyte-rich organs (liver, kidney, spleen, and testicles) in which they were phagocytized by monocytes/macrophages. In the interstitial lung, only parts of Fe_2O_3 NPs were deposited. The plasma elimination half-life of the NPs was calculated to be 22.8 days and the lung clearance rate was 3.06 μg/day. The results conclude that the inhaled NPs were not confined to the lungs but also went through the systemic circulation system and were redistributed (Zhu et al., 2009). Therefore, to counteract the occupational and other long term exposure of the inhaled NPs, systemic toxicity has to be cautioned and monitored.

To understand the mechanism of toxicity of NPs, a detailed comparative study of toxic responses of nano- and micron-sized Fe_2O_3 NPs is required. The impact of the Fe_2O_3 NPs on the pulmonary and coagulation system has been explored, and was followed by further investigation into toxicity potential and important toxicity-related factors (the size, dosage, and exposure time). Fe_2O_3 NPs of a size between 22 and 280 nm were intratracheally instilled to male Sprague-Dawley rats at low (0.8 mg/kg bw) and high (20 mg/kg bw) doses, respectively. Then toxic effects were monitored at one, seven, and 30 days post-instillation. Oxidative stress-related lung injury was induced with nano-sized Fe_2O_3 particles, which are more potent than submicron-sized ones. Alveolar macrophages were overloaded with the phagocytized NPs and the pro-sign of lung fibrosis was triggered. Moreover, Fe_2O_3 NPs affected the coagulation process, as shown by the prolongation of prothrombin time (PT) and activated partial thromboplastin time (APTT). These two are the typical coagulation parameters measured at the thirtieth day post-installation (Zhu et al., 2008). Weak pulmonary fibrosis was developed at the thirtieth day post-exposure in adult male Wister rats when a single dose of 5 mg/kg of Fe_2O_3 NPs was instilled (Szalay et al., 2012).

Other organ toxicity of nano-Fe_2O_3 has also been widely reported. After a single intravenous administration of γ-Fe_2O_3 NPs (0.8 mg/kg) into rats, 40% and 75% of the administered γ-Fe_2O_3 NPs were excreted from the body via urine within 24 hours and 72 hours, respectively. However, the remaining γ-Fe_2O_3 NPs could induce inflammation in the liver, kidney, and lungs (Hanini et al., 2011). The metabolic analysis of the injected γ-Fe_2O_3 NPs (at a single dose of 25 mmol/L) in rats showed NP-induced metabolic abnormality in the spleen and kidneys of the rats, with alterations in their metabolism-associated small molecules and enzymes such as triglycerides, phospholipids, N- and O-acetyl glycoprotein, glucose, and glycogen, as observed 48 hours post-injection (Feng et al., 2011). The question then arises as to whether Fe_2O_3 NPs cross the biological membrane during biodistribution in the body. The first thing to be concerned with is the natural barrier, that is the blood–brain barrier (BBB), which directly reflects the neurotoxicity. Actually, few results indicated that NPs can directly cross through the BBB and induce brain damage. Intranasally instilled Fe_2O_3 NPs transferred to animal brains through the olfactory pathway. The microdistribution of Fe in mice brains was detected by synchrotron radiation X-ray fluorescence analysis after intranasally instilled with Fe_2O_3 NPs (280 ± 80 nm) at a single dose of 40 mg/kg. Fe contents were significantly increased in the olfactory nerve and the triage minus of the brain stem at two weeks post-exposure (Wang et al., 2007). The ratios of Fe (III)/Fe (II) were increased in the olfactory bulb and brain stem, as analyzed by X-ray absorption of the near-edge structure. Concomitantly, the neuron fatty degeneration occurred in the CA3 area of the hippocampus. This result demonstrated that Fe_2O_3 NPs transferred via the olfactory nerve into the brain poses a risk to the neurological system (Wang et al., 2007). Fe_2O_3 NPs were observed to be mainly deposited in the olfactory bulb, hippocampus, and striatum, where neurological damage happens. Microglial, the main participator in the immune system of the central nervous system (CNS), was injured. Meanwhile, substantial ROS and nitric oxide (NO) was produced. In the brains of the exposed mice, neuropathological changes were observed, including the irregular arrangement of neuron cells in the olfactory bulb, cellular swelling, nuclear chromatin condensation, fragmentation in the striatum and cerebral cortex, and damage towards the integrity of neuronal cells in hippocampal CA1 (Wang et al., 2011c). The intranasal exposure of nano- and submicron-sized Fe_2O_3 particles (21 nm and 280 nm) to mice at repeatedly low doses (130 μg) were found to be located in the hippocampus and olfactory bulb. Size-dependent, oxidative stress-related nerve cell damage was induced. In the olfactory bulb and hippocampus, alterations in the activity of oxidative stress-related nerve cell damages were observed in the exposed mice. In the olfactory bulb and hippocampus of the exposed mice, alterations in the activity of the oxidative stress-related biomarkers were detected with a significant increase in the activity of GSH-Px, Cu, zinc-superoxide dismutase (Zn-SOD), and nitrus oxide system (NOS) and cause obvious decrease in total GSH and reduced GSH/oxidized glutathione (GSSG) ratio. Coupled with oxidative stress, ultrastructural alterations in nerve cells, membranous disruption, and lysosome increases in the olfactory bulb also occur (Wang et al., 2009). These studies have suggested that the inhaled Fe_2O_3 NPs may cross the BBB and enter the brain where neurotoxicity is augmented.

It was reported that in the incubation of human aortic endothelial cells (HAECs) with nano-Fe_2O_3 NPs, the NPs penetrated the cell membrane and localized into the cells, thereby leading to cytoplasmic vacuolation, mitochondrial swelling, and cell death. Additionally, they further increased the expression and secretion of various inflammatory factors such as ICAM-1 and IL-8 (Zhu et al., 2011). In another cell line, bPC12, Fe_2O_3 NPs were also internalized and resulted in dose-dependent decreases in cell viability and cell response to nerve growth factor (Pisanic et al., 2007).

Associated with reproductive and developmental toxicity, genotoxicity is also of high concern. Although the data supplied is still limited so far, some evidence has shown that genotoxicity probably depends upon particle size and exposure dosage. For example, in human broncho epithelial cells (BEAS-2B), α-Fe_2O_3 NPs were internalized and induced significant genotoxic effects at certain particle concentrations higher than 50 μg/mL. Size-dependent genotoxicity was also observed in nano-scale Fe_2O_3 particles and was found to be more potent than their micro-scale counterparts (Bhattacharya et al., 2012). This possible reproductive and developmental toxicity was found in the new model organism, zebrafish. When zebrafish in their early life stages were exposed to nano-Fe_2O_3 of increasing doses, nano-Fe_2O_3 at does higher than 50 mg/L produced significant developmental toxicity, including hatching delay, increased mortality rate, and structural deformation (Zhu et al., 2012).

12.9 Conclusion

Use of NPs for commercial and biomedical purpose has been increasing exponentially; therefore, evaluation of safety and toxicity of NPs has become an issue of interest to the public. To date, the question remains whether metal and metal oxide NPs are safe to be used for biomedical purpose. As discussed in the present chapter, different physico-chemical properties of NPs have significant impact on humans and ecological environments. Therefore, careful characterization of NPs in their biologically relevant environment and large scale comparative studies could be a first step in increasing our understanding in this field. Elucidation of detailed biochemical mechanism of NPs-induced ROS formation leading to toxicity is become challenging. The hydroxyl radicals can serve as an ideal representative ROS and is reactive with almost all types of biomolecules including lipids, proteins, and nucleic acids. A lot of research needs to be done to understand the mechanism of NPs-induced toxicity for hazard prevention. On the other hand, the evaluation of ecological impact of metal and metal oxide NPs is an important field. A great number of NPs are widely distributed into the air or soil and can be ingested by bacteria or enter our food chain. Therefore, a careful assessment of the possible negative effects associated with these NPs in biomedical field is necessary in order to allow them to be used in a safe and well-controlled manner for the benefit of mankind. More and more data are becoming available regarding NP toxicity, but a lot of effort is still required in order to truly advance our knowledge in this field.

REFERENCES

Abdelhalim MA. Exposure to gold NPs produces cardiac tissue damage that depends on the size and duration of exposure. Lipids Health Dis. 2011;10:205–14.

Abdelhalim MA, Jarrar BM. Renal tissue alterations were size-dependent with smaller ones induced more effects and related with time exposure of gold NPs. Lipids Health Dis. 2011;10:163–9.

Adams LK, Lyon DY, Alvarez PJJ. Comparative eco-toxicity of nanoscale TiO_2, SiO_2, and ZnO water suspensions. Water Sci Technol. 2006;54:327–34.

Ahamed M, Karns M, Goodson M, Rowe J, Hussain SM, Schlager JJ, Hong Y. DNA damage response to different surface chemistry of silver NPs in mammalian cells. Toxicol Appl Pharmacol. 2008;233:404–10.

Ahamed M, Posgai R, Gorey TJ, Nielsen M, Hussain SM, Rowe JJ. Silver NPs induced heat shock protein 70, oxidative stress and apoptosis in *Drosophila melanogaster*. Toxicol Appl Pharmacol. 2009;242:264–9.

Apopa P, Qian Y, Shao R, Guo N, Schwegler-Berry D, Pacurari M, Porter D, Shi X, Vallyathan V, Castranova V, Flynn DC. Iron oxide NPs induce human microvascular endothelial cell permeability through reactive oxygen species production and microtubule remodeling. Part Fibre Toxicol. 2009;6:1. DOI:10.1186/1743-8977-6-1.

Ariano P, Zamburlin P, Gilardino A, Mortera R, Onida B, Tomatis M, Ghiazza M, Fubini B, Lovisolo D. Interaction of spherical silica NPs with neuronal cells: Size-dependent toxicity and perturbation of calcium homeostasis. Small. 2011;7:766–74.

Aruoja V, Dubourguier HC, Kasemets K, Kahru A. Toxicity of NPs of CuO, ZnO and TiO_2 to microalgae *Pseudokirchneriella subcapitata*. Sci Total Environ. 2009;407:1461–8.

AshaRani PV, Wu YL, Gong ZY, Valiyaveettil S. Toxicity of silver NPs in zebrafish models. Nanotechnology. 2008;19:1–8.

Asli S, Neumann PM. Colloidal suspensions of clay or titanium dioxide NPs can inhibit leaf growth and transpiration via physical effects on root water transport. Plant Cell Environ. 2009;32:577–84.

Auffan M, Rose J, Wiesner MR, Bottero JY. Chemical stability of metallic NPs: A parameter controlling their potential cellular toxicity *in vitro*. Environ Pollut. 2009a;157:1127–33.

Auffan M, Rose J, Bottero J, Lowry GV, Jolivet J, Wiesner MR. Towards a definition of inorganic NPs from an environmental, health and safety perspective. Nat Nanotechnol. 2009b;4:634–41.

Badireddy AR, Hotze EM, Chellam S, Alvarez P, Wiesner MR. Inactivation of bacteriophages *via* photosensitization of fullerol NPs. Environ Sci Technol. 2007;41:6627–32.

Bai W, Zhang Z, Tian W, He X, Ma Y, Zhao Y, Chai Z. Toxicity of zinc oxide NPs to zebrafish embryo: A physicochemical study of toxicity mechanism. J Nanopart Res. 2010;12:1645–54.

Balasubramanyam A, Sailaja N, Mahboob M, Rahman MF, Misra S, Hussain SM, Grover P. Evaluation of genotoxic effects of oral exposure to aluminium oxide NPs in rat bone marrow. Mutat Res-Gen Tox En. 2009;676:41–7.

Bar-Ilan O, Albrecht RM, Fako VE, Furgeson DY. Toxicity assessments of multisized gold and silver NPs in zebrafish embryos. Small. 2009;5:1897–910.

Baveye P, Laba M. Aggregation and toxicology of titanium dioxide NPs. Environ Health Perspect. 2008;116:A152.

Beer C, Foldbjerg R, Hayashi Y, Sutherland DS, Autrup H. Toxicity of silver NPs – Nanoparticle or silver ion? Toxicol Lett. 2012;208:286–92.

Behra R, Krug H. NPs at large. Nat Nanotechnol. 2008;3:253–4.

Benn TM, Westerhoff P. Nanoparticle silver released into water from commercially available sock fabrics. Environ Sci Technol. 2008;42:4133–9.

Bhattacharya K, Hoffmann E, Schins RF, Boertz J, Prantl EM, Alink GM, Byrne HJ, Kuhlbusch TJ, Rahman Q, Wiggers H, Schulz C, Dopp E. Comparison of micro- and nanoscale Fe^{3+}-containing (hematite) particles for their toxicological properties in human lung cells *in vitro*. Toxicol Sci. 2012;126:173–82.

Blaser SA, Scheringer M, Macleod M, Hungerbuhler K. Estimation of cumulative aquatic exposure and risk due to silver: Contribution of nanofunctionalized plastics and textiles. Sci Total Environ. 2008;390: 396–409.

Borm P, Klaessig FC, Landry TD, Moudgil B, Pauluhn J, Thomas K, Trottier R, Wood S. Research strategies for safety evaluation of NPs, Part V: Role of dissolution in biological fate and effects of nanoscale particles. Toxicol Sci. 2006;90:23–32.

Botta C, Labille J, Auffan M, Borschneck D, Miche H, Cabie M, Masion A, Rose J, Bottero JY. TiO_2-based NPs released in water from commercialized sunscreens in a life-cycle perspective: Structures and quantities. Environ Pollut. 2011;159:1543–8.

Boudreau M, Courtenay SC, MacLatchy DL, Berube CH, Hewitt LM, Van Der Kraak GJ. Morphological abnormalities during early-life development of the estuarine mummichog *Fundulus heteroclitus*, as an indicator of androgenic and anti-androgenic endocrine disruption. Aquat Toxicol. 2005;71:357–69.

Boyoglu C, He Q, Willing G, Boyoglu-Barnum S, Dennis VA, Pillai S, Singh SR. Microscopic studies of various sizes of gold NPs and their cellular localizations. ISRN Nanotechnol. 2013. DOI:10.1155/2013/123838.

Braydich-Stolle L, Hussain S, Schlager JJ, Hofmann MC. *In vitro* cytotoxicity of NPs in mammalian germline stem cells. Toxicol Sci. 2005;88:412–9.

Braydich-Stolle LK, Schaeublin NM, Murdock RC, Jiang J, Biswas P, Schlager JJ, Hussain SM. Crystal structure mediates mode of cell death in TiO_2 nanotoxicity. J Nanopart Res. 2009;11:1361–74.

Braydich-Stolle LK, Speshock JL, Castle A, Smith M, Murdock RC, Hussain SM. Nanosized aluminum altered immune function. ACS Nano. 2010;4:3661–70.

Brunner TJ, Wick P, Manser P, Spohn P, Grass RN, Limbach LK, Bruinink A, Stark WJ. *In vitro* cytotoxicity of oxide NPs: Comparison to asbestos, silica, and the effect of particle solubility. Environ Sci Technol. 2006;40:4374–81.

Cabiscol E, Tamarit J, Ros J. Oxidative stress in bacteria and protein damage by reactive oxygen species. Int Microb. 2000;3:3–8.

Chaudière J, Ferrari-Iliou R. Intracellular antioxidants: From chemical to biochemical mechanisms. Food Chem Toxicol. 1999;37:949–62.

Chen X, Schluesener HJ. Nanosilver: A nanoproduct in medical application. Toxicol Lett. 2008;176:1–12.

Chen L, Yokel RA, Hennig B, Toborek M. Manufactured aluminum oxide NPs decrease expression of tight junction proteins in brain vasculature. J Neuroimmune Pharmacol. 2008;3:286–95.

Chen JY, Zhou HJ, Santulli AC, Wong SS. Evaluating cytotoxicity and cellular uptake from the presence of variously processed TiO_2 nanostructured morphologies. Chem Res Toxicol. 2010;23:871–9.

Chen J, Dong X, Xin Y, Zhao M. Effects of titanium dioxide NPs on growth and some histological parameters of zebrafish (*Danio retio*) after long-term exposure. Aquat Toxicol. 2011;101:493–9.

Chiang HM, Xia Q, Zou X, Wang C, Wang S, Miller BJ, et al. Nanoscale ZnO induces cytotoxicity and DNA damage in human cell lines and rat primary neuronal cells. J Nanosci Nanotechnol. 2012;12:2126–35.

Chithrani BD, Ghazani AA, Chan WC. Determining the size and shape dependence of gold nanoparticle uptake into mammalian cells. Nano Lett. 2006;6:662–8.

Cho M, Cho WS, Choi M, Kim SJ, Han BS, Kim SH, Kim HO, Sheen YY, Jeong J. The impact of size on tissue distribution and elimination by single intravenous injection of silica NPs. Toxicol Lett. 2009;189:177–83.

Choi O, Hu ZQ. Size dependent and reactive oxygen species related nanosilver toxicity to nitrifying bacteria. Environ Sci Technol. 2008;42:4583–8.

Choi O, Deng KK, Kim NJ, Ross JL, Surampalli RY, Hu Z. The inhibitory effects of silver NPs, silver ions, and silver chloride colloids on microbial growth. Water Res. 2008;42:3066–74.

Choi JE, Kim S, Ahn JH, Youn P, Kang JS, Park K, Yi J, Ryu DY. Induction of oxidative stress and apoptosis by silver NPs in the liver of adult zebrafish. Aquat Toxicol. 2010;100:151–9.

Christen V, Fent K. Silica NPs and silver-doped silica NPs induce endoplasmatic reticulum stress response and alter cytochrome P4501A activity. Chemosphere. 2012;87:423–34.

Christensen FM, Johnston HJ, Stone V, Aitken RJ, Hankin S, Peters S, Aschberger K. Nano-silver – Feasibility and challenges for human health risk assessment based on open literature. Nanotoxicology. 2010;4:284–95.

Coleman JG, Johnson DR, Stanley JK, Bednar AJ, Weiss CA, Boyd RE, Steevens JA. Assessing the fate and effects of nano aluminum oxide in the terrestrial earthworm *Eisenia fetida*. Environ Toxicol Chem. 2010;29:1575–80.

Colvin VL. The potential environmental impact of engineered NPs. Nat Biotechnol. 2003;21:1166–70.

Daimon T, Nosaka Y. Formation and behavior of singlet molecular oxygen in TiO_2 photocatalysis studied by detection of near-infrared phosphorescence. J Phys Chem C. 2007;111:4420–4.

Daughton CG, Ternes TA. Pharmaceuticals and personal care products in the environment: Agents of subtle change? Environ Health Perspect. 1999;107:907–38.

De Jong WH, Hagens WI, Krystek P, Burger MC, Sips AJ, Geertsma RE. Particle size-dependent organ distribution of gold NPs after intravenous administration. Biomaterials. 2008;29:1912–9.

Dunford R, Salinaro A, Cai LZ, Serpone N, Horikoshi S, Hidaka H, Knowland J. Chemical oxidation and DNA damage catalysed by inorganic sunscreen ingredients. FEBS Lett. 1997;418:87–90.

Edwards-Jones V. The benefits of silver in hygiene, personal care and healthcare. Lett Appl Microbiol. 2009;49:147–52.

Ekstrand-Hammarstrom B, Akfur CM, Andersson PO, Lejon C, Osterlund L, Bucht A. Human primary bronchial epithelial cells respond differently to titanium dioxide NPs than the lung epithelial cell lines A549 and BEAS-2B. Nanotoxicology. 2012;6:623–34.

Eom HJ, Choi J. P38 MAPK activation, DNA damage, cell cycle arrest and apoptosis as mechanisms of toxicity of silver NPs in Jurkat T cells. Environ Sci Technol. 2010;44:8337–42.

Fahmy B, Cormier SA. Copper oxide NPs induce oxidative stress and cytotoxicity in airway epithelial cells. Toxicol In Vitro 2009;23:1365–71.

Faisal M, Saquib Q, Alatar AA, Al-Khedhairy AA, Hegazy AK, Musarrat J. Phytotoxic hazards of NiO-NPs in tomato: A study on mechanism of cell death. J Hazard Mater. 2013;250–251:318–32.

Farkas J, Christian P, Gallego-Urrea JA, Roos N, Hassellov M, Tollefsen KE, Thomas KV. Uptake and effects of manufactured silver NPs in rainbow trout (*Oncorhynchus mykiss*) gill cells. Aquat Toxicol. 2011;101:117–25.

Farré M, Gajda-Schrantz K, Kantiani L, Barceló D. Ecotoxicity and analysis of NPs in the aquatic environment. Anal Bioanal Chem. 2009;393:81–95.

Federici G, Shaw B, Handy R. Toxicity of titanium dioxide NPs to rainbow trout (*Oncorhynchus mykiss*): Gill injury, oxidative stress, and other physiological effects. Aquat Toxicol. 2007;84:415–30.

Feng JH, Liu HL, Bhakoo KK, Lu LH, Chen Z. A metabonomic analysis of organ specific response to USPIO administration. Biomaterials. 2011;32:6558–69.

Fent K, Weisbrod CJ, Wirth-Heller A, Pieles U. Assessment of uptake and toxicity of fluorescent silica NPs in zebrafish (*Danio rerio*) early life stages. Aquat Toxicol. 2010;100:218–28.

Foldbjerg R, Olesen P, Hougaard M, Dang DA, Hoffmann HJ, Autrup H. PVP-coated silver NPs and silver ion induce reactive oxygen species, apoptosis and necrosis in THP-1 monocytes. Toxicol Lett. 2009;190:156–62.

Foldbjerg R, Dang DA, Autrup H. Cytotoxicity and genotoxicity of silver NPs in the human lung cancer cell line, A549. Arch Toxicol. 2011;85:743–50.

Franklin NM, Rogers NJ, Apte SC, Batley GE, Gadd GE, Casey PS. Comparative toxicity of nanoparticulate ZnO, bulk ZnO, and $ZnCl_2$ to a freshwater microalga (*Pseudokirchneriella subcapitata*): The importance of particle solubility. Environ Sci Technol. 2007;41:8484–90.

Gao J, Youn S, Hovsepyan A, Llaneza VL, Wang Y, Bitton G, Bonzongo JCJ. Dispersion and toxicity of selected manufactured NPs in natural river water samples: Effect of water chemical composition. Environ Sci Technol. 2009;43:3322–8.

Gardes-Albert M, Bonnefont-Rousselot D, Abedinzadeh Z, Jore D. Especes reactives de l'oxygène: Comment l'oxygène peut-il devenir toxique? Mécanismes Biochimiques. 2003;91–6.

Gerloff K, Fenoglio I, Carella E, Kolling J, Albrecht C, Boots AW, Förster I, Schins RPF. Distinctive toxicity of TiO_2 rutile/anatase mixed phases NPs on Caco-2 cells. Chem Res Toxicol. 2012;25:646–55.

Ghosh M, Bandyopadhyay M, Mukherjee A. Genotoxicity of titanium dioxide (TiO_2) NPs at two trophic levels: Plant and human lymphocytes. Chemosphere. 2010;81:1253–62.

Gonzalez L, Lison D, Kirsch-Volders M. Genotoxicity of engineered NPs: A critical review. Nanotoxicology. 2008;2:252–73.

Gottschalk F, Nowack B. The release of engineered NPs to the environment. J Environ Monit. 2011;13:1145–55.

Gottschalk F, Sonderer T, Scholz RW, Nowack B. Modelled environmental concentrations of engineered NPs (TiO_2, ZnO, Ag, CNT, fullerenes) for different regions. Environ Sci Technol. 2009;43:9216–22.

Griffitt RJ, Luo J, Gao J, Bonzongo JC, Barber, DS. Effects of particle composition and species on toxicity of metallic NPs in aquatic organisms. Environ Toxicol Chem. 2008;27:1972–8.

Griffitt RJ, Hyndman K, Denslow ND, Barber DS. Comparison of molecular and histological changes in zebrafish gills exposed to metallic NPs. Toxicol Sci. 2009;107:404–15.

Guzman KAD, Taylor MR, Banfield JF. Environmental risks of nanotechnology: National nanotechnology initiative funding, 2000–2004. Environ Sci Technol. 2006;40:1401–7.

Handy RD, Owen R, Valsami-Jones E. The ecotoxicology of NPs and NPs: Current status, knowledge gaps, challenges, and future needs. Ecotoxicology. 2008;17:315–25.

Hardman R. A toxicologic review of quantum dots: Toxicity depends on physicochemical and environmental factors. Environ Health Perspect. 2006;114:165–72.

Hauck TS, Anderson RE, Fischer HC, Newbigging S, Chan WC. *In vivo* quantum dot toxicity assessment. Small. 2010;6:138–44.

He W, Zhou YT, Wamer WG, Boudreau MD, Yin JJ. Mechanisms of the pH dependent generation of hydroxyl radicals and oxygen induced by Ag NPs. Biomaterials. 2012;33:7547–55.

Heinlaan M, Ivask A, Bilnova I, Dubourguier HC, Kahru A. Toxicity of nanosized and bulk ZnO, CuO and TiO_2 to bacteria *Vibrio fischeri* and crustaceans *Daphina magna* and *Thamnocephalus platyurus*. Chemosphere. 2008;71:1308–16.

Hill AJ, Bello SM, Prasch AL, Peterson RE, Heideman W. Water permeability and TCDD-induced edema in zebrafish early-life stages. Toxicol Sci. 2004;78:78–87.

Hirakawa K, Hirano T. Singlet oxygen generation photocatalyzed by TiO_2 particles and its contribution to biomolecule damage. Chem Lett. 2006;35:832–3.

Hirakawa K, Mori M, Yoshida M, Oikawa S, Kawanishi S. Photoirradiated titanium dioxide catalyzes site specific DNA damage via generation of hydrogen peroxide. Free Radical Res. 2004;38:439–47.

Hogstrand C, Galvez F, Wood CM. Toxicity, silver accumulation and metallothionein induction in freshwater rainbow trout during exposure to different silver salts. Environ Toxicol Chem. 1996;15:1102–8.

Hoshino A, Hanada S, Yamamoto K. Toxicity of nanocrystal quantum dots: The relevance of surface modifications. Arch Toxicol. 2011;85:707–20.

Hoskins C, Cuschieri A, Wang L. The cytotoxicity of polycationic iron oxide NPs: Common endpoint assays and alternative approaches for improved understanding of cellular response mechanism. J Nanobiotechnol. 2012;10:15.

Huang CC, Aronstam RS, Chen DR, Huang YW. Oxidative stress, calcium homeostasis, and altered gene expression in human lung epithelial cells exposed to ZnO NPs. Toxicol In Vitro. 2010;24:45–55.

Hund-Rinke K, Simon M. Ecotoxic effect of photocatalytic active NPs TiO_2 on algae and daphnids. Environ Sci Pollut Res. 2006;13:225–32.

Hussain SM, Hess KL, Gearhart JM, Geiss KT, Schlager JJ. *In vitro* toxicity of NPs in BRL 3A rat liver cells. Toxicol In Vitro. 2005;19:975–83.

Hwang ET, Lee JH, Chae YJ, Kim YS, Kim BC, Sang BI, Gu MB. Analysis of the toxic mode of action of silver NPs using stress-specific bioluminescent bacteria. Small. 2008;4:746–50.

Ispas C, Andreescu D, Patel A, Goia DV, Andreescu S, Wallace KN. Toxicity and developmental defects of different sizes and shape nickel NPs in zebrafish. Environ Sci Technol. 2009;43:6349–56.

Ivankovic S, Music S, Gotic M, Ljubesic N. Cytotoxicity of nanosize V2O5 particles to selected fibroblast and tumor cells. Toxicol In Vitro. 2006;20:286–94.

Jang H, Kim S, Kim S. Effect of particle size and phase composition of titanium dioxide NPs on the photo-catalytic properties. J Nanopart Res. 2007;3:141–7.

Ji J, Long ZF, Lin DH. Toxicity of oxide NPs to the green algae Chlorella sp. Chem Eng J. 2011;170:525–30.

Jiang GX, Shen ZY, Niu JF, Bao YP, Chen J, He TD. Toxicological assessment of TiO_2 NPs by recombinant *Escherichia coli* bacteria. J Environ Monit. 2011;13:42–8.

Jin CY, Zhu BS, Wang XF, Lu QH. Cytotoxicity of titanium dioxide NPs in mouse fibroblast cells. Chem Res Toxicol. 2008;21:1871–7.

Kang SG, Brown AL, Chung JH. Oxygen tension regulates the stability of insulin receptor substrate-1 (IRS-1) through caspase-mediated cleavage. J Biol Chem. 2007;282:6090–7.

Kangwansupamonkon W, Lauruengtana V, Surassmo S, Ruktanonchai U. Antibacterial effect of apatite-coated titanium dioxide for textiles applications. Nanomedicine. 2009;5:240–9.

Kappus H. Oxidative stress in chemical toxicity. Arch Toxicol. 1987;60:144–9.

Kashiwada S. Distribution of NPs in the see-through medaka (*Oryzias latipes*). Environ Health Perspect. 2006;14:1697–702.

Kawata K, Osawa M, Okabe S. *In vitro* toxicity of silver NPs at noncytotoxic doses to HepG2 human hepatoma cells. Environ Sci Technol. 2009;43:6046–51.

Khare P, Sonane M, Pandey R, Ali S, Gupta KC, Satish A. Adverse effects of TiO_2 and ZnO NPs in soil nematode *Caenorhabditis elegans*. J Biomed Nanotechnol. 2011;7:116–7.

Khus M, Gernjak W, Ibanez PF, Rodriguez SM, Galvez JB, Icli S. A comparative study of supported TiO_2 as photocatalyst in water decontamination at solar pilot plant scale. J Sol Energy Eng Trans ASME. 2006;128:331–7.

Kim S, Ryu DY. Silver nanoparticle-induced oxidative stress, genotoxicity and apoptosis in cultured cells and animal tissues. J Appl Toxicol. 2013;33:78–89.

Kim YJ, Choi HS, Song MK, Youk DY, Kim JH, Ryu JC. Genotoxicity of aluminum oxide (Al_2O_3) nanoparticle in mammalian cell lines. Mol Cell Toxicol. 2009;5:172–8.

Klaine SJ, Alvarez PJJ, Batley GE, Fernandes TF, Handy RD, Lyon DY, Mahendra S, McLaughlin MJ, Lead JR. NPs in the environment: Behavior, fate, bioavailability, and effects. Environ Toxicol Chem. 2008;27:1825–51.

Komatsu T, Tabata M, Kubo-Irie M, Shimizu T, Suzuki K, Nihei Y, Takeda K. The effects of NPs on mouse testis Leydig cells *in vitro*. Toxicol In Vitro. 2008;22:1825–31.

Laban G, Nies L, Turco R, Bickham J, Sepulveda MS. The effects of silver NPs on fathead minnow (*Pimephales promelas*) embryos. Ecotoxicology. 2010;19:185–95.

Lai JC, Lai MB, Jandhyam S, Dukhande VV, Bhushan A, Daniels CK, Leung SW. Exposure to titanium dioxide and other metallic oxide NPs induces cytotoxicity on human neural cells and fibroblasts. Int J Nanomed. 2008;3:533–45.

Lankveld DP, Oomen AG, Krystek P, Neigh A, Troost-de Jong A, Noorlander CW, Van Eijkeren JC, Geertsma RE, De Jong WH. The kinetics of the tissue distribution of silver NPs of different sizes. Biomaterials. 2010;31:8350–61.

Lee KJ, Nallathamby PD, Browning LM, Osgood CJ, Xu XH. *In vivo* imaging of transport and biocompatibility of single silver NPs in early development of zebrafish embryos. ACS Nano. 2007;1:133–43.

Lee S, Yun HS, Kim SH. The comparative effects of mesoporous silica NPs and colloidal silica on inflammation and apoptosis. Biomaterials. 2011;32:9434–43.

Leonard A, Lauwerys RR. Carcinogenicity and mutagenicity of chromium. Mutat Res-Rev Genet. 1980;76:227–39.

Lewinski N, Colvin V, Drezek R. Cytotoxicity of NPs. Small. 2008;4:26–49.

Li XB, Zheng H, Zhang ZR, Li M, Huang ZY, Schluesener HJ, Li YY, Xu SQ. Glia activation induced by peripheral administration of aluminium oxide NPs in rat brains. Nanomed Nanotechnol. 2009;5:473–9.

Li Y, Tian X, Lu Z, Yang C, Yang G, Zhou X, Yao H, Zhu Z, Xi Z, Yang X. Mechanism for alpha-MnO_2 nanowire-induced cytotoxicity in HeLa cells. J Nanosci Nanotechnol. 2010;10:397–404.

Liao KH, Lin YS, Macosko CW, Haynes CL. Cytotoxicity of graphene oxide and graphene in human erythrocytes and skin fibroblasts. ACS Appl Mater Interfaces. 2011;3:2607–15.

Limbach LK, Wick P, Manser P, Grass RN, Bruinink A, Stark WJ. Exposure of engineered NPs to human lung epithelial cells: Influence of chemical composition and catalytic activity on oxidative stress. Environ Sci Technol. 2007;41:4158–63.

Lin W, Huang YW, Zhou XD, Ma Y. *In vitro* toxicity of silica NPs in human lung cancer cells. Toxicol Appl Pharmacol. 2006a;217:252–9.

Lin W, Huang YW, Zhou XD, Ma Y. Toxicity of cerium oxide NPs in human lung cancer cells. Int J Toxicol. 2006b;25:451–7.

Lin W, Stayton I, Huang YW, Zhou XD, Ma Y. Cytotoxicity and cell membrane depolarization induced by aluminum oxide NPs in human lung epithelial cells A549. Toxicol Environ Chem. 2008;90:983–96.

Lin W, Xu Y, Huang CC, Ma Y, Shannon KB, Chen DR, Huang YW. Toxicity of nano- and micro-sized ZnO particles in human lung epithelial cells. J Nanopart Res. 2009;11:25–39.

Liu T, Li L, Teng X, Huang X, Liu H, Chen D, Ren J, He J, Tang F. Single and repeated dose toxicity of mesoporous hollow silica NPs in intravenously exposed mice. Biomaterials. 2011;32:1657–68.

Livingstone DR. Contaminant-stimulated reactive oxygen species production and oxidative damage in aquatic organisms. Mar Pollut Bull. 2001;42:656–66.

Lok CN, Ho CM, Chen R, He QY, Yu WY, Sun HZ, Tam PKH, Chiu JF, Che CM. Proteomic analysis of the mode of antibacterial action of silver NPs. J Proteome Res. 2006;5:916–24.

Lu W, Senapati D, Wang S, Tovmachenko O, Singh AK, Yu H, Ray PC. Effect of surface coating on the toxicity of silver NPs on human skin keratinocytes. Chem Phys Lett. 2010;487:92–6.

Lu X, Tian Y, Zhao Q, Jin T, Xiao S, Fan X. Integrated metabonomics analysis of the size-response relationship of silica NPs-induced toxicity in mice. Nanotechnology. 2011;22:055101.

Lubick N. Nanosilver toxicity: Ions, NPs, or both? Environ Sci Technol. 2008;42:8617.

Luoma SN, Ho YB, Bryan GW. Fate, bioavailability and toxicity of silver in estuarine environments. Mar Pollut Bull. 1995;31:44–54.

Ma H, Bertsch PM, Glenn TC, Kabengi NJ, Williams PL. Toxicity of manufactured zinc oxide NPs in the nematode *Caenorhabditis elegans*. Environ Toxicol Chem. 2009;28:1324–30.

Ma H, Kabengi NJ, Bertsch PM, Unrine JM, Glenn TC, Williams, PL. Comparative phototoxicity of nanoparticulate and bulk ZnO to a free-living nematode *Caenorhabditis elegans*: The importance of illumination mode and primary particle size. Environ Pollut. 2011;159:1473–80.

Maccormack TJ, Clark RJ, Dang MK, Ma G, Kelly JA, Veinot JG, Goss GG. Inhibition of enzyme activity by NPs: Potential mechanisms and implications for nanotoxicity testing. Nanotoxicology. 2012;6:514–25.

Mahto SK, Yoon TH, Rhee SW. A new perspective on *in vitro* assessment method for evaluating quantum dot toxicity by using microfluidics technology. Biomicrofluidics. 2010;4. DOI:10.1063/1.3486610.

Maness PC, Huang Z, Smolinsky S, Jacoby W, Blake D, Wolfrum E. Photosterilization and photomineralization of microbial cells with titanium dioxide. Photochem Photobiol. 1999;69:64S–5S.

Maynard A. Nanotechnology and safety. 2006. Available from: http://www.cleanroom-technology.co.uk/story.asp?storyCode=44919.

McCarthy J, Inkielewicz-Stepniak I, Corbalan JJ, Radomski MW. Mechanisms of toxicity of amorphous silica NPs on human lung submucosal cells *in vitro*: Protective effects of fisetin. Chem Res Toxicol. 2012;25:2227–35.

McLaren A, Valdes-Solis T, Li G, Tsang SC. Shape and size effects of ZnO nanocrystals on photocatalytic activity. J Am Chem Soc. 2009;131:12540–1.

Menard A, Drobne D, Jemec A. Ecotoxicity of nanosized TiO_2: Review of *in vivo* data. Environ Pollut. 2011;159:677–84.

Miao AJ, Zhang XY, Luo Z, Chen CS, Chin WC, Santschi PH, Quigg A. Zinc oxide engineered NPs: Dissolution and toxicity to marine phytoplankton. Environ Toxicol Chem. 2010;29:2814–22.

Miller RJ, Lenihan HS, Muller EB, Tseng N, Hanna SK, Keller AA. Impact of metal oxide NPs on marine phytoplankton. Environ Sci Technol. 2010;44:7329–34.

Mohammed Sadiq I, Dalai S, Chandrasekaran N, Mukerjee A. Ecotoxicity study of titania (TiO_2) NPs on two microalgae species: *Scenedesmus* sp. and *Chlorella* sp. Ecotoxicol Environ Saf. 2011;74:1180–7.

Moore MN. So NPs present ecotoxicological risks for the health of the aquatic environment? Environ Int. 2006;32:967–76.

Morones JR, Elechiguerra JL, Camacho A, Holt K, Kouri JB, Ramirez JT, Yacaman MJ. The bactericidal effect of silver NPs. Nanotechnology. 2005;16:2346–53.

Mukherjee SG, O'Claonadh N, Casey A, Chambers G. Comparative *in vitro* cytotoxicity study of silver nanoparticle on two mammalian cell lines. Toxicol In Vitro. 2012;26:238–51.

Mura GM, Ganadu ML, Lubinu G, Maida V. Photodegradation of organic waste coupling hydrogenase and titanium dioxide. Ann NY Acad Sci. 1999;879:267–75.

Napierska D, Thomassen LC, Lison D, Martens JA, Hoet PH. The nanosilica hazard: Another variable entity. Part Fibre Toxicol. 2010;3:39.

Nations S, Long M, Wages M, Canas J, Maul JD, Theodorakis C, Cobb GP. Effects of ZnO NPs on *Xenopus laevis* growth and development. Ecotoxicol Environ Saf. 2011;74:203–10.

Navarro E, Baun A, Behra R, Hartmann N, Filser J, Miao AJ, Quigg A, Santschi P, Sigg L. Environmental behavior and ecotoxicity of engineered NPs to algae, plants, and fungi. Ecotoxicology. 2008;17:372–86.

Nebeker AV, McAuliffe CK, Mshar R, Stevens DG. Toxicity of Ag to steelhead and rainbow trout, fathead minnows and *Daphnia magna*. Environ Toxicol Chem. 1983;2:95–104.

Nel A, Xia T, Mädler L, Li N. Toxic potential of materials at the nanolevel. Science. 2006;311:622–7.

Ngwa HA, Kanthasamy A, Gu Y, Fang N, Anantharam V, Kanthasamy AG. Manganese nanoparticle activates mitochondrial dependent apoptotic signaling and autophagy in dopaminergic neuronal cells. Toxicol Appl Pharmacol. 2011;256:227–40.

Nishanth RP, Jyotsna RG, Schlager JJ, Hussain SM, Reddanna P. Inflammatory responses of RAW 264.7 macrophages upon exposure to NPs: Role of ROS-NFkB signalling pathway. Nanotoxicology. 2011;5:502–16.

Nowrouzi A, Meghrazi K, Golmohammadi T, Golestani A, Ahmadian S, Shafiezadeh M, Shajary Z, Khaghani S, Amiri AN. Cytotoxicity of subtoxic AgNP in human hepatoma cell line (HepG2) after long-term exposure. Iran Biomed J. 2010;14(1–2):23–32.

Oberdörster G, Oberdörster E, Oberdörster J. Nanotoxicology: An emerging discipline evolving from studies of ultrafine particles. Environ Health Perspect. 2005;113:823–39.

Oh WK, Kim S, Choi M, Kim C, Jeong YS, Cho B, Hahn J, Jang J. Cellular uptake, cytotoxicity, and innate immune response of silica-titania hollow NPs based on size and surface functionality. ACS Nano. 2010;4:5301–13.

Ohno T, Sarukawa K, Matsumura M. Crystal faces of rutile and anatase TiO_2 particles and their roles in photocatalytic reactions. New J Chem. 2002;26:1167–70.

Pal S, Tak YK, Song JM. Does the antibacterial activity of silver NPs depend on the shape of the nanoparticle? A study of the Gram-negative bacterium *Escherichia coli*. Appl Environ Microbiol. 2007;73:1712–20.

Papageorgiou I, Brown C, Schins R, Singh S, Newson R, Davis S, Fisher J, Ingham E, Case CP. The effect of nano- and micron-sized particles of cobalt–chromium alloy on human fibroblasts *in vitro*. Biomaterials. 2007;28:2946–58.

Park EJ, Yi J, Kim Y, Choi K, Park K. Silver NPs induce cytotoxicity by a Trojan-horse type mechanism. Toxicol In Vitro. 2010;24:872–8.

Persson H, Kobler C, Molhave K, Samuelson L, Tegenfeldt JO, Oredsson S, Prinz CN. Fibroblasts cultured on nanowires exhibit low motility, impaired cell division, and DNA damage. Small. 2013;9:4006–16.

Petersen EJ, Nelson BC. Mechanisms and measurements of nanomaterial-induced oxidative damage to DNA. Anal Bioanal Chem. 2010;398:613–50.

Piao MJ, Kang KA, Lee IK, Kim HS, Kim S, Choi JY, Choi J, Hyun JW. Silver NPs induce oxidative cell damage in human liver cells through inhibition of reduced glutathione and induction of mitochondria-involved apoptosis. Toxicol Lett. 2011;201:92–100.

Pisanic TR, Blackwell JD, Shubayev VI, Finones RR, Jin S. Nanotoxicity of iron oxide nanoparticle internalization in growing neurons. Biomaterials. 2007;28:2572–81.

Pourbaix M. Atlas of Electrochemical Equilibria in Aqueous Solutions. Houston, TX: National Association of Corrosion Engineers; 1974. p. 644.

Powers CM, Slotkin TA, Seidler FJ, Badireddy AR, Padilla S. Silver NPs alter zebrafish development and larval behavior: Distinct roles for particle size, coating and composition. Neurotoxicol Teratol. 2011;33:708–14.

Prabhakar PV, Reddy UA, Singh SP, Balasubramanyam A, Rahman MF, Kumari SI, Agawane SB, Murty USN, Grover P, Mahboob M. Oxidative stress induced by aluminum oxide NPs after acute oral treatment in Wistar rats. J Appl Toxicol. 2012;32:436–45.

Przybyszewska M, Zaborski M. The effect of zinc oxide nanoparticle morphology on activity in crosslinking of carboxylated nitrile elastomer. Express Polym Lett. 2009;3:542–52.

Qiu Y, Liu Y, Wang L, Xu L, Bai R, Ji Y, Wu X, Zhao Y, Li Y, Chen C. Surface chemistry and aspect ratio mediated cellular uptake of Au nanorods. Biomaterials. 2010;31:7606–19.

Ray PC, Yu H, Fu PP. Toxicity and environmental risks of NPs: Challenges and future needs. J Environ Sci Health C Environ Carcinog Ecotoxicol Rev. 2009;27:1–35.

Ringwood AH, McCarthy M, Bates TC, Carroll DL. The effects of silver NPs on oyster embryos. Mar Environ Res. 2009;69:S49–51.

Roh J, Park Y, Park K, Choi J. Ecotoxicological investigation of CeO$_2$ and TiO$_2$ NPs on the soil nematode *Caenorhabditis elegans* using gene expression, growth, fertility, and survival as endpoints. Environ Toxicol Pharmacol. 2010;29:167–72.

Sadiq IM, Pakrashi S, Chandrasekaran N, Mukherjee A. Studies on toxicity of aluminum oxide (Al$_2$O$_3$) NPs to microalgae species: *Scenedesmus* sp. and *Chlorella* sp. J Nanopart Res. 2011;13:3287–99.

Sakai N, Matsui Y, Nakayama A, Tsuda A, Yoneda M. Functional dependent and size-dependent uptake of NPs in pc12. J Phys Conf Ser. 2011;304:012049.

Saquib Q, Al-Khedhairy AA, Siddiqui MA, Abou-Tarboush FM, Azam A, Musarrat J. Titanium dioxide NPs induced cytotoxicity, oxidative stress and DNA damage in human amnion epithelial (WISH) cells. Toxicol In Vitro. 2012;26:351–61.

Sarkar S, Sharma C, Yog R, Periakaruppan A, Jejelowo O, Thomas R, Barrera EV, Rice-Ficht AC, Wilson BL, Ramesh GT. Analysis of stress responsive genes induced by single-walled carbon nanotubes in BJ Foreskin cells. J Nanosci Nanotechnol. 2007;7:584–92.

Sato T, Taya M. Enhancement of phage inactivation using photocatalytic titanium dioxide particles with different crystalline structures. Biochem Eng J. 2006;28:303–8.

Schoonen MA, Cohn CA, Roemer E, Laffers R, Simon SR, O'Riordan T. Mineral-induced formation of reactive oxygen species. Rev Mineral Geochem. 2006;64:179–221.

Schrand AM, Rahman MF, Hussain SM, Schlager JJ, Smith DA, Syed AF. Metal based NPs and their toxicity assessment. Wiley Interdiscip Rev Nanomed Nanobiotechnol. 2010;2:544–68.

Scown TM, Santos EM, Johnston BD, Gaiser B, Baalousha M, Mitov S, Lead JR, Stone V, Fernandes TF, Jepson M, Van Aerle R, Tyler CR. Effects of aqueous exposure to silver NPs of different sizes in rainbow trout. Toxicol Sci. 2010;115:521–34.

Shaligram S, Campbell A. Toxicity of copper salts is dependent on solubility profile and cell type tested. Toxicol In Vitro. 2013;27:844–51.

Sharma VK. Aggregation and toxicity of titanium dioxide NPs in aquatic environment a review. J Environ Sci Health Part A. 2009;44:1485–95.

Shavandi Z, Ghazanfari T, Moghaddam KN. *In vitro* toxicity of silver NPs on murine peritoneal macrophages. Immunopharmacol Immunotoxicol. 2011;33:135–40.

Shen C, James SA, de Jonge MD, Turney TW, Wright PF, Feltis BN. Relating cytotoxicity, zinc ions and reactive oxygen in ZnO nanoparticle exposed human immune cells. Toxicol Sci. 2013;136:120–30.

Shi M, Kwon HS, Peng Z, Elder A, Yang H. Effects of surface chemistry on the generation of reactive oxygen species by copper NPs. ACS Nano. 2012;6:2157–64.

Siddiqi NJ, Abdelhalim MA, El-Ansary AK, Alhomida AS, Ong WY. Identification of potential biomarkers of gold nanoparticle toxicity in rat brains. J Neuroinflammation. 2012;9:123–30.

Sondi I, Salopek-Sondi B. Silver NPs as antimicrobial agent: A case study on *E. coli* as a model for Gram-negative bacteria. J Colloid Interface Sci. 2004;275:177–82.

Soto K, Garza KM, Murr LE. Cytotoxic effects of aggregated NPs. Acta Biomater. 2007;3:351–8.

Stanley JK, Coleman JG, Weiss CA, Steevens JA. Sediment toxicity and bioaccumulation of nano and micronsized aluminum oxide. Environ Toxicol Chem. 2010;29:422–9.

Studer AM, Limbach LK, Van Duc L, Krumeich F, Athanassiou EK, Gerber LC, Moch H, Stark WJ. Nanoparticle cytotoxicity depends on intracellular solubility: Comparison of stabilized copper metal and degradable copper oxide NPs. Toxicol Lett. 2010;197:169–74.

Sycheva LP, Zhurkov VS, Iurchenko VV, Daugel-Dauge NO, Kovalenko MA, Krivtsova EK, Durnev AD. Investigation of genotoxic and cytotoxic effects of micro- and nanosized titanium dioxide in six organs of mice *in vivo*. Mutat Res. 2011;726:8–14.

Szalay B, Tatrai E, Nyiro G, Vezer T, Dura G. Potential toxic effects of iron oxide NPs in *in vivo* and *in vitro* experiments. J Appl Toxicol. 2012;32:446–53.

Tarimala S, Kothari N, Abidi N, Hequet E, Fralick J, Dai LL. New approach to antibacterial treatment of cotton fabric with silver nanoparticle-doped silica using sol–gel process. J Appl Polym Sci. 2006;101:2938–43.

Thomas KV. Uptake and effects of manufactured silver NPs in rainbow trout (*Oncorhynchus mykiss*) gill cells. Aquat Toxicol. 2011;101:117–25.

Thomas DJ, Styblo M, Lin S. The cellular metabolism and systemic toxicity of arsenic. Toxicol Appl Pharm. 2001;176:127–44.

Vamanu CI, Cimpan MR, Hol PJ, Sornes S, Lie SA, Gjerdet NR. Induction of cell death by TiO$_2$ NPs: Studies on a human monoblastoid cell line. Toxicol In Vitro. 2008;22:1689–96.

Vevers WF, Jha AN. Genotoxic and cytotoxic potential of titanium dioxide (TiO$_2$) NPs on fish cells *in vitro*. Ecotoxicology. 2008;17:410–20.

Voinov MA, Sosa Pagan JO, Morrison E, Smirnova TI, Smirnov AI. Surface-mediated production of hydroxyl radicals as a mechanism of iron oxide nanoparticle biotoxicity. J Am Chem Soc. 2011;133:35–41.

Wadhwa S, Rea C, O'Hare P, Mathur A, Roy SS, Dunlop PSM, Byrne JA, Burke G, Meenan B, McLaughlin JA. Comparative *in vitro* cytotoxicity study of carbon nanotubes and titania nanostructures on human lung epithelial cells. J Hazard Mater. 2011;191:56–61.

Wang B, Feng WY, Wang M, Shi JW, Zhang F, Ouyang H, Zhao YL, Chai ZF, Huang YY, Xie YN, Wang HF, Wang J. Transportation of intranasal instillation of fine Fe$_2$O$_3$ particles in brain: Micro-distribution, chemical states, and histopathological observation. Biol Trace Elem Res. 2007;118:233–43.

Wang H, Wick RL, Xing B. Toxicity of nanoparticulate and bulk ZnO, Al$_2$O$_3$ and TiO$_2$ to the nematode *Caenorhabditis elegans*. Environ Pollut. 2009;157:1171–7.

Wang SH, Lee CW, Chiou A, Wei P. Size-dependent endocytosis of gold NPs studied by three dimensional mapping of plasmonic scattering images. J Nanobiotechnol. 2010;8:33–45.

Wang B, Yin JJ, Zhou X, Kurash I, Chai Z, Zhao Y, Feng W. Physicochemical origin for free radical generation of iron oxide NPs in biomicroenvironment: Catalytic activities mediated by surface chemical states. J Phys Chem C. 2012;117:383–92.

Wang CC, Wang S, Xia Q, He W, Yin J, Fu PP, Li J. Phototoxicity of zinc oxide NPs in HaCaT keratinocytes – Generation of oxidative DNA damage during UVA and visible light irradiation. J Nanosci Nanotechnol. 2013;13:3880–8.

Warheit DB, Web TR, Reed KL, Frerichs S, Sayes CM. Pulmonary toxicity study in rats with three forms of ultrafine-TiO$_2$ particles: Differential responses related to surface properties. Toxicology. 2007;230:90–104.

Wiesner MR, Lowry GV, Alvarez PJJ. Assessing the risks of manufactured NPs. Environ Sci Technol. 2006;40:4337–45.

Wiesner MR, Bottero JY. Nanotechnology and the environment. In: Wiesner MR, Bottero JY, editors. Environmental nanotechnology – Applications and impacts of nano-materials. New York: McGraw Hill; 2007.

Wiesner MR, Lowry GV, Jones KL, Hochella Jr. MF, Di Giulio RT, Casman E, Bernhardt ES. Decreasing uncertainties in assessing environmental exposure, risk, and ecological implications of NPs. Environ Sci Technol. 2009;43:6458–62.

Wijnhoven SWP, Peijnenburg WJGM, Herberts CA, Hagens WI, Oomen AG, Heugens EHW, Roszek B, Bisschops J, Gosens I, Meent DVD, Dekkers S, Jong WHD, Zijverden MV, Sips AJAM, Geertsma RE. Nano-silver – A review of available data and knowledge gaps in human and environmental risk assessment. Nanotoxicology. 2009;3:109–38.

Winnik FM, Maysinger D. Quantum dot cytotoxicity and ways to reduce it. Acc Chem Res. 2013;46:672–80.

Wong SWY, Leung PTY, Djurišić AB, Leung KMY. Toxicities of nano zinc oxide to five marine organisms: Influences of aggregate size and ion solubility. Anal Bioanal Chem. 2010;396:609–18.

Wu Y, Zhou QF, Li HC, Liu W, Wang T, Jiang GB. Effects of silver NPs on the development and histopathology biomarkers of Japanese medaka (*Oryzias latipes*) using the partial-life test. Aquat Toxicol. 2010;100:160–7.

Xia T, Kovochich M, Brant J, Hotze M, Sempf J, Oberley T, Sioutas C, Yeh JI, Wiesner MR, Nel AE. Comparison of the abilities of ambient and manufactured NPs to induce cellular toxicity according to an oxidative stress paradigm. Nano Lett. 2006;6:1794–807.

Xia T, Kovochich M, Liong M, Madler L, Gilbert B, Shi H, Yeh JI, Zink JI, Nel AE. Comparison of the mechanism of toxicity of zinc oxide and cerium oxide NPs based on dissolution and oxidative stress properties. ACS Nano. 2008;2:2121–34.

Yang L, Watts DJ. Particle surface characteristics may play an important role in phytotoxicity of alumina NPs. Toxicol Lett. 2005;158:122–32.

Yang X, Gondikas AP, Marinakos SM, Auffan M, Liu J, Hsu-Kim H, Meyer JN. Mechanism of silver nanoparticle toxicity is dependent on dissolved silver and surface coating in *Caenorhabditis elegans*. Environ Sci Technol. 2011;46:1119–27.

Yeo MK, Yoon JW. Comparison of the effects of nano-silver antibacterial coatings and silver ions on zebrafish embryogenesis. Mol Cell Toxicol. 2009;5:23–31.

Yin JJ, Lao F, Fu PP, Wamer WG, Zhao Y, Wang PC, Qiu Y, Sun B, Xing G, Dong J, Liang XJ, Chen C. The scavenging of reactive oxygen species and the potential for cell protection by functionalized fullerene materials. Biomaterials. 2009;30:611–21.

Yin JJ, Liu J, Ehrenshaft M, Roberts JE, Fu PP, Mason RP, Zhao B. Phototoxicity of nano titanium dioxides in HaCaT keratinocytes e generation of reactive oxygen species and cell damage. Toxicol Appl Pharmacol. 2012;263:81–8.

Yoon KY, Hoon Byeon J, Park JH, Hwang J. Susceptibility constants of *Escherichia coli* and *Bacillus subtilis* to silver and copper NPs. Sci Total Environ. 2007;373:572–5.

Yoshida T, Yoshikawa T, Nabeshi H, Tsutsumi Y. Relation analysis between intracellular distribution of NPs, ROS generation and DNA damage. Yakugaku Zasshi. 2012;132:295–300.

Zhang H, Banfield JF. Phase stability in the nanocrystalline TiO_2 system. In: Ma E, Bellon P, Atzmon M, Trivedi R, editors. Phase transformations and systems driven far from equilibrium. MRS; 1998. p. 619–24.

Zhang T, Qian L, Tang M, Xue YY, Kong L, Zhang S, Pu Y. Evaluation on cytotoxicity and genotoxicity of the L-glutamic acid coated iron oxide NPs. J Nanosci Nanotechnol. 2012a;12:2866–73.

Zhang XD, Wu D, Shen X, Liu PX, Fan FY, Fan SJ. *In vivo* renal clearance, biodistribution, toxicity of gold nanoclusters. Biomaterials. 2012b;33:4628–38.

Zhao X, Striolo A, Cummings PT. C60 binds to and deforms nucleotides. Biophys J. 2005;89:3856–62.

Zhao K, Zhang L, Wang J, Li Q, He W, Yin JJ. Surface structure-dependent molecular oxygen activation of BiOCl single-crystalline nanosheets. J Am Chem Soc. 2013;135:15750–3.

Zhu X, Zhu L, Duan Z, Qi R, Li Y, Lang Y. Comparative toxicity of several metal oxide nanoparticle aqueous suspensions to zebrafish (*Danio retio*) early developmental stage. J Environ Sci Health Part A. 2008;43:278–84.

Zhu MT, Feng WY, Wang Y, Wang B, Wang M, Ouyang H, Zhao YL, Chai ZF. Particokinetics and extrapulmonary translocation of intratracheally instilled ferric oxide NPs in rats and the potential health risk assessment. Toxicol Sci. 2009;107:342–51.

Zhu MT, Wang Y, Feng WY, Wang B, Wang M, Ouyang H, Chai ZF. Oxidative stress and apoptosis induced by iron oxide NPs in cultured human umbilical endothelial cells. J Nanosci Nanotechnol. 2010;10:8584–90.

Zhu ZJ, Yeh YC, Tang R, Yan B, Tamayo J, Vachet RW, Rotello VM. Stability of quantum dots in live cells. Nat Chem. 2011;3:963–8.

Zhu XS, Tian S, Cai Z. Toxicity assessment of iron oxide nanoparticles in zebrafish (Danio rerio) early life stages. PloS One. 2012;7:e46286. http://dx.doi.org/10.1371/journal.pone.0046286.

Zhu X, Hondroulis E, Liu W, Li CZ. Biosensing approaches for rapid genotoxicity and cytotoxicity assays upon nanomaterial exposure. Small. 2013;9:1821–30.

13

Functional Foods and Nutraceuticals: An Overview of the Clinical Outcomes and Evidence-Based Archive

Manjir Sarma Kataki, Ananya Rajkumari, and Bibhuti Bhusan Kakoti

CONTENTS

13.1 Defining Functional Foods and Nutraceuticals

The two terms "nutraceuticals" and "functional foods" are used interchangeably in numerous situations. However, there are some differences in their definitions by various regulatory agencies. One of the most feasible and worthwhile definitions was provided by Health Canada. According to Health Canada, "a nutraceutical is a product isolated or purified from foods that is generally sold in medicinal forms not usually associated with foods. A nutraceutical is demonstrated to have a physiological benefit or provide protection against chronic disease". On the other hand, "a functional food is similar in appearance to, or may be, a conventional food that is consumed as part of a usual diet, and is demonstrated to have physiological benefits and/or reduce the risk of chronic disease beyond basic nutritional functions".

The term "functional foods" was first coined in Japan in the mid-1980s and denotes processed foods containing ingredients which augment specific bodily functions, apart from being nutritious. The only country which has prescribed a particular regulatory process for the approval of functional foods is Japan. This regulation is known as "Foods for Specified Health Use (FOSHU)" (Shimizu, 2003). The United States did not legalize the functional foods category, but it is becoming a booming area of research in the US food industry. Other countries, including India, are not providing any clear guidelines for functional foods. However, there are many guidelines and recommendations for nutraceuticals and more amendments are coming with the better understanding and developments happening in food science and technology. The health promoting possibilities, disease modulating capabilities, and cost effectiveness for healthcare are the modalities that led to publicizing the concept of functional foods (Action, 1999).

13.2 Food and Health

Being healthy is the most promising and satisfying way of being alive. Food cannot be separated from health. Food and health are two sides of the same coin and cannot be neglected at any cost. As a primary and basic need, food has been positioned in a very significant and indispensable place in life. Food gives us the energy to live life and keep us thinking, moving, and working in just the way we do. Human food and feeding behavior has evolved tremendously since the beginning of human kind. In the long run, the concept of a "balanced diet" comes into light as a result of some kind of abuse of foods by human kind, alongside a sedentary lifestyle with little or no physical activity. The knowledge of a balanced diet is therefore indispensable to good health. A balanced diet has also been mentioned in ancient texts and in yoga science (World Health Organization, 2004).

13.3 Evolution of the Human Diet

The human diet has evolved a lot over the past decades. Human feeding behavior has been changing since time immortal and has been influenced by many factors, including environmental changes, behavioral changes in the human race, and the tremendous developments in science and technology. The modern pattern of food intake resulted from a 10,000 year evolution of the human diet since the Paleolithic era (Jew et al., 2009). The tremendous development in food processing, preservation techniques, and food additives shifted the human diet that was originally rich in fiber, vegetables, fruits, lean meat, and seafood to processed food engorged with a high sodium content, hydrogenated fats, and chemical preservatives, and that was low in fiber. This shifting of the diet pattern evoked many heath related concerns affecting the dietary input of health parameters leading to various chronic diseases, including cardiovascular disease (CVD), diabetes, and cancer (Link and Phelan, 1995). Obesity is another primary concern directly related to diet patterns now-a-days and is creating an alarming situation worldwide. Research also identified the beneficial effect of the older ancestral dietary pattern and the newer diet regimes incorporating the elements from ancestral diets. However, the evidence of calcium and vitamin D deficiency is of prime concern in our traditional ancestral dietary patterns.

With the significant advances in food science and technology, food and nutrition has become a mainstream research field as big as science itself. The concept of managing disease conditions and their pathophysiologic hallmarks has been found to be associated with the functional capabilities of food or food ingredients. In most of the prevailing diseases, oxidative stress has been identified as a crucial pathophysiologic hallmark sign underlying the initiation or progression of the disease condition. A plethora of scientific literature demonstrated many significantly effective neutralizers of oxidative stress from traditional foods and these food components or nutraceuticals have been well studied (Aruoma, 1998; Maritim et al., 2003).

Recent research and growing scientific evidence has revealed the health benefits of various functional foods and nutraceuticals. The most studied food ingredients include omega-3 fatty acids, polyphenols, flavonoids, carotenoids, fiber, and plant sterols. The utilization of the functional abilities of certain foods in the prophylaxis, cure, or management of various diseases has become a novel trend supported by significant positive clinical outcomes (Jew et al., 2009).

13.4 Dietary Patterns and Their Relation to Disorders

Dietary patterns and habits can make a significant difference when it comes to healthy lifestyles and disease occurrences. Optimum physical activity, energy expenditure, and calorie intake must be balanced for a healthy life with a lower possibility of various nutrition-related chronic diseases. Various disease conditions including diabetes, CVDs, osteoporosis, dental diseases, and cancer, etc., are found to be linked strongly to nutrition and lifestyle. Obesity is another risk factor associated with these above-mentioned disease conditions and has become a worldwide health-related concern (Kant, 1996; Nicklas et al., 2001).

13.4.1 Obesity

Physical inactivity and excess intake of foods rich in sugar and fat are the main culprits for the obesity epidemic. Obesity further contributes to various lifestyle diseases including diabetes, hypertension, other

CVDs, and osteoporosis. Many other diseases are also found to be linked with obesity. The prevention of obesity is therefore a must for a healthy lifestyle. But dealing with obesity is not easy; curbing obesity needs many diverse configurations which include increasing physical activity, reducing excess calorie intake, reducing foods high in sugar and fats, and many other social and environmental changes. Making these lifestyle changes is the only way to prevent or reduce obesity effectively (Bellisle et al., 1988).

13.4.2 Diabetes

Diabetes is clinically proven to be linked with excessive weight, obesity, and lack of physical activity in daily life. The increasing occurrence of type 2 diabetes is the outcome of these detrimental lifestyle changes. Diabetes can further lead to an increased risk of coronary diseases, CVDs, stroke, nephrological disorders, and various infections.

13.4.3 Cardiovascular Diseases (CVDs)

CVDs are a major cause of death worldwide. Dysfunctional sedentary lifestyles along with imbalanced dietary practices are concerned with an alarming increment in CVD. Heart disease and stroke are the main CVDs risking life in millions worldwide. Positive lifestyle modifications incorporating physical activity to avoid obesity or weight gain can be an initiating factor to be considered for reducing the incidence of fatal cardiovascular outcomes. Alongside this, balanced dietary practices need to be applied to reduce the intake of saturated trans fats and increase polyunsaturated fats, fruits, and vegetables. Reducing salt intake is also a very important factor that is well supported by clinical evidence for a healthy cardiovascular system (Ghafoorunissa, 2009).

13.4.4 Cancer

Studies and growing literature references indicate tobacco use as the prime cause of cancer, however dietary practices contribute significantly to some types of cancer. Maintaining a normal healthy weight and incorporating various functional nutritional food components into diets can be a therapeutic or prophylactic intervention in cancer therapeutics or management.

13.4.5 Osteoporosis and Bone Fractures

Healthy aging is a graceful gift and can be earned to a significant extent by maintaining a physically active lifestyle, healthy dietary practices, and a healthy weight. In the older population, fragile bones prone to fractures are a common problem. Dietary fortification of calcium and vitamin D helps to manage this issue. Osteoporotic bones can be significantly prevented by adequate dietary practices, physical exercise, and adequate sun exposure from an early age.

13.4.6 Dental Disease

Acidic foods and sugars are always linked with dental caries and tooth decay. Optimizing the consumption of acidic foods, sugars, and alcoholic beverages and increasing fluoride exposure might be a way to prevent or reverse tooth decay and dental issues (Kim et al., 2002; Sen Savara and Suher, 1955).

13.5 Dietary Patterns and Their Reduced Risk of Chronic Diseases

13.5.1 Mediterranean Diet

The traditional diet of countries around the Mediterranean Sea varies to some extent. Within several countries adjacent to the Mediterranean Sea, a number of different versions of the Mediterranean diet have evolved. The Mediterranean diet is basically the traditional diet of these Mediterranean countries, characterized by a high consumption of vegetables with olive oil, alongside moderate protein intake. This particular diet regime is very popular worldwide and has been prescribed by doctors and dieticians in many chronic conditions, including heart disease, high blood pressure, depression, and dementia as an adjunct to mainstream therapies. The Mediterranean diet is first and foremost a plant-based intake plan that embraces the daily intake of whole grains, olive oil, fruits, vegetables, beans and other legumes, nuts, herbs, and spices (Figure 13.1). The diet includes animal protein in smaller quantities. In 1993, the World Health Organization, Harvard School of Public Health, and Oldways Preservation and Exchange Trust designed a "diet pyramid" introducing the most common foods of Mediterranean regions. The pyramid incorporates the traditional dietary components of Greece, Crete, and Southern Italy during the mid-twentieth century. This was a promising step forward for familiarizing the Mediterranean diet to the world front in a formal manner. The Mediterranean food pyramid not only emphasizes the food ingredients but also regular physical exercise and the habit of eating socially. The Mediterranean diet is not just a diet plan, rather it is a way of eating, an eating plan consisting of mainly fruits and vegetables, beans, nuts, whole grains, fish, olive oil, small amounts of dairy, and red wine. The portion size was also indicated in the food pyramid, with the shape of the pyramid indicating the portion size of the food component. However, the amount of physical activity and size of the individual is an important factor in deciding the portion size of meals. During the mid-twentieth century, the Mediterranean countries demonstrated a significantly higher life expectancy along with a low incidence of chronic diseases, which further strengthens the existing popularity of the Mediterranean diet (De Lorgeril et al., 1999; Willett et al., 1995).

Scientific literature has demonstrated the effectiveness of the Mediterranean diet in a plethora of clinical studies both in animals and humans. The studies indicated this diet as a risk minimizer in CVD and overall mortality. The PREDIMED trial studied the effectiveness of the Mediterranean diet, fortified with extra virgin olive oil along with nuts, avoiding any kind of fat and calorie restrictions, in a diabetic patient population (7,400 patients) with CVD risk factors. This was a large primary prevention trial which demonstrated a significant 30% reduction of death rate from stroke. This trial also demolished a myth about low or no fat consumption by patients at risk for CVD. The total fat consumption in the trial ranges from 39 to 42% of the total daily calorie intake and most all of the fats were healthy fats from olive oil, fish, and nuts. The risk of type 2 diabetes was also found to be reduced in the PREDIMED trial (Guasch-Ferre et al., 2017; Toledo et al., 2013).

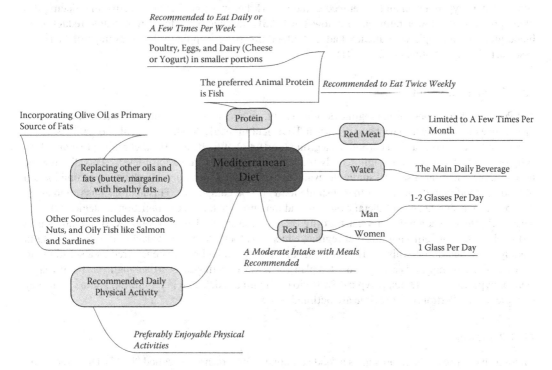

FIGURE 13.1   The unique eating plan of the Mediterranean diet.

Research has identified and supported the significant health benefits of the Mediterranean diet, but owing to the ambiguity in portion size of certain foods, the possibility of calorie over-intake is likely, which needs to be re-elucidated. Research hold ups the utilization of the Mediterranean diet as a healthy eating prototype for the avoidance of CVD, increasing lifespan, and healthy aging. When applied in combination with caloric restraint, the diet may also sustain healthy weight loss (Fitó et al., 2014; Hu et al., 2013; Toledo et al., 2013).

13.5.2 Dietary Approaches to Stop Hypertension Diet

The Dietary Approaches to Stop Hypertension (DASH) diet exists to help manage hypertension or high blood pressure by means of dietary intervention along with lifestyle modification and other therapeutic intervention. Being a risk factor for CVD, hypertension needs to be treated and managed effectively not only with pharmacologic interventions but also with lifestyle and dietary interventions. In these cases, the DASH diet regime is recommended for treating and managing hypertension effectively. The DASH diet emphasizes the consumption of fruits and vegetables, whole grains, and lean meat as part of the daily diet. Basically, the DASH diet is a plant-based diet regime with low sodium intake on a daily basis. However, nutritional requirements are primarily focused in this diet regime to avoid all possible deficiencies. Scientific reports revealed that vegans and vegetarians taking primarily plant-based diets were found to be less prone to hypertension, which in turn ignited researchers to design a diet rich in nutrients (DASH diet) to protect individuals from hypertension. The DASH diet comprises fruits and vegetables along with lean protein sources including fish, chicken, and beans. This diet is low in sodium (salt), extra added sugars, fats, and red meat. The reduction in salt intake in the DASH diet is one of the hallmark factors associated with the effective reduction of hypertension. The DASH diet has a recommended intake of sodium (1 tsp) not exceeding more than 2,300 mg per day. However, a lower version of this salt regime has been recommended for a more precise control of blood pressure, which indicates the minimization of sodium intake to 1,500 mg per day. The DASH diet has been documented as successful in lowering blood pressure in clinical trials, provided strict and persistent compliance to the diet is maintained. A blood pressure reading of more or equal to 140/90 (Systolic 140/Diastolic 90) can be considered as hypertension and adherence to the DASH diet was found to be effective in reducing this blood pressure to a lower number. This low-salt DASH diet demonstrated a remarkable reduction in blood pressure in people who already had high blood pressure, dropping it by an average of 11 points (Sacks et al., 2001; Vollmer et al., 2001).

13.5.3 Portfolio Diet

The Portfolio diet is another diet regime designed to target a reduction in serum cholesterol levels to further reduce the risk of coronary heart disease (CHD) (Jenkins et al., 2008; Keith et al., 2015). This diet is called the Portfolio diet as it incorporates a group (portfolio) of foods into the diet regime that are clinically proven to be effective in significantly reducing serum-raised cholesterol levels. This diet regime provoked the initiation of several studies to evaluate the CHD risk minimization claims by a combination of foods or food ingredients rather than a single food or food component. This approach has augmented the prospective significance of dietary therapy and may provide the link in nutrition strategies between what is regarded as a good diet versus a drug therapy. Research identified a noteworthy reduction in serum cholesterol by utilizing this diet regime, which was found to be equivalent to the standard statin therapy. However, despite this equivalency, this diet regime should not be considered as a substitute to pharmacologic therapy. It is a very good diet plan and a way of eating that further augments the management of hypercholesterolemia. A typical Portfolio diet regime is depicted in Figure 13.2. Some of the key ingredients in a Portfolio diet regime are outlined below.

13.5.3.1 Nuts

Almonds have already been proven as a food component for reducing so-called "bad" LDL cholesterol. Studies demonstrated other that nuts, including peanuts, cashews, pecans, pistachios, walnuts, and Brazil

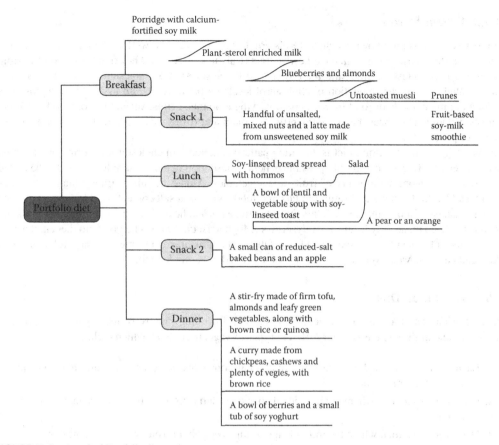

FIGURE 13.2 A typical Portfolio diet regime.

nuts, also possess a similar therapeutic effect. Nuts are low in detrimental saturated fat and are great sources of appetite-busting fiber and protein. However, being energy-dense, nuts should be taken in moderation to avoid excessive calorie intake (Kris-Etherton et al., 2001).

13.5.3.2 Soy

Soybeans are rich in soluble fiber which augments the absorption of cholesterol in the blood. Soybeans are also rich in phytoestrogen, which is considered beneficial for heart health. Studies in populations consuming soy in their staples indicated a decrease in occurrence of heart related diseases. The retrieved studies compared the heart health in populations eating soy on a daily basis against populations eating western, meat-based diets. The results were interesting as they revealed the beneficial effect of a soy diet compared to a meat-based diet in terms of heart health (Chen et al., 2004).

13.5.3.3 Soluble Fiber

Soluble fiber is an important food component leading to various beneficial effects in humans in terms of a reduction in blood cholesterol levels. Soluble fiber is profusely found in foods including barley, kidney beans, oats, chickpeas, and lentils. Certain fruits are also rich in soluble fiber, which includes pears, oranges, and grapefruit. Soluble fibers absorb water, forming a gel-like consistency in the gastrointestinal tract, which in turn removes some amounts of cholesterol from systemic absorption. These also work for evacuating the stomach by improving digestion and the microbial flora of the gastrointestinal tract (Abdul-Hamid and Luan, 2000).

13.5.3.4 Plant Sterols

Plant-derived sterols are known as phytosterols. Phytosterols have been well studied and are known to reduce cholesterol absorption from the digestive tract (Raicht et al., 1980). They trap the cholesterol from food in the gastrointestinal tract and thereby reduce the absorption of cholesterol from food. Studies demonstrated a significant reduction in cholesterol levels by phytosterols, up to 10–15%. Phytosterols are found in many foods or food ingredients in varying amounts. Fruits, vegetables, olive oil, different legumes including chickpeas, beans, and lentils, and plant and fruit seeds are known to contain a small amount of phytosterol.

In the long run, the Portfolio diet is an eating pattern functional in cholesterol lowering and keeping the heart healthy. To get the optimum possible benefits, these food items or ingredients need to be added to the daily diet, along with plenty of fruits and vegetables. Alongside this, salmon, tuna, or sardines are to be added to the menu and saturated fats and foods with high salt content need to be cut from the diet. Oily fish have important omega-3 fats that encourage normal heart function, control blood pressure, and assist to keep up healthy levels of triglycerides (Moghadasian, 2000). The Portfolio diet emphasizes the avoidance of unhealthy fats that include the fatty parts of meat, pasties, pies, sausage rolls, biscuits, cakes, and pastries. Moreover, full-fat dairy products need to be moderated.

13.5.4 Vegetarian Diet

The vegetarian diet is a plant and vegetable-based diet, including mainly fruits, vegetables, legumes, nuts, seeds, and grains. There are four different types of vegetarian diets, which include:

1. Lacto-ovo-vegetarian diet, where the individual consumes dairy products and eggs, but no meat, poultry, or seafood
2. Lactovegetarian diet, where the individual consumes dairy products but no eggs, meat, poultry, or seafood
3. Ovo-vegetarian diet, where the individual consumes eggs but no dairy products, meat, poultry, or seafood
4. Vegan diet, where the individual does not consume or eat any animal products, including meat, fish, poultry, eggs, and dairy products. Many vegans will also avoid honey. A vegan diet is the most pure form of a vegetarian diet that restricts all animal-derived or related foods or food ingredients.

A 2009 countrywide survey accomplished by the Vegetarian Resource Group estimated that approximately 3% of US adults are vegetarian (indicating that they never eat meat, poultry, fish, or seafood), and around 1% are vegan (they also never eat dairy, eggs, or honey) (Fraser et al., 2009). Individuals choose vegetarian diets considering the possibilities of improved health, as many studies demonstrate the significant health benefits of a vegetarian diet. A plethora of research also indicates the low incidence of health issues such as obesity, CVD, hypertension, type 2 diabetes, some cancers, gallstones, kidney stones, constipation, and diverticular disease among vegetarians. The low obesity incidence among vegetarians is also a prime factor that supports the utilization of the vegetarian diet as a tool to prevent various health problems and also to adjunct the management of various disease conditions (Szeto et al., 2004).

13.5.5 Okinawan Diet

Ageing and lifespan are intensely related to nutrition. Historically in many animal models, lifespan and/or health span is influenced by caloric restriction through nutritional intervention. Reducing protein consumption also showed an increase in lifespan in many studies, although not as noticeably as caloric constraint. Additional current research on nutritional geometry has focused on describing the influence of nutrition on ageing to a wide backdrop of dietary macronutrients and energy sources. The longest lifespans were observed in those animals which were kept with libitum admittance to low-protein, high-carbohydrate diets in studies in insects and mice. Interestingly, the most favorable content and ratio of dietary protein to carbohydrates for ageing in experimental animals was approximately the same as those in the habitual diets of the long-lived individuals on the island of Okinawa (Le Couteur et al., 2016).

The inhabitants of Okinawa, in the southernmost region of Japan, are well-known for their elongated normal life expectancy, elevated figures of centenarians, and associated low risk of age-related diseases. A great deal of the long-life benefit in Okinawa is thought to be interrelated to a healthy way of life, particularly the traditional food habits, which is low in calories yet nutritionally dense, mainly with regards to phytonutrients in the form of antioxidants and flavonoids. Research revealed that diets linked with a lower risk of chronic diseases are comparable to the traditional Okinawan diet, that is, vegetable and fruit heavy (and therefore phytonutrient and antioxidant rich) but restricted in meat, refined grains, saturated fat, sugar, salt, and full-fat dairy products. Several of the features of the diet in Okinawa are pooled with other healthy dietary types, such as the traditional Mediterranean diet or the modern DASH diet. Characteristics like the low levels of saturated fat, high antioxidant consumption, and lower glycemic content in these diets are possibly related to a lower threat of CVD, certain cancers, and other chronic diseases via multiple mechanisms, together with decreased oxidative stress. Comparing the nutrient profiles of the three dietary types revealed that the traditional Okinawan diet is the least possible in fat intake, mainly in terms of saturated fat, and the maximum in carbohydrate consumption, in observance with the extremely elevated consumption of antioxidant-rich, calorie-poor, orange-yellow root vegetables, like sweet potatoes, and green leafy vegetables. Extensive studies of the individual components of the Okinawan diet disclose that a lot of the traditional foods, herbs, or spices taken on a day-to-day basis could be labeled "functional foods" and are at present being studied for their impending health-improving properties (Willcox et al., 2009).

The traditional diet in Okinawa is occupied by root vegetables (principally sweet potatoes), green and yellow vegetables, soybean-based foods, and medicinal plants. Marine foods, lean meats, fruit, medicinal garnishes and spices, tea, and alcohol are also consumed. The healthy fat consumption is one probable mechanism for optimizing cholesterol and decreasing inflammation and other threatening factors. Moreover, the lower caloric concentration of plant-rich diets marks a lower caloric ingestion while simultaneously elevating the consumption of phytonutrients and antioxidants. Some other characteristics comprise low glycemic content, low inflammation and oxidative stress, and probable inflection of aging-related biological pathways. This may perhaps decrease the possibility of chronic age-related diseases and encourage healthy aging and long life (Willcox et al., 2014).

Long-term caloric restriction (CR) is a powerful way of minimizing age-associated diseases and expanding life span in multiple species, but the impacts in humans are unidentified. The low caloric consumption, long life expectation, and the elevated occurrence of centenarians in Okinawa have been considered as an argument to uphold the CR hypothesis in humans. Nevertheless, no long-term, epidemiologic analysis has been performed on the traditional dietary habits, energy balance, and potential CR phenotypes for the particular cohort of Okinawans who are supposed to have had a calorically limited diet. Nor has this cohort's following mortality incidence been thoroughly studied. Hence, six decades of archived population data on the elderly cohort of Okinawans (aged 65 plus) for support of CR was studied. Analyses incorporated traditional diet constituent, energy consumption, energy outflow, anthropometry, plasma DHEA, mortality from age-related diseases, and existing survival patterns. The results comprise lower caloric consumption and negative energy balance at younger ages, lower weight addition with age, life-long low body mass index (BMI), comparatively high plasma DHEA levels at older ages, less risk of mortality from age-associated diseases, and survival characteristics similar with the extended mean and maximum life span. The study showed the epidemiologic patterns hold up for the phenotypic benefits of CR in humans and are similar with the renowned literature on animals with regard to CR phenotypes and healthy aging (Willcox et al., 2007).

13.6 Dietary Bioactive Agents and Functional Foods

13.6.1 Curcumin

Curcumin (diferuloylmethane) is the main phytocomponent of turmeric, a key spice in Indian cuisine and an ancient Ayurvedic medicine. In India it is known as Haldi, with curcuminoids as

phytocompounds in it. It is a very popular perennial spice and medicinal herb, known for its significant medicinal properties. Curcumin has been used for various diseases including inflammation, liver
disorders, respiratory and other hepatic disorders, obesity, diabetic wounds, rheumatism, and certain
malignant tumors and cancers. Curcumin demonstrated a significant safety profile in many animal and
human clinical studies. A significantly higher dose can be safe for curcumin, as evident from many
toxicological studies. The preventive and curative properties of curcumin have been established by a
huge pile of published clinical reports. Recently there was a significant therapeutic search for natural
bioactives for the intervention of carcinogenesis and curcumin has been proven to be preventive in
carcinogenesis and mutagenesis in many studies. There was observed significant inhibition of cyclo-
oxygenase-2 (COX-2) expression in 12-O-tetradecanoylphorbol-13-acetate (TPA) induced tumors in
a mice skin model. The mechanism of carcinogenesis intervention includes the blocking of ERK1/2
and p38 MAPK pathways resulting in inactivation of the eukaryotic transcription factor NF-kB (Chun
et al., 2003). In human multiple myeloma and murine melanoma cells, curcumin has been found to
inhibit IkBα phosphorylation, leading to proapoptotic, antiproliferative, and antimetastatic activities
(Bharti et al., 2003; Philip and Kundu, 2003). Another study has demonstrated the protective effect
of curcumin via the genesis of the pro-inflammatory cytokines TNF-α and IFN-γ in a chronic colitis-
associated colorectal carcinoma (CRC) model (Villegas et al., 2011). A plethora of research activities
and reports have demonstrated the significant therapeutic profile of curcumin as a chemo preventive,
antioxidant, and protective.

13.6.2 Isoflavones

Isoflavones are plant-derived polyphenolic compounds and have been known for various protective and
therapeutic mechanisms. A huge array of health benefits have been observed for isoflavones, which
include a protective effect in heart diseases, prostate cancer, breast cancer, menopausal symptoms, and
osteoporosis. Chemically, isoflavones are similar in structure to physiological estrogen (17-β-estradiol),
which indicates modulation of estrogen activities. This mechanism has justified the estrogenic or anti-
estrogenic properties of isoflavones as observed in many literature works. These are called phytoestro-
gens and are responsible for balancing hormone modulation and imbalances (Andres et al., 2011). Soy
and soy foods are the best source of isoflavones, along with red clover as another source. Many different
kinds of isoflavones are found in soy, however genistein and daidzein are the most beneficial ones. Soy
products are mainly a prime part of Asian diets, indicating a significant intake of polyphenols in terms
of isoflavones.

Genistein is an isoflavone that has been found to decrease myeloperoxidase (MPO) activity and COX-2
messenger RNA (mRNA) protein expression, thereby producing significant anti-inflammatory activity in
TNBS-induced chronic colitis rodent models (Eric and Jordan, 2006). Genistein, along with other polyphenols including quercetin and apigenein, has also been demonstrated to produce induction of apoptosis
in various cancer cells as revealed by studies (Kuo, 1996; Pagliacci et al., 1994). Isoflavones are also
found to be potent antioxidants, reducing the free radical damage caused to DNA. Genistein is primarily
indicated as the most potent antioxidant among all the isoflavones under study.

13.6.3 Resveratrol

Resveratrol is a stilbene compound chemically termed *trans*-3,5,4'-trihydroxystilbene. It is a polyphenol
phytoalexin compound, plentiful in nature in a diversity of plant species, together with white helebore
(*Veratrum grandiflorum O. Loes*), *Polygonum cuspidatum*, blueberries, mulberries, wine, grapes, and
peanuts (Borrelli et al., 2007; Spanier et al., 2009). Resveratrol appeared in publicity following the
finding of cardioprotective properties in red wine (Vidavalur et al., 2006). This has evoked innovative
research actions concerning the clinical health benefits of resveratrol. Many research reports subsequently showed that resveratrol possibly would perk up a therapeutic improvisation in a variety of disease
conditions, including CVD, ischemic disorders, neurodegenerative disorders, and cancer (Bradamante
et al., 2004; Li et al., 2012; Patel et al., 2011; Petrovski et al., 2011; Smoliga et al., 2011). Resveratrol has
also been proven to be a powerful antioxidant as well as a free radical scavenger. It can modulate diverse

intracellular signal transduction pathways and exhibit therapeutic properties together with the modulation of apoptosis and angiogenesis and the survival of cells (Nie et al., 2007; Niemeyer, 2001). A plethora of scientific studies publicized the anticancer potential of resveratrol (Aggarwal et al., 2004; Aluyen et al., 2012; Cal et al., 2003; Sun et al., 2008).

13.6.4 Carotenoids

Carotenoids are a different class of the natural, chemo-preventive compounds that have received growing consideration as a consequence of the lower incidence of cancers associated with eating them (Seddon et al., 1994). Carotenoids are lipidic in nature and found naturally in varying quantities in vegetables, fruits, algae, some animals, fungi, and bacteria. Among the many carotenoid compounds, human plasma contains six, namely β-carotene, α-carotene, lycopene, lutein/zeaxanthin, astaxanthin, and β-cryptoxanthin. The first three belong to the carotene sub-group, while the last three are xanthophylls. These carotenoids and xanthophylls are found to be potent free radical scavengers as well as antioxidants. Due to these mechanisms, carotenoids have been found to possess chemotherapeutic efficacy in various intervention studies. Carotenoids showed their ability to induce cellular protective responses via the Nrf2/ARE pathway that further justifies the protective effect of carotenoids at a cellular level (Kumar et al., 2014). Earlier *in vitro* studies also demonstrated the growth inhibitory and pro-apoptotic effect of β-carotene in human colon adenocarcinoma cells by a reduction in the expression of COX-2 (Palozza et al., 2001).

13.6.5 Fatty Acids

Fatty acids are an important part of any diet worldwide, irrespective of culture and food habits. They are known to be therapeutic in numerous pathological conditions, including various inflammatory disorders, inflammatory bowel disease (IBD), atherosclerosis, Parkinson's and Alzheimer's diseases, and cancer (Simopoulos, 1999). They are mainly long-chain fatty acids, which fall under two classes, namely saturated or unsaturated fatty acids. The polyunsaturated fatty acids (PUFAs) are the most studied and significant fatty acid, with tremendous clinical potential for health benefits, as evident by previous and growing research. Fish oils are a major source of N-3 PUFAs owing to their high levels of docosahexaenoic acid (DHA), eicosapentaenoic acid (EPA), and docosapentaenoic acid (DPA) (Connor, 2000). These components have been found to be significantly cardio-protective and have also demonstrated noteworthy, documented health benefits, apart from this preventive effect, in CVDs. Linoleic acid (LNA), α-linolenic acid (ALA), EPA, and DHA are the fatty acids with the most beneficial effects and are profusely abundant in marine sources, including various microalgae. Among the various fatty acids, PUFAs are known to bind PPAR-γ, thereby modulating inflammation via the reduction of TNF-α production.

13.6.6 Green Tea

Green tea (*Camellia sinensis*) is one of the most extensively consumed beverages worldwide (Chacko et al., 2010). Cancer therapeutics have been found to be significantly influenced by various epidemiological studies on green tea (Cabrera et al., 2006; Jankun et al., 1997). Green tea possesses many compounds, of which epigallocatechin gallate (EGCG) is the most important, a biologically active compound with anticancer properties as evident by a growing body of clinical studies (Elbling et al., 2005). The well-studied anti-inflammatory and antioxidant properties of EGCG indicate a justification of the anticancer efficacy demonstrated by EGCG. The various mechanisms revealed by EGCG include a significant reduction of NF-kB and AP-1 activation, as well as an inhibition of TNF-α and IFN-γ production. In an experimental colitis model, EGCG has also been found to reduce NO and malondialdehyde (MDA) alongside an increase in superoxide dismutase (SOD) in colonic mucosa. *In vitro* studies have also reported that EGCG induces apoptosis in multiple cancer cells. EGCG has been publicized to induce apoptosis by the suppression of COX-2 expression and the subsequent production of PGE2 in diverse colon cancer cells, such as HT-29, SW837, and HCA-7 cells (Ahmad et al., 1997; Fang et al., 2003; Hong et al., 2002).

13.6.7 Vegetables (Cruciferous)

Cruciferous vegetables are vegetables of the family *Brassicaceae* (*Cruciferae*) and include Brussels sprouts, cabbage, Japanese radish, broccoli, and cauliflower. These are commonly consumed vegetables and form part of every diet regime worldwide, depending upon availability. The vegetables were studied and linked to a reduction in the risk of cancer. Glucosinolates are the main phytocomponent found in these vegetables, but another class of phytochemical known as isothiocyanates (ITCs) are responsible for the observed chemo-preventive effect of these cruciferous vegetables. The glucosinolates are not bioactive naturally, but upon hydrolysis by myrosinase get converted to ITCs and indole-3 carbinols, which are found to be active and chemo-preventive. Some ITCs obtained from cruciferous vegetables, such as benzyl isothiocyanates (BITC), phenethyl isothiocyanates (PEITC), and sulforaphane (SFN) have been established to be incredibly potent chemo-preventive moieties, as evident in various animal carcinogenesis models as well as cell culture models (Abdull et al., 2013; Higdon et al., 2007).

13.6.8 Dithiolethiones

Cruciferous vegetables are the source of another class of compound showing chemo-preventive efficacy in various clinical settings, which includes cabbage and Brussels sprouts. This class of compound is known as dithiolethiones. There are several compounds from this class which have been well studied; among them, 3H-1, 2-dithiole-3-thione (D3T) is a noteworthy compound representative of dithiolethiones that have been studied extensively as chemo-preventives in cancer therapeutics (Kwak et al., 2001; Talalay and Fahey, 2001).

13.6.9 Garlic and Other Allium Vegetables

Garlic and other allium vegetables (e.g., onions) are known for their medicinal properties in various illness conditions. These vegetables are reported to be protective against diseases including CVD, diabetes, infections, and cancer. Several organosulfur compounds (OSCs) are found in these vegetables which are known to be responsible for their medicinal properties. Mostly all these phyto-compounds are water soluble, which further enhances their actions. These compounds include allicin, diallyl sulfide (DAS), diallyl disulfide (DADS), diallyl trisulfide (DATS), diallyl tetrasulfide, as well as S-allylcysteine (SAC), or S-allylmercaptocysteine (SAMC). DATS is the most promising and medicinally active constituent among all the allium phyto-compounds, as suggested by a plethora of research studies. DATS demonstrated significant chemo-preventive efficacy via various mechanisms, as elucidated in many clinical settings which include cell cycle arrest, inhibition of angiogenesis, induction of apoptosis, and suppression of oncogenic signaling (Pinto and Rivlin, 1999; You et al., 1989).

13.7 Functional Components from Foods

The functional components of foods include numerous phytochemicals and chemically active moieties present in the food or food ingredients. These phytochemicals may be non-nutritive but active pharmacologically. These phytochemicals are abundant in foods or food ingredients and are also found to be pharmacologically active in various pathophysiological conditions. More than 900 phytochemicals have been identified in various foods (Srividya et al., 2010). Functional food components are not only found in plant-based foods but are also found in various animal-derived foods or food ingredients. Probiotics, conjugated linolenic acid, long-chain omega-3, -6, and -9 polyunsaturated fatty acids, and bioactive peptides are the animal-derived chemical entities equally active biologically and found abundantly in milk, fermented milk products, and cold-water fish. These chemical moieties are not only therapeutic in various disease conditions but also modulate various physiological metabolic reactions to maintain human health. Functional components generally occur in different forms, such as esterified, glycosylated, thiolated, or hydroxylated materials in food. Due to the diverse formation of these functional components of foods, they modulate multiple target physiological phenomena which in turn affect the disease mechanisms. Studies revealed several key functional components in food or food ingredients that have been found to be significantly prophylactic and therapeutic in various disease pathophysiologies. Some of the key functional food components are summarized as follows.

13.7.1 Dietary Fibers

Dietary fibers (DF) are an integral part of food ingredients and can be categorized as soluble or insoluble fiber. They are basically the structural components of cell walls of microorganisms or cereals and are non-starchy polysaccharides, indigestible in nature. The indigestibly of these dietary fibers makes them significantly functional and are reported to be beneficial for gastrointestinal tract functions. They are the indigestible element of plant-derived foods compiled of long, straight, and branched chains of carbohydrate molecules packed together by bonds that cannot be hydrolyzed by human digestive enzymes. The insoluble dietary fibers include lignin, cellulose, and hemicelluloses, etc., and the soluble fibers include gums, pectin, mucilage, β-glucans, and arabinoxylans (Andlauer and Fürst, 2002). These dietary fibers demonstrate gastro protective actions and improve digestion. They have also been found to entrap harmful toxicants as well as carcinogens from the gastrointestinal tract, thereby removing these harmful moieties from the body. They are also reported to be functional in terms of reducing cholesterol plasma levels by hindering the excess absorption of cholesterol from foods.

13.7.2 Antioxidants, Anti-Cancerous Agents, and Immune-Modulating Agents

Antioxidants are a diverse class of compounds with the ability to trap and neutralize free radicals and reactive oxygen species (ROS) generated in the physiological system. Oxidative stress is one highly focused topic of research and is reported to be a very important pathophysiological contributor in many disease conditions. Antioxidants have been studied extensively in this regard and have been reported to be a clinically-relevant intervention in the management of various disease conditions. Some of the functional components reported to be potent antioxidant are summarized as follows.

13.7.2.1 Carotenoids

Carotenoids (e.g., lycopene, lutein) are potent antioxidants and are basically plant-derived, oxygenated or non-oxygenated pigments that are lipid-soluble in nature. These compounds are extensively conjugated structurally and possess unique capabilities in neutralizing ROS. Lycopene and lutein are the most studied and endorsed antioxidant carotenoids. Other important carotenoids include α-carotene and β-carotene (Brush, 1990; Nelis and De Leenheer, 1991).

13.7.2.2 Polyphenols

Polyphenols are the most abundant group of functional compounds, diversely distributed in food items. These are naturally occurring compounds that structurally contain more than one benzene ring along with a varying number of hydroxyl (OH), carbonyl (CO), and carboxylic acid (COOH) groups. Polyphenols are found largely in fruits, vegetables, cereals, and beverages. Fruits like apples, pears, grapes, cherries, and berries contains up to 200–300 mg polyphenols per 100 g fresh fruit weight. Flavonoids and isoflavones are the most common, therapeutically active class of polyphenols found in foods. There are thousands of polyphenolic compounds abundant in foods derived from plants, including catechins, thearubigens, and theaflavins as the most common ones (Lobo et al., 2010). Epidemiological studies and numerous published meta-analyses strongly recommend the long term use of diets engorged with plant polyphenols present some level of protection against the development of cancers, CVD, diabetes, osteoporosis, and neurodegenerative diseases (Pandey and Rizvi, 2009).

13.7.2.3 Phytosterols

Phytosterols are plant sterols chemically and biologically equivalent to cholesterol. Mostly, they are similar in structure to cholesterol with the only differences being the additional double bonds present in the side chain structure of plant sterols along with methyl and/or ethyl groups. Stigmasterol, β-sitosterol, and campesterol are the commonly found bioactive phytosterols. The Mediterranean diet, Portfolio diet, Okinawa diet, and vegetarian diet are the diet regimes that contain high amounts of phytosterols. The vegetarian diet contains a higher amount of phytosterols than non-vegetarian diets (Awad and Fink, 2000; Moreau et al., 2002). In a vegetarian diet and a non-vegetarian diet approximately 500 mg and 250 mg respectively of unsaturated phytosterols are consumed on a daily basis. Sitostanol is an example of a saturated derivative of plant sterols that has been studied extensively.

13.7.2.4 Tocopherols and Tocotrienols

Tocopherols and tocotrienols are lipid-soluble monophenols which exist as four homologues (alpha, beta, delta, and gamma). These homologous tocopherols and tocotrienols differ from each other by the number and location of the methyl groups in their chemical structures. They contain a phenolic-chromanol ring linked to an isoprenoid side chain which is either saturated (tocopherols) or unsaturated (tocotrienols) (Kamal-Eldin and Appelqvist, 1996). The health benefits of tocopherols and tocotrienols are owed to the regulation of gene expression, signal transduction, and modulation of other cellular functions. The potential health benefits of tocopherols and tocotrienols incorporate the prevention of heart disease, certain types of cancer, and other chronic ailments (Shahidi and de Camargo, 2016).

13.7.2.5 Organosulfur Compounds

The organosulfur compounds are a group of functional compounds derived from plant-based foods, including various vegetables like broccoli, cauliflower, onion, garlic, and Brussels sprouts. These compounds are mainly found in cruciferous vegetables as discussed earlier in this chapter. These compounds possess sulfur atoms in their structure linked with a cyanate group and can also be found in cyclic configurations (Bellamy, 1975; Sparnins et al., 1988). These compounds are mostly volatile and pungent. Crushing or chewing them releases the compounds leading to exposure to the pungent smell.

13.8 Nutraceuticals and Plant-Derived Therapeutics

13.8.1 Coconut

The coconut is a topical fruit from the coconut palm tree (*Cocos nucifera*) and has been cultivated since time immemorial. A large quantity of products are produced from coconuts with nutritional and medicinal significance. Coconut oil is the most used product from coconuts, with significant health benefits. Studies suggested a great degree of superiority of coconut oil over other oils in a daily diet. Coconut oil is rich in medium-chain fatty acids (MCFA), making it absorbable and easily metabolized by the liver. Metabolic products of coconut oil include ketone bodies which are reported to be functional and therapeutic in Alzheimer's disease. Coconuts are also classified as a greatly nutritious functional food, rich in dietary fiber, vitamins, and minerals. Growing evidence from scientific literature demonstrates the beneficial effect of coconut in the treatment and management of dyslipidaemia, elevated LDL cholesterol, insulin resistance, hypertension, obesity, and type 2 diabetes (Fernando et al., 2015).

13.8.2 Buckwheat

Buckwheat (*Fagopyrum esculentum*) is a gluten-free functional food known as a pseudo cereal for its grain-like seeds. Buckwheat flour is used for its great nutritional value. It contains high levels of protein, polyphenols, and minerals. Quercetin is also abundant in buckwheat, which shows potent antioxidant activity. Studies also suggest the therapeutic and preventive effect of buckwheat in terms of the reduction in risk of hyperlipidaemia and hypertension (Li and Zhang, 2001).

13.8.3 Litchi

The litchi (*Litchi chinensis*) is a very popular and relished fruit in India during the summers. It belongs to the *Sapindaceae* family and is well known in the Indian traditional system for its medicinal and nutritive uses. Various phytochemicals are present in all parts of the *Litchi chinensis*, which include procyanidin A2, procyanidin B2, epicatechin, leucocyanidin, malvidin glycoside, cyanidin glycoside, saponins, rutin, butylated hydroxytoluene, isolariciresinol, kaempferol, and stigmasterol. These phytochemicals impart potent antioxidant activity to the litchi. Research also suggested the protective effect of litchi in neuro-degenerative diseases, arthritis, inflammation, asthma, and cancer (Wall, 2006; Zhang et al., 2000).

13.8.4 Wild Apple Fruit

Wild apple fruit (*Malus sylvestris*) is a good source of polyphenolic compounds and has been found to be an antioxidant. Due to the presence of polyphenols, wild apple fruit has demonstrated beneficial health benefits against many diseases caused by ROS and oxidative stress, including degenerative diseases, CVD, atherosclerosis, osteoporosis, diabetes, dermatitis, phototoxity, and cancer (Guyot et al., 1997).

13.8.5 Seaweed

Seaweeds are marine plants known as algae and are considered to be nutrient dense when compared to any kind of land vegetables. They are a rich source of micronutrients, including folate, calcium, magnesium, iron, zinc, and selenium. They are consumed as sea vegetables and are regarded as a functional as well as nutritious delicacy worldwide (MacArtain et al., 2007). However, this delicacy has been most popular among the Asian countries. A cholesterol compound called fucosterol, or 24-ethylidene cholesterol, is primarily found in seaweeds, marine algae, and some diatoms. This compound exhibits significant biological and therapeutic efficacy in various pathological conditions, including diabetes, oxidative damages, liver pathologies, fungal infections, allergy conditions, and various cancers. Seaweeds have been found to be anti-histaminic, anti-cholinergic, anti-adipogenic, anti-photodamaging, and anti-osteoporotic. Studies also suggest they have a plasma cholesterol lowering effect and anti-platelet and cholinesterase inhibition activities (Chakraborty et al., 2013; Fleurence, 1999; Kılınç et al., 2013).

13.8.6 Grapes

Grapes are a delicious fruit consumed in the entire worldwide diet regime. They are considered a functional and nutritious food, rich in the antioxidants, vitamins, and minerals essential for healthy status. They are a natural source of polyphenols and are known to reverse oxidative damage. Resveratrol is known to be the most widely studied, polyphenolic compound isolated from grapes and red wine. Due to the polyphenolic content and antioxidant properties, several grape products have been formulated which include grape seed oil, grape seed and skin powder, and pomace extract. These lead to the development of grape-based food additives, dietary supplements, and bulk nutraceuticals. These grape-based nutraceuticals have been found to possess several biological activities, including anti-inflammatory and antimicrobial properties (Sano et al., 2007; Sovak, 2001).

13.8.7 Oryzanol

Oryzanol is another compound that has recently come under the radar of the health paradigm. It is the main chemical constituent of rice bran oil. The heart health aspect of oryzanol has been scientifically validated in several clinical settings. The beneficial effects of oryzanol include an improvement of capillary action of blood vessels and a reduction of cholesterol in the blood (Rong et al., 1997). Anti-aging properties have also been reported for oryzanol. These beneficial effects paved a new range of cooking oils fortified with oryzanol in the food market (Cicero and Gaddi, 2001; Xu et al., 2001).

13.8.8 Ginger

Ginger root/rhizome (*Zingiber officinale* Roscoe) is well known as a spice vegetable and a traditional medicine in India and other countries. Ayurveda, the Indian system of traditional medicine, also documented the medicinal properties of ginger in various disease conditions. Ginger has been used as a ready remedy for stomach problems, indigestion, dyspepsia (bloating, heartburn, flatulence), colic, irritable bowel, loss of appetite, travel sickness, nausea and chills, cold, flu, poor circulation, and menstrual cramps (Ghayur and Gilani, 2005; Rasool Hassan, 2012; Vutyavanich et al., 2001). Clinical trials also suggest ginger has anti-inflammatory and antiemetic properties (Bliddal et al., 2000). Several phytochemicals such as 6-gingerol, 8-gingerol, 10-gingerol, and 6-shogaol were indicated as responsible biological and functional modalities. Some of the various pharmacological and therapeutic mechanisms of ginger include NK1 antagonism, 5-hydroxytryptamine (5-HT3) antagonism, and pyrokinesis. Among the various phytoconstituents, 6-gingerol possessed na acceptable safety profile and exhibited chemopreventive as well as chemotherapeutic activities in oral and cervical tumors through cell cycle arrest and apoptosis induction (Hsu et al., 2015; Shukla and Singh, 2007; Zorn et al., 2005).

13.9 The Changing Nutrition/Health Paradigm

The development and supply of safe nutraceuticals and functional foods is the new emerging area in global food industries as it tends to improve healthcare and boost health status. The Indian continent produces a huge number of medicinal spices, herbs, fruits, and trees, therefore India has become a major producer as well as supplier of nutraceuticals and functional foods. Exporting extracts of traditional herbs and medicinal plants is becoming a trend in India and further newer research has been initiated focusing upon the standardization and optimization of herbal extracts and functional ingredients according to global needs. Research also emphasized the evaluation of safety and efficacy of potential herbal extracts in various chronic diseases. The function of nutraceuticals in the prevention of various diseases is not essentially owing to a single compound in the herbs, fruits, or foods, rather the observed therapeutic actions are found to be due to the cumulative effect of several components present in the product. However, several isolated single compounds were reported to be highly functional in curbing various disease pathologies, which include curcumin, lycopene, and quercetin, etc. Therefore, it is necessary to carry out biomarker research for the comparison of defensive mechanisms for different kinds of foods. Industrial levels of supplies need reproducibility of the exact extract every time it gets isolated, which further focuses on the standardization parameters of the extracts. New merchandise can be developed by the acknowledgment of variation in the composition of nutraceuticals and functional foods. The approach of standardizing the products can target specific populations as well as markets and generate produce with biochemical uniformity for reproducible and predictable health outcomes. The industrial application of isolation, purification, characterization, and formulation of nutraceuticals and functional foods can be a costly affair and measures need to be initiated to devise newer, more cost effective strategies. Modern approaches such as proteomics, genomics, metabolomics, and biotechnological tools can be explored to further characterize, analyze, and formulate nutraceutical and functional foods for optimal health benefits. Nutraceuticals and foods need to be evaluated on a par with western medicinal drugs; only then will better optimization and safer therapeutics be achieved. The conceptual trend prevailing in frontier drug evaluation and delivery technologies has been started and incorporated in the nutraceutical and food industries. Newer, more novel nutraceutical and functional food formulations with high safety profiles are paving their way into the global market. In this context, nanotechnology can be stated as the new emerging scientific front in the food industry.

Nanotechnology handles matter in the nanoscale range of 1–100 nm, which includes the ability to control, measure, image, and manipulate in this nanoscale range, which further creates enormous possibilities and functionalities. The emergence of nanotechnological interventions in the food industry changes every aspect of the science and technology prevailing earlier. The tremendous possibilities imparted by nanotechnology have created newer horizons in the way we evaluate and present matter, and various government as well as private sectors have launched initiatives to deploy this technology to improve the quality of human life and health. Exploiting matter in the nanoscale can impart newer and more novel functionalities to nutraceuticals and functional foods. This nanoscale exploitation has been employed in the formulation, manufacturing, and other processes of nutraceuticals and functional foods. The main outputs derived from these nanotechnology-based explorations involve the improvisation of functionalities of the final product as well as especially improved delivery efficiency (Cho et al., 2008; Scott and Chen, 2012). In nanometer resolution, improved visualization and a better understanding of the functionalities of nutraceuticals and food components can be achieved. Accordingly, nanotechnology has revealed enormous potential for improving the usefulness and efficiency of the delivery of bioactive compounds and nutraceuticals in functional foods to improve human health. Nanotechnology application in nutraceuticals and functional foods can solve several issues, including solubility, bioavailability, and stability. It can impressively augment solubility, assist controlled release, perk up bioavailability, and secure the stability of micronutrients and bioactive compounds during various stages of production, including processing, storage, and distribution. Other aspects of application of nanotechnology in nutraceuticals and functional foods include the various possibilities for the development of new flavors, colors, and functionalities. Potent functional ingredients and nutraceuticals can be targeted and delivered using nanotechnology applications. Controlling release can be a new approach for nutraceuticals that can be explored for a better therapeutic and pharmacokinetic profile. This will finally provide a platform for the better understanding of the mechanisms involved in the delivery of bioactives from nutraceuticals, leading to the design of smart food systems.

13.9.1 Nutrients

13.9.1.1 Nutrients of Concern

As the intake of vegetables, fruits, whole grains, milk and milk products, and seafood are generally less than recommended, some nutrients are consumed less and have become a public health concern. These nutrients include potassium, dietary fiber, calcium, and vitamin D. Additionally, the consumption of iron, folate, and vitamin B12 is alarming in specific population groups.

13.9.1.2 Dietary Fiber

The non-digestible variety of carbohydrates and lignin constitutes dietary fiber. Dietary fiber occurs naturally in plants, which aids in providing a feeling of fullness and enhances healthy laxation. The best sources of dietary fiber are beans, like navy beans, split peas, lentils, pinto beans, black beans, and peas. Vegetables, fruits, whole grains, and nuts are additional sources of dietary fiber. Another excellent source of dietary fiber is bran, even though it is not a whole grain. The risk of CVD, obesity, and type 2 diabetes might be reduced by the consumption of dietary fiber that occurs naturally in foods. The consumption by children and adults of foods rich in dietary fiber increases the nutrient density, enhances healthy lipid profiles and glucose tolerance, and assures normal gastrointestinal function.

In the typical American diet these foods are consumed less than the recommended levels, with the normal intake range only at an average of 15 g per day. Occasionally, fiber could be incorporated into foods, however it is unclear if added fiber provides the same health benefits as naturally occurring sources. Even though breads, rolls, buns, and pizza crusts made of refined flour are not the best sources of dietary fiber, they are known to make a significant contribution to dietary fiber consumption as they are included in almost all typical American diets. In order to attain the recommended level of dietary fiber, Americans should elevate the intake of vegetables, fruits, whole grains, and other foods with naturally occurring fiber, including beans and peas. The fiber content of whole grains differ and therefore the nutrition facts label can be used to evaluate whole-grain products and find options that are higher in dietary fiber.

13.9.1.3 Additional Nutrients of Concern for Specific Groups

13.9.1.3.1 Iron

Iron deficiency is very common among women now-a-days worldwide. In India, the deficiency is alarming. A significant number of women, including teenagers, females of child bearing age, and adolescent girls, have been found to be deficient in iron. This alarming deficiency of iron has been causing various problems and health issues in women, especially in those who are pregnant or trying to conceive. It is the need of the hour to emphasize the adoption of necessary measures to correct or manage this iron deficiency among women. Iron status can be improved by consuming foods rich in heme iron, which is the most easily absorbed form of iron in foods. Alongside this, other good sources of iron need to be added into the daily diet with vitamin C enriched foods. Vitamin C can correct the absorption issues related to iron bioavailability. As a source of heme iron, lean meat as well as poultry should be consumed. White beans, lentils, and spinach are the non-heme iron sources which are plant-based and readily available. Pregnant women are recommended to take an iron supplement, as suggested by an obstetrician or other healthcare provider.

13.9.1.3.2 Folate

Folate or folic acid is another crucial nutrient which has been considered essential in pregnancy. Folic acid fortification has been victorious in dropping the incidence of neural tube defects during child birth. Earlier, this issue was a grave concern during pregnancy and child birth. Folic acid along with iron supplementation is a must in conceiving women. However, many women capable of becoming pregnant still do not meet the recommended intake for folic acid. A 400 mcg of folic acid is highly recommended for women planning to become pregnant. They can fulfil this folic acid requirement either by taking fortified foods or by supplements. During pregnancy this requirement tends to be 600 mcg daily from all sources. Sources of food folate include beans and peas, oranges and orange juice, and dark-green, leafy vegetables such as spinach and mustard greens.

13.9.2 Minerals

13.9.2.1 Potassium

Potassium is a mineral considered crucial to the physiological system around the clock, along with sodium. Sodium excess can precipitate or exacerbate high blood pressure, which can be nullified or reversed by an adequate potassium intake. Adequate potassium intake also improves bone health and prevents kidney stone formation. The adequate intake (AI) for potassium for adults is 4,700 mg per day. Scientific evidence suggests that African Americans and individuals with hypertension especially benefit from increasing their intake of potassium. A low intake of potassium below the AI level for long periods can cause health related issues. Therefore, it is recommended to take potassium fortified foods to fulfil the AI level of required potassium. However, individuals with kidney disease and those who have been taking drugs like ACE inhibitors must consult their physician before taking any potassium supplements. Fruits and vegetables, along with dairy products, are a good natural source of potassium.

13.9.2.2 Calcium

Calcium is necessary for an endless number of physiological phenomena, starting with muscle contraction through to bone health. Adequate calcium levels are of the utmost importance for optimal bone health status. This is more important in old age. Calcium governs multiple vital phenomena in the physiological system, including nerve transmission, blood vessel constriction cycles, and receptor activation. Individuals in old age are like to suffer from osteoporosis as a result of chronic calcium deficiency. Osteoporosis is becoming a very widespread health concern worldwide. This has increased the risk of bone fracture in old age which in turn is difficult to manage in older individuals. Calcium also plays a vital role during the growth phase in children. Therefore, all ages of individuals are advised to fulfil the daily requirement of calcium to maintain an optimum health, especially optimum bone health. Dairy products are considered the main source of calcium worldwide. Individuals with no milk or dairy products in their diet need to be cautious in terms of replacements with some reliable sources of calcium. They can include supplements of calcium along with vitamin D or other calcium fortified foods.

13.9.3 Vitamins

13.9.3.1 Vitamin B₁₂

The average American population of age 50 years and older have a sufficient vitamin B_{12} intake. However, a considerable percentage of that population might have less potential to take up the naturally occurring vitamin B_{12}. Despite this, the crystalline form of the vitamin is absorbed well. Hence, it is recommended for this population to consume foods fortified with vitamin B_{12}, such as fortified cereals, or take dietary supplements.

13.9.3.2 Vitamin D

For a healthy life a sufficient amount of vitamin D is essential. Vitamin D deficiency may cause rickets in children and osteomalacia (softening of bones) in adults. The risk of bone fractures can be reduced by an optimum intake of vitamin D. In Americans, the consumption of dietary vitamin D is below recommendations, however recent data from the National Health and Nutrition Examination Survey (NHANES) shows that in 80% of Americans the blood levels of vitamin D are adequate. Sunlight is a unique source that allows the body to produce vitamin D. Most dietary vitamin D in the United States is obtained from fortified foods like milk and some yogurts. Vitamin D is also rich in other foods and beverages like breakfast cereals, margarine, orange juice, and soy beverages. Fish like salmon, herring, mackerel, and tuna are natural sources of vitamin D and egg yolks have a small amount of vitamin D. Vitamin D is also available as a dietary supplement. The RDAs of vitamin D for children and most adults are 600 IU (15 mcg) per day, assuming minimal sun exposure, and for adults older than 70 years it is 800 IU (20 mcg). The potential risk of adverse effects increases when intake increases above 4,000 IU (100 mcg) per day.

13.9.4 Ayurceuticals

The term "Ayurceuticals" is a relatively new term for the ancient Indian system of medicine renowned as "Ayurveda". The botanicals and medicinal plants along with other medicinal modalities from the Ayurveda, Unani, and Siddha systems are included and described under the term Ayurceuticals. Some of the most popular functional components from this ancient system of medicine include aloe vera, arjun (*Terminalia arjuna*), karela (*Momordica charantia*), brahmi (*Centella asiatica*), vasa (*Adhatoda vasica*), ginger (*Zingiber officinale*), garlic (*Allium sativum*), yashtimadhu (*Glycyrrhiza glabra*), shallaki (*Boswellia serrata*), bilwa (*Aegle marmelos*), amalaki (*Phyllanthus emblica*), guggul (*Commiphora wightii*), kutki (*Picrorhiza kurroa*), methi (*Trigonella foenum-graecum*), ashwagandha (*Withania somnifera*), and pomegranate (*Punica granatum*), etc. Research has already demonstrated significant clinical and therapeutic benefits of these Ayurceuticals in various pathologies and disease conditions, including CVD, liver disorders, disorders related to immunity, nephrological and urological disorders, and various infections, etc.

13.10 Food Safety

Safety is always a prime concern in medicine, foods, and healthcare. Securing the safety of food products and ingredients is a highly attributed aspect of recent research and development. With the development of newer formulation technologies and the design of smart foods, safety issues are becoming a growing concern. The nanotechnological exploration in food science further changes the traditional aspect of safety in nanoscale dimensions. Moreover, plant-based products and botanicals need to be screened at multiple levels, including collection, post-harvesting conditions, and processing, before formulation. Human kind always regards food as safe compared to medicine. Therefore, the safety issues must be of supreme concern for food scientists. It is the need of the hour and it is critical to deploy science-based strategies that concentrate on food safety.

13.11 Health Supplements: Science versus Gimmicks

With the development of technology and the subsequent significant lifestyle changes evolved the pattern of human feeding behavior to a different level, with the incorporation of processed foods, packed foods, fast foods, and refrigerated food into daily diets. Moreover, the enormous projected advantages of nutraceuticals and functional foods created a wave of newer food products in the health supplement market. The launches are high as there is a lack of proper regulation for food supplements. The norms in most countries are not up to the mark and are incapable of justification as to the efficacy and safety of these nutraceuticals and functional foods. Traditional nutraceutical supplements are also in the lime light. These are crucial to human health, but only if taken in the form of real foods and in moderation. Newer studies raised several question marks as to the efficacy of these nutraceuticals. The Harvard-led Physicians Health Study II (PHS II) recently revealed that taking a multivitamin slightly lowers the risk of being diagnosed with cancer. This was a large study involving 15,000 physicians, and was the first of its kind where multivitamin therapy has been put to the test for disease prevention. Of the other studies retrieved, most covered only a single vitamin or nutrient. The significance of taking multivitamins as part of a daily routine as a supplement for functional benefits like protection from cancer-like disease is questionable. The PHS II study also demonstrated the failure of multivitamin therapy in reducing the risk of CVD. Therefore, it is recommended to fulfil the daily requirement of these nutrients via taking real foods including fruits, vegetables, and other foods. A plethora of studies dealing with multivitamin therapy for disease prevention have been disappointing clinically. Convincing evidence is less and questionable. However, the functionalities of traditional nutrients for health have been found to be intact and it is always a good idea to take a multivitamin supplement for the fulfilment of daily requirements and for correcting deficiencies. For other newer nutraceuticals and functional foods, the health beneficial projections are very promising but clinical studies need to be conducted in multiple to justify their functionalities.

13.12 Clinical Outcomes: Functional Foods and Nutraceuticals

Functional foods and nutraceuticals are beneficial for various disease conditions as suggested by numerous published scientific literature works. The food industry is growing very fast and a new generation of smart foods with a great deal of functionalities are likely in the future. Functional foods and nutraceuticals have been demonstrated to lessen the risk of CVD, modulate weight, reduce osteoporosis, reduce the risk of cancer, improve foetal health, improve memory, and reduce the risk of other many diseases. Functional foods and nutraceuticals will be encouraged for superior health in the future, and have been persuasively established to be advantageous for their proposed uses when consumed as part of a commonly well-balanced and healthy diet. Well-studied, key nutraceuticals and functional foods and their clinical outcomes are summarized in Table 13.1.

13.13 Conclusion

In conclusion, it can be summarized that nutraceuticals and functional foods are becoming an integral part of the current protocols for various disease management. Without the intervention of diet therapy, no disease or disorder can be managed effectively. Diets fortified with nutraceutical supplements or functional dietary elements become an effective adjunct in the therapy and management of different diseases. Moreover, specific populations need specific dietary requirements and this needs to be addressed while dealing with disorders in these populations. With the advent of newer functional components or foods coming up to the present scenario, it is becoming necessary to revisit the traditional disease management protocols for a better disease management approach. More research activities are warranted to evaluate the efficacy and safety of the numerous nutraceuticals and functional foods for a better understanding of their functionalities.

TABLE 13.1

Clinical Outcomes of Some of the Key Functional Components/Foods

Functional Component/Food	Formulation	Clinical Outcome	References
Curcumin	Curcumin nanomicelle	Anti-infertility, improves semen quality	Alizadeh et al. (2018)
	Curcuminoid complex extract from turmeric rhizome with turmeric volatile oil	Beneficial in osteoarthritis treatment along with *Boswellia serrata* extracts	Haroyan et al. (2018)
	Curcumin	Improves chronic inflammation and prevention of carcinogenic changes in patients with chronic gastritis associated by Helicobacter pylori	Judaki et al. (2017)
	Phytosomal curcumin supplement	Improves fatty liver and liver transaminase levels in patients with non-alcoholic fatty liver disease (NAFLD)	Panahi et al. (2017)
	Curcumin supplement	Reduces oxidative stress and improves resistance artery endothelial function by increasing vascular nitric oxide bioavailability	Santos-Parker et al. (2017)
Resveratrol	Intranasal resveratrol	Improves nasal symptoms in adults with allergic rhinitis (AR)	Lv et al. (2018)
	Resveratrol	Modulates neuro-inflammation, induces adaptive immunity	Moussa et al. (2017)
	Resveratrol supplement (800 mg/day)	Antioxidant effect in the blood and peripheral blood mononuclear cell smear of patients with type 2 diabetes	Seyyedebrahimi et al. (2018)
	Resveratrol	Reduces chronic pain in age-related osteoarthritis in post-menopausal women	Wong et al. (2017)
Oryzanol	Oryzanol supplement (600 mg/day)	Increases muscular strength in young healthy males during exercise	Eslami et al. (2014)
Lycopene	Lycopene supplement (29.4 mg/day)	Slows the progression of heart failure	Biddle et al. (2015)
	Lycopene supplement	Improves endothelial function in cardiovascular disease patients	Gajendragadkar et al. (2014)
Isoflavones	Flavanols, flavonols, isoflavones, flavanones, anthocyanins, pro-anthocyanidins	Anti-obesity	Akhlaghi et al. (2018)
	Dietary flavonoids	Reduces oxidative stress and thereby prevents retinal damage in diabetic retinopathy	Ola et al. (2018)
	Soy protein with isoflavones	Improves cardiovascular disease risk (CVR) markers	Sathyapalan et al. (2018)
Carotenoids	13-Cis retinoic acid (13-CRA)	Reduces the incidence of second primary tumors in patients with squamous cell cancers of the head and neck	Bhatia et al. (2017)
	Lutein, zeaxanthin	Increases the density of macular pigment	Korobelnik et al. (2017)
Polyphenols	Apple and blackcurrant polyphenol-rich drinks	Reduces postprandial glucose, insulin, and incretin response to a high-carbohydrate meal in healthy people, both male and female, inhibits intestinal glucose transport	Castro-Acosta et al. (2017)

(Continued)

TABLE 13.1 (CONTINUED)

Clinical Outcomes of Some of the Key Functional Components/Foods

Functional Component/Food	Formulation	Clinical Outcome	References
	A polyphenol-rich diet	Pleiotropic effect on cardiometabolic risk factors	Vetrani et al. (2018)
Phytosterols	Sitosterol	Suppresses obesity-related chronic inflammation	Kurano et al. (2018)
	Phytosterols (PhyS)-milk	Induces attenuation of the pro-inflammatory pathways	Lambert et al. (2017)
Fatty acids	Omega-3 fatty acids	Improves cancer cachexia in patients with unresectable pancreatic and bile duct cancer	Abe et al. (2018)
	n-3 fatty acids (3,000 mg) (n-3 eicosapentaenoic and docosahexaenoic acids)	Slightly improves dry eye symptoms if consumed for a long period of time	Asbell et al. (2018)
Green tea	Theanine	Suppresses the excessive stress response	Habibagahi et al. (2017)
	Epigallocatechin-3-gallate (EGCG)	Treatment of dystrophic epidermolysis bullosa, a rare genodermatosis with severe blistering for which no curative treatments are available	Chiaverini et al. (2016)
	Green tea	Improves systemic lupus erythematosus disease activity, anti-inflammatory, immunomodulatory benefits	Shamekhi and Amani (2017)
	Green tea, capsaicin, and ginger extracts	Beneficial effects on weight, BMI, markers of insulin metabolism, and plasma GSH levels in overweight women	Taghizadeh et al. (2017)
	Theanine	Reduces stress in middle-aged individuals, improves quality of sleep	Unno et al. (2017)
	Epigallocatechin-3-gallate (EGCG)	Reduces the pain, burning-feeling, itching, pulling, and tenderness in radiation-induced dermatitis in breast cancer patients undergoing radiotherapy	Zhu et al. (2016)

REFERENCES

Abdul-Hamid A, Luan YS. Functional properties of dietary fibre prepared from defatted rice bran. Food Chem. 2000;68(1):15–9.

Abdull R, Ahmad F, Noor NM. Cruciferous vegetables: Dietary phytochemicals for cancer prevention. Asian Pac J Cancer Prev. 2013;14(3):1565–70.

Abe K, Uwagawa T, Haruki K, Takano Y, Onda S, Sakamoto T, et al. Effects of omega-3 fatty acid supplementation in patients with bile duct or pancreatic cancer undergoing chemotherapy. Anticancer Res. 2018;38(4):2369–75. DOI:10.21873/anticanres.12485.

Action E. Scientific concepts of functional foods in Europe: Consensus document. Br J Nutr. 1999;81(1):1–27.

Aggarwal BB, Bhardwaj A, Aggarwal RS, Seeram NP, Shishodia S, Takada Y. Role of resveratrol in prevention and therapy of cancer: Preclinical and clinical studies. Anticancer Res. 2004;24(5A):2783–840.

Ahmad N, Feyes DK, Agarwal R, Mukhtar H, Nieminen A-L. Green tea constituent epigallocatechin-3-gallate and induction of apoptosis and cell cycle arrest in human carcinoma cells. J Natl Cancer Inst. 1997;89(24):1881–6.

Akhlaghi M, Ghobadi S, Mohammad Hosseini M, Gholami Z, Mohammadian F. Flavanols are potential anti-obesity agents, a systematic review and meta-analysis of controlled clinical trials. Nutr Metab Cardiovasc Dis. 2018. DOI:10.1016/j.numecd.2018.04.001.

Alizadeh F, Javadi M, Karami AA, Gholaminejad F, Kavianpour M, Haghighian HK. Curcumin nanomicelle improves semen parameters, oxidative stress, inflammatory biomarkers, and reproductive hormones in infertile men: A randomized clinical trial. 2018;32(3):514–21. DOI:10.1002/ptr.5998.

Aluyen JK, Ton QN, Tran T, Yang AE, Gottlieb HB, Bellanger RA. Resveratrol: Potential as anticancer agent. J Diet Suppl. 2012;9(1):45–56.

Andlauer W, Fürst P. Nutraceuticals: A piece of history, present status and outlook. Food Res Int. 2002;35(2–3):171–6.

Andres S, Abraham K, Appel KE, Lampen A. Risks and benefits of dietary isoflavones for cancer. Crit Rev Toxicol. 2011;41(6):463–506. DOI:10.3109/10408444.2010.541900.

Aruoma OI. Free radicals, oxidative stress, and antioxidants in human health and disease. J Am Oil Chem' Soc. 1998;75(2):199–212.

Asbell PA, Maguire MG, Pistilli M, Ying GS, Szczotka-Flynn LB, Hardten DR, et al. n-3 fatty acid supplementation for the treatment of dry eye disease. N Engl J Med. 2018;378(18):1681–90. DOI:10.1056/NEJMoa1709691.

Awad AB, Fink CS. Phytosterols as anticancer dietary components: Evidence and mechanism of action. J Nutr. 2000;130(9):2127–30.

Bellamy L. Organo-sulphur compounds. In: The infra-red spectra of complex molecules. Springer; 1975:394–418.

Bellisle F, Rolland-Cachera M-F, Deheeger M, Guilloud-Bataille M. Obesity and food intake in children: Evidence for a role of metabolic and/or behavioral daily rhythms. Appetite. 1988;11(2):111–8.

Bharti AC, Donato N, Singh S, Aggarwal BB. Curcumin (diferuloylmethane) down-regulates the constitutive activation of nuclear factor-kappa B and IkappaBalpha kinase in human multiple myeloma cells, leading to suppression of proliferation and induction of apoptosis. Blood. 2003;101(3):1053–62. DOI:10.1182/blood-2002-05-1320.

Bhatia AK, Lee JW, Pinto HA, Jacobs CD, Limburg PJ, Rubin P, et al. Double-blind, randomized phase 3 trial of low-dose 13-cis retinoic acid in the prevention of second primaries in head and neck cancer: Longterm follow-up of a trial of the Eastern Cooperative Oncology Group-ACRIN Cancer Research Group (C0590). Cancer. 2017;123(23):4653–62. DOI:10.1002/cncr.30920.

Biddle MJ, Lennie TA, Bricker GV, Kopec RE, Schwartz SJ, Moser DK. Lycopene dietary intervention: A pilot study in patients with heart failure. J Cardiovasc Nurs. 2015;30(3):205–12. DOI:10.1097/jcn.0000000000000108.

Bliddal H, Rosetzsky A, Schlichting P, Weidner M, Andersen L, Ibfelt H-H, et al. A randomized, placebo-controlled, cross-over study of ginger extracts and ibuprofen in osteoarthritis. Osteoarthr Cartilage. 2000;8(1):9–12.

Borrelli F, Capasso R, Izzo AA. Garlic (Allium sativum L.): Adverse effects and drug interactions in humans. Mol Nutr Food Res. 2007;51(11):1386–97. DOI:10.1002/mnfr.200700072.

Bradamante S, Barenghi L, Villa A. Cardiovascular protective effects of resveratrol. Cardiovasc Drug Rev. 2004;22(3):169–88.

Brush AH. Metabolism of carotenoid pigments in birds. FASEB J. 1990;4(12):2969–77.

Cabrera C, Artacho R, Giménez R. Beneficial effects of green tea– A review. J Am Coll Nutr. 2006;25(2):79–99.

Cal C, Garban H, Jazirehi A, Yeh C, Mizutani Y, Bonavida B. Resveratrol and cancer: Chemoprevention, apoptosis, and chemo-immunosensitizing activities. Curr Med Chem Anticancer Agent. 2003;3(2):77–93.

Castro-Acosta ML, Stone SG, Mok JE, Mhajan RK, Fu CI, Lenihan-Geels GN, et al. Apple and blackcurrant polyphenol-rich drinks decrease postprandial glucose, insulin and incretin response to a high-carbohydrate meal in healthy men and women. J Nutr Biochem. 2017;49:53–62. DOI:10.1016/j.jnutbio.2017.07.013.

Chacko SM, Thambi PT, Kuttan R, Nishigaki I. Beneficial effects of green tea: A literature review. Chin Med. 2010;5(1):13.

Chakraborty K, Praveen NK, Vijayan KK, Rao GS. Evaluation of phenolic contents and antioxidant activities of brown seaweeds belonging to Turbinaria spp. (Phaeophyta, Sargassaceae) collected from Gulf of Mannar. Asian Pac J Trop Biomed. 2013;3(1):8–16.

Chen Y-M, Ho SC, Lam SS, Ho SS, Woo JL. Beneficial effect of soy isoflavones on bone mineral content was modified by years since menopause, body weight, and calcium intake: A double-blind, randomized, controlled trial. Menopause. 2004;11(3):246–54.

Chiaverini C, Roger C, Fontas E, Bourrat E, Bourdon-Lanoy E, Labreze C, et al. Oral epigallocatechin-3-gallate for treatment of dystrophic epidermolysis bullosa: A multicentre, randomized, crossover, double-blind, placebo-controlled clinical trial. Orphanet J Rare Dis. 2016;11:31. DOI:10.1186/s13023-016-0411-5.

Cho Y-J, Kim C-J, Kim N, Kim C-T, Park B. Some cases in applications and agricultural systems of nanotechnology to food. Biochip J. 2008;2(3):183–5.

Chun KS, Keum YS, Han SS, Song YS, Kim SH, Surh YJ. Curcumin inhibits phorbol ester-induced expression of cyclooxygenase-2 in mouse skin through suppression of extracellular signal-regulated kinase activity and NF-kappaB activation. Carcinogenesis. 2003;24(9):1515–24. DOI:10.1093/carcin/bgg107.

Cicero A, Gaddi A. Rice bran oil and γ-oryzanol in the treatment of hyperlipoproteinaemias and other conditions. Phytother Res. 2001;15(4):277–89.

Connor WE. Importance of n-3 fatty acids in health and disease. Am J Clin Nutr. 2000;71(1):171S–5S.

De Lorgeril M, Salen P, Martin J-L, Monjaud I, Delaye J, Mamelle N. Mediterranean diet, traditional risk factors, and the rate of cardiovascular complications after myocardial infarction: Final report of the Lyon Diet Heart Study. Circulation. 1999;99(6):779–85.

Elbling L, Weiss R-M, Teufelhofer O, Uhl M, Knasmueller S, Schulte-Hermann R, et al. Green tea extract and (–)-epigallocatechin-3-gallate, the major tea catechin, exert oxidant but lack antioxidant activities. FASEB J. 2005;19(7):807–9.

Eric AA, Jordan VC. Estrogen-related receptors as emerging targets in cancer and metabolic disorders. Curr Top Med Chem. 2006;6(3):203–15. DOI:10.2174/156802661060603020303.

Eslami S, Esa NM, Marandi SM, Ghasemi G, Eslami S. Effects of gamma oryzanol supplementation on anthropometric measurements and muscular strength in healthy males following chronic resistance training. Indian J Med Res. 2014;139(6):857–63.

Fang MZ, Wang Y, Ai N, Hou Z, Sun Y, Lu H, et al. Tea polyphenol (–)-epigallocatechin-3-gallate inhibits DNA methyltransferase and reactivates methylation-silenced genes in cancer cell lines. Cancer Res. 2003;63(22):7563–70.

Fernando WM, Martins IJ, Goozee KG, Brennan CS, Jayasena V, Martins RN. The role of dietary coconut for the prevention and treatment of Alzheimer's disease: Potential mechanisms of action. Br J Nutr. 2015;114(1):1–14. DOI:10.1017/s0007114515001452.

Fitó M, Estruch R, Salas-Salvadó J, Martínez-Gonzalez MA, Arós F, Vila J, et al. Effect of the Mediterranean diet on heart failure biomarkers: A randomized sample from the PREDIMED trial. Eur J Heart Fail. 2014;16(5):543–50.

Fleurence J. Seaweed proteins: Biochemical, nutritional aspects and potential uses. Trends Food Sci Technol. 1999;10(1):25–8.

Fraser AJ, Webster TF, McClean MD. Diet contributes significantly to the body burden of PBDEs in the general US population. Environ Health Perspect. 2009;117(10):1520.

Gajendragadkar PR, Hubsch A, Maki-Petaja KM, Serg M, Wilkinson IB, Cheriyan J. Effects of oral lycopene supplementation on vascular function in patients with cardiovascular disease and healthy volunteers: A randomised controlled trial. PLoS One. 2014;9(6):e99070. DOI:10.1371/journal.pone.0099070.

Ghafoorunissa. Impact of quality of dietary fat on serum cholesterol and coronary heart disease: Focus on plant sterols and other non-glyceride components. Natl Med J India. 2009;22(3):126–32.

Ghayur MN, Gilani AH. Pharmacological basis for the medicinal use of ginger in gastrointestinal disorders. Dig Dis Sci. 2005;50(10):1889–97.

Guasch-Ferre M, Salas-Salvado J, Ros E, Estruch R, Corella D, Fito M, Martinez-Gonzalez MA. The PREDIMED trial, Mediterranean diet and health outcomes: How strong is the evidence? Nutr Metab Cardiovasc Dis. 2017;27(7):624–32. DOI:10.1016/j.numecd.2017.05.004.

Guyot S, Doco T, Souquet J-M, Moutounet M, Drilleau J-F. Characterization of highly polymerized procyanidins in cider apple (Malus sylvestris var. Kermerrien) skin and pulp. Phytochemistry. 1997;44(2):351–7.

Habibagahi Z, Namjoyan F, Ghadiri A, Saki Malehi A, Unno K, Yamada H, et al. Anti-stress effect of green tea with lowered caffeine on humans: A pilot study. Phytother Res. 2017;40(6):902–9. DOI:10.1002/ptr.5827 10.1248/bpb.b17-00141.

Haroyan A, Mukuchyan V, Mkrtchyan N, Minasyan N, Gasparyan S, Sargsyan A, et al. Efficacy and safety of curcumin and its combination with boswellic acid in osteoarthritis: A comparative, randomized, double-blind, placebo-controlled study. BMC Complement Altern Med. 2018;18(1):7. DOI:10.1186/s12906-017-2062-z.

Higdon JV, Delage B, Williams DE, Dashwood RH. Cruciferous vegetables and human cancer risk: Epidemiologic evidence and mechanistic basis. Pharmacol Res. 2007;55(3):224–36.

Hong J, Lu H, Meng X, Ryu J-H, Hara Y, Yang CS. Stability, cellular uptake, biotransformation, and efflux of tea polyphenol (–)-epigallocatechin-3-gallate in HT-29 human colon adenocarcinoma cells. Cancer Res. 2002;62(24):7241–6.

Hsu Y-L, Hung J-Y, Tsai Y-M, Tsai E-M, Huang M-S, Hou M-F, Kuo P-L. 6-shogaol, an active constituent of dietary ginger, impairs cancer development and lung metastasis by inhibiting the secretion of CC-chemokine ligand 2 (CCL2) in tumor-associated dendritic cells. J Agric Food Chem. 2015;63(6):1730–8.

Hu EA, Toledo E, Diez-Espino J, Estruch R, Corella D, Salas-Salvado J, et al. Lifestyles and risk factors associated with adherence to the Mediterranean diet: A baseline assessment of the PREDIMED trial. PLoS One. 2013;8(4):e60166.

Jankun J, Selman SH, Swiercz R, Skrzypczak-Jankun E. Why drinking green tea could prevent cancer. Nature. 1997;387(6633):561.

Jenkins D, Kendall C, Faulkner D, Kemp T, Marchie A, Nguyen T, et al. Long-term effects of a plant-based dietary portfolio of cholesterol-lowering foods on blood pressure. Eur J Clin Nutr. 2008;62(6):781.

Jew S, AbuMweis SS, Jones PJ. Evolution of the human diet: Linking our ancestral diet to modern functional foods as a means of chronic disease prevention. J Med Food. 2009;12(5):925–34.

Judaki A, Rahmani A, Feizi J, Asadollahi K, Hafezi Ahmadi MR. Curcumin in combination with triple therapy regimes ameliorates oxidative stress and histopathologic changes in chronic gastritis-associated helicobacter pylori infection. Phytother Res. 2017;54(3):177–82. DOI:10.1002/ptr.5998 10.1590/s0004-2803.201700000-18.

Kamal-Eldin A, Appelqvist LÅ. The chemistry and antioxidant properties of tocopherols and tocotrienols. Lipids. 1996;31(7):671–701.

Kant AK. Indexes of overall diet quality: A review. J Am Diet Assoc. 1996;96(8):785–91.

Keith M, Kuliszewski MA, Liao C, Peeva V, Ahmed M, Tran S, et al. A modified portfolio diet complements medical management to reduce cardiovascular risk factors in diabetic patients with coronary artery disease. Clin Nutr. 2015;34(3):541–8.

Kılınç B, Cirik S, Turan G, Tekogul H, Koru E. Seaweeds for food and industrial applications. In: Food industry. InTech; 2013.

Kim YK, Lee HO, Chang R, Choue R. A study on the food habits, nutrient intake and the disease distribution in the elderly (aged over 65 years)(I). Korean J Community Nutr. 2002;7(4):516–26.

Korobelnik JF, Rougier MB, Delyfer MN, Bron A, Merle BMJ, Savel H, et al. Effect of dietary supplementation with lutein, zeaxanthin, and omega-3 on macular pigment: A randomized clinical trial. JAMA Ophthalmol. 2017;135(11):1259–66. DOI:10.1001/jamaophthalmol.2017.3398.

Kris-Etherton PM, Zhao G, Binkoski AE, Coval SM, Etherton TD. The effects of nuts on coronary heart disease risk. Nutr Rev. 2001;59(4):103–11.

Kumar H, Kim I-S, More SV, Kim B-W, Choi D-K. Natural product-derived pharmacological modulators of Nrf2/ARE pathway for chronic diseases. Nat Prod Rep. 2014;31(1):109–39.

Kuo S-M. Antiproliferative potency of structurally distinct dietary flavonoids on human colon cancer cells. Cancer Lett. 1996;110(1):41–8. DOI:10.1016/S0304-3835(96)04458-8.

Kurano M, Hasegawa K, Kunimi M, Hara M, Yatomi Y, Teramoto T, Tsukamoto K. Sitosterol prevents obesity-related chronic inflammation. Biochim Biophys Acta. 2018;1863(2):191–8. DOI:10.1016/j.bbalip.2017.12.004.

Kwak M-K, Itoh K, Yamamoto M, Sutter TR, Kensler TW. Role of transcription factor Nrf2 in the induction of hepatic phase 2 and antioxidative enzymes in vivo by the cancer chemoprotective agent, 3H-1,2-dimethiole-3-thione. Mol Med. 2001;7(2):135.

Lambert C, Cubedo J, Padro T, Sanchez-Hernandez J, Antonijoan RM, Perez A, Badimon L. Phytosterols and omega 3 supplementation exert novel regulatory effects on metabolic and inflammatory pathways: A proteomic study. Nutrients. 2017;9(6). DOI:10.3390/nu9060599.

Le Couteur DG, Solon-Biet S, Wahl D, Cogger VC, Willcox BJ, Willcox DC, et al. New horizons: Dietary protein, ageing and the Okinawan ratio. Age Ageing. 2016;45(4):443–7. DOI:10.1093/ageing/afw069.

Li SQ, Zhang QH. Advances in the development of functional foods from buckwheat. Crit Rev Food Sci Nutr. 2001;41(6):451–64. DOI:10.1080/20014091091887.

Li F, Gong Q, Dong H, Shi J. Resveratrol, a neuroprotective supplement for Alzheimer's disease. Curr Pharm Des. 2012;18(1):27–33.

Link BG, Phelan J. Social conditions as fundamental causes of disease. J Health Soc Behav. 1995;80–94.

Lobo V, Patil A, Phatak A, Chandra N. Free radicals, antioxidants and functional foods: Impact on human health. Pharmacogn Rev. 2010;4(8):118.

Lv C, Zhang Y, Shen L. Preliminary clinical effect evaluation of resveratrol in adults with allergic rhinitis. Int Arch Allergy Immunol. 2018;175(4):231–6. DOI:10.1159/000486959.

MacArtain P, Gill CI, Brooks M, Campbell R, Rowland IR. Nutritional value of edible seaweeds. Nutr Rev. 2007;65(12):535–43.

Maritim A, Sanders A, Watkins J 3rd. Diabetes, oxidative stress, and antioxidants: A review. J Biochem Mol Toxicol. 2003;17(1):24–38.

Moghadasian MH. Pharmacological properties of plant sterols: In vivo and in vitro observations. Life Sci. 2000;67(6):605–15.

Moreau RA, Whitaker BD, Hicks KB. Phytosterols, phytostanols, and their conjugates in foods: Structural diversity, quantitative analysis, and health-promoting uses. Prog Lipid Res. 2002;41(6):457–500.

Moussa C, Hebron M, Huang X, Ahn J, Rissman RA, Aisen PS, Turner RS. Resveratrol regulates neuro-inflammation and induces adaptive immunity in Alzheimer's disease. J Neuroinflammation. 2017;14(1):1. DOI:10.1186/s12974-016-0779-0.

Nelis HJ, De Leenheer A. Microbial sources of carotenoid pigments used in foods and feeds. J Appl Microbiol. 1991;70(3):181–91.

Nicklas TA, Baranowski T, Cullen KW, Berenson G. Eating patterns, dietary quality and obesity. J Am Coll Nutr. 2001;20(6):599–608.

Nie S, Xing Y, Kim GJ, Simons JW. Nanotechnology applications in cancer. Annu Rev Biomed Eng. 2007;9:257–88.

Niemeyer CM. Semi-synthetic nucleic acid-protein conjugates: Applications in life sciences and nanobiotechnology. J Biotechnol. 2001;82:47–66.

Ola MS, Al-Dosari D, Alhomida AS. Role of oxidative stress in diabetic retinopathy and the beneficial effects of flavonoids. Curr Pharm Des. 2018. DOI:10.2174/1381612824666180515151043.

Pagliacci MC, Smacchia M, Migliorati G, Grignani F, Riccardi C, Nicoletti I. Growth-inhibitory effects of the natural phyto-oestrogen genistein in MCF-7 human breast cancer cells. Eur J Cancer. 1994;30(11):1675–82. DOI:10.1016/0959-8049(94)00262-4.

Palozza P, Calviello G, Serini S, Maggiano N, Lanza P, Ranelletti FO, Bartoli GM. β-Carotene at high concentrations induces apoptosis by enhancing oxy-radical production in human adenocarcinoma cells. Free Radic Biol Med. 2001;30(9):1000–7.

Panahi Y, Kianpour P, Mohtashami R, Jafari R, Simental-Mendia LE, Sahebkar A. Efficacy and safety of phytosomal curcumin in non-alcoholic fatty liver disease: A randomized controlled trial. Drug Res (Stuttg). 2017;67(4):244–51. DOI:10.1055/s-0043-100019.

Pandey KB, Rizvi SI. Plant polyphenols as dietary antioxidants in human health and disease. Oxid Med Cell Longev. 2009;2(5):270–8.

Patel KR, Scott E, Brown VA, Gescher AJ, Steward WP, Brown K. Clinical trials of resveratrol. Ann N Y Acad Sci. 2011.

Petrovski G, Gurusamy N, Das DK. Resveratrol in cardiovascular health and disease. Ann N Y Acad Sci. 2011.

Philip S, Kundu GC. Osteopontin induces nuclear factor kappa B-mediated promatrix metalloproteinase-2 activation through I kappa B alpha /IKK signaling pathways, and curcumin (diferulolylmethane) down-regulates these pathways. J Biol Chem. 2003;278(16):14487–97. DOI:10.1074/jbc.M207309200.

Pinto J, Rivlin R. Garlic and other allium vegetables in cancer prevention. In: Nutritional oncology. San Diego: Academic Press; 1999:393–403.

Raicht RF, Cohen BI, Fazzini EP, Sarwal AN, Takahashi M. Protective effect of plant sterols against chemically induced colon tumors in rats. Cancer Res. 1980;40(2):403–5.

Rasool Hassan B. Medicinal plants (importance and uses). Pharmaceut Anal Acta. 2012;3:e139.

Rong N, Ausman LM, Nicolosi RJ. Oryzanol decreases cholesterol absorption and aortic fatty streaks in hamsters. Lipids. 1997;32(3):303–9.

Sacks FM, Svetkey LP, Vollmer WM, Appel LJ, Bray GA, Harsha D, et al. Effects on blood pressure of reduced dietary sodium and the Dietary Approaches to Stop Hypertension (DASH) diet. N Engl J Med. 2001;344(1):3–10.

Sano A, Uchida R, Saito M, Shioya N, Komori Y, Tho Y, Hashizume N. Beneficial effects of grape seed extract on malondialdehyde-modified LDL. J Nutr Sci Vitaminol. 2007;53(2):174–82.

Santos-Parker JR, Strahler TR, Bassett CJ, Bispham NZ, Chonchol MB, Seals DR. Curcumin supplementation improves vascular endothelial function in healthy middle-aged and older adults by increasing nitric oxide bioavailability and reducing oxidative stress. Aging. 2017;9(1):187–208. DOI:10.18632/aging.101149.

Sathyapalan T, Aye M, Rigby AS, Thatcher NJ, Dargham SR, Kilpatrick ES, Atkin SL. Soy isoflavones improve cardiovascular disease risk markers in women during the early menopause. Nutr Metab Cardiovasc Dis. 2018. DOI:10.1016/j.numecd.2018.03.007.

Scott N, Chen H. Nanoscale science and engineering for agriculture and food systems. Ind Biotechnol. 2012;8(6):340–3.

Seddon JM, Ajani UA, Sperduto RD, Hiller R, Blair N, Burton TC, et al. Dietary carotenoids, vitamins A, C, and E, and advanced age-related macular degeneration. JAMA. 1994;272(18):1413–20.

Sen Savara B, Suher T. Dental caries in children one to six years of age as related to socioeconomic level, food habits, and toothbrushing. J Dent Res. 1955;34(6):870–5.

Seyyedebrahimi S, Khodabandehloo H, Nasli Esfahani E, Meshkani R. The effects of resveratrol on markers of oxidative stress in patients with type 2 diabetes: A randomized, double-blind, placebo-controlled clinical trial. 2018;55(4):341–53. DOI:10.1007/s00592-017-1098-3.

Shahidi F, de Camargo AC. Tocopherols and tocotrienols in common and emerging dietary sources: Occurrence, applications, and health benefits. Int J Mol Sci. 2016;17(10):1745. DOI:10.3390/ijms17101745.

Shamekhi Z, Amani R. A randomized, double-blind, placebo-controlled clinical trial examining the effects of green tea extract on systemic lupus erythematosus disease activity and quality of life. 2017;31(7):1063–71. DOI:10.1002/ptr.5827.

Shimizu T. Health claims on functional foods: The Japanese regulations and an international comparison. Nutr Res Rev. 2003;16(2):241–52.

Shukla Y, Singh M. Cancer preventive properties of ginger: A brief review. Food Chem Toxicol. 2007;45(5):683–90.

Simopoulos AP. Essential fatty acids in health and chronic disease. Am J Clin Nutr. 1999;70(3):560s–9s.

Smoliga JM, Baur JA, Hausenblas HA. Resveratrol and health – A comprehensive review of human clinical trials. Mol Nutr Food Res. 2011;55(8):1129–41.

Sovak M. Grape extract, resveratrol, and its analogs: A review. J Med Food. 2001;4(2):93–105.

Spanier G, Xu H, Xia N, Tobias S, Deng S, Wojnowski L, et al. Resveratrol reduces endothelial oxidative stress by modulating the gene expression of Superoxide Dismutase 1 (SOD1), Glutathione Peroxidase 1 (GPX1) and NADPH Oxidase Subunit (NOX4). J Physiol Pharmacol. 2009;60(Suppl 4):111–6.

Sparnins VL, Barany G, Wattenberg LW. Effects of organosulfur compounds from garlic and onions on benzo [a] pyrene-induced neoplasia and glutathione S-transferase activity in the mouse. Carcinogenesis. 1988;9(1):131–4.

Srividya A, Dhanabal S, Misra V, Suja G. Antioxidant and antimicrobial activity of Alpinia officinarum. Indian J Pharm Sci. 2010;72(1):145.

Sun W, Wang W, Kim J, Keng P, Yang S, Zhang H, et al. Anti-cancer effect of resveratrol is associated with induction of apoptosis via a mitochondrial pathway alignment. Adv Exp Med Biol. 2008;614:179–86.

Szeto Y, Kwok TC, Benzie IF. Effects of a long-term vegetarian diet on biomarkers of antioxidant status and cardiovascular disease risk. Nutrition. 2004;20(10):863–6.

Taghizadeh M, Farzin N, Taheri S, Mahlouji M, Akbari H, Karamali F, Asemi Z. The effect of dietary supplements containing green tea, capsaicin and ginger extracts on weight loss and metabolic profiles in overweight women: A randomized double-blind placebo-controlled clinical trial. Ann Nutr Metab. 2017;70(4):277–85. DOI:10.1159/000471889.

Talalay P, Fahey JW. Phytochemicals from cruciferous plants protect against cancer by modulating carcinogen metabolism. J Nutr. 2001;131(11):3027S–33S.

Toledo E, Hu FB, Estruch R, Buil-Cosiales P, Corella D, Salas-Salvadó J, et al. Effect of the Mediterranean diet on blood pressure in the PREDIMED trial: Results from a randomized controlled trial. BMC Med. 2013;11(1):207.

Unno K, Noda S, Kawasaki Y, Yamada H, Morita A, Iguchi K, Nakamura Y. Reduced stress and improved sleep quality caused by green tea are associated with a reduced caffeine content. Nutrients. 2017;9(7). DOI:10.3390/nu9070777.

Vetrani C, Vitale M, Bozzetto L, Della Pepa G, Cocozza S, Costabile G, et al. Association between different dietary polyphenol subclasses and the improvement in cardiometabolic risk factors: Evidence from a randomized controlled clinical trial. Acta Diabetol. 2018;55(2):149–53. DOI:10.1007/s00592-017-1075-x.

Vidavalur R, Otani H, Singal PK, Maulik N. Significance of wine and resveratrol in cardiovascular disease: French paradox revisited. Exp Clin Cardiol. 2006;11(3):217–25.

Villegas I, Sanchez-Fidalgo S, de la Lastra CA. Chemopreventive effect of dietary curcumin on inflammation-induced colorectal carcinogenesis in mice. Mol Nutr Food Res. 2011;55(2):259–67. DOI:10.1002/mnfr.201000225.

Vollmer WM, Sacks FM, Ard J, Appel LJ, Bray GA, Simons-Morton DG, et al. Effects of diet and sodium intake on blood pressure: Subgroup analysis of the DASH-sodium trial. Ann Intern Med. 2001;135(12):1019–28.

Vutyavanich T, Kraisarin T, Ruangsri R-A. Ginger for nausea and vomiting in pregnancy: Randomized, double-masked, placebo-controlled trial. Obstet Gynecol. 2001;97(4):577–82.

Wall MM. Ascorbic acid and mineral composition of longan (Dimocarpus longan), lychee (Litchi chinensis) and rambutan (Nephelium lappaceum) cultivars grown in Hawaii. J Food Compos Anal. 2006;19(6–7):655–63.

Willcox BJ, Willcox DC, Todoriki H, Fujiyoshi A, Yano K, He Q, et al. Caloric restriction, the traditional Okinawan diet, and healthy aging: The diet of the world's longest-lived people and its potential impact on morbidity and life span. Ann N Y Acad Sci. 2007;1114:434–55. DOI:10.1196/annals.1396.037.

Willcox DC, Willcox BJ, Todoriki H, Suzuki M. The Okinawan diet: Health implications of a low-calorie, nutrient-dense, antioxidant-rich dietary pattern low in glycemic load. J Am Coll Nutr. 2009;28(Suppl):500s–16s.

Willcox DC, Scapagnini G, Willcox BJ. Healthy aging diets other than the Mediterranean: A focus on the Okinawan diet. Mech Ageing Dev. 2014;136–7:148–62. DOI:10.1016/j.mad.2014.01.002.

Willett WC, Sacks F, Trichopoulou A, Drescher G, Ferro-Luzzi A, Helsing E, Trichopoulos D. Mediterranean diet pyramid: A cultural model for healthy eating. Am J Clin Nutr. 1995;61(6):1402S–6S.

Wong RHX, Evans HM, Howe PRC. Resveratrol supplementation reduces pain experience by postmenopausal women. Acta Diabetol. 2017;24(8):916–22. DOI:10.1007/s00592-017-1098-3 10.1097/gme.0000000000000861.

World Health Organization . Food and health in Europe: A new basis for action. World Health Organization Regional Office for Europe; 2004.

Xu Z, Hua N, Godber JS. Antioxidant activity of tocopherols, tocotrienols, and γ-Oryzanol components from rice bran against cholesterol oxidation accelerated by 2, 2 '-Azobis (2-methylpropionamidine) dihydrochloride. J Agric Food Chem. 2001;49(4):2077–81.

You W-C, Blot WJ, Chang Y-S, Ershow A, Yang ZT, An Q, et al. Allium vegetables and reduced risk of stomach cancer. J Natl Cancer Inst. 1989;81(2):162–4.

Zhang D, Quantick PC, Grigor JM. Changes in phenolic compounds in Litchi (Litchi chinensis Sonn.) fruit during postharvest storage. Postharvest Biol Tec. 2000;19(2):165–72.

Zhu W, Jia L, Chen G, Zhao H, Sun X, Meng X, et al. Epigallocatechin-3-gallate ameliorates radiation-induced acute skin damage in breast cancer patients undergoing adjuvant radiotherapy. Oncotarget. 2016;7(30):48607–13. DOI:10.18632/oncotarget.9495.

Zorn KK, Bonome T, Gangi L, Chandramouli GV, Awtrey CS, Gardner GJ, et al. Gene expression profiles of serous, endometrioid, and clear cell subtypes of ovarian and endometrial cancer. Clin Cancer Res. 2005;11(18):6422–30.

14

Emulsification Technology for Cosmetics Preparation

Kazi Asraf Ali and Sanjit Roy

CONTENTS

ABSTRACT: Cosmetics are products that can decorate our bodies by means of cleansing, beautifying, or altering appearance. Emulsions technology is widely used for the preparation of cosmetics for the topical administration of hydrophilic and lipophilic active ingredients. Recently, it has been observed that there is an increasing tendency for pharmaceutical companies to incorporate a number of ingredients such as olive oil, almond, fruit extracts, and essential oils in the form of emulsions due to their beneficial and therapeutic properties. Various types of emulsion technology, like microemulsion, multiple emulsions, non-aqueous emulsion, liposome emulsion, emulsion polymerization, and nanoemulsion, have great magnitude as effective vehicles for the continuous release of cosmetic agents and for the uniform distribution of cosmetic agents in skin layers. Oil-in-water (o/w) emulsions play an important role in cosmetics: they are fundamental in the formulation of some products like body lotions, skin creams, sunscreens, hair conditioner, hair oil, shampoo, etc.

   This chapter deals with the theories behind emulsification technology in cosmetics, the types of emulsion utilized in cosmetics, the advantages of the emulsification technique over other methods, the method of preparation and evaluation of the formulation, the risk of hazards and the effect on the environment, etc.

KEY WORDS: *Cosmetics, emulsions, emulsion technology, microemulsion, nanoemulsion*

14.1 Introduction

The Food and Drug Administration (FDA) defines cosmetics as "articles intended to be applied to the human body for cleansing, beautifying, promoting attractiveness, or altering the appearance without affecting the body's structure or functions" (Durej et al., 2005; Millikan, 2001). Cosmetics include skin care, (Butler, 2013) hair care, nail care, and care of teeth. Emulsion technology is widely used for the preparation of cosmetics for the topical administration of hydrophilic and lipophilic active ingredients. Recently, it has been observed that there is an increasing tendency for pharmaceutical companies to incorporate a number of ingredients such as olive oil, almond, fruit extracts, and essential oils in the form of emulsions due to their beneficial and therapeutic properties (Mihranyan et al., 2012). Various types of emulsion technology, like microemulsion, multiple emulsion, non-aqueous emulsion, liposome emulsion, emulsion polymerization, and nanoemulsion, have great magnitude as effective vehicles for the continuous release of cosmetic agents and for the uniform distribution of cosmetic agents in skin layers. Oil-in-water (o/w) emulsions play an important role in cosmetics: they are fundamental in the formulation of some products like body lotions, skin creams, sunscreens, hair conditioner, hair oil, shampoo, etc (Matsumoto et al., 1976).

14.2 Emulsion and Its Properties

Emulsions are biphasic, thermodynamically-unstable but physically-stable liquid dosage form. It consists of two immiscible liquids in which one liquid is uniformly distributed into another by the action of a third substance known as an emulsifying agent. Here the first liquid is called the dispersed phase, or internal phase, and the second liquid is called the dispersion medium, or continuous medium/phase. The basic requirement to prepare an emulsion is oils, surfactants/emulgents, co-surfactants, and water. Table 14.1 gives a list of the ingredients in emulsions (Sharma and Sarangdevot, 2012).

   The ideal properties of an emulsion are as follows:

1. Droplet size of the internal phase should be uniform in size.
2. The droplets should not aggregate or coalesce to each other.
3. Rancidity of oils in emulsion must be prevented using antioxidants.
4. Emulsions should not be prone to microbial contaminations.

5. Emulsions must be stable in a variety of temperature ranges.
6. Creaming, cracking, and phase inversion of emulsions must be avoided by applying suitable measures (Even and Gregory, 1994; Madaan et al., 2014).

14.3 Theories of Emulsification

There are various theories which can be utilized to form stable emulsions by the emulsifiers but no single theory is universally applicable to all types of emulsions. A few theories, in brief, are illustrated below (Bancroft, 1913; Becher, 1965; Finkle et al., 1923; Haanai, 1968).

1. **The phase volume theory:** According to this theory, spheres of the identical diameter will touch 12 others, and these can occupy about 74% of the total volume. So, 74 parts of the spherical particles can be dispersed into the continuous phase.
2. **Hydration theory:** This theory is postulated by Fischer (Bancroft, 1913). According to this theory, emulsions can only be produced when the liquid which is to become the continuous phase is all used in the formation of a hydrated compound of the emulsifying agent employed.
3. **The electrical double layer theory:** In a o/w emulsion, oil globules bear a negative charge. The negative charge on the oil may come from adsorption of the hydroxyl (OH) ions, which is obtained by ionization of water. A layer of adsorbed hydroxyl ions is formed around the oil globules. Outside the layer of negative ions, another layer of positively-charged ions forms in the liquid. These two layers of oppositely charged ions are known as the Helmholtz double layer. The electrical double layers are not restrained to emulsions but escort all border phenomena. The electric charge is a factor in all emulsions, even those stabilized with emulsifying agents.
4. **Oriented wedge theory:** Langmuir and Harkins developed this theory (Finkle et al., 1923), which deals with the stability of emulsion. According to this theory, the emulgent molecules orient themselves in such a manner in the interface between the dispersed phase and dispersion medium that a wedge is formed, which limits the size of the dispersed particles.
5. **Adsorbed film and interfacial tension theory:** According to this theory (Haanai, 1968), Clayton emphasized that emulsification is influenced by (a) the mass of the emulsifier present in the emulsion, (b) the ease with which this agent is adsorbed at the interfacial separating surface, and (c) the nature of the ions adsorbed by the resultant film (Figure 14.1).

14.4 Development of Ternary Phase Diagram

Through the development of the ternary phase diagram (Figure 14.2), the microemulsion region is predicted. The relative amounts of the basic ingredients (oil, water, and surfactant/fixed ratio of surfactant and co-surfactant) required to produce a microemulsion are utilized to construct the ternary phase diagram. At the angular point of the triangle, each component represents 100% by volume fraction. Within the triangle, every point describes the possible composition of the three components (Date and Nagarsenke, 2007; Singh et al., 2012).

TABLE 14.1

Components of Emulsion (Oils, Surfactants and Co-Surfactants)

Oils	Surfactants	Co-surfactants
Sun flower oil, sesame oil, soybean oil, castor oil, jojoba oil, olive oil, peanut oil, fish oil, cotton seed oil, coconut oil, Captex® 200, Captex® 8000, Witepsol®, Myritol® 318	Gelucire® 44/14, Cremophor® RH 40, Capryol® 90, Softigen® 701, Tween® 80, Poloxamer™ 188, Plurol® Oleique CS 497, Labrafil® M 1944 CS	Propanol, ethanol, glycerol, Transcutol® P, ethylene glycol, propylene glycol

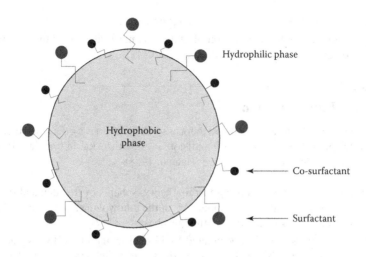

FIGURE 14.1   Interfacial film of emulsion.

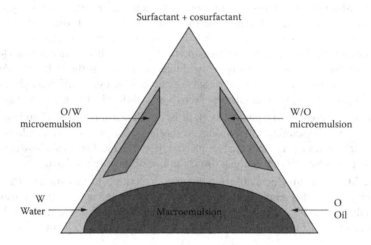

FIGURE 14.2   Ternary phase diagram consisting of oil, water and a mixture of surfactant and co-surfactant.

14.5 Types of Emulsion

There are two main types of emulsions, namely o/w and water-in-oil (w/o) emulsions. In o/w emulsions, oil globules are distributed in a continuous aqueous phase, whereas in w/o emulsions water droplets are distributed in the dispersion medium, i.e., oil phase. Another type of emulsion is multiple emulsions which is of two types, i.e., water-in-oil-in-water (w/o/w) and oil-in-water-in-oil (o/w/o). Self-emulsion is the newest form of emulsion which is the isotropic mixture of oil, surfactant, and co-surfactant which, when they come in contact with the aqueous phase, form fine o/w micro/nanoemulsion (Devarajan and Ravichandran, 2011; Jansen et al., 2006; Kumar et al., 2013).

14.6 Advantages of Emulsion Technology in Cosmetics

1. Cosmetic emulsions are thermodynamically stable and applicable to self-emulsification of the system.
2. The energy requirement for emulsification is less when compared to other preparations.

3. Emulsions can act as the good solvents for cosmetic agents. Hydrophilic as well as lipophilic drugs are soluble in emulsions due to the presence of microdomains of different polarity within the same single-phase solution (Shaji and Reddy, 2004).

4. The dispersed phase of either type of emulsion (o/w or w/o) is useful for the storage of the lipid-soluble or water-soluble agents, respectively.

5. The smaller droplet size (below 0.22 mm) in the case of microemulsion-based cosmetics provides a larger interfacial area of application (Vyas and Khar, 2002).

6. Emulsion-based cosmetics can reduce the use of volatile organic components.

7. Emulsion-based cosmetic preparations are free of greasiness and offer more compliance to the customers.

14.7 Disadvantages of Emulsion Technology in Cosmetics

1. Higher concentrations of surfactant and co-surfactant are requisite for stabilizing nano-globules (Wu et al., 2010).

2. The solubilizing capacity is limited.

3. The surfactant should have harmless properties for use in pharmaceutical applications.

4. Microemulsion has an instability problem because it is affected by pH and temperature.

5. Sometimes a specific device is needed for the preparation of the emulsion.

14.8 Preparation Methods of Emulsion

14.8.1 General Method of Preparation of Cosmetic Emulsion

There are lots of procedures that have been suggested for the preparation of emulsions in literature. Generally, o/w emulsions are prepared by separating the oily phase into minute globules surrounded by an envelope of emulsifying agent. Then the oil globules are suspended in the aqueous phase. For the preparation of w/o emulsions (Figure 14.3), aqueous phases are separated into minute globules surrounded by an envelope of emulsifying agent and the globules are suspended in the oily phase (Boom, 2008).

14.8.2 Method of Preparation for Multiple Emulsions in Cosmetics

Multiple emulsions can be prepared by the re-emulsification of a primary emulsion when an emulsion is inversed from one type to another.

14.8.2.1 Two-Step Emulsification

This is the most common and easiest method that gives a high yield with reproducibility. Generally, multiple emulsions can be formulated by the two-step emulsification method by using conventional rotor-stator or high-pressure valve homogenizers. The secondary emulsification step is carried out with fewer shears to avoid rupture of the liquid membrane between the outermost and innermost phase, while the primary emulsions are formulated under high-shear conditions to obtained small inner droplets. The methods are divided into two steps.

1. Primary emulsions (w/o or o/w) are obtained by using appropriate emulsifier.

2. The freshly prepared primary emulsions are re-emulsified with an excess of aqueous phase or oil phase. The final prepared emulsion could be w/o/w or o/w/o respectively.

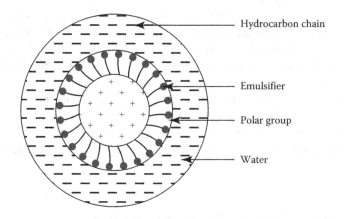

FIGURE 14.3 A typical structure of o/w emulsion.

14.8.2.2 Modified Two-Step Emulsification

This method is different from the conventional two-step technique in two ways.

1. Sonication and stirring are used to obtain fine, homogenous, and stable w/o emulsion.
2. A continuous phase is poured into a dispersed phase for preparing w/o/w emulsion.

Moreover, the composition of internal aqueous phase – oily phase – external phase is fixed at 1:4:5, which produces the most stable formulation as reported for most w/o/w emulsions.

14.8.2.3 Phase Inversion Method

Phase inversion means the change of one type of emulsion into another type, that is o/w emulsion into w/o type and vice versa (Klucker et al., 2012). The temperature at which the inversion occurs depends on the emulsifier concentration, which is called phase inversion temperature (PIT). In the phase inversion method, the aqueous phase is added to the oil phase in response to temperature so as to form a w/o emulsion. Emulsion can be stabilized by adding surfactant that contains oil-swollen micelles of the surfactant and emulsifying oil. At the time of the phase inversion process, drastic physical changes occur, including changes in particle size. This technique is based on the principle of altering the continuous curvature of the surfactant. PIT can be attained by altering the temperature from the o/w system with a lower temperature to the w/o emulsion at a higher temperature. During cooling, the system crosses a point of spontaneous curvature and minimal surface tension that promotes the formation of finally dispersed oil droplets. This method is also referred to as the PIT method. Figure 14.4 describes the steps related to the preparation of multiple emulsions by two-step methods.

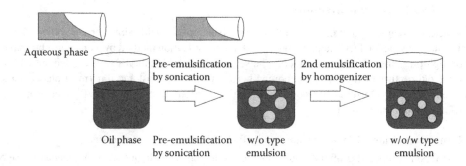

FIGURE 14.4 Multiple emulsion (w/o/w) formulation by the modified two-step emulsification method.

14.8.2.4 Membrane Emulsification Method

This method is based on the novel concept of generating droplets "drop-by-drop" to produce emulsion. Here, w/o emulsion is extruded into an external aqueous phase with constant pressure through a porous glass membrane that should bear controlled homogenous pores. The particle size of the resulting emulsion will be controlled through the proper selection of the porous glass membrane. The relation between particle size and membrane pore size of w/o/w emulsion exhibits a good co-relation that can be described by the following relationship: $Y = 5.03X + 0.19$ (where Y is the particle size and X is the pore size of the multiple emulsions).

14.9 Microemulsions in Cosmetics

Microemulsion-based cosmetics are becoming more and more important at the present time due to the changing behaviour of the consumer. Good appearance is considered a crucial component of quality of life today. It gives a good conditioning and hair feel effect when it is wet.

14.9.1 Microemulsions in Hair Conditioner

Hair is one of the vital fragments of the body derived from the ectoderm of the skin and contains a protein called keratin that is produced in hair follicles in the outer layer of skin (Adhirajan et al., 2001; Purwal et al., 2008; Thorat et al., 2009). Hair loss, or alopecia, is one of the most common problems of many communities, causing many economical and physiological problems. The treatment of hair loss is sometimes difficult because of the deficient efficacy and limited options in treatment (Ohyama, 2010). Microemulsions appear to be a promising carrier system for cosmetic active ingredients due to their numerous advantages over the existing conventional formulations. They are capable of encapsulating both hydrophilic and lipophilic molecules. Microemulsions can be used successfully in hair care compositions to ensure not only good conditioning of the hair but also good hair feel and hair gloss (Azeem et al., 2008). Through emulsification, materials can be diluted to an appropriate level to deposit on hair without negative side effects. Shahtalebi et al. (2016) formulated clove oil in emu oil self-emulsion to enhance routine hair care such as conditioning, cleaning, and grooming hair and stimulating hair follicles. Nazir et al. (2011) prepared uniform-sized silicone oil microemulsions that were added into model shampoos and conditioners to investigate the effects of size, uniformity, and storage stability of silicone oil deposition on the hair surface. They observed more storage stability in uniform-sized microemulsion than non-uniform-sized microemulsions prepared by a conventional homogenizer (Nazir et al., 2011).

14.9.2 Antiperspirant and Deodorant

The terms "antiperspirant" and "deodorant" are frequently used interchangeably, though they have quite marked distinctions. Antiperspirant actively reduces the amount of underarm perspiration while deodorant masks or reduces odor through the use of an antimicrobial agent or fragrance. Antiperspirant and deodorant products constitute a major segment in the health and beauty aids industry of today. Now most people use antiperspirant and/or deodorant products frequently to prevent or mask body odor. Aluminium chlorohydrate, aluminium sesquichlorohydrate, and aluminium zirconium tetrachlorohydrate are common examples of antiperspirants.

14.9.3 Hair Bleaches and Hair Colorants

Cosmetics including deodorants, shampoo, skin care, and perfumes are an integral part of modern life. Today's customers are trying to get skin, hair, and sun products to a value-added perspective and are willing to pay for additional benefits. Microemulsions appear as a tool to add multiple benefits that can improve appearance, elegance, and stability, are easy to apply, and are useful to control the release of active agents. Emulsion bleaches have the advantage of being particularly suitable for full-head bleaches because they have a cooling agent added to them. They also have a brilliant consistency that prevents them from dripping and drying out. Emulsion-form hydrogen peroxide compositions, in the form of an o/w emulsion containing oil or fatty components, emulsifiers, and hydrogen peroxide, are particularly suitable as an oxidizing component in processes for dyeing or lightening hair, resulting in improved depth of color and brightness. Yang et al. (2017) developed a novel microemulsion to deliver phenyletha-noid glycoside (PG) for use in skin lighteners and sunscreens. Here, the oil phase was selected on the basis of drug solubility, while the surfactant and co-surfactant were screened and selected on the basis of their solubilizing capacity. They proved that the permeating ability of microemulsion-carried PG was significantly increased compared with saline solution (Yang et al., 2017).

14.9.4 Depilatories

A depilatory is also known as hair removal cream. It is used to remove hair from the skin on the human body (in particular superfluous hair occurring on the face, legs, axilla, etc.) without causing any injury to the skin (Tortora and Grabowski, 1996). It is a totally different method of hair removal from others such as (1) mechanical removal of hair, (2) destruction of hair by use of laser energy, and (3) shaving. Chemical depilatories have existed from ancient times as a temporary hair removal method. At present, different types of depilatory agents are available in the market. They are frequently used, simple to apply, inexpensive, and readily available. The qualities of an ideal depilatory should be (1) non-toxic and non-irritant to skin, (2) fast and efficient in action, preferably causing depilation within five minutes, (3) preferably odorless, (4) stable upon storage, (5) non-staining/damaging to clothing, and (6) cosmetically elegant. The use of depilatories is rapidly growing and becoming widely accepted as a temporary hair removal method. Their widespread availability, ease of use, and broad product range has improved their popularity with customers. At present, this is the most used method of hair removal in the world.

14.9.5 Shaving Preparations

Shaving preparations like creams or lotions are a group of male-oriented toiletry products. They are one of the most important requirements for a smooth and comfortable shave. The cream is mainly an emulsion consisting of oil, soap, or surfactants. The soaps are basically composed of sodium and potassium stearate, mixed with water and glycerol to give a creamy texture. The application of microemulsion technology in after-shave lotions can provide reduced stinging and irritation to the skin and a comfortable feeling without tackiness.

14.10 Nanoemulsions in Cosmetics

14.10.1 Nanogels as Skin Care Products

Nanogels are defined as nano-sized, three-dimensional networks that are formed by physically or chemically crosslinked polymer chains (Figure 14.5). 'Nanogel' is nothing but the cross linked polymer of polyion and non-ionic polymer. It has a bi-functional network and able to polynucleotides (cross-linked poly ethylene imine (PEI) and polyethylene glycol (PEG), or PEG-*cl*-PEI) (Biancoet al., 2013). Nanotechnology has introduced the need for developing nanogel systems which prove their potential efficacy in a wide range of skin care products. They can even be used successfully for the treatment of skin cancer (Figure 14.6).

A nanogel is a unique nanoemulsion carrier system that has been designed around easy formulation. The formula is particularly suited to minimize transepidermal water loss and enhance the penetration of active ingredients, which is particularly useful for sun care products as well as moisturizing and anti-aging creams (Table 14.2).

In the field of cosmetics, nanogels have a wide range of application in sunscreens. Some inorganic components, like titanium dioxide (TiO_2) and zinc oxide (ZnO), in sunscreen act as ultraviolet (UV) filters and have been shown to inhibit the production of free radicals. It has been shown that the addition of these inorganic compounds into sunscreens can effectively reduce the absorption of UVA and UVB radiation

High HLB surfactant

Coated nanoemulsion

FIGURE 14.5 A typical structure of nanoemulsion.

and give protection against the harmful effects of UV radiation. Nanogels are very useful for UV protection due to their small particle size. TiO_2 is most widely used due to its brightness and high refractive index. They have a significantly higher effectiveness in blocking UV light compared to natural material due to their large surface area to volume ratio. TiO_2 and ZnO are the main inorganic components of sunscreen that can be used in sensitive skin and baby products and in daily-wear skin lotion (Luther, 2004). Nanogels are advantageous because they approach a zero-order release profile which means that they release less of the sunscreen formulation over time than o/w emulsions. Thus, the sunscreen remains on the surface of the skin longer and provides better protection against UV rays.

14.10.2 PEG-Free Nanoemulsions in Baby Care and Makeup Removal

O/w emulsions can play a significant role in cosmetics: they are essential for the formulation of products such as skin creams and sunscreens that are free from emulsifiers and based on PEG. These types of blends are highly attractive in the growing market for impregnating emulsions for moisturized tissue. In this somewhat up-to-date but quickly developing field of application, emulsion-based wet wipes can be

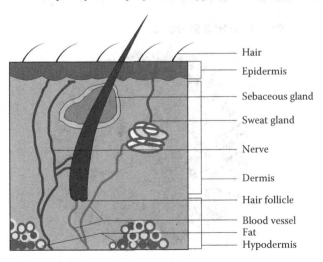

Hair
Epidermis
Sebaceous gland
Sweat gland
Nerve
Dermis
Hair follicle
Blood vessel
Fat
Hypodermis

FIGURE 14.6 A typical structure of human skin.

TABLE 14.2

Some Consumer Needs in Skin Care

Protection from	Repair of
Ultraviolet radiation	Dry skin
Infrared radiation	Sunburn
Wind	Acne
Central heating chemicals	Trauma stress, smoking burns, illness
Cold	Wrinkles
Pollution	Puffy eyes
Insects	Razor bumps, stretch marks, cellulite, skin cancer, patchy pigmentation, freckles, age spots, scars

applicable to baby care and makeup removal. The key components in these products are low viscosity o/w emulsions with good storage stability. Classical emulsions have a typical particle size of between 0.5 and 10 μm, which causes their typical white appearance, and usually show viscosities of over 1,000 mPa. They are kinetically stable and can be formulated with the help of a homogenizer. On the other hand, o/w microemulsions are easy to produce due to their thermodynamic stability. They are translucent and their typical particle radii range between 10 and 40 nm. Microemulsions form spontaneously on mixing and the order in which the components are added makes no difference. However, microemulsion formation usually requires large quantities of emulsifiers and surfactants. In terms of their properties, nanoemulsions are positioned between microemulsions and traditional emulsions. Their particle size range is between 30 and 100 nm, which causes their typical shiny blue appearance. At these small particle sizes, the Brownian motion prevents creaming and, as a result, nanoemulsions often have good long-term stability. Like classical emulsions, nanoemulsions are kinetically stable (Chaudhri et al., 2015).

14.11 Multiple Emulsions in Cosmetics

Multiple emulsions can be defined as emulsions in which both types of emulsions, i.e., w/o and o/w, are present simultaneously. The properties of both types of emulsions are seen in multiple emulsions. The system can be described as a heterogeneous system of one immiscible liquid dispersed in another in the form of droplets, which usually have diameters greater than 1 μm (Akhtar et al., 2010). These two liquids form a system that is characterized by its low thermodynamic stability (Figure 14.7).

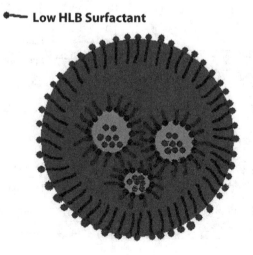

FIGURE 14.7   A typical structure of multiple emulsion.

Multiple emulsions are very complex systems as the dispersed phase droplets themselves contain even smaller droplets, which normally consist of a liquid miscible which is, in most cases, identical with the continuous phase. Both hydrophilic and lipophilic emulsifiers are used for the formation of multiple emulsions. Multiple emulsions were determined to be promising in many fields, particularly in pharmaceutics and in separation science. Their potential biopharmaceutical applications (Sinha and Kumar 2002) include their use as adjuvant vaccines (Sonakpuriya et al., 2013), prolonged drug delivery systems (Asuman and Ongun, 2008; Nisisako, 2008; Omotosho et al., 1986), sorbent reservoirs in drug overdose treatments (Matsumoto et al., 1976), and in mobilization of enzymes (Akhtar et al., 2010; Goto et al., 1995). Multiple emulsions were also investigated for cosmetics for their potential advantages of prolonged release of active agents, incorporation of incompatible materials, and protection of active ingredients by dispersion in the internal phase (Carlotti et al., 2005; Dahms and Tagawa, 1996; Faure et al., 2013).

14.12 Liposomes Emulsion

Liposomes are small, spherical-shaped, artificial vesicles that can be formulated from cholesterol and other natural, non-toxic phospholipids (Müller et al., 2002). At the present time, they are a promising system for drug delivery and skin-care treatment due to their nano-size and hydrophilic/lipophilic nature. The system can ensure the effectiveness of the cosmetic formulation by transporting the active ingredients to the target site, mostly the epidermis.

Liposomes are commonly used in dermal applications as a protectant for the active ingredients and for their moisturizing properties. The spherical vesicles are composed of phospholipids with an aqueous core. Either lipophilic or hydrophilic active ingredients can be incorporated in these vesicles. Liposomes which will be applied as dermal carriers should be small-sized, unilamellar, and equipped with a flexible membrane (Figure 14.8). On the other hand, flexible liposomes are the only choice to carry active molecules into the deeper skin layers (via biological syringe) as they stabilize the active ingredients and even act as a pharmaceutical or cosmetic ingredient due to their chemical nature.

Liposomes are used for transferring drugs safely to their target site in a precise time period to have a controlled release and attain the maximum therapeutic effect. Liposomes offer several advantages in delivering genes to cells: they can complex both with negatively and positively charged molecules, offer a degree of protection to the deoxyribonucleic acid (DNA) from derivative processes, carry large pieces of DNA, potentially as large as a chromosome, and can be targeted to specific cells or tissues. Generally, liposomes are prepared in the following ways.

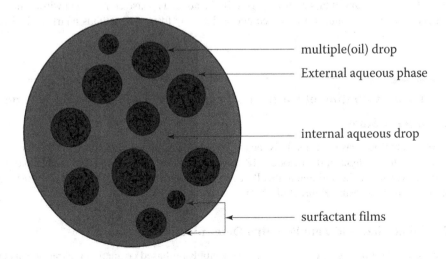

FIGURE 14.8 A typical structure of liposome emulsion.

14.12.1 Solvent Emulsification–Evaporation Technique

The hydrophobic drug and lipophilic material are dissolved in a water-immiscible, organic solvent (e.g., cyclohexane, dichloromethane, toluene, and/or chloroform) and then emulsified in an aqueous phase by using a high-speed homogenizer (Lemos-Senna et al., 1998). To improve the efficiency of fine emulsification, the coarse emulsion is directly passed through the microfluidizer. Thereafter, the organic solvent is evaporated by mechanical stirring at room temperature and reduced pressure (e.g., using a rotary evaporator), leaving lipid precipitates of solid lipid nanoparticles (SLNs). The mean particle size depends on the concentration of lipid in the organic phase. A very small particle size can be obtained with a low lipid load (5%) related to organic solvent. The major advantage of this method is the prevention of any thermal stress, which makes it appropriate for the incorporation of highly thermolabile drugs. The use of organic solvent, which may interact with drug molecules and limit the solubility of the lipid in the organic solvent, is another side of the coin.

14.12.2 Solvent Emulsification–Diffusion Technique

In the solvent emulsification–diffusion technique, the solvent used (e.g., benzyl alcohol, butyl lactate, ethyl acetate, isopropyl acetate, methyl acetate) must be partially miscible with water and this technique can be carried out either in aqueous phase or in oil phase. For example, cyclosporine SLNs were prepared by the emulsification–diffusion method to improve absorption and bioavailability. Initially, both the solvent and water were mutually saturated in order to ensure the initial thermodynamic equilibrium of both liquids. When heating was required to solubilize the lipid, the saturation step was performed at that temperature. Then the lipid and drug were dissolved in water saturated solvent and this organic phase (internal phase) was emulsified with solvent saturated aqueous solution containing stabilizer (dispersed phase) using a mechanical stirrer. After the formation of o/w emulsion, water (dilution medium) in typical ratio ranges of 1:5 to 1:10 was added to the system in order to allow solvent diffusion into the continuous phase, thus forming aggregation of the lipid in the nanoparticles. Here both the phases were maintained at the same elevated temperature and the diffusion step was performed either at room temperature or at the temperature under which the lipid was dissolved. Throughout the process, constant stirring was maintained. Finally, the diffused solvent was eliminated by vacuum distillation or lyophilization (Mora-Huertas et al., 2011; Trotta et al., 2003).

14.12.3 Double Emulsion Technique

In the double emulsion technique, the drug is dissolved in the aqueous phase and then it is emulsified in melted lipids. This primary emulsion is stabilized by adding stabilizer such as gelatin or Poloxamer™ 407. Then this stabilized primary emulsion is dispersed in the aqueous phase containing hydrophilic emulsifier. After that, the double emulsion is stirred and isolated by filtration methods (Garti and Bisperink, 1998; Masmoudi et al., 2005).

14.13 *In Vitro* Evaluation of Cosmetics Prepared by Emulsification Technology

14.13.1 Turbidity Study

Turbidity of an emulsion-based cosmetic is measured using a turbidity meter at a temperature of 68°F. The apparatus utilizes a light in the infrared (IR) region (860 nm). There is a continuous monitoring of turbidity data, which is collected automatically every second throughout a duration of 1,000 seconds, and then in 5 second intervals (Kabri et al., 2011).

14.13.2 Globule Size and Zeta Potential Determination

The particle diameter ranging 3 to 10,000 nm in o/w emulsion-based cosmetic preparation was carried out through the Dynamic Light Scattering (DLS) method, using the Zetasizer Nano ZS instrument (from

Malvern Instruments Ltd., UK) coupled with its Diversified Technical Systems (DTS) NANO software. The detection signal of the DLS was integrated at an angle of 173° in room temperature (Bachhav and Patravale, 2009; Negi et al., 2016).

The charge of the droplets of the microemulsion was measured by a Zeta-Meter which was monitored at 77°F at a scattering angle of 173° (Zetasizer Nano ZS, Malvern Instruments Ltd., UK). The value of zeta potential may vary from −30 to +30 mV (Negi et al., 2016; Patil et al., 2007).

14.13.3 Stability Study

The physical stability of an emulsion-based cosmetic formulation is vital to its performance, otherwise it can produce a negative effect by precipitating the active constituents in the excipient matrix. In addition, the reduced physical stability of the formulation can cause phase separation of the excipients, which affects formulation performance as well as the visual appearance of the formulation. The following steps are performed for thermodynamic stability testing.

1. Heating–cooling cycle

   In this phase, six cycles between refrigerator temperature (39.2°F) and 113°F with storage at each temperature of not less than 48 hours were studied. Those formulations which were stable at these temperatures were selected for centrifugation testing (Patel et al., 2013; Shafiq-un-Nabi et al., 2007).

2. Centrifugation

   The formulations passed in the heating–cooling cycle were centrifuged at 3,500 rpm for 30 minutes. Those formulations that did not show any phase separation were taken for the freeze–thaw stress test (Palanisamy and Raichur, 2009; Sawant et al., 2007).

3. Freeze–thaw cycle

   In this test, selected formulations were exposed for three freeze–thaw cycles which included freezing at 24.8°F for 24 hours followed by thawing at 77°F (Ghosh and Coupland, 2008; Vavia et al., 2007; Shafiq et al., 2007).

14.13.4 Differential Scanning Calorimetry

Differential Scanning Calorimetry (DSC) technique was applied with a DSC apparatus. About 10 ml of an emulsion sample was placed in an aluminium pan, hermetically sealed, before being placed in the calorimeter thermocouples. The samples were then heated from 68°F to 140°F in 41°F/min heating increments to eliminate initial thermal history (equilibrated at 140°F), and then cooled down to −76°F following the same process (equilibrated at −76°F). The samples were then reheated to 140°F in 41°F/min heating increments as well (Kabri et al., 2011).

14.13.5 Polydispersity Study

The size, shape and dynamics of the emulsion particles can be determined by small-angle X-ray scattering, small-angle neutron scattering and static and DLS techniques. Modification of the structure and composition of the pseudophase due to dilution can be overcome by measuring the intensity of the scattered light at different angles. In DLS the size distribution of molecules or particles is the property of interest. Here, the distribution describes how much material is present of the different size "slices". Traditionally, this overall polydispersity has also been converted into an overall polydispersity index (PDI) which is the square of the light scattering polydispersity. For a perfectly uniform sample, the PDI would be 0.0 (Kale and Deore, 2017).

14.13.6 Phase Analysis

To determine the type of nanoemulsion that has formed, the phase system (o/w or w/o) of the nanoemulsions is determined by measuring the electrical conductivity using a conductometer (Mishra et al., 2014).

14.13.7 Viscosity Measurement

The rheological properties of the microemulsions can be evaluated by viscometer. The viscosity determination confirms whether the system is w/o or o/w. A low viscosity system shows o/w type emulsion whereas a high viscosity system describes the w/o type (Ho Ching and Gao, 2009; Vavia et al., 2007).

14.13.8 Transmission Electron Microscopy Study

The morphology and structure of the nanoemulsion can be studied using transmission electron microscopy. A combination of bright field imaging at increasing magnifications and diffraction modes can be used to reveal the form and size of the nanoemulsion droplets. Observations were performed as a drop of the nanoemulsion was directly deposited on the holey film grid and again after drying (Mishra et al., 2014).

14.14 Risk of Health Hazards and Effect on the Environment

The toxic surfactants used to formulate cosmetic emulsion may cause health hazards. Prior to the production of emulsion-based cosmetics, it is a worthwhile to verify the list of generally recognized as safe (GRAS) materials published by the United States (US) FDA. There are several alternative measures that can be taken to abolish the harmful properties of surfactants. Polymeric surface-active agents, for example acrylic acid copolymer and hydroxypropyl cellulose (solid form), are the most desirable nontoxic alternatives (Ozgün, 2013; Sowjanya and Bandhavi, 2012).

14.15 Future Perspective

Cosmetic manufacturers are facing more and more challenges day-to-day in searching for new environmental and health hazard-free surfactants to increase product performance and functionality. Emulsification technology creates a good delivery mechanism for the immiscible cosmetic agents. Recently, there was a trend to prepare alcohol-free, water-based perfumes and fragrances which were more environmentally-friendly and reduced the use of volatile organic chemicals. In growing emulsion technology, the introduction of new w/o emulsifiers make products acceptable to the users and restrict greasiness. Silicone copolymers and modified carbomer copolymers can be incorporated into the cosmetic formulation to enhance skin moisturization and product performance.

14.16 Conclusion

Microemulsion-based cosmetics are highly expedient and up-to-standard for formulating skin care, including sunscreen formulation. They are capable of attaining a high sun protection factor (SPF) and have the ability to penetrate skin and increase skin retention. Hair care products prepared by microemulsion technology exert good comprehension, which was lacking in comparison with the earlier, conventional formulations. Furthermore, clear emulsions are useful with regard to consumer compliance.

REFERENCES

Adhirajan N, Dixit VK, Chandrakasan G. Development and evaluation of herbal formulations for hair growth. Indian Drugs. 2001;38(11):559–63.

Akhtar N, Ahmad M, Shoaib Khan HM, Akram J, Gulfishan G, Mahmood A, Uzair M. Formulation and characterization of a multiple emulsion containing 1% L-ascorbic acid. Bull Chem Soc Ethiop. 2010;24(1): 1–10.

Asuman B, Ongun MS. Multiple emulsions https://www.ajol.info/index.php/bcse/article/view/52955/41554. Hoboken, NJ: John Wiley and Sons, Inc.; 2008. 293–306 p. Available from: eu.Wiley.com.

Azeem A, Rizwan M, Ahmad FJ, Khan ZI, Khar RK, Aqil M, Talegaonkar S. Emerging role of microemulsions in cosmetics. Recent Pat Drug Deliv Formul. 2008;2(3):275–89.

Bachhav YG, Patravale VB. Microemulsion based vaginal gel of fluconazole: Formulation, in vitro and in vivo evaluation. Int J Pharm. 2009;365(1):175–9.

Bancroft WD. The theory of emulsification, V. J Phys Chem. 1913;17(6):501–19.

Becher P, Hamor WA. Emulsions: theory and practice. New York: Reinhold; 1957.

Bianco A. Graphene: safe or toxic? The two faces of the medal. Angewandte Chemie International Edition. 2013 May 3;52(19):4986-97.

Boom RM. Emulsions: Principles and preparation. J Food Sci. 2008:305–39, Springer, New York, NY.

Carlotti M, Gallarate M, Sapino S, Ugazio E, Morel S. W/O/W multiple emulsions for dermatological and cosmetic use, obtained with ethylene oxide free emulsifiers. J Dispers Sci Technol. 2005;26(2):183–92.

Chaudhri N, Soni GC, Prajapati SK. Nanotechnology: An advance tool for nano-cosmetics preparation. IJPRR. 2015;4(4):28–40.

Dahms GH, Tagawa M. Novel multiple phase emulsions for stable incorporation of vitamin C derivatives and enzymes. In Proceedings of the 19th IFSCC Congress,Society of Cosmetic Chemists, Sydney; 1996. pp. 79–90. Available from: https://books.google.co.in/books?id=Im4oDwAAQBAJ&pg=PA424&lpg= PA424&dq=GH+Dahms,+M+Tagawa+-+Proceedings+of+the+19th+IFSCC+Congress,+Sydney,+1996 &source=bl&ots=yXto-9H08B&sig=3pSjdttNUjXlE2YT8F2Ekck8FLM&hl=en&sa=X&ved=2ahUK EwiQjr7jpOvfAhUQfSsKHZwIAT4Q6AEwAHoECAEQAQ#v=onepage&q=GH%20Dahms%2C%20 M%20Tagawa%20-%20Proceedings%20of%20the%2019th%20IFSCC%20Congress%2C%20 Sydney%2C%201996&f=.

Date AA, Nagarsenker MS. Design and evaluation of self-nanoemulsifying drug delivery systems (SNEDDS) for cefpodoxime proxetil. Int J Pharm. 2007;329(1):166–72.

Devarajan V, Ravichandran V. Nanoemulsions: As modified drug delivery tool. IJCP. 2011;4(1):1–6.

Dureja H, Kaushik D, Gupta M, Kumar V, Lather V. Cosmeceuticals: An emerging concept. Indian J Pharmacol. 2005;37(3):155.

Even WR, Gregory DP. Emulsion-derived foams preparation, properties, and application. Mrs Bulletin. 1994;19(4):29–33.

Faure B, Salazar-Alvarez G, Ahniyaz A, Villaluenga I, Berriozabal G, De Miguel YR, Bergström L. Dispersion and surface functionalization of oxide nanoparticles for transparent photocatalytic and UV-protecting coatings and sunscreens. Sci Technol Adv Mater. 2013;14(2):23001.

Finkle P, Draper HD, Hildebrand JH. The theory of emulsification 1. J Am Chem Soc. 1923;45(12):2780–8.

Garti N, Bisperink C. Double emulsions: Progress and applications. Curr Opin Colloid Interface Sci. 1998;3(6):657–667.

Ghosh S, Coupland JN. Factors affecting the freeze–thaw stability of emulsions. Food Hydrocoll. 2008;22(1):105–11.

Goto M, Miyata M, Kamiya N, Nakashio F. Novel surfactant-coated enzymes immobilized in poly (ethylene glycol) microcapsules. Biotechnol Tech. 1995;9(2):81–4.

Hanai T. Electrical Properties of Emulsions. Emulsion Sci. 1968: 354–377. Edited by Philip Sherman, London; New York: Academic Press.

Ho CJ, Gao JY. Preparation and thermophysical properties of nanoparticle-in-paraffin emulsion as phase change material. Int J Heat Mass. 2009;36(5):467–70.

Jansen T, Hofmans MPM, Theelen MJG, Manders F, Schijns VEJC. Structure-and oil type-based efficacy of emulsion adjuvants. Vaccine. 2006;24(26):5400–5.

Kabri TH, Arab-Tehrany E, Belhaj N, Linder M. Physico-chemical characterization of nano-emulsions in cosmetic matrix enriched on omega-3. J Nanobiotechnol. 2011;9(1):41.

Kale SN, Deore SL. Emulsion micro emulsion and nano emulsion: A review. Sys Rev Pharm. 2017;8(1):39.

Klucker M-F, Dalençon F, Probeck P, Haensler J. AF03, an alternative squalene emulsion-based vaccine adjuvant prepared by a phase inversion temperature method. J Pharm Sci. 2012;101(12):4490–500.

Kumar S, Kim DW, Lee HJ, Changez M, Yoon TH, Lee JS. Exploration of the mechanism for self-emulsion polymerization of amphiphilic vinylpyridine. Macromolecules. 2013;46(18):7166–72.

Lemos-Senna E, Wouessidjewe D, Lesieur S, Duchene D. Preparation of amphiphilic cyclodextrin nanospheres using the emulsification solvent evaporation method. Influence of the surfactant on preparation and hydrophobic drug loading. Int J Pharm. 1998;170(1):119–28.

Luther W. Industrial application of nanomaterials: Chances and risks: Technology analysis. Dusseldorf: Future Technologies Division of VDI Technologiezentrum; 2004. 1–119 p.

Madaan V, Chanana K, Kataria MK, Bilandi A. Emulsion technology and recent trends in emulsion applications. Int Res J Pharm. 2014;5(7):533–42.

Masmoudi H, Le Dréau Y, Piccerelle P, Kister J. The evaluation of cosmetic and pharmaceutical emulsions aging process using classical techniques and a new method: FTIR. Int J Pharm. 2005;289(1):117–31.

Matsumoto S, Kita Y, Yonezawa D. An attempt at preparing w/o/w multiple-phase emulsions. J Colloid Interface Sci. 1976;57(2):353–61.

Mihranyan A, Ferraz N, Strømme M. Current status and future prospects of nanotechnology in cosmetics. Prog Mater Sci. 2012;57(5):875–910.

Millikan LE. Cosmetology, cosmetics, cosmeceuticals: Definitions and regulations. Clin Dermatol. 2001;19(4):371–4.

Mishra RK, Soni GC, Mishra RP. A review article: On nanoemulsion. World J Pharm Pharm Sci. 2014;3(9):259.

Mora-Huertas CE, Fessi H, Elaissari A. Influence of process and formulation parameters on the formation of submicron particles by solvent displacement and emulsification–diffusion methods: Critical comparison. Adv Colloid Interface Sci. 2011;163(2):90–122.

Müller RH, Radtke M, Wissing SA. Solid lipid nanoparticles (SLN) and nanostructured lipid carriers (NLC) in cosmetic and dermatological preparations. Adv Drug Deliv Rev. 2002;54:S131–55.

Nazir H, Lv P, Wang L, Lian G, Zhu S, Ma G. Uniform-sized silicone oil microemulsions: Preparation, investigation of stability and deposition on hair surface. J Colloid Interface Sci. 2011;364(1):56–64.

Negi P, Singh B, Sharma G, Beg S, Raza K, Katare OP. Phospholipid microemulsion-based hydrogel for enhanced topical delivery of lidocaine and prilocaine: QbD-based development and evaluation. Drug Deliv. 2016;23(3):941–57.

Nisisako T. Microstructured devices for preparing controlled multiple emulsions. Chem Eng Technol. 2008;31(8):1091–8.

Ohyama M. Management of hair loss diseases. Dermatol Sin. 2010;28(4):139–45.

Omotosho JA, Whateley TL, Law TK, Florence AT. The nature of the oil phase and the release of solutes from multiple (w/o/w) emulsions. J Pharm Pharmacol. 1986;38(12):865–70.

Özgün S. Nanoemulsions in cosmetics. Anadolu Univ. 2013;1(6):3–11.

Palanisamy P, Raichur AM. Synthesis of spherical NiO nanoparticles through a novel biosurfactant mediated emulsion technique. Mater Sci Eng C. 2009;29(1):199–204.

Patel AR, Vavia PR. Preparation and in vivo evaluation of SMEDDS (self-microemulsifying drug delivery system) containing fenofibrate. AAPS J. 2007;9(3):E344–52.

Patel AR, Schatteman D, De Vos WH, Dewettinck K. Shellac as a natural material to structure a liquid oil-based thermo reversible soft matter system. RSC Adv. 2013;3(16):5324–7.

Patel D, Sawant KK. Oral bioavailability enhancement of acyclovir by self-microemulsifying drug delivery systems (SMEDDS). Drug Dev Ind Pharm. 2007;33(12):1318–26.

Patil S, Sandberg A, Heckert E, Self W, Seal S. Protein adsorption and cellular uptake of cerium oxide nanoparticles as a function of zeta potential. Biomaterials. 2007;28(31):4600–7.

Purwal L, Gupta SPBN, Pande SM. Development and evaluation of herbal formulations for hair growth. J Chem. 2008;5(1):34–8.

Shafiq S, Shakeel F, Talegaonkar S, Ahmad FJ, Khar RK, Ali M. Development and bioavailability assessment of ramipril nanoemulsion formulation. Eur J Pharm Biopharm. 2007;66(2):227–43.

Shafiq-un-Nabi S, Shakeel F, Talegaonkar S, Ali J, Baboota S, Ahuja A, Khar RK, Ali M. Formulation development and optimization using nanoemulsion technique: A technical note. AAPS PharmSciTech. 2007;8(2):E12–7.

Shaji J, Reddy MS. Microemulsions as drug delivery systems. Pharma Times. 2004;36(7):17–24.

Sharma S, Sarangdevot K. Nanoemulsions for cosmetics. IJARPB. 2012;1(3):408–15.

Singh V, Kataria MK, Bilandi A, Sachdeva V. Recent advances in pharmaceutical emulsion technology. J Pharm Res. 2012;5(8):4250–8.

Sinha VR, Kumar A. Multiple emulsions: An overview of formulation, characterization, stability and applications. Indian J Pharm Sci. 2002;64(3):191.

Sonakpuriya P, Bhowmick M, Pandey GK, Joshi A, Dubey B. Formulation and evaluation of multiple emulsion of valsartan. Int J Pharmtech Res. 2013;5(1):132–46.

Sowjanya GN, Bandhavi P. Nanoemulsions an emerging trend: A review. IJPRD. 2012;4(6):137–52.

Thorat RM, Jadhav VM, Kadam VJ. Development and evaluation of polyherbal formulations for hair growth-promoting activity. Int J PharmTech Res. 2009;1(4):1251–4.

Tortora GJ, Grabowski SR. Principles of anatomy and physiology. 8th ed: Harper Collins College Publishers, John Wiley & Sons, Inc; 1996.

Trotta M, Debernardi F, Caputo O. Preparation of solid lipid nanoparticles by a solvent emulsification–diffusion technique. Int J Pharm. 2003;257(1):153–60.

Vyas SP, Khar RK. Submicron emulsions, eds. Targeted and controlled drug delivery novel carrier systems. New Delhi: CBS Publishers; 2002. 282–2 p.

Wu J, Lu J, Wilson C, Lin Y, Lu H. Effective liquid–liquid extraction method for analysis of pyrethroid and phenylpyrazole pesticides in emulsion-prone surface water samples. J Chromatogr A. 2010;1217(41):6327–33.

Yang J, Xu H, Wu S, Ju B, Zhu D, Yan Y, Wang M, Hu J. Preparation and evaluation of microemulsionbased transdermal delivery of Cistanche tubulosa phenylethanoid glycosides. Mol Med Rep. 2017;15(3):1109–16.

15

Nanotechnology in Cosmetics: Safety Evaluation and Assessment

Sanjoy Kumar Das, Rajan Rajabalaya, and Sheba Rani N. David

CONTENTS

ABSTRACT: Nanotechnology in cosmetics assists consumers in looking more attractive through enhanced delivery with intradermal penetration. The application of materials in the nanometric range is responsible for the enhancement of physical, chemical, and biological properties as compared to that of coarse particles. The use of nanoparticles in cosmetic products needs a lot of attention in terms of its assessment and evaluation as cosmetics are in direct contact with the skin. This chapter discusses the determination of the physicochemical properties of nanoparticles for safety assessments and toxicological studies, along with international test guidelines. During the preparation of cosmetics, the assessment of the ingredient/s should be done very carefully so that the final products do not produce unintended harmful effects. After launching cosmetic products in the market, reevaluation is also necessary as safety assessment is a continuous process. Raw materials and other ingredients used in the cosmetic preparation should be properly selected based on characterizations of these materials using different analytical techniques. Finally, this chapter discusses international regulatory guidelines and the post-marketing analysis of cosmetic products in different countries. It will be useful for the cosmetic industry and their product safety evaluations during development, manufacturing, and submission of the products for approval.

KEY WORDS: *Nanotechnology, cosmetics, safety assessment, safety evaluation, toxicological studies*

15.1 Introduction

Nanotechnology uses the materials within 1 nm and 100 nm in size. The branch of nanotechnology reveals novel and considerably enhanced physical, chemical, and biological properties, phenomena, and procedures due to applications of materials in the nanometric range (Wang, 2018). The application of nanotechnology in the cosmetic field and in health products started nearly 40 years ago with liposomal moisturizing creams (Chaudhri et al., 2015). Cosmetic products containing nanoparticles exhibit a more attractive color and an enhanced intradermal penetration profile (Montenegro et al., 2016; Morganti, 2010). Cosmetic manufacturers' have incorporated nanoparticles in shampoos, toothpastes, anti-aging and anti-cellulite creams, fairness creams, moisturizers, face powders, aftershave lotions, body perfumes, sunscreens, makeup products, and nail polishes (Melo et al., 2015). The various kinds of nanomaterials used in cosmetics include liposomes, nanoemulsions, solid lipid nanoparticles, nanocrystals, nanosilver and nanogold, dendrimers, cubosomes, hydrogels, and fullerenes such as buckyballs (Raj et al., 2012). Among various industries in personal and health care, cosmetic industry recognized the prospects of utilizing nanotechnology in their products, in an advanced way than others (Ajazuddin et al., 2015). According to a market analysis done by the Woodrow Wilson Project on emerging nanotechnology, as at October 2013, 1628 products or product lines are produced. Among various categories of these products, health and fitness was considered as the main and largest category containing 788 products. This category was further divided into the sub-categories like personal care, clothing, cosmetics, sporting goods, filtration and sunscreen. It was observed that cosmetics and sunscreen products sub-category contains 154 and 40 products respectively. Personal care was the largest sub-category with 292 products (Nanotechproject, 2018). More than sixteen hundred products had been produced. Of these, health and personal care is the main category, covering almost 50% of the products produced (788 products). Of these 788 items, products are classified as nanotechnology products for cosmetic use and an additional 40 products under the sunscreen category. The total number of cosmetic nanotechnology products increased by nearly 5.16% between the years 2006–2013 (Ajazuddin et al., 2015). The Federal Food, Drug and Cosmetic Act defines cosmetics as "articles intended to be rubbed, poured, sprinkled, or sprayed on, introduced into, or otherwise applied to the human body for cleansing, beautifying, promoting attractiveness, or altering the appearance" (FDA, 2017). This formalization covers various products such as skin moisturizers, perfumes, lipsticks, fingernail polishes, eye and facial makeup preparations, shampoos, permanent hair waves, hair colors, toothpaste, deodorants, and the materials used in the production of cosmetics. Except for the organoleptic additives imparting color, pre-approval from the Food and Drug Administration (FDA) is not available for other products or ingredients used in cosmetic product manufacturing. However, cosmetic manufacturers bear the responsibility of marketing the products, maintaining proper safety and labelling, avoiding prohibited ingredients, and staying within the limits on restricted ingredients (FDA, 2017). Product safety is considered a very important issue in the cosmetic and personal care manufacturing industries (Cosmeticsinfo, 2016). Mathematically it is difficult to define the link between benefit and risk, possibly due to the difficulties associated with the measurement of benefits obtained from cosmetics products. It is not adequate to measure the proportion of risk and benefits based on adverse effects only; instead, a qualitative investigation of risk is required. For proper risk assessment it is important to consider the severity of the adverse effects associated with a particular cosmetic product, rather than just expressing it in terms of numerical incidence. For proper consumer use, the cosmetic product must be sufficiently authentic in terms of not being harmful to human health in both specified usage and foreseeable condition (Masuda and Harada, 2017). However, one can never claim that the cosmetic products are free from adverse effects during prolonged use (Kojima, 2017).

All the products whose labels contain mention of "nanoparticles" impart questions about the safety of those particular products. Unlike pharmaceutical products, there is a lack of regulation regarding the proper evaluation and testing of cosmetic products before they are introduced into the market to be sold. Although nowadays, according to the claim of several cosmetic manufacturers, safety-oriented research has been started in cosmetic field companies, but the outcomes of this research is kept under a high degree of privacy and is not available to the public (Buzea et al., 2007). This chapter discusses the safety evaluation and assessment of nanotechnology in cosmetics.

15.2 Safety Evaluation of Nanomaterials in Cosmetic Products

The ingredients intended to be used in the manufacture of cosmetic preparations should undergo a thorough assessment procedure to ensure that they are suitable for consumer safety upon the use of that particular product. It is difficult to evaluate possible intercomponent interactions by only performing assessments of single components alone. Performing the safety evaluation of the ingredients and finished products both are variable processes based on the intended use of the product and the consumer exposure condition (Masuda and Harada, 2017). The characterization of nanomaterials is considered a vital part of the safety assessment of those nanomaterials involved in cosmetic products. Essential characterization should be performed, including particle size, shape, surface area, surface charge, surface morphology, rheology, porosity, crystallinity, and amorphocity on primary nanoparticles, agglomerates, and/or aggregates (SCCP, 2007a). Nanoparticle characterization in the finished products is a complex task due to their opaqueness in nature, and therefore, during characterization, pre-treatment of the substances present in the products is necessary, which may produce changes in the viscosity, aggregation/agglomeration, and pH of the finished products (Lu et al., 2015). Therefore, a few instrumental techniques are briefly outlined for the determination of the physicochemical properties and impurities present in nanocosmetics.

15.3 Determination of Physicochemical Properties

15.3.1 Physical Properties

15.3.1.1 Particle Size Measurement

Different methods are used for the measurement of particle size, including transmission electron microscopy (TEM), scanning transmission electron microscopy (STEM), X-ray diffraction (XRD), extended X-ray absorption fine structures (EXAFS), probe molecule absorption, magnetic measurements, optical absorption, and small angle neutron scattering (SANS) (Pyrz and Buttrey, 2008). TEM has been employed for the determination of nanoparticle size measurements (Bonevich and Haller, 2010) and it provides a quantitative analysis of shape, size, and size distribution of the particles (Bantz et al., 2014). The disadvantage of TEM is that the data is always based on single particle measurements. Apart from this, TEM needs vacuum conditions to reach a reasonable resolution (Bantz et al., 2014). Lu et al. (2015) measured the particle size of three commercial sunscreen sprays using conventional TEM, and it was observed that the sizes of titanium dioxide (TiO_2) nanoparticles in the commercial sunscreen sprays were in the range of 23.9–27.9 nm. STEM has been proven to be more advantageous than traditional TEM as it offers the chance of obtaining both bright and dark field images, along with the ease of regulating the range of detectors during analysis (Kempen et al., 2013). XRD analysis is most frequently used to measure particle size and crystallite-structure in polycrystals in the nano-range. This analysis additionally gives information about the nanostructures, because the width and shape of the reflections yield information on the substructure of the nanomaterials (Dorofeeva et al., 2012). EXAFS is also one of the best types of equipment to study both the atomic and electronic structure of small ensembles, due to its local structure sensitivity and spatial resolution. It is useful for the measurement of coordination numbers, bond lengths, disorder up to the fifth coordination shell, size, and shape, such as the different shapes of oblate, raft-like, or truncated polyhedral in nanoparticles. It also useful to measure the surface morphology and disorder of 1–2 nm clusters, which can be determined in various conditions such as temperature, alloy composition, and substrate material (Frenkel et al., 2011). Conventional optical microscopy has been proven to be inadequate for size determination of nanometric materials due to its inability to resolve particles situated closer than 200 nm spatial distance. Near-field scanning optical microscopy (NSOM) can impart better performance for the size analysis of such kinds of nanoparticles. It is a surface probe microscopy technique which adopts a combination of principles from both STEM and optical microscopy (Lin et al., 2014). NSOM provides the opportunity for optical imaging of the particles separated by less than 2 nm spatial distance in a non-destructive manner (Wabuyele et al., 2005) that increases detection sensitivity approximately 4,000 fold (Hamann et al., 1998). Generally, the current practice used to determine the average particle size, particle size distribution, and polydispersity index (PDI) of a nanoparticle sample is dynamic light scattering (DLS), as it is an economical and fast methodology. DLS can be used to examine whether acceptable nanoparticle dispersion has been obtained or whether particle aggregates/agglomerates exist in the system (Lu et al., 2018). Particle size analysis of nanoparticles, colloids, emulsions, and submicron suspensions were carried out using this instrument. It is able to demonstrate, based on the physical properties of the sample, the dynamic range from 0.3 nm to 8 µm (Horiba, 2018). The particle size of the ethosomal vesicle (Iizhar et al., 2016) cosmetic nanoemulsion was measured by using the DLS method (Riberio et al., 2015). The separation and quantification of nanoparticles with different particle sizes can be measured using high-performance liquid chromatography (HPLC) with size exclusion chromatography (Mu and Sprando, 2010).

15.3.1.2 Shape

Variability in the shape of nanoparticles, like spheres, tubes, and sheets, can be considered a major health hazard in nanoparticles (Raj et al., 2012). The various shapes of particles are used to describe the geometry of the nanoparticles. The correlation between width and height can be used as an aspect ratio. The smallest possible aspect ratio of a circle is 1:1. Lu et al. (2015) worked on three commercial samples to observe the shape of the nanoparticles present in them; needle-shaped TiO_2 particles were detected in sample one and roundish TiO_2 particles were detected in the other two samples. TiO_2 nanoparticles with an aspect ratio of 4.09 ± 1.04, 1.72 ± 0.62, and 1.70 ± 0.54 were observed in samples one to three, respectively. Roundish zinc oxide (ZnO) particles with an aspect ratio of 1.35 ± 0.36 were also found in sample one. The shapes of the TiO_2 structures were almost like those of circles, needles, and lances. The shapes of the ZnO structures included rod-like and isometrically-shaped particles (Lu et al., 2015). The shape of the nanostructures can affect the cellular response towards them (Banerjee, 2017). The cellular uptake of rod-shaped, gold nanoparticles is lower than spherical, gold nanoparticles (Chithrani et al., 2006). Tak et al. (2015) reported that the shape (spherical, rod-shaped, and triangular-shaped) was the dependent variable in skin penetration and the comparative bactericidal property of the silver nanoparticles. The study was conducted by administering the nanoparticles through two different pathways; one was the follicular penetration pathway (entering through the hair route) and another one was the intercellular penetration pathway (entering through the gaps between two consecutive corneocytes). An *in vitro* skin permeability study was conducted employing Franz diffusion cells (containing phosphate buffer with pH 7.0 as the receptor fluid), which revealed that an abnormal lag time of 8 hours was observed in the case of rod-shaped nanoparticles and triangular-shaped nanoparticles. Furthermore, the amount of silver that penetrated from differently shaped nanoparticles was in decreasing order from rod-shaped nanoparticles (1.82 µg/cm^2) to spherical-shaped nanoparticles (1.17 µg/cm^2) to triangular-shaped nanoparticles (0.52 µg/cm^2). The comparative study of shape-dependent bactericidal action of silver nanoparticles exposed that truncated triangular, silver nanoparticles showed maximum bactericidal action as compared to the other two shapes. Spectrophotometric analysis was carried out that exhibited the existence of three distinct pictures at different wavelengths, indicating the presence of three differently shaped silver nanoparticles. The TEM study showed that the mean diameter of the spherical and rod-shaped nanoparticles was 50 nm and 20 nm respectively, whereas the average length of the rod-shaped nanoparticles was 50 nm. Triangular nanoparticles (TNPs) were observed as 2 nm thick, equilateral, triangular plates with an average side length of 50 nm (Tak et al., 2015). The summary of their studies and results are shown in Figure 15.1.

15.3.1.3 Surface Area

The application of nanoparticles in cosmetics as dispersion aids work well due to their small diameter and high surface area, exhibiting enhanced catalytic activity (Hosokawa et al., 2007). The Brunauer, Emmett, and Teller (BET) method is usually employed to determine the surface area of solid nanoparticles. The principle of this method is based on the physisorption of a single layer made up of the gas molecules on the surface of the nanoparticles at a definite pressure and temperature. The gas molecules used may be either liquid nitrogen or argon (Contado, 2015). High vacuum conditions are a prerequisite criterion for the sample preparation prior to the measurement. Apart from nitrogen (which is most abundantly used), several other gases such as argon, carbon dioxide, or krypton can also be employed.

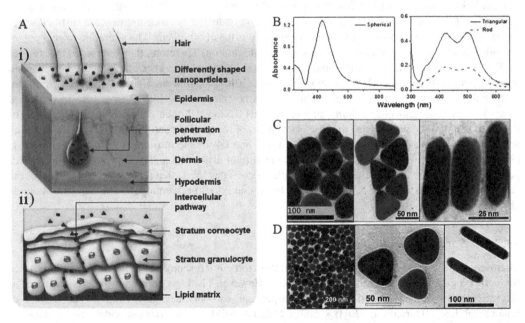

FIGURE 15.1 (A) Skin penetration pathways: (i) enters via hair follicles, and (ii) the intercellular penetration pathway. (B) Absorption spectra of solutions containing chemically synthesized spherical nanoparticles, triangular nanoparticles, and rod-shaped nanoparticles. (C) TEM images of synthesized AgNPs. (D) TEM images of differently shaped AgNPs after dispersing them in phosphate buffer (pH 7.0). (Adapted from Tak et al. (2015).)

In some situations, it is observed that the surface area of the particles to be measured are present in agglomerated condition, which affects the outcome of the particle surface area measurement procedure to a small extent, probably due to their very small dimensions. It is also found that the small size of these adsorbate molecules is marginally affected by the degree of agglomeration of the measured surface area (Powers et al., 2006). The smaller silver nanoparticles (<10 nm) can penetrate into the cell to prevent the growth of microbes. Owing to their higher surface area the minute and uncharged silver nanoparticles possess the ability to directly interact with the cell membrane. As a consequence, this eases the penetration of a higher number of nanoparticles in entering the cell. Choi and Hu (2008) observed that particles of a size less than 5 nm are more subjected to microbial inhibition, but this outcome was not same for the other sizes (10, 15, and 20 nm) (Choi and Hu, 2008). Instinctively, the outer surface area of a nanoparticle exhibits a more significant role than the inner surface in stimulating a definite cell response (Mihranyan et al., 2012).

15.3.1.4 Surface Charge

The zeta (ζ) potential indicates the charge possessed by the particle surface for all materials, especially in the case of particles dispersed in a fluid medium which is related to the surface charge the materials possess or acquire when suspended in that fluid medium (Particle Sciences, 2012). The zeta potential and the surface charge density are the characteristics of the particles that can be directly associated with the stability of the colloidal system and *in vivo* performance of the particles (Badri et al., 2015). High zeta potential values impart significant steric stability towards the particles and aid them in maintaining the individual existence of the nanoparticles (Bhatia, 2016). Emulsion systems of various types, like oil-in-water (o/w), water-in-oil (w/o), oil-in-water-in-oil (o/w/o), and water-in-oil-in-water (w/o/w), are used by cosmetic product developers (Gasparelo et al., 2014). Currently, a technique has been developed with employing the combination strategy of acoustic spectroscopy and electro-acoustics spectroscopy, which provides an opportunity to assess the surface charge of the cosmetic emulsions (applicable for o/w and w/o emulsion) without diluting the emulsion, giving stability and the ability to performance assess cosmetic emulsions without (Fairhurst et al., 2013). Freitas and Muller (1998) determined the surface charge of solid lipid nanoparticles (SLNs) as zeta potential using a zeta-sizer. They took an SLN formulation comprising 10% compritol and 1.2% pluronic F68 and conducted the measurements in distilled water adjusted with sodium chloride to obtain a conductivity value of 50 µS/cm. The alteration in zeta potential value is specific to the alteration in structural configuration. Therefore, they concluded that further studies regarding the structure of the lipid phase need to be executed to check the proposed mechanism of gel formation (Freitas and Muller, 1998). Kanazawa (2010) prepared casein nanoparticles with a mean particle diameter of 10 nm to 300 nm, wherein zeta potential is positive. At first, they incorporated casein into an acidic aqueous medium ranging from pH 0.5 to pH 7, then gradually increased the pH of the solution above the pH value of the isoelectric point of the casein, continuously stirring the solution, and finally added an active substance (ionic substance or a fat-soluble substance). Zeta potential value is an important parameter for the assessment of the quality of dispersion and/or aggregation properties of particles, their interaction, and their surface modification. They found that the zeta potential value of the casein nanoparticle was positive, and they concluded that 3 to 30 mV zeta potential value was suitable (Kanazawa, 2010).

15.3.1.5 Surface Morphology

Microscopic techniques like TEM, SEM, and atomic force microscopy (AFM) are commonly utilized for morphological characterization of nanoparticles. However, TEM has greater advantages compared to other techniques, which is related to the use of suspensions without the need for additional purification to remove stabilizers (Durán et al., 2011). Usually, TEM and SEM are the two most widely used techniques to study the surface morphology (using SEM) as well as internal structure (using TEM) of the nanoparticles. TEM is based on the measurement of the scattering induced alteration of the electron beam when it is passed through the sample, whereas in SEM analysis, electrons beams are utilized for the purpose of scanning and consequent imaging of the particle surface in a raster scan arrangement. When electrons pass through the sample they interact with atoms resulting in the generation of the signals carrying information regarding the sample's external features, composition, and additional properties, such as electrical conductivity (Liu et al., 2010). X-ray photoelectron spectroscopy (XPS), XRD, and Fourier transform infrared spectroscopy (FTIR) are employed to examine the particle's surface characteristics (Mu and Sprando, 2010). The hydrophobic nature of the surface can be measured by using different techniques, for example, hydrophobic interaction chromatography, biphasic partitioning, adsorption of probes, and contact angle measurements, etc. Currently, X-ray photon correlation spectroscopy has been introduced, which provides the relevant data regarding surface hydrophobicity along with the identification of specific chemical moieties situated in the particle's surface (Bhatia, 2016).

15.3.1.6 Rheology of Nanoparticles in Cosmetics

Generally, rheological studies on pharmaceutical and cosmetic constituents provides information regarding the basic nature of a system for the quality control of raw ingredients, finished products, and manufacturing processes such as mixing, pumping, packaging, and filling, and to observe the influence of different factors such as formulation, storage period, and temperature on the quality and acceptance of a finished product. The consistency of the materials used in cosmetic preparation may lie in a variable range, from fluid to solid. Rheological characterization of the semisolid products is mostly difficult, owing to the combination of both the properties of solid and liquid (Herh et al., 1998). Rotational viscometers are most commonly used for the determination of flow curves expressing the relationship of shear stress to rate of shear between suitable limits (Barry and Warburton, 1968). The stability of a product depends upon the alteration in the consistency with the consequent changes in shear rate. The selection of appropriate rheological additives as well as temperature regulation is an important parameter to give the system suitable flow properties and thermostable rheology in the product. Rheological additives from Elementis Specialties are classified as aqueous-phase thickeners (e.g., BENTONE® EW, BENTONE® EW CE, and BENTONE® MA, etc.) and non-aqueous-phase thickeners (e.g., BENTONE®27 V, BENTONE® 27 V CG, and BENTONE® 38 V, etc.). BENTONE® (hydrophilic clay) is used as a rheological additive (e.g., hectorite) because it is non-abrasive in nature, making the skin feel pleasantly silky, quite different from the majority of cellulose, polymer, and polysaccharide-based thickeners. The highest degree of thixotropy in aliphatic liquids makes THIXCIN® additive (e.g., trihydroxy stearin) the product to use. In addition to its effective rheological properties, it acts as a water repelling agent, emulsion stabilizer, and stiffening agent in lipsticks and ointments (Elementis-specialities, 2018). Nanoemulsions are commonly considered to be in the size range of less than 100 nm in diameter (Sarker, 2005). In the cosmetic industry, nanoemulsions are widely used in the preparation of various products (e.g., bath oils, body creams, and anti-wrinkle and anti-aging preparations) owing to their higher spreading coefficient value in comparison with conventional emulsions. Higher spreading coefficient value leads to the proper spreading of the formulation containing the nanoemulsion in a better way. Besides the above-mentioned facts, transparent visual aspect, hydrating power, and good skin feel make nanoemulsions one of the most popular choices in the cosmetic industry (Wu and Guy, 2009). The rheological study is commonly done using a stress rheometer (Alves et al., 2011). Aswal et al. (2013) prepared and evaluated a cosmetic cream containing polyherbal extracts from natural products such as *Aloe vera*, *Cucumis sativus*, and *Daucus carota* to obtain a multipurpose effect including whitening, anti-wrinkle, anti-aging and sunscreening of the skin. The viscosity of the cream was measured by a Brookfield viscometer, working at 100 rpm, employing spindle number seven. It was observed that the measured viscosity lay in the range of 28,001–27,025 cps, which indicates the easy spreadability of the cream upon the application of a small amount of shear (Aswal et al., 2013). A Ford viscosity cup or Brookfield viscometer are considered to be useful in the measurement of the viscosity of creams and lotions (Nema et al., 2009). In the case of non-Newtonian materials, the Ford viscosity cup can be employed in accordance to test method ASTM D2196 (ASTM International, 2015). Measurement of the viscosity was performed at $77 \pm 35.6°F$ with a Ford viscometer (Viswanath et al., 2007). The apparent viscosity, shear thinning property, and thixotropic behavior in the case of non-Newtonian materials can be estimated using a rotational viscometer operating at a shear rate ranging from 0.1 to 50 s^{-1} in a fluid of "infinite" dimensions (ASTM International, 2015). Kabri et al. (2011) prepared a cosmetic matrix made up of nanoemulsion with long-chain, polyunsaturated fatty acids

(LC-PUFA) and evaluated the rheological properties of the prepared formulation using a StressTech rheometer equipped with a C40.4 cup. They observed that all the formulations showed a Newtonian flow property. The coefficient of viscosity was measured by plotting shear rate against shear stress (Kabri et al., 2011).

15.3.1.7 Porosity, Crystallinity, and Amorphocity

The porosity, crystallinity, and amorphocity of the nanoparticles should be characterized. Silica can be classified into two main forms: crystalline and amorphous. Crystalline micron-sized silica is a principal component of soil, sand, granite, and many other minerals, while amorphous silica is synthetic, except for biogenic kieselguhr which is composed of particles or structural units in the nanometric size range (<1 μm in diameter). Synthetic amorphous silica particles are found in different forms, three of which can be used in skin care treatments and cosmetics. These include pyrogenic or fumed silica, non-porous silica nanoparticles, and mesoporous silica nanoparticles (Nafisi et al., 2015).

15.3.2 Chemical Properties

15.3.2.1 Impurities

Impurities present in cosmetic products comprised of nanosubstances are not the same as may be established in traditional formulations or materials. Nanoscale impurities may arise from the altered manufacturing procedure (Nanda et al., 2016). The type of impurity present in the cosmetic final product may vary depending upon the different solvents used, time/temperature settings, and variations to the preliminary substances. Sometimes dispersing agents and surface modifying agents used in nanoparticle production may be considered an impurity and should be taken into account during a safety test. In June 2014, guidance for the industry safety of nanomaterials for cosmetic products was issued by the United States Department of Health and Human Services FDA Center for Food Safety and Applied Nutrition (FDA, 2014). This guidance document was proposed to aid in supporting the nanomaterial production industries and other investors involved in the findings of safety profiles associated with nanocosmetics and developing an outline for the assessment of finished products. They also provided information so that anybody can contact the FDA regarding this guidance. Physicochemical properties of any cosmetic ingredients used in nanoscale should be completely described, including the name of the nanomaterial, chemical abstracts service (CAS) CAS number, structural formula, the elemental and molecular composition, including the degree of purity, and any known impurities or other additives that are used (FDA, 2014). Sometimes, due to the recurrent unattainability of the pure constituents for cosmetic raw materials, it is also necessary to determine the existence of impurities and their probable byproducts (Masuda and Harada, 2017). The cosmetic manufacturer uses a range of synthetic substances together with amidoamines. Amidoamines can be prepared as a result of reacting fatty acids or triglycerides with 3-dimethylaminopropylamine (DMAPA). Amidoamines are employed in personal care products and shampoos to impart reduced static electricity and to improve hair combability, shine, and texture. It also possesses surfactant properties, particularly for some individual materials (e.g., betaines). The incorporation of such materials may induce several abnormalities, like dermatitis. Normally DMAPA quantification is done by liquid chromatography-mass spectrometry (LC-MS) LC-MS, but it has several disadvantages like the specific detection limit ($0.5 \ \mu g \ g^{-1}$), the requirement for a large amount of solvent, and the consumption of more time. Zamora et al. (2011) evaluated the ability of ion mobility spectrometry (IMS) in the finding and quantification of DMAPA residues in stearamidopropyl dimethylamine (SAPDA) production samples. It possesses the potential advantages of less time consumption (less than 1 min) and sample requirement. They concluded that the proposed technique is simpler and more cost-effective, with good precision in comparison to LC-MS (Zamora et al., 2011). A reverse-phase, ion-pair HPLC method can be employed to estimate the qualitative determination of dequalinium chloride and its associated impurities in pharmaceuticals and cosmetic products (Gagliardi et al., 1998). Non-ionic surfactants are widely used in cosmetic products and 1,4-dioxane is present in the non-ionic surfactants as an impurity. The detection of 1,4-dioxane in cosmetic products is usually done by employing solid-phase microextraction (SPME) coupled with gas chromatography (GC) and gas chromatography-mass spectrometry (GC-MS). In comparison to the large time consumption of traditional extraction methods (1 hour), this SPME method takes about 10 minutes only. SPME also gives better analyses of 1,4-dioxane in non-ionic surfactants and cosmetics than traditional solvent extraction (Fuha et al., 2005). The maximum limits of impurity present in polyethylene glycol (PEG) and its derivatives, according to the recommendations prescribed by the German Cosmetic, Toiletry, Perfumery, and Detergent Association (IKW), are mentioned below (Fruijtier-Pölloth, 2005):

- 1,4-dioxane: Not more than 10 mg/kg (10 ppm)
- Ethylene and diethylene glycol: Not more than 0.3% (by weight)
- Ethylene oxide: Not more than 1 mg/kg (1 ppm)
- Heavy metals (combined): Not more than 10 mg/kg (10 ppm).

15.3.2.2 Chemical Composition

Information regarding the complete chemical composition of the nanoparticle must be presented. This information must also contain the purity, the nature of the impurities, coatings, or surface moieties, the doping material, the encapsulating materials, the processing chemicals, the dispersing agents, and any other additives or formulants, e.g., stabilizers. Different analytical methods such as ultraviolet (UV) visible spectroscopy (UV-Vis), HPLC, GC/LC-MS, atomic absorption spectrometry (AAS), inductively coupled plasma-mass spectroscopy (ICP-MS), FTIR, nuclear magnetic resonance (NMR) spectroscopy, and XRD, etc., are used for the determination of the chemical composition of the nanoparticles (SCCS, 2012). The United States FDA has approved safe concentration levels of lead (not more than 20 ppm) in mica intended for good manufacturing practice (GMP). The presence of Pb, Cd, As, and Hg was studied in a clear digested solution using an air-acetylene flame atomic absorption spectrophotometer by the standard calibration technique (Al-Qahtani et al., 2016). The usual toiletry formulations and cosmetics, such as deodorants, shampoos, hair conditioners, hair dyes and dye removers, liquid soaps, detergents, and other cleaning products, contain tolerable amount ranges between 1–10% of urea.

A less time consuming, relatively sensitive, and relatively inexpensive method for the quantitative estimation of water-soluble urea content in cosmetic products intended for use in dermatological complications was developed using a new spectrophotometric assay that employs water as the only extraction solvent (Bojic et al., 2008). The color additives used in the cosmetic formulation must obey the FDA's regulations and their prescribed limit of heavy metal impurities. The maximum limits of heavy metals in color additives, according to the FDA, are listed below (FDA, 2018):

- Arsenic: Not more than 3 ppm
- Lead: Not more than 20 ppm
- Mercury: Not more than 1 ppm.

Although there are not any regulations regarding chromium, the listing regulation for the color additive FD&C Blue No. 1 limits chromium contamination to 50 ppm. The presence of chromium as an impurity in cosmetics will not be recorded on the label (FDA, 2018).

15.3.2.3 Solubility

The solubility of a nanomaterial in suitable solvents and its partition coefficient is regarded as an important formulation factor in the case of cosmetic preparation. The rate of dissolution of nanomaterials of different degrees of solubility in a particular solvent system is also regarded as significant information. Proper studies regarding the hygroscopicity, solubility, and dissolution behavior in both aqueous and non-aqueous mediums must be performed (SCCS, 2012). TiO_2 nanoparticles can be regarded as insoluble even under physiological pH conditions due to its low degree of aqueous solubility, as well as significantly poor disassociation constant in the case of aqueous mediums. There is not a single method available for measuring the partition coefficient of nanoparticles containing a coating of organic materials yet. The distribution of these between polar as well as nonpolar substances should also be studied (SCCS, 2014). The absorption, distribution, and elimination behaviors of the particles are governed by their existence in molecular form, while the particles possessing delayed disassociation properties exist as particulate matter. A similar incident is observed in the case of ZnO and cadmium selenide particles (quantum dots) where biological properties of the particles are regulated by their disassociated ions. The dissolution behavior of the nanoparticles is dependent upon several factors, such as the composition of the medium, ionic strength, temperature, pH, and the presence of ligands, e.g., proteins, lipids, or formed elements in the blood, containing preferential susceptibility to binding of the nanomaterial and the ability to alter the clearance. The structural configuration of the particle surface along with the surface coating and agglomeration of nanoparticles are also considered important factors in regulating the dissolution profile. Surface architecture, coating, and agglomeration condition also affect the dissolution. There is a lack of proper standards to define solubility with respect to the dissolution rate. In the case of inhaled particles, the rate of clearance is dependent on their solubility in the lung fluid with a directly proportional relationship. Very often it is observed that the larger-scale, soluble, inhaled particles also exhibit biological as well as toxicological profiles similar to nanoparticles, possibly due to the gradual dissolution of the larger particles in the lung fluid and the consequent alteration of particle size to the nanometric range (Boverhof et al., 2015).

15.3.2.4 Surface Coating

Generally, surface coatings of nanomaterials are applied with the purpose of bringing alterations to distinct particle properties. The surface coating of nanoparticles can be achieved by covering the surface with a wide variety of materials in either single or multiple layers. The layer can be complete or sometimes may also be incomplete (Nanopartikel, 2018). Usually, the generation of an interface between the surface of the nanoparticle and the use matrix is favored in most of the cases. For example, to obtain a uniform dispersion of naturally occurring, hydrophilic nanoparticles in a hydrophobic, nonpolar dispersion medium such as cosmetic oils, surface coating of the nanoparticles with a hydrophobic coating material is necessary. Nanophase employs both patented and trademarked particle coating technology to change the surface of the nanoparticles through means of the encapsulation of the individual particles. This process employs versatile functionality in the nanoparticles which in turn improves the ease of application as well as patient compliance (Nanophase, 2018). For example, surface coating agents for silver nanoparticles retard the oxidation of the surface silver atoms to silver oxide, while the incorporation of coating agents for ZnO nanoparticles helps to prevent the breakdown of ZnO in acidic mediums. Charged organic surface coating compounds, by means of inducing electrostatic repulsion as well as a steric hindrance among the particles, prevents cluster-related instability in the nanoparticles and keeps their separate existence (Zhang et al., 2015).

15.4 Reevaluation of Cosmetic Products after Launch

According to the European Union, an in-market control system (also known as a post-market control) is considered preferable over a pre-market approval procedure for cosmetics to ensure their safety and that the manufacturer meets the required standards as prescribed in regulation. The whole procedure of cosmetic product development is regulated by the European Union Cosmetics Regulation (Cosmetics Europe, 2018b). Various sources are available regarding the safety information of cosmetic products, among which the Scientific Committee on Consumer Safety in the European Union and the Cosmetic Ingredient Review in the United States provide information regarding safety assessments, especially in case of active ingredient use in cosmetics (Masuda and Harada, 2017). The eChemPortal was introduced by Organisation for Economic Co-operation and Development (OECD) in 2004 to provide important information regarding chemicals. From 12 June 2015, the current version of the portal was made available. This version allows users to get detailed information on a chemical, based on their globally harmonized system (GHS) classification (OECD, 2018 OECD, 2018a). According to the new European Union Cosmetics Regulation, there is a requirement for cosmetic products to include post-marketing surveillance and reporting of serious undesirable effects (SUEs). The regulation suggests a pre-marketing responsible person designation and the maintenance of this throughout its lifecycle. Companies must sincerely present the small information. Efficient planning and training is necessary for facilitation of product launches, as well as to prevent compliance issues and support to control cosmetics in the marketplace (Bhatt, 2014).

15.5 Toxicological Concerns

15.5.1 Route and Extent of Exposure

Exposure of nanomaterials to the human body can pose serious health risks. Nanomaterials enter the body through three main routes, as outlined below (Raj et al., 2012).

15.5.1.1 Inhalation

Inhalation may be the route of exposure to nanomaterials for workers during the production of cosmetics if appropriate safety devices are not available. According to the National Institute of Health, after entering the pulmonary tract the inhaled particles enter the brain through nasal nerves and also get distributed to the blood, nervous system, and other organs (Raj et al., 2012). For the equivalent mass of inhaled nanoparticles, due to their greater surface area, the capacity of adsorbing potentially toxic agents is greater than in larger-sized particles, leading to a greater chance of the generation of free radicals and reactive oxygen species. Thus, ultrafine particles (diameter <0.1 μm) have been observed to possess a greater reactivity with lung cells (Schwab, 2011), which ultimately results in a greater extent of inflammation in the lungs when compared to fine particles (Yah et al., 2011). Depending upon the size, different inhaled particles exhibit different region-specific fractional deposition in the human respiratory tract. Ultrafine particles with diameters <100 nm show deposition in all segments, whereas particles <10 nm deposit specifically in the tracheobronchial region, and particles between 10 and 20 nm deposit in the alveolar region (Gatoo et al., 2014). Bermudez et al. (2004) conducted a study to evaluate the pulmonary reactions of laboratory rodent species upon inhalation of TiO_2 particles, and also compared the pulmonary responses of the laboratory rodent species with mice and hamsters. It was observed that there was a significant difference in species-specific pulmonary responses on the inhalation of ultrafine-TiO_2 particles. When the existence of ultrafine particles in the lungs of a mouse and a rat was equivalent, it was observed that a more severe inflammatory response was obtained in the case of the rat, which subsequently leads to the development of progressive epithelial and fibroproliferative alterations. The main cause behind this result was probably the impaired clearance of the particles in the case of the mouse (Bermudez et al., 2004).

15.5.1.2 Ingestion

The probability of consumers touching their mouths when using cosmetic products leads to the risk of exposure to cosmetic products by ingestion. This route accounts for exposure to only a minute extent and is hence considered an insignificant route (Wiechers and Musee, 2010).

15.5.1.3 Through the Skin

TiO_2 can be used in sunscreen cosmetic products and different studies have shown that TiO_2 is unable to permeate into the healthy stratum corneum. However, a few other nanoparticles do not have information regarding their penetration capacity into the stratum corneum (SC) of the skin. The dose of nanoparticles to the intact organ and the dose in secondary target organs should be considered for their toxicological issues (SCCP, 2007b).

15.5.2 Uptake and Absorption

Uptake and absorption of cosmetic products is based on the different types of nanoparticles. There are two types of nanoparticles and these are (1) soluble or biodegradable nanoparticles which break down upon application to the skin into molecular species (e.g., liposomes, microemulsions, nanoemulsions), and (2) insoluble or biopersistent particles (e.g., TiO_2, fullerenes, quantum dots). Insoluble nanoparticles are the primary health concern when the possible uptake of these particles occurs. If these particles are systemically available, it may create a lot of health problems. Sometimes transportation and final accumulation in secondary target organs may happen. Due to the repeated use of cosmetic products which contain insoluble nanoparticles, this may become a more problematic issue in relation to environmental and health concerns. Therefore, the safety evaluation of nanoparticles in cosmetics should be done from batch to batch (SCCP, 2007c).

15.5.3 Toxicological Studies

15.5.3.1 Acute Toxicity Study

Studies of the acute toxicity of a cosmetic product and its data are not required for submission with the dossiers to the Scientific Committee on Consumer Safety (SCCS), however, usually it's desirable to include according to the industry. These acute toxicity studies are usually carried out in rats, mice, and rabbits by oral, topical, and inhalation routes. All established methods for acute oral toxicity are based on *in vivo* models after the single dose of administration and an assessment of the dose inducing death in 50% of the animals (Vinardell and Mitjans, 2017). Acute dermal toxicity is defined as the adverse effect triggered by a material following a single dermal application in 24 hours or less. The fixed dose procedure for acute dermal toxicity (Test Guideline (TG) 434) was mentioned in detail as per the relevant OECD guideline (OECD, 2004). An acute inhalation toxicity–acute toxic class method (TG 436) was adopted on 7 September 2009 by the OECD (OECD, 2009a). Furthermore, an acute inhalation toxicity–fixed concentration procedure (TG 433) was drafted as a proposal for a new guideline by the OECD on 18 December 2015 (OECD, 2015a). Grassian et al. (2007) purchased the smallest commercially available TiO_2 nanoparticles with an average primary particle size of 5 nm and a surface area of $210 \pm 10\ m^2/g$. They performed acute studies and sub-acute toxicity studies in exposed mice in a group of six animals. They exposed the mice to TiO_2 nanoparticles in a whole-body exposure chamber acutely (4 h) or sub-acutely (4 h/day for 10 days). Mice unexposed to nebulized water were kept for control mice for the comparative analysis. In the sub-acute study, a necropsied group was used as soon as possible after the last day of the exposure (week 0), and the other remaining animal groups were euthanized at weeks one, two, and three post exposure. According to their studies, they found that an individualistic characterization of nanoparticles in a toxicity study was very important. They found that in the exposure chamber nanoparticles form aggregates with a geometric mean diameter between 120 and 130 nm. They found that acute exposure confirmed no adverse effects 4 hours after the exposure began. The lung response of the mice after sub-acute exposure of the aggregate was studied and it was observed that a significant but modest inflammation was developed among animals' necropsied at week zero, one, or two after the last exposure. It was observed that the inflammatory response was recovered after three weeks of exposure. The relatively modest response was observed in the case of TiO_2 nanoparticles with a primary particle size of less than 10 nm, which can be considered as a benchmark against which other particles can be compared (Grassian et al., 2007).

15.5.3.2 Skin Irritation

The OECD guidelines for the testing of chemicals are periodically reviewed to safeguard and generate improvements by animal welfare organizations. The updated Guideline 404 (originally adopted in 1981, revised in 1992, 2002, and 2015) contains a reference to the Guidance Document on Integrated Approaches to Testing and Assessment (IATA) for Skin Irritation/Corrosion, proposing a modular approach for skin irritation and skin corrosion testing. This updated guideline clearly states that *in vivo* studies will be carried out to obtain all available data, such as potential dermal corrosivity/irritation, as per the guidelines for Skin Irritation/Corrosion. The *in vivo* test principle, preparation, test procedures, observation period, and evaluation of results are mentioned in detail in the OECD test guideline (TG 404) for the testing of chemicals (OECD, 2015b). The acute skin irritation test is usually accomplished using rabbit skin (OECD, 2002). This test also provides information about erythema and eschar formation and edema after single application on the skin (Cotovio et al., 2007; OECD, 2002). Table 15.1 shows the different grades of erythema and oedema.

Acute irritation is categorized by local and reversible nonimmunological inflammatory response in living skin subsequent to an injury or single contact with an irritant (Cotovio et al., 2007). The European Centre for the Validation of Alternative Methods (ECVAM) sets performance standards for *in vitro* skin irritation test methods based on reconstructed human epidermis (RhE). Two methods, namely the EpiSkin™ test and the EpiDerm™test methods, were part of the first skin irritation validation study (ECVAM, 2009). Therefore, the accessibility of these test methods is dependent on the commercial interest of the manufacturer (Groeber et al., 2016). Contri et al. (2014) studied the effect of nanoencapsulated capsaicinoids on the skin to produce dermal irritation, and also evaluated the dermal tolerance against a novel vehicle composed of chitosan hydrogel and polymeric nanocapsules. The study was conducted on human volunteers (n=13, males and females aged 18–45 years) and the product was applied (0.01 mg/cm^2 of skin area) on the arms (three circles of 30 mm in diameter each) of the volunteers within a specified area and the product was allowed to remain in contact with the skin for 30 minutes. Evaluation of the skin irritation was conducted by a double-blind study with the help of an electronic probe and an arbitrary visual scale from 0 to 5, as presented in Table 15.2.

Skin transepidermal water loss and pH values were also determined as they are considered an indicator of dermal irritation. The measurements (taken using a Tewameter TM30M® and a Skin-pH-Meter® from CK Electronic) were done before application and 60 and 180 minutes after application of the test formulations. The study showed that after one hour of application of the test formulations about 31% of volunteers felt slight irritation from the CH-NC-CP formulation, while moderate (46% for CH-ET-CP and 23% for the commercial product) and severe (8% for CH-ET-CP and 69% for the commercial product) irritation was obtained from the preparations comprising free capsaicinoids. In the case of dermal application of CH-NC, erythema was not observed and irritation was observed for only 8% of the volunteers (Contri et al., 2014). The U-SENS™ method for the purpose of detection and prediction of dermal allergy was developed in L'Oréal's laboratories. The method is mainly based on *in vitro* testing of human cells resulting in the expression of a specific marker of immunity. The method was evaluated by the European Union Reference Laboratory (EURL)-ECVAM and was declared to be efficient and reproducible (L'Oréal, 2017).

TABLE 15.1

Different Grades in the Case of Erythema and Escher (OECD, 2015b)

Erythema	Grade	Oedema	Grade
No erythema	0	No oedema	0
Very slight erythema (barely perceptible)	1	Very slight oedema (barely perceptible)	1
Well defined erythema	2	Slight oedema (edges of area well defined by definite raising)	2
Moderate to severe erythema-3	3		
Severe erythema (beef redness) to eschar formation	4		

TABLE 15.2

Evaluation of the Skin Irritation Grades (Contri et al., 2014)

Skin Irritation	Grade
Redness	0
Slight red points	1
Bigger red points	2
Concrete regions of redness	3
Redness encompassing the whole circle where the formulation was applied	4
Redness extending beyond the drawn circle	5

15.5.3.3 Ocular Irritation

Usually, ocular irritation is studied by performing a Draize test on rabbits. A single dose of the formulation is applied to one eye of three adult rabbits whereas the other eye is allowed to remain unaltered. The degree of irritation is measured in different parts of the eye by the direct comparison of pictures over a period of three days. To evaluate the reversible nature of the effect, the testing period may be extended to three weeks (United States Congress, 1986). The OECD TG 437 covering the bovine corneal opacity and permeability (BCOP) test method allows the conduct of *in vitro* studies without employing any animals. This test procedure is carried out under specified conditions, but possesses limitations for eye hazard classification and the labeling of chemicals (OECD, 2013a). The recently adopted OECD Test Guideline 492 suggests the human cornea-like epithelium test method allows the *in vitro* identification of chemicals, free from the ability to cause eye irritation or serious eye damage, irrespective of their classification (OECD, 2016a). Nigam (2009) performed the Draize test by dripping chemical substances into the eyes of six to nine immobilized, conscious albino rabbits. The gradual damage was recorded as per protocol for 72 hours and was sometimes extended for up to seven to 18 days (Nigam, 2009). Ema et al. (2013), as per OECD Guideline 405, conducted an "Acute Eye Irritation/Corrosion" test by incorporating fullerenes into the conjunctival sac of a single eye of a rabbit. In the initial test using one rabbit, no corrosive effects were observed one hour after the ocular application of the fullerenes. After 72 hours of application in the eye, sodium fluorescein was applied to stain the corneal epithelium. The imparted ocular damage was evaluated according to the McDonald and Shadduck method (Ema et al., 2013). The isolated chicken eye (ICE) test method (TG 438) is an organotypic model that provides short-term maintenance of the chicken eye *in vitro*. In this test method, the test chemical mediated ocular damage was measured by corneal swelling, opacity, and fluorescein retention (OECD, 2013b). Fluorescence leakage test methods (OECD TG 460) provide the opportunity for identification of ocular corrosives and severe irritants (OECD, 2012). OECD TG 492 suggests a short time exposure *in vitro* test method for the identification of chemicals causing serious eye damage as well as chemicals not requiring classification for eye irritation or serious eye damage (OECD, 2016a).

15.5.3.4 Phototoxicity/Photoirritation Testing

Acute phototoxic, or photo*irritant* in contrast to photo*allergic*, skin reactions are usually developed by some chemical species on the absorption of light (predominantly in the UV region 280–400 nm) (Pape, 1997; Thong and Maibach, 2008). This study is usually conducted for products intended to be in prolonged contact with the skin (Intertek, 2018). The 3T3 neutral red uptake phototoxicity assay (3T3-NRU-PT) can be performed to evaluate the cytotoxic potency of a test substance after exposure to UV rays. The result is compared with the cytotoxic effect of a test substance evaluated in the presence or absence of a non-cytotoxic dose of long wavelength UV rays (UVA) as well as visible light. The cytotoxic response is recorded as a concentration-dependent reduction of the uptake of the vital dye, neutral red. This test can be utilized for the qualitative estimation of substances exhibiting *in vivo* phototoxicity after oral application and distribution to the skin, as well as for compounds behaving in the same way after topical application as well (MB Research Labs, 2018). TiO_2 nanoparticles are photoactive and generate reactive oxygen species (ROS) under normal sunlight. These ROS can be damaging to numerous organisms, inducing oxidative damage, cell injury, and death. Most of the studies for the determination TiO_2 nanoparticle toxicity are not considered necessary since this nanoparticle did not simulate natural irradiation either in dark conditions or under artificial lighting. The phototoxicity of TiO_2 nanoparticles studied showed the ratio between experiments performed in the absence of sunlight and those conducted under solar or simulated solar radiation for aquatic species. This phototoxicity ratio can be used to correct the endpoints of the toxicity tests with nano-TiO_2 that were performed in the absence of sunlight. This correction of endpoints may be important for regulators and risk assessors (Jovanovic, 2015). Yin et al. (2012) determined the phototoxicity of TiO_2 nanoparticles with four different molecular sizes (<25 nm, 31 nm, <100 nm, and 325 nm) and two crystal forms (<25 nm anatase and <100 nm rutile) in human skin keratinocytes under UVA irradiation. The nano-TiO_2 with the smaller particle size produces a higher level of cell damage. The rutile form of nano-TiO_2 showed less phototoxicity when compared to the anatase form of nano-TiO_2 (Yin et al., 2012).

According to the SCCS, the established safe amount of TiO$_2$ nanoparticles for healthy human skin is up to a concentration of 25% as a UV protector in sunscreen formulations. Though the labeling may not contain any information regarding the concentration of TiO$_2$ nanoparticles, during manufacturing, manufacturers of cosmetic products containing TiO$_2$ nanoparticles have to follow the limit according to European legislation (EC, 2009; EC, 2018).

15.5.3.5 Skin Sensitization

Cosmetic ingredients need to be evaluated within the frame of reference of present and upcoming regulations for all chemicals for skin sensitization (Corsini et al., 2014). Local lymph node assay (LLNA) is utilized for the estimation of skin sensitization in mice. The process was updated based on scientific data and experience, and this is known as the second test guidelines for the assessment of skin sensitization potential of chemicals in animals (OECD, 2010). The other TG (i.e., TG 406) suggests guinea pig tests, most notably the guinea pig maximization test and the Buehler test (OECD, 1992). The *in vitro* and *in chemico* methods for predicting skin sensitization are: (1) OECD TG 442C: *In Chemico* Skin Sensitization Direct Peptide Reactivity Assay (DPRA) (OECD, 2015c), (2) OECD TG 442D: *In Vitro* Skin Sensitization ARE-Nrf2 Luciferase Test Method (OECD, 2017a; OECD, 2017), and (3) OECD TG 442E: *In Vitro* Skin Sensitization Human Cell Line Activation Test (h-CLAT) (PETA International Science Consortium, 2017). For evaluating the molecular initiating event of the skin sensitization adverse outcome pathway (AOP), DPRA is the proposed method. Cysteine and lysine percent peptide depletion values are also important to determine whether a substance is a skin sensitizer or non-sensitizer (OECD, 2015c). For obtaining keratinocyte activation, the cell line containing the luciferase reporter gene (KeratinoSens™ ARE-Nrf2 luciferase test method) is usually employed (OECD, 2017a; OECD, 2017). The h-CLAT method is an *in vitro* assay that is used to evaluate changes in cell surface marker expression (i.e., CD86 and CD54) on a human monocytic leukemia cell line (THP-1 cells) after 24 hours exposure to the test chemical (OECD, 2017b; OECD, 2018b), including fragrance ingredients, preservatives, and dyes as human allergens (Sakaguchi et al., 2010). OECD TG 442E has been approved on 9 October 2017 (PETA International Science Consortium, 2017). Previously, the h-CLAT was evaluated in an EURL-ECVAM for alternatives to animal testing and was considered scientifically valid to be used as part of an IATA to differentiate between sensitizers and non-sensitizers for hazard classification and labeling (PETA International Science Consortium, 2017; MB Research, 2017).

15.5.3.6 Mutagenicity/Genotoxicity

Mutagenicity indicates permanent transmissible cellular changes in the genetic level (both in number and structure). Genotoxicity is used in a larger sense and indicates the harmful impact on genetic material, not always associated with mutagenicity. Thus, tests for genotoxicity usually help to evaluate any possible damages to the number and structure of deoxyribonucleic acids (DNA), including mutation-induced damages such as unscheduled DNA synthesis (UDS), sister chromatid exchange (SCE), DNA strand breaks, DNA adduct formation, or mitotic recombination (SCCNFP, 2003). The different tests are discussed below.

- TG 471: Evaluation of bacterial reverse mutation

    A bacterial reverse mutation test (often simply referred to as an "Ames test") is a principle safety assessment test performed by employing amino acid-requiring strains of *Salmonella typhimurium* or *Escherichia coli*. Here the bacterial growth on a selective agar medium undergoes a scoring procedure after exposing the bacterial stains to that agar medium either by direct incorporation or by pre-incubation (GLP OECD, 2014). OECD TG 471 gives the detailed procedure for the bacterial reverse mutation test. This test involves prokaryotic cells, different from mammalian cells in several parameters such as uptake, metabolism, chromosomal structure, and DNA repair processes. An external source for the metabolic activation is required for conducting *in vitro* tests. *In vitro*, metabolic activation systems are not identical to mammalian *in vivo* conditions in the true sense. Thus, this test is unable to provide direct information regarding the mutagenic and carcinogenic potency of a substance in mammals (OECD,1997).

- TG 473: Chromosomal aberration test by *in vitro*

     The *in vitro* chromosomal aberration test is performed for the purpose of identification of the substances imparting structural chromosomal aberrations of the mammalian cells in cell culture, either of established cell lines (e.g., Chinese hamster ovary (CHO), Chinese hamster lung (CHL) V79, CHL/IU, TK6) or primary cell cultures (human or other mammalian peripheral blood lymphocytes) (OECD, 2014).

- TG 476: Mammalian cell gene mutation test by *in vitro*

     The *in vitro* mammalian cell gene mutation test provides information regarding chemical substance induced mutations. In this test, the induced mutation in hypoxanthine-guanine phosphoribosyltransferase (HPRT) and the transgene of xanthine-guanine phosphoribosyl transferase (XPRT) is usually evaluated (OECD, 2015d; OECD, 2016b).

- TG 490: *In vitro* mammalian cell gene mutation test

     The *in vitro* mammalian cell gene mutation test can be performed in two different cell lines (L5178Y tk+/-3.7.2C cells for the mouse lymphoma assay (MLA) and TK6 tk+/- cells for the TK6 assay) as per OECD TG 490 (OECD, 2015e; OECD, 2016c).

- TG 487: Micronucleus test *in vitro* (MNvit)

     The *in vitro* micronucleus (MNvit) test is a genotoxicity test for the identification of micronuclei (MN) in the cytoplasm of interphase cells (OECD, 2016b; OECD, 2016d).

15.5.3.7 Repeated Dose Toxicity

Repeated dose toxicity involves the development of common adverse effects associated with repetitive daily dosing with, or exposure to, a material for a definite time span in the test species (CHMP, 2008). Different types of *in vivo* repeated dose toxicity tests along with used animals as per the OECD guidelines are given in Table 15.3.

15.5.3.8 Cellular Toxicity

Engineered TiO_2 and ZnO nanoparticles (TiO_2- and ZnO-ENP) are on the market and are 70% and 30%, respectively. The TiO_2- and ZnO-ENP have been reported to have various toxicological effects such as oxidative stress, the release of toxic ions, or the adsorption of pollutants. These toxicological effects not only affect the cellular mass uptake but are also dependent on the particle's diameter, shape, agglomeration/aggregation status, and/or surface (Detoni et al., 2011; Lorenz et al., 2010). To evaluate the safety of Ag nanoparticles, assessment with skin transcutaneous passage assay is usually employed. In the study by Kokura et al. (2010), it was observed that the silver nanoparticles could not penetrate the skin but may penetrate if there is a disruption in the skin's barrier function, e.g., a rupture. It was also observed that in allergic conditions there was skin penetration of nanoparticles, but to a much lesser extent (0.2–2%) (Kokura et al., 2010).

15.5.3.9 Reproductive Toxicity

For existence of a species, fertility, reproduction, and fetal development are essential, and this is what indicates the importance of the discussion related to the toxicity of nanoparticles on the reproductive system. Only 400 follicles in a woman reach maturity and undergo ovulation during their lifetime, indicating the opportunity for reproduction. The hormonal control system regulating the functions of the uterus and ovaries, exhibiting periodic growth and regeneration, is seriously affected by foreign particles, and any stress in female reproduction leads to fetal abnormalities (Brohi et al., 2017). The embryonic stem cell test (EST) has been adopted to evaluate the embryotoxic potential of nanoparticles in animals. The test can be performed only in the case of those chemicals which do not require any metabolic conversion and act in the early stages of embryonic development (Riebeling et al., 2012).

TABLE 15.3

Different Types of *In Vivo* Repeated Dose Toxicity Tests as per OECD Guidelines

In Vivo **Repeated Dose Toxicity Tests**	**Used Animals**	**Test No.**
Repeated dose (28 days) toxicity (oral)	Rat (preferred)	TG 407 (OECD, 2008)
Repeated dose (21/28 days) toxicity (dermal)	Rats (200 to 300 g), rabbits (2 to 3 kg), guinea pigs (350 to 450 g), other species (use requires justification)	TG 410 (OECD, 1981a)
Repeated dose (28 days) toxicity (inhalation)	Rat	TG 412 (OECD, 2009b)
Sub-chronic oral toxicity test: repeated dose 90-day oral toxicity study in rodents	Rat (preferred), mouse	TG 408 (OECD, 1998a)
Sub-chronic oral toxicity test: repeated dose 90-day oral toxicity study in non-rodents	Dog (commonly used), which should be of a defined breed, i.e., the beagle is frequently used. Other species, e.g., swine, mini-pigs, and primates, are not recommended and their use should be justified.	TG 409 (OECD, 1998b)
Sub-chronic dermal toxicity study: repeated dose 90-day dermal toxicity study using rodent species	Rats (200 to 300 g), rabbits (2 to 3 kg), guinea pigs (350 to 450 g), other species (use requires justification).	TG 411 (OECD, 1981b)
Sub-chronic inhalation toxicity study: repeated dose 90-day inhalation toxicity study using rodent species	Rat (preferred), other species (use requires justification)	TG 413 (OECD, 2009c)
Chronic toxicity test	Rat (preferred), other rodent species, i.e., mice, may be used	TG 452 (OECD, 2018c)

15.5.3.10 Toxicokinetic Study

Toxicokinetics act to establish a correlation between the kinetic behaviors of a toxic agent and the occurrence of toxic events by means of pharmacokinetic tools. Both the pharmacokinetic and toxicokinetic profiles of nanoparticles are dependent upon the route of exposure, as well as the physicochemical properties of the nanoparticles. Biokinetic profiles depend on physicochemical properties, dissolution behavior in biological fluids, and nanoparticle–protein interaction. It is not yet clear what the exact form (particles or ions) is in which ZnO nanoparticles are absorbed into systemic circulation. The binding of ZnO nanoparticles with vascular proteins by means of electrostatic interactions helps to regulate its absorption. Therefore, an *in vivo* study is necessary to evaluate the effects of nanoparticle–protein interaction on biokinetics, absorption, tissue distribution, bioavailability, and potential toxicity (Choi and Choy, 2014). There is no available, fully validated, *in vitro* test procedure that replaces *in vivo* toxicokinetic studies for regulatory use (Kojima, 2017).

15.5.3.11 Finished Products

The *in vitro* assay of the finished product is usually performed based on the in-house experience of the cosmetic product manufacturing companies, although there is not a successful *in vitro* alternative test method available in validation studies. According to the Scientific Committee on Cosmetic Products and Non-food Products Intended for Consumers (SCCNFP), the finished product evaluation in the cosmetic industry can be performed dependent upon the gathered knowledge regarding ingredient toxicity. Apart from these, the vehicles used in the preparation and in the toxicity assay must not be identical. Proper monitoring of the finished product to avoid the generation of toxic substances through the interaction of ingredients must be performed (SCCNFP, 2000). Suitable tests for confirming the physical stability of the finished product should be carried out to ensure that the product remains stable during transport, storage, and handling of the product. Various factors like temperature variation, humidity, UV light, and mechanical stress can alter the physical stability of the product, leading to an effect on consumer safety. Evaluation tests should also be conducted to ensure that the type of container, i.e., its material composition, has no impact on the physical stability of the product. The safety profile of the finished product generated by these tests should be clearly expressed in the safety evaluation report (SCCP, 2006).

15.6 Global Trends in Regulatory Guidelines for the Use of Cosmetics

15.6.1 International Cooperation on Cosmetics Regulation (ICCR)

The International Cooperation on Cosmetics Regulation (ICCR) is an internationally-working organization composed of regulatory authorities from different countries like the United States, Japan, the European Union, and Canada, with an objective to secure consumer safety globally to maximum extent and to maintain the worldwide trade in cosmetic products (ICCR, 2018). ICCR acts on the necessity of reducing animal studies and replacing them with different proposed alternatives from different industries and validation centers. To collaborate and communicate regarding the design, execution, and peer review of validation studies, ICCR invites interagency coordinating committee on the validation of alternative methods (ICCVAM), ECVAM, the Japanese Center for the Validation of Alternative Methods (JaCVAM), and a competent representative from the Government of Canada to provide input regarding these issues (ICCR, 2007).

15.6.2 Regulatory Bodies for Cosmetics in Europe

The European Union Cosmetics Regulation governs how cosmetics and personal care products are made and placed on the market (Cosmetics Europe Cosmetics, 2017). A globally-acting organization of 39 different partner organizations will gain funds granted by the European Commission during the Horizon 2020 program, a program initiated by the European Union for proper innovation and research, to gather and integrate new ideas to govern chemical safety assessments. These new concepts include cutting-edge, human-relevant, *in vitro* non-animal methods and *in silico* computational technologies to convert toxicity in the perspective of the molecular mechanics into different strategies for safety evaluation, ultimately providing opportunities for testing the safety of chemicals without the utilization of animals and to ensure the delivery of a safe and reliable product (EU-ToxRisk, 2015).

The Long-range Science Strategy (LRSS) research program represents the research materials that are gaining maximum priority in cosmetics in Europe from 2016 to 2020. The main thrust areas of this program are repeat dose toxicity, bioavailability (ADME), and systemic toxicity. The principle objectives are:

1. The prediction of toxicity in the human body without animal utilization as well as innovation of new substances with proper safety profiles and the maintenance of safety in the materials already existing in the industry.

2. A wide spectrum of acceptance in both the scientific and regulatory aspects (Cosmetics Europe, 2018a).

EU-ToxRisk, an integrated European "flagship" program with the objective of studying toxicity from the mechanistic view and perform risk assessments for the twenty-first century, is a European collaborative project with funding from Horizon 2020. The project is of six years duration (from 1 January 2016) with a budget of over 30 million euros (EU-ToxRisk, 2018). The goal of Safety Evaluation Ultimately Replacing Animal Testing-1 (SEURAT-1) ("towards the replacement of *in vivo* repeated dose systemic toxicity testing") is to properly design a pathway for the long-term research and development of procedures for the assessment of human safety in the field of repeated dose systemic toxicity testing of chemicals (SEURAT-1, 2014).

15.6.3 Regulatory Bodies for Cosmetics in Japan

The JaCVAM was set up at the National Center for Biological Safety and Research, affiliated with the National Institute of Health Sciences (NIHS) in Japan. On 4 February 2011, the Ministry of Health, Labor, and Welfare of Japan was notified that the data generated from different alternative testing methods approved by the JaCVAM Steering Committee could be employed for the submission of quasi-drug applications or for petitions to bring cosmetic ingredients under the proper regulation of standards. The different nationally as well as globally peer-reviewed methods included Bhas cell transformation assay and the short time exposure (STE) assay for eye irritation testing. Additionally, JaCVAM is working in collaboration with several other international agencies in ongoing validation studies, which include studies into the h-CLAT, *in vivo/in vitro* Comet assays, the stably transfected transactivation assay (STTA) antagonist test for screening of endocrine disruptors, and an assay for phototoxicity (Kojima, 2018).

15.6.4 Regulatory Bodies for Cosmetics in Australia

Cosmetics in Australia are regarded as a branch of industrial chemicals. Therefore, the Industrial Chemicals (Notification and Assessment) Regulations 1990 are application to all cosmetics. The Australian Government Department of Health empower the "National Industrial Chemicals Notification and Assessment Scheme" (NICNAS), which is completely in charge of industrial chemicals (including cosmetics). Various regulatory reform is ongoing within the NICNAS and it will be replaced by a new scheme called the "Australian Industrial Chemical Introduction Scheme (AICIS)" as of 1 July 2018. The AICIS will develop a method for the complete safety assessment of chemicals, comprising a wider monitoring of both lower and higher risk chemicals, and post-market surveillance (ChemLinked, 2018).

15.7 Conclusion

In this safety evaluation of nanocosmetics, materials must be evaluated on par with pharmaceutical compounds, such as physiochemical characterizations, size, shape, surface area, charge, morphology, rheology including porosity, and crystallinity and amorphocity. This chapter also elaborated on how impurities, chemical compositions, solubility, and surface coating leads to different methodology and safety evaluation requirements and their approaches were presented. Toxicological evaluations were studied and their methodologies, including oral, dermal, inhalation, and cellular update, were discussed within various parameters. This chapter has also covered acute toxicity studies, skin and ocular irritation, phototoxicity, and skin sensitization. Also, this chapter briefly discussed mutagenicity and genotoxicity by giving various examples of cell-based *in vitro* testing methods. Repeated dose toxicity is an important parameter in cosmetic products, hence this chapter has provided a detailed methodology as well as a reproductive and cellular toxicity analysis of various cosmetic products. This safety evaluation and assessment chapter also discussed details in terms of research and development, as well as international regulatory guidelines, including the post-marketing analysis of cosmetic products in different countries.

REFERENCES

Ajazzuddin M, Jeswani G, Jha AK. Nanocosmetics: Past, present and future trends. Recent Pat Nanomed. 2015;5:3–11.

Al-Qahtani KMA, Ahmed HAM, Al-Otaibi MB. Detection of toxic metals in lipsticks products in Riyadh, Saudi Arabia. Orient J Chem. 2016;32:1929–36.

Alves MP, Raffin RP, Fagan SB. Rheological behavior of semisolid formulations containing nanostructured systems. In: Beck R, Guterres S, Pohlmann A, editors. Nanocosmetics and nanomedicines: New approaches for skin care. Heidelberg: Springer-Verlag Berlin; 2011. p. 37–48.

ASTM International. Standard D2196-15 – Standard test methods for rheological properties of non-newtonian materials by rotational viscometer. 2015. Available from: www.astm.org.

Aswal A, Kalra M, Rout A. Preparation and evaluation of polyherbal cosmetic cream. Der Pharmacia Lettre. 2013;5:83–8.

Badri W, Miladi K, Eddabra R, Fessi H, Elaissari A. Elaboration of nanoparticles containing indomethacin: Argan oil for transdermal local and cosmetic application. J Nanomater. 2015 2015:1-9.

Banerjee R. Nanocosmetics: The good, the bad and the beautiful. TCOJ. 2017;1:e9–e11.

Bantz C, Koshkina O, Lang T, Galla HJ, Kirkpatrick CJ, Stauber RH, Maskos M. The surface properties of nanoparticles determine the agglomeration state and the size of the particles under physiological conditions. Beilstein J Nanotechnol. 2014;5:1774–86.

Barry BW, Warburton B. Rheological aspects of cosmetics. J Soc Cosmetic Chemists. 1968;19:725–44.

Bermudez E, Mangum JB, Wong BA, Asgharian B, Hext PM, Warheit DB, Everitt JI. Pulmonary responses of mice, rats, and hamsters to subchronic inhalation of ultrafine titanium dioxide particles. Toxicol Sci. 2004;77:347–57.

Bhatia S. Natural polymer drug delivery systems. Switzerland: Springer International Publishing; 2016.

Bhatt D. New EU regulation on post marketing surveillance of cosmetic products. Regulatory Affairs Professionals Society; 2014 [cited 2018 April 2]. Available from: https://www.sciformix.com/wp-conten t/uploads/RAPS-RF-2014-02-Cosmetic-Products.pdf.

Bojic J, Radovanovic B, Dimitrijevic J. Spectrophotometric determination of urea in dermatologic formulations and cosmetics. Anal Sci. 2008;24:769–74.

Bonevich JE, Haller WK. Measuring the size of nanoparticles using transmission electron microscopy (TEM) [Internet]. 2010 [cited 2018 February 2]. Available from: http://www.metallurgy.nist.gov/.

Boverhof DR, Bramante CM, Butala JH, Clancy SF, Lafranconi M, West J, Gordon SC. Comparative assessment of nanomaterial definitions and safety evaluation considerations. Regul Toxicol Pharmacol. 2015;73:137–50.

Brohi RD, Wang L, Talpur HS, Wu D, Khan FA, Bhattarai D, Rehman ZU, Farmanullah F, Huo LH. Toxicity of nanoparticles on the reproductive system in animal models: A review. Front Pharmacol. 2017; 8:606.

Buzea C, Pacheco II, Robbie K. Nanomaterials and nanoparticles: Sources and toxicity. Biointerphases. 2007;2:MR17–71.

Chaudhri N, Soni GC, Prajapati SK. Nanotechnology: An advance tool for nano-cosmetics preparation. IJPRR. 2015;4:28–40.

ChemLinked. Australia Cosmetics Regulation [Internet]. Beijing: ChemLinked; 2018 [cited 2018 April 4]. Available from: https://cosmetic.chemlinked.com/countrypage/australia-cosmetics-regulation.

Chithrani BD, Ghazani AA, Chan WCW. Determining the size and shape dependence of gold nanoparticle uptake into mammalian cells. Nano Lett. 2006;6:662–8.

Committee for Medicinal Products for Human Use (CHMP). Doc. Ref. EMEA/CHMP/SWP/488313/2007: Guideline on repeated dose toxicity (Draft). London: European Medicines Agency; 2008. p. 1–9.

Choi O, Hu Z. Size dependent and reactive oxygen species related nanosilver toxicity to nitrifying bacteria. Environ Sci Technol. 2008;42:4583–8.

Choi SJ, Choy JH. Biokinetics of zinc oxide nanoparticles: Toxicokinetics, biological fates, and protein interaction. Int J Nanomed. 2014;9:261–9.

Contado C. Nanomaterials in consumer products: A challenging analytical problem. Front Chem. 2015;3:48.

Contri RV, Frank LA, Kaiser M, Pohlmann AR, Guterres SS. The use of nanoencapsulation to decrease human skin irritation caused by capsaicinoids. Int J Nanomed. 2014;9:951–62.

Corsini E, Papale A, Galbiati V, Roggen EL. Safety evaluation of cosmetic ingredients: In vitro opportunities for the identification of contact allergens. Cosmetics. 2014;1:61–74.

Cosmetics Europe. Non-animal approaches to safety assessment of cosmetic products, cutting-edge science and constant innovation: The keys to success [Internet]. 2017 [cited 2018 April 2]. Available from: https:// www.cosmeticseurope.eu/files/1215/0245/3923/Non-animal_approaches_to_safety_assessment_of_ cosmetic_products.pdf.

Cosmetics Europe. Promoting science and research [Internet]. Brussels: Cosmetics Europe; 2018a [cited 2018 April 4]. Available from: https://www.cosmeticseurope.eu/how-we-take-action/promoting-science-re search/.

Cosmetics Europe. Understanding the cosmetics regulation [Internet]. Brussels: Cosmetics Europe; 2018b [cited 2018 April 22]. Available from: https://www.cosmeticseurope.eu/cosmetics-industry/understand ing-cosmetics-regulation/.

Cosmeticsinfo. Cosmetic & personal care product safety [Internet]. 2016 [cited 2018 March 11]. Available from: http://www.cosmeticsinfo.org/Cosmetics-safety.

Cotovio J, Grandidier MH, Lelièvre D, Roguet R, Tinois-Tessonneaud E, Leclaire J. *In vitro* acute skin irritancy of chemicals using the validated EPISKIN model in a tiered strategy results and performances with 184 cosmetic ingredients. In: Proceedings of the 6th World Congress on Alternatives & Animal Use in the Life Sciences; 2007 August 21–25; Tokyo. Tokyo: AATEX 14; 2007. p. 351–8.

Detoni CB, Paese K, Beck RCR, Pohlmann AR, Guterres SS. Nanosized and nanoencapsulated sunscreens. In: Beck R, Guterres S, Pohlmann A, editors. Nanocosmetics and nanomedicines: New approaches for skin care. Heidelberg: Springer-Verlag Berlin; 2011. p. 333–62.

Dorofeeva GA, Streletskiib AN, Povstugara IV, Protasova AV, Elsukova EP. Determination of nanoparticle sizes by the X-ray diffraction method. Colloid J. 2012;74:678–88.

Durán N, Teixeira Z, Marcato PD. Topical application of nanostructures: Solid lipid, polymeric and metallic nanoparticles. In: Beck R, Guterres S, Pohlmann A, editors. Nanocosmetics and nanomedicines: New approaches for skin care. Heidelberg: Spinger-Verlag Berlin; 2011. p. 69–100.

European Commission (EC). Regulation (EC) No 1223/2009 of the European Parliament and of the Council of 30 November 2009 on cosmetic products. Off J Eur Union (EN). 2009;52:59–209.

European Commission (EC). Sunscreens with titanium dioxide as nanoparticles: Health risks? [Internet]. GreenFact publications; 2018 [cited 2018 April 2]. Available from: https://copublications.greenfacts.org/en/titanium-dioxide-nanoparticles/citizens-summary-titanium-dioxide-en.pdf.

ECVAM. ECVAM performance standards for skin irritation testing (updated)-24.8.2009 [Internet]. EU Science Hub2009 [cited 2018 April 2]. Available from: https://eurl-ecvam.jrc.ec.europa.eu/validation-regulatory-acceptance/docs-skin-irritation-1/DOC8-updated_ECVAM_PS_2009.pdf.

Elementis-specialities. Rheological additives in cosmetics, elimentis specialities [Internet]. 2018 [cited 2018 April 4]. Available from: http://www.elementis-specialties.com/esweb/webprodliterature.nsf/allbydocid/3BCB996F45DEBB93852575FB004C77BA/$FILE/Rheology%20cosmetics%20brochure%20Dec-2013.pdf.

Ema M, Matsuda A, Kobayashi N, Naya M, Nakanishi J. Dermal and ocular irritation and skin sensitization studies of fullerene C60 nanoparticles. Cutan Ocul Toxicol. 2013;32:128–34.

EU-ToxRisk. EU-ToxRisk: An integrated European Flagship Program driving mechanism-based toxicity testing and risk assessment [Internet]. EU-ToxRisk; 2015 [cited 2018 April 4]. Available from: https://www.upf.edu/en/web/e-noticies/home_upf?_101_assetEntryId=3011567&_101_groupId=10193&_101_struts_action=%2Fasset_publisher%2Fview_content&_101_type=content&_101_urlTitle=eu-toxrisk-un-programa-europeu-lider-en-conduccio-d-assajos-de-toxicitat-i-avaluacio-de-riscos-per-al-segle-21&inheritRedirect=truc&p_p_id=101&p_p lifecycle=0&p_p_state=maximized#.XA4Fqdszbcc .

EU-ToxRisk. EU-ToxRisk: An integrated European Flagship Programme driving mechanism based toxicity testing and risk assessment for the 21st century [Internet]. EU-ToxRisk 2018. Available from: http://www.eu-toxrisk.eu/.

Fairhurst D, Dukhin A, Klein K. A new way to characterize stability and performance of cosmetic emulsions and suspensions. Mount Kisco, NY: Dispersion Technology, Inc.; 2013 [cited 2018 March 11]. Available from: http://www.dispersion.com/a-new-way-to-characterize-stability-and-performance-of-cosmetic-emulsions-and-suspensions.

Food and Drugs Administration (FDA). Contains nonbinding recommendations, guidance for industry safety of nanomaterials in cosmetic products, 2014 [Internet]. Rockville: U.S. Department of Health and Human Services Food and Drug Administration, Center for Food Safety and Applied Nutrition 2014 [cited 2018 April 22]. Available from: https://www.fda.gov/downloads/Cosmetics/GuidanceRegulation/GuidanceDocuments/UCM300932.pdf.

Food and Drugs Administration (FDA). Cosmetics & US law [Internet]. Silver Spring: U.S. Food and Drug Administration 2017 [cited 2018 March 11]. Available from: https://www.fda.gov/Cosmetics/GuidanceRegulation/LawsRegulations/ucm2005209.htm.

Food and Drugs Administration (FDA). FDA's testing of cosmetics for arsenic, cadmium, chromium, cobalt, lead, mercury, and nickel [Internet]. Silver Spring: U.S. Food and Drug Administration 2018 [cited 2018 April 22]. Available from: https://www.fda.gov/Cosmetics/ProductsIngredients/PotentialContaminants/ucm452836.htm.

Freitas C, Muller RH. Effect of light and temperature on zeta potential and physical stability in solid lipid nanoparticle (SLN™) dispersions. Int J Pharm. 1998;168:221–9.

Frenkel AI, Yevick A, Cooper C, Vasic R. Modeling the structure and composition of nanoparticles by extended X-Ray absorption fine-structure spectroscopy. Annu Rev Anal Chem. 2011;4:23–39.

Fruijtier-Pölloth C. Safety assessment on polyethylene glycols (PEGs) and their derivatives as used in cosmetic products. Toxicology. 2005;214:1–38.

Fuha CB, Laib M, Tsaic HY, Chang CM. Impurity analysis of 1,4-dioxane in nonionic surfactants and cosmetics using headspace solid-phase microextraction coupled with gas chromatography and gas chromatography-mass spectrometry. J Chromatogr A. 2005;1071:141–5.

Gagliardi L, Cavazzutti G, Tonelli D. Determination of dequalinium chloride and related impurities in cosmetics and pharmaceuticals by reversed-phase HPLC. Anal Lett. 1998;31:829–39.

Gasparelo AP, Pizzol CD, Fernanda Campos de Menezes P, Knapik RV, Costa MAT, Machado Prado MR, de Oliveira Praes CE. Zeta potential and particle size to predict emulsion stability [Internet]. Gundersen Drive Allured Business Media 2014 [cited 2018 March 11]. Available from: http://www.cosmeticsandto iletries.com/testing/invitro/Zeta-Potential-and-Particle-Size-to-Predict-Emulsion-Stability-premium-283256431.html.

Gatoo MA, Naseem S, Yasir Arfat M, Mahmood Dar A, Qasim K, Zubair S. Physicochemical properties of nanomaterials: Implication in associated toxic manifestations. BioMed Res Int. 2014;2014:8.

GLP OECD. AM-FL03 V1: Ames bacterial reverse mutation screening assay, Guideline-471 [Internet]. Cheshire; Gentronix Ltd; 2014 [cited 2018 April 1]. Available from: http://www.gentronix.co.uk/wp-content/uploads/2014/07/AM-FL03.pdf.

Grassian H, O'Shaughnessy PT, Adamcakova-Dodd A, Pettibone JM, Thorne PS. Inhalation exposure study of titanium dioxide nanoparticles with a primary particle size of 2 to 5 nm. Environ Health Perspect. 2007;115:397–402.

Groeber F, Schober L, Schmid FF, Traube A, Kolbus-Hernandez S, Daton K, Hoffmann S, Petersohn D, Schäfer-Korting M, Walles H, Mewes KR. Catch-up validation study of an in vitro skin irritation test method based on an open source reconstructed epidermis (phase II). Toxicol in vitro. 2016;36:254–61.

Hamann HF, Gallagher A, Nesbitt DJ. Enhanced sensitivity near-field scanning optical microscopy at high spatial resolution. Appl Phys Lett. 1998;73:1469.

Herh P, Tkachuk J, Wu S, Bernzen M, Rudolph B. The rheology of pharmaceutical and cosmetic semisolids [Internet]. USA: American Laboratory; 1998 [cited 2018 March 1]. Available from: https://www.can noninstrument.com/en/Image/GetDocument/455.

Horiba. Particle size analysis [Internet]. Horiba Scientific; 2018 [cited 2018 March 1]. Available from: http://www.horiba.com/scientific/products/particle-characterization/particle-size-analysis/details/sz-100-7245/.

Hosokawa M, Nogi K, Naito M, Yokoyama T. Nanoparticle technology handbook. Amsterdam: Elsevier; 2007.

ICCR. International cooperation on cosmetic regulation: Outcome of meeting. September 26–28, 2007. Silver Spring: U.S. Food and Drug Administration; [cited 2018 March 1]. Available from: https://www.fda.gov/Cosmetics/InternationalActivities/ICCR/ucm126530.htm

ICCR. International cooperation on cosmetics regulation, last update: April 23, 2018. Silver Spring: U.S. Food and Drug Administration; [cited 2018 March 1]. Available from: https://www.fda.gov/cosmetics/inter nationalactivities/iccr/default.htm.

Iizhar SA, Syed IA, Satar R, Ansari SA. In vitro assessment of pharmaceutical potential of ethosomes entrapped with terbinafine hydrochloride. J Adv Res. 2016;7:453–61.

Intertek. Dermal safety studies for cosmetics and skincare products [Internet]. Intertek Group plc 2018 [cited 2018 April 2]. Available from: http://www.intertek.com/consumer-healthcare-trials/dermal-safety/.

Jovanovic B. Review of titanium dioxide nanoparticle phototoxicity: Developing a phototoxicity ratio to correct the endpoint values of toxicity tests. Environ Toxicol Chem. 2015;34:1070–7.

Kabri TH, Arab-Tehrany E, Belhaj N, Linder M. Physico-chemical characterization of nanoemulsions in cosmetic matrix enriched on omega-3. J Nanobiotechnol. 2011;9:41.

Kanazawa K. Casein nanoparticle. US Patent Application 20100143424. 2010.

Kempen PJ, Thakor AS, Zavaleta C, Gambhir SS, Sinclair R. A scanning transmission electron microscopy (STEM) approach to analyzing large volumes of tissue to detect nanoparticles. Microsc Microanal. 2013;19:1290–7.

Kojima H. The Japanese Center for the Validation of Alternative Methods (JaCVAM): Recent ICATM contributions and future plans [Internet]. Japan Altex; 2018 [cited 2018 April 9]. p. 337–8. Available from: http://www.altex.ch/resources/337338_Kojima31.pdf.

Kojima H. Safety assessment of cosmetic ingredients. In: Sakamoto K, Lochhead RY, Maibach H, Yamashita Y, editors. Cosmetic science and technology: Theoretical principles and applications. Amsterdam: Elsevier Inc.; 2017. p. 793–803.

Kokura S, Handa O, Takagi T, Ishikawa T, Naito Y, Yoshikawa T. Silver nanoparticles as a safe preservative for use in cosmetics. NBM. 2010;6:570–4.

L'Oréal. 2017. Research and innovation, [Internet]. L'Oréal 2017 Annual Report; 11.23.2017; [cited 2018 April 9]. Available from: https://www.loreal-finance.com/en/annual-report-2017/research-innovation.

Lin PC, Lin S, Wang PC, Sridhar R. Techniques for physicochemical characterization of nanomaterials. Biotechnol Adv. 2014;32:711–26.

Liu F, Wu J, Chen K, Xue D. Morphology study by using scanning electron microscopy. In: Méndez-Vilas A, Díaz J, editors. Microscopy: Science, technology, applications and education. Spain: FORMATEX; 2010. p. 1781–92. Available from: http://formatex.info/microscopy4/1781-1792.pdf.

Lorenz C, Tiede K, Tear S, Boxall A, Goetz NV, Hungerbühler K. Imaging and characterization of engineered nanoparticles in sunscreens by electron microscopy, under wet and dry conditions. Int J Occup Environ Health. 2010;16:406–28.

Lu PJ, Cheng WL, Huang SC, Chen YP, Chou HK, Chen HF. Characterizing titanium dioxide and zinc oxide nanoparticles in sunscreen spray. Int J Cosmet Sci. 2015;37:620–6.

Lu PJ, Fu WE, Huang SC, Lin CY, Ho ML, Chen YP, Cheng HF. Methodology for sample preparation and size measurement of commercial ZnO nanoparticles. J Food DrugAnal. 2018;26:628–36.

Masuda M, Harada F. Safety evaluation. In: Sakamoto K, Lochhead RY, Maibach H, Yamashita Y, editors. Cosmetic science and technology: Theoretical principles and applications. Amsterdam: Elsevier Inc.; 2017. p. 785–92.

MB Research. Human cell line activation test (hclat): OECD 442E-*in vitro* sensitization test [Internet]. Spinnerstown: MB Research Laboratories; 2017 [cited 2018 April 1]. Available from: http://www.mbresearch.com/hclat.htm.

Melo A, Amadeu MS, Lancellotti M, Maria de Hollanda L, Machado D. The role of nanomaterials in cosmetics: National and international legislative aspects. Química Nova. 2015;38:599–603.

Mihranyan A, Ferraz N, Strømme M. Current status and future prospects of nanotechnology in cosmetics. Prog Mater Sci. 2012;57:875–910.

Montenegro L, Lai F, Offerta A, Sarpietro MG, Micicche L, Maccioni AM, Valenti D, Fadda AM. From nanoemulsions to nanostructured lipid carriers: A relevant development in dermal delivery of drugs and cosmetics. J Drug Deliv Sci Technol. 2016;32:100–12.

Morganti P. Use and potential of nanotechnology in cosmetic dermatology. Clin Cosmet Investig Dermatol. 2010;3.5–13.

Mu L, Sprando RL. Application of nanotechnology in cosmetics. Pharm Res. 2010;27.1746–9.

Nafisi S, Maibach HI. Silica nanoparticles for increased cosmetic ingredient efficacy [Internet]. Cosmetics and Toiletries; 2015 [cited 2018 March 1]. Available from: http://www.cosmeticsandtoiletries.com/research/chemistry/Silica-Nanoparticles-for-Increased-Cosmetic-Ingredient-Efficacy--300987651.html.

Nanda S, Nanda A, Lohan S, Kaur R, Singh B. Nanocosmetics: Performance enhancement and safety assurance. In: Grumezescu AM, editor. Nanobiomaterials in Galenic Formulations and Cosmetics. Vol 10. Amsterdam: William Andrew Publishing ; 2016. p. 47–67. Available from: http://www.sciencedirect.com/science/article/pii/B9780323428682000036.

Nanopartikel. Coatings for nanomaterials [Internet]. Francfurt 2018 [cited 2018 April 22]. Available from: https://nanopartikel.info/en/nanoinfo/cross-cutting/993-coatings-cross-cutting-section.

Nanophase. Nanoparticle surface treatment, nanophase technologies corporation [Internet]. Romeoville: Nanophase Technologies Corporation; 2018 [cited 2018 April 22]. Available from: http://nanophase.com/technology/nanoparticle-surface-treatment/.

Nanotechproject. The project on emerging nanotechnologies [Internet]. Pennsylvania, 2018 [cited 2018 Dec 8]. Available from: www.nanotechproject.org/cbi/about/analysis/.

Nema RK, Rathore KS, Dubey BK. Textbook of cosmetics. New Delhi: CBS Publishers and Distributors; 2009.

MB Research Labs. Neutral Red Phototoxicity (3T3 NRU PT) Testing Services [Internet]. Spinnerstown: MB Research Labs; 2018 [cited 2018 April 2]. Available from: http://3t3nru.mbresearchlabs.com/.

Nigam PK. Adverse reactions to cosmetics and methods of testing. Indian J Dermatol Venereol Leprol. 2009;75:10–9.

OECD. OECD test guidelines for the testing of chemicals – Guideline 410: Repeated dose dermal toxic-ity: 21/28-day study, 12 May 1981. [Internet]. Paris: OECD Publishing; 1981a. [cited 2018 April 2]. Available from: https://doi.org/10.1787/9789264070745-en.

OECD. OECD test guideline for the testing of chemicals – Guideline 411: Subchronic dermal toxicity: 90-day study, 12 May 1981. [Internet]. Paris: OECD Publishing; 1981b. [cited 2018 April 2]. Available from: http://doi.org/10.1787/9789264070769-en.

OECD. OECD guideline for the testing of chemicals – Guideline 406: Skin sensitisation, adopted 17th July 1992. [Internet]. Paris: OECD Publishing; 1992. [cited 2018 April 2]. Available from: https://ntp.niehs.nih.gov/iccvam/suppdocs/feddocs/oecd/oecdtg406.pdf.

OECD. OECD guideline for the testing of chemicals – Guideline 471: Bacterial reverse mutation test, adopted 21st July 1997.[Internet]. Paris: OECD Publishing; 1997. [cited 2018 April 9]. Available from: https://www.oecd.org/chemicalsafety/risk-assessment/1948418.pdf.

OECD. OECD guideline for the testing of chemicals – Guideline 408: Repeated dose 90-day oral toxicity study in rodents, B.26 [Internet]. Paris: OECD Publishing;1998a [cited 2018 April 9]. Available from: http://www.intermed.it/istbiotech/reach/B26web2001.pdf.

OECD. OECD guideline for the testing of chemicals – Guideline 409: Repeated dose 90-day oral toxicity study in non-rodents, adopted 21st September 1998. [Internet]. Paris: OECD Publishing; 1998b. [cited 2018 April 9].

OECD. OECD guideline for the testing of chemicals – Guideline 404: Acute dermal irritation/corrosion, adopted 24th April 2002. [Internet]. Paris: OECD Publishing; 2002. [cited 2018 April 9]. Available from: http://www.nihs.go.jp/hse/chem-info/oecd/160603_sec4.pdf.

OECD. OECD guideline for testing of chemicals: Proposal for a new draft guideline 434: Acute dermal toxicity-fixed dose procedure, draft guideline 14 May 2004 (1st Version). [Internet]. Paris: OECD Publishing; 2004. [cited 2018 April 9]. Available from: http://www.oecd.org/chemicalsafety/testing/32037747.pdf.

OECD. OECD guidelines for the testing of chemicals – Guideline 407: Repeated dose 28-day oral toxicity study in rodents. Paris: OECD Publishing; [Internet]. Paris: OECD Publishing; 2008. [cited 2018 April 9]. Available from: https://doi.org/10.1787/9789264070684-en.

OECD. OECD guideline for the testing of chemicals – Guideline 436: Acute inhalation toxicity-acute toxic class method, adopted 7 September 2009. [Internet]. Paris: OECD Publishing; 2009a. [cited 2018 April 9]. Available from: https://ntp.niehs.nih.gov/iccvam/suppdocs/feddocs/oecd/oecd-tg436.pdf.

OECD. OECD guideline for the testing of chemicals – Guideline 412: Subacute inhalation toxicity: 28-day study, adopted 7 September 2009. [Internet]. Paris: OECD Publishing; 2009b. [cited 2018 April 9]. Available from: https://ntp.niehs.nih.gov/iccvam/suppdocs/feddocs/oecd/oecd-tg412.pdf.

OECD. OECD guideline for the testing of chemicals – Guideline 413: Subchronic inhalation toxicity: 90-day study, adopted 7 September 2009. [Internet]. Paris: OECD Publishing; 2009c. [cited 2018 April 9]. Available from: https://doi.org/10.1787/9789264070806-en.

OECD. OECD guideline for the testing of chemicals – Guideline 452: Chronic toxicity studies, adopted 25 June 2018. [Internet]. Paris: OECD Publishing; 2018c. [cited 2018 Dec 10]. Available from: https://doi.org/10.1787/9789264071209-en

OECD. OECD guideline for the testing of chemicals – Guidelines 429: Skin sensitization: Local lymph node assay, adopted 22 July 2010. [Internet]. Paris: OECD Publishing; 2010. [cited 2018 April 9]. Available from: https://ntp.niehs.nih.gov/iccvam/suppdocs/feddocs/oecd/oecd-tg429-2010.pdf.

OECD. OECD guideline for the testing of chemicals – Guidelines 460: Fluorescein leakage test method for identifying ocular corrosives and severe irritants, adopted 2 October 2012.[Internet]. Paris: OECD Publishing; 2012. [cited 2018 April 9]. Available from: https://ntp.niehs.nih.gov/iccvam/suppdocs/feddocs/oecd/oecd-tg460-508.pdf.

OECD. OECD guidelines for the testing of chemicals – Guideline 437: Bovine corneal opacity and permeabil-ity test method for identifying i) chemicals inducing serious eye damage and ii) chemicals not requiring classification for eye irritation or serious eye damage, adopted 26 July 2013.[Internet]. Paris: OECD Publishing; 2013a. [cited 2018 April 9]. Available from: https://ntp.niehs.nih.gov/iccvam/suppdocs/feddocs/oecd/oecd-tg437-2013-508.pdf.

OECD. OECD guidelines for the testing of chemicals – Guideline 438, adopted 26 July 2013. [Internet]. Paris: OECD Publishing; 2013b. [cited 2018 April 9]. Available from: http://www.efcc.eu/media/2950/2013-07-oecd-tg-438-in-vitro-eyeirritation.pdf.

OECD. OECD guideline for the testing of chemicals – Guideline 473: In vitro mammalian chromosomal aberration test, adopted 26 September 2014. [Internet]. Paris: OECD Publishing; 2014. [cited 2018 April 9]. Available from: https://ntp.niehs.nih.gov/iccvam/suppdocs/feddocs/oecd/oecd-tg473-2014-508.pdf.

OECD. Draft proposal for a new guideline 433: Acute inhalation toxicity-fixed concentration procedure, 18th December 2015. [Internet]. Paris: OECD Publishing; 2015a. [cited 2018 April 9]. Available from: https://www.oecd.org/env/ehs/testing/Revised%20TG433%20(2015)%20(clean%20version)_18Dec%202015.pdf.

OECD. OECD guideline for testing of chemicals – Guideline 404: Acute dermal irritation/corrosion, adopted 28 July 2015. Paris: OECD Publishing;2015b. [cited 2018 April 9]. Available from: https://dx.doi.org/1 0.1787/9789264242678-en.

OECD. OECD guideline for the testing of chemicals – Guideline 442c: InChemico skin sensitisation: Direct Peptide Reactivity Assay (DPRA), adopted 4 February 2015. Paris: OECD Publishing; 2015c. [cited 2018 April 9]. Available from: https://ntp.niehs.nih.gov/iccvam/suppdocs/feddocs/oecd/oecd-tg442c-508.pdf.

OECD. OECD guideline for the testing of chemicals – Guideline 476: In vitro mammalian cell gene mutation tests using the Hprt and xprt genes. Adopted 29 July 2016. Paris: OECD Publishing; 2016b. [cited 2018 April 9]. Available from: https://doi.org/10.1787/9789264264809-en.

OECD. OECD guideline for the testing of chemicals– Guideline 490: In vitro mammalian cell gene mutation tests using the thymidine kinase gene. Adopted 29 July 2016. Paris: OECD Publishing; 2016c. [cited 2018 April 9]. Available from: https://dx.doi.org/10.1787/9789264264908-en.

OECD. OECD guideline for the testing of chemicals – Guideline 492: Reconstructed Human Cornea-like Epithelium (RhCE) test method for identifying chemicals not requiring classification and labelling for eye irritation or serious eye damage. Draft revised TG 492, December 2016. Paris: OECD Publishing; 2016a. [cited 2018 April 9]. Available from: http://www.oecd.org/env/ehs/testing/TG%20492_SkinE thic%20HCE%20revision_Dec%202016_clean.pdf.

OECD. OECD guideline for the testing of chemicals -- Guideline 487: In vitro mammalian cell micronucleus test, adopted 29 July 2016. Paris: OECD Publishing; 2016d. [cited 2018 April 9]. Available from: https://www.oecd-ilibrary.org/docserver/9789264264861-en.pdf?expires=1544504698&id=id&accname=g uest&checksum=FD18A481438B23EAEEBB1026548E6CA1.

OECD. Draft key event based test guidelines 442D: In vitro skin sensitisation assays addressing the AOP key event on keratinocyte activation, revised TG 442D–Draft v. 7 July 2017. Paris: OECD Publishing; 2017. [cited 2018 April 9]. Available from: https://www.oecd.org/env/ehs/testing/latestdocuments/2017-07-07_Draft%20Revised_TG_442D.pdf.

OECD. Test guideline for the testing of chemicals based on key events – Guideline 442e, adopted 25 June 2018. Paris: OECD Publishing; 2018b. [cited 2018 Dec 9]. Available from: https://www.oecd-ilibrary. org/docserver/9789264264359-en.pdf?expires=1544505159&id=id&accname=guest&checksum=B CD7F3253FF5A25B1BFD3DB3FDA768CC.

OECD. eChemPortal: Global portal to information on chemical substances [Internet]. Paris: OECD Publishing; 2018a [cited 2018 April 9]. Available from: http://www.oecd.org/chemicalsafety/risk-assessment/eche mportalglobalportaltoinformationonchemicalsubstances.htm.

Pape WJW. Validation of in vitro methods to single out photoirritants using mechanistically based tests. In: Seilver JP, Vilanova E, editors. Applied toxicology: Approaches through basic science, proceedings of the 1996 EUROTOX congress meeting, held in Alicante, Spain, September 22–25, 1996. Heidelberg: Springer-Verlag Berlin; 1997. p. 239–48.

Particle Sciences. An Overview of the Zeta Potential. Bethlehem: Particle Sciences, Inc.; 2012 [cited 2018 March 11]. Available from: http://www.particlesciences.com/docs/technical_briefs/TB_2012_2-Overvi ew-of-Zeta-Potential.pdf.

PETA International Science Consortium. . In vitro and in chemico methods to predict skin sensitisation: The adverse outcome pathway for skin sensitisation initiated by covalent binding to proteins [Internet]. Saints Street: PETA International Science Consortium; 2017 [cited 2018 April 1]. Available from: https://www.piscltd.org.uk/skin-sensitisation/.

Powers KW, Brown SC, Krishna VB, Wasdo SC, Moudgil BM, Roberts SM. Research strategies for safety evaluation of nanomaterials. Part VI. Characterization of nanoscale particles for toxicological evaluation. Toxicol Sci. 2006;90:296–303.

Pyrz WD, Buttrey DJ. Particle size determination using TEM: A discussion of image acquisition and analysis for the novice microscopist. Langmuir. 2008;24:11350–60.

Raj S, Jose S, Sumod US, Sabitha M. Nanotechnology in cosmetics: Opportunities and challenges. J Pharm Bioallied Sci. 2012;4:186–93.

Ribeiro RC, Barreto SM, Ostrosky EA, da Rocha-Filho PA, Veríssimo LM, Ferrari M. Production and characterization of cosmetic nanoemulsions containing *Opuntiaficus-indica* (L.) mill extract as moisturizing agent. Molecules. 2015;20:2492–509.

Riebeling C, Hayess K, Peters AK, Steemans M, Spielmann H, Luch A, Seiler AEM. Assaying embryotoxicity in the test tube: Current limitations of the embryonic stem cell test (EST) challenging its applicability domain. Crit Rev Toxicol. 2012;42:443–64.

Sakaguchi H, Ryan C, Ovigne JM, Schroeder KR, Ashikaga T. Predicting skin sensitization potential and inter-laboratory reproducibility of a human Cell Line Activation Test (h-CLAT) in the European Cosmetics Association (COLIPA) ring trials. Toxicol in Vitro. 2010;24:1810–20.

Sarker DK. Engineering of nanoemulsions for drug delivery. Curr Drug Deliv. 2005;2:297–310.

The Scientific Committee on Cosmetic Products and Non-food Products Intended for Consumers (SCCNFP). SCCNFP/0321/00 Final: Notes of guidance for testing of cosmetic ingredients for their safety evaluation. In: Proceedings of the SCCNFP Plenary Meeting; 24 October 2000.

The Scientific Committee on Cosmetic Products and Non-food Products Intended for Consumers (SCCNFP). SCCNFP/0755/03: Proposal for recommended mutagenicity/genotoxicity tests for the safety testing of cosmetic ingredients to be included in the annexes to council directive 76/768/EEC. In: Proceedings of the SCCNFP Plenary Meeting; 9 December 2003.

Scientific Committee on consumer products (SCCP). The SCCP's notes of guidance for the testing of cosmetic ingredients and their safety evaluation 6th revision. In: Proceedings of the SCCP 10th Plenary Meeting; 19 December 2006.

SCCP. SCCP/1147/07: Opinion on safety of nanomaterials in cosmetic products. Adopted by the SCCP after the public consultation on the 14th plenary of 18 December 2007. 2007a.

SCCP. Preliminary opinion on safety of nanomaterials in cosmetic products, approved by the SCCP for public consultation 12th plenary of 19 June 2007. 2007b.

SCCP. SCCP/1147/07: SCCP opinion on safety of nanomaterials in cosmetic products, adopted by the SCCP after the public consultation on the 14th plenary of 18 December 2007. 2007c.

SCCS. Guidance on the safety assessment of nanomaterials in cosmetics, the SCCS adopted this opinion at its 15th plenary meeting of 26–27 June 2012. 2012.

SCCS. Opinion on titanium dioxide (nano form) COLIPA n° S75. Revision of 22 April 2014. 2014.

Schwab M. Encyclopaedia of cancer. 3rd ed. Heidelberg: Springer-Verlag Berlin; 2011.

SEURAT-1. SEURAT-1 annual meeting: Alternative methods for repeated dose systemic toxicity testing of chemicals [Internet]. 2014 [cited 2018 April 9]. Available at: https://ec.europa.eu/growth/content/seurat-1-annual-meeting-alternative-methods-repeated-dose-systemic-toxicity-testing-0_en.

Tak YK, Pal S, Naoghare PK, Rangasamy S, Song JM. Shape-dependent skin penetration of silver nanoparticles: Does it really matter? Sci Rep. 2015;5: 1-11.

Thong HY, Maibach HI. Photosensitivity induced by exogenous agents: Phototoxicity and photoallergy. In: Roberts MS, Walters KA, editors. Dermal absorption and toxicity assessment. Boca Raton, FL: CRC Press; 2008. p. 391–404.

United States Congress. Office of technology assessment, alternatives to animal use in research, testing, and education. Washington, DC: United States Government Publishing Office; 1986. Report No.: OTA-BA-273, February 1986. 144.

Vinardell PM, Mitjans M. Alternative methods to animal testing for the safety evaluation of cosmetic ingredients: An overview. Cosmetics. 2017;4:1–14.

Viswanath DS, Ghosh TK, Prasad DHL, Dutta NVK, Rani KY. Viscosity of liquids: Theory, estimation, experiment, and data. Dordrecht: Springer; 2007.

Wabuyele MB, Culha M, Griffin GD, Viallet PM, Vo-Dinh T. Near-field scanning optical microscopy for bioanalysis at nanometer resolution. Methods Mol Biol. 2005;300:437–52.

Wang ZL. What is nanotechnology? [Internet]. Atlanta: Prof. Zhong L. Wang's Nano Research Group; 2018 [cited 2018 March 14]. Available from: http://www.nanoscience.gatech.edu/zlwang/research/nano.html.

Wiechers JW, Musee N. Engineered inorganic nanoparticles and cosmetics: Facts, issues, knowledge gaps and challenges. J Biomed Nanotechnol. 2010;6:408–31.

Wu X, Guy RH. Applications of nanoparticles in topical drug delivery and in cosmetics. J Drug Deliv Sci Technol. 2009;19:371–84.

Yah CS, Iyuke SE, Simate GS. A review of nanoparticles toxicity and their routes of exposures. IJPS Winter. 2011;8:299–314.

Yin JJ, Liu J, Ehrenshaft M, Roberts JE, Fu PP, Mason RP, Zhao B. Phototoxicity of nano titanium dioxides in HaCaT keratinocytes-generation of reactive oxygen species and cell damage. Toxicol Appl Pharmacol. 2012;263:81–8.

Zamora D, Alcalà M, Blanco M. Determination of trace impurities in cosmetic intermediates by ion mobility spectrometry. Anal Chim Acta. 2011;708:69–74.

Zhang Y, Newton B, Lewis E, Fu PP, Kafoury R, Ray PC, Yu H. Cytotoxicity of organic surface coating agents used for nanoparticles synthesis and stability. Toxicol In Vitro. 2015;29:762–8.

Index

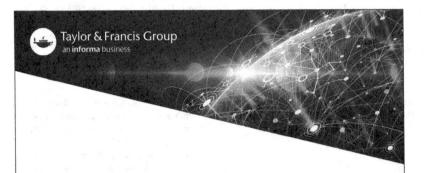

9781032338552